ANTIVIRAL CHEMOTHERAPY 5

New Directions for Clinical Application and Research

ADVANCES IN EXPERIMENTAL MEDICINE AND BIOLOGY

Recent Volumes in this Series

Volume 450
ADVANCES IN MODELING AND CONTROL OF VENTILATION
Edited by Richard L. Hughson, David A. Cunningham, and James Duffin

Volume 451
GENE THERAPY OF CANCER
Edited by Peter Walden, Uwe Trefzer, Wolfram Sterry, and Farzin Farzaneh

Volume 452
MECHANISMS OF LYMPHOCYTE ACTIVATION AND IMMUNE REGULATION VII: Molecular Determinants of Microbial Immunity
Edited by Sudhir Gupta, Alan Sher, and Rafi Ahmed

Volume 453
MECHANISMS OF WORK PRODUCTION AND WORK ABSORPTION IN MUSCLE
Edited by Haruo Sugi and Gerald H. Pollack

Volume 454
OXYGEN TRANSPORT TO TISSUE XX
Edited by Antal G. Hudetz and Duane F. Bruley

Volume 455
RHEUMADERM: Current Issues in Rheumatology and Dermatology
Edited by Carmel Mallia and Jouni Uitto

Volume 456
RESOLVING THE ANTIBIOTIC PARADOX: Progress in Understanding Drug Resistance and Development of New Antibiotics
Edited by Barry P. Rosen and Shahriar Mobashery

Volume 457
DRUG RESISTANCE IN LEUKEMIA AND LYMPHOMA III
Edited by G. J. L. Kaspers, R. Pieters, and A. J. P. Veerman

Volume 458
ANTIVIRAL CHEMOTHERAPY 5: New Directions for Clinical Application and Research
Edited by John Mills, Paul A. Volberding, and Lawrence Corey

Volume 459
IMPACT OF PROCESSING ON FOOD SAFETY
Edited by Lauren S. Jackson, Mark G. Knize, and Jeffrey N. Morgan

---

A Continuation Order Plan is available for this series. A continuation order will bring delivery of each new volume immediately upon publication. Volumes are billed only upon actual shipment. For further information please contact the publisher.

ANTIVIRAL CHEMOTHERAPY 5

New Directions for Clinical Application and Research

Edited by

John Mills

Macfarlane Burnet Centre for Medical Research
Fairfield, Victoria, Australia

Paul A. Volberding

University of California, San Francisco
San Francisco, California

and

Lawrence Corey

University of Washington
Seattle, Washington

Kluwer Academic / Plenum Publishers
New York, Boston, Dordrecht, London, Moscow

Proceedings of the Fifth Triennial Symposium on New Directions in Antiviral Chemotherapy,
held November 13 – 15, 1997, in San Francisco, California

ISBN 0-306-46107-2

233 Spring Street, New York, N.Y. 10013

10 9 8 7 6 5 4 3 2 1

A C.I.P. record for this book is available from the Library of Congress.

Printed in the United States of America

PREFACE

Scientists and clinicians attending the last "New Directions in Antiviral Therapy" conference in late 1994 could hardly have predicted the revolution in the management of patients with HIV infection that has occurred since. Two new classes of antiretrovirals have been licensed, the second-site RT inhibitors and the protease inhibitors; the long incubation period of active HIV infection, when the infection is clinically latent, is now understood to be a period of intense viral replication and turnover of CD4 lymphocytes; measurements of HIV RNA concentration in plasma have been shown to be essential tools for monitoring the course of HIV infection, deciding when to treat, and assessing the results of treatment; and finally, combinations of antiretrovirals, particularly combinations including protease inhibitors, have been shown to have dramatically beneficial effects on patients with HIV infection. These advances, coupled with new drugs for the management of herpesvirus infections, have made dramatic differences in the quality and length of life of HIV-infected patients.

Additional advances have been made since 1994 in the prevention or management of influenza virus (zanamavir), respiratory syncytial virus (palvizumab), hepatitis B virus (lamivudine and famciclovir), and enterovirus infections (pleconaril). It is difficult to remember that only slightly more than a decade ago there were only a handful of antiviral agents available (none of which were antiretrovirals), and a number of those were either highly toxic, of dubious efficacy, or both.

This book is the edited, expanded, and updated proceedings of the "Fifth Triennial Conference on Antiviral Chemotherapy," held in San Francisco in November, 1997. The conference was sponsored by the University of California, San Francisco, and co-sponsored by the Macfarlane Burnet Centre for Medical Research in Melbourne, Australia. The conference speakers and contributors to this volume are all experts in their respective areas, and much unpublished data were included in the presentations and discussions. This text should serve as a topical reference on antiviral chemotherapy, at least until the proceedings of the next meeting, the Sixth Triennial Conference, are published in 2001!

The editors wish to thank Abbott Laboratories, Amgen Inc., AMRAD Corporation, Boehringer Ingelheim Diagnostics, Bristol Myers Squibb, Gilead Sciences, Glaxo Wellcome Inc., Roche Diagnostics, and SmithKline Beecham for their extraordinarily generous and unencumbered support without which neither the symposium nor this publication would be possible, and the speakers and authors for their ungrudging preparation of high-quality lectures and manuscripts.

Adrienne Williams and Kathy Mello of UCSF Postgraduate Programs and Valerie Jessop of Macfarlane Burnet Centre are thanked for their assistance with organizing this conference. Mardi Malone and Adam Chapman are thanked for their rigorous attention to editing and formatting the text of this monograph.

John Mills
Paul Volberding
Lawrence Corey

CONTENTS

OVERVIEW OF HBV THERAPY

Mario G. Pessoa and Teresa L. Wright

Department of Veterans Affairs Medical Center
4150 Clement Street
San Francisco, California 94121

The natural history of infection in patients with active HBV viral replication (HBsAg, HBeAg and HBV DNA positive) tends to be progressive, with development of long-term complications such as portal hypertension and hepatic decompensation. As well, the risk of developing hepatocellular carcinoma (HCC) is increased 10 to 390-fold in patients with chronic HBV infection,[1] and in regions where HBV is endemic, HCC is the leading cause of cancer related deaths. Evolving therapies, discussed in this chapter, may be changing the natural history of HBV infection and even reducing the risk of HCC development.

Therefore, in patients with chronic HBV infection, the main goals of treatment are (i) to reduce the rate of progression of liver disease, hence preventing the development of complications such as decompensated liver disease, hepatocellular carcinoma and liver-related death; and (ii) to eradicate the virus, thus reducing the number of chronic carriers who serve as a reservoir for HBV transmission. Because complications may take many years to develop, most treatment trials have used inhibition of viral replication and improvement in liver histology as intermediate endpoints to evaluate efficacy. Loss of HBV DNA and seroconversion from HBeAg to anti-HBe are usually associated with normalization of liver enzymes and decreasing inflammation on liver biopsy.[2] HBeAg seroconversion, whether induced by therapy or occurring spontaneously, has been shown to be associated with improved long-term survival.[3]

Prospective clinical studies are needed to evaluate whether therapeutic interventions which inhibit viral replication (see below) lead to improved survival or decreased rates of hepatocellular carcinoma.

1. HBV REPLICATION

Knowledge of the mechanism of HBV replication has been essential in the identification and testing of potential new drugs. Understanding the host immune response to in-

Antiviral Chemotherapy 5, edited by Mills *et al.*
Kluwer Academic / Plenum Publishers, New York, 1999.

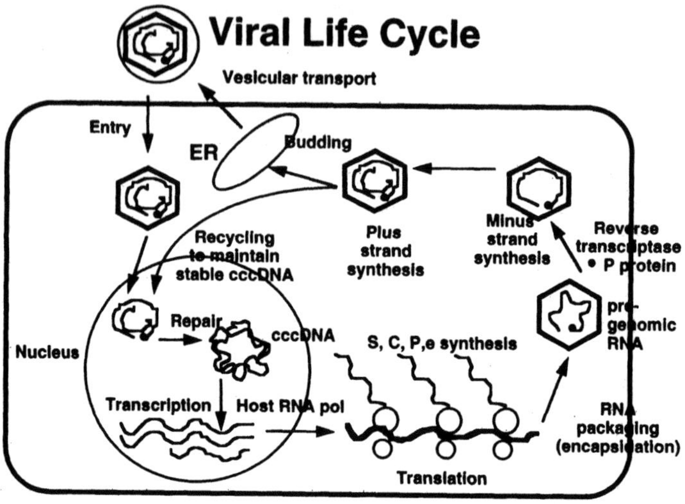

Figure 1. The viral life cycle of HBV (see text). ER = endoplasmic reticulum; cccDNA = covalently closed circular DNA; pol, P = polymerase; S = surface protein; C = core protein; e = e antigen.

fection is also critical to understanding the mechanism of clearance of virus with acute infection, the failure of viral clearance in persistent infection and the pathogenesis of liver disease associated with chronic infection.

HBV polymerase catalyzes by reverse transcription both the generation of DNA complementary to the viral RNA template, and the synthesis of positive strand viral DNA from the negative strand DNA template of the virus. The error rate during replication is low, even though the viral cycle includes an intermediate RNA step, with the production of viral DNA catalyzed by a reverse transcriptase/RNA-dependent DNA polymerase.[4] The high fidelity during replication is probably a function of the small size and the compact organization of the genome, which renders most new mutations non-viable. Despite this relatively low mutation rate, up to 12% of nucleotides may vary between isolates of HBV.[4]

The life cycle of HBV includes the following steps: (i) Viral binding and entry; (ii) viral uncoating in the cytoplasm; (iii) synthesis of complete double-stranded DNA (ccc DNA) in the nucleus; (iv) synthesis of genomic or pregenomic (pg) RNA which forms the template for viral DNA synthesis, and synthesis from cccDNA of viral transcripts necessary for viral protein production; (v) translation of viral transcripts in the cytoplasm; (vi) encapsidation or packaging of genomic RNA in the cytoplasm with production of viral cores; (vii) synthesis of minus strand DNA from pregenomic RNA, by reverse transcription followed by synthesis of the partially complete plus strand; (vii) envelopment of viral cores with excretion of viral progeny from the cell or transport of viral cores back to the nucleus (Figure 1). Knowledge of these steps is important in understanding which steps in replication are inhibited by specific drugs.

2. THERAPY

All drugs currently under investigation against hepatitis B virus are listed in Table 1.

Table 1

*Drugs in clinical development**
Lamivudine (phase III complete, FDA approval likely in 1998)
Famiciclovir (phase III under analysis)
Lobucavir (phase II complete, phase III underway)
Adefovir (phase II complete)
FTC (phase I)
Monoconal HBIg (phase I)
Tucarasol (phase I)
Drugs in pre-clinical development
Pre-S vaccine by Chiron
BMS 200475
LMFAU
PMPA
FLG
Medeva vaccine

*Interferon remains the only FDA approved therapy

2.1. Interferon

Alpha interferon has been most extensively studied, and is still the only antiviral drug currently FDA-approved in the U.S. for clinical use in chronic hepatitis. In the largest U.S. trial, 169 patients were randomized to receive interferon α-2b 5 million units (MU) daily, 1 MU daily, 6 weeks of oral prednisone followed by 5 MU daily, or observation for a period of 16 weeks.[5] Disappearance of HBV DNA and loss of HBeAg occurred in 37% of patients receiving 5 MU daily, a significantly higher rate than observed in patients receiving 1 MU daily (17% responded), or in patients receiving no therapy (7%). A decrease in inflammation was evident on the liver biopsies of patients who responded. Loss of HBV DNA was typically preceded by an increase in serum ALT activity, believed to result from interferon-induced, immune-mediated hepatocyte necrosis. This 'flare', defined as a greater than two fold rise in ALT, occurred more commonly in responders (63%) than in non-responders (27%). Ten percent of all treated patients lost HBsAg. The majority of patients experienced minor side effects during interferon treatment, but these rarely led to cessation of therapy. A meta-analysis[6] from 15 randomized controlled trials including 837 HBV carriers (HBsAg and HBeAg positive, biopsy-proven chronic hepatitis) has confirmed the efficacy of alpha-interferon. Interferon at doses of 5–10 MU tiw (ranging from 7 to 30 MU/m^2 weekly) for at least 3 months was associated with significantly greater loss of HBV DNA (37% versus 17%), HBeAg (33% versus 21%) and HBsAg (7.8% versus 1.8%) in treated patients compared to controls. Treatment was also associated with normalization of serum ALT levels. The meta-analysis excluded studies involving children, patients who were delta-coinfected, or patients who were predominantly anti-HBe positive. Loss of HBeAg and HBV DNA occurred more frequently when patients received doses greater than 15 MU/m^2/-week, or total doses greater than 200 MU/m^2. Three times weekly administration is equivalent in efficacy to treatment for 6 months, and as a consensus, treatment for four months is currently recommended.

Long-term follow-up of patients treated with interferon indicates that remission is maintained in the majority of those who initially respond to therapy and that additional seroconversion occurs after treatment is complete. After a follow-up of 5 to 7 years, responders will continue to lose HbsAg.[3] HBV DNA is absent in the serum of the majority of patients who cleared HBsAg and loss of HBeAg and HBV DNA in serum is usually associated with disappearance of replicative forms of HBV DNA in liver.[7] HBeAg seroconversion is associated with an improved long-term natural history, whether due to interferon treatment or spontaneous seroconversion.

Thus, patients with chronic HBV appear to obtain long-term benefit from interferon therapy. While prospective studies have not established this benefit, recent data strongly suggest that HBeAg seroconversion with interferon therapy is associated with improved survival and reduced complications. One study from Eurohep looked at long-term outcome of hepatitis B e antigen positive patients with compensated cirrhosis treated with alpha interferon. The authors had shown improved survival in the treated group despite the same rate of HBeAg seroconversion. They observed higher rates of ALT normalization and HBsAg clearance in the treated group.[8] The recommended regimen of alpha interferon in patients with biopsy-proven hepatitis, persistent elevations of serum aminotransferase levels and detectable levels of HBsAg, HBeAg and HBV DNA in serum is 5 million units daily or 10 million units thrice weekly for 4 months.

The patients included in most clinical trials of alpha-interferon represent a highly select group of chronic HBV carriers. In particular, patients with decompensated liver disease have been excluded. These patients frequently have leukopenia and thrombocytopenia due to hypersplenism, which limits the dose of interferon that can be administered. Therapy with low doses of interferon (0.5 MU/day to 3 MU tiw) adjusted according to tolerability is an option for patients with mildly decompensated cirrhosis,[9] but in general, such patients should be treated with caution. Limited studies involving patients with decompensated disease suggest that interferon therapy may be of benefit and produce sustained inhibition of viral replication and clinical stabilization, which could in theory postpone the need for liver transplantation. Treating marginally compensated patients, however, must be undertaken with caution since side effects are common and potentially life threatening infections and exacerbation of liver disease may occur. Liver transplantation should be available for those patients who deteriorate with therapy.

Pre-treatment variables which are predictive of a sustained loss of HBeAg and HBV DNA are high serum aminotransferase levels, low serum HBV DNA levels, short duration of infection, wild-type (HBeAg-positive) virus, older age of infection, absence of immunosuppression and a hystological picture of active hepatitis. Patients with HBV DNA level less than 200 pg/mL (by liquid hybridization assay) are more likely to respond than those with higher levels, but due to different methods of measuring levels of viremia, quantitation of HBV DNA by the liquid hybridization assay is not directly comparable to quantitation by other assays such as dot blot hybridization or branched DNA assays. The age at the time of infection, as well as the duration of infection, are likely to be important in determining the rate of response. Patients who acquired their infection in the perinatal period typically have low aminotransferase levels, and are less likely to respond to interferon. The age of infection is more important than ethnicity in determining response. HBeAg-positive Asians with high pre-treatment ALT levels have a similar response rate to other ethnic groups. Interferon is not recommended in HBsAg-positive patients with normal serum ALT, since these patients usually respond poorly to therapy and long-term follow-up is benign.[10] A randomized controlled trial of 24 month course of 6 MU of alpha IFN in patients infected with precore variant virus who were HBV DNA positive and HBeAg negative in serum, led to virological sustained response in 33% of the treated group, compared to none of the untreated controls.[11] Patients coinfected with HIV have lower rates of response to interferon than immunocompetent patients.[12]

As only a minority of patients achieves loss of HBV DNA, HBeAg and HBsAg with interferon treatment, analyses of cost-effectiveness are useful. Wong and colleagues utilized a hypothetical cohort of 35 year old chronic HBV carriers (HBsAg and HBeAg positive) with chronic hepatitis (without cirrhosis) on liver biopsy, to examine the cost-benefits of a 16 week course of alpha-interferon (10 MU three times weekly) compared to standard management of chronic liver disease. In the analysis, interferon treatment was associated with an increase in life expectancy and quality-adjusted life-expectancy, as well as a substantial decrease in the total healthcare costs. In patients receiving interferon therapy, the absolute cumulative lifetime risk of developing cirrhosis, decompensated liver disease and hepatocellular carcinoma was decreased by 13%, 9%, and 4%, respectively. Based upon the predictors of this model, 7 patients would need to be treated with interferon therapy to prevent 1 case of cirrhosis, 11 patients would need to be treated to prevent 1 case of decompensated liver disease, and 26 patients would need to be treated to prevent 1 case of HCC.[13] In a second analysis by Dusheiko and Roberts, a cohort of 25–35 year old HBV carriers who were HBV DNA positive was used for modeling costs and benefits. This analysis also showed that therapy with interferon is cost-effective.[14]

Alpha interferon has been used in combination with other anti-viral (gamma-interferon, ara-AMP, azidothymidine) and immunomodulatory drugs (levamisole, thymosin, interleukin-2). However combination therapy of α-interferon with other agents has shown no clear advantage over treatment with interferon alone, yet is associated with increased side effects.

2.2. Nucleoside and Nucleotide Analogs

Four agents which specifically inhibit HBV replication are in clinical development for the treatment of chronic hepatitis B infection. These nucleoside and nucleotide analogs are incorporated into the growing DNA chain and terminate replication. Lamivudine or 3-thiacytadine is a cytadine analog which inhibits the reverse transcription step, synthesis of the HBV negative strand DNA from pre-genomic messenger RNA.[15] This agent predictably inhibits replication in patients who are both HBe antigen and HBV DNA positive (with an initial 2 to 3 log reduction in HBV DNA), improving liver enzymes as well as liver histology. However, both with short (3 months) and long (12 months) term therapy, clearance of virus occurs in the minority and relapse following cessation of treatment is common.[16, 17] In a randomized study of 42 HBV carriers[17] in which 53% had previously failed interferon, HBV DNA became undetectable by hybridization assay (= 1.5 pg/mL) in 100% of patients receiving lamivudine at doses of 100 or 300 mg daily and in 70% of those receiving 25 mg daily for 3 months. After stopping therapy, only 19% of the patients, including 5 of the non-responders to interferon, maintained a sustained response with suppression of HBV DNA and normalization of ALT. A flare during treatment (rise in serum aminotransferase levels greater than twice baseline values) occurred in 36% of patients treated with both doses. No serious adverse effects were observed. In a more recent randomized placebo-controlled study of lamivudine[18] at doses of 25 mg or 100 mg per day versus placebo for 12 months, conducted in 358 Chinese chronic HBV carriers, loss of circulating HBV DNA was almost universal in the 100 mg group (Figure 2). Prolonged therapy with both doses was also associated with sustained normalization of serum ALT and improved histology when compared with placebo. Relapse was observed in the majority when treatment was stopped, yet therapy was significantly associated with off-treatment loss of HBV DNA, HBeAg seroconversion and normalization of serum ALT. The antiviral

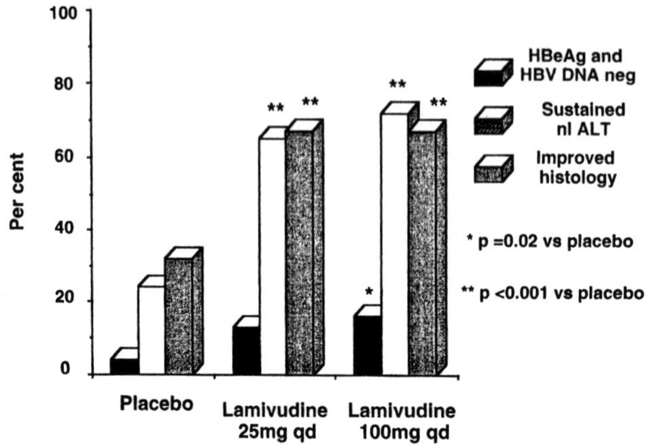

Figure 2. Phase III, double-blind, placebo-controlled study of 2 doses of lamivudine given for 12 months in 358 Chinese HBV carriers. Loss of circulating HBV DNA was almost universal in the 100 mg group. Prolonged therapy with both doses was also associated with sustained normalization of serum ALT and improved histology when compared to placebo. Adapted from CL Lai, EASL 1997.

effects produced by lamivudine treatment appeared to be similar in vertically acquired or horizontally acquired patients. More recent data with prolonged therapy for up to 18 months have shown that the drug is well tolerated and leads to HBV DNA suppression in 87% of patients during treatment, with a sustained loss of HBeAg in 39% of the patients treated.[19] The major concern with long-term suppressive antiviral therapy is the development of viral resistance with DNA breakthrough, which occurs in approximately 15% of the patients treated for 12 months[16] and has also been described in immunocompromised patients following liver transplantation.[20, 15] This resistance is associated with specific mutations in the YMDD locus (methionine to isoleucine or less commonly to valine), the same region of the viral polymerase which has been associated with resistance to lamivudine in the therapy of HIV infection.[15] Clinical consequences of variant viruses are under investigation, although in a phase III trial, the development of YMDD-mutant HBV was not associated with any consistent clinical signs or symptoms, and these patients tended to maintain lower HBV DNA levels and ALT values than their pre-treatment levels.[16]

Acyclovir, ganciclovir, famciclovir and penciclovir are a family of related guanosine analogs, which inhibit viral DNA and protein synthesis. These drugs differ in their oral bioavailability and anti-HBV potency. Famciclovir has good oral bioavailability and is well-tolerated at a dose of 500 mg three times daily.[21] In a randomized placebo-controlled multicenter study, 333 HBeAg positive patients were randomized to receive 16 weeks treatment with famciclovir 500 mg, 250 mg, 125 mg or placebo three times a day. In all famciclovir treatment groups the effect on HBV DNA levels was evident within 1 week of therapy and was maintained throughout the 16 week treatment period. There was no difference in the efficacy of famciclovir in patients who had previously received interferon compared to those who had not.[22] Also, preliminary study has demonstrated a beneficial effect of famciclovir in reducing HBV replication in the post-transplantation period.[23] Phase III studies using famciclovir for the treatment of chronic HBV are in progress.

Agents in phase I/II trials include lobucavir, a novel nucleoside analogue, and adefovir dipivoxil, a nucleotide analog. Preliminary results from phase I/II trial of both these new drugs have demonstrated substantial, although transient suppression of viral replication. Lobucavir-triphosphate is a potent inhibitor of HBV DNA polymerase. In a phase I/II placebo-controlled study in 22 patients with chronic HBV infection, oral lobucavir at doses of 200 mg twice daily or four times daily was given for 4 weeks. In both treatment groups, HBV DNA levels were reduced by 2.8 logs (bDNA assay - limit of detection 2.5

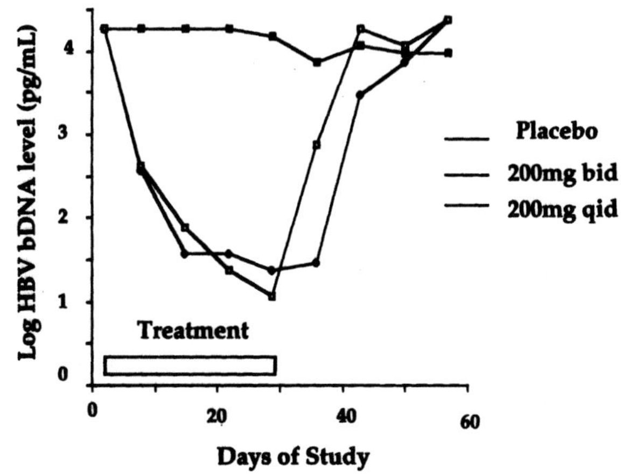

Figure 3. Phase I/II, double-blind, placebo-controlled study of 2 doses of oral lobucavir given for 28 days in patients with chronic hepatitis B infection. Both doses (200 mg bid and 200 mg qid) were equally effective in reducing the serum HBV DNA titer by ~2.8 logs during treatment and were well tolerated. Adapted from J. Bloomer *et al.* Second International Conference on Therapies for Viral Hepatitis. Hawaii, Dec. 1997.

pg/mL). Twenty-four percent of treated patients had undetectable HBV DNA levels by the end of the treatment period (Figure 3). Lobucavir was well tolerated with no discontinuations due to adverse events. The most common adverse events were headache, fatigue, insomnia, infection, diarrhea, abdominal pain and flu syndrome.[24,25] A current randomized, double-blind, placebo-controlled trial is studying several once daily dosing regimens of oral lobucavir given for 3 months. BMS-200475, a cyclopentyl guanosine analogue, is a potent inhibitor of HBV DNA polymerase in animal studies. The safety and pharmacokinetics of BMS-200475 have been studied in both single and multiple oral dose studies in healthy volunteers. The most common adverse events were headaches and visual disturbances. Dose-ranging studies of oral BMS-200475 in chronic HBV infection are planned.[25]

Nucleotide analogs, such as adefovir dipivoxil, in contrast to nucleoside analogs, do not require the nucleoside kinase enzyme for initial phosphorylation, an enzyme which is absent in 1% to 2% of hepatocytes. Thus nucleotide analogs may be effective against HBV infection in cell types in which nucleoside analogs such as lobucavir cannot be activated. In one phase I/II placebo-controlled trial, 20 patients with chronic hepatitis B were treated with adefovir at a single oral dose of 125 mg/day for 1 month. The drug was well tolerated, producing >90% suppression of HBV DNA after only one week of therapy. The initial mean log reduction of HBV DNA was 1.8, however HBV DNA levels returned to baseline or even above baseline values after stopping therapy (Figure 4). A rise in serum aminotransferase levels, also seen with other anti-HBV drugs, such as lamivudine or interferon, in association with seroconversion to anti-HBe and declines in HBV DNA levels, was the only side effect observed.[26] The ALT rise is thought to be related to an immune-mediated response of the host to drug-induced viral suppression. Two parallel phase II trials studying lower doses (5, 30 and 60 mg daily) and a longer duration of therapy in HIV-1 negative patients, with normal or abnormal ALTs, are in progress. *In vitro* assays have shown that the YMDD motif (M552I and M552V) mutant HBV DNA polymerases remain sensitive to adefovir.[27] Thus, adefovir may be beneficial in combination with lamivudine or as monotherapy for the treatment of chronic HBV infection, including those patients who have failed lamivudine monotherapy.

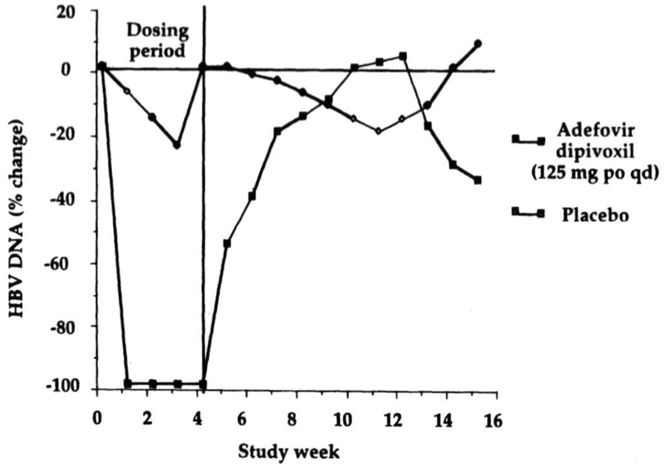

Figure 4. Phase I/II, double-blind, placebo-controlled study of 125 mg qd oral adefovir dipivoxil given for 4 weeks in patients with chronic hepatitis B infection. There was a rapid fall in HBV DNA in all treated patients (mean reduction of 1.8 log). HBV DNA returned to baseline levels in all patients after discontinuation of the 4 week treatment. Extracted from RJC Gilson *et al.* Second International Conference on Therapies for Viral Hepatitis. Hawaii, Dec. 1997.

2.3. Immunotherapy

Theradigm-HBV™ is a therapeutic vaccine consisting of three components, the viral protein (HBV core antigen peptide), a T helper peptide (tetanus toxoid peptide) which enhances immunogenicity, and two palmitic acid molecules. The vaccine is capable of inducing an HBV-specific MHC class I-restricted CTL response in healthy volunteers. In a phase I dosage escalation study, the vaccine was well-tolerated and response rates, defined as detectable cytotoxic T-lymphocyte activity, were dose-related.[28] Phase I/II studies in patients with chronic HBV infection have demonstrated that the peptide can induce a specific cytotoxic T cell response against HBV, although the antiviral effects appear to be minor.[29]

2.4. Future Directions

The effective therapy against HBV infection will likely include combination agents with one or more nucleoside/nucleotide analogs[30] and immune stimulants such as interferon and/or therapeutic vaccines. Issues which must be considered when planning studies to optimize combination therapy include synergistic effects of using two drugs against HBV replication, as has been shown with lamivudine and penciclovir *in vitro*,[30] lack of cross resistance, as described in sequential famciclovir and lamivudine therapy,[31] and additive side effects of two or more agents. Evaluating how to optimize therapy with two or more of the currently available antiviral agents, will be an exciting task over the next few years.

REFERENCES

1. Beasley R, Ling C, Hwang L, *et al.* Hepatocellular carcinoma and hepatitis B virus: a prospective study of 22,707 men in Taiwan. Lancet 1981; 2:1129–33.
2. Wright T, Lau J. Clinical aspects of hepatitis B virus infection. Lancet 1993; 342:1340–5.
3. Niederau C, Heinges T, Lange S, *et al.* Long-term follow-up of the HBeAg-positive patients treated with interferon alfa for chronic hepatitis B. New England Journal of Medicine 1996; 334:1422–1427.
4. Lau J, Wright T. Molecular virology and pathogenesis of hepatitis B. Lancet 1993; 342:1335–40.
5. Perrillo R, Schiff E, Davis G, *et al.* A randomized, controlled trial of interferon alfa-2b alone and after prednisone withdrawal for the treatment of chronic hepatitis B. N Engl J Med 1990; 323:295–301.
6. Wong D, Cheung A, O'Rourke K, Naylor C, Detsky A, Heathcote J. Effect of alpha-interferon treatment in patients with hepatitis B e antigen-positive chronic hepatitis B. A meta-analysis. Annals of Internal Medicine 1993; 119(4):312–23.
7. Lok A, Lai C, Lau J. Interferon alfa therapy in patients with chronic hepatitis B virus infection. Effects on hepatitis B virus DNA in the liver. Gastroenterology 1991; 100:756–61.
8. Fattovich G, Giustina G, Realdi G, Corrocher R, Schalm S. Long term outcome of hepatitis B e antigen-positive patients with compensated cirrhosis treated with interferon alpha. Hepatology 1997; 26(5):1338–1342.
9. Perrillo RP, Tamburron C, Regenstein F, *et al.* Low-dose, titratable interferon alfa in decompensated liver disease caused by chronic infection with hepatitis B virus. Gastroenterology 1995; 109:908–916.
10. Villeneuve J, Desrochers M, Infante-Rivard L, *et al.* A long-term follow-up study of asymptomatic hepatitis B surface antigen-positive carriers in Montreal. Gastroenterology 1994; 106(1000–5.):577–80.
11. Lampertico P, Ninno E, Manzin A, *et al.* A randomized, controlled trial of a 24 month course of interferon alfa 2b in patients with chronic hepatitis B who had hepatitis B virus DNA without hepatitis B e antigen in serum. Hepatology 1997; 26(6):1621–1625.
12. Davis G. Interferon treatment of viral hepatitis in immunocompromised patients. Semin Liver Dis 1989; 9(4):267–272.
13. Wong JB, Koff RS, Tine F, Pauker SG. Cost-effectiveness of interferon-alpha 2b treatment for hepatitis B e antigen-positive hepatitis B. Ann Intern Med 1995; 122(9):664–675.
14. Dusheiko G, Roberts J. Treatment of chronic type B and C hepatitis with interferon alfa: an economic appraisal. Hepatology 1995; 22(6):1863–73.
15. Ling R, Multimer D, Ahmed M, *et al.* Selection of mutations in hepatitis B virus polymerase during therapy of transplant recipients with lamivudine. Hepatology 1996; 24:711–713.
16. Lai C, Ching C, Ting A, *et al.* Lamivudine is effective in suppressing Hepatitis B virus DNA in Chinese hepatitis B surface antigen carriers: a placebo controlled trial. Hepatology 1997; 25:241–244.
17. Dienstag JL, Perrillo RP, Schiff ER, *et al.* A preliminary trial of lamivudine for chronic hepatitis B infection. N Engl J Med 1995; 333(25):1657–61.
18. Lai C, Chien R, Leung N *et al.* A one-year trial of lamivudine for chronic hepatitis B. N Engl J Med 1998; 339(2):61–68.
19. Dienstag J, Schiff E, Gitlin N, *et al.* Extended lamivudine retreatment for chronic hepatitis B. Hepatology 1996; 24(Suppl 1):188A.
20. Bartholomew M, Jansen R, Jeffers L, *et al.* Hepatitis B virus resistance to lamivudine given for recurrent infection after orthotopic liver transplantation. Lancet 1997; 349:20–22.
21. Main J, Brown J, Howells C, *et al.* A double-blind, placebo controlled study to assess the effect of famciclovir on virus replication in patients with chronic hepatitis B virus infection. J Viral Hepat 1996; 3:211–215.
22. Trepo C, Jezek P, Atkinson G, Boon R. Efficacy of famciclovir in chronic hepatitis B: Results of a dose finding study. Hepatology 1996; 24(4):188A.
23. Kruger M, Tillmann H, Trautwein C, *et al.* Famciclovir treatment of hepatitis B virus recurrence after liver transplantation: a pilot study. Liver Transplantation and Surgery 1996; 2:253–262.
24. Bloomer J, Chan R, Sherman M, Ingraham P, DeHertogh D, *et al.* Phase I/II study of oral lobucavir in adults with chronic hepatitis B infection. Second International Conference on Therapies for Viral Hepatitis 1997:2.
25. DeHertogh D, Colonno R, Grasela D. Lobucavir and BMS-200475: two potent new agents for hepatitis B. Second International Conference of Therapies for Viral Hepatitis 1997.
26. Gilson R, Chopra K, Murray-Lyon I, *et al.* A placebo-controlled phase I/II study of Adefovir dipixovil (BIS-POM PMEA) in patients with chronic hepatitis B infection. Hepatology 1996; 24:620A.

27. Virani-Ketter N. Adefovir dipivoxil for the treatment of chronic hepatitis B. Second International Conference on Therapies for Viral Hepatitis 1997:22.
28. Vitiello A, Ishioka G, Grey H, *et al.* Development of a lipopeptide-based therapeutic vaccine to treat chronic HBV infection. J Clin Invest 1995; 95:341–9.
29. Heathcote J, McHutchison J, Benner K, *et al.* CY-1899: a therapeutic vaccine for chronic hepatitis B. Hepatology 1996; 24:628A.
30. Colledge D, Locarnini S, Shaw T. Synergistic inhibition of hepadnaviral replication by lamivudine in combination with penciclovir *in vitro*. Hepatology 1997; 26:216–225.
31. Locarnini S, Aye T, Shaw T, De Man R, *et al.* The emergence of famciclovir resistant mutations in the hepatitis B virus polymerase during therapy in patients following liver transplantation. Hepatology 1997; 26(4):368A.

2

LAMIVUDINE THERAPY OF CHRONIC HEPATITIS B

Christine M. Hunt, Nathaniel A. Brown, and Marc Rubin

Department of Infectious Diseases
Glaxo Wellcome Incorporated
5 Moore Drive
Research Triangle Park, North Carolina 27709

1. OVERVIEW OF HEPATITIS B

In the U.S., approximately 5–10% of patients with chronic liver disease and/or cirrhosis exhibit chronic hepatitis B.[1] Globally, approximately one third of the world's population has been infected with hepatitis B virus (HBV) and it is estimated that 5% of the world's population (approximately 350 million people) are chronic carriers of HBV.[2,3] Of chronic carriers, 25–40% of patients will ultimately develop cirrhosis or hepatocellular carcinoma.[1,4] The World Health Organization cites hepatitis B as the ninth leading cause of death worldwide.[5]

Approximately half of chronic hepatitis B viral (HBV) carriers exhibit active viral replication. These patients are at increased risk for progressive inflammatory changes in the liver, and 15–20% develop cirrhosis over a 5 year period.[6] As only 1–2% of patients annually will spontaneously clear hepatitis B surface antigen (HBsAg), most patients remain chronically infected with HBV.

Alpha-interferon has been well studied in the treatment of chronic hepatitis B. A meta-analysis of multiple trials which evaluated alpha-interferon therapy of 4–6 months duration for HBeAg-positive chronic hepatitis B revealed a 33% hepatitis B e antigen (HbeAg) loss in interferon-treated patients, in comparison to a 12% HBeAg loss in patients receiving placebo.[7] Following adult-acquired hepatitis B, HBeAg seroconversion, with the loss of HBeAg and appearance of HBeAb, is the harbinger of eventual HBV clearance and generally reflects a durable transition to a state of lower HBV replication. HBeAg loss appears to be associated with significant improvements in survival and clinical outcome in patients with chronic hepatitis B responding to interferon-alfa therapy.[8,9] HBsAg seroconversion may be delayed by several years following HBeAg seroconversion.[9,10] Unfortunately, alpha-interferon therapy is effective in only a minority of patients

Antiviral Chemotherapy 5, edited by Mills *et al.*
Kluwer Academic / Plenum Publishers, New York, 1999.

and is further complicated by its numerous side effects, cost, parenteral use,[11] and decreased efficacy in patients with vertically-acquired HBV infection (the largest patient subgroup globally).[12] Therefore, other antiviral agents and immunomodulators are being investigated for patients with chronic hepatitis B.

2. NUCLEOSIDE THERAPY OF CHRONIC HEPATITIS B

Although the hepatitis B virus (HBV) is a DNA virus, it replicates via an RNA intermediate, requiring the activity of a reverse transcriptase (i.e. RNA-dependent DNA polymerase).[13] Nucleosides which inhibit the activity of the HBV polymerase/reverse-transcriptase have been examined as antiviral agents for the treatment of chronic hepatitis B. Other than lamivudine, nucleoside analogues that have been investigated in hepatitis B patients include vidarabine (adenine arabinoside), ganciclovir, acyclovir, several HIV-directed nucleosides (e.g. zidovudine, zalcitabine, and didanosine), famciclovir, adefovir, lobucavir, and fialuridine (FIAU). Most of these agents have been found to be clinically inactive or to have relatively weak antiviral activity in hepatitis B patients, while several continue to be evaluated in clinical trials. Vidarabine and FIAU both appeared to have clinically relevant antiviral activity in HBV-infected patients, but were associated with significant adverse effects (e.g. anemia and neurotoxicity with vidarabine, and a fatal lactic acidosis and multi-organ failure syndrome with FIAU, putatively mediated by mitochondrial toxicity).[14,15]

The pre-clinical toxicologic evaluation of lamivudine predicted a good clinical safety profile, and a low probability of mitochondrial toxicity in view of its lack of a 3'-hydroxyl group, which prevents incorporation into mitochondrial DNA. Lamivudine's lack of mitochondrial toxicity has since been established in a number of preclinical and clinical studies.[19–30] Lamivudine exhibited potent inhibitory activity against HBV replication *in vitro* ($IC_{50} \approx 4$–20 micromoles in HepG2 cells), and was highly active in the duck hepadnavirus model as well as in HBV-infected chimpanzees.

As lamivudine demonstrated a favorable safety profile in large HIV trials, its efficacy in HBV was assessed in Phase II dose-ranging trials in immunocompetent HIV negative patients with chronic hepatitis B. Retrospective observations in a cohort of patients coinfected with HIV and HBV had suggested efficacy in these patients.[19] In patients with advanced HIV infection and high HBV replication, lamivudine 300 to 600 mg daily had reduced HBV DNA concentrations to less than 5 pg/ml in 26/27 (96%) of those treated at 12 months. Despite the immunocompromised status of these patients (baseline mean CD4 lymphocyte count = 113± 88 cells/ml), 3 of 27 patients with high HBV replication at baseline underwent HBeAg seroconversion during lamivudine therapy. Mean alanine aminotransferase (ALT) levels also declined after 10–12 months of therapy. However, within 4 months of lamivudine cessation, ALT values returned to pretreatment levels. Lamivudine was well tolerated in this study.

3. SHORT-TERM DOSE-RANGING STUDIES OF LAMIVUDINE THERAPY IN CHRONIC HEPATITIS B (1–6 MONTHS)

Phase II dose-ranging studies of lamivudine therapy in chronic hepatitis B patients were designed to meet four objectives: to determine the safety and toxicity of lamivudine when administered orally for one to six months in diverse populations, to examine the

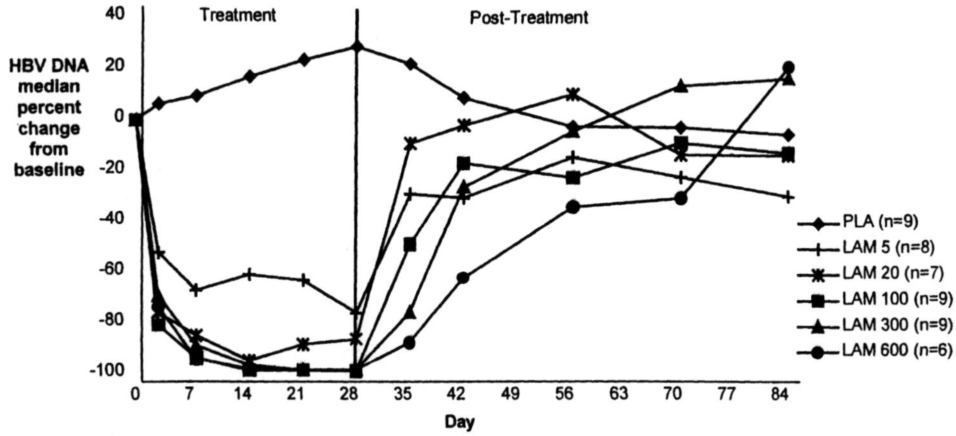

Figure 1. Four week dose-ranging study of lamivudine therapy in HBV. The median percentage change from baseline HBV DNA is portrayed during 28 day treatment and 56 day follow-up of patients with chronic hepatitis B receiving placebo (PLA), lamivudine 5 mg/day (LAM 5), lamivudine 20 mg/day (LAM 20), lamivudine 100 mg/day (LAM 100), lamivudine 300 mg/day (LAM 300), and lamivudine 600 mg/day (LAM600).

pharmacokinetic profile of lamivudine at differing doses and in combination with alpha-interferon, to conduct a preliminary evaluation of the clinical antiviral effects of lamivudine in patients with chronic hepatitis B, and to determine the optimal dose of lamivudine for Phase III controlled trials. A decrease in serum HBV DNA was the primary measurement of efficacy in the Phase II trials; secondary efficacy measurements included improvement in baseline ALT values and HBeAg status.

Lamivudine doses of 5–600mg orally daily were administered to patients with chronic hepatitis B in multinational Phase II studies.[20–22] All doses of lamivudine dramatically reduced serum HBV DNA values during treatment, compared with placebo. Lamivudine doses as low as 5 mg daily decreased median HBV DNA by 70%. Reduction in HBV DNA levels were dose-proportional in the 5mg-100mg dosing ranges, with little additional antiviral effect at doses above 100mg/day (Figure 1).

Serum HBV DNA values returned toward baseline values during the follow-up period. There was no change in HBeAg status during this short treatment regimen. The majority of patients in all treatment groups entered the study with elevated ALT values. Few ALT normalizations were seen in either the treatment or follow-up period and none of the patients experienced a sustained ALT normalization. Lamivudine was generally well tolerated throughout the study and did not exhibit dose-related toxicity; similar adverse event and laboratory profiles were noted among the varying dosing cohorts.

To explore the efficacy of lamivudine in patient populations with predominantly vertically transmitted chronic hepatitis B, lamivudine was administered for 4 weeks to 36 Chinese HbsAg-positive carriers in dosages of 25, 100, or 300 mg daily.[20] Patients receiving lamivudine 25 mg daily exhibited a 94% suppression of HBV DNA at Week 4. More potent HBV DNA suppression was evident with the use of lamivudine 100 or 300 mg daily. No substantive change in ALT was noted during treatment; however, baseline ALT was low (< 2 fold ULN). Lamivudine was well tolerated in this Asian population.

Lamivudine therapy was then examined in predominantly Western patients with presumed horizontally transmitted hepatitis B.[21] In a randomized double-blind, multicenter, dose-ranging U.S. trial, 32 patients with chronic hepatitis B (of whom 17 were prior inter-

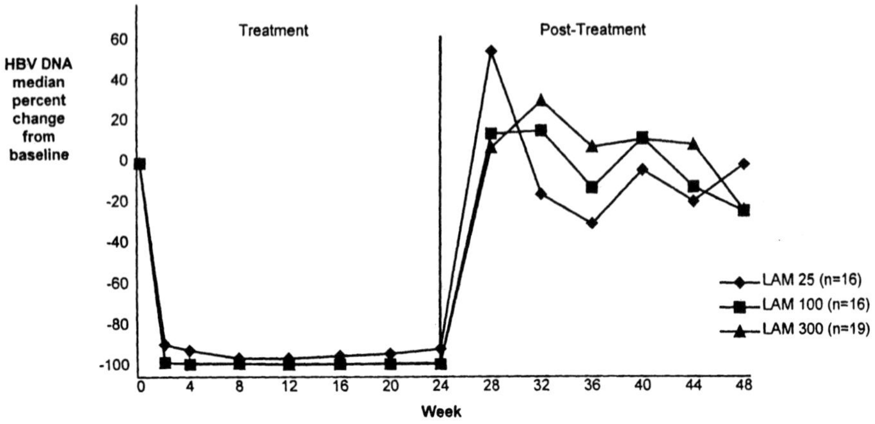

Figure 2. Twenty-four week dose-ranging study of lamivudine therapy in HBV. The median percentage change from baseline HBV DNA is portrayed during 24 week treatment and 24 week follow-up of patients with chronic hepatitis B receiving lamivudine 25 mg/day (LAM 25), lamivudine 100 mg/day (LAM 100), and lamivudine 300 mg/day (LAM 300).

feron nonresponders) received lamivudine doses of 25, 100, or 300 mg daily for 3 months. Lamivudine produced marked, rapid reductions in serum HBV DNA levels in all dosing cohorts which lasted until the end of treatment, with patients receiving lamivudine at doses ≥100mg/day exhibiting the greatest HBV DNA suppression. In addition, a greater proportion of patients in the LAM (100mg) group experienced undetectable (<1.6pg/mL) HBV DNA levels during the 6 month follow-up period, compared with the other treatment groups. Post-treatment, HBV DNA levels returned toward pretreatment levels.

Serum ALT and AST concentrations did not change appreciably during treatment in the majority of patients, although a minority of patients exhibited 2–3 fold elevations of serum ALT. During the trial, 6/32 patients (19%), of whom 5 of 6 were prior interferon nonresponders, exhibited sustained suppression of HBV DNA; 4 of these 6 patients also exhibited HBeAg loss. The lamivudine antiviral effect did not vary appreciably with previous treatment with interferon. Lamivudine was generally well tolerated throughout the study, with similar adverse event and clinical laboratory safety profiles noted among the three treatment groups.

To examine efficacy of chronic hepatitis B over longer treatment periods, lamivudine 25, 100 and 300 mg daily was administered for 6 months.[22] Median HBV DNA was closely followed during lamivudine treatment; by Week 2, median HBV DNA was ≥ 90% reduced from pretreatment values in patients receiving lamivudine 25 mg daily and 94–99% in patients receiving lamivudine 100 mg daily (Figure 2). HBV DNA suppression was maintained throughout the six month treatment period.

Median serum ALT and AST also decreased during lamivudine treatment in all dosing cohorts with a trend to normalization; by the end of treatment, 70% of patients who began treatment with chronically elevated ALT levels had achieved ALT normalization. A significant difference was detected between the treatment groups for both the normalized AUC of serum HBV DNA adjusted for baseline serum HBV DNA concentration (p=0.017), and time to suppression of serum HBV DNA concentration to below detection by the solution hybridization assay (p=0.022). The greatest difference was seen in the comparison of the lamivudine 100mg to the 25mg cohorts, with a plateau of antiviral activity observed at lamivudine 100mg and 300mg/day. Lamivudine was well tolerated with no increase in the incidence or nature of adverse events with increasing dose.

Post-treatment, serum concentrations of HBV DNA returned toward pre-treatment levels, with a more gradual return of serum HBeAg and HBsAg levels and elevated aminotransferase (ALT/AST) in almost all patients. Four patients (8%) underwent HBeAg seroconversion during the study, with two events occuring during treatment and two observed during the six month post-treatment observation period.

A subsequent study evaluated the effect of lamivudine on the safety and tolerability of alpha-interferon and potential interactions of lamivudine with alpha-interferon pharmacokinetics. This double-blind, placebo-controlled (for one month), randomized trial was conducted in 20 patients with chronic hepatitis B. Patients were randomized to receive either lamivudine 100mg daily for 16 weeks or placebo daily for 4 weeks followed by lamivudine 100mg daily for 12 weeks in combination with a 16-week course of 10MU alpha-interferon administered subcutaneously three times weekly. Treatment efficacy was evaluated by analysis of change of serum HBV DNA and HBeAg status.

Serum HBV DNA values were dramatically reduced during treatment, similar to results with lamivudine monotherapy. In both treatment groups, quantitative serum HBeAg declined. One patient underwent HBeAg seroconversion following treatment. Co-administration of lamivudine did not alter the pharmacokinetics of alpha-interferon.

The combination of lamivudine and alpha-interferon appeared safe; the safety profile was consistent with published reports with alpha-interferon monotherapy and without unexpected toxicity.

4. EXTENDED DURATION LAMIVUDINE THERAPY IN CHRONIC HEPATITIS B (12–18 MONTHS)

Persistent HBV DNA suppression and sustained improvements in ALT levels with prolonged lamivudine therapy would be predicted to be associated with improved liver histology. Histologic and clinical improvement has been demonstrated with HBV DNA suppression and HBeAg seroconversion following alpha-interferon therapy.[8,9,31,32] With alpha-interferon, only a minority of patients experience sustained HBV DNA suppression and HBeAg seroconversion,[7–9,31,32] while the majority of patients treated with lamivudine exhibit marked suppression of HBV DNA and ALT normalization;[22–27] hence, a greater proportion of lamivudine-treated patients are likely to experience histologic improvement, compared to results achievable with interferon therapy. Overall, Phase II results suggested that long-term treatment with lamivudine was likely to produce significant clinical benefits (ALT normalization, histologic improvements), and it was also apparent that large-scale controlled studies utilizing longer dosing regimens would be needed to definitively assess the potential HBeAg-seroconverting effects of lamivudine therapy.

As a prelude to large-scale Phase III trials, a U.S. multicenter, open-label trial was conducted to assess the efficacy and safety of extended lamivudine treatment 100 mg daily (up to 18 months) in patients with chronic hepatitis B who had previously participated in Phase II dose-ranging trials;[26] the final results of this trial will be forthcoming this year.

A large, double-blind, placebo-controlled trial examining the use of lamivudine for 12 months was performed in Southeast Asia (Hong Kong/Singapore/Taiwan) to determine the dose, safety, and efficacy of prolonged lamivudine therapy in patients with predominantly vertically transmitted chronic hepatitis B.[27] In this study, 358 Asian patients were

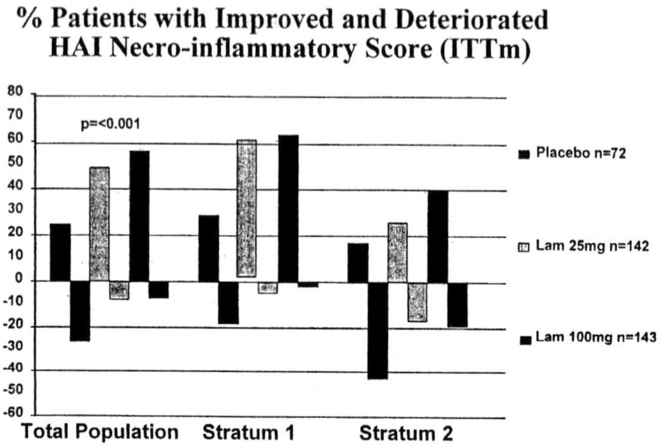

Figure 3. Liver histologic response after 52 weeks of lamivudine therapy in HBV. The percent improvement or deterioration of necroinflammatory histologic activity index (HAI) score (33) is quantified for the modified intent to treat (MITT) population, Stratum 1 subpopulation (patients with moderate to severe necroinflammatory activity on hepatic histologic analysis), and Stratum 2 subpopulation (patients with mild necroinflammatory activity on hepatic histologic analysis) by comparing liver histology at baseline and following 52 week therapy with placebo, lamivudine 25 mg/day (LAM 25), or lamivudine 100 mg/day (LAM 100).

randomized to receive one of three treatments: placebo, lamivudine 25mg or lamivudine 100mg for 12 months. Before randomization, patients were assigned to one of two strata according to the necroinflammatory activity of a liver biopsy; Stratum 1 patients exhibited moderate/severe hepatitis on histologic evaluation, while Stratum 2 patients exhibited mild hepatitis. The primary efficacy endpoint was improvement in liver histology after one year (52 weeks) of treatment.

Pre-treatment and Week 52 liver biopsy slides were assessed in a blinded fashion by a central histopathologist, by use of the Knodell Histologic Activity Index (HAI).[33] The primary efficacy parameter was reduction in the sum of the first three components of the Knodell HAI score, i.e., the so-called "necroinflammatory" components. Reduction in the Knodell HAI necroinflammatory score of at least 2 points occurred in 49 and 56% of patients treated with lamivudine 25mg and 100mg, respectively, compared to 25% of patients receiving placebo ($p<0.05$)(Figure 3). When patients with moderate to severe hepatitis (Stratum 1) were examined, more striking improvements in Knodell HAI necroinflammatory score were noted, with a 2 point reduction noted in 59 and 63% of patients treated with lamivudine 25mg and 100mg, respectively, compared to 29% of patients receiving placebo. Patients with mild hepatitis (Stratum 2) exhibited less notable improvements in necroinflammatory score with lamivudine treatment (Figure 3).

When patients with missing biopsies and those with a baseline HAI score <2 were excluded, reductions in Knodell HAI necro-inflammatory activity in Stratum 1 patients of at least 2 points occurred in 70–75% of patients treated with lamivudine 25mg and 100mg, respectively, compared to 36% of patients receiving placebo. The reductions in HAI necroinflammatory activity were confirmed in a separate, blinded "ranked" analysis, which revealed that lamivudine 100mg significantly prevented progression of hepatic fibrosis compared to placebo (3% v 15% ; $p=0.009$).

Ninety-six percent of patients receiving lamivudine 100mg became HBV DNA negative during the Phase III Asian trial compared with 23% of placebo patients. Importantly, HBeAg seroconversion was significantly more common in patients receiving lamivudine 100mg/day (16%) compared to patients receiving placebo (4%) ($p<0.04$).

Of patients with elevated ALT at baseline, 72% of those treated with lamivudine 100mg/day achieved ALT normalization and maintained their ALT values within the normal range during treatment compared to 24% of those receiving placebo. The overall incidence of clinical adverse events and laboratory abnormalities was comparable in patients receiving either lamivudine or placebo. In summary, lamivudine treatment for one year was well tolerated in doses of 100mg and 25mg daily resulting in significant histologic improvement in the majority of patients and a significant enhancement of HBeAg seroconversion and ALT normalization, compared to placebo treatment.

5. LAMIVUDINE TREATMENT IN RECURRENT HEPATITIS B FOLLOWING LIVER TRANSPLANTATION

Lamivudine has also been used in open clinical trials involving liver transplant patients, in which treatment is used for pre- and post-transplant suppression of HBV replication,[23,24] or for treatment of post-transplant recurrences of hepatitis B (OLT).[25,34] Grellier *et al* reported that 10 of 17 European patients with chronic HBV treated with lamivudine peri-transplantation survived greater than 1 month.[23]

A North American study sought to evaluate the use of prolonged lamivudine treatment for pre- and post-transplant HBV suppression in patients with HBV-related cirrhosis.[24] Thirty-seven of 41 post-transplant patients received lamivudine for a median duration of 1 month pre-transplantation and greater than 6 months post-transplantation. Of 34 patients with data at 6 months post-transplantation, no patients exhibited detectable HBV DNA (by solution hybridization assay), 5/31 (16%) exhibited HBsAg, 2/32 (6%) exhibited HbeAg, and 4/31 (13%) exhibited HBsAb. Fourteen patients withdrew, of whom 11 had adverse events (none of these were attributed to lamivudine therapy). Prolonged lamivudine therapy was well tolerated in this patient population.

An additional North American study enrolled 64 patients with established post-transplant recurrences of hepatitis B, a clinical entity which often has a severe or fatal course and for which there is currently no available treatment.[25] Of 44 patients with data after 12 months of lamivudine therapy, mean ALT was 55 IU/l and mean bilirubin was 1.0 mg/dl; 11/44 (25%) patients exhibited detectable HBV DNA (by solution hybridization method). Eleven patients have withdrawn, of whom 8 had adverse events (one of these was attributed to lamivudine therapy). In patients treated for 12 months, serum HBV DNA elevations in concert with YMDD mutations (or lamivudine genotypic resistance) were noted in 14 patients; 7/14 (50%) patients exhibited ALT elevations at the onset of HBV DNA elevations. Five of these had progression of liver disease.

6. YMDD MUTANT HBV

HBV replication requires the activity of a reverse transcriptase (i.e. RNA-dependent DNA polymerase). This polymerase replicates with a fidelity intermediate between that of a retroviral reverse transcriptase and the DNA-dependent DNA polymerase of a DNA virus.[13, 35] Upon exposure to prolonged lamivudine therapy, point mutations in the HBV polymerase gene result in amino acid changes in the YMDD motif, which decrease the sensitivity of the HBV polymerase to the chain terminating activity of lamivudine triphosphate *in vitro.*[26] Similiar mutations have been described for the YMDD motif of the HIV reverse transcriptase (at codon 184), resulting in diminished efficacy of lamivudine therapy.[27]

To date, the development of YMDD mutants during lamivudine treatment of hepatitis B patients has been most closely assessed in a large Asian trial examining lamivudine therapy in immunocompetent patients with chronic hepatitis B.[28] In this trial, the incidence of detectable YMDD-mutant HBV was 4% following 36 weeks of lamivudine therapy and increased to 15% at 52 weeks of therapy.[28]

To further elucidate the clinical relevance of YMDD mutant HBV, correlation and regression analyses of HBV genotypic data from this trial were performed. Despite the development of YMDD mutant HBV, patients receiving lamivudine therapy exhibited significantly improved total HAI scores and sustained ALT response in comparison to those receiving placebo. In addition, despite the development of YMDD mutants, lamivudine-treated patients exhibited no significant differences in ALT or bilirubin at Week 52 when compared to lamivudine-treated patients without YMDD-mutant HBV. However, upon development of YMDD mutants, median HBV DNA values rose to a value intermediate between that seen in lamivudine-treated and placebo-treated patients with non-mutant HBV at Week 52. In addition, three of six patients harboring YMDD mutant HBV exhibited ALT elevations within 8 weeks of developing YMDD mutant HBV.

Correlation analyses have identified baseline HAI score (total or necro-inflammatory), baseline ALT, stratum, body mass index, gender, weight, age, and cirrhosis to be significantly ($p<0.08$) correlated with the presence of YMDD mutants. Regression analyses (along with correlation analysis and cluster analysis) have identified three factors that appear to be pre-treatment factors associated with an increased probability of developing YMDD mutants: (a) disease severity, represented by a high baseline total histologic activity index (HAI) score (other variables related to this factor are baseline ALT, stratum, presence/absence of cirrhosis, and age); (b) high body mass index (a factor based on weight and height); and (c) high baseline HBV DNA. High baseline total HAI is a significant ($p=0.005$) factor associated with YMDD mutants, while body mass index and baseline DNA are marginally significant ($p=0.06$ and 0.09, respectively) factors associated with the development of YMDD-mutant HBV.

7. YMDD MUTANT HBV ARISING POST-TRANSPLANTATION

Accruing data suggest that the YMDD mutant HBV is less replication competent than the wild type HBV DNA.[36,37] Data from a case series of three HBV patients treated with lamivudine post-transplantation revealed the development of YMDD mutants at 9–10 months; all three patients remained clinically asymptomatic with mild hepatitis for 11 months following the detection of the mutants.[38] Despite the development of YMDD mutant HBV, lamivudine was reported to have improved the clinical course of HBV reinfection post-transplantation in four patients who were followed for 6–16 months after the development of YMDD mutant HBV.[39] Two of the four patients received famciclovir in combination with lamivudine treatment for marked elevations of HBV DNA (exceeding 2500 pg/mL); HBV DNA subsequently decreased 50–70%. No patient experienced an episode of allograft rejection or a diminution in quality of life during the 6–16 month follow-up. Similarly, Ma *et al* describe two patients harbouring YMDD mutant HBV who have evidence of HBV reinfection on liver biopsy, but remain clinically well. These investigators suggested that the disease course of patients with HBV reinfection with YMDD mutant virus post-transplantation may be less 'aggressive'.[40] These findings are suggestive that continuation of lamivudine in patients with YMDD mutants may result in continued suppression of native or "wild type" HBV and amelioration of hepatic injury compared to that which would be seen in untreated patients.

8. CONCLUSIONS

In summary, multiple Phase II and Phase III international clinical trials examining the use of lamivudine in patients with chronic hepatitis B have been completed or are nearing completion. In the first completed Phase III trial, lamivudine therapy in patients with HBeAg-positive chronic hepatitis B resulted in significant histologic improvement, decrease in inflammation and retardation of fibrosis, and in enhanced rate of HBeAg seroconversion, compared to placebo treatment. In all studies, lamivudine profoundly suppressed HBV DNA replication, and resulted in sustained ALT normalization in the majority of patients receiving treatment for 6–12 months or greater.

Prolonged lamivudine therapy has been associated with the development of YMDD mutant HBV in a minority of patients. The majority of patients developping YMDD mutants remain clinically stable, with serum ALT and HBV DNA levels persisting below pretreatment values. The findings of various clinical trials, to date, suggest that continuation of lamivudine in patients harboring YMDD-mutant HBV results in continued benefit, probably due to suppression of residual "wild type" HBV infection.

ACKNOWLEDGMENTS

The authors wish to gratefully acknowledge the collaborative efforts of their coworkers at Glaxo Wellcome Inc, (especially those of Lynn Crowther, Fraser Gray, Mark Atkins, Lynn Condreay, Mary Woessner, Gaya Anschuetz, Lilly Sanathanan, Judy Barber, and Randy Davis) as well as a large group of international clinical investigators, without whom these studies would not have been possible.

REFERENCES

1. Hoofnagle JH, Di Bisceglie AM. The treatment of chronic viral hepatitis. N Engl J Med 1997; 336:347–56.
2. Maynard JE. Hepatitis B: global importance and need for control. Vaccine 1990; 8(Suppl):S18–20.
3. Zuckerman AJ. Progress towards the comprehensive control of hepatitis B - introduction. Gut 1996; 38(Suppl 2):S1.
4. Beasley RP, Ling CC, Hwang LY, *et al.* Hepatocellular carcinoma and hepatitis B virus: a prospective study of 22,707 men in Taiwan. Lancet 1981; 2:1129–33.
5. Hoofnagle JH. Chronic hepatitis B. N Engl J Med 1990; 323:337–9.
6. Liaw YF, Tai DI, Chu CM, Chen TJ. The development of cirrhosis in patients with chronic type B hepatitis: a prospective study. Hepatology 1988; 8:493–6.
7. Wong DKH, Cheung AM, O'Rourke K, *et al.* Effect of alpha-interferon treatment in patients with hepatitis B e antigen-positive chronic hepatitis B: a meta-analysis. Ann Intern Med 1993; 119:312–23.
8. Niederau C, Heintges T, Lange S, *et al.* Long-term follow-up of HBeAg-positive patients treated with interferon alfa for chronic hepatitis B. N Engl J Med 1996; 334:1422–1427.
9. Lau DTY, Everhart J, Kleiner DE, *et al.* Long-term follow-up of patients with chronic hepatitis B treated with interferon alfa. Gastroenterol 1997; 113:1660–1667.
10. Korenman J, Baker B, Waggoner J, *et al.* Long-term remission of chronic hepatitis B after alpha-interferon therapy. Ann Intern Med 1991; 114:629–634.

11. Renault PF, Hoofnagle JH. Side Effects of Alpha Interferon. Seminars in Liver Disease 1989; 9(4):273–277.
12. Lok ASF, Wu PC, Lai CL, *et al.* Long-term follow-up of chronic hepatitis B patients treated with interferon alfa. Gastroenterology 1993; 105:1833–1838.
13. Blum HE. Variants of hepatitis B, C, and D viruses: molecular biology and clinical significance. Digestion 1995; 56:85–95.
14. McKenzie R, Fried MW, Sallie R *et al.* Hepatic failure and lactic acidosis due to fialuridine (FIAU), an investigational nucleoside analogue for chronic hepatitis B. N Engl J Med 1995; 333:1099–1105.
15. Cui L. Yoon S. Schinazi RF. Sommadossi JP. Cellular and molecular events leading to mitochondrial toxicity of 1-(2-deoxy-2-fluoro-1-beta-D-arabinofuranosyl)-5-iodouracil in human liver cells. Journal of Clinical Investigation. 1995; 95:555–63.
16. Dalakas MC, Illa I, Pezeshkpour GH, *et al.* Mitochondrial myopathy caused by long-term zidovudine therapy. N Engl J Med 1990; 322:1098–1105.
17. Chen CH, Cheng YC. Delayed cytotoxicity and selective loss of mitochondrial DNA in cells treated with the anti-human immunodeficiency virus compound 2',3'-dideoxycytidine. J Biol Chem 1989; 264:11934–11937.
18. Faulds D, Brogden RN. Didanosine-a review of its antiviral activity, pharmacokinetic properties and therapeutic potential in human immunodeficiency virus infection. Drugs 1992; 44:94–116.
19. Benhamou Y, Katlama C, Lunel F, *et al.* Effects of lamivudine on replication of hepatitis B virus in HIV-infected men. Ann Intern Med 1996; 125:705–12.
20. Lai CL, Ching CK, Tung AKM, *et al.* Lamivudine is effective in suppressing hepatitis B virus DNA in Chinese hepatitis B surface antigen carriers: a placebo-controlled trial. Hepatol 1997; 25:241–44.
21. Dienstag JL, Perrillo RP, Schiff ER, *et al.* A preliminary trial of lamivudine for chronic hepatitis B infection. N Engl J Med 1995; 333:1657–61.
22. Nevens F, Main J, Honkoop P, *et al.* Lamivudine therapy for chronic hepatitis B: a six months randomized dose ranging study. Gastroenterology 1997; 113:1258–1263.
23. Grellier L, Mutimer D, Ahmed M, *et al.* Lamivudine prophylaxis against reinfection in liver transplantation for hepatitis B cirrhosis. Lancet 1996; 348:1212–5.
24. Perrillo R, Rakela , Wright T, *et al.* Lamivudine for suppression and/or prevention of hepatitis B given pre/post liver transplantation. Hepatol 1997; 26:260A.
25. Perrillo R, Rakela J, Martin P, *et al.* Long term lamivudine therapy of patients with recurrent hepatitis B post liver transplantation. Hepatol 1997; 26:177A.
26. Dienstag JL, Schiff ER, Mitchell M, *et al.* Extended lamivudine retreatment for chronic hepatitis B. Hepatology 1996; 24:188A (Abstract 245).
27. Leung NWY, Lai CL, Liaw YF, *et al.* Lamivudine (100 mg od) for 1 year significantly improves necro-inflammatory activity and reduces progression in fibrosis stage: Results of a placebo-controlled multicentre study in Asia of lamivudine for chronic hepatitis B infection. Hepatology 1997; 26:357A (Abstract 915).
28. Lai CL, Liaw YF, Leung NWY, *et al.* Genotypic resistance to lamivudine in a prospective, placebo-controlled multicentre study in Asia of lamivudine therapy for chronic hepatitis B infection: Incidence, kinetics of emergence, and correlation with disease parameters. Hepatology 1997; 26:259A (Abstract 522).
29. Swartz MN. Mitochondrial toxicity - new adverse drug effects. N Engl J Med 1995; 333: 1146–8.
30. Honkoop P, De Man RA, Scholte HR, *et al.* Effect of lamivudine on morphology and function of mitochondria in patients with chronic hepatitis B. Hepatol 1997; 26:211–215.
31. Perrillo RP, Schiff ER, Davis GL, *et al.* A randomized, controlled trial of interferon alfa-2b alone and after prednisone withdrawal for the treatment of chronic hepatitis B. New England Journal of Medicine 1990; 323:295–301.
32. Wong JB, Koff RS, Tine F, Pauker SG. Cost-effectiveness of interferon-alpha 2b treatment for hepatitis B e antigen-positive chronic hepatitis B. Ann Intern Med 1995; 122:664–75.
33. Knodell RG, Ishak KG, Black WC, *et al.* Formulation and application of a numerical scoring system for assessing histological activity in asymptomatic chronic active hepatitis. Hepatology 1981; 1:431–5.
34. Bartholomew MM, Jansen RW, Jeffers LJ, *et al.* Hepatitis-B-virus resistance to lamivudine given for recurrent infection after orthotopic liver transplantation. Lancet 1997; 349:20–2.
35. Gunther S, Sommer G, Plikat U, *et al.* Naturally occurring hepatitis B virus genomes bearing the hallmarks of retroviral G to A hypermutation. Virol 1997; 235:104–8.
36. Melegari M, Scaglioni PP, Wands JR. Hepatitis B virus mutants associated with 3TC and famciclovir administration are replication defective. Hepatology 1998; 27:628–633.
37. Allen, MI, Deslauriers M, Andrews CW, *et al.* Identification and characterization of mutations in hepatitis B virus resistant to lamivudine. Hepatol. 1998; 27:1670–77.

38. Ben-Ari Z, Scmeuli D, Mor E, *et al.* Beneficial effects of lamivudine in recurrent hepatitis B after liver transplantation. Transplantation 1997; 63:393–396.
39. Rodriguez MJ, Martinez EJ, Magill A, *et al.* Clinical management of hepatitis B recurrence after liver transplantation with lamivudine resistant strains. Hepatology 1997; 26:590A.
40. Ma MM, Tipples GA, Fischer KP, *et al.* Lamivudine resistant HBV infection of hepatic allografts: One year follow-up. Can J Gastroenterology 1997; 11:68A.

3

FAMCICLOVIR UPDATE

Chronic Hepatitis B

Clarence L. Young

SmithKline Beecham Pharmaceuticals
1250 South Collegeville Road
Collegeville, Pennsylvania 19426-0989

Famciclovir is the well-absorbed (77% bioavailability) oral formulation of penciclovir. It was first licensed in 1994[1] and since then has been approved in a number of countries, including the United States, for the treatment of acute herpes zoster, acute recurrent and primary genital herpes, and suppression of genital herpes recurrences. In addition, *Denavir*, a topical formulation of penciclovir, has been approved for treatment of herpes labialis cold sores. Famciclovir and the intravenous formulation, penciclovir, have recently been granted regulatory approval in Europe for treatment of herpesvirus (i.e. varicella zoster virus, herpes simplex virus) infections in immunocompromised hosts.

1. PENCICLOVIR ACTIVITY AGAINST HEPATITIS B VIRUS (HBV)

During the famciclovir clinical development program for the treatment of herpesvirus infections, it was observed that penciclovir inhibited hepatitis B virus (HBV) in *in vitro* cell culture systems.[2] This observation gave rise to a clinical development program which today consists of studies in immunocompetent patients with chronic hepatitis B, as well as programs in liver transplant recipients for the prevention or treatment of HBV recurrences.

The initial observation of penciclovir activity in an *in vivo* model of hepatitis B was made by Tsiquaye *et al.*, (1994)[3]. These workers observed complete suppression of duck hepatitis B virus (DHBV) replication in Pekin ducks treated with penciclovir for 18 days. When penciclovir therapy was discontinued, replication of duck HBV resumed within three days.[4]

Studies *in vitro*, as well as *in vivo* studies in animal models of chronic hepatitis B, have helped to elucidate the mechanism of action and activity of penciclovir in the treat-

Antiviral Chemotherapy 5, edited by Mills *et al.*
Kluwer Academic / Plenum Publishers, New York, 1999.

ment of HBV infection. Penciclovir inhibits HBV replication by inhibition of the chain elongation step mediated by the HBV DNA polymerase.[2] Penciclovir also inhibits the priming step in which single-stranded DNA synthesis is covalently linked to the DNA polymerase.[5] In contrast to its anti-herpesvirus activity, the R-enantiomer of penciclovir triphosphate is a more efficient inhibitor of HBV polymerase than the S-enantiomer.[6] Once taken up by hepatitis B-infected cells, penciclovir is phosphorylated by (as yet undefined) host enzymes which are thought to be localised in the cytoplasm.[6] Deoxycytidine kinase is not thought to play an important role in the phosphorylation of penciclovir in these cells (A. Fridland, personal communication).

Phosphorylation of penciclovir in cells of hepatic origin does not require the presence of HBV infection. Following *in vitro* exposure to the same concentration of penciclovir, intracellular levels of penciclovir triphosphate were nearly identical in 2.2.15 cells expressing hepatitis B virions and in uninfected HepG2 cells of liver cell origin.[6] The low peak levels of penciclovir triphosphate measured in these systems, approximately 0.4 μM, is far in excess of the penciclovir triphosphate concentrations required for inhibition of specific cellular polymerases. The half life of penciclovir triphosphate within cells of liver origin has been measured as approximately 18 hours.[6]

Further studies in the Pekin duck have demonstrated inhibition of the production of hepatitis B pre-S1 and core antigens, as well as significant inhibition of covalently closed circular (CCC) DNA, the replicative intermediate responsible for maintenance of HBV replication within infected cells.[7] Alterations in mitochondrial morphology have not been observed in Pekin ducks treated with either penciclovir or famciclovir (S. Locarnini, unpublished data). Work is in progress to further elucidate the mechanism of activation of penciclovir within human hepatocytes, including identification of the enzyme(s) responsible for phosphorylation of penciclovir, confirmation of penciclovir triphosphate levels within the cells, determination of the half life of penciclovir triphosphate in human hepatocytes as well as determination of the ratio of R to S enantiomers of penciclovir triphosphate formed in these cells.

2. FAMCICLOVIR IN THE TREATMENT OF CHRONIC HEPATITIS B

Following the encouraging findings of a small, pilot phase IIa study in patients with chronic hepatitis B,[8] a phase IIb dose-ranging study was undertaken to define the efficacy and safety of penciclovir for the treatment of chronic hepatitis B in immunocompetent patients.[9] In this study, 333 patients who were positive for hepatitis B e antigen (HBeAg) and surface (HBsAg) antigens at least 6 months prior to study entry, were randomized to receive treatment with either famciclovir at 125 mg, 250 mg, or 500 mg or placebo three times daily for up to 16 weeks. The study was conducted in Europe and Australia. Seventy-three percent of the patients were male; 74% were Caucasian and the mean age was 36 years. Patients enrolled in this study represented a traditionally difficult-to-treat population, with relatively low baseline alanine transaminase (ALT) levels (median, 1.7 times upper limit of normal [ULN]) and relatively high baseline HBV DNA levels (median, 3900 pg/ml); 41% of patients had previously been treated with interferon-α without clearance of HBV infection. Following an 8-week screening period, patients received 16 weeks of double-blind study medication and were then followed for 8 months to assess the durability of treatment. At the end of the follow-up period, all patients were given the option to receive 16 weeks of treatment with open-label famciclovir at 250 mg three times a day. In the intent-to-treat analysis, famciclovir treatment resulted in a rapid, dose-dependent suppression of HBV DNA which was apparent after the first week of study medication and was sustained as long as therapy with famciclovir was continued. Following discontinuation of famciclovir, HBV DNA levels promptly increased, approaching baseline values.[9]

Nearly a quarter of the patients in the 500 mg tid treatment group experienced complete loss of HBV DNA, as well as a reduction in median ALT levels, compared with placebo, both at the end of the 16-week treatment period and at the end of the 8-month follow-up period. In addition, there was evidence that anti-HBe seroconversion over the study period was increased in the 500 mg tid treatment group compared with placebo ($p=0.04$).[10]

There was no evidence for any influence of age, gender, race or prior interferon treatment on normalization of ALT. Patients with normalization of ALT were found to have relatively high levels of ALT at baseline (median ALT level, 3.5 × ULN) ($p < 0.05$). A trend was also observed which suggested that patients with lower baseline HBV DNA levels were more likely to experience normalization of ALT, compared with those patients who had relatively high HBV DNA levels at the start of treatment.

Tolerance to hepatitis B infection is usually characterised by high levels of viral replication (HBV DNA) and minimal disease activity i.e. low levels of ALT.[11] Asian patients represent a population which is traditionally difficult to treat by virtue of the development of tolerance to hepatitis B virus, a result of acquisition of the infection at birth or during early childhood. A subgroup analysis of Asian patients confirmed potent activity for famciclovir in the 500 mg tid treatment group. Famciclovir was generally well tolerated with an adverse event profile comparable to placebo. There was no evidence of severe hepatic decompensation following discontinuation of famciclovir treatment at the end of 16 weeks treatment, and up to 8 months following the conclusion of famciclovir therapy.

In summary, in this study in otherwise healthy adults with chronic hepatitis B, famciclovir was found to be effective in reducing HBV DNA levels as well as levels of ALT. Famciclovir 500 mg tid resulted in a significant and sustained normalization of ALT, and significant anti-HBe seroconversion, compared with placebo. Pivotal studies are underway in North America, Europe, Asia, and Australia to evaluate famciclovir in the treatment of chronic hepatitis B in HBeAg-positive populations, as well as in patients in Southeastern Europe who are HBeAg negative and HBeAb positive.

3. FAMCICLOVIR FOR TREATMENT OF HEPATITIS B RECURRENCES IN LIVER TRANSPLANT RECIPIENTS

In 1994, Böker *et al.* reported a dramatic effect of famciclovir treatment in a liver transplant recipient who had developed fulminant hepatitis following recurrence of the hepatitis B infection.[12] Following this observation, a compassionate use program was initiated to provide famciclovir 500 mg three times a day to those patients who developed hepatitis B recurrences after liver transplantation. The dosage of famciclovir administered to patients was adjusted according to the extent of their underlying renal impairment. In addition to safety information, investigators were also requested to provide information regarding HBV replication and transaminase levels in these patients.

In the four years from 1993 to 1997, a total of 122 patients were treated. Of these, 70 patients are continuing to receive famciclovir. As of December 1997, more than 100 patients had received > 6 months therapy, 34 patients had received > 2 years of treatment and 14 patients had been treated for at least 3 years. Patients involved in this program were predominantly male with a median age of 48 years (range, 12–71 years). Half of the patients had developed hepatitis B recurrence approximately 6 months after undergoing liver transplantation, and the median time from recognition of hepatitis B recurrence until administration of the first dose of famciclovir was 42 weeks. Median HBV DNA levels at the time of starting treatment were 432 pg/ml (range, 0–18,480 pg/ml) and median ALT baseline levels were 2.9 times upper limit of normal (range, 0–94). Ten (8%) patients were infected with hepatitis C virus and nine (7%) had evidence of co-infection with delta hepatitis virus.[13]

In patients for whom baseline HBV DNA and ALT levels were available, and who had been treated for at least 6 months, famciclovir treatment resulted in an 80% decrease in HBV DNA levels 6 months after the onset of treatment. Median HBV DNA levels continued to decline by >90% at 18 months and by approximately 99% at 24 months after the onset of famciclovir treatment. ALT levels decreased by 40% in the first 6 months of treatment, with a decrease of 70% compared with baseline observed at 24 months after initiation of famciclovir treatment.[13]

Suppression of HBV DNA replication was more profound in those patients who had lower levels of HBV DNA (< 300 pg/ml) at the time of onset of famciclovir treatment, with a > 90% decrease in HBV DNA replication noted at 6 months. Patients with baseline ALT levels > 3 times upper limit of normal experienced more rapid decreases in HBV DNA levels compared to the overall study population, with a > 95% reduction in HBV DNA levels occurring within 6 months following the start of therapy. These patients were also more likely to experience a profound decrease (approximately 70%) in ALT levels compared with baseline, whereas patients whose ALT levels at baseline were < 3 times upper limit of normal experienced little or no decrease in ALT on famciclovir treatment up to 30 months. Famciclovir was generally well tolerated, with an adverse event profile that was not unexpected in this group of severely ill, immunosuppressed patients, many of whom developed complications of their liver transplant in the course of the study period. Of the 46 patients who were followed for at least six months, 17 had negative HBV DNA levels noted at the time of their last assessment.[13]

Genotypic analysis of HBV breakthroughs in lamivudine treated patients has revealed structural changes in the catalytic site of HBV DNA polymerase within the YMDD locus (methionine to valine or isoleucine at position 550). Alterations in this motif in the human immunodeficiency virus (HIV) reverse transcriptase are associated with the rapid development of lamivudine resistance by HIV.[14] Information on HBV resistance to penciclovir is currently based on case reports of liver transplant recipients who developed breakthroughs while on treatment. Preliminary findings suggest an association of breakthroughs with genotypic changes in domain B (template binding site) of the HBV DNA polymerase.[14] In contrast with lamivudine, famciclovir breakthroughs have not been associated with genotypic changes in the catalytic site of the HBV DNA polymerase at the YMDD locus. *In vitro* studies have indicated that development of mutations in domain B is associated with a 3-fold decrease in susceptibility of the HBV DNA polymerase to inhibition by penciclovir triphosphate.[14] This contrasts with the > 40-fold decrease in lamivudine susceptibility associated with methionine to valine or isoleucine changes in the YMDD locus.[15]

4. SUMMARY AND CONCLUSIONS

The growing body of preclinical data, as well as clinical data from the phase II dose-ranging study in immunocompetent patients and from liver transplant recipients participating in the compassionate use program, indicates that famciclovir offers the promise of safe and convenient oral treatment of chronic hepatitis B infection. The results of phase III pivotal studies are eagerly awaited.

Denavir is a trademark of SmithKline Beecham plc.

REFERENCES

1. Pue MA & Benet LZ Pharmacokinetics of famciclovir in man. Antiviral Chem. Chemother. 1993; 4(Suppl 1):47–55.
2. Shaw T, Amor P, Civitico G, Boyd M, Locarnini S. *In vitro* antiviral activity of penciclovir, a novel purine nucleoside, against duck hepatitis B virus. Antimicrob Agents Chemother. 1994; 38:719–723.
3. Tsiquaye KN, Slomka MJ, Maung M. Oral famciclovir against duck hepatitis B virus replication in hepatic and nonhepatic tissues of ducklings infected in ovo. J Med Virol. 1994; 42:306–310.
4. Tsiquaye KN, Sutton D, Maung M, Boyd MR. Antiviral activities and pharmacokinetics of penciclovir and famciclovir in Pekin ducks chronically infected with duck hepatitis B virus. Antiviral Chem. Chemother. 1996; 7(3):153–159.
5. Dannaoui E, Trepo C, Zoulim F. Inhibitory effect of penciclovir-triphosphate on duck hepatitis B virus reverse transcription. Antiviral Chem Chemother. 1997; 8(1):38–46.
6. Shaw T, Mok S, Locarnini S. Inhibition of hepatitis B virus DNA polymerase by enantiomers of penciclovir Triphosphate and metabolic basis for selective inhibition of HBV replication by penciclovir. Hepatology. 1996; 24(5):996–1002.
7. Lin E, Luscombe C, Wang Y, *et al.* The Guanine Nucleoside Analog Penciclovir Is Active against Chronic Duck Hepatitis B Virus Infection *In Vivo*. Antimicrobial Agents Chemother. 1996;40(2):413–418.
8. Main J, Brown J, Karayiannis P, Georgiou P, Boyd M, Prince W, Thomas H. A double-blind, placebo-controlled study to assess the effect of famciclovir on virus replication in patients with chronic hepatitis B virus infection. (Poster P1C1/38). J Hepatol. Suppl 1994; S32A.
9. Trepo C, Jezek P, Atkinson GF, Boon RJ. Efficacy of famciclovir in chronic hepatitis B: results of a dose finding study. Hepatology. 1996; 24(Suppl 4, Pt 2): 188A (Abstract 247).
10. Trepo C, Jezek P, Atkinson GF, Boon RJ. Long term efficacy of famciclovir (FCV) in chronic hepatitis B: results of a Phase IIB study. J Hepatol. 1997; 26(Suppl 1):74 (Abstract WP3/22).
11. Lok ASF. Treatment of chronic hepatitis B. J Vir Hepatitis. 1994; 1:105–124.
12. Böker KH, Ringe B, Krüger M, Pichlmayr R, Manns MP. Prostaglandin E plus famciclovir - a new concept for the treatment of severe hepatitis B after liver transplantation. Transplantation. 1994; 57:1706–1708.
13. Neuhaus P, Manns M, Atkinson G. Safety and efficacy of Famciclovir for the treatment of recurrent Hepatitis B in liver transplant recipients. Hepatology. 1997; 26(4, Pt. 2):260a.
14. Aye TT, Bartholomeusz A, Shaw T, Bowden S, Breschkin A, McMillan J, *et al.* Hepatitis B virus polymerase mutations during antiviral therapy in a patient following liver transplantation. J Hepatol. 1997; 26(5):1148–1153.
15. Bartholomew, M, Jansen R, Jeffers, L *et al.* Hepatitis-B-virus resistance to lamivudine given for recurrent infection after orthotopic liver transplantation. Lancet 1997; 349:20–22.

4

CURRENT STATUS OF ANTIVIRAL THERAPY FOR CHRONIC HEPATITIS C

Stanley M. Lemon

Department of Microbiology and Immunology
The University of Texas Medical Branch at Galveston
Galveston, Texas 77555-1019

Hepatitis C virus (HCV) was first identified in the late 1980s, and soon thereafter was shown to be the cause of almost all cases of non-B posttransfusion hepatitis.[1,2] Today, in the United States, it is by far the most common cause of chronic viral hepatitis, and is thought to be responsible for approximately 10,000 deaths annually due to chronic liver failure or hepatocellular carcinoma.[3] Rather than a single virus, HCV is actually a complex of genetically related flaviviruses with unique genome features and a strong predilection for establishing persistent hepatic infections in humans. These viruses can be classified into six major genotypes based on nucleotide sequence comparisons,[4] and have been placed taxonomically within the recently established genus *Hepacivirus* of the family *Flaviviridae*. It is likely that different HCV genotypes may also be distinguished by differences in their antigenic characteristics, but no tests have been developed for the measurement of neutralizing antibodies to the virus. There are no practical cell culture-based systems allowing propagation of the virus. Both genetically and in terms of multiple other features of the viral genome, the most closely related other member of the *Flaviviridae* is GB virus B (GBV-B),[5] an hepatotropic virus that may be of either human or simian origin.

1. MOLECULAR VIROLOGY OF HEPATITIS C VIRUS

The single-stranded, positive-sense RNA genome of HCV can be functionally divided into three major domains. These include: the 5' nontranslated region (5' NTR) which controls expression of the viral polyprotein and likely also controls initiation of positive-strand RNA synthesis; the open-reading frame (ORF) encoding the viral polyprotein; and, the 3' NTR which controls initiation of minus-strand RNA synthesis (Figure 1). The functions of the 5' NTR and 3' NTR are dependent upon well mapped, higher-ordered RNA secondary and tertiary structures.[6] It is likely that the functional (and structural) features of the 5' NTR extend into the well-conserved 5' segment of the ORF, within which

Antiviral Chemotherapy 5, edited by Mills *et al.*
Kluwer Academic / Plenum Publishers, New York, 1999.

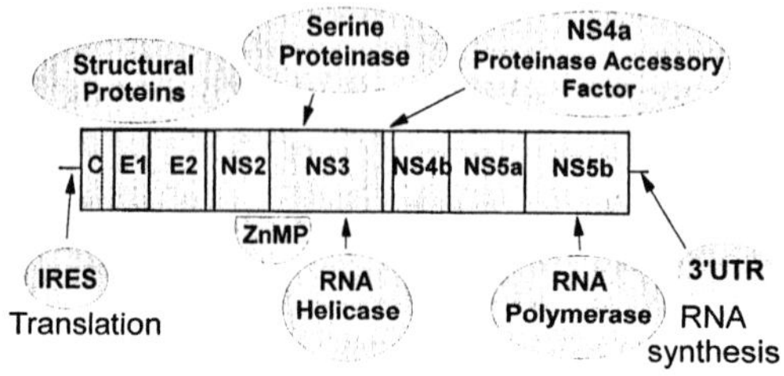

Figure 1. The genetic organization of the HCV genome. Major protein products derived from proteolytic processing of the polyprotein are indicated. E1 and E2 are secreted glycoproteins that form the envelope of the virus.

two large stem-loop structures have been defined.[7] The functional significance of these latter structures is unknown.

Translation of the polyprotein occurs by internal entry of the 40S ribosome subunit on the viral RNA.[8] Unique among all eukaryotic RNAs, the internal ribosome entry site (IRES) which is comprised of nucleotides ~42–345 of the 5' NTR binds directly to 40S ribosome subunits to form a binary complex in the absence of any eukaryotic translation initiation factor (eIF).[9] The ribosome forms a primary contact with the RNA at the site of the initiator AUG; there is no scanning.[6] Thus, hepatitis C translation initiation parallels events in prokaryotes. The eIF2/Met-tRNA/GTP ternary complex binds at the AUG, while eIF3 binds to the upstream stem-loop III of the IRES; these factors enhance the stability of the complex by extending contacts along the length of the RNA.

The polyprotein encoded by the ORF undergoes post-translational cleavages directed by both host cell and virus-encoded proteinases.[10] Signal sequences within the amino terminal third of the polyprotein direct its secretion into the endoplasmic reticulum (ER).[11,12] Several cleavages directed by host cell signalase produce a series of structural proteins which include the nucleocapsid protein, two envelope glycoproteins, E1 and E2, and a small membrane-associated protein, p7 or NS2A. The nucleocapsid protein remains within the cytoplasm, but some data suggest it may undergo further cleavage with the result that part of the molecule is transported to the nucleus. E1 and E2 are glycosylated in the ER and the Golgi, but details of the viral assembly and secretion process remain obscure. E2 contains a highly variable domain near its amino terminus (HVR-1 domain), which is likely to form an immunogenic loop on the surface of the virion and which may interact with some neutralizing antibodies.[13] As such, it might be quite analogous to the V3 loop of HIV-1.

At least 6 nonstructural proteins are derived from the remainder of the polyprotein. The functions of these are only partly understood. They include NS2B (which with the adjacent NS3 sequence functions as a *cis*-active metalloproteinase at the NS2B/NS3 cleavage site),[14] NS3 (a serine proteinase/NTPase/RNA helicase),[15–17] NS4A (serine proteinase accessory factor, which is an integral part of the mature NS3 proteinase as it forms one strand of a β-barrel structure which flanks the proteinase active site),[16] NS4B, NS5A (both of unknown function), and NS5B (RNA-dependent RNA polymerase).[18] Efforts to develop effective inhibitors of HCV replication are being based on structural and functional characterization of these proteins (and the IRES), but good *in vitro* systems have been difficult to develop for many of these viral activities.

The positive-strand genome of HCV replicates through a minus-strand intermediate. Thus, at one point in the replication cycle, it is likely that there is a duplex RNA molecule that serves as template for the production of new positive RNA strands that are packaged to make new virus. These details are not known, but almost certainly the 3' UTR figures prominently in the early events in the replication cycle. The 3' UTR can be subdivided into a variable domain, a poly(U) segment of variable length, and a highly-conserved series of stem-loops at the extreme 3' end of the genome (3'X region).[19,20] The function of the poly(U) is unknown, but somewhat similar polypyrimidine tracts are found in the 5' nontranslated RNA segments of other positive-strand RNA viruses. The RNA replicase is probably a complex of viral and perhaps host cell proteins; it is likely to assemble on the 3'X structure which has recently been shown to be essential for virus replication. These events have yet to be replicated in any *in vitro* system. However, since the virus genome is messenger-sense, it has been possible to develop cDNA clones that transcribe full-length RNAs that are infectious when injected directly into chimpanzees.[21,22] This represents a very important tool that will help unravel the details of the replication cycle and will aid enormously in dissecting the structure/function relationships of HCV proteins. Before such studies can be contemplated, however, it will be necessary to develop cell culture systems that will permit the reliable recovery of virus from synthetic infectious RNA.

2. INTERFERON TREATMENT OF CHRONIC HEPATITIS C

A considerable number of controlled clinical trials have proven that treatment with several different formulations of interferon may be beneficial in a small proportion of patients with chronic hepatitis C.[3] This beneficial response is marked by normalization of serum alanine aminotransferase (ALT) values, elimination of viral RNA, and/or improvements in hepatic histology. A full discussion of these reports is well beyond the scope of this review. However, several points are worth emphasizing. First of all, the mechanism by which interferon exerts its therapeutic benefit is not well understood. There is an apparent antiviral effect, with relatively rapid declines in HCV viremia on institution of interferon therapy. However, it is not clear whether this reflects a direct suppression of viral replication, or enhanced immunologic suppression of viral replication due to the immunomodulatory action of interferon. The latter includes the upregulation of class I markers on the surface of hepatocytes, which would enhance the recognition of infected cells by virus-specific cytotoxic lymphocytes.

It is interesting to note, however, that improved liver histology is the measure of response to interferon that occurs with the greatest frequency in treated patients.[23–25] This significantly exceeds the frequency with which ALT levels are normalized. In four early, randomized, placebo-controlled, prospective clinical trials, improvements in histology occurred in an average of 69% of all treated patients (2–3 million units interferon-α t.i.w. x 6 months). In contrast, ALT levels were normal in only 43% of patients at the end of therapy. Thus, whatever the underlying therapeutic mechanism, a reduction in hepatic inflammation frequently accompanies interferon therapy. An important question that is now under study is whether there is a measurable clinical benefit for those patients who improve histologically with interferon therapy, but who do not normalize their ALT levels or clear virus from the blood.

Since chronic inflammation is an important factor in activating stellate cells within the liver to lay down an abnormal collagen matrix, documented improvements in inflammation suggest that interferon may have a significant impact on one of the primary pathogenic events that leads to disease, fibrogenesis. In addition, it is likely that one of the important proximate causes of hepatocellular carcinoma in patients with chronic hepatitis C is oxidative stress related to chronic inflammation, with resultant damage to cellular DNA. However, data supporting a tangible improvement in disease outcome or quality of life following interferon therapy are sparse. One randomized, controlled, prospective

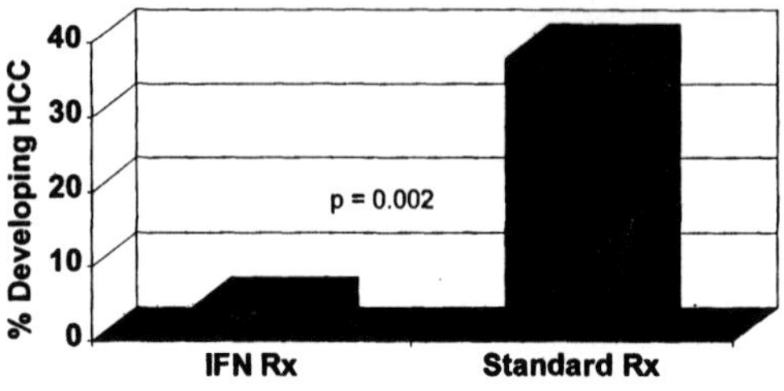

Figure 2. Summary of a Japanese trial that demonstrated that interferon therapy may prevent or delay the onset of hepatocellular carcinoma (HCC) in patients with chronic hepatitis C and well compensated cirrhosis.[26]

study has demonstrated a reduction in the incidence of hepatocellular carcinoma in Japanese patients with well compensated cirrhosis who were given relatively high dose interferon therapy (6 million units of interferon-α t.i.w.) over a period of 12–24 weeks[26] (Figure 2). Hepatologists in the U.S. have been slow to accept these results due to the extraordinarily high incidence of liver cancer in the control group in this study, which approached 40% after 2–7 years of post-treatment follow-up. However, it is likely that the high incidence of cancer in the untreated patients may simply reflect the natural history of hepatitis C in Japan, which appears to differ from that in the U.S.[27,28] A second, retrospective study also suggests that patients who respond to interferon may have a lower risk of developing hepatocellular carcinoma than interferon nonresponders.[29] There are no data available that address the issue of whether interferon therapy prevents or delays the onset of clinically evident cirrhosis.

In addition to the relatively low overall response rate, there are other major impediments to successful treatment of chronic hepatitis C with interferon. These include the extraordinarily high cost of the drug, its administration and attendant patient monitoring. Moreover, interferon therapy for chronic hepatitis C is marked by a relatively high frequency of adverse side effects such as cytopenia, depression, autoimmunity, and increased frequency of bacterial infections, among others. Most notable, however, is the high rate of relapse that typically follows completion of interferon therapy. This generally occurs in about half of all treated patients whose ALT levels have been rendered normal by the end of 6 months of therapy, although it may occur in a slightly smaller proportion of patients if therapy is continued for a year.[3] The low overall frequency of sustained response to interferon is a major reason for the current intensive search for better therapeutic regimens. Nonetheless, the results of a recent NIH Consensus Development Conference on interferon therapy and hepatitis C were a clear endorsement of interferon treatment for patients with active inflammatory disease and no major contraindications to treatment.[3] Further details regarding indications, treatment regimens, and contraindications should be sought from that source.

3. APPROACH TO ANTIVIRAL THERAPY—THE CASE FOR COMBINATION THERAPY

Recent successes in the treatment of HIV-1 infections with combinations of antiviral drugs argue strongly for a similar approach to the treatment of chronic HCV infection. This is particularly so since HCV infections share a number of features in common with persistent HIV-1 infection, including the capacity for substantial quasispecies variation and the potential for selection of resistant viruses in patients receiving monotherapy. Combination therapy with antimicrobial agents has a long history in the treatment of chronic

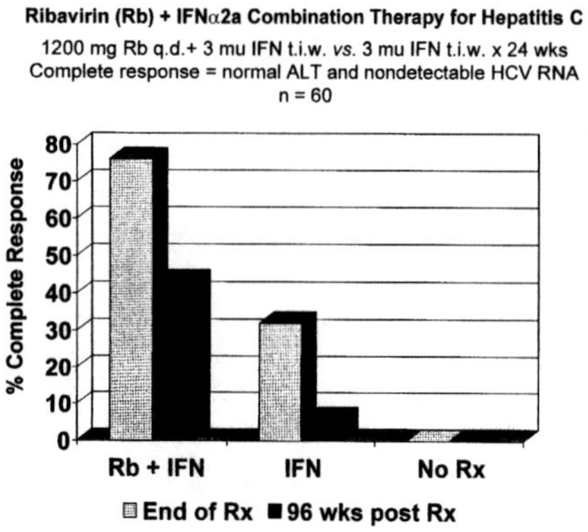

Figure 3. Summary of results from a prospective, randomized, controlled clinical trial of ribavirin-interferon combination therapy vs interferon monotherapy, as reported by Lai *et al.*[33]

infectious diseases. It has been used for decades in the treatment of tuberculosis for very similar reasons. In the case of hepatitis C, recent data strongly support a synergistic effect when ribavirin is combined with interferon, both in interferon-naïve patients as well as in the retreatment of interferon nonresponders.[30–33]

Ribavirin is a synthetic, oral guanosine nucleoside analog that has been used clinically for a number of years as treatment for several different viral infections. It has demonstrated efficacy in the treatment of neonatal respiratory syncytial virus infections, as well as in life-threatening arenavirus infections.[34,35] The compound is unusual in that it possesses a broad range of antiviral activity against viruses of vastly different type. It has been suggested to block the synthesis of functional rhabdovirus mRNAs,[36] possibly by interfering with capping of the 5' ends of the RNA (a phenomenon that is likely to be irrelevant in the case of HCV as its translation occurs by a cap-independent process). Other data suggest that it may suppress transcription of double-stranded reovirus RNAs, possibly by interfering with viral RNA helicase activity.[37] However, its mode of broad antiviral action has never been satisfactorily explained and, as indicated below, it may have important immunomodulatory activities.

Early studies of ribavirin monotherapy in patients with chronic hepatitis C were prompted by its known activity against other RNA virus infections. For the most part, these studies were disappointing. Although slight improvement was noted in serum ALT values, there was very little suppression of HCV viremia even after prolonged therapy. This indicates that the antiviral effect of ribavirin against HCV, if any, is extremely limited.[38–41]

Despite the early results with ribavirin monotherapy, recent clinical trials indicate that the addition of ribavirin to standard courses of interferon results in significantly better response to therapy. This is marked by increases both in the proportion of patients with end-of-treatment responses, as well as in the proportion of patients with sustained responses at 6–18 months following cessation of therapy. For example, in a prospective randomized trial, Lai *et al.* [33] treated interferon-naïve patients with the combination of ribavirin (1200 mg q.d.) and interferon-α2a (3 million units t.i.w.) for 24 weeks. This combination treatment regimen resulted in a complete response (normalization of ALT and elimination of detectable HCV RNA) in 76% of patients by the end of therapy, and a sustained response at 96 weeks after therapy in 43% (Figure 3). In contrast, comparable response rates in patients receiving a similar course of interferon without ribavirin were 32% and 6%, respectively.

Similar findings were reported recently by Reichard *et al.*[42] in a larger prospective study examining the combination of ribavirin and interferon-α2b (3 million units t.i.w.), given for a total of 24 weeks. Altogether, 42% of the patients had a sustained virologic re-

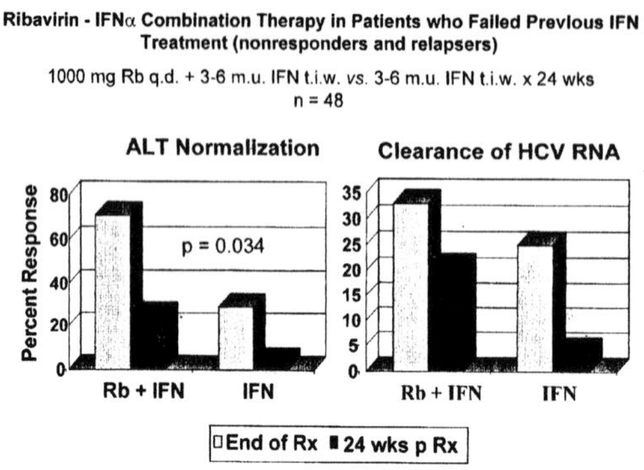

Figure 4. Summary of results from a prospective, randomized, controlled clinical trial of ribavirin-interferon combination therapy vs interferon monotherapy, as reported by Bellobuono *et al.*[30]

sponse after 1 year of follow-up, compared to only 20% of patients receiving interferon alone. Importantly, a retrospective analysis of the patients included in this study indicated that the combination was beneficial only in patients with high levels of serum HCV RNA (> 3 x 10^6 genome equivalents/ml). The combination offered no advantage over interferon monotherapy in those patients with lower levels of HCV viremia, who overall have a greater chance of a favorable response to interferon alone.

Although the response to interferon-ribavirin combination therapy has not been as impressive in those patients who have previously failed interferon monotherapy, Bellobuono *et al.* [30] reported both greater end-of-treatment and sustained ALT responses in patients undergoing retreatment with combination therapy, compared with interferon alone (Figure 4). In all studies, the addition of ribavirin to interferon regimens has been relatively well tolerated, with few patients forced to withdraw from therapy due to adverse reactions. However, low grade hemolysis has been a frequent side-effect, and may lead to both anemia and increased intrahepatic iron stores.[43] The latter is a worrisome finding, because it could be detrimental to liver function if allowed to reach a significant level.

The studies described above all contain only relatively small numbers of patients. Data from large international, multicenter studies comparing the combination of interferon and ribavirin with interferon monotherapy are now under review, and a clearer idea of the benefits to be gained by this combination therapy should soon be available. Nonetheless, even though the overall sustained virologic response rates remain less than satisfactory, the benefit that appears to accrue from the use of interferon-ribavirin combination therapy highlights the general potential for combination therapy against this disease.

The specific mechanisms underlying the synergy between interferon and ribavirin remain unknown. However, it is interesting that the combination of ribavirin and interferon also shows synergy in the treatment of experimental subacute sclerosing panencephalitis (SSPE) virus infections in hamsters.[44] This suggests a broad effect of ribavirin on the host response to chronic viral infections treated with interferon. Furthermore, as indicated above, the mechanism of action of ribavirin in the setting of hepatitis C is not likely to be related to direct suppression of viral replication. This is suggested by the failure of ribavirin to suppress HCV viremia when administered as monotherapy,[39,40] as well as a series of observations concerning the effects of ribavirin on various aspects of the host immune response.

In animal studies, the course of disease in mice experiencing fulminant hepatitis due to the coronavirus murine hepatitis virus (strain 3) can be attenuated by treatment with ribavirin.[45] This has been shown to be associated with inhibition of macrophage production of TNF and the procoagulant fgl2 prothrombinase. Ribavirin inhibited Th2 cytokine responses, but preserved Th1 cytokine production. IL6 production has also been shown to be modulated by ribavirin in human pulmonary epithelial cells that are infected with respiratory syncytial virus.[46] However, the relevance of these findings to the mechanism of action of ribavirin in the setting of combination therapy for hepatitis C is not known.

Ribavirin inhibits cellular inosine monophosphate (IMP) dehydrogenase,[47] and this may be the central mechanism through which it exerts its broad influence on intracellular events in a variety of viral infections. IMP dehydrogenase plays a central role in cellular metabolic processes, and its suppression may influence the ability of cells either to support viral replication or to respond to appropriate immunologic stimuli. Both ribavirin and the related compound, tiazofurin, also an inhibitor of cellular IMP dehydrogenase, may influence various signal transduction pathways. These compounds have both been shown to suppress the replication of yellow fever virus in cell culture, a flavivirus that is distantly related to HCV.[47,48] Thus, it is possible that ribavirin could act both to modulate the host immune response and to suppress replication through interaction with common cellular metabolic pathways involving IMP dehydrogenase.

Whatever the mechanism of action of ribavirin when used in combination with interferon for treatment of hepatitis C, the success of this strategy has increased the urgency of the search for additional, specific inhibitors of HCV replication. Although this search is hampered by the absence of tractable cell culture systems that would allow the development of broad-based screens for novel antiviral compounds, significant progress is being made through the use of more sophisticated biochemical screens. This is certain to be a field that will become complicated by the availability of a number of new antiviral drugs in the future.

REFERENCES

1. Choo Q-L, Kuo G, Weiner AJ, Overby LR, Bradley DW, Houghton M. Isolation of a cDNA clone derived from a blood-borne non-A, non-B viral hepatitis genome. Science 1989; 244:359–362.
2. Kuo G, Choo Q-L, Alter HJ, Gitnick GL, Redeker AG, Purcell RH, *et al.* An assay for circulating antibodies to a major etiologic virus of human non-A, non-B hepatitis. Science 1989; 244:362–364.
3. Consensus Development Panel. National Institutes of Health Consensus Development Conference Panel statement: Management of hepatitis C. Hepatology 1997; 26, Supplement 1:2S-10S.
4. Simmonds P, Holmes EC, Cha T-A, Chan S-W, McOmish F, Irvine B, *et al.* Classification of hepatitis C virus into six major genotypes and a series of subtypes by phylogenetic analysis of the NS-5 region. J.Gen.Virol. 1993; 74:2391–2399.
5. Simons JN, Pilot-Matias TJ, Leary TP, Dawson GJ, Desai SM, Schlauder GG, *et al.* Identification of two flavivirus-like genomes in the GB hepatitis agent. Proc.Natl.Acad.Sci.USA 1995; 92:3401–3405.
6. Honda M, Brown EA, Lemon SM. Stability of a stem-loop involving the initiator AUG controls the efficiency of internal initiation of translation on hepatitis C virus RNA. RNA 1996; 2:955–968.
7. Smith DB, Simmonds P. Characteristics of nucleotide substitutions in the hepatitis C virus genome: Constraints on sequence change in coding regions at both ends of the genome. J.Mol.Evol. 1997; In press.
8. Tsukiyama-Kohara K, Iizuka N, Kohara M, Nomoto A. Internal ribosome entry site within hepatitis C virus RNA . J.Virol. 1992; 66:1476–1483.
9. Pestova TV, Shatsky IN, Fletcher SP, Jackson RJ, Hellen CU. A prokaryotic-like mode of cytoplasmic eukaryotic ribosome binding to the initiation codon during internal translation initiation of hepatitis C and classical swine fever virus RNAs. Genes Dev. 1998; 12:67–83.
10. Major ME, Feinstone SM. The molecular virology of hepatitis C. Hepatology 1997; 25:1527–1538.
11. Dubuisson J, Hsu HH, Cheung RC, Greenberg HB, Russell DG, Rice CM. Formation and intracellular localization of hepatitis C virus envelope glycoprotein complexes expressed by recombinant vaccinia and Sindbis viruses. J.Virol. 1994; 68:6147–6160.
12. Hijikata M, Mizushima H, Tanji Y, Komoda Y, Hirowatari Y, Akagi T, *et al.* Proteolytic processing and membrane association of putative nonstructural proteins of hepatitis C virus. Proc.Natl.Acad.Sci.USA 1993; 90:10773–10777.

13. Kato N, Seikya H, Ootsuyama Y, Nakazawa T, Hijikata M, Ohkoshi S, *et al.* Humoral immune response to hypervariable region 1 of the putative envelope glycoprotein (gp70) of hepatitis C virus. J.of Virology 1993; 67:3923–3930.
14. Grakoui A, McCourt DW, Wychowski C, Feinstone SM, Rice CM. A second hepatitis C virus-encoded proteinase. Proc.Natl.Acad.Sci.USA 1993; 90:10583–10587.
15. Love RA, Parge HE, Wickersham JA, Hostomsky Z, Habuka N, Moomraw EW, *et al.* The crystal structure of hepatitis C virus NS3 proteinase reveals a trypsin-like fold and a structural zinc binding site. Cell 1996; 87:331–342.
16. Kim JL, Morgenstern KA, Lin C, Fox T, Dwyer MD, Landro JA, *et al.* Crystal structure of the hepatitis C virus NS3 protease domain complexed with a synthetic NS4A cofactor peptide. Cell 1996; 87:343–355.
17. Kim JL, Morgenstern KA, Griffith JP, Dwyer MD, Thomson JA, Murcko MA, *et al.* Hepatitis C virus NS3 RNA helicase domain with a bound oligonucleotide: the crystal structure provides insights into the mode of unwinding. Structure. 1998;6:89–100.
18. De Francesco R, Behrens SE, Tomei L, Altamura S, Jiricny J. RNA-dependent RNA polymerase of hepatitis C virus. Meth.Enzymol. 1996; 275:58–67.
19. Tanaka T, Kato N, Cho M-J, Shimotohno K. A novel sequence found at the 3' terminus of the hepatitis C virus genome. Biochem.Biophys.Res.Commun. 1995; 215:744–749.
20. Kolykhalov AA, Feinstone SM, Rice CM. Identification of a highly conserved sequence element at the 3' terminus of hepatitis C virus genome RNA. J.Virol. 1996; 70:3363–3371.
21. Kolykhalov AA, Agapov AA, Blight KJ, Mihalik K, Feinstone SM, Rice CM. Transmission of hepatitis C by intrahepatic inoculation with transcribed RNA. Science 1997; 277:570–574.
22. Yanagi M, Purcell RH, Emerson SU, Bukh J. Transcripts from a single full-length cDNA clone of hepatitis C virus are infectious when directly transfected into liver of a chimpanzee. Proc.Natl.Acad.Sci.USA 1997; 97:8738–8743.
23. Di Bisceglie AM, Martin P, Kassianides C, Lisker-Melman M, Murray L, Waggoner J, *et al.* Recombinant interferon alpha therapy for chronic hepatitis C: A randomized, double-blind, placebo-controlled trial. N.Engl.J.Med. 1989; 321:1506–1510.
24. Davis GL, Balart LA, Schiff ER, Lindsay K, Bodenheimer HC, Jr., Perillo RP, *et al.* Treatment of chronic hepatitis C with recombinant interferon alpha: A multicenter randomized, controlled trial. N.Engl.J.Med. 1989; 321:1501–1506.
25. Sáez-Royuela F, Porres JC, Moreno A, Castillo I, Martinez G, Galiana F, *et al.* High doses of recombinant alpha-interferon or gamma-interferon for chronic hepatitis C: A randomized, controlled trial. Hepatology 1991; 13:327–331.
26. Nishiguchi S, Kuroki T, Nakatani S, Morimoto H, Takeda T, Nakajima S, *et al.* Randomised trial of effects of interferon-α on incidence of hepatocellular carcinoma in chronic active hepatitis C with cirrhosis. Lancet 1995; 346:1051–1055.
27. Lemon SM, Chisari FV, Lai MMC, Mishiro S, Nishioka K, Johnson L. Meeting report: The 19th Joint U.S.-Japan Hepatitis Panel Meeting. Submitted 1998.
28. Kiyosawa K, Furuta S. Hepatitis C virus and hepatocellular carcinoma. In: Hepatitis C Virus. Reesink HW, editor. Basel: Karger, 1994:98–120.
29. Mazzella G, Accogli E, Sottili S, Festi D, Orsini M, Salzetta A, *et al.* Alpha interferon treatment may prevent hepatocellular carcinoma in HCV-related liver cirrhosis. J.Hepatol. 1996; 24:141–147.
30. Bellobuono A, Mondazzi L, Tempini S, Silini E, Vicari F, Ideo G. Ribavirin and interferon-alpha combination therapy vs interferon- alpha alone in the retreatment of chronic hepatitis C: a randomized clinical trial. J.Viral.Hepat. 1997; 4:185–191.
31. Schvarcz R, Ando Y, Sönnerborg A, Weiland O. Combination treatment with interferon alpha-2b and ribavirin for chronic hepatitis C in patients who have failed to achieve sustained response to interferon alone: Swedish experience. J.Hepatol. 1995; 23:17–21.
32. Schalm SW, Brouwer JT, Chemello L, Alberti A, Bellobuono A, Ideo G, *et al.* Interferon-ribavirin combination therapy for chronic hepatitis C . Dig.Dis.Sci. 1996; 41:131S-134S.
33. Lai MY, Kao JH, Yang PM, Wang JT, Chen PJ, Chan KW, *et al.* Long-term efficacy of ribavirin plus interferon alpha in the treatment of chronic hepatitis C. Gastroenterology 1996; 111:1307–1312.
34. Smith DW, Frankel LR, Mathers LH, Tang AT, Ariagno RL, Prober CG. A controlled trial of aerosolized ribavirin in infants receiving mechanical ventilation for severe respiratory syncytial virus infection. N.Engl.J.Med. 1991; 325:24–29.
35. McCormick JB, King IJ, Webb PA, Scribner CL, Craven RB, Johnson KM, *et al.* Lassa fever. Effective therapy with ribavirin. N.Engl.J.Med. 1986; 314:20–26.
36. Toltzis P, Huang AS. Effect of ribavirin on macromolecular synthesis in vesicular stomatitis virus-infected cells. Antimicrob.Agents Chemother. 1986; 29:1010–1016.

37. Rankin JT, Jr., Eppes SB, Antczak JB, Joklik WK. Studies on the mechanism of the antiviral activity of ribavirin against reovirus. Virology 1989; 168:147–158.
38. Di Bisceglie AM, Shindo M, Fong T-L, Fried MW, Swain MG, Bergasa NV, *et al.* A pilot study of ribavirin therapy for chronic hepatitis C. Hepatology 1992; 16:649–654.
39. Di Bisceglie AM, Conjeevaram HS, Fried MW, Sallie R, Park Y, Yurdaydin C, *et al.* Ribavirin as therapy for chronic hepatitis C - A randomized, double-blind, placebo-controlled trial. Ann.Intern.Med. 1995; 123:897–903.
40. Bodenheimer HC, Jr., Lindsay KL, Davis GL, Lewis JH, Thung SN, Seeff LB. Tolerance and efficacy of oral ribavirin treatment of chronic hepatitis C: a multicenter trial. Hepatology 1997; 26:473–477.
41. Dusheiko G, Main J, Thomas H, Reichard O, Lee C, Dhillon A, *et al.* Ribavirin treatment for patients with chronic hepatitis C: results of a placebo-controlled study. J.Hepatol. 1996; 25:591–598.
42. Reichard O, Norkrans G, Fryden A, Braconier JH, Sonnerborg A, Weiland O. Randomised, double-blind, placebo-controlled trial of interferon alpha- 2b with and without ribavirin for chronic hepatitis C. The Swedish Study Group. Lancet 1998; 351:83–87.
43. Di Bisceglie AM, Bacon BR, Kleiner DE, Hoofnagle JH. Increase in hepatic iron stores following prolonged therapy with ribavirin in patients with chronic hepatitis C. J.Hepatol. 1994; 21:1109–1112.
44. Takahashi T, Hosoya M, Kimura K, Ohno K, Mori S, Takahashi K, *et al.* The cooperative effect of interferon-alpha and ribavirin on subacute sclerosing panencephalitis (SSPE) virus infections, *in vitro* and *in vivo*. Antiviral Res. 1998; 37:29–35.
45. Ning Q, Brown D, Parodo J, Cattral M, Gorczynski R, Cole E, *et al.* Ribavirin inhibits viral-induced macrophage production of TNF, IL-1, the procoagulant fgl2 prothrombinase and preserves Th1 cytokine production but inhibits Th2 cytokine response. J.Immunol. 1998; 160:3487–3493.
46. Jiang Z, Kunimoto M, Patel JA. Autocrine regulation and experimental modulation of interleukin- 6 expression by human pulmonary epithelial cells infected with respiratory syncytial virus. J.Virol. 1998; 72:2496–2499.
47. Yamada Y, Natsumeda Y, Weber G. Action of the active metabolites of tiazofurin and ribavirin on purified IMP dehydrogenase. Biochemistry 1988; 27:2193–2196.
48. Neyts J, Meerbach A, McKenna P, de Clercq E. Use of the yellow fever virus vaccine strain 17D for the study of strategies for the treatment of yellow fever virus infections. Antiviral Res. 1996; 30:125–132.

5

PREVENTION AND TREATMENT OF RESPIRATORY SYNCYTIAL VIRUS INFECTIONS

John Mills

Macfarlane Burnet Centre for Medical Research
Yarra Bend Road
Fairfield, Victoria 3078, Australia

Respiratory syncytial virus (RSV) infections remain a major cause of morbidity and mortality in infants and young children. Based on a recent Canadian study, it can be estimated that the annual cost of RSV-related illness in the USA is over $300 million per year, and the bulk of this cost is related to hospitalization.[1] Over the past decade or so it has become apparent that RSV is also a major pathogen of elderly and immunosuppressed adults.[2–7] RSV disease has proven to be a major problem in institutionalized elderly (for example, those living in nursing homes or long-term convalescent hospitals),[4,6,7] and in bone marrow transplant recipients. In the latter group, mortality rates in excess of 50% have been reported.[5, 8–11]

Antiviral chemotherapy of RSV infections has largely been less than impressive. Although aerosolized ribavirin was approved over a decade ago for treatment of RSV infections in infants, based upon a single randomized, placebo-controlled trial,[12] subsequent evaluations of ribavirin therapy have been relatively disappointing, and the present reality is that ribavirin is rarely used except in the most severely-affected infants.[13–15] Even in the case of RSV pneumonia in the setting of bone marrow transplantation, the results of ribavirin therapy have been mixed but in no instance has impressive benefit been documented.[16–18] Although further antivirals with activity against RSV may progress to clinical development,[19–23] it seems unlikely that an expensive and potentially toxic antiviral compound represents the best long-term, overall solution to RSV disease in either children or adults.

For these reasons, there has been increasing focus on immunotherapeutic and immunoprophylactic strategies for management of RSV infections. This chapter reviews the current data on the use of ribavirin and some investigational compounds as therapy for RSV infections, but focuses on immune control of RSV infection.

Antiviral Chemotherapy 5, edited by Mills *et al.*
Kluwer Academic / Plenum Publishers, New York, 1999.

1. TREATMENT OF RSV INFECTION

1.1. Ribavirin and Other Antivirals

Ribavirin remains the only licenced antiviral for RSV infections. The seminal studies documenting its efficacy in RSV infections in infants were done more than a decade ago, and showed modest clinical benefit when compared with placebo or no treatment.[24, 25] However, these studies did not look for consequential effects on disease–that is, reduction in the length of hospitalization, need for intensive care, or mortality. These measures are now considered critical to antimicrobial development and licencing.

More recent studies of ribavirin therapy in infants not receiving mechanical ventilation have generally confirmed the efficacy seen in earlier studies, but have suggested that the benefits of ribavirin on the consequential effects of RSV infection, if any, were relatively modest. Groothius conducted a prospective, placebo-controlled trial of early ribavirin therapy for mild RSV bronchiolitis in 47 infants with bronchopulmonary dysplasia or congenital heart disease.[26] Although the results again suggested a beneficial effect of ribavirin on clinical findings such as oxygen saturation, no beneficial effects were seen on duration of hospitalization or overall outcome.[26] A meta-analysis of eight double-blind, randomized, placebo-controlled trials of ribavirin for RSV pneumonitis failed to show a beneficial effect of ribavirin therapy on the duration of hospitalization, the risk of respiratory failure or mortality, and the drug's effects on other clinical variables were conflicting.[14] Similar results were found in a review of patients from two hospitals, one using ribavirin in virtually every case of RSV bronchiolitis and the other never using ribavirin,[27] and in a retrospective review from a hospital with mixed ribavirin usage patterns.[28] A retrospective review of infants with RSV bronchiolitis, 19 of whom were treated and 22 of whom received only supportive care, suggested that after one year of follow-up the incidence of reactive airway disease was significantly less in the ribavirin treated group.[29] However, it is difficult to draw firm conclusions from this study considering the small numbers of patients studied and the short duration of follow-up. Supporting this concern, in a 6–8 year followup study by Krilov and collaborators of 100 patients hospitalized with RSV bronchiolitis, the incidence of reactive airway disease (as determined by formal pulmonary function testing) was virtually identical in the ribavirin-treated and conservatively-treated groups, although there was less "bronchitis" by history in the ribavirin-treated patients.[30] Perhaps as the result of these studies, ribavirin utilization for RSV infections in infants has been decreasing, and in some locations, such as Australia, the drug is virtually never utilized.[28] The infants most in need of therapeutic benefit from ribavirin are those who develop respiratory failure due to RSV infection, and therefore require monitoring in the intensive care unit, or also intubation and mechanical ventilation. Although the initial study which showed ribavirin to have both clinical and consequential benefits (specifically, reduced intensive care unit utilization) in this context was soundly designed,[31] the number of patients studied (28) was relatively small and the authors were criticized for using a water placebo rather than a saline placebo. The reasoning behind this criticism was that aerosolization of water may cause bronchospasm, which might have worsened the clinical condition of the placebo group relative to the ribavirin group, despite the fact that the amount of water aerosolized is thought to be insufficient to be physiologically significant.[32] A recent, large (223 patients) multi-centre study of ribavirin therapy of previously-well infants with respiratory failure secondary to RSV infection failed to show therapeutic benefit, and there was a statistically-significant *prolongation* of mechanical ventilation in the ribavirin-treated patients.[33] Another study in 41 ventilated in-

Table 1. Clinical situations in which ribavirin therapy "may be considered" for infants with RSV disease, as promulgated by the American Academy of Pediatrics

- Infants with clinically significant, complicated congenital heart disease or chronic lung disease.
- Premature infants (<37 weeks gestation) and mature infants <6 weeks of age, without underlying disease.
- Infants with underlying immune compromise, such as AIDS, SCID, or organ transplantation requiring immunosuppressive therapy.
- Infants who are severely ill with RSV lower respiratory tract disease, with or without the need for mechanical ventilation.
- Nosocomially-transmitted RSV infections, especially in infants with severe underlying disease (e.g. multiple congenital abnormalities, cerebral palsy, myasthenia gravis).

Adapted from reference [15].

fants used a saline placebo but which otherwise was of comparable design to the Smith study also failed to demonstrate any benefit to ribavirin therapy; however, this latter study was too small to detect even modest clinical benefit.[34]

As a result of these and other data, the American Academy of Pediatrics has recently re-evaluated the indications for ribavirin therapy in infants with RSV infection (Table 1).[15] The most significant change in the Academy's recommendations is from the statement that, "...ribavirin should be used..." to, "...ribavirin may be considered [for use]..." in specific clinical situations (Table 1). As DeVincenzo has commented, however, the AAP guidelines fail to address some practical issues which are relevant to the practising clinician.[32] The principal issue is that, like any antimicrobial, ribavirin is more effective when treatment is begun early; hence, a decision to treat or not should be made within the first 24 hours of hospitalization, and in immunocompetant individuals, treatment should seldom be begun after the sixth day of illness.[32] Further, illness severity is best assessed by the degree of hypoxemia, and the decision to treat should be a function of the severity of illness at baseline as well as the presence of risk factors increasing the severity of disease (e.g. bronchopulmonary displasia).[32]

The efficacy of ribavirin for bone marrow transplant recipients with RSV lower respiratory infections is difficult to assess. The available data suggest that RSV infection has a high mortality in this setting, whether in adults or children.[8,11] However, results of the small, uncontrolled studies of ribavirin therapy have wildly discrepant results, varying from full recovery in most patients[35] to 80% mortality[17] and intermediate results.[9,18] However, given the high mortality from RSV pneumonia in this context, and the lack of alternative therapeutic options or substantive toxicity from ribavirin, empiric therapy seems warranted.

Despite the well-recognized disadvantages of ribavirin as an RSV antiviral, few other therapeutic agents are under development. Although preclinical studies with some novel antivirals, including antisense oligonucleotides and ribozymes, have been promising, clinical trials have not been initiated with any of these agents [19–22 and P. Young, personal communication 1998].

1.2. Immunotherapy

Intravenous or aerosol administration of neutralizing antibodies to animals with established, experimental RSV pneumonitis dramatically reduces lung viral titers and hastens resolution of pulmonary pathology.[36–41] For this reason, clinical trials of immunotherapy of RSV infections in infants and adults were undertaken, either alone or in combination with ribavirin.

The first therapeutic trial was conducted over a decade ago by Hemming and collaborators, using intravenously-administered, commercially-available IVIG, a preparation which had not been selected for high RSV neutralizing antibody titers. This small study showed some evidence of therapeutic benefit which did not reach statistical significance.[42] As the result of this salutory experience, a larger randomized prospective controlled trial was conducted using RSV-IG (human IVIG prepared from donors selected for high titers of neutralizing antibodies to RSV) in 107 infants less than two years old with risk factors for severe RSV disease.[43] This study showed absolutely no evidence of therapeutic benefit from the immunoglobulin therapy. Two clinical trials of aerosolized IVIG (not selected for high RSV antibody titers) for RSV infections in Swiss infants were recently completed, and likewise failed to show any clinical benefit despite causing considerable toxicity.[44,45] However, the design of this study was flawed in many respects, including the investigator's failure to use IVIG with high titers of RSV neutralizing antibody (i.e. RSV-IG). Whimbey and collaborators reported treating bone marrow transplant recipients with RSV pneumonia in a non-comparative fashion with the combination of ribavirin and RSV-IG, with good results.[16] However, the small numbers of patients studied and the lack of a comparison group make the results of this study virtually impossible to interpret.

In conclusion, although RSV immunoglobulin therapy has shown benefit in experimental animal infections, the clinical experience to date with RSV-IG has been very disappointing. It remains to be seen whether the newly-developed humanized or fully human neutralizing monoclonal antibodies, which can achieve astronomically high neutralizing antibody titers in patients following a single injection, will be effective for therapy.[39,46–53]

1.2.1. Other Treatment Modalities. Although infants with RSV bronchiolitis experience air trapping, they do not have bronchospasm in the usual sense. It is therefore not surprising that trials of bronchodilator therapy have uniformly been disappointing, and their use is generally not recommended, especially in very young infants where reactive airway disease which has been worsened by concurrent RSV infection is unlikely.[32] Glucocorticoid therapy has likewise not shown benefit in randomized prospective trials.[32] Two prospective trials of vitamin A for RSV bronchiolitis, one conducted in the USA and the other in Chile, came to divergent conclusions.[54,55] The US study found no therapeutic benefit from vitamin A therapy, and treatment was associated with a significant prolongation of hospitalization in infants > 1 year of age.[55] In contrast, although the Chilean study showed now statistically-significant beneficial effect of vitamin A therapy overall, vitamin A-treated patients with significant hypoxemia on admission had more rapid resolution of tachypnea and shorter hospitalization than controls.[54] It seems likely that the therapeutic benefit of vitamin A, if any, is marginal, and there is at least the potential of clinically-significant adverse effects. Two cases of respiratory failure due to RSV bronchiolitis responded to empiric therapy with pulmonary surfactant,[56] but no confirmatory controlled trials have been performed.

2. PREVENTION OF RSV INFECTION

2.1. Passive Immunoprophylaxis

The past several years have seen the concept of passive immunoprophylaxis of RSV infections finally and completely validated by controlled clinical trials, and these trials have set the direction of future development and clinical research.

Passive immunoprophylaxis for RSV infection has been shown to be effective in a wide variety of animal models.[37,40,41,57–60] The protective antibodies are those directed against the major membrane glycoproteins, the F (fusion) protein and the G (attachment) protein.[59] In addition, prospective studies by Glezen and Wang and their coworkers have shown that passively-transferred maternal antibodies could protect infants from severe RSV disease,[61,62] further supporting the important protective role of neutralizing antibodies to the F and G proteins.

2.1.1. Polyclonal RSV Antibodies. The first study attempting to prevent RSV infection by passive immunoprophylaxis utilized unscreened human IVIG administered on a monthly basis, and failed to show prophylactic efficacy, presumably because the antibody titers achieved in the recipients were insufficiently high to be protective.[63] However, as there were no adverse effects, there was an impetus to refine this strategy by screening donors for RSV neutralizing antibody titers and preparing pooled IVIG only from those donors with high titers—so-called RSV-IG.[64] The initial clinical trial with RSV-IG showed a significant (>50%) reduction in RSV disease, as measured by episodes of lower respiratory infection, need for hospitalization or intensive care, and by duration of stay once hospitalized.[65] This study also documented a clear dose-response relationship between placebo, and 250mg/kg/month or 750mg/kg/month infusions of the RSV-IG, and also showed that the polyclonal immunoglobulin which was administered offered some protection against infections other than RSV.[65] Efficacy was also demonstrated in a subgroup of patients with prematurity.[66]

Concern by the US Food and Drug Administration about some aspects of this study prompted two further controlled clinical trials, one in infants with pulmonary disease (the PREVENT study)[67] and the other in infants with cardiopulmonary disease (the CARDIAC study).[68] The PREVENT study largely confirmed the earlier Groothuis study, and as a result RSV-IG was licenced by the FDA in January 1996 for prevention of RSV infections in premature infants with or without bronchopulmonary dysplasia.

The outcome of the CARDIAC trial of RSV-IG, however, was quite different, although the results to date have only been presented in abstract form.[68] Although RSV-IG appeared to reduce the impact of RSV disease, none of the reductions (in comparison to the placebo group) were statistically significant, despite the fact that the study enrolled over 400 patients.[68] Further, the CARDIAC trial also suggested that there was an increase in deaths in the RSV-IG group compared with the placebo controls. These findings were of interest as the Groothuis study also showed that RSV-IG was least beneficial in patients with cardiac disease, and likewise experienced an excess of deaths in RSV-IG-treated patients with cyanotic congenital heart disease (albeit statistically-insignificant).[65] Although none of these adverse events seemed clearly related to the RSV-IG infusions either temporally or pathophysiologically, caution is generally recommended in using any form of IVIG prophylaxis in infants with cyanotic congenital heart disease.[69]

RSV is a recognized cause of otitis media,[70,71] and RSV-IG has also been shown to reduce the incidence of otitis media associated with RSV infection.[72] A sub-analysis of the Groothuis study of RSV-IG[65] showed that the high dose of RSV-IG (750 mg/kg/month) resulted in an approximately 4-fold reduction in otitis media when compared with placebo recipients.[72] This is an interesting observation considering that one recent study suggested that many of the episodes of otitis media associated with bronchiolitis had a bacterial etiology.[73] The results of the recent Phase III study of an RSV monoclonal antibody, which failed to show an effect on otitis media,[51] suggest that the efficacy of RSV-IG in prevent-

Table 2. Recommendations for use of RSV antibody-enriched IVIG (RSV-IG) for prevention of RSV infections in high-risk infants*

- Infants ≤2 years of age (including premature infants) with bronchopulmonary dysplasia who have received or are receiving oxygen therapy within six months of the RSV season are likely to benefit from RSV-IG. Depending on the severity of the disease, prophylaxis may continue for up to two RSV seasons.†
- Premature infants ≤32 weeks gestational age, otherwise healthy, may benefit from RSV-IG. In this group, benefit depends on the risk of disease, which is related to gestational age and chronologic age at the onset of the RSV season, and to the presence of other risk factors for RSV disease. In general infants of gestational age ≤28 weeks may need prophylaxis through two RSV seasons, while those between 28 and 32 weeks may need prophylaxis through only one season.† Other risk factors that may contribute to the cost-effectiveness of prophylaxis include multiple births, discharge from the neonatal intensive care unit just prior to the RSV season, crowding in the home environment, and school-age siblings.‡
- Some premature infants between 32 and 35 weeks gestational age may also benefit from RSV-IG prophylaxis if other risk factors for severe RSV disease are present.‡

The benefits of RSV-IG in other groups of patients thought to be at risk of severe RSV disease, such as those with chemotherapy-induced myelosuppression or immunosuppression related to bone marrow or organ transplantation,†‡ or those with nosocomially-acquired RSV infection, are unknown.†

RSV-IG should not be used in infants with symptomatic cyanotic congenital heart disease (asymptomatic heart disease is not a contraindication if another clear indication for use of RSV-IG is present).‡

*Adapted from references [69, 75, 118]
† - from the AAP [69]
‡ - from [75] and [118]

ing otitis media may have been due to antibodies in the preparation to pathogens other than RSV.

The principal difficulty with RSV-IG is, of course, that it is very expensive (ca US$5000 per season) and troublesome to administer (since it requires intravenous access monthly during the RSV season). For these reasons, an economic benefit could only be demonstrated in those infants at high risk of requiring hospitalisation as a result of RSV infection, especially if intensive care is also required.[50,74]

The American Academy of Pediatrics has recently put forward recommendations for the use of RSV-IG for prevention of RSV infections in selected high-risk infants,[69] which have been followed by a slightly different set of recommendations from investigators familiar with the relevant clinical trials;[75] these two sets of recommendations have been combined in Table 2.

2.1.2. Monoclonal RSV Antibodies. The neutralizing antibody response to RSV is exclusively directed against the major RSV surface proteins, F and G.[59] Administration of monoclonal neutralizing antibodies to the RSV F and G proteins offers significant practical advantages over RSV-IG. Since the monoclonal antibodies are highly potent and purified, they can be administered intramuscularly in small volumes, possibly less frequently than monthly. On the other hand, since the antibodies bind to only one epitope on the F or G protein, the opportunity for escape mutants (either pre-existing in the community, or developing under the selection pressure of the antibodies) is higher.[76–80] Further, a monoclonal antibody directed solely against RSV would not be expected to reduce the incidence of other respiratory infections, as was noted in some of the RSV-IG trials.[65] However, the ability to deliver preventative (or even therapeutic) antibody titers conveniently clearly outweighs any other disadvantages of monoclonal antibodies.

As with polyclonal RSV-IG and IVIG, parenterally-administered monoclonal antibodies have been highly effective in preventing RSV infections in experimental animals.[39,47,59] For clinical trials, manufacturers have all selected monoclonals to the F protein of RSV, as the F protein demonstrates significantly less sequence and antigenic diversity than the G glycoprotein, and neutralizing F monoclonals reactive with all known clinical isolates have been relatively easy to obtain.[59,81–83] These monoclonals have either been reshaped ("humanized") murine monoclonal antibodies, or fully-human monoclonal Fab fragments derived using recombinant DNA technology (none of the latter are in clinical trials).[39,46,47,84]

However, despite these promising preclinical results, the clinical research path to developing monoclonal antibodies which can protect against RSV infection has been decidedly bumpy. A humanized monoclonal developed by SmithKline Beecham[46–48,84] failed to prevent RSV infections when studied in a large randomized prospective trial [Everitt et al, unpublished data]. The reasons for this failure are unclear, since the antibody works extremely well in animal models, but possible reasons include inadequate dosage, low binding affinity of the antibody itself, a generally low attack rate in the study population during the RSV season studied, and a dosing schedule which did not begin soon enough before the RSV season.[50] Further, a murine IgA neutralizing RSV monoclonal antibody (HNK20) which prevented RSV infection when administered intranasally to a variety of experimental animals, including primates,[49,85] also failed to protect infants in a large Phase III study conducted in the Southern Hemisphere, despite promising Phase II clinical data [Thomas Monath, OraVax Corporation, personal communication]. Although the data showed a trend toward protection, with an 11% decrease in the incidence of RSV lower respiratory infections overall and a 39–47% reduction in infants ≤3 months of age, the results were not statistically significant [Thomas Monath, OraVax Corporation, personal communication]. Although the reasons for the failure of HNK20 are not known, it was thought that the dose was too low and a repeat study is being considered with a larger dose.

The results of the recent Medimmune trial of a humanized RSV antibody represents a breakthrough in passive immunoprophylaxis of RSV infections in infants. The Medimmune antibody, known as MEDI-493 or palivizumab, is a humanized murine monoclonal to the A epitope of the RSV fusion protein.[52,86] The antibody has a high affinity for the F protein, neutralizes all RSV isolates tested (including A and B types) and, in contrast to some F monoclonals, does not readily facilitate the development of neutralization escape mutants.[52] Similar to the results with other neutralizing monoclonals to the F or G proteins,[37,40,41,57–60] prophylaxis with palivizumab reduced pulmonary virus titers in experimentally-infected mice by more than two logs at serum concentrations of ≥30 μg/ml.[52] A phase I/II study in at-risk infants showed that the antibody had human IgG pharmacokinetics and safety, and that monthly infusions of 15mg/kg resulted in mean trough serum concentrations of 60–70 μg/ml, with all patients having levels above 40μg/ml, well above the levels shown to be protective in the aforementioned animal studies.[87] Although this Phase I/II study was not powered to show efficacy, the results were somewhat encouraging. Four of 20 placebo-treated patients had RSV infections versus only six of 42 palizumab-treated patients, and only two of 22 patients treated with high-dose (15mg/kg) palivizumab had RSV infections (all comparisons not significant).[87] A subsequent Phase I/II also suggested prophylactic efficacy.[53] Two of 11 children in a low-dose (5 mg/kg/month) palivizumab group had RSV infections, compared with none of 54 in the two higher-dose groups (10mg/kg and 15 mg/kg per month) ($p = 0.026$).[53] Further, the incidence of otitis media also appeared to be reduced in the high-dose group (3 {6%} of 48 patients) compared with that in the two lower-dose groups (4 {24%} of 17 patients) ($p = 0.13$).[53]

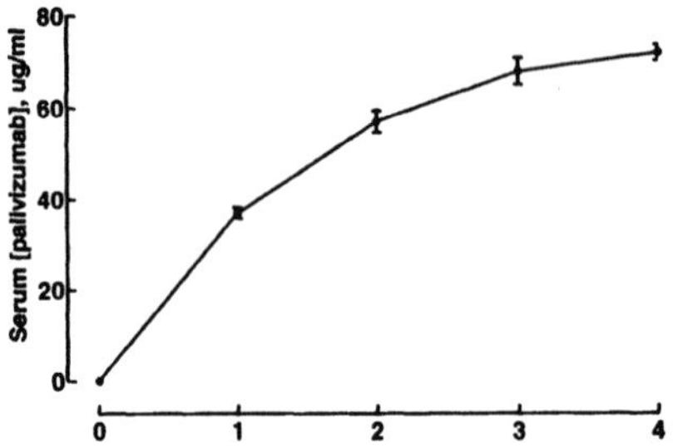

Figure 1. Trough serum concentrations of a humanized, neutralizing RSV F protein monoclonal antibody (palivizumab) after the indicated number of monthly doses of 15 mg/kg in high-risk infants. Data adapted from [51], with permission.

The conclusive Phase III study of Palivizumab (study MI-CP018) was conducted in about 1500 high-risk infants (those with bronchopulmonary dysplasia or prematurity) at 139 centers in the Northern Hemisphere. The study randomized subjects to either palivizumab or placebo, in a 2:1 ratio, during one RSV season. The study material had a palivizumab concentration of 100mg/ml, and it was given intramuscularly at a dose of 15mg/kg/month. Trough serum palivizumab concentrations in treated infants were above the protective level established in experimental animal infection[87] after the first dose, and rose progressively throughout the duration of the study (Figure 1).[51] As previously reported,[88] palivizumab reduced hospitalization due to RSV infection by 40–80%; the reduction was seen in all age and weight groups (Figure 2).[51] Palivizumab also reduced morbidity from RSV infection as measured by duration of hospitalization and need for supplemental oxygen or intensive care (Figure 3).[51] As a result of these data palivizumab was licenced by the US Food and Drug Administration in June 1998 (brand name, "Synagis"). Because of the difficulties with RSV-IG in infants with cyanotic cardiovascular disease discussed above, a Phase III study of palivizumab for such patients is planned to begin shortly [Franklin H. Top, Jr., personal communication].

2.1.3. Conclusions—Passive Immunoprophylaxis. The data from both experimental animals and humans show clearly that parenterally administered RSV neutralizing antibodies will dramatically reduce the severity of RSV infections, resulting in a decreased need for medical care, hospitalization, and intensive care. Passive immunoprophylaxis with polyclonal human IVIG will also prevent RSV-related otitis media. The type of antibody, the dosage and the dosing regimen are important to the success of the immunoprophylaxis regimen. The logistical difficulties of delivering human RSV antibody-enriched immunoglobulins (e.g. RSV-IG) are such that it is highly likely that they will be replaced by monoclonal antibodies such as palivizumab in the near future. Passive immunoprophylaxis is certainly a practical alternative for infants at high risk of severe RSV disease (e.g.

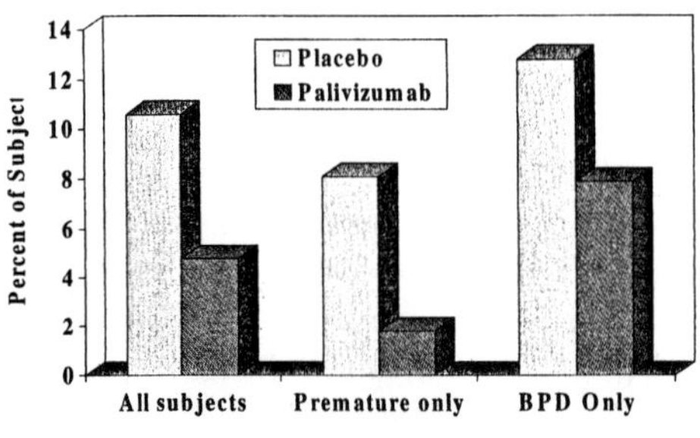

Figure 2. Palivizumab reduces hospitalization from RSV infection in high-risk infants. Data adapted from [51], with permission.

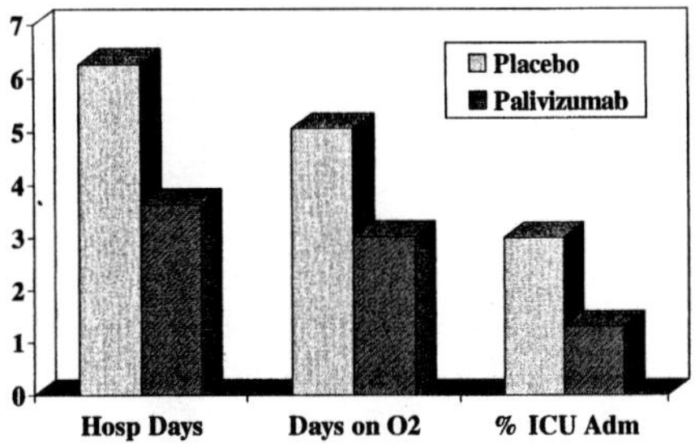

Figure 3. Palivizumab reduces morbidity from RSV pneumonitis in high-risk infants. Data shown are hospital days or days requiring supplemental oxygen per 10 infants, and the proportion of subjects requiring intensive care admission, because of RSV infection. Data adapted from [51], with permission.

those with bronchopulmonary dysplasia) and doubtless will be widely implemented now that the monoclonals are licenced. Whether the cost-benefit analysis will make it a sensible strategy for all infants remains to be seen.

2.2. Active Immunoprophylaxis (Immunization)

Development of an RSV vaccine has been one of infectious disease's "holy grails", and this goal has been actively pursued through research and development for well over thirty years, so far without success. However, the recent success of a live attenuated influenza vaccine,[89] and the development of vector- and DNA-based immunization strategies which have been highly successful for inducing broad protective antibody- and CTL-based immune responses against influenza in experimental animals,[90–93] suggest that there is some prospect that an effective RSV vaccine may be developed in the intermediate term. Further, DNA immunization has been shown to induce antibodies and cytotoxic T lymphocytes in neonates with passively-acquired maternal antibody,[94] a major consideration with RSV vaccines which will need to be given within weeks of birth. Development efforts using more conventional strategies are also continuing.

The development of live attenuated RSV vaccines has been a long-standing goal, particularly driven by the work of Bob Chanock and his collaborators.[95,96] Although the recent animal studies of temperature-sensitive and/or cold-adapted RSV strains as vaccines were promising,[97,98] the early human trials have been somewhat disappointing.[99] Specifically, vaccine strains have been transmitted to 20–25% of household members, it was difficult to achieve a degree of attenuation which provided adequate immunogenicity without causing disease; and mutants which were sufficiently attenuated for use in very young infants tended to be over-attenuated for older infants, while those suitable for older infants tended to retain some pathogenicity in seronegative infants.[99] The recent development of an RSV reverse genetics system by Collins and his collaborators in Chanock's laboratory,[100] and the identification of the mutations responsible for attenuation and their transfer to other strains of RSV using this technique, may allow rational design of improved live attenuated RSV vaccines.[97,101–104] Failing this goal, it may be possible to implement a two-stage strategy, in which young infants are given highly-attenuated strains of RSV, and are then re-immunized at a later date with a less attenuated variant.

The classical approach of developing subunit RSV vaccines using the F (or F and G) protein subunits as immunogens has been suspect because of adverse outcomes in the first Phase III clinical trial of an inactivated RSV vaccine conducted in the 1960's.[105–107] In this study, infants receiving a formalin-inactivated vaccine not only were not protected from RSV infection, but tended to get more severe infections, and at an older age, than placebo recipients.[105–107] The mechanism of this paradoxical vaccine effect has never been proven, but based on a variety of studies is thought to be due to selective inactivation of virion antigens by the formalin resulting in an inappropriate antibody response.[108–110] Furthermore, it is unlikely that very young infants—those most at risk for severe RSV disease—will respond to conventional purified protein immunogens. However, clinical trials by Lederle-Praxis of a purified F protein (PFP-2) vaccine have been promising in older infants at risk for severe RSV disease[111–114] as well as in at-risk adults.[115,116] Recent randomized trials of the PFP-2 vaccine in children with cystic fibrosis did not show protection against RSV infection *per se*, but did show a significant reduction in number of lower respiratory illnesses, antibiotic courses, and days of illness in the vaccine group.[111,117] Trials of the PFP-2 vaccine in elderly adults have shown it to be satisfactorily immunogenic, and the very preliminary protection data are promising.[115,116] Further studies with the fusion protein vaccine will be of interest.

It is clear that the long-term goal of preventing RSV disease will depend on immunoprophylaxis. Whether control will be achieved solely by passive immunoprophylaxis (allowing natural immunization by wild-type RSV infections "attenuated" by the passively administered antibody), by a combination of passive and active immunoprophylaxis, or whether vaccine technology will allow an active immunization strategy to be effective on its own, is unknown.

REFERENCES

1. Langley JM, *et al.* Economic evaluation of respiratory syncytial virus infection in Canadian children: a Pediatric Investigators Collaborative Network on Infections in Canada (PICNIC) study. Journal of Pediatrics 1997; 131:113–117.
2. Spelman DW, *et al.* Respiratory syncytial virus pneumonitis in adults. Medical Journal of Australia 1983; 1:430–431.
3. Dowell SF, *et al.* Respiratory syncytial virus is an important cause of community-acquired lower respiratory infection among hospitalized adults. Journal of Infectious Diseases 1996; 174:456–462.
4. Mlinaric-Galinovic G, *et al.* Respiratory syncytial virus infection in the elderly. European Journal of Clinical Microbiology & Infectious Diseases 1996; 15:777–781.
5. Whimbey E, *et al.* Respiratory syncytial virus pneumonia in hospitalized adult patients with leukemia. Clinical Infectious Diseases 1995; 21:376–379.
6. Falsey AR, *et al.* Respiratory syncytial virus and influenza A infections in the hospitalized elderly. Journal of Infectious Diseases 1995; 172:389–394.
7. Falsey AR, *et al.* Acute respiratory tract infection in daycare centers for older persons. Journal of the American Geriatrics Society 1995; 43:30–36.
8. Harrington RD, *et al.* An outbreak of respiratory syncytial virus in a bone marrow transplant center. Journal of Infectious Diseases 1992; 165:987–993.
9. Englund JA, *et al.* Respiratory syncytial virus infection in immunocompromised adults. Annals of Internal Medicine 1988; 109:203–208.
10. Fouillard L, *et al.* Severe respiratory syncytial virus pneumonia after autologous bone marrow transplantation: a report of three cases and review. Bone Marrow Transplantation 1992; 9:97–100.
11. Whimbey E, *et al.* Community respiratory virus infections among hospitalized adult bone marrow transplant recipients. Clinical Infectious Diseases 1996; 22:778–782.
12. Hall CB, *et al.* Aerosolized ribavirin treatment of infants with respiratory syncytial viral infection. A randomized double-blind study. New England Journal of Medicine 1983; 308:1443–1447.
13. Mills J. Management of respiratory syncytial virus infections. Advances in Experimental Medicine & Biology 1996; 394:163–174.
14. Randolph AG, *et al.* Ribavirin for respiratory syncytial virus lower respiratory tract infection. A systematic overview. Archives of Pediatrics & Adolescent Medicine 1996; 150:942–947.
15. Anonymous. Reassessment of the indications for ribavirin therapy in respiratory syncytial virus infections. American Academy of Pediatrics Committee on Infectious Diseases. Pediatrics 1996; 97:137–140.

16. Whimbey E, *et al.* Combination therapy with aerosolized ribavirin and intravenous immunoglobulin for respiratory syncytial virus disease in adult bone marrow transplant recipients. Bone Marrow Transplantation 1995; 16:393–399.
17. Lewinsohn DM, *et al.* Phase I study of intravenous ribavirin treatment of respiratory syncytial virus pneumonia after marrow transplantation. Antimicrobial Agents & Chemotherapy 1996; 40:2555–2557.
18. Sparrelid E, *et al.* Ribavirin therapy in bone marrow transplant recipients with viral respiratory tract infections. Bone Marrow Transplantation 1997; 19:905–908.
19. Cirino NM, *et al.* Targeting RNA decay with 2',5' oligoadenylate-antisense in respiratory syncytial virus-infected cells. Proceedings of the National Academy of Sciences of the United States of America 1997; 94:1937–1942.
20. Jairath S, *et al.* Inhibition of respiratory syncytial virus replication by antisense oligodeoxyribonucleotides. Antiviral Research 1997; 33:201–213.
21. Barnard DL, *et al.* Potent inhibition of respiratory syncytial virus by polyoxometalates of several structural classes. Antiviral Research 1997; 34:27–37.
22. Wyde PR, *et al.* Evaluation of the antiviral activity of N-(phosphonoacetyl)-L-aspartate against paramyxoviruses in tissue culture and against respiratory syncytial virus in cotton rats. Antiviral Research 1995; 27:59–69.
23. Villanueva N, *et al.* Antiviral effects of xanthate D609 on the human respiratory syncytial virus growth cycle. Virology 1991; 181:101–108.
24. Barry W, *et al.* Ribavirin aerosol for acute bronchiolitis. Archives of Disease in Childhood 1986; 61:593–597.
25. Hall CB, *et al.* Respiratory syncytial viral infection in children with compromised immune function. New England Journal of Medicine 1986; 315:77–81.
26. Groothuis JR, *et al.* Early ribavirin treatment of respiratory syncytial viral infection in high-risk children. Journal of Pediatrics 1990; 117:792–798.
27. Wheeler JG, *et al.* Historical cohort evaluation of ribavirin efficacy in respiratory syncytial virus infection. Pediatric Infectious Disease Journal 1993; 12:209–213.
28. Ohmit SE, *et al.* Ribavirin utilization and clinical effectiveness in children hospitalized with respiratory syncytial virus infection. Journal of Clinical Epidemiology 1996; 49:963–967.
29. Edell D, *et al.* Reduced long-term respiratory morbidity after treatment of respiratory syncytial virus bronchiolitis with ribavirin in previously healthy infants: a preliminary report. Pediatric Pulmonology 1998; 25:154–158.
30. Krilov LR, *et al.* Follow-up of children with respiratory syncytial virus bronchiolitis in 1986 and 1987: potential effect of ribavirin on long term pulmonary function. The Bronchiolitis Study Group. Pediatric Infectious Disease Journal 1997; 16:273–276.
31. Smith DW, *et al.* A controlled trial of aerosolized ribavirin in infants receiving mechanical ventilation for severe respiratory syncytial virus infection. New England Journal of Medicine 1991; 325:24–29.
32. DeVincenzo J. Prevention and treatment of respiratory syncytial virus infections. Advances in Pediatric Infectious Diseases 1997; 13:1–47.
33. Moler FW, *et al.* Effectiveness of ribavirin in otherwise well infants with respiratory syncytial virus-associated respiratory failure. Pediatric Critical Study Group. Journal of Pediatrics 1996; 128:422–428.
34. Meert KL, *et al.* Aerosolized ribavirin in mechanically ventilated children with respiratory syncytial virus lower respiratory tract disease: a prospective, double-blind, randomized trial. Critical Care Medicine 1994; 22:566–572.
35. McColl MD, *et al.* Respiratory syncytial virus infection in adult BMT recipients: effective therapy with short duration nebulised ribavirin. Bone Marrow Transplantation 1998; 21:423–425.
36. Groothuis JR, *et al.* Immunoprophylaxis and immunotherapy: role in the prevention and treatment of respiratory syncytial virus. Internat J Antimicrob Agents 1993; 2:97–104.
37. Graham BS, *et al.* Immunoprophylaxis and immunotherapy of respiratory syncytial virus-infected mice with respiratory syncytial virus-specific immune serum. Pediatric Research 1993; 34:167–172.
38. Hemming VG, *et al.* Topically administered immunoglobulin reduces pulmonary respiratory syncytial virus shedding in owl monkeys. Antimicrobial Agents & Chemotherapy 1988; 32:1269–1270.
39. Crowe JE, Jr., *et al.* Recombinant human respiratory syncytial virus (RSV) monoclonal antibody Fab is effective therapeutically when introduced directly into the lungs of RSV-infected mice. Proceedings of the National Academy of Sciences of the United States of America 1994; 91:1386–1390.
40. Prince GA, *et al.* Immunoprophylaxis and immunotherapy of respiratory syncytial virus infection in the cotton rat. Virus Research 1985; 3:193–206.
41. Prince GA, *et al.* Mechanism of antibody-mediated viral clearance in immunotherapy of respiratory syncytial virus infection of cotton rats. Journal of Virology 1990; 64:3091–3092.

42. Hemming VG, *et al.* Intravenous immunoglobulin treatment of respiratory syncytial virus infections in infants and young children. Antimicrobial Agents & Chemotherapy 1987; 31:1882–1886.
43. Rodriguez WJ, *et al.* Respiratory syncytial virus (RSV) immune globulin intravenous therapy for RSV lower respiratory tract infection in infants and young children at high risk for severe RSV infections: Respiratory Syncytial Virus Immune Globulin Study Group. Pediatrics 1997; 99:454–461.
44. Rimensberger PC, *et al.* Clinical experience with aerosolized immunoglobulin treatment of respiratory syncytial virus infection in infants. Pediatric Infectious Disease Journal 1994; 13:328–330.
45. Rimensberger PC, *et al.* Aerosolized immunoglobulin treatment of respiratory syncytial virus infection in infants. Pediatric Infectious Disease Journal 1996; 15:209–216.
46. Davis CB, *et al.* Preclinical pharmacokinetic evaluation of the respiratory syncytial virus-specific reshaped human monoclonal antibody RSHZ19. Drug Metabolism & Disposition 1995; 23:1028–1036.
47. Wyde PR, *et al.* Evaluation of the protective efficacy of reshaped human monoclonal antibody RSHZ19 against respiratory syncytial virus in cotton rats. Pediatric Research 1995; 38:543–550.
48. Everitt DE, *et al.* The pharmacokinetics, antigenicity, and fusion-inhibition activity of RSHZ19, a humanized monoclonal antibody to respiratory syncytial virus, in healthy volunteers. Journal of Infectious Diseases 1996; 174:463–469.
49. Weltzin R, *et al.* Intranasal monoclonal IgA antibody to respiratory syncytial virus protects rhesus monkeys against upper and lower respiratory tract infection. Journal of Infectious Diseases 1996; 174:256–261.
50. Welliver RC. Respiratory syncytial virus immunoglobulin and monoclonal antibodies in the prevention and treatment of respiratory syncytial virus infection. Seminars in Perinatology 1998; 22:87–95.
51. The IMpact-RSV Study Group. Palivizumab, a humanized respiratory syncytial virus monoclonal antibody, reduces hospitalization from respiratory syncytial virus infection in high-risk infants. Pediatrics 1998; 102:531–537.
52. Johnson S, *et al.* Development of a humanized monoclonal antibody (MEDI-493) with potent in vitro and in vivo activity against respiratory syncytial virus. Journal of Infectious Diseases 1997; 176:1215–1224.
53. Saez-Llorens X, *et al.* Respiratory syncytial virus monoclonal antibody (MEDI-493) administered intramuscularly in high-risk children. Presented at the 37th Interscience Conference on Antimicrobial Agents and Chemotherapy (ICAAC), 1997. Toronto, Canada. Abstract 138 H
54. Dowell SF, *et al.* Treatment of respiratory syncytial virus infection with vitamin A: a randomized, placebo-controlled trial in Santiago, Chile. Pediatric Infectious Disease Journal 1996; 15:782–786.
55. Bresee JS, *et al.* Vitamin A therapy for children with respiratory syncytial virus infection: a multicenter trial in the United States. Pediatric Infectious Disease Journal 1996; 15:777–782.
56. Vos GD, *et al.* Treatment of respiratory failure due to respiratory syncytial virus pneumonia with natural surfactant. Pediatric Pulmonology 1996; 22:412–415.
57. Graham BS, *et al.* Topical immunoprophylaxis of respiratory syncytial virus (RSV)-challenged mice with RSV-specific immune globulin. Journal of Infectious Diseases 1995; 171:1468–1474.
58. Sami IR, *et al.* Systemic immunoprophylaxis of nasal respiratory syncytial virus infection in cotton rats. Journal of Infectious Diseases 1995; 171:440–443.
59. Taylor G, *et al.* Monoclonal antibodies protect against respiratory syncytial virus infection in mice. Immunology 1984; 52:137–142.
60. Walsh EE, *et al.* Protection from respiratory syncytial virus infection in cotton rats by passive transfer of monoclonal antibodies. Infection & Immunity 1984; 43:756–758.
61. Wang E, *et al.* PICNIC (Pediatric Investigators Collaborative Network on Infections in Canada) study of the role of age and respiratory syncytial virus neutralizing antibody on respiratory syncytial virus illness in patients with underlying heart or lung disease. Pediatrics (*Electronic Pages*) 1997; 99:9–15.
62. Glezen WP, *et al.* Risk of respiratory syncytial virus infection for infants from low-income families in relationship to age, sex, ethnic group, and maternal antibody level. Journal of Pediatrics 1981; 98:708–715.
63. Meissner HC, *et al.* Controlled trial to evaluate protection of high-risk infants against respiratory syncytial virus disease by using standard intravenous immune globulin. Antimicrob Agents Chemother 1993; 37:1655–1658.
64. Siber GR, *et al.* Protective activity of a human respiratory syncytial virus immune globulin prepared from donors screened by microneutralization assay. Journal of Infectious Diseases 1992; 165:456–463.
65. Groothuis JR, *et al.* Prophylactic administration of respiratory syncytial virus immune globulin to high-risk infants and young children. The Respiratory Syncytial Virus Immune Globulin Study Group. New England Journal of Medicine 1993; 329:1524–1530.
66. Groothuis JR, *et al.* Respiratory syncytial virus (RSV) infection in preterm infants and the protective effects of RSV immune globulin (RSVIG). Respiratory Syncytial Virus Immune Globulin Study Group. Pediatrics 1995; 95:463–467.

67. Anonymous. Reduction of respiratory syncytial virus hospitalization among premature infants and infants with bronchopulmonary dysplasia using respiratory syncytial virus immune globulin prophylaxis. The PREVENT Study Group. Pediatrics 1997; 99:93–99.
68. Simoes E, *et al.* Respiratory syncytial virus immunoglbulin as prophylaxis against respiratory syncytial virus in children with congential heart disease [Abstract]. Pediatric Research 1996; 39:113A.
69. Anonymous. Respiratory syncytial virus immune globulin intravenous: indications for use. American Academy of Pediatrics Committee on Infectious Diseases, Committee on Fetus and Newborn. Pediatrics 1997; 99:645–650.
70. Okamoto Y, *et al.* Presence of respiratory syncytial virus genomic sequences in middle ear fluid and its relationship to expression of cytokines and cell adhesion molecules. Journal of Infectious Diseases 1993; 168:1277–1281.
71. Shaw CB, *et al.* Incidence of adenovirus and respiratory syncytial virus in chronic otitis media with effusion using the polymerase chain reaction. Otolaryngology - Head & Neck Surgery 1995; 113:234–241.
72. Simoes EA, *et al.* Respiratory syncytial virus-enriched globulin for the prevention of acute otitis media in high risk children. Journal of Pediatrics 1996; 129:214–219.
73. Andrade MA, *et al.* Acute otitis media in children with bronchiolitis. Pediatrics 1998; 101:617–619.
74. Hay J, *et al.* Respiratory syncytial virus immune globulin: a cost-effectiveness analysis. Am J Managed Care 1996; 2:851–861.
75. Meissner HC, *et al.* Immunoprophylaxis and the control of respiratory syncytial virus disease. Pediatrics 1997; 100:260–263.
76. Garcia-Barreno B, *et al.* Frame shift mutations as a novel mechanism for the generation of neutralization resistant mutants of human respiratory syncytial virus. EMBO Journal 1990; 9:4181–4187.
77. Rueda P, *et al.* Premature stop codons in the G glycoprotein of human respiratory syncytial viruses resistant to neutralization by monoclonal antibodies. Journal of Virology 1991; 65:3374–3378.
78. Rueda P, *et al.* Loss of conserved cysteine residues in the attachment (G) glycoprotein of two human respiratory syncytial virus escape mutants that contain multiple A-G substitutions (hypermutations). Virology 1994; 198:653–662.
79. Garcia-Barreno B, *et al.* Oligo(A) sequences of human respiratory syncytial virus G protein gene: assessment of their genetic stability in frameshift mutants. Journal of Virology 1994; 68:5460–5468.
80. Rueda P, *et al.* The three C-terminal residues of human respiratory syncytial virus G glycoprotein (Long strain) are essential for integrity of multiple epitopes distinguishable by antiidiotypic antibodies. Viral Immunology 1995; 8:37–46.
81. Duncan RJ, *et al.* Antigenic diversity of respiratory syncytial viruses and its implication for immunoprophylaxis in ruminants. Veterinary Microbiology 1993; 37:319–341.
82. Orvell C, *et al.* Preparation and characterization of monoclonal antibodies directed against five structural components of human respiratory syncytial virus subgroup B. Journal of General Virology 1987; 68:3125–3135.
83. Collins PL, *et al.* Respiratory syncytial viruses (Chapter 44), in: Fields BN, Knipe DM, Howley PM, Editors. Virology, Lippincott-Raven Publishers: Philadelphia. 1996. p. 1313–1352.
84. Taylor G, *et al.* Humanised monoclonal antibody to respiratory syncytial virus. Lancet 1991; 337:1411–1412.
85. Weltzin R, *et al.* Intranasal monoclonal immunoglobulin A against respiratory syncytial virus protects against upper and lower respiratory tract infections in mice. Antimicrobial Agents & Chemotherapy 1994; 38:2785–2791.
86. Beeler JA, *et al.* Neutralization epitopes of the F glycoprotein of respiratory syncytial virus: effect of mutation upon fusion function. Journal of Virology 1989; 63:2941–2950.
87. Subramanian KN, *et al.* Safety, tolerance and pharmacokinetics of a humanized monoclonal antibody to respiratory syncytial virus in premature infants and infants with bronchopulmonary dysplasia. MEDI-493 Study Group. Pediatric Infectious Disease Journal 1998; 17:110–115.
88. Connor E, *et al.* Reduction of respiratory syncytial virus (RSV) hospitalization in children with prematurigy or bronchopulmonary sdysplasia (BPD) using intramuscular (IM) humanized monoclonal antibody to the RSV F protein (MEDI-493; palivizumab). Presented at the Society for Pediatric Research Annual Meeting, 1998. San Francisco. CA
89. Belshe RB, *et al.* The efficacy of live attenuated, cold-adapted, trivalent, intranasal influenza virus vaccine in children. New England Journal of Medicine 1998; 338:1405–1412.
90. Ulmer JB, *et al.* Heterologous protection against influenza by injection of DNA encoding a viral protein. Science 1993; 259:1745–1749.
91. Webster RG, *et al.* Protection of ferrets against influenza challenge with a DNA vaccine to the haemagglutinin. Vaccine 1994; 12:1495–1498.

92. Kodihalli S, *et al.* Cross-protection among lethal H5N2 influenza viruses induced by DNA vaccine to the hemagglutinin. Journal of Virology 1997; 71:3391–3396.
93. Donnelly JJ, *et al.* Preclinical efficacy of a prototype DNA vaccine: enhanced protection against antigenic drift in influenza virus. Nature Medicine 1995; 1:583–587.
94. Hassett DE, *et al.* Neonatal DNA immunization with a plasmid encoding an internal viral protein is effective in the presence of maternal antibodies and protects against subsequent viral challenge. Journal of Virology 1997; 71:7881–7888.
95. Wright PF, *et al.* Evaluation of a temperature-sensitive mutant of respiratory syncytial virus in adults. Journal of Infectious Diseases 1971; 124:505–511.
96. Gharpure MA, *et al.* Temperature-sensitive mutants of respiratory syncytial virus. Journal of Virology 1969; 3:414–421.
97. Juhasz K, *et al.* The temperature-sensitive (*ts*) phenotype of a cold-passaged (*cp*) live attenuated respiratory syncytial virus vaccine candidate, designated cpts530, results from a single amino acid substitution in the L protein. Journal of Virology 1997; 71:5814–5819.
98. Crowe JE, Jr, *et al.* A comparison in chimpanzees of the immunogenicity and efficacy of live attenuated respiratory syncytial virus (RSV) temperature-sensitive mutant vaccines and vaccinia virus recombinants that express the surface glycoproteins of RSV. Vaccine 1993; 11:1395–1404.
99. Karron RA, *et al.* Evaluation of two live, cold-passaged, temperature-sensitive respiratory syncytial virus vaccines in chimpanzees and in human adults, infants, and children. Journal of Infectious Diseases 1997; 176:1428–1436.
100. Collins PL, *et al.* Production of infectious human respiratory syncytial virus from cloned cDNA confirms an essential role for the transcription elongation factor from the 5' proximal open reading frame of the M2 mRNA in gene expression and provides a capability for vaccine development. Proceedings of the National Academy of Sciences of the United States of America 1995; 92:11563–11567.
101. Whitehead SS, *et al.* Recombinant respiratory syncytial virus (RSV) bearing a set of mutations from cold-passaged RSV is attenuated in chimpanzees. Journal of Virology 1998; 72:4467–4471.
102. Karron RA, *et al.* Respiratory syncytial virus (RSV) SH and G proteins are not essential for viral replication in vitro: clinical evaluation and molecular characterization of a cold-passaged, attenuated RSV subgroup B mutant. Proceedings of the National Academy of Sciences of the United States of America 1997; 94:13961–13966.
103. Bukreyev A, *et al.* Recombinant respiratory syncytial virus from which the entire SH gene has been deleted grows efficiently in cell culture and exhibits site-specific attenuation in the respiratory tract of the mouse. Journal of Virology 1997; 71:8973–8982.
104. Connors M, *et al.* A cold-passaged, attenuated strain of human respiratory syncytial virus contains mutations in the F and L genes. Virology 1995; 208:478–484.
105. Fulginiti VA, *et al.* Respiratory virus immunization. I. A field trial of two inactivated respiratory virus vaccines; an aqueous trivalent parainfluenza virus vaccine and an alum-precipitated respiratory syncytial virus vaccine. American Journal of Epidemiology 1969; 89:435–448.
106. Chin J, *et al.* Field evaluation of a respiratory syncytial virus vaccine and a trivalent parainfluenza virus vaccine in a pediatric population. American Journal of Epidemiology 1969; 89:449–463.
107. Kim HW, *et al.* Respiratory syncytial virus disease in infants despite prior administration of antigenic inactivated vaccine. American Journal of Epidemiology 1969; 89:422–434.
108. Murphy BR, *et al.* Current status of respiratory syncytial virus (RSV) and parainfluenza virus type 3 (PIV3) vaccine development: memorandum from a joint WHO/NIAID meeting. Bulletin of the World Health Organization 1997; 75:307–313.
109. Waris ME, *et al.* Priming with live respiratory syncytial virus (RSV) prevents the enhanced pulmonary inflammatory response seen after RSV challenge in BALB/c mice immunized with formalin-inactivated RSV. Journal of Virology 1997; 71:6935–6939.
110. Tang YW, *et al.* T cell source of type 1 cytokines determines illness patterns in respiratory syncytial virus-infected mice. Journal of Clinical Investigation 1997; 99:2183–2191.
111. Piedra PA, *et al.* Sequential annual administration of purified fusion protein vaccine against respiratory syncytial virus in children with cystic fibrosis. Pediatric Infectious Disease Journal 1998; 17:217–224.
112. Tristram DA, *et al.* Second-year surveillance of recipients of a respiratory syncytial virus (RSV) F protein subunit vaccine, PFP-1: evaluation of antibody persistence and possible disease enhancement. Vaccine 1994; 12:551–556.
113. Belshe RB, *et al.* Immunogenicity of purified F glycoprotein of respiratory syncytial virus: clinical and immune responses to subsequent natural infection in children. Journal of Infectious Diseases 1993; 168:1024–1029.

114. Tristram DA, *et al.* Immunogenicity and safety of respiratory syncytial virus subunit vaccine in seropositive children 18–36 months old. J Infect Dis 1993; 167:191–195.
115. Falsey AR, *et al.* Safety and immunogenicity of a respiratory syncytial virus subunit vaccine (PFP-2) in the institutionalized elderly. Vaccine 1997; 15:1130–1132.
116. Falsey AR, *et al.* Safety and immunogenicity of a respiratory syncytial virus subunit vaccine (PFP-2) in ambulatory adults over age 60. Vaccine 1996; 14:1214–1218.
117. Piedra PA, *et al.* Purified fusion protein vaccine protects against lower respiratory tract illness during respiratory syncytial virus season in children with cystic fibrosis. Pediatric Infectious Disease Journal 1996; 15:23–31.
118. Meissner HC, *et al.* Prevention of respiratory syncytial virus infection in high risk infants: consensus opinion on the role of immunoprophylaxis with respiratory syncytial virus hyperimmune globulin. Pediatric Infectious Disease Journal 1996; 15:1059–1068.

6

UPDATE ON INFLUENZA AND RHINOVIRUS INFECTIONS

Frederick G. Hayden

University of Virginia School of Medicine
Box 473, Health Sciences Center
Charlottesville, Virginia 22908

Since the 4th Triennial Conference at Antiviral Chemotherapy held in November 1994, no new antiviral drugs or vaccines have been approved for use in influenza or rhinovirus infections in the United States. The only newly approved therapy for symptomatic management of upper respiratory illnesses is topical ipratropium. Intranasal administration of this quaternary anticholinergic reduces colds-associated rhinorrhea to a moderate extent.[1] However, considerable clinical investigation has led to important advances in regard to prevention and treatment of influenza virus infections and provided new insights in regard to possible measures for management of rhinovirus infections. This paper will update previous reviews from this conference[2,3,4,5] and summarize recent developments with agents that are in active clinical development.

1. NEURAMINIDASE INHIBITORS

1.1. Historical Background

In 1942 Hirst reported the discovery of an enzyme activity on the surface of influenza viruses that removed virus receptors from erythrocytes.[6] Five years later Burnet and his colleagues showed that a similar activity, which they termed receptor destroying enzyme or RDE, was present in a culture filtrate of *Vibrio cholera*. This group established that the actions of the influenza virus enzyme were reproduced by the bacterial one and that the substrates of both enzymes were components of a variety of mucopolysaccharides.[7] These workers also showed that various mucins and mucoid substances, including crude human respiratory tract mucus, inhibited hemagglutination by influenza virus and that this inhibition was reversible upon addition of fresh virus or receptor destroying enzyme.[8] They speculated "an effective 'competitive poison' for the virus enzyme might be similarly administered which, when deposited on the mucus film lining the respiratory

Antiviral Chemotherapy 5, edited by Mills *et al.*
Kluwer Academic / Plenum Publishers, New York, 1999.

tract, would render this an effective area against infection, both initial infection from without and the spreading surface infection mucosa which follows the initiation of infection".[7] Subsequently Gottschalk[9] characterized the chemical structure of *N*-acetylneuraminic acid (sialic acid), its mode of linkage to neuraminidase-cleavable molecules, and the specificity of the enzyme toward terminal sialic acid residues. A number of neuraminidase inhibitors have been described but most have lacked potency and selectivity.[10] Following definition of the crystal structure of neuraminidase in 1983 by Colman and his colleagues,[11] a rational design approach led to the synthesis of sialic acid derivatives with potent and selective antiviral activity.[12] Currently, two agents (zanamivir or GG167 and GS4104 or Ro 64–0796) have undergone clinical testing and are being taken forward in large-scale field trials.

1.2. Neuraminidase Actions

The neuraminidase of influenza A and B viruses exerts multiple actions and is essential for replication *in vitro* but not for viral entry, replication, assembly, or budding.[13] Cleavage of the terminal sialic acid residues eliminates the receptor for viral hemagglutinin, and this action allows release of budded virus from infected cells, such that virions do not aggregate at the cell surface or with each other. Influenza aggregates have been observed by electron microscopy when the activity of neuraminidase has been reduced by antibody, selective inhibitors, or mutation.[13] Replication of neuraminidase-deficient virus occurs at lower levels in mice. In addition, neuraminidase may cleave sialic acid residues from respiratory tract mucus and prevent viral inactivation and indirectly potentiate viral penetration to the respiratory mucosa. The neuraminidase also appears to influence viral virulence. For certain strains it appears to enhance viral pathogenicity by removal of carbohydrate from hemagglutinin to enhance its cleavability.[14] In addition, recent evidence suggests that neuraminidase directly activates latent transforming growth factor B *in vitro* and in experimentally infected mice, which may contribute to apoptosis in cells.[15] Neuraminidase and/or hemagglutinin have been implicated in the pyrogenic response to influenza infection in ferrets[16] and purified neuraminidase, but not hemagglutinin, has been shown to increase production of interleukin-1 and tumor necrosis factor in macrophage cultures.[17] Because of these diverse actions and because the essential amino acids that define the active enzyme site are highly conserved among influenza A virus subtypes and influenza B viruses, neuraminidase is an important target for antiviral drug development.

1.3. Zanamivir

The neuraminidase inhibitor which has undergone the most extensive clinical testing is zanamivir, also termed 4-guanidino-Neu5Ac2en or GG167 (Figure 1). Mechanistic studies have shown that its antiviral activity is influenza neuraminidase specific. It has no effect on hemagglutinin function or on the neuraminidases from other pathogenic organisms or mammalian cells except at concentrations that are approximately 10^6 higher than those for influenza.[12] The interaction of the guanidino group with two conserved framework residues, Glu 119 and Glu 227, gives zanamivir a tight affinity for the active enzyme site. Zanamivir is inhibitory for all nine influenza A neuraminidase subtypes and influenza B neuraminidases studied to date. Inhibition occurs at nanomolar concentrations (0.64–7.9 nM) and varies over approximately 10-fold range in enzyme inhibition assays. In contrast, inhibitory concentrations for growth of influenza A and B viruses in cell culture systems show wide variability. The 50% inhibitory concentrations in cell cultures systems vary up

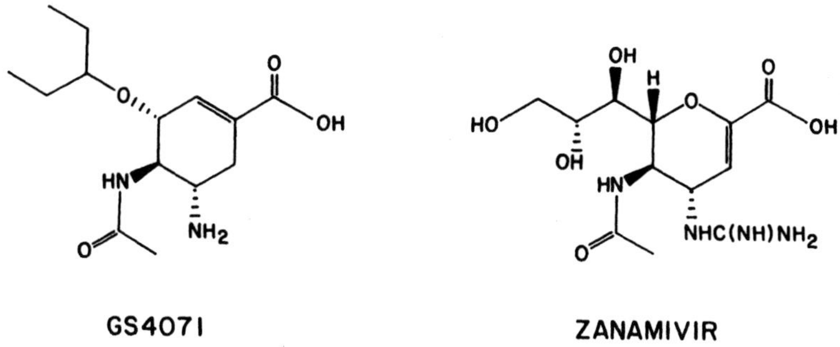

Figure 1. Structure of GS4071 and zanamivir.

to 4000-fold (0.005–16 uM).[18,19] The activity of the inhibitor is dependent on the cell culture system used and it is likely that the relative importance of hemagglutinin and neuraminidase in allowing release of virus from cells and in forming plaques varies among virus stains in cell culture systems. A reliable and simple cell culture system for assessing the susceptibility of clinical isolates of influenza viruses has not been established. MDCK cells do not appear to provide good correlations between inhibition in plaque assays and inhibition observed in assays of enzymatic activity.

Experiments in ferret models of influenza have shown that topically applied zanamivir inhibits viral replication at relatively low doses given on a twice-daily basis.[12,20] When administered intranasally in mice, zanamivir is about 100-fold more potent in inhibiting viral replication on a weight-adjusted basis than amantadine and 1000-fold more active than ribavirin. Delayed administration has been shown to reduce mortality and viral replication in such models, and topical zanamivir is also active in ferrets at low doses.[21] Inhibitory doses after intraperitoneal administration are over 1000-fold higher in murine systems. The compound is poorly absorbed after oral administration in multiple species (generally <10%) and rapidly cleared by renal elimination (plasma elimination $T_{1/2}$ <1 hour). There was a small volume of distribution, generally limited to the extracellular space, and apparently poor penetration into respiratory secretions. The pharmacologic profile of zanamivir is consistent with a extracellular cell surface effect.[20] Despite these limitations, Sidwell and coworkers have found recently that oral zanamivir does have dose-related antiviral activity for certain influenza viruses in experimentally infected mice.[22]

The human pharmacokinetics of zanamivir have been determined in uninfected subjects following intravenous, topical, and oral administration.[23] The oral bioavailability of zanamivir has been estimated to be 5% or less in humans. Humans also show rapid renal elimination of zanamivir with no significant metabolism and a volume of distribution limited to extracellular body water. The plasma elimination $T_{1/2}$ ranges from 1.5–1.8 hours. Because of these reasons, clinical studies to date have used direct application to the upper and/or lower respiratory tract. For studies in natural influenza zanamivir has been formulated as a micronized dry powder blended with lactose for inhalation. Bioavailability after inhalation is approximately 15–20%. Gamma scintigraphy studies in uninfected volunteers have found that an average of 13% (8–21%) of a zanamivir dose from the inhaler device was deposited in the lungs compared to an average of 78% deposited in the oropharynx.[24] The distribution in the lung appears to be relatively uniform. Such results are indicative of a relatively inefficient delivery of drug to the tracheobronchial tree and lower respiratory tract. The durability of antiviral effects after administration to the respiratory tract remains to be determined but most studies have employed twice daily dosing regimens.

In humans prophylaxis with intranasal doses ranging from 3.6–16 mg per application and frequencies of 2 or 6 times per day all showed protection against experimental influenza A(H1N1) infection.[25] For the combined zanamivir recipients, the protective efficacy was 96% against viral shedding, 82% against laboratory documented infection, and 95% against febrile illness compared to placebo. In similar studies conducted in influenza B virus inoculated volunteers, intranasal zanamivir did not significantly reduce overall infection rates but was associated with 60% protection against viral shedding. Those receiving twice-daily dosing also had reduced frequencies of upper respiratory illness compared to placebo or to recipients of once-daily doses in studies involving small numbers of subjects.[26]

When early treatment with zanamivir was initiated approximately one day after experimental influenza A(H1N1) infection, significant antiviral and clinical effects were observed.[25] Peak viral titers were reduced by over 100-fold and the median duration of shedding reduced by 3 days compared to placebo. The risk of febrile illness was reduced by 84% and other illness measures were lessened, including use of acetaminophen. To assess the potential of antiviral interventions to reduce upper respiratory tract complications of influenza, middle ear pressures were monitored during these studies.[27] With prophylaxis the frequency of middle ear pressure abnormalities, principally marked under-pressures, was reduced from 61% of placebo recipients to 15% of zanamivir recipients. Similarly, with early treatment only 32% of zanamivir recipients compared to 73% of placebo recipients developed such abnormalities. The frequency of individuals with earache or pressure symptoms was also reduced by 2–3 fold compared to placebo. These findings indicate that zanamivir is effective not only in reducing experimental influenza illness but also in reducing the risk of middle ear pressure abnormalities. Whether antiviral therapy might reduce the risk of otitis media in a high-risk group like young children remains to be determined.

Because of these encouraging results, studies were undertaken to assess the efficacy and safety of zanamivir in treating naturally occurring influenza virus infections in adults. During the 1994–95 influenza season 70 centers in North America and Europe enrolled a total of 417 adults with influenza-like illness of <48 hours duration.[28] These individuals were randomly assigned to one of three treatment groups: inhaled zanamivir (10 mg /dose), inhaled zanamivir and intranasal spray zanamivir (6.4 mg/dose), or placebo by both routes, each administered twice daily for 5 days. Among the 63% of patients with laboratory documented influenza virus infection (56% influenza A, 44% influenza B), the median time to alleviation of all major influenza symptoms was one day shorter in either of the zanamivir groups (4 days) compared to placebo (5 days). Among those who were febrile at enrollment (Figure 2) or who were enrolled within 30 hours of symptom onset, the duration of illness was more protracted in the placebo group (7 days) but the time to alleviation remained at 4 days in the zanamivir groups. The latter zanamivir groups also experienced 1–2 day shortening in the time to resumption of their usual activities compared to placebo. No significant benefit was observed in those who lacked fever at enrollment or were treated at 30 hours after symptom onset. As predicted by the antiviral spectrum of zanamivir, effectiveness was also observed in both influenza A and B virus infections. No local intolerance or other significant side effects were recognized during the trial. Reduced titers of virus in upper respiratory tract secretions were observed in those who received intranasal zanamivir (Figure 3). Whether the observed reductions in viral titers and the duration of cough might reduce the risk of transmitting influenza remains to be determined.

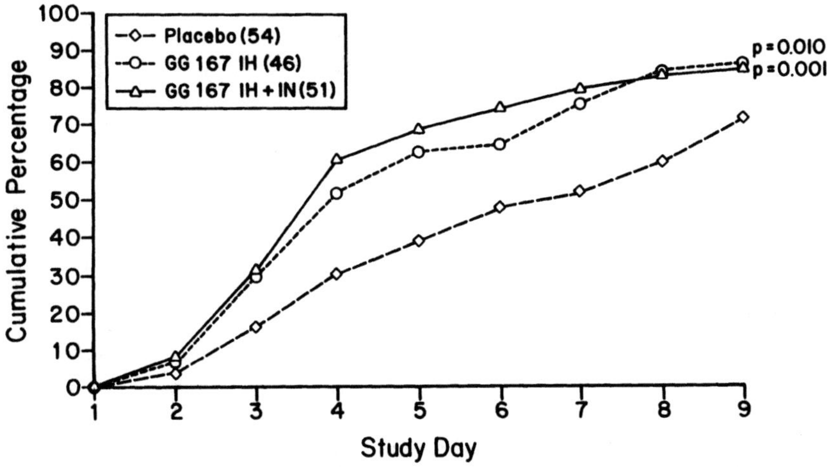

Figure 2. Duration of illness in patients with fever at enrollment and laboratory documented influenza treated with (◇) placebo, (O) inhaled zanamivir (GG167 IH), (△) inhaled and intranasal spray zanamivir (GG167 IH + IN).

A subsequent placebo-controlled trial conducted during the 1995–96 influenza season compared twice and four times daily treatments with zanamivir using both inhaled (10 mg/dose) and intranasal (6.4 mg/dose) routes of administration for 5 days.[29] The study also found zanamivir-related reductions in the time to alleviation of major symptoms of 1–2 days. This study also assessed the impact of influenza on patient quality of life, productivity, and health status assessed by a standard questionnaire. Significantly less sleep disturbance, inability to engage in work or leisure activities, work time lost, and impact on overall health status were found in zanamivir recipients. Unscheduled contacts with health care providers were significantly reduced from 14% in the placebo group to 8% in each of the zanamivir groups. No significant differences were detected between the twice and four times daily zanamivir groups. These results highlight the adverse consequences of acute influenza infections and show that antiviral treatment can provide meaningful functional and quality of life benefits. Further studies of zanamivir for both prophylaxis and for treating acute influ-

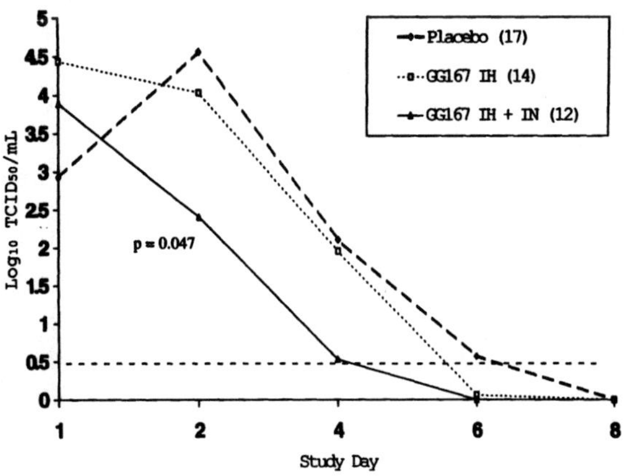

Figure 3. Influenza virus titres in upper respiratory tract secretions (nasal washings) in patients with laboratory documented influenza treated with (◇) placebo, (□) inhaled zanamavir (GG167 IH), (△) inhaled and intranasal zanamavir (GG167 IH + IN).

enza in outpatients and serious influenza in hospitalized patients are planned for the 1997–98 influenza season in the northern hemisphere. In addition, a placebo controlled trial of nebulized zanamivir is being initiated under the auspices of the NIH Collaborative Antiviral Study Group for treating serious influenza in hospitalized patients.

2. GS4104

In an effort to develop orally bioavailable influenza virus neuraminidase inhibitors, a series of carbocyclic transition-state analogs of sialic acid have been synthesized.[30] Replacement of the polar glycerol of zanamivir with a lipophilic 3-pently group and elimination of the guanidinium moiety has resulted in a molecule designated GS4107 (Figure 1). This compound inhibits the neuraminidase activity and replication in cell culture systems of various influenza viruses at concentrations similar to those for zanamivir.[31,32] Activity has been demonstrated against all nine influenza A neuraminidase subtypes at concentrations similar to those observed for zanamivir. Although clinical isolates of influenza virus exhibit a wide range (> 100-fold) in inhibitory concentrations that vary depending on the cell culture system and virus, inhibition of enzyme activity occurs over a narrow range (approximately 10-fold). GS4071 is similar to zanamivir in that it lacks significant inhibitory effect against the neuraminidases from human liver, parainfluenza virus, or bacterial sources. The crystal structure of GS4071 complexed into the neuraminidase active site reveals that the lipophilic side chain of the molecule occupies a new hydrophobic pocket in the enzyme within the region corresponding to the glycerol side chain of the natural substrate. The antiviral potency of GS4071 has been attributed to the highly favorable interaction in this pocket.[33]

GS4071 has poor oral bioavailability itself, but its ethyl ester derivative, designated GS4104, provides good oral biolavailability of the active parent drug (30–73%) in multiple species.[34] After oral administration of GS4104 in the rat, levels of the parent compound GS4071 in bronchioalveolar lavage fluid reached maximum concentrations similar to those observed in plasma but persisted for a longer period of time.[35]

In experimental murine influenza, GS4104 has shown protection after twice daily oral administration against mortality due to influenza A(H1N1), A(H3N2), and B strains.[22] The 50% protective doses against mortality depend on virus type and inoculum but range from approximately 0.1–3.2 mg/kg/day. Therapeutic effects with reductions in mortality have been seen with administration delayed up to 60 hours post inoculation. Interestingly, both zanamivir and the parent GS4071 are active at approximately 10-fold greater doses. GS4104 (5 or 25 mg/kg) is also effective at reducing the pyrogenic and nasal inflammatory responses to influenza in ferrets, although oral bioavailability is relatively low in this species.[36]

Single and multiple ascending dose studies of the tolerability and pharmacology of GS4104 have been conducted in uninfected humans.[37] GS4104 was generally well tolerated over doses ranges from 20 to 1000 mg, although gastrointestinal upset with nausea, and less often emesis, have been observed with doses of 500 mg and above. Hydrolysis of the ester groups of GS4104 occurs in the gut mucosa and bloodstream to yield the parent drug GS4071. Peak plasma concentrations of the GS4071 occur 2.5–6 hours after administration and there is generally good dose proportionality in observed plasma concentrations. The plasma elimination half- life of the parent drug is approximately 7–8 hours, and it is eliminated unchanged by renal excretory mechanisms. Administration of GS4104 with food is associated with modest effects on pharmacokinetics, manifested by a slight delay in time to maximum plasma concentration and a slight increase in the overall exposure to GS4071. Plasma concentrations of the the prodrug are low, and the estimated exposure is less than 5% of that of the active. These results indicate that high and sustained levels of the active drug are achievable. Preliminary analysis of studies using oral GS4104 for prevention and early treatment of experimental influenza A(H1N1) infection in humans have found significant clinical and antiviral effects.[38] Consequently, this agent will also be tested for its prophylactic and therapeutic activities during the 1997–98 influenza season.

2.1. Resistance

Serial *in vitro* passage of influenza virus in the presence of zanamivir has yielded phenotypic variants with 1000-fold or greater reductions in susceptibility in cell culture assays. However, this *in vitro* resistance has not been associated consistently with loss of enzyme inhibition or of *in vivo* inhibitory activity of the compound.[39,40,41,42,43] Mutations in or near the receptor binding site of the hemagglutinin decrease the affinity of hemagglutinin for sialic acid. Such variants have decreased dependence on neuraminidase activity for cleavage of sialic acid and release from infected cells and are able to replicate in cell culture despite neuraminidase inhibition by zanamivir. Although these variants replicate and cause disease in experimentally infected animals like their wild-type parental viruses, the replication of one of these variants is inhibited in murine models of influenza at zanamivir doses that are comparable to those inhibitory for wild-type virus.[43] This result suggests that the hemagglutinin mutants may not be likely to be of major clinical significance, but study of additional variants is needed.

The second type of resistant variant has mutations in the active enzyme site of the neuraminidase and shows decreased affinity for zanamivir. These mutations have been difficult to select in most laboratories. The more frequently recognized neuraminidase mutation is a framework mutation at amino acid 119 in which a glutamine is replaced by a glycine, alanine or less often asparagine residue. This decreases the affinity of neuraminidase for zanamivir resulting from a loss of interaction between the guanidinium of the molecule and the carboxylate of glutamine 119. More recently, a catalytic site mutation with substitution of asparagine 292 by lysine has been reported.[39] The 119 mutations have loss of zanamivir inhibition of neuraminidase enzymatic activity of up to 500-fold, whereas the 292 mutation is associated with only about a 10-fold reduction in inhibition. These neuraminidase mutations are usually associated with decreased activity and/or stability of the neuraminidase.[41,42] *In vivo* studies have found that the replication and pathogenicity of the neuraminidase variants is comparable in ferrets to the parental wild-type virus, whereas infectivity in the mouse is reduced 4–8 fold for the 119 variants and 500-fold for the 292 mutants. The 119 mutants are inhibited by zanamivir in mice at approximately 10-fold higher doses than required for wild-type virus. The difficulty in generating neuraminidase variants, their reduced enzymatic activity in some assay systems, lower infectivity for mice, and their continued inhibition by zanamivir *in vitro* at higher drug doses suggests reduced fitness of these viruses. Of note, GS4071 retains activity against viruses which are resistant to zanamivir on the basis of neuraminidase mutations at position 119 but is highly cross-resistant for 292 mutants.

Limited data are available regarding resistance emergence during clinical use. Analysis of small numbers of isolates collected from volunteers participating in the experimental challenge studies and those participating in the natural influenza studies have found no important differences in *in vitro* susceptibility between first and last day or when compared to isolates from placebo recipients.[26,44] However, natural clinical isolates of influenza virus show wide differences in *in vitro* susceptibility, the significance of which remains to be clarified. Neuraminidase assays have found that isolates from patients taken before and after zanamivir treatment were comparably sensitive to the drug, which indicated that no neuraminidase resistance had emerged during therapy.[44]

3. RHINOVIRUS INFECTIONS

Increasing epidemiologic evidence demonstrated the importance of rhinovirus infections in causing a variety of clinical problems in addition to common colds. The application of nucleic acid amplification techniques has allowed more sensitive detection of these infections compared to virus isolation in cell culture. During the usual fall peak of rhinovirus activity in temperate climates, approximately 80% of common colds in adults are due to rhinovirus infections.[45] In addition, rhinovirus infections have been implicated in between one-third and two-thirds of asthma exacerbations in children and adults, approximately one-third of acute otitis media episodes in children, and up to one-half of acute sinusitis episodes in adults.[46,47,48,49] They have been linked to lower airways disease in infants, the elderly, and immunocompromised patients. The impact of these infections clearly indicates a need for effective methods for prevention and treatment of rhinovirus infections.

3.1. Soluble Intercellular Adhesion Molecule-1

The cellular receptor for the major human rhinovirus (HRV) receptor group, which contains approximately 90% of the numbered serotypes, is intercellular adhesion molecule-1 (ICAM-1).[50] This receptor interacts with amino acids lining the base of the canyon or cleft that encircles each of the 5-fold vertices of the viral capsid. A cellular receptor for the minor HRV receptor groups belongs to the low density lipoprotein receptor (LDLR) family. ICAM-1 is a single chain membrane glycoprotein which has five extracellular immunoglobulin-like domains and a molecular weight of 80–100 Kd depending on the extent of glycosylation. The natural function of ICAM-1 is to serve as a ligand for the β_2 integrins, CD11A/CD18 or lymphocyte-related antigen-1(LFA-1) and CD11B/CD18 (MAC-1). The binding sites for LFA-1 and HRV overlap and involve the distal two domains. ICAM-1 functions as an adhesion molecule and mediates multiple immune interactions including migration of neutrophils and eosinophils into tissues, activation of CD4 and B lymphocytes, and T cell and NK cell cytotoxicity. During inflammatory responses ICAM-1 expression is increased in vascular endothelium and epithelium. Antibody to ICAM blocks cell adhesion to vascular endothelium and also inhibits mucosal eosinophilia and airway reactivity in primates after antigen exposure.[51]

Soluble or truncated forms of recombinant ICAM-1 consisting of the extracellular portion of the molecule specifically inhibit the replication of major HRV serotypes *in vitro*,[52] and inhibition of other respiratory viruses has not been recognized. The inhibitory concentrations of sICAM-1 for the numbered rhinovirus serotypes range over 400-fold in cultures of human embryonic lung cells.[53] The mean 50% inhibitory concentration is approximately 2 μg/ml but the range extends from 0.1 to 41 μg/ml. Inhibition of rhinovirus replication has also been observed in explants of human respiratory epithelium. A similar spectrum of activity has been reported for nontyped field isolates of rhinovirus.[54] The antiviral mechanism of sICAM-1's action appears to be competitive extracellular inhibition of rhinoviral binding.[55] Inhibition is reversible and is lost upon sICAM-1 removal from the medium. Direct virus exposure to sICAM-1 can also trigger extracellular receptor-mediated uncoating of the virion, but the extent of this phenomenon depends on the virus serotype and temperature.[56] Variants with reduced *in vitro* susceptibity to sICAM-1 have been selected by serial passage in cell culture.[57] Depending on the serotype under study, the loss of *in vitro* susceptibility ranges from 10 to approximately 60 fold. The mechanism of resistance and its effect on the biologic properties of rhinovirus are under study.

Limited studies in chimpanzees experimentally inoculated with rhinovirus type 16 have found evidence of protection with a truncated ICAM-1 molecule.[58] When nasal sprays of tICAM-1 (doses of 5 or 10 mg) were given immediately (within 5–10 minutes) before virus exposure, the rates of infection in chimpanzees were reduced from 100% in controls to 13%. A single dose given 10 minutes before virus challenge was sufficient to protect against infection. These experiments did not establish the duration of antiviral action in the upper airway, which one would anticipate to be influenced substantially by mucociliary clearance mechanisms. Recently, studies of sICAM for prevention of experimental rhinovirus colds have been completed with HRV type 39 challenge of susceptible adults.[59] These placebo-controlled, blinded studies used intranasal sICAM-1 doses of approximately 0.7 mg given 6 times daily for 7 days beginning at either 4 hours before or 12 hours after viral inoculation. Both dry powder and aqueous solution forms were tested, but no important differences were found between these formulations or between those starting sICAM-1 before virus exposure or 12 hours later. No significant reductions in infection rates between the sICAM-1 and placebo groups, and the overall frequency of virus recovery from nasal washings was only modestly reduced (92% vs 78%). However, the frequency of clinical colds was reduced by 34% overall, and there were significant reductions in symptom burden and nasal mucus weights. Early therapeutic administration of ICAM was associated with significant reductions in nasal wash viral titers. The lack of difference between the prophylaxis and early treatment groups suggests that the observed effects were not simply due to prevention of initial viral infection but rather to sustained antiviral effects within the upper respiratory tract. These results establish that receptor blockade may offer a realistic approach to prevention and possibly treatment of rhinoviral infections. However, formulations of ICAM that could be used less frequently would be needed to make this a practical approach.

3.2. Pleconaril

A variety of molecules have been described which bind to a hydrophobic pocket within VP-1 of the rhinovirus capsid and interfere with early events of in viral replication.[60] This binding induces conformational changes in the protein shell which interferes with virus-receptor interactions and/or uncoating of the virus. Most of these agents have been abandoned because of lack of efficacy, intolerance, or unfavorable pharmacokinetic properties. For example one methylisoxazole compound, designated WIN 54954, shows broad spectrum anti-picornoviral activity *in vitro*. Oral WIN 54954 (600 mg tid) provided partial protection against experimental coxsackie A21 virus infection,[61] but it was associated with skin rash and possibly abnormal liver functions, perhaps related to one of its metabolic breakdown products. Furthermore, the agent was not significantly protective against rhinovirus infection when given in a comparable dose.[62] Almost all WIN 54954 recipients had trough plasma levels that exceeded the inhibitory concentration for the challenge virus under *in vitro* conditions, but detectable levels of the drug were found only in a small portion of nasal washings. The observed lack of antiviral and clinical efficacy in

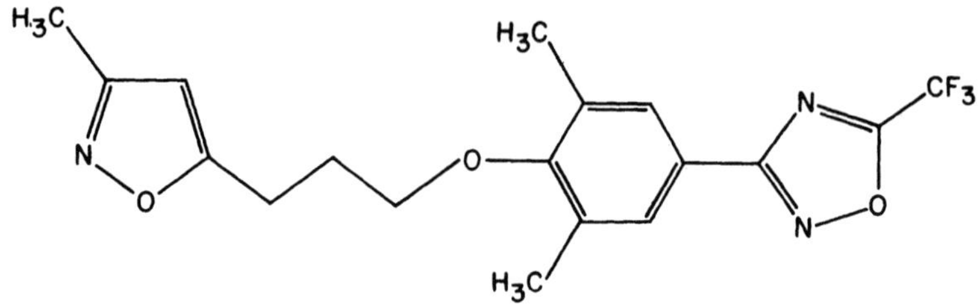

PLECONARIL

Figure 4. Structure of pleconaril.

these studies likely resulted from inability to achieve sufficient drug levels at the nasal epithelium.

Another compound of this series has been found to have greater metabolic stability, improved pharmacokinetic properties, and retain excellent anti-picornovirus activity (Figure 4). Pleconaril (VP63843) inhibits *in vitro* replication of more than 90% of clinical isolates of common enterovirus serotypes at concentrations less than 0.1 μg/ml. Pleconaril is also active in protecting mice infected with several different coxsackie virus serotypes. with 50% protective doses ranging from 12–88 mg/kg/day depending on virus.[63] Pleconaril is also active against most rhinovirus serotypes and inhibits 90 of 101 HRV serotypes tested at concentrations from 0.01 to 1 μg/ml. Concentrations of 0.02–0.09 μg/ml are inhibitory for a range of rhinoviruses by yield reduction assay in human cells. In human pharmacokinetic studies, dose proportional increases in plasma concentrations and drug exposure are seen with doses ranging from 50–1000 mg. The time to reach maximum concentration ranges from 1.5–5 hours, and the plasma elimination half-life averages approximately 25 hours. The prolonged half-life means that infrequent dosing schedules are possible. Dosages of 200 mg twice daily provide maximum plasma concentrations of 2.3 μg/ml which exceeds *in vitro* inhibitory concentrations for most picornoviruses. Administration in the fed state increases drug exposure approximately two-fold.

The efficacy of oral pleconaril has been tested in experimental coxsackie A21 virus infection.[64] Pleconaril therapy (400 mg loading dose followed by 200 mg twice daily for 6 days) was begun 14 hours before inoculation with coxsackie virus A21. Pleconaril significantly reduced the frequency of fever (44% vs 6% control), as well as rhinorrhea, peak viral titers, and overall symptom burden. These findings provide the impetus to further characterize the potential anti- rhinoviral activity of this capsid-binding agent. In addition to the trials being conducted for enteroviral syndromes, studies are currently in place to examine its possible therapeutic value in treating common colds in asthmatic patients.

REFERENCES

1. Hayden FG, Diamond L, Wood PB, Korts DC, Wecker MT. Effectiveness and safety of intranasal ipratropium bromide in common colds. A randomized, double-blind, placebo-controlled trial. Annals of Internal Medicine 1996; 125:89–97.
2. Hayden FG. Use of interferons for prevention and treatment of respiratory viral infections. In: Antiviral Chemotherapy, New Directions for Clinical Application and Research. Mills J, Corey L, eds. New York: Elsevier, 1986; 28–39.

3. Hayden FG. Update on antiviral chemotherapy of respiratory viral infection. In: Antiviral Chemotherapy, New Directions for Clinical Application and Research. Mills J, Corey L, eds. New York: Elsevier, 1989; 117–142.
4. Hayden FG. Update on treatment of respiratory viral infections. In: Antiviral Chemotherapy, New Directions for Clinical Application and Research. Mills J, Corey L, eds. Englewood Cliffs, New Jersey: PTR Prentice-Hall, 1993; 207–228.
5. Arruda E., Hayden FG. Update on therapy of influenza and rhinovirus infections. In: Antiviral Chemotherapy 4 New Directions for Clinical Application and Research. Mills J, Volberding PA, Corey L, eds. New York: Plenum Press, 1996; 175–187.
6. Hirst GK. Absorption of influenza hemagglutinins and virus by red blood cells. Journal of Experimental Medicine 1942; 76:195–209.
7. Anderson SG, Burnet FM, Fazekas deSt Groths. Mucins and mucoids in relation to influenza virus action, VI. general discussion. Aust J Exp Biol Med Sci 1948; 26:403–411.
8. Burnet FM. Mucins and mucoids in relation to influenza virus action, III: inhibition of virus haemagglutination by glandular mucins. Aust J Exp Biol Med Sci 1948; 26: 371–379.
9. Gotschalk A. Neuraminidase: its substrate and mode of action. Advanced Enzymology 1958; 20:135–146.
10. Bamford MJ. Neuraminidase inhibitors as potential anti-influenza drugs. J Enzyme Inhibitiion 1995; 10:1–16.
11. Colman PM, Varghese JN, Laver WG. Structure of the catalytic and antigenic sites in influenza virus neuraminidase. Nature 1983; 303:41–44.
12. von Itzstein M, Wu WY, Kok GB, *et al.* Rational design of potent sialidase-based inhibitors of influenza virus replication. Nature 1993; 363:418–423.
13. Liu C, Eichelberger MC, Compans RW, Air GM. Influenza type A virus neuraminidase does not play a role in viral entry, replication, assembly, or budding. Journal of Virology 1995; 69:1099–1106.
14. Li S, Schulman J, Itamura S, Palese P. Glycosylation of neuraminidase determines the neurovirulence of influenza A/WSN/33 virus. Journal of Virology 1993; 67:6667–6673.
15. Schultz-Cherry S, Hinshaw VS. Influenza virus neuraminidase activates latent transforming growth factor beta. Journal of Virology 1996; 70:8624–8629.
16. Alluwaimi AM, Smith H, Sweet C. Role of surface glycoproteins in influenza virus pyrogenicity. Journal of General Virology 1994; 75:2835–2840.
17. Houde M, Arora DJ. Stimulation of tumor necrosis factor secretion by purified influenza virus neuraminidase. Cellular Immunology 1990; 129:104–111.
18. Woods JM, Bethell RC, Coates JA, *et al.* 4-Guanidino-2,4-dideoxy-2,3-dehydro-N-acetylneuraminic acid is a highly effective inhibitor both of the sialidase (neuraminidase) and of growth of a wide range of influenza A and B viruses *in vitro*. Antimicrobial Agents & Chemotherapy 1993; 37:1473–1479.
19. Hayden FG, Rollins BS, Madren LK. Anti-influenza virus activity of the neuraminidase inhibitor 4-guanidino-Neu5Ac2en in cell culture and in human respiratory epithelium [published erratum appears in Antiviral Res 1994 Dec; 25(3–4):287]. Antiviral Res 1994; 25:123–131.
20. Ryan DM, Ticehurst J, Dempsey MH, Penn CR. Inhibition of influenza virus replication in mice by GG167 (4-guanidino-2,4-dideoxy-2,3-dehydro-N-acetylneuraminic acid) is consistent with extracellular activity of viral neuraminidase (sialidase). Antimicrobial Agents & Chemotherapy 1994; 38:2270–2275.
21. Ryan DM, Ticehurst J, Dempsey MH. GG167 (4-guanidino-2,4-dideoxy-2,3-dehydro-N-acetylneuraminic acid) is a potent inhibitor of influenza virus in ferrets. Antimicrobial Agents & Chemotherapy 1995; 39:2583–2584.
22. Sidwell RW, Huffman JH, Bailey K.W., Wong MH, Kim CU. *In vivo* influenza virus-inhibitory activity of an orally administered neuraminidase inhibitor. Antiviral Research 1997; 34:A74, Abstr #113.
23. Efthymiopoulos C, Barrington P, Patel JA. Pharmacokinetics of the neuraminidase inhibitor 4-guanidino Neu5Ac2en (GG167) following intravenous, intranasal, and inhaled administration in man. Abstracts of the 34th Interscience Conference on Antimicrobial Agents and Chemotherapy American Society of Microbiology 1994; 265, Abstr #H70.
24. Newman S, Brown J, Pickford M, Fayinka S, Cass L, Bye A. Desposition pattern in the respiratory tract of the neuraminidase inhibitor zanamivir; a gamma scintigraphic study. Abstracts of the 37th Interscience Conference on Antimicrobial Agents and Chemotherapy 1997; 237, Abstr #H-134.
25. Hayden FG, Treanor JJ, Betts RF, Lobo M, Esinhart JD, Hussey EK. Safety and efficacy of the neuraminidase inhibitor GG167 in experimental human influenza. JAMA 1996; 275:295–299.
26. Hayden FG, Lobo M, Hussey EK, Eason CU. Efficacy of intranasal GG167 in experimental human influenza A and B virus infection. In: Options for the control of influenza III. Brown LE, Hampson AW, Webster RG, eds. Elsevier Science B.V., 1996; 718–725.

27. Walker JB, Hussey EK, Treanor JJ, Montalvo A, Hayden FG. Effects of the neuraminidase inhibitor zanamivir on otologic manifestations of experimental human influenza. Journal of Infectious Diseases 1997; 176:1417–1422.
28. Hayden FG, Osterhaus ADME, Treavor JJ, Fleming DM *et al.* Efficacy and safety of the neuraminidase inhibitor zanamivir in the treatment of influenza virus infections. New England Journal of Medicine 1997;337:874–880.
29. Aoki FY, Fleming DM, Lacey L, Keene O, Keech M, Macleod P. Impact of treatment of influenza with zanamivir on patients' health status, sleep quality, productivity and healthcare use. Abstracts of the 37th Interscience Conference on Antimicrobial Agents and Chemotherapy 1997; 384, Abstr #N-15.
30. Kim CU, Lew W, Williams MA, *et al.* Influenza Neuraminidase Inhibitors Possessing a Novel Hydrophobic Interaction in the Enzyme Active Site: Design, synthesis, and Structural Analysis of Carbocyclic Sialic Acid Analogues with Potent Anti-Influenza Activity. J Am Chem Soc 1997; 119:681–690.
31. Hayden FG, Rollins BS. *In vitro* activity of the neuraminidase inhibitor GS4071 against influenza viruses. Antiviral Research 1997; 34: A86, Abstr #159.
32. Mendel DB, Tai CY, Escarpe PA, *et al.* GS4104 is a potent and selective inhibitor of the growth and neuraminidase activity of influenza A and B viruses *in vitro*. Antiviral Research 1997; 34:A73, Abstr # 111.
33. Kim CU, Lew W, Williams MA, *et al.* Rational drug design approach to an orally bioabvailable neuraminidase inhibitor. Abstracts of the 37th Interscience Conference on Antimicrobial Agents and Chemotherapy 1997; 388, Abstr #S-10.
34. Cundy K, Eisenberg G, Bidgood A, *et al.* The novel influenza neuraminidase inhibitor prodrug GS4104 is highly bioavailable in animals. Abstracts of the 37th Interscience Conference on Antimicrobial Agents and Chemotherapy 1997; 237, Abstr #H-135.
35. Eisenberg G, Bidgood A, Lynch G, Lee WA, Cundy K. Penetration of GS4071, a novel influenza neuraminidase inhibitor, into rat bronchoalveolar lining fluid following oral administration of the prodrug GS4104. Antimicrob Agents Chemother 1997; 41: 19491952.
36. Kim C.U., Bischofberger N, Williams MA, Lew W, Merson J, Sweet C. Efficacy of GS4104 in ferrets infected with influenza a virus. Antiviral Research 1997; 34:A74, Abstr #114.
37. Wood ND, Aitken M, Sharp S, Evison H. Tolerability and pharmacokinetics of the influenza neuraminidase inhibitor Ro-64–0802 ((GS4071) following oral administration of the prodrug Ro-64–0796 (GS4104) to health male volunteers. Abstracts of the 37th Interscience Conference on Antimicrobial Agents and Chemotherapy 1997; 25, Abstr #A-123.
38. Hayden FG, Lobo M, Treanor JJ, Miller M, Mills RG. Efficacy and tolerability of oral GS4104 for early treatment of experimental influenza in humans. Abstracts of the 37th Interscience Conference on Antimicrobial Agents and Chemotherapy 1997; 14, Abstr #LB-26.
39. Gubareva LV, Robinson MJ, Bethell RC, Webster RG. Catalytic and framework mutations in the neuraminidase active site of influenza viruses that one resistant to 4-guanidino-Neu5Ac2en. Journal of Virology 1997; 71:3385–3390.
40. Staschke KA, Colacino JM, Baxter AJ, Air GM *et al.* Molecular basis for the resistance of influenza viruses to 4-guanidino-Neu5Ac2en. Virology 1995; 214:642–646.
41. McKimm-Breschkin JL, Blick TJ, Sahasrabudhe A, *et al.* Generation and characterization of variants of NWS/G70C influenza virus after *in vitro* passage in 4-amino-Neu5Ac2en and 4-guanidino-Neu5Ac2en. Antimicrobial Agents & Chemotherapy 1996; 40:40–46.
42. Gubareva LV, Bethell R, Hart GJ, Murti KG, Penn CR, Webster RG. Characterization of mutants of influenza A virus selected with the neuraminidase inhibitor 4-guanidino-Neu5Ac2en. Journal of Virology 1996; 70:1818–1827.
43. Penn CR, Barnett J, Bethell R, *et al.* Selection of influenza virus with reduced sensitivity *in vitro* to the neuraminidase inhibitor GG167 (4-guanidino-Neu5Ac2en): changes in haemagglutinin may compensate for loss of neuraminidase activity. In: Options For The Control of Influenza III. Brown LE, Hampson AW, Webster RG, eds. Amsterdam, The Netherlands: Elsevier Science B.V., 1996; 735–740.
44. Barnett J, Dempsey M, Rothbarth PH, DeGroot R, Osterhaus ADME. Susceptibility monitoring of influenza virus clinical isolates to the neuraminidase inhibitor zanamivir (GG167) during phase II clinical efficacy trials. Abstracts of the 37th Interscience Conference on Antimicrobial Agents and Chemotherapy 1997; 230, Abstr #H-93.
45. Arruda E, Pitkaranta A, Witek TJ, Doyle CA, Hayden FG. Frequency and natural history of rhinovirus infections in adults during autumn. Journal of Clinical Microbiology 1997; 35:2864–2868.
46. Pitkaranta A, Arruda E, Malmberg H, Hayden FG. Detection of rhinovirus in sinus brushings of patients with acute community-acquired sinusitis by reverse transcription-PCR. Journal of Clinical Microbiology 1997; 35:1791–1793.

47. Johnston SL, Pattemore PK, Sanderson G, *et al.* Community study of role of viral infections in exacerbations of asthma in 9–11 year old children. British Medical Journal 1995; 310:1225–1229.
48. Nicholson KG, Kent J, Ireland DC. Respiratory viruses and exacerbations of asthma in adults. British Medical Journal 1993; 307:982–986.
49. Nicholson KG, Kent J, Hammersley V, Esperanza C. Risk factors for lower respiratory complications of rhinovirus infections in elderly people living in the community: prospective cohort study. British Medical Journal 1996; 313: 1119–1123.
50. Norkin LC: Virus receptors. Implications for pathogenesis and the design of antiviral agents. Clinical Microbiology Reviews 1995; 8:293–315.
51. Wegner CD, Gundel RH, Reilly P, Haynes N, Letts LG, Rothlein R. Intercellular adhesion molecule-1 (ICAM-1) in the pathogenesis of asthma. Science 1990; 247:456–459.
52. Marlin SD, Staunton DE, Springer TA, Stratowa C, Sommergruber W, Merluzzi VJ. A soluble form of intercellular adhesion molecule-1 inhibits rhinovirus infection. Nature 1990; 344:70–72.
53. Crump C, Arruda E., Hayden FG. *In vitro* inhibitory activity of soluble ICAM-1 for the numbered serotypes of human rhinovirus. Antiviral Chemistry & Chemotherapy 1993; 4: 323–327.
54. Ohlin A, Hoover-Litty H, Sanderson G, *et al.* Spectrum of activity of soluble intercellular adhesion molecule-1 against rhinovirus reference strains and field isolates. Antimicrob Agents Chemother 1994; 38:1413–1415.
55. Arruda E, Crump CE, Marlin SD, Merluzzi VJ, Hayden FG. *In vitro* studies of the antirhinovirus activity of soluble intercellular adhesion molecule-1. Antimicrob Agents Chemother 1992; 36:1186–1191.
56. Hoover-Litty H, Greve JM. Formation of rhinovirus-soluble ICAM-1 complexes and conformational changes in the virion. Journal of Virology 1993; 67:390–397.
57. Arruda E, Crump CE, Hayden FG. *In vitro* selection of human rhinovirus relatively resistant to soluble intercellular adhesion molecule-1. Antimicrob Agents Chemother 1994; 38:66–70.
58. Huguenel ED, Cohn D, Dockum DP, *et al.* Prevention of rhinovirus infection in chimpanzees by soluble intercellular adhesion molecule-1. American Journal of Respiratory & Critical Care Medicine 1997; 155:1206–1210.
59. Turner RB, Wecker MT, Polk BF, *et al.* Efficacy of Soluble ICAM-1 (sICAM) for Prevention of Rhinovirus Infection and Illness. Abstracts of the 37th Interscience Conference on Antimicrobial Agents and Chemotherapy 1997; 229, Abstr #H-85.
60. Arruda E, Hayden FG. Clinical Studies of Antiviral Agents for Picornaviral Infections. In: Antiviral Chemotherapy. Jeffries DJ, De Clercq E, eds. John Wiley & Sons Ltd, 1995; 321–355.
61. Schiff GM, Sherwood J, Young EC, Mason L. Prophylactic efficacy of WIN 54954 in prevention of experimental human coxsackievirus A21 infection and illness. Program and Abst 5th International Conference on Antiviral Research, Vancouver, BC March 8–13, 1992. Antiviral Research 1992; Suppl 1: 92, Abstr #92.
62. Turner RB, Dutko FJ, Goldstein NH, Lockwood G, Hayden FG. Efficacy of oral WIN 54954 for prophylaxis of experimental rhinovirus infection. Antimicrob Agents Chemother 1993; 37:297–300.
63. Viropharma Inc. Clinical Investigators Brochure VP63843. 1997.
64. Schiff GM, Sherwood JR, Sherwood. Oral Efficacy of VP 63843. Abstracts of the 36th Interscience Conference on Antimicrobial Agents and Chemotherapy 1996; 171, Abstr #H43.

7

PLECONARIL

A Broad Spectrum Antipicornaviral Agent

Jon M. Rogers, Guy D. Diana, and Mark A. McKinlay

Viropharma Incorporated
405 Eagleview Boulevard
Exton, Pennsylvania 19341

Members of the virus family Picornaviridae are ubiquitous in nature and responsible for a broad spectrum of clinical syndromes ranging from the common cold to neonatal enteroviral sepsis. The two genuses of classification of picornaviridae are the enteroviridae and rhinoviridae. Rhinoviruses express human parisitization clinically as a nasopharyngeal syndrome (the "common cold") which occurs in all populations and age groups and can take up to two weeks to resolve. Exceptions to the uncomplicated course of rhinovirus nasopharyngeal infection include special patient sets at risk for infectious complications including patients with asthma, congestive heart failure, bronchiectasis and cystic fibrosis.[1] In contrast, the pathology of enterovirus disease in humans relies principally upon the somatotopic localization of infection after primary replication in the gastrointestinal tract and subsequent bloodborne dissemination.[2] Myocarditis, meningitis, encephalitis, respiratory syndromes (ranging from nasopharyngitis to lobar pneumonia), and mucocutaneous manifestations such as hand, foot and mouth disease and herpangina characterize the multifaceted presentations of enterovirus disease in humans.[3] The absence of precise localization of seeded bloodborne infection by enterovirus may eventuate in an ill-defined "transition" syndrome characterized by fever, malaise, myalgias, headache and mild respiratory complaints, often referred to as "summer flu", given the predilection of enteroviruses to express disease during the summer and early fall months. The high prevalence of asymptomatic or mildly symptomatic infections has led to the analogy of the human populations representing an 'ocean' of enterovirus infections, with 'islands' in the ocean representing discrete and well recognized organ-specific clinically expressed syndromes.

No FDA approved antiviral therapies are available for the treatment of picornavirus infections. The mainstay of treatment continues to be directed toward symptomatic relief of the most prominent symptoms of each clinical syndrome. However, the characterization of the three dimensional structure of human picornaviruses by x-ray crystallography has been

Antiviral Chemotherapy 5, edited by Mills *et al.*
Kluwer Academic / Plenum Publishers, New York, 1999.

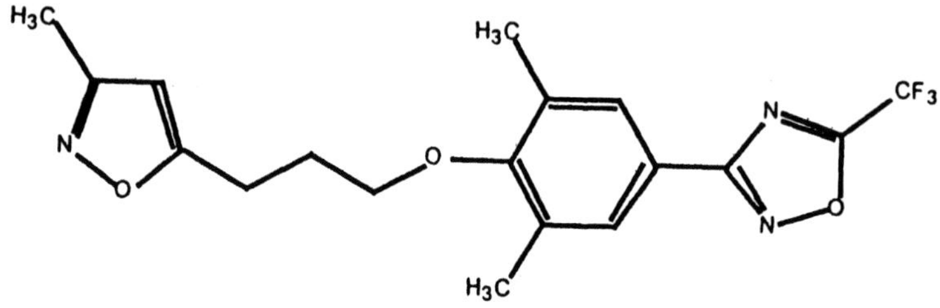

Figure 1. The Lewis structure of pleconaril (VP 63843).

useful in guiding the design of drugs to inhibit picornavirus replication.[4] One such molecular target is a hydrophobic pocket beneath the canyon floor of the VP1 capsid protein of enteroviruses and rhinoviruses. The functionality of the hydrophobic pocket in the pathogenesis of picornavirus disease is unknown but it is thought to serve an important role in the virion uncoating process. It has been shown that the canyon floor on the virion capsid is a putative binding site for intercellular adhesion molecule-I (ICAM-I), a glycoprotein responsible for cellular recognition. Antiviral agents such as pleconaril, which bind to the pocket increase conformational stability of the virion capsid protein thereby preventing uncoating of the virus after ingress into the intracellular space.[5] Activity of these agents, particularly those of the isoxazole series, has been extensively characterized in both *in vitro* and *in vivo* systems, including viral sensitivity of clinical isolates in cell culture and animal models of disease using well characterized pathogenic enterovirus strains.

A large variety of molecular structures bind to the hydrophobic pocket to stabilize the virus capsid.[6, 7] Several agents have been advanced into the clinic and have undergone preliminary evaluation in established rhinovirus and enterovirus challenge models.[8–10] Despite encouraging results with predecessor molecules in some classical challenge models, complex drug metabolism or lack of broad spectrum antipicornavirus activity has posed a significant impediment to further development of earlier agents. Fluorination of the isoxazole ring of predecessor molecules in the structure of pleconaril (Figure 1) (formerly WIN 63843, Sterling Winthrop; VP 63843, Viropharma Incorporated) has resulted in a moiety of relative metabolic stability.[11]

The medical literature is generally lacking in detailed characterizations of the natural history or health economic cost of picornavirus diseases. Longitudinal information on disease course is essential in designing clinical trials of therapeutic agents in order to craft robust and meaningful clinical trial endpoints. An estimate of the total health economic burden of any disease permits the appropriate contextual framing of the utility of a therapeutic agent for treatment of that disease. Recently the natural history and health economic cost of enterovirus meningitis in adults was evaluated in a longitudinal study by Rotbart and coworkers.[12] Subjects with enterovirus meningitis, confirmed by reverse transcriptase-polymerase chain reaction (RT-PCR) testing of cerebrospinal fluid (CSF) were followed with daily telephone contacts until the clinical course of the acute illness had resolved. In the study cohort, 80% of subjects were hospitalized for a mean hospital duration of four days. This high percentage of subjects undergoing hospitalization is consistent with other large published studies.[13–15] Subjects were usually hospitalized to receive broad spectrum parenteral antibiotic coverage until the results of bacterial CSF cultures were made available. In addition, the majority of subjects required in-hospital parenteral fluid replacement therapy and two thirds of the survey subjects required parenteral narcotic

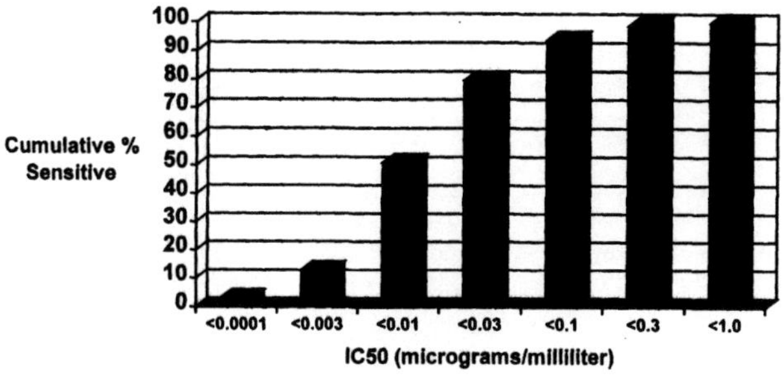

Figure 2. Cumulative sensitivity of 215 clinical enterovirus isolates to pleconaril (Pevear, 1996).

therapy for headache pain relief. The average duration of illness was 16 days (range 4–52 days), with subjects absent from work or school for a mean of 6 days. Based on estimates of the prevalence of enterovirus meningitis in the United States, the total cost of inpatient hospitalization was estimated to be $650 million dollars annually. This health economic assessment was consistent with that of Opal and colleagues who estimated the hospital costs associated with an outbreak of enterovirus meningitis in Rhode Island in 1991.[15] In addition, based on the data from the survey by Rotbart, an estimated 700,000 work weeks of productivity are lost in the United States each year from adult enterovirus meningitis. These data provide insight into the annual economic and societal burden of one enterovirus disease in the United States. The Rotbart study also provided longitudinal information on the natural history of the resolution of symptoms associated with enterovirus meningitis. Employing an empiric disease scoring system (Total Morbidity Score; TMS) for the most common symptoms, a linear decrease in that multidimensional symptom score was observed over the course of the meningitis illness. A minimal score correlated with functionality and return to work by approximately nine days after presentation.

The preclinical profile of pleconaril has been extensively characterized. The oral bioavaility is 66% in beagle dogs. Acute toxicity testing demonstrated no mortality when single doses of up to 2,000 milligrams per kilogram were administered orally to rats. One month toxicology studies in rats and dogs demonstrated excellent tolerability with mild, reversible increases in liver weights consistent with induction of cytochrome P-450 superfamily enzymes. *In vitro* sensitivity testing against commonly occurring clinical enterovirus isolates from the United States Centers for Disease Control and Prevention, and comprising both echoviruses and coxsackieviruses has uniformly demonstrated sensitivity to pleconaril at submicromolar levels (Figure 2).[16] Pleconaril is also active against human rhinovirus strains. Figure 3 demonstrates the cumulative sensitivity of pleconaril to the 101 rhinovirus serotypes from the American Tissue Type Culture Collection. While somewhat less sensitive, pleconaril is also active against wild-type poliovirus serotypes, primarily serotypes II and III. In radiolabel studies with rats, the drug is widely distributed in tissues, with 2- to 6-fold higher concentrations in brain, spinal cord and meninges compared with plasma. Nasal epithelial concentrations of pleconaril are 11- to 13-fold higher than corresponding plasma levels. These highly desirable features are due to the lipophilicity of the molecule and make it an attractive candidate for penetration into the meninges, central nervous system and respiratory epithelium, the primary sites of picornavirus replication in man. Mass balance studies in vertebrate species also demonstrate that pleconaril undergoes extensive enterohepatic recirculation, with approximately 80% of the orally administered dose excreted in feces within 48 hours and the reminder of the drug (largely unmetabolized) being excreted in

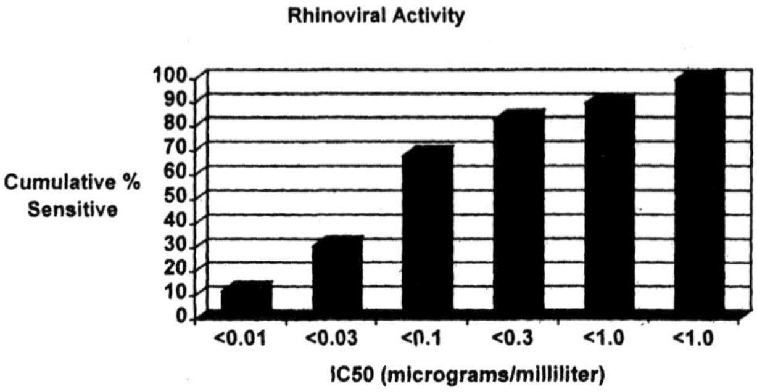

Figure 3. Cumulative rhinovirus sensitivity to pleconaril.

urine. Administered either therapeutically or prophylactically, pleconaril is protective against mortality in *in vivo* models of enterovirus infections using suckling or weanling mice. The potential for the development of resistance is low, with resistant strains exhibiting dramatically attenuated virulence in these models.[17]

The pharmacokinetics of pleconaril in adults, children and neonates are consistent and predictable.[18–19] Single oral doses of pleconaril of up to 1000 milligrams in healthy adult volunteers have demonstrated dose proportionality and linearity of maximal plasma concentration and area under the plasma concentration-time curve. Repeated dosing with pleconaril in volunteer subjects has demonstrated modest accumulation of pleconaril (accumulation ratio of less than 2) with trough steady state plasma concentrations of pleconaril in the range of 1.0 - 1.5 micrograms per milliliter. These concentrations are 10 to 40 times higher than the range of *in vitro* drug concentrations required to inhibit the majority of enterovirus serotypes from *in vitro* $TCID_{50}$ assays. Prepubescent children exhibit similar pharmacokinetics to adults, which eliminates the need for a dosage adjustment in this important group of patients with picornavirus infections. Investigations in newborns are ongoing, but preliminary evidence exists for adequate absorption and predictable pharmacokinetics in this special population as well.

Pleconaril has been evaluated in an experimental model of picornavirus infection in adult volunteers.[20] The clinical characteristics of the respiratory and systemic syndrome produced by intranasal inoculation with a multiply passaged safety-tested strain of coxsackievirus A-21 have been previously described.[21–22] Inoculation produces respiratory (nasal congestion, rhinorrhea, increased nasal mucous production, cough) and systemic (fever, myalgias, headache, malaise) symptoms for 24–72 hours after incubation with a predictable attack rate of approximately 70%. Coxsackievirus A-21 is unique among the enterovirus serotypes because it, like rhinoviruses, requires ICAM-1 for respiratory cell membrane attachment in the early phase of cellular infectivity. Thirty-three adult subjects received a loading dose of 400 milligrams of pleconaril or placebo, followed 12 hours later by virus inoculation in this double blind study. Dosing was then continued with placebo or pleconaril 200 mg BID for 6.5 days. Subjects were evaluated by a multidimensional symptom score, a visual analogue self-reported score, quantitative daily mucous weight, fever and viral load from nasal respiratory secretions. Pleconaril is active against coxsackievirus A-21 with a sensitivity of 0.03 micrograms per milliliter. A summary of the clinical results in subjects with a peak respiratory symptom score of greater than 6 is presented in Figure 4. Pleconaril produced greater than a two $\log_{10}$ reduction in nasal secretion viral titer during the treatment period when compared with placebo which was highly

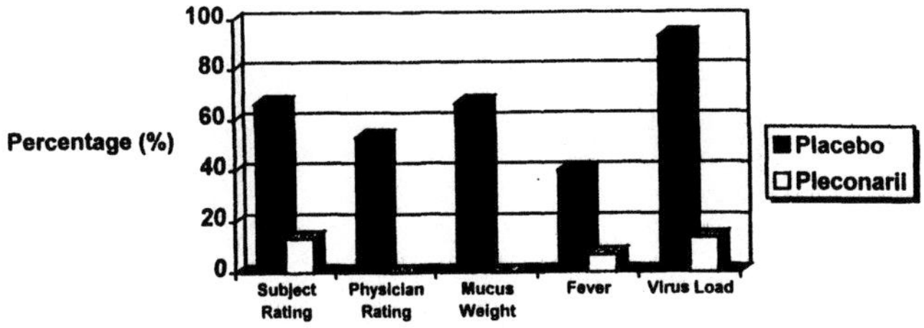

Figure 4. Disease measure reductions in the coxsackievirus A-21 challenge study. Subjects with a peak multimensional cold score of greater than 6 are included.

statistically significant (p=0.0001). This reduction in viral load correlated with prevention of clinical expression of post-inoculation illness in pleconaril-treated subjects. Daily nasal mucous production of greater than 5 or 10 grams per day was present in 69% and 44% of placebo volunteers respectively, compared to 13% and 6% of those treated with pleconaril (p=0.006). The percentage of subjects with peak daily oral temperature greater than 100 degrees Fahrenheit was 44% in the placebo group compared to 6% in the pleconaril group (p=0.026), and physician and subject self-reported illness scores were also significantly lower in the pleconaril-treated group than to placebo. These results suggest that pleconaril has the potential for therapeutic and curative activity in a variety of clinically important human infections caused by susceptible picornavirus strains. Clinical treatment trials are currently underway to determine if pleconaril can reduce the severity and duration of respiratory and systemic symptoms of picornavirus infections in ambulatory patients.

In light of its high prevalence, substantial health economic costs and significant short term morbidity, enterovirus meningitis represents a rational disease target for broad spectrum antipicornoviral agents such as pleconaril. A placebo controlled exploratory study in 39 adults with laboratory and clinical evidence of enterovirus meningitis was conducted using two different oral doses of pleconaril - 200 mg or 400 mg TID for seven days.[23] The primary endpoint was the time required to achieve a predefined minimal value for a multidimensional meningitis disease severity index (Total Morbidiy Score - TMS) which correlated with hospital discharge and return to work, school or normal recreational activity.[11] Secondary endpoints included the time to complete absence of headache, the number of days of analgesic use for headache pain, nuchal rigidity and myalgias, and an index of total analgesic consumption during the study observation period (Figures 5 and 6). Compared to placebo, subjects treated with pleconaril demonstrated a 45% reduction in meningitis disease duration (from 11.5 days to 5 days; p=0.005), as well as a 50% reduction in the time to complete absence of headache (14.2 days to 7 days; p=0.014). The duration of analgesic use and total analgesic medication use were significantly reduced with pleconaril compared to placebo; 50% reduction in duration of analgesic use (12 days to 6 days; p=0.021), 49% reduction in total analgesic use using an analgesic potency normalization formula (p=0.002), and 50% reduction in time to a patient global assessment score of ≤ 3 (12 days to 6 days; p=0.022). These results suggest a substantial clinical benefit of pleconaril with prominent reductions observed in both meningitis disease duration and severity and have stimulated the conduct of several large scale controlled clinical evaluations of pleconaril in enterovirus meningitis in infants, children and adults.

The development of chronic meningoencephalitis due to persistent enterovirus infections in patients with congenital or acquired immune deficiency represents a daunting challenge for the clinical immunologist.[24–25] These infections are often progressive, lead-

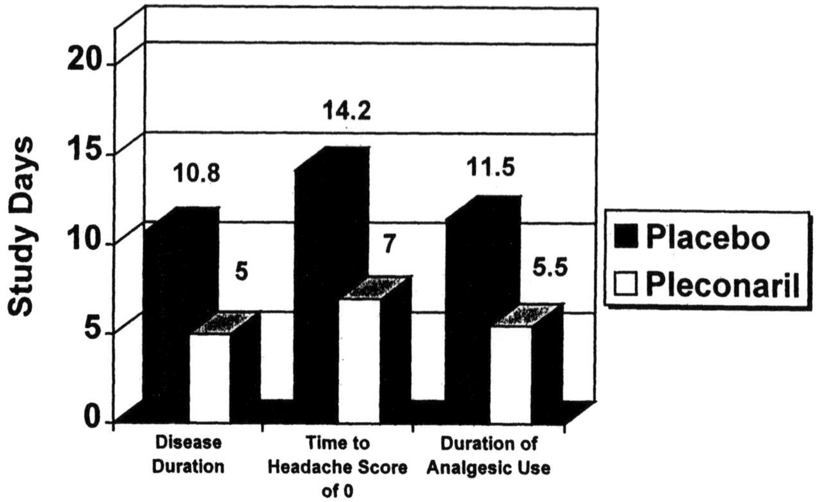

Figure 5. Primary and selected secondary meningitis endpoint measures for pleconaril (Weiner, 1997).

ing to significant disability, progressive neurological deficits, and death. These patients may represent a model of stable chronic enterovirus infection, as cerebrospinal fluid (CSF) findings (cell count, protein, viral culture, examination of the cerebrospinal fluid for enterovirus ribonucleic acid by reverse transcriptase polymerase chain reaction testing, or any combination) are persistently abnormal after the onset of clinically expressed infection. Short courses of pleconaril have been administered to 20 adult and pediatric patients with confirmed chronic enterovirus meningoencephalitis. The preliminary results[26] of this treatment program suggest that improvement or normalization of abnormal CSF findings, stabilization or reversal of abnormal neurological symptoms and overall symptomatic improvement are linked to pleconaril treatment. Included in this cohort is one four month old white female with adenosine deaminase-dependent severe combined immunodeficiency who developed a vaccine-associated neurological syndrome due to confirmed infection with a neurovirulent revertant of a Sabin poliovirus serotype II strain. Treatment with ple-

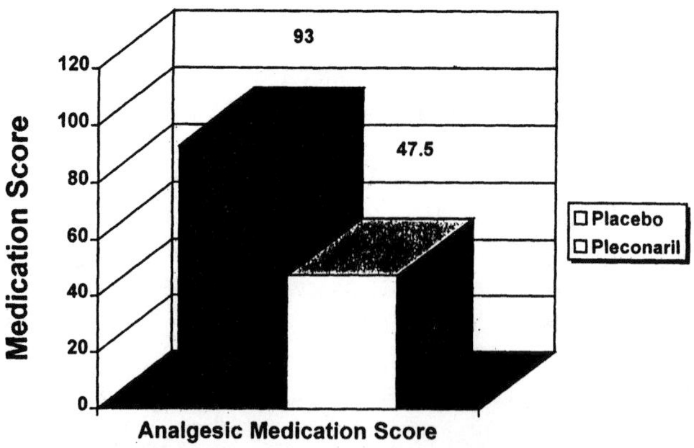

Figure 6. Reduction in total analgesic use in enterovirus meningitis patients with pleconaril (Weiner, 1997).

conaril aborted the clinical infection and neurological syndrome. Although the optimal pleconaril dosage regimen for chronic enterovirus infection has not been defined, short courses of therapy appear to be associated with a durable remission or cure from chronic enterovirus infection in some patients with immune deficiency. Other patient sets which have also been treated in an expanded access program include enterovirus infections in transplant recipients, acute enterovirus myocarditis and neonates with multiorgan systemic involvement due to overwhelming enterovirus infections.

Patient groups at high risk for untoward complications from rhinovirus infections are important considerations in the development of an antipicornaviral agent. Persistent rhinovirus replication in the lower airways of patients with asthma may reduce responsiveness to asthma medications and can eventuate in urgent outpatient treatment, hospitalization or death.[26] An exploratory placebo-controlled evaluation is currently underway to determine if pleconaril can reduce the frequency or severity of exacerbations of asthma following rhinovirus nasopharyngeal infections. Children with recurrent viral nasopharyngeal infections are prone to the development of otitis media, either through resultant Eustatian tube dysfunction or direct viral infection of the ear.[27–28] Pleconaril is currently being considered for evaluation in children at high risk for the development of otitis media after rhinoviral upper respiratory tract infections. Other rhinovirus-infection populations at high risk for complications, and which are being considered for evaluation, are patients with bronchopulmonary dysplasia, congestive heart failure, chronic obstructive pulmonary disease and diffuse alveolar pneumonia due to rhinovirus infections in bone marrow transplant recipients.

With approximately 800 single and mutiple dose exposures in infants, children and adults, the safety profile of pleconaril continues to be favorable. Since pleconaril has a viral-specific mechanism of action, there is negligible potential liability through extension of pharmacologic action by binding to mammalian receptor systems, membranes, or intracellular enzymatic machinery. The serious and nonserious adverse event profile of pleconaril continues to be similar to placebo treatment in identical populations. The safety profile observed thus far suggests that broad use of pleconaril in populations with self-limited infections such as upper respiratory disease may be accompanied by a favorable risk:benefit ratio.

Pleconaril has demonstrated broad activity through a unique and specific mechanism of action against enteroviruses and rhinovirus in both *in vitro* and *in vivo* systems. Confirmation of the antiviral activity of pleconaril in several exploratory clinical trials has been encouraging, stimulating the conduct of large scale clinical investigations in viral meningitis, respiratory disease and other allied clinical syndromes caused by picornaviruses. Pleconaril has the potential to provide significant clinical benefit to patients afflicted with disease caused by the ubiquitous Picornaviridae family.

REFERENCES

1. Gwaltney GM, Ruckert RR. Rhinovirus. DD Richman, RJ Whitley and FG Hayden (ed.) Clinical Virology. Churchill Livingstone. New York. 1997; 1025–1047.
2. Rotbart HA.. Enteroviruses. DD Richman, RJ Whitley and FG Hayden (ed.) Clinical Virology. Churchill Livingstone. New York. 1997; 997–1023.
3. Rotbart HA (ed.) Human Enterovirus Infections. ASM Press. Washington, D.C.1994.
4. McKinlay MA, Pevear DC, Rossmann MG. Treatment of the picornavirus common cold by inhibitors of viral uncoating and attachment. Annual Reviews of Microbiology 1992; 46:635–654.

5. Smith TJ, Kremer MJ, Luo M, *et al.* The site of attachment of human rhinovirus 14 and antiviral agents that inhibit uncoating. Science. 1986; 233:1286–1293.
6. Diana GD, Pevear DC. Antipicornavirus drugs: current status. Antiviral Chemistry and Chemotherapy. 1997; 8:412–418.
7. O'Connell J, Albin A, Blum D, *et al.* Development of antiviral agents for picornavirus infections. In HA Rotbart (ed.) Human Enterovirus Infections. ASM Press. Washington, DC.1995; 419–434.
8. Hayden FG, Andries A, Janssen PAJ. Safety and efficacy of oral pirodavir (R 77975) in experimental rhinovirus infection. Antimicrobial Agents and Chemotherapy. 1992; 36:727–732.
9. Schiff GM, Sherwood JR, Young EC, *et al.* Prophylactic efficacy of WIN 54954 in prevention of experimental human coxsackievirus A21 infection and illness. Antiviral Research. 1992; 17 (Suppl 1):92.
10. Turner RB, Dutko FJ, Goldstein NH, *et al.* Efficacy of oral WIN 54954 for the prophylaxis of experimental rhinovirus infection. Antimicrobial Agents and Chemotherapy. 1993; 37: 297–300.
11. Diana GD, Rudewicz P, Pevear DC, *et al.* Picornavirus inhibitors: triflouromethyl substitution provides a global protective effect against hepatic metabolism. Journal of Medicinal Chemistry 1995; 38: 1355–1371.
12. Rotbart HA, Brennan PJ, Fife KH, *et al.* Enterovirus meningitis in adults. Clinical Infectious Diseases. 1998; 27(4) (in press).
13. Rotbart HA. Enteroviral infections of the central nervous system. Clinical Infectious Diseases. 1995; 20: 971–981.
14. Elmore, JG, Horwitz, RI, Quagliarello, VJ. Acute Meningitis With a Negative Gram's Stain: Clinical and Management Outcomes in 171 Episodes. The American Journal of Medicine. 1996; 100: 78–84.
15. Rice, SK, Heinl, RE, Thornton, *et al.* Clinical Characteristics, Management Strategies, and Cost Implications of a Statewide Outbreak of Enterovirus Meningitis. Clinical Infectious Diseases. 1995; 20: 931–937.
16. Pevear DC, Seipel ME, Pallansch MA. *In vitro* activity of VP 63843 against field isolates of non-polio enteroviruses. H-99 Accepted for presentation at the 36th Interscience Conference on Antimicrobial Agents and Chemotherapy, New Orleans, LA September 15–18, 1996
17. Groarke JM and Pevear DC. Attenuated virulence of pleconaril-resistent coxsackievirus B3 variants. (submitted for publication).
18. Kearns GL. Pharmacokinetics of pleconaril in adults. Abstract H-21. Accepted for presentation at the 37th Interscience Conference on Antimicrobial Agents and Chemotherapy. New Orleans, LA September 27 - October 1, 1997
19. Kearns GL, Wells TG *et al.* Pharmacokinetics of pleconaril in children. Abstract A-206. Accepted for presentation at the 99th Annual Meeting of the American Society of Clinical Pharmacology and Therapeutics. New Orleans, LA March 30 - April 1, 1998.
20. Schiff GM *et al.* Oral efficacy of VP 63843 in coxsackievirus A21 infected volunteers (Abstract H42). Accepted for presentation at the 36th Interscience Conference on Antimicrobial Agents and Chemotherapy, New Orleans, LA. September 15–18, 1996.
21. Couch RB, Knight VR, Gerone PR, *et al.* Factors influencing response of volunteers to inoculation with coxsackievirus A type 21. American Review of Respiratory Disease. 1969; 99:34–69.
22. Couch RB, Cate TR, Gerone PJ, *et al.* Production of illness with a small-particle aerosol of coxsackie A21. Journal of Clinical Investigation. 1965; 44:535–542.
23. Weiner LB, Rotbart HA, Gilbert DL, *et al.* Treatment of enterovirus meningitis with pleconaril (VP 63843) an antipiconaviral agent. Accepted for presentation at the 37th Interscience Conference on Antimicrobial Agents and Chemotherapy, Toronto, Ontario, Canada, September 28-October 1, 1997.
24. McKinney RA, Katz SL, Wilfert KM. Chronic enteroviral meningoencephalitis in agammagolbulinemic patients. Reviews of Infectious Diseases. 1987; 9:334–356.
25. Webster ADB, Rotbart HA, Warner T, *et al.* Diaignosis of enterovirus brain disease in hypogammaglobulinemic patients by polymerase chain reaction. Clinical Infectious Diseases. 1993; 17:657–661.
26. Rotbart HA, Jones A, Novelli V, *et al.* Pleconaril treatment of chronic enteroviral central nervous system infections in children with agammaglobulinemia. Accepted for presentation at the 37th Interscience Conference on Antimicrobial Agents and Chemotherapy, Toronto, Ontario, Canada, September 28-October 1, 1997.
27. Sokhandan, M, *et al.* The contribution of respiratory viruses to severe exacerbations of asthma in adults. Chest. 1995; 107: 1570–1575.
28. Sung BS, Chonmaitree T, Broemeling LD, *et al.* Association of rhinovirus infection with poor bacteriologic outcome of bacterial viral otitis media. Clinical Infectious Diseases. 1993; 17:38–42.
29. Pitkaranta A, Virolainen A, Jero J, *et al.* Detection of rhinovirus, respiratory syncytial virus, and coronavirus infections in acute otitis media by reverse transcriptase polymerase chain reaction. Pediatrics . 1998;102:291–295.

8

MANAGEMENT OF CMV INFECTIONS IN HIV-INFECTED PATIENTS

Jennifer Hoy

Clinical Research Section
Infectious Diseases
Alfred Hospital
Melbourne, Victoria 3000, Australia

CMV disease is an important complication of advanced HIV infection. CMV disease was the initial AIDS defining illness (ADI) in 8% of patients, and occurred during the lifetime of 29% of a large Australian AIDS cohort. In the 10 years from 1985 to 1994, the cumulative risk for CMV disease in the 2 years following an AIDS diagnosis rose to 40%.[1] After the introduction of highly active antiretroviral therapy (HAART), marked reductions in the incidence of CMV disease were noted by several investigators.[2–4] CMV retinitis accounts for 90% of cases of CMV disease, and occurs late in the course of HIV disease, as determined by the mean CD4 cell count at the time of diagnosis of 50 cells/μl in 1989, and 10 cells/μl in 1995. The median time to development of CMV retinitis after the CD4 cell count falls below 50 cells/μl was 13 months in one study.[5] A report of 5 patients diagnosed with CMV retinitis with CD4 cell counts greater than 195 cells/μl, revealed that 4 patients had CD4 cell counts below 50 cells/μl, and the fifth had a count below 85 cells/μl 5–24 weeks earlier when they initiated HAART. This suggests that individuals with advanced HIV immunodeficiency remain susceptible to CMV retinitis for at least 6 months after initiation of successful HAART.[6] Improved outcomes for individuals with CMV retinitis have been noted with apparently longer times to relapse of CMV disease in those who are treated with HAART as well as specific anti-CMV therapy[7]. Selected patients with stable CMV retinitis successfully treated with HAART have not had reactivation of the CMV retinitis on discontinuation of the maintenance CMV treatment.[8–9]

1. CMV RETINITIS

The diagnosis of CMV retinitis remains a clinical/ophthalmological one. Between 13–19% of HIV-infected patients with less than 50 CD4 cells/μl were found to have as-

Antiviral Chemotherapy 5, edited by Mills *et al.*
Kluwer Academic / Plenum Publishers, New York, 1999.

Table 1. CMV PCR as a predictor of end-organ disease in ganciclovir and placebo treated patients

	Oral ganciclovir	Placebo
Baseline CMV PCR +ve	26%	43%
CMV PCR -ve	1%	14%

ymptomatic CMV retinitis,[10,11] while other individuals presented with a range of symptoms including blurred vision, visual loss, floaters, blind spots or flashing lights.

Currently, assays to identify those individuals at risk of reactivation of CMV disease, are under development with interest in prophylaxis or pre-emptive therapy. Surveillance urine or leucocyte cultures for CMV have limited, if any value in predicting those at highest risk of future development of CMV disease,[12] but recently, the use of PCR, branched chain DNA methods and antigen detection (pp65) have shown promise in identifying these individuals. The qualitative CMV PCR assay which detects the presence or absence of CMV DNA in plasma or leucocytes was used in the Roche oral ganciclovir prophylaxis study. The baseline CMV PCR and rate of CMV end-organ disease at 12 months are noted below[13] (Table 1).

Individuals who were CMV PCR positive at entry to the study (45% of the cohort) had a 3.4 fold greater risk of developing CMV disease and a 2.5 fold greater risk of death. Of those patients who were CMV PCR negative at study entry, 88% were PCR positive prior to the development of CMV disease.[14]

A number of assays have been used in retrospective and prospective studies to identify those individuals at greatest risk of development of CMV end organ disease, with variable success. Some assays may identify CMV positive individuals who should receive pre-emptive treatment to prevent disease and others may select individuals who would benefit from prophylaxis. The assays used include the CMV pp65 antigen assay, quantitative CMV PCR assay on plasma or white blood cells and the branched chain DNA assay on plasma, as well as a quantitative hybrid capture assay on whole blood. Results from these studies have not been conclusive, and only two small studies have compared the assays in the same patient population.[12,15,16] Both studies were performed prior to the introduction of protease inhibitors with 19–25% of the patients developing CMV disease over 12–20 months of follow up. One revealed that pp65 antigenemia was able to predict the development of CMV retinitis a median of 4 months prior to disease, while the qualitative CMV PCR and quantitative CMV DNA hybridisation assays predicted disease 0.4 and 0.1 months prior to disease.[16] The second study revealed the CMV PCR to be superior to the antigenemia test detecting CMV disease a median of 46 days prior to its subsequent development compared with 34 days for the antigenemia test.[17]

Identification of a threshold value of CMV burden which is predictive of subsequent development of CMV end organ disease is dependent on the assay methodology used, and the population studied. Two studies have suggested that there is only marginal additional value in quantitation of CMV burden to predict progression to retinitis once a threshold CMV DNA copy number was reached.[12,15] In contrast, others have shown that quantitative CMV PCR in plasma appears to have the ability to predict the development of CMV retinitis.[18] In one study there was an overlap of CMV burden in those with CMV disease and those who remained CMV disease free.

Table 2. The 12 month risk of CMV retinitis

	Overall risk	Risk reduction with oral ganciclovir
PCR negative	14%	90%
< 2,500 copies/ml	22%	84%
2,500-50,000 copies/ml	50%	42%
50,000-100,000 copies/ml	75%	100%

In a subset of 281 patients enrolled in the Roche oral ganciclovir prophylaxis study, the prophylactic benefit of ganciclovir appeared to be related to the baseline quantitative CMV viral load (Table 2).

There was no benefit from oral ganciclovir for those with greater than 50,000 copies/ml at baseline.[14] The PCR can therefore identify those patients most likely to benefit from CMV prophylaxis. Of note, between 15–25% of individuals may be CMV PCR negative at the time of diagnosis of CMV retinitis.[15]

All CMV antibody positive patients whose CD4 cell count is less than 50/µl or patients who have reached a nadir of less than 50/µl in the previous 3–6 months, should be screened by an experienced ophthalmologist using dilated indirect funduscopy every 3 months, as should any patient who develops any symptoms of visual disturbance. Individuals with extraocular CMV disease should also be screened.

2. THERAPEUTIC OPTIONS FOR CMV RETINITIS

Treatment for CMV retinitis involves an induction phase using higher daily doses of the virustatic anti-CMV agents. The induction phase lasts 14–21 days and is complete when the retinal inflammation has resolved or markedly decreased. Maintenance therapy is the second phase using the same agent at lower daily doses to suppress CMV replication and prevent relapse of CMV retinitis.

2.1. Parenteral Therapy

There are several agents available for the treatment of CMV disease; ganciclovir, foscarnet, and cidofovir (Table 3).

The duration of time between relapses of CMV retinitis progressively shorten from a median of 48 days to first relapse, 41 days to second relapse and 31 days to third relapse (approaching the time to disease progression in untreated patients).[24]

The Cytomegalovirus Retreatment Trial sought to determine the most effective regimen (induction and maintenance therapy) for either persistently active or relapsed CMV retinitis.[25] (Table 4)

Table 3. Antiviral options for CMV disease

	Ganciclovir	Foscarnet	Cidofovir
Induction dose	5 mg/kg bd	90 mg/kg bd	5 mg/kg weekly
Major toxicity	leukopenia thrombocytopenia	nephrotoxicity hypocalcemia mucosal ulceration	nephrotoxicity proteinuria neutropenia
Maintenance dose	30 mg/kg per week	120 mg/kg 5 times per week	5 mg/kg once every 2nd week

Table 4. Ganciclovir and forscarnet regimen efficacy in treating CMV retitinitis

	GCV	FCN
Median time to relapse (days)	56	59
Switched treatment		
toxicity	1%	22%
treatment failure	7%	8%
Survival (months)	8.5	12.6 ($p<0.001$)

2.1.1. Ganciclovir. A nucleoside analogue which requires intracellular phosphorylation to form the active ganciclovir triphosphate which inhibits CMV DNA polymerase. The first phosphorylation to the monophosphate form occurs by a protein kinase produced by the viral UL97 gene, and the other phosphorylation steps are facilitated by cellular enzymes.

2.1.2. Foscarnet. A pyrophosphate analogue that does not require intracellular activation to inhibit the CMV DNA polymerase.

2.1.3. Cidofovir. A nucleotide analogue which is phosphorylated to the active form (cidofovir diphosphate) by cellular enzymes rather than viral encoded enzymes. Cidofovir diphosphate inhibits the CMV DNA polymerase enzyme. It has a long intracellular half-life enabling once to second weekly dosing. Its major toxicities are neutropenia, iritis/uveitis, ocular hypotonia and renal impairment (occasionally irreversible); probenecid has been shown to reduce the renal toxicity by blocking secretion of cidofovir into the distal renal tubule. In one study cidofovir was discontinued because of renal toxicity in 24% of participants.[19]

In studies evaluating treatment for CMV retinitis, the median time to progression of retinitis in those untreated was 15–22 days, compared with 51 days for ganciclovir, 53 days for foscarnet and 120 days for cidofovir.

A randomised, comparative study of ganciclovir (GCV) and foscarnet (FCN) for induction and maintenance treatment for the initial episode of CMV retinitis in 234 patients with AIDS revealed there was no difference in the time to relapse, however there was a significant survival advantage for those randomised to foscarnet.[23] (Table 5)

Combination maintenance therapy using the recommended maintenance doses of ganciclovir and foscarnet achieved significantly lower rates of relapse and longer median times to retinitis progression than ganciclovir at twice the usual maintenance dose and foscarnet at the usual maintenance dose. In addition, there was no significant difference in the rate of visual acuity loss between arms, but there was a significantly lower rate of visual field loss in the combination arm. Quality of life measures revealed that the combination arm had the greatest negative impact on the quality of life.

Table 5. Comparison of ganciclovir and foscarnet treatment for CMV retinitis in AIDS patients

	GCV	FCN	GCV+FCN
Induction dose	5 mg/kg bd	90 mg/kg	5 mg/kg GCV or 90 mg/kg FCN
Survival (months)	9.0	8.4	8.6
Relapse photographic median time (months)	2.0	1.3	4.3 ($p<0.001$)
% patients	81.9	82	58.6
Clinical funduscopy median time (months)	3.6	2.0	5.4
% patients	67.0	78.3	49.4

2.2. Intraocular Therapy

The penetration of ganciclovir and foscarnet into the eye is limited when the drugs are administered systemically, and the concentrations achieved may not be sufficient to effectively inhibit CMV replication, especially when maintenance doses are used. The ability to deliver high doses of these agents directly to the site of disease without associated systemic toxicity is the attraction of intraocular therapy. The downside to this approach is that it does not provide systemic cover for the development of extraocular CMV disease. In addition, there is a paucity of clinical trial data comparing systemic therapy with local therapy provided by intravitreal administration of the drugs.

2.2.1. Intravitreal Ganciclovir Injection. After anaesthetising the eye with topical drops and the instillation of a drop of 5% betadine, 0.2 mg (200μg) ganciclovir in a volume of 0.05–0.1 ml is injected into the vitreous. Induction treatment involves twice weekly injections for 3 weeks, followed by weekly injections. Some individuals undergo induction treatment with systemic anti-CMV therapy and continue maintenance treatment with intravitreal therapy. In one study of relapsed CMV retinitis, there were no relapses recorded after a mean 21 weeks treatment.[26] About 75% patients can be maintained on intravitreal therapy without relapse.[27]

2.2.2. Intravitreal Foscarnet Injection. The standard dose of intravitreal foscarnet (2.4 mg) is administered by the same technique as for intravitreal ganciclovir, and given weekly for maintenance suppression of CMV retinitis, or twice weekly for 3 weeks as an induction course of treatment for active disease. In one study, complete resolution of CMV retinitis occurred in 63% cases and no patient failed to respond or developed progressive disease. During maintenance therapy, 33% relapsed by 20 weeks.[28]

2.2.3. Intravitreal Cidofovir Injection. Cidofovir injections of 20 μg administered every 6 weeks were shown to have some effect with a median time to CMV retinitis progression of 55 days after one injection.[29] Adverse effects included ocular inflammation causing iritis or uveitis, and ocular hypotonia especially in those receiving high dose (100μg) cidofovir. These usually occur in the first 3 months of therapy. It is apparent that this complication is much more common than the rate of 6% reported in clinical trial participants. Although the use of ophthalmic steroids allows the continuation of cidofovir in most patients, a reduction in intraocular pressure (ocular hypotonia) is an indication for cessation of the drug.

2.2.4. Intravitreal Sustained-Release Ganciclovir Implant. This device contains 4.5 mg of ganciclovir in a 4 mm pellet that allows for slow diffusion of the drug out of the pellet. Drug is released at approximately 1 μg per hour over 6–8 months achieving constant intraocular drug levels. Implantation of the pellet is an outpatient procedure, but visual acuity may not return to baseline for 2–3 weeks. A randomised trial evaluating the efficacy of the ganciclovir implant for immediate or deferred treatment of peripheral CMV retinitis found the median time to progression for immediate treatment was 226 days.[30] A comparison of the therapeutic effect of the implant with conventional intravenous ganciclovir therapy revealed a statistically significant superior treatment effect for the implants; median time to retinitis progression was 220 days for the 1 μg/hr implant, 182 days for the 2 μg/hr implant and 72 days for intravenous ganciclovir. But the development of CMV disease in the contralateral eye (40%) or other sites (11%) in those who received the implant but no systemic

Table 6. Comparison of oral and IV ganciclovir in preventing CMV retinitis progression

	Oral GCV 1 gm tds	IV GCV 5 mg/kg daily
ICM 1653 (newly diagnosed CMV)		
Mean time to progression		
Funduscopy	68 days	96 days p = 0.03
Photography	57 days	62 days p = 0.63
ICM 1774 (stable CMV 4 weeks to 4 mths after induction)		
Mean time to progression		
Funduscopy	59 days	106 days p = 0.05
Photography	54 days	66 days p = 0.1
EACG (both newly diagnosed and stable CMV disease)		
Mean time to progression		
Funduscopy	86 days	109 days p = 0.02
Photography	51 days	62 days p = 0.15

therapy compared with 16% contralateral eye CMV retinitis and 0% extraocular disease in the group randomised to IV ganciclovir (31). If the implant is inserted in the setting of successful HAART, it may not require replacement at 6–8 months (8–9).

Complications of intraocular therapy include endophthalmitis, and retinal detachment (20%), as well as the development of extraocular disease.

2.3. Oral Therapy- Maintenance Therapy after Successful Induction Treatment

Three studies have compared oral ganciclovir with standard intravenous ganciclovir for the prevention of CMV retinitis progression (Table 6).

There was a significantly shorter time to retinitis progression when clinical funduscopy was used as the measurement tool, but no significant difference in the mean time to progression was evident on masked photography.[32–34] However, in one study there was a significant difference in the median time to progression judged on photography (29 days for oral GCV and 49 days for IV GCV).[32] Oral ganciclovir has poor oral bioavailability (6–9%) and many clinicians are reticent to use this therapy as the sole maintenance therapy for prolonged periods of time.

2.4. Antiviral Resistance in CMV

Resistance to ganciclovir results from mutations in either the UL97 gene or the polymerase genes. Foscarnet resistance is associated with mutations in the polymerase gene only. Mutations conferring resistance to cidofovir are predicted to also be in the polymerase gene. Ganciclovir-resistant isolates fall into 2 groups: low level resistance (<6 fold increase in IC50, 8.5–30μM) and high level resistance (>6 fold increase in IC50, >30μM). Low level ganciclovir resistance is associated with mutations in the UL97 gene only and these isolates retain susceptibility to cidofovir. High level resistant isolates carry mutations in both the UL97 and the DNA polymerase genes, and these isolates have reduced susceptibility to cidofovir.[35] Treatment with ganciclovir for less than 3 months does not appear to be associated with resistance, but about 50% isolates are resistant after 3 months and 76% are resistant by 7 months treatment. Clinical failure can be associated with low level ganciclovir resistance.

Foscarnet resistance did not develop in the first 3 months of treatment in individuals not previously treated with ganciclovir, however resistance did develop by 4 months in 60% of patients with at least 12 months prior treatment with ganciclovir. Foscarnet resistance can emerge within 1–2 months of therapy in those with ganciclovir resistant isolates. There does not appear to be cross resistance between foscarnet resistant ganciclovir susceptible isolates and cidofovir resistance isolates. However, 80% of isolates resistant to both foscarnet and ganciclovir are also resistant to cidofovir.

Cidofovir resistance also does not appear to develop by 3–4 months in those with no prior anti-CMV treatment, but does occur in 50% of those extensively pre-treated with ganciclovir. Pre-therapy isolates revealed high level resistance to ganciclovir and *in vitro* cross resistance to cidofovir.

Clinical failure and relapse of CMV retinitis can be due either to the inadequate trough concentrations of the drugs used in the maintenance phase of treatment or to the development of resistance. Relapse of CMV retinitis during the first 3 months of treatment is likely to respond to re-induction treatment with the same agent. However, the time to relapse becomes shorter as resistance develops.

3. RECOMMENDATIONS FOR THE TREATMENT OF CMV RETINITIS

a. Initial therapy CMV retinitis

- Induction ganciclovir 5 mg/kg bd for 2–3 weeks, followed by
- Maintenance ganciclovir 5 mg/kg daily or 10 mg/kg administered 3 times per week (30 mg/kg/week)
- Sustained release ganciclovir implant, replaced every 6–8 months. Concurrent systemic therapy recommended.

b. Therapy for first relapse CMV disease

- Re-induction with ganciclovir 5 mg/kg bd for 2–3 weeks
- Maintenance ganciclovir with 5 mg/kg daily or 10 mg/kg 3 times weekly
- Induction foscarnet 90 mg/kg bd for 2–3 weeks
- Maintenance foscarnet 100–120 mg/kg administered 5 days of each week.
- Consider combination ganciclovir and foscarnet maintenance therapy if severe/sight threatening disease.

c. Therapy for second or subsequent relapses

- Because over 50% of isolates will be ganciclovir resistant at the second or subsequent relapse
- Reinduction with foscarnet if this agent has not been used, or foscarnet and ganciclovir combination
- Maintenance with combination ganciclovir 5 mg/kg daily or 10 mg/kg 3 times weekly plus foscarnet 120 mg/kg 5 times weekly.

d. Salvage therapy

- Indicated for those with progressive disease on conventional therapy or those unable to tolerate the toxicity of foscarnet. Ganciclovir toxicity is generally manageable with the use of G-CSF.
- Intraocular therapy
  Foscarnet 2.4 mg in 0.1 ml, administered twice weekly for induction and twice weekly for maintenance therapy. Concurrent systemic foscarnet is recom-

mended, or alternatively maintenance ganciclovir with intraocular foscarnet. Rarely, ganciclovir 0.2 mg in 0.1 ml, administered twice weekly for induction and weekly for maintenance treatment. Patients should also receive concurrent foscarnet systemic therapy if tolerated, to prevent the up to 37% development of extraocular disease, and because the CMV is likely to be resistant to ganciclovir by this time.

- Cidofovir - induction with 5 mg/kg, 2 doses administered IV 1 week apart, with probenecid and prehydration, followed by maintenance treatment with IV cidofovir 5 mg/kg every 2 weeks.

4. GASTROINTESTINAL CMV DISEASE

The treatment of CMV colitis has been evaluated in a placebo controlled trial of ganciclovir and a comparative study of ganciclovir and foscarnet. Greater improvement in histological abnormalities, endoscopic findings and CMV cultures occurred in the ganciclovir treated group compared with placebo.[36] In the comparative study there was no difference in efficacy between foscarnet and ganciclovir,[37] however foscarnet has been shown to be effective in 80% of those who failed induction therapy with ganciclovir.[38] The duration of CMV therapy is often determined by endoscopic findings (usually 3 weeks), and the question of maintenance therapy in this setting is controversial. There is no clinical trial data to support routine maintenance therapy, however, the analogy with CMV retinitis and the need for maintenance therapy to suppress CMV replication and prevent relapse of CMV disease is hard to ignore. Uncontrolled, noncomparative studies suggest that both ganciclovir and foscarnet are effective treatments for CMV gastritis and oesophagitis with 75–80% of patients responding in 2–3 weeks of treatment.[39,40]

5. CMV DISEASE OF THE CENTRAL NERVOUS SYSTEM

CMV encephalitis usually occurs in the setting of pre-existing CMV disease and often in patients already receiving ganciclovir. The prognosis is poor with the mean time to death just 9 weeks.[41] Some patients may improve or have stabilisation of the disease process after treatment with ganciclovir or foscarnet. Foscarnet is indicated for those patients who have received more than 3 months of prior ganciclovir therapy because of the likelihood of ganciclovir resistance. In contrast to encephalitis, therapy with ganciclovir for CMV polyradiculomyelitis is likely to be successful with neurological improvement over several months.[42] If the patient has extensive pre-treatment with ganciclovir, foscarnet should be instituted. The role of combination ganciclovir-foscarnet therapy has not been evaluated in this setting, but may be indicated for severe disease not responding to monotherapy.

6. PRIMARY PROPHYLAXIS OF CMV INFECTION

6.1. Oral Ganciclovir

Two randomized placebo-controlled studies have been performed. Ganciclovir 1 gm tds was shown to significantly reduce the incidence of CMV disease from 39% in the placebo arm to 20% in the oral ganciclovir arm (49% risk reduction, p=0.0001) in one study[43] but did not show a significant effect on prevention of CMV retinitis in the CPCRA study.[44] The risk of CMV disease appeared related to concomitant use of didanosine, with a decreased risk noted for those individuals not taking didanosine at baseline.

6.2. Oral Valaciclovir

Valaciclovir is the valine ester of aciclovir which achieves plasma aciclovir concentrations after oral administrations equivalent to those achieved after intravenous administration of aciclovir 5 mg/kg. A randomized controlled study comparing the efficacy of valaciclovir 2 gm qid with high dose aciclovir (800 mg qid) and low dose aciclovir (400 mg bd) in the prevention of CMV disease revealed a significantly longer time to development of CMV disease in the valciclovir recipients and a reduction in the incidence of CMV disease (11.7% compared with 17.5% in the combined aciclovir arms).[45]

At this time, primary CMV prophylaxis for all individuals with CD4 cell counts less than 50/μl is not recommended because of the lack of supportive data, the pill burden and potential side-effects. The commercial availability of CMV PCR assays will help to determine those individuals who would benefit from prophylaxis or pre-emptive therapy.

REFERENCES

1. Dore GJ, Hoy JF, Mallal SA, *et al.* Trends in the incidence of AIDS illnesses in Australia from 1983 to 1994: The Australian AIDS Cohort. J Acquir Immune Defic Syndr. 1997; 16:39–43.
2. Pallela FJ, Delaney KM, Moorman AC *et al.* Declining morbidity and mortality among patients with advanced human immunodeficiency virus infection. N Engl J Med. 1988; 338:853–860.
3. Michaels S, Clark R, Kissinger P. Differences in the incidence rates of opportunistic processes before and after the availability of protease inhibitors. In; Abstracts of the 5th Conference on Retroviruses and Opportunistic Infections, Chicago, February 2–5, 1998, Abstract No. 180.
4. Moore RD, Keruly JC, Chaisson RE. Decline in CMV and other opportunistic disease with combination antiretroviral therapy. In; Abstracts of the 5th Conference on Retroviruses and Opportunistic Infections, Chicago, February 2–5, 1998, Abstract No. 184.
5. Pertel P, Hirschtick R, Phair J, *et al.* Risk of developing cytomegalovirus retinitis in persons infected with the human immunodeficiency virus. J Acquir Immune Defic Syndr. 1992; 5:1069–1074.
6. Jacobson MA, Zegans M, Pavan PR, *et al.* Cytomegalovirus retinitis after initiation of highly active antiretroviral therapy. Lancet. 1997; 349:1443–1445.
7. Mallolas J, Arrizabalaga J, Lonca M, *et al.* Cytomegalovirus disease in HIV-1 infected patients treated with protease inhibitors. AIDS. 1997; 11:1785–1787.
8. Tural C, Sirera D, Andreu D, *et al.* Long lasting remission of cytomegalovirus retinitis without maintenance therapy in HIV+ patients. In: Programs and Abstracts of the 37th International Conference on Antimicrobial Agents and Chemotherapy, Toronto, Canada, September 28-October 1, 1997.
9. Torriani FJ, MacDonald JC, Karevellas M, *et al.* Lack of progression after discontinuation of maintenance therapy (MT) for cytomegalovirus retinitis (CMVR) in AIDS patients responding to Highly Active Antiretroviral Therapy (HAART). In: Programs and Abstracts of the 37th International Conference on Antimicrobial Agents and Chemotherapy, Toronto, Canada, September 28-October 1, 1997.
10. Baldassano V, Dunn JP, Feinberg J, *et al.* Cytomegalovirus retinitis and low CD4+ T-lymphocyte counts. N Engl J Med. 1995; 333:670.
11. Kuppermann BD, Petty JG, Richman DD, *et al.* Correlation between CD4+ counts and prevalence of cytomegalovirus retinitis and human immunodeficiency virus-related non-infectious retinal vasculopathy in patients with acquired immunodeficiency syndrome. Am J Ophthalmol. 1993; 115:575–582.
12. Shinkai M, Bozzette SA, Powderly W, *et al.* Utility of urine and leukocyte cultures and plasma DNA polymerase chain reaction for identification of AIDS patients at risk for developing human cytomegalovirus disease. J Infect Dis. 1997; 175:302–308.

13. Spector S, Pilcher M, Lamy P, *et al.* PCR of plasma for cytomegalovirus DNA identifies HIV-infected persons most likely to benefit from oral ganciclovir prophylaxis. XI International Conference on AIDS, Vancouver, Canada, July 7–12 1996. Abstract ThB.302
14. Spector SA, Hsia K, Crager M, *et al.* Impact of oral ganciclovir (GCV) on plasma cytomegalovirus DNA and development of CMV disease in advanced AIDS. In: Programs and Abstracts of the 37th International Conference on Antimicrobial Agents and Chemotherapy, Toronto, Canada, September 28-October 1, 1997. Abstract No. I-232.
15. Rasmussen L, Zipeto D, Wolitz RA, *et al.* Risk for retinitis in patients with AIDS can be assessed by quantitation of threshold levels of cytomegalovirus DNA burden in blood. J Infect Dis. 1997;176:1146–1155.
16. Riss JM, Zandotti C, Petit N, *et al.* Prospective evaluation of CMV antigenemia, PCR and DNA hybridization as predictors of CMV retinitis in HIV infected patients. In: Abstracts of the 4th Conference on Retroviruses and Opportunistic Infections. Washington DC January 1997. Abstract No. 316.
17. Dodt KK, Jacobsen PH, Hofmann B, *et al.* Development of cytomegalovirus (CMV) disease may be predicted in HIV-infected patients by CMV polymerase chain reaction and the antigenemia test. AIDS. 1997;11:F21–8.
18. Bowen EF, Sabin, CA, Wilson P, *et al.* Cytomegalovirus (CMV) viraemia detected by polymerase chain reaction identifies a group of HIV-positive patients at high risk of CMV disease. AIDS, 1997; 11: 889–893.
19. Lalezari JP, Stagg RJ, Kuppermann BD, *et al.* Intravenous cidofovir for peripheral cytomegalovirus retinitis in patients with AIDS. A randomized, controlled trial. Ann Intern Med 1997; 126:257–263.
20. Spector SA, Weingeist T, Pollard RB, *et al.* A randomized, controlled study of intravenous ganciclovir therapy for cytomegalovirus peripheral retinitis in patients with AIDS. J Infect Dis. 1993; 168:557–563.
21. Palestine AG, Polis MA, De Smet MD, *et al.* A randomized, controlled trial of foscarnet in the treatment of cytomegalovirus retinitis in patients with AIDS. Ann Intern Med. 1991; 115:665–673.
22. Studies of Ocular Complications of AIDS Research Group in collaboration with the AIDS Clinical Trial Group. Parenteral cidofovir for cytomegalovirus retinitis in patients with AIDS: The HPMPC Peripheral Cytomegalovirus Retinitis Trial. A randomized, controlled trial. Ann Intern Med. 1997; 126:264–274.
23. Studies of Ocular Complications of AIDS Research Group. AIDS Clinical Trial Group. Mortality in patients with the acquired immunodeficiency syndrome treated with either foscarnet or ganciclovir for cytomegalovirus retinitis. N Engl J Med. 1992; 326:213–220.
24. Studies of Ocular Complications of AIDS Research Group in collaboration with the AIDS Clinical Trial Group. Foscarnet-ganciclovir cytomegalovirus retinitis trial, IV: visual outcomes. Ophthalmol. 1994; 101:1250–1261.
25. Studies of Ocular Complications of AIDS Research Group in collaboration with the AIDS Clinical Trials Group. Combination foscarnet and ganciclovir therapy vs monotherapy for the treatment of relapsed cytomegalovirus retinitis in patients with AIDS. Arch Ophthalmol 1996; 114:23–33.
26. Young SH, Morlet N, Heery S, *et al.* High dose intravitreal ganciclovir in the treatment of cytomegalovirus retinitis. Med J Aust. 1992; 157:370–373.
27. Heinemann MH. Long-term intravitreal ganciclovir therapy for cytomegalovirus retinopathy. Arch Ophthalmol. 1989; 107:1767–1772.
28. Diaz-Llopis M, Espana E, Munoz G, *et al.* High dose intravitreal foscarnet in the treatment of cytomegalovirus retinitis in AIDS. Br J Ophthalmol. 1994; 78:120–124.
29. Kirsch LS, Arevalo JF, de la Paz EC, *et al.* Intravitreal cidofovir (HPMPC) treatment of cytomegalovirus retinitis in patients with the acquired immunodeficiency syndrome. Ophthalmol. 1995; 102:533–542.
30. Martin DF, Parks DJ, Mellow SD, *et al.* Treatment of cytomegalovirus retinitis with an intraocular sustained-release ganciclovir implant. A randomized controlled clinical trial. Arch Ophthalmol. 1994; 112:1531–1539.
31. Musch DC, Martin DF, Gordon JF *et al.* Treatment of AIDS related CMV retinitis with Ganciclovir implant: a multicenter randomized controlled trial. N Engl J Med. 1997; 115:665–673.
32. Drew WL, Ives D, Lalezari JP, *et al.* Oral ganciclovir as maintenance treatment for cytomegalovirus retinitis in patients with AIDS. N Engl J Med. 1995; 333:615–620.
33. The Oral Ganciclovir European and Australian Cooperative Study Group. Intravenous versus oral ganciclovir: European/Australian comparative study of efficacy and safety in the prevention of cytomegalovirus retinitis recurrence in patients with AIDS. AIDS. 1995; 9:471–477.
34. Squires KE, Stempien MJ, Shadman A, *et al.* Oral ganciclovir (POG) versus intravenous ganciclovir (IVG) maintenance therapy for cytomegalovirus retinitis in patients with AIDS: preliminary results of a phase III study. First National Conference on Human Retroviruses and Related Infections, Washington DC 1993 Abstract no. 540.
35. Smith IL, Cherrington JM, Jiles RE, *et al.* High-level resistance of cytomegalovirus to ganciclovir is associated with mutations in both the UL97 and DNA polymerase genes. J Infect Dis. 1997; 176:69–77.

36. Dieterich DT, Kotler DP, Busch DF, *et al.* Ganciclovir treatment of cytomegalovirus colitis in AIDS: A randomized, double-blind, placebo-controlled multicenter study. J Infect Dis. 1993;167:278–282.
37. Blanshard C, Benhamou Y, Dohin E, *et al.* Treatment of AIDS-associated gastrointestinal infection with foscarnet and ganciclovir: a randomized comparison. J Infect Dis. 1995; 172:622–628.
38. Dieterich DT, Poles MA, Dicker M, *et al.* Foscarnet treatment of cytomegalovirus gastrointestinal infections in acquired immunodeficiency syndrome patients who have failed ganciclovir induction. Am J Gastroenterol. 1993; 88:542–548.
39. Nelson MR, Connolly GM, Hawkins DA, *et al.* Foscarnet in the treatment of cytomegalovirus infection of the esophagus and colon in patients with the acquired immunodeficiency syndrome. Am J Gastroenterol 1991; 86:876–881.
40. Dieterich DT, Chachoua A, Lafleur F, *et al.* Ganciclovir treatment of gastrointestinal infections caused by cytomegalovirus in patients with AIDS. Rev Infect Dis. 1988; 10(Suppl. 3):S532–537.
41. Holland NR, Power C, Matthews VP, *et al.* Cytomegalovirus encephalitis in acquired immunodeficiency syndrome (AIDS): Clinical features and course. Neurology. 1994; 44:507–514.
42. Cohen BA, McArthur JC, Grohman S, *et al.* Neurologic prognosis of cytomegalovirus polyradiculomyelopathy in AIDS. Neurology. 1993; 43:493–499.
43. Spector SA, McKinley GF, Lalezari JP, et al. Oral ganciclovir for the prevention of cytomegalovirus disease in persons with AIDS. N Engl J Med. 1996; 334:1491–1497.
44. Brosgart CL, Louis TA, Hillman DW, et al. A randomized, placebo-controlled trial of the safety and efficacy of oral ganciclovir for prophylaxis of cytomegalovirus disease in HIV-infected individuals. AIDS. 1998; 12:269–277.
45. Feinberg J, Hurwitz S, Cooper DA, et al. A randomized, double-blind trial of valaciclovir (VACV) prophylaxis for cytomegalovirus (CMV) end-organ disease in patients with AIDS. J Inf Dis. 1998; in press.

9

MANAGEMENT OF CYTOMEGALOVIRUS INFECTIONS IN BLOOD AND MARROW TRANSPLANT RECIPIENTS

Michael Boeckh

Fred Hutchinson Cancer Research Center and
The University of Washington
Program in Infectious Diseases
1100 Fairview Ave. North
Seattle, Washington 98109-1024

1. HOST-VIRUS RELATIONSHIP AND IMMUNOBIOLOGY OF CMV

As one of the 8 human herpes viruses (designated human herpes virus 5), CMV shares characteristics with other herpes viruses including virion and genome structure and the ability to establish lifelong persistent and latent infections after primary exposure.[1] The basic biology and pathogenesis of CMV disease go beyond the scope of this overview and are reviewed elsewhere.[1–4] CMV infection in immunocompromised individuals is acquired exogenously or results from reactivation of latent virus. Reactivation of latent endogenous virus appears to be the dominant mechanism of infection in immunocompromised patients. The frequency at which CMV in seropositive patients can be superinfected by exogenous CMV is uncertain, but co-infection with different strains can occur.[5,6] In seronegative individuals, CMV is virtually always acquired from blood products or from transplanted organs.[7] Transplant of an organ from a seropositive donor into a seronegative recipient results in a high risk of infection (80–90%), higher levels of CMV replication, and a higher incidence of disease than reactivation of latent virus.[8] In contrast, transplantation of marrow or peripheral blood stem cells (PBSC) from a seropositive donor into a seronegative recipient results in CMV infection in only 20–25% of patients.[9] The reasons for this difference are only poorly understood.

Both host and viral factors appear to be responsible for the progression from CMV infection to disease in immunocompromised patients. In the transplant setting, host factors that have been associated with CMV disease after transplant include CMV serostatus before transplant, patient's age, post-transplant immunosuppression, development of acute

Antiviral Chemotherapy 5, edited by Mills *et al.*
Kluwer Academic / Plenum Publishers, New York, 1999.

graft versus host disease (GvHD) and its treatment, and the use of total body irradiation (TBI) pre transplantation.[10]

Hematogenous dissemination of CMV is an important factor in the pathogenesis of disease, and hematogenous dissemination of CMV occurs in most immunocompromised patients with active infection. In allogeneic marrow transplant recipients, viremia and CMV viral load are highly predictive for the development of CMV disease.[11–13a] Studies using sensitive techniques for detection of CMV such as the antigenemia assay or PCR for CMV DNA show that CMV can be detected in the blood of almost all patients with CMV disease. However, due to the high sensitivity of these assays, CMV is also detectable in a substantial number of patients with asymptomatic infection who never progress to disease,[14,15] although patients with disease often have a higher systemic viral load (measured by quantitative pp65 antigenemia or quantitative PCR in PBL or plasma) than those who remain asymptomatic.[16,17] CMV DNA can be detected in both mononuclear (MN) and polymorphonuclear leukocytes (PMNL).[18,19] While the major amount of DNA in PMNL seems to originate from phagocytosis,[20] there is also evidence of active replication.[18] Although CMV is a cell-associated virus, CMV DNA can also be detected in plasma or serum.[17,21] Longitudinal studies indicate that detection in plasma occurs coincident or after detection in PMNL.[15,17] It has been hypothesized that cell-free viral DNA in plasma could be derived from lysis of CMV-infected cells such as PMNL and endothelial cells. Additional support for endothelial cells as an important source of cell-free CMV DNA comes from the observation that CMV DNA can be detected in plasma from patients with absolute neutropenia early after marrow transplantation.[15] CMV-infected endothelial cells have also been detected in the peripheral blood during active infection in both transplant and HIV-infected patients.[22–24]

Studies in the murine model and in human transplant recipients suggest that HLA-restricted CMV specific cytotoxic T cell (CTL) responses play an important role in protective immunity and the elimination of active infection.[25,26] $CD4^+$ T helper (Th) responses are responsible for maintaining CTL responses and clearing CMV from specific sites.[27,28] In allogeneic marrow transplant recipients who develop CMV disease, both CMV-specific $CD8^+$ CTL and $CD4^+$ Th responses are usually undetectable, although in an occasional patient, weak $CD4^+$ Th reactivity is observed.[3,29] There is also a direct correlation between absence of CMV-specific CTL and Th responses and high CMV viral load in allogeneic marrow transplant recipients.[30] Recovery of CMV-specific T cell immunity is delayed in recipients of ganciclovir prophylaxis.[31] In autologous peripheral blood and marrow transplant recipients, recovery of $CD8^+$ CTL responses was associated with subsequent protection from CMV infection.[32] Current data suggest that antibodies to CMV do not protect from CMV diesease.

There is increasing evidence that strain differences play an important role in the pathogenesis of CMV disease. Recent studies suggest that strain differences based on the gB envelope protein may contribute to the tropism and pathogenicity of CMV in both marrow transplant and AIDS patients.[5,33,34]

There is also an interesting association between CMV infection and acute and chronic GvHD. Patients with acute GvHD are at an increased risk for CMV disease.[10,11] Whether CMV increases the risk for acute and chronic GvHD remains controversial.[35–43]

1.1. Clinical Manifestations, Epidemiology, and Risk Factors

The most common manifestations of CMV infection after blood and marrow transplantation (BMT) are pneumonia and gastrointestinal disease.[44] Hepatitis, retinitis, marrow suppression, and encephalitis are rare manifestations. CMV infection can also cause fever but the association is often difficult to ascertain after BMT because of the multiple causes of fever in this setting. Without antiviral prophylaxis, CMV infection and disease are more common after allogeneic transplant than after autologous transplant.[44] CMV seropositive recipients are at highest risk of disease.[10]

The peak incidence of CMV disease traditionally occurred approximately 45 to 60 days after BMT, but during the last few years CMV disease has been increasingly diagnosed late after transplantation.[30,31,45] This change has been associated with the introduction of ganciclovir prophylaxis,[30,46] which by effectively suppressing CMV replication, delays recovery of CMV-specific T cell immunity.[31,47] Other factors associated with late CMV disease are acute and chronic GvHD[46,48] and a high CMV viral load in plasma during the first three months after transplant[4] as well as later.[30] At the Fred Hutchinson Cancer Research Center in Seattle, the incidence of disease before day 100 in CMV seropositive allogeneic recipients was 35% in 1987 and 6% in 1994. The corresponding figures for disease occurring later than 100 days after BMT were 4% and 15%, respectively.[30] The incidence of CMV infection after allogeneic unmodified G-CSF-mobilized PBSC transplant appears similar to that after marrow transplantation.[49] However, CD34+ selected PBSC transplantation is associated with a higher incidence of CMV infection and higher levels of quantitative pp65 antigenemia than marrow transplantation, possibly due to the T cell depletion which is part of the CD34+ selection process.[50] There are no reports on the incidence of CMV infection after umbilical cord blood transplantation, but there is conflicting data as to whether CMV seropositivity of the recipient is a negative predictor for survival and a risk factor for acute GvHD.[51,52]

Several studies have shown an incidence of CMV pneumonia after autologous BMT of between 1 and 6%.[12,53–56] The outcome of CMV pneumonia after autologous BMT is similar to that found after allogeneic BMT.[12,54–56] However, these reports included mostly patients with hematological malignancies many of whom were treated with TBI or with toxic combination chemotherapy regimens. CMV disease did not occur in one series of patients undergoing autologous PBSC transplant for breast cancer.[57] However, there are no large studies which have analyzed the impact of the underlying diagnosis and the conditioning regimens on the risk for CMV disease after autologous transplantation. Using the antigenemia assay or PCR, CMV infection rates of 39% and 42% have been reported,[58,59] with no apparent difference between marrow and PBSC with regard to the incidence of CMV infection.[58] There has been some controversy in the literature regarding the influence of CMV on marrow engraftment after marrow transplant,[53,54] but no data exists regarding CMV and engraftment after autologous PBSC transplantation.

1.2. Diagnosis

Detection and quantification of CMV DNA by PCR amplification and quantification of CMV antigen in peripheral blood leukocytes are assays which are being used increasingly to monitor CMV infection after BMT.[15,60–62] These assays are more sensitive than culture-based assays (i.e. shell vial centrifugation cultures), resulting in earlier detection of CMV infection after transplant. In addition, both assays can be quantified.[63,64] (Details of quantitative assays for detection of CMV have been reviewed elsewhere.[65]) CMV DNA is easier to detect in peripheral blood leukocytes than in plasma.[15,17] Because early detection may be critical for early institution of antiviral therapy peripheral blood leukocytes (PBL) are preferable to plasma for detection of CMV DNA.[15]

Diagnosis of CMV pneumonia requires documentation of CMV in the bronchoalveolar lavage (BAL) or biopsy (including autopsy specimens) by culture, immunohistology or direct fluorescent antibody methods (in BAL specimens) in conjunction with radiographic signs of pneumonia.[66] The symptoms of pneumonia are initially uncharacteristic with dry cough, hypoxemia, and low degree fever, but it can progress rapidly to respiratory failure. The radiographic picture often shows an interstitial pattern but alveolar infiltrates or nodular changes can also be seen. Severe hypoxemia is a poor prognostic sign and most patients who require ventilator assistance will succumb.[12]

Because BAL has largely replaced open lung biopsy for the diagnosis of CMV pneumonia,[67] there is a risk of overdiagnosis, as it has been shown that CMV can be isolated from BAL in patients without any signs or symptoms of respiratory disease.[13,68] The presence of CMV in BAL from asymptomatic patients at day 35 after BMT is highly predictive for later development of CMV pneumonia.[13] However, approximately one third of asymptomatic patients with CMV in BAL do not subsequently develop CMV pneumonia. During the last few years even more sensitive techniques such as PCR have been introduced for the detection of CMV DNA in BAL. In a study by Cathomas *et al.*, the absence of CMV DNA in BAL using the PCR assay had a very high negative predictive value, however, the predictive value of a positive PCR was low due to the high sensitivity of PCR assay.[69] Thus, qualitative PCR for CMV DNA in BAL can be used to rule out CMV as a cause of pneumonia in BMT recipients.

Quantitation of CMV from the lung or BAL specimens has been explored as a way of differentiating asymptomatic virus shedding from CMV pneumonia. In BMT recipients, there was no relation between the virus load in centrifugation culture and asymptomatic CMV excretion or CMV pneumonia.[70] More recently, Boivin *et al.* measured CMV viral load by quantitative PCR and CMV gene expression in 18 patients with asymptomatic pulmonary shedding and 19 patients with definite or probable CMV pneumonitis.[71] There was $> 2 \log_{10}$ difference in the amount of DNA between patients with asymptomatic shedding and definite or probable CMV pneumonia. In addition, all patients with definite or probable CMV pneumonitis had detectable CMV glycoprotein H mRNA compared to none of the asymptomatic shedders.[71] These data suggest that the amount of CMV in the lung is related to the pathogenesis of CMV pneumonia, and that quantitative PCR may also be a useful method to diagnose CMV pneumonia.

CMV gastrointestinal disease can occur from the esophagus to the colon, thus, symptoms will vary depending on the location of the disease. These may include epigastric pains, vomiting, abdominal cramps, and diarrhea. At endoscopy, ulcerations are frequently seen. CMV gastrointestinal disease is diagnosed by culture or immunohistology examination of biospy specimens. Shell vial centrifugation cultures are an excellent tool to diagnose CMV rapidly in biopsy and autopsy specimens, but two diagnostic tests should be performed on each specimen for optimal sensitivity.[72] CMV DNA can also be detected by PCR in these samples but the predictive values of this test in tissue specimens have only been poorly charcterized in the marrow transplant setting. Therefore, PCR for CMV DNA is presently not used to diagnose CMV gastrointestional disease.

CMV hepatitis occurs in BMT recipients but is rarely severe. Evaluation of liver tissue is required to differentiate from other causes of liver function abnormalities.[73] Encephalitis and retinitis have been rare complications of CMV infection, but recent reports indicate that these manifestations might become more common among those patients developing CMV disease late after transplantation. As the symptoms of CMV encephalitis are frequently non-specific (headache, confusion, and fatigue), the diagnosis depends on detection of CMV in the cerebrospinal fluid. Recent data document that PCR for CMV

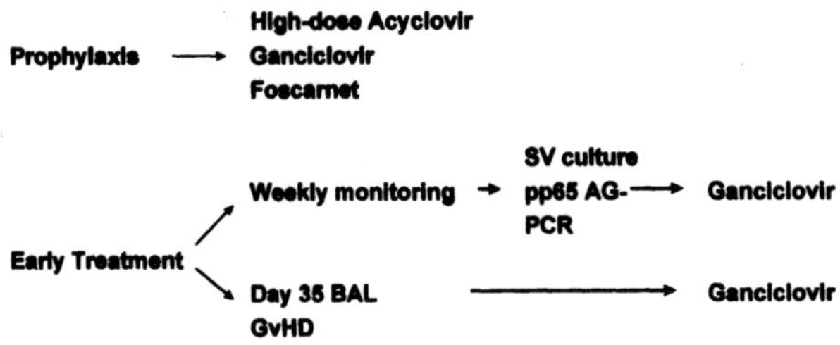

Figure 1. Options for prevention of CMV in seropositive allogeneic recipients. Abbreviations: BAL: bronchoalveolar lavage; GvHD: graft vs host disease; SV: shell-vial centrifugation culture; AG: antigenemia; PCR: polymerase chain reaction amplification to quantify or detect CMV DNA (see text)

DNA is both sensitive and highly specific for the diagnosis of CMV CNS disease, as it is for HSV encephalitis.[74] The diagnosis of CMV retinitis is based on typical retinal lesions. The so called CMV syndrome with fever and bone marrow suppression is difficult to differentiate from several other etiologies such as other viral infections, drug toxicity and marrow rejection, and therefore it is currently not accepted as a separate disease entity in the current consensus definition of CMV disease.[73] However, *in vitro* studies support that CMV has an inhibitory effect on hematopoiesis,[75] and there are well documented cases of late marrow failures after BMT which can also occur with low or undetectable systemic viral load.[76]

2. MANAGEMENT

2.1. Priniciples of Prevention

Two basic ganciclovir administration strategies are used to prevent CMV disease in transplant recipients (Figure 1). Marrow transplant patients with evidence of early CMV infection without disease (as indicated by a positive culture from blood, urine, throat or BAL or CMV antigens or DNA in blood) can be treated with ganciclovir, in an 'early treatment' or 'preemptive therapy' strategy. Alternatively, all patients at risk of CMV infection or disease, based on pretransplant serologic status, are given prophylactic ganciclovir regardless of whether or not there is post-transplant CMV excretion, CMV antigenemia or CMV DNA in blood (an 'early prophylaxis' or 'universal treatment' strategy).

2.2. Antiviral Prophylaxis

2.2.1. Ganciclovir. Three randomized double-blind studies using an early prophylaxis strategy have been published.[44,80,81] Winston *et al.* randomized patients before transplant to either ganciclovir or placebo while Goodrich *et al.* and Boeckh *et al.* started prophylaxis at engraftment. Also, Goodrich *et al* compared prophylaxis with ganciclovir given for CMV excretion from blood, urine or throat while Boeckh *et al* compared prophylaxis with a strategy based on quantitative antigenemia. In all 3 studies, ganciclovir prophylaxis decreased CMV infection and/or disease (Table 1). Indeed, 2 studies showed an almost complete elimination of CMV disease while ganciclovir was given.[45,80] However, severe neutropenia was a limiting factor in all 3 studies, and there was no benefit in overall survival in any of the studies. Salzberger *et al* analyzed risk factors for neutropenia in 278 patients who received ganciclovir at engraftment and found that early liver failure,

Table 1. Results of randomized ganciclovir prophylaxis studies

Author	Ganciclovir Regimen	No. of patients	CMV infection (day 100)	CMV disease (day 100)	Adverse events
Goodrich et al.[80]	At engraftment: 5 mg/kg BID for 5 days, followed by 5 mg/kg/day until day 100	33	3% (culture)	0%	33% neutropenia[1]
Winston et al.[81]	Day -7 to -1: 2.5mg/kg TID; at engraftment: 6 mg/kg/day, Monday through Friday until day 120	65	20% (culture)	9%	58% neutropenia[2]
Boeckh et al.[45]	At engraftment: 5 mg/kg BID for 5 days, followed by 5 mg/kg/day, Monday through Saturday until day 100	112	0% (viremia) 41% (antigenemia) 50% (CMV-DNA by PCR)	2.7%	31% neutropenia[1] 4% CNS side effects

[1] Defined as absolute neutrophil count of < 750/mm^3 for 2 days
[2] Defined as absolute neutrophil count of <1000/mm^3

renal insufficiency after engraftment, and a low marrow cellularity at day 28 were significantly associated with the development of neutropenia.[82]

A pretransplant induction course of ganciclovir during the week prior to transplantation, followed by lower maintenance doses of ganciclovir (i.e. 5 mg/kg, 3 times a week) starting at engraftment, has been evaluated in non-randomized trials.[83–87] Two of these studies, which were performed in unrelated marrow transplant recipients and recipients of T-cell depleted marrow, respectively, showed unacceptably high rates of CMV disease.[84,85]

Although early ganciclovir prophylaxis appears to be highly effective in preventing CMV disease, there are significant disadvantages associated with this strategy (Table 2). Ganciclovir given at engraftment causes prolonged neutropenia leading to an increase in invasive bacterial and fungal infections.[45,80,82] In addition, a substantial number of patients not at risk for disease (i.e. 60–65%) will unnecessarily receive a potentially marrow-toxic drug. Both factors contribute substantially to morbidity and financial cost. Further, ganciclovir interferes with the recovery of CMV-specific immune responses.[31] Finally, ganciclovir given at engraftment appears to increase the risk of late onset CMV disease (ie after day 100) to as high as 17%.[45]

2.2.2. Aciclovir. Two earlier prospective studies have shown a survival benefit of high-dose aciclovir in allogeneic marrow transplant recipients.[77,78] What the role of high-dose aciclovir is in current prevention strategies consisting of prophylactic or preemptive use of ganciclovir can not be determined from these studies, but a recent retrospective analysis suggests that there is no additional survival benefit of giving high-dose aciclovir when ganciclovir was given either as prophylaxis or preemptively for CMV antigenemia.[79]

2.2.3. Foscarnet. The role of foscarnet for prevention of CMV disease remains undefined since no controlled studies have been published. Two small uncontrolled studies suggest that nephrotoxicity is the major side effect,[88,89] but foscarnet's antiviral efficacy relative to ganciclovir could not be determined from these studies. Because foscarnet does

Table 2. Options for prevention of CMV disease in allogeneic BMT recipients

Strategy	Advantages	Disadvantages	References
Prophylaxis			
Intravenous Immunoglobulin	Low risk for side effects	Low efficacy High cost Over treatment	7, 101-111
High-dose aciclovir	Low risk for side effects Survival benefit in patients not receiving ganciclovir prophylaxis or preemptive therapy	Low effectiveness for prevention of CMV disease High cost Over treatment	77, 78, 79,100
Ganciclovir at engraftment	Highly effective	High risk of neutropenia High risk of invasive fungal infections Delay of recovery of CMV-specific T cell function Over treatment	30, 31, 80-82
Preemptive Therapy			
Based on PCR for CMV-DNA	Effective Targeted treatment	Requirement for close monitoring Occasionally missed cases of CMV disease	91
Based on pp65 antigenemia	Effective Targeted treatment Less invasive fungal infections (compared to ganciclovir at engraftment)	Requirement for close monitoring Increased risk of CMV disease (compared to ganciclovir at engraftment) when ganciclovir is delayed until ≥ 2 positive cells per slide and discontinued based on a negative test	45

not cause marrow toxicity it is often used as second line drug in patients with neutropenia who require antiviral prophylaxis or treatment.

2.3. Preemptive Antiviral Therapy

Several studies have shown that CMV viremia is predictive for the development of CMV disease.[11,13,48] However, the rapid or standard isolation techniques used in the early studies were not sensitive enough to allow initiation of antiviral therapy prior to the onset of CMV disease in a significant proportion of the patients (12–13%).[13,90] More recent studies which included a high number of seropositive unrelated and HLA-mismatched recipients showed disease rates of as high as 30% with shell vial-guided early treatment.[80,91] Therefore, more sensitive techniques such as the CMV antigenemia assay and detection of CMV DNA by PCR have recently replaced shell vial centrifugation cultures in early treatment strategies.

PCR for detection of CMV DNA has been evaluated in several studies. Einsele *et al* showed that PCR of leukocyte DNA could detect CMV infection earlier than rapid isolation,[61] and similar data were presented by Wolf *et al.*, who used plasma for detection of CMV DNA.[21] Direct comparisons of the two tests showed that PCR of plasma was less sensitive than PCR of leukocytes.[15,62] Several laboratories are now evaluating quantitative CMV PCR and also PCR for CMV RNA.[63,92,93] Recently, Einsele *et al.* showed in a randomized trial that a preemptive treatment strategy in which ganciclovir was given to patients who had CMV DNA detected in PBL reduced the incidence of CMV disease and CMV associated mortality compared to preemptive ganciclovir initiated when CMV was recovered using the shell vial culture system.[91] (Table 2) Whether quantitative CMV DNA PCR can further improve the outcome of preemptive therapy strategies is presently unknown. However, because even patients with low systemic CMV viral load can develop CMV disease after allogeneic BMT, a strategy based in viral load might fail in patients who progress rapidly from low viral load to overt disease.[45]

The CMV antigenemia assay is based on the detection of the CMV lower matrix protein pp65 in PBL by immunostaining with monoclonal antibodies.[60,94,95] Results can be quantified, with patients who have a high number (≥ 2 positive cells per slide, made from 150,000 PBL) of antigen-positive leukocytes being at increased risk for developing CMV disease.[60] In a randomized, double-blind study, a preemptive therapy strategy based on CMV viral load (measured by quantitative antigenemia) was compared with ganciclovir prophylaxis given at engraftment (Table 2). The results of this study indicated that antigenemia-guided early ganciclovir treatment, based on a pre-determined level of CMV antigenemia, was as effective at preventing CMV disease as empiric ganciclovir prophylaxis.[45] However, ganciclovir prophylaxis was more effective in preventing CMV disease during the time it was given (i.e. the first 100 days after BMT) while the risk for late CMV disease was higher in the empiric ganciclovir prophylaxis group than in the antigen-guided prophylaxis group, with the risk for CMV disease equalizing at 180 days after transplantation. Survival was similar in both groups throughout the first 400 days after transplant. The higher incidence of CMV disease before day 100 in the antigen-guided, preemptive therapy group was most likely due to ganciclovir treatment being delayed until the levels of CMV antigen in PBL were ≥ 2 per slide, and discontinuation of treatment being based on a negative test in patients with severe GvHD. Thus, the authors recommend treatment of any level of antigenemia in patients with severe acute GvHD and continuation of ganciclovir until day 100.[45] After institution of a such a strategy, the incidence of early disease decreased during the study from 14.1% to 6%.[50]

Another risk-adapted strategy that combines both immunologic and virologic risk factors is to give a short-course of ganciclovir to patients who receive high-dose steroids for treatment of acute GvHD or for CMV antigenemia.[96] In an uncontrolled trial that included forty-one HLA matched-related allogeneic transplant recipients, there were no cases of CMV disease within the first 180 days after transplant.[96] Whether such an approach is useful in unrelated and HLA mismatched patients and whether it is superior to early treatment strategies based on virologic markers only has not been studied in a randomized fashion.

The duration of preemptive therapy has varied greatly in published studies. In early studies, once ganciclovir therapy was started, it was continued until day 100 after transplantation, resulting in 6–8 weeks of therapy in most patients.[13,90] Shorter periods of therapy have been used in the past few years. For example, Einsele *et al.* gave ganciclovir for a mean of three weeks[91] and Ljungman *et al.* gave a mean of two weeks of therapy.[97] Drawbacks with shorter courses of therapy are that treatment has to be reinstituted in up to 30% of patients and that occasional cases of CMV disease occur shortly after discontinuation of ganciclovir.[14,45] However, the advantages of short-course preemptive therapy are lower cost, a lower risk of side-effects, and potentially, a better reconstitution of the specific immune response to CMV, which might thereby reduce the risk of late CMV disease.[45]

Foscarnet has been used in two small uncontrolled studies of preemptive therapy which showed efficacy similar to ganciclovir.[98,99] A randomized trial comparing ganciclovir and foscarnet is currently ongoing within the European Group for Blood and Marrow Transplantation.

2.4. Critique of Current Prevention Strategies in Seropositive Allograft Recipients

Whether preemptive therapy or universal prophylaxis should be used in seropositive allograft recipients is being intensely debated in the marrow transplant community. An attractive feature of any universal prophylaxis strategy is that it is simple and does not require virologic monitoring. Presently, intravenous ganciclovir prophylaxis appears to be the most effective way of preventing CMV disease. This strategy should be used in CMV seropositive allograft recipients when no virologic monitoring by PCR or the antigenemia assay is available. However, ganciclovir prophylaxis has significant disadvantages (Table 2) which can be overcome, in part, by PCR- or antigenemia-guided preemptive treatment strategies. Whether empiric reduction of the maintenance dose of ganciclovir can reduce the incidence of neutropenia has not been studied in a controlled fashion. However, there is good evidence that such a strategy is not safe in high-risk patients, including those with severe GvHD or after T cell depletion.[84,85] Prophylaxis with high-dose aciclovir, although it has only a limited effect on CMV disease, is associated with a survival benefit in low-risk patients (i.e. those with a low risk of acute GvHD) when no ganciclovir prophylaxis or antigenemia- or PCR-guided therapy is given.[77,78,100] However, there does not seem to be a survival advantage when either of these ganciclovir strategies is used.[79] Due to high-dose aciclovir's high cost, only moderate effect on CMV, and the lack of data demonstrating that it adds additional benefit when ganciclovir is given as prophylaxis at engraftment or for antigenemia or PCR positivity, most centers do not use aciclovir for prevention of CMV disease.

Rapid culture-guided preemptive treatment strategies are not optimal for prevention of CMV disease.[80,91] Based on two randomized trials, antigenemia- and PCR-guided strategies currently provide the best results (Table 3), however there were important differences in the design and outcome of these two trials. While Einsele *et al.* performed a superiority trial that was designed to show an advantage of PCR-guided preemptive therapy compared to that guided by rapid culture, Boeckh *et al.* conducted an equivalence trial comparing short-term ganciclovir based on quantitative antigenemia with ganciclovir at engraftment, presently the most effective way of preventing CMV disease. Consequently, these two trials cannot really be compared directly (Table 3). Einsele *et al.* found a reduction of CMV disease and improved survival with PCR-guided ganciclovir while the antigenemia-guided treatment as used by Boeckh *et al.* was associated with more CMV disease by day 100. However, the actual incidence of CMV disease before day 100 was similar, especially if one considers that in the Boeckh study only seropositive individuals were included. Because the control group in the Boeckh study had received the best available prophylaxis for CMV disease, there was no apparent difference in survival between the two study groups. Although both studies used a short-term ganciclovir treatment regimen, the incidence of severe neutropenia seemed lower in patients who received PCR-guided treated[91] than in patients treated based on antigenemia[45] (Table 3). However, Einsele *et al* did not report the incidence of neutropenia in high-risk patients (i.e. seropositive recipients) and the use of hematopoietic growth factors was not standardized in either study. Also, patients in the Boeckh study may have included a patient population at higher risk for ganciclovir-related neutropenia.[82]

Table 3. Design and results of randomized studies using PCR- or antigenemia-guided early treatment

Characteristic/Endpoint	Einsele et al.[91]		Boeckh et al.[45, 50]
Recipient pretransplant CMV	• Seropositive or seronegative		• Seropositive
Treatment study group	• Ganciclovir and CMV-Ig for 2 consecutive positive PCR results		• Ganciclovir for antigenemia (≥ 2 positive cells/slide)
	• Discontinuation after 2 weeks or when PCR negative		• Discontinuation after 3 weeks or 6 days after negative antigenemia assay
	• Repeated treatment if PCR positivity recurred		• Repeated treatment if antigenemia recurred
Treatment control group	• Ganciclovir for 2 weeks for detection of CMV by shell vial cultures from blood, urine or throat wash		• Ganciclovir from engraftment until day 100
	• Thereafter, same PCR-guided ganciclovir as above		
Design	Randomized, open-label		Randomized, double-blind
Time of randomization	Before transplant		At engraftment
Primary endpoints	CMV disease, CMV-related mortality, survival		CMV disease, neutropenia
Sample size	71[1]		226
Incidence of CMV disease[2]	*All randomized patients*	*High-risk engrafted patients*[4]	
Day 100	5.4% vs 23.5% P = 0.004	7.7% vs 32% P = 0.03	14.1%[5] vs 2.7% P = 0.002
Day 180	8.1% vs 32.4% P = NR	11.5% vs 44% P = NR	15.8% vs 9.8% P = 0.17
Day 400	13.5% vs 38.2% P = NR	19.2% vs 52% P = NR	20.2% vs 16.1% P = 0.42
CMV-related mortality[2,3]			
Day 100	0% vs 14.7% P = 0.02	NR	6.1% vs 0.9% P = 0.07
Day 180	NR	NR	7.0% vs 3.6% P = 0.21
Day 400	NR	NR	11.4% vs 11.6% P = 1.0
Neutropenia[1] < $500/mm^3$	5.4% vs 23.5% P = 0.02	NR	26.0% vs 20.0% P = 0.27
Invasive infections[2]	2.7% vs 17.6% P = 0.03	NR	6% vs 16% P = 0.03
Fungal Transplant survival[2]			
Day 100	87% vs 67% P = 0.05	92% vs 72% P = 0.05	84% vs 87% P = 0.51
Day 180	84% vs 59% P = 0.02	88% vs 60% P = 0.02	73% vs 71% P = 0.91
Day 400	NR	NR	61% vs 59% P = 0.80

NR: not reported

[1] Study included 14 seronegative recipients with seronegative donors and 18 seronegative recipients with seropositive donors

[2] Results show study group versus control group

[3] Defined as histopathologic evidence at autopsy (Einsele et al.) or death within 6 weeks after diagnosis of CMV disease (Boeckh et al.)

[4] To compare both studies, results from high-risk patients who have engrafted are presented (excluding seronegative recipients of seropositive marrow)

[5] Disease before or following antigenemia of ≥ 2 positive cells per slide 8.8%; disease shortly after discontinuation of ganciclovir based on a negative test 5.3%

2.5. Prophylaxis with Intravenous Immunoglobulin (IVIG)

The use of IVIG or hyperimmune globulin for the prophylaxis of CMV infection and disease after allogeneic transplants remains controversial. Although the prophylactic use of IVIG is associated with virtually no toxicity, the regimens proposed are very costly in most countries and controlled studies show conflicting results with regard to the prevention of CMV disease.[7,101–107] In addition, some studies showed a reduction of bacteremia, interstitial pneumoniae and/or acute GVHD, while others did not. An improvement of survival has not been reported in any of the studies. There are several factors that may explain the inconsistent results of IVIG prophylaxis studies, including (1) use of nonspecific IVIG versus hyperimmune globulin, (2) differing doses, dose schedules, duration of administration after transplant and preparation of the product, (3) mixed patient populations with different risk for CMV disease (i.e. inclusion of autologous and allogeneic patients with different CMV serostatus), (4) varying supportive care techniques, (5) and different GvHD prophylaxis regimens. Recently, the available studies have been analyzed in 3 reviews which also came to conflicting conclusions.[108–110] Furthermore, there is no randomized trial which evaluates IVIG when ganciclovir prophylaxis or antigenemia- or PCR-based preemptive treatment is given. Therefore, most transplant centers do not use IVIG for prevention of CMV disease after allogeneic transplant. IVIG is not effective for prevention of CMV disease in autologous transplant recipients.[111]

2.6. Prophylaxis with CMV-Specific Monoclonal Antibodies

MSL-109 is a human monoclonal antibody specific for CMV glycoprotein H. Preliminary results from a multicenter prospective randomized, placebo-controlled, double blind study of 2 doses of MSL-109 suggest that MSL-109 is safe but has no effect on CMV infection or disease in CMV seropositive allogeneic recipients or seronegative recipients with a seropositive donor. However, based on a subgroup analysis, there seemed to be a survival advantage to seronegative recipients with a seropositive donor who received high doses of MSL-109. This survival advantage was not due to a reduction in CMV-related deaths.[112]

2.7. Prevention of Primary CMV Infection

Although it is preferable that patients who are CMV seronegative before transplantation should receive their transplant from a CMV seronegative donor, this is rarely possible due to the limited number of available donors. CMV infection of CMV seronegative patients receiving marrow from seronegative donors is mainly due to blood products.[7] Two options exist for reducing this risk of CMV transmission, restricting administration of blood products to those from CMV seronegative donors or attempting to remove CMV from unscreened blood products by the use of leukocyte-filters. A recent randomized trial suggested that both strategies result in similar rates of CMV infection.[113] The risk of transmission of CMV via the marrow or stem cell product from a seropositive donor to the recipient is approximately 20–25%.[114] Thus, these patients should be considered at risk for CMV disease and antigenemia- or PCR-guided antiviral therapy should be given. IVIG prophylaxis is not effective in preventing CMV disease in this setting.[105,107]

2.8. Prevention in Autologous Transplantation

High-dose aciclovir does not seem to be effective in prevention of CMV disease in CMV seropositive autograft recipients.[56] Autologous marrow and stem cell transplant patients have a lower risk for development of CMV than allogeneic marrow transplant recipients and few studies have studied preemptive therapy after autologous transplantation. Boeckh *et al.* investigated the use of the CMV antigenemia assay to survey autologous marrow and stem cell transplant patients and found it to be more sensitive than CMV viremia. The antigemia assay had the additional advantage of allowing quantification of viral load.[58] Hebart *et al*, using semi-quantitative PCR for CMV DNA in whole blood, reported similar findings.[59] However, since CMV disease after autologous transplantation is rare, it is uncertain whether monitoring for CMV with any technique is cost-effective in all seropositive patients. More research is needed to define a high-risk group for CMV disease after autologous transplant that might benefit most from virologic monitoring.

2.9. Therapy of Established CMV Disease

The need to treat established CMV disease in transplant recipients has decreased markedly with the advent of more effective prevention strategies. However, in some high risk patient populations the incidence of CMV disease, especially late onset CMV disease, is still substantial.

2.9.1. CMV Pneumonia. There have been no randomized trials evaluating treatment for CMV pneumonia. The currently-used regimen of ganciclovir in combination with IVIG is based on several non-randomized trials which showed improved survival in patients so treated compared to historical controls.[115–118] Most transplant centers use ganciclovir at a dose of 5 mg/kg BID for 14–21 days followed by maintenance therapy (5 mg/kg/day) for at least two weeks. (Table 4). IVIG dosages were quite variable in early studies and none was clearly superior. A dosage frequently used today is 0.5 g / kg bodyweight of standard intravenous immune globulin given every other day during the first two weeks of ganciclovir therapy. There does not seem to be an advantage for the use of CMV-hyperimmune globulin as compared to standard IVIG.[50,118]

Although foscarnet has similar efficacy to ganciclovir against CMV *in vitro* and in the treatment of CMV retinitis in HIV-infected patients,[119] only anecdotal cases of the use of foscarnet combined with immune globulin for therapy for CMV pneumonia in allogeneic BMT recipients have been reported. Combination antiviral therapy (e.g. ganciclovir and foscarnet) has not been evaluated in BMT recipients with CMV disease. Because results of treatment of established CMV pneumonia clearly remains unsatisfactory despite combination therapy of ganciclovir and IVIG (i.e. fatality rate > 50%),[30] better treatments for established CMV pneumonia are clearly needed.

2.9.2. CMV Gastrointestinal Disease. At present, there is no established therapy for CMV gastrointestinal disease. A double-blind, randomized study suggested that ganciclovir, 7.5 mg/kg/day for two weeks, was no better than placebo in reducing the ulcer size but eliminated CMV in cultures from biopsy specimens.[120] Possible reasons for the failure of this study to show a beneficial effect of ganciclovir therapy include the short duration of administration, inadequate dosing of ganciclovir, other concomitant causes of gastrointestinal symptoms such as GvHD, or ineffectiveness of ganciclovir alone. Due to the latter hypothesis, some centers have added IVIG to ganciclovir, as is currently done for treatment of CMV pneumonia. However, a recent retrospective study showed no advantage with the addition of immune globulin.[121] Foscarnet has been used in a small uncontrolled trial with clinical response in 2 of 5 patients treated for CMV enteritis.[122] Because of the unsatisfactory experience with using two weeks of antiviral therapy, many centers now give longer courses of therapy, e.g. at least two weeks of induction therapy (5 mg/kg twice daily) followed by 2–4 weeks maintenance (Table 4). Supportive treatment should be given (e.g. omeprazole). If CMV large-sized (> 1cm) or deep ulcers (mainly in stomach or duodenum) are seen at endoscopy or if symptoms persist, longer maintenance courses may

Table 4. Management of CMV infection after BMT: recommendations

Indication	Strategy	Comment
Prevention		
Allogeneic transplant		
• seropositive recipient	Antigenemia- or PCR-guided early ganciclovir treatment: 5 mg/kg BID for 7-14 days, followed by 5 mg/kg/day until day 100 (or until negative PCR or antigenemia) or ganciclovir prophylaxis at engraftment: 5 mg/kg BID for 5 days followed by 5 mg/kg/day on 5-6 day per week until day 100	Some cases of CMV disease may occur shortly after discontinuation based on negative PCR or antigenemia.[45] Recommended if neither PCR nor antigenemia testing is available
• seronegative recipient/ seropositive donor	Antigenemia- or PCR-guided early ganciclovir treatment: 5 mg/kg BID for 7-14 days, followed by 5 mg/kg/day until day 100 (or until negative PCR or antigenemia) and seronegative or filtered blood products	Prophylaxis at engraftment not recommended because of low incidence of posttransplant infection
• seronegative recipient/ seronegative donor	Seronegative or filtered blood products	
Autologous transplant		
• seropositive recipient	Antigenemia- or PCR-guided early ganciclovir treatment at >5 positive cells/slide: 5 mg/kg BID for 7 days, followed by 5 mg/kg/day for 14 days	Monitoring not uniformly advocated due to very low disease risk in some settings
• seronegative recipient	seronegative or filtered blood products	
Treatment of disease		
• CMV pneumonia	Ganciclovir (5 mg/kg BID) for 14-21 days followed by 5 mg/kg/day for at least 3-4 weeks plus IVIG or CMV-Ig	Extended maintenance throughout the period of severe immunosuppression (ie.GvHD treatment) may be considered
• Gastrointestinal disease	Ganciclovir (5 mg/kg BID) for 14-21 days followed by 5 mg/kg/day for at least 3-4 weeks	If deep ulcerations are present, longer maintenance may be required
• Marrow failure	Foscarnet (90 mg/kg BID) for 14 days followed by 90 mg/kg/day for 2 weeks plus G-CSF	Ganciclovir plus IVIG has also been used
• Retinitis	Ganciclovir (5 mg/kg BID) for 14-21 days followed by 5 mg/kg/day for at least 3-4 weeks	Extended maintenance may be required

be required. Antiviral combination therapy for CMV gastrointestinal disease has not been evaluated in BMT recipients.

2.9.3. Other Manifestations of CMV Diseases. There have been no treatment controlled trials for any of the infrequent manifestations of CMV diseases in BMT recipients (i.e. hepatitis, retinitis, marrow failure, encephalitis, sinusitis). In the experience of many centers, however, either ganciclovir or foscarnet can be used for therapy. This is consistent with studies in HIV infected patients which show, for example, that these agents are equally effective in the treatment of CMV retinitis.[119] In patients with marrow failure, foscarnet is the preferred drug; however, ganciclovir in combination with IVIG has been used in a small number of patients. Hematopoietic growth factors may be useful as well.

2.10. Antiviral Resistance

Strains of CMV resistant to ganciclovir or foscarnet have been reported in HIV-infected patients[123] but there are only anecdotal reports of such strains in BMT patients to date.[124,125] However, with the increased use of repeated courses of therapy needed for controlling CMV in patients with severe acute and chronic GvHD, it is possible that more resistant CMV mutants will be found also in BMT recipients.

3. CONCLUSIONS AND FUTURE PERSPECTIVE

Major progress has been made in the prevention of CMV disease after allogeneic BMT. In seropositive recipients, CMV disease before day 100 can be effectively prevented by ganciclovir prophylaxis given at engraftment or by antigenemia- or PCR-guided preemptive treatment (Table 4). In seronegative recipients of marrow or stem cells from a seropositive donor, seronegative or filtered blood in combination with a PCR-or antigenemia-guided early treatment should be given due to the low incidence of CMV infection following transplantation (20–25%).[114] Seronegative recipients with a seronegative donor should also receive seronegative or filtered blood products to prevent CMV infection altogether.[113] The strategies for preventing CMV infection in autologous transplant recipients are less clearly defined. At the Fred Hutchinson Cancer Research Center CMV-seropositive autograft recipients are tested weekly for CMV antigenemia and we give a 3 week course of ganciclovir if antigenemia exceeds 5 positive cells per slide.[58] Seronegative autograft recipients receive screened or filtered blood. Clearly, more research is needed to define the autograft recipients who will benefit most from prevention strategies.

Several new antivirals active against CMV are currently in various stages of clinical evaluation. (See also other chapters in this monograph). Oral ganciclovir has been evaluated in a phase I/II study in allogenic marrow transplant recipients and is absorbed as well as it is in other patients.[126] However, with the current formulation gastrointestinal intolerance seems to be a limiting factor in the early post-transplant period.[126] Whether oral ganciclovir can prevent CMV disease, especially in patients with acute GvHD, has not been studied. A prodrug of oral ganciclovir (valganciclovir) is currently being studied in HIV-infected subjects. This prodrug has a much higher bioavailability than ganciclovir which may allow once or twice a day dosing. Valaciclovir, the pro-drug of aciclovir, is currently being tested in two different phase III trials. Cidofovir (HPMPC), a nucleotide analogue with a long half-life allowing once-a-week administration, has been used successfully for treatment of CMV retinitis in AIDS patients. Cidofovir's principal toxicity, renal insufficiency, can be ameliorated by concomitant use of probenecid.[127] Whether or not cidofovir is safe and effective after BMT has not been studied. Lobucavir, a cyclobutyl analog of guanine with broad *in vitro* activity against most herpes viruses, and 1263W94, a benzimidazole riboside with potent anti-CMV activity, have entered clinical trials in HIV-infected patients.

Combination antiviral therapy has been evaluated in a non-randomized trial[128] and may offer an effective strategy with acceptable toxicity in selected situations after BMT, e.g. in patients with very high viral load or CMV disease in whom morbidity and mortality remains high. A randomized trial is needed to evaluate such strategies.

It is clear that one of the most important tasks for the future is to develop prevention strategies for late CMV disease. Most cases of late CMV disease occur between 100 and

365 days after transplantation.[50] Either antiviral or immunologic interventions may be useful for preventing late CMV disease. With applications for BMT expanding and more high-risk patients being transplanted (e.g. older recipient age, HLA mismatch), it can be expected that this problem will increase in the future. Possible future antiviral strategies include extended prophylaxis or virologic monitoring combined with short-term courses of preemptive antiviral therapy in high-risk patients. Both strategies are potentially harmful with currently available antiviral agents (due to drug toxicity or emergence of antiviral resistance) and should therefore be evaluated in controlled clinical trials. Clearly, new compounds with an improved side effect profile, a long serum half life, and a high oral bioavailability which allow long-term prophylaxis in the outpatient setting are needed. Because the principal immunologic defect in the host defense against CMV seems to be a delayed recovery of HLA-restricted CMV-specific CD8+ cytotoxic T lymphocyte (CTL) function, studies are currently being conducted to adoptively transfer donor-derived CTL after transplant. In a pioneering study, Riddell and coworkers demonstrated that donor CMV-specific cloned CTL can be expanded *in vitro* and safely infused after transplant. The infused lymphocytes were detectable for up to 12 weeks.[129] A phase II study of CTL infusion is currently underway. Another possible approach that might be useful in the future is to augment the donor immunity before transplant by donor vaccination, possibly combined with recipient vaccination. The most promising candidate antigen for vaccine development is pp65 (UL83) which has been shown to be an important target antigen for CMV-specific CTL.[130]

ACKNOWLEDGMENTS

Grant support: Michael Boeckh was supported by the National Institutes of Health (CA 18029).

REFERENCES

1. Mocarski ES. Cytomegaloviruses and their replication. In: Fields BN, Knipe DM, Howley PM, eds. Virology. Philadelphia: Lippincott-Raven, 1996:2447–2492.
2. Britt WJ, Alford CA. Cytomegalovirus. In: Fields BN, Knipe DM, Howley PM, eds. Virology. Philadelphia: Lippincott-Raven, 1996:2493–2523.
3. Riddell SR. Pathogenesis of cytomegalovirus pneumonia in immunocompromised hosts. Seminars in Respiratory Infections 1995; 10:199–208.
4. Zaia JA, Forman SJ. Cytomegalovirus infection in the bone marrow transplant recipient. In: Rubin RH, ed. Infectious Disease Clinics of North America: Infection in transplantation. Philadelphia and London: WB Saunders Co., 1995:879–900.
5. Fries BC, Chou S, Boeckh M, Torok-Storb B. Frequency distribution of cytomegalovirus envelope glycoprotein genotypes in bone marrow transplant recipients. Journal of Infectious Diseases 1994; 169:769–74.
6. Chou SW. Acquisition of donor strains of cytomegalovirus by renal-transplant recipients. New England Journal of Medicine 1986; 314:1418–23.
7. Bowden RA, Sayers M, Flournoy N, *et al.* Cytomegalovirus immune globulin and seronegative blood products to prevent primary cytomegalovirus infection after marrow transplantation. New England Journal of Medicine 1986; 314:1006–10.
8. Grossi P, Minoli L, Percivalle E, Irish W, Vigano M, Gerna G. Clinical and virological monitoring of human cytomegalovirus infection in 294 heart transplant recipients. Transplantation 1995; 59:847–51.
9. Goodrich JM, Boeckh M, Bowden R. Strategies for the prevention of cytomegalovirus disease after marrow transplantation. Clinical Infectious Diseases 1994; 19:287–98.

10. Meyers JD, Flournoy N, Thomas ED. Risk factors for cytomegalovirus infection after human marrow transplantation. Journal of Infectious Diseases 1986; 153:478–88.
11. Meyers JD, Ljungman P, Fisher LD. Cytomegalovirus excretion as a predictor of cytomegalovirus disease after marrow transplantation: importance of cytomegalovirus viremia. Journal of Infectious Diseases 1990; 162:373–80.
12. Enright H, Haake R, Weisdorf D, *et al.* Cytomegalovirus pneumonia after bone marrow transplantation. Risk factors and response to therapy. Transplantation 1993; 55:1339–46.
13. Schmidt GM, Horak DA, Niland JC, Duncan SR, Forman SJ, Zaia JA. A randomized, controlled trial of prophylactic ganciclovir for cytomegalovirus pulmonary infection in recipients of allogeneic bone marrow transplants; The City of Hope-Stanford-Syntex CMV Study Group. New England Journal of Medicine 1991; 324:1005–11.
13a. Gor D, Sabin C, Prentice H *et al.* Lougitudinal fluctuations in cytomegalovirus load in bone marrow transplant patients - relationship between peak virus load, donor/recipent sero status, acute GvHD and CMV disease. Bone Marrow Transplantation 1998;21:597–605.
14. Vlieger AM, Boland GJ, Jiwa NM, *et al.* Cytomegalovirus antigenemia assay or PCR can be used to monitor ganciclovir treatment in bone marrow transplant recipients. Bone Marrow Transplantation 1992; 9:247–53.
15. Boeckh M, Hawkins G, Myerson D, Zaia J, Bowden RA. Plasma PCR for cytomegalovirus DNA after allogeneic marrow transplantation: comparison with PCR using peripheral blood leukocytes, pp65 antigenemia, and viral culture. Transplantation 1997; 64:108–113.
16. The TH, van der Ploeg M, van den Berg AP, Vlieger AM, van der Giessen M, van Son WJ. Direct detection of cytomegalovirus in peripheral blood leukocytes--a review of the antigenemia assay and polymerase chain reaction. Transplantation 1992; 54:193–8.
17. Gerna G, Furione M, Baldanti F, Sarasini A. Comparative quantitation of human cytomegalovirus DNA in blood leukocytes and plasma of transplant and AIDS patients. Journal of Clinical Microbiology 1994; 32:2709–17.
18. Gerna G, Zipeto D, Percivalle E, *et al.* Human cytomegalovirus infection of the major leukocyte subpopulations and evidence for initial viral replication in polymorphonuclear leukocytes from viremic patients. Journal of Infectious Diseases 1992; 166:1236–44.
19. Grefte JM, van der Gun BT, Schmolke S, *et al.* The lower matrix protein pp65 is the principal viral antigen present in peripheral blood leukocytes during an active cytomegalovirus infection. Journal of General Virology 1992; 73:2923–32.
20. Grefte A, Harmsen MC, van der Giessen M, Knollema S, van Son WJ, The TH. Presence of human cytomegalovirus (HCMV) immediate early mRNA but not ppUL83 (lower matrix protein pp65) mRNA in polymorphonuclear and mononuclear leukocytes during active HCMV infection. Journal of General Virology 1994; 75:1989–98.
21. Wolf DG, Spector SA. Early diagnosis of human cytomegalovirus disease in transplant recipients by DNA amplification in plasma. Transplantation 1993; 56:330–4.
22. Grefte A, van der Giessen M, van Son W, The TH. Circulating cytomegalovirus (CMV)-infected endothelial cells in patients with an active CMV infection. Journal of Infectious Diseases 1993; 167:270–7.
23. Percivalle E, Revello MG, Vago L, Morini F, Gerna G. Circulating endothelial giant cells permissive for human cytomegalovirus (HCMV) are detected in disseminated HCMV infections with organ involvement. Journal of Clinical Investigation 1993; 92:663–70.
24. Salzberger B, Myerson D, Boeckh M. Circulating CMV-infected endothelial cells in marrow transplant patients with CMV disease and CMV infection. Journal of Infectious Diseases 1997; 176:778–781.
25. Quinnan GV, Jr., Kirmani N, Rook AH, *et al.* Cytotoxic t cells in cytomegalovirus infection: HLA-restricted T-lymphocyte and non-T-lymphocyte cytotoxic responses correlate with recovery from cytomegalovirus infection in bone-marrow-transplant recipients. New England Journal of Medicine 1982; 307:7–13.
26. Reddehase MJ, Mutter W, Munch K, Buhring HJ, Koszinowski UH. CD8-positive T lymphocytes specific for murine cytomegalovirus immediate-early antigens mediate protective immunity. Journal of Virology 1987; 61:3102–8.
27. Matloubian M, Concepcion RJ, Ahmed R. CD4+ T cells are required to sustain CD8+ cytotoxic T-cell responses during chronic viral infection. Journal of Virology 1994; 68:8056–63.
28. Jonjic S, Mutter W, Weiland F, Reddehase MJ, Koszinowski UH. Site-restricted persistent cytomegalovirus infection after selective long-term depletion of CD4+ T lymphocytes. Journal of Experimental Medicine 1989; 169:1199–212.
29. Reusser P, Riddell SR, Meyers JD, Greenberg PD. Cytotoxic T-lymphocyte response to cytomegalovirus after human allogeneic bone marrow transplantation: pattern of recovery and correlation with cytomegalovirus infection and disease. Blood 1991; 78:1373–80.

30. Boeckh M, Riddell SR, Cunningham T, Myerson D, Flowers M, Bowden RA. Increased incidence of late CMV disease in allogeneic marrow transplant recipients after ganciclovir prophylaxis is due to a lack of CMV-specific T cell responses. Blood 1996; 88 Supplement 1:302a.
31. Li CR, Greenberg PD, Gilbert MJ, Goodrich JM, Riddell SR. Recovery of HLA-restricted cytomegalovirus (CMV)-specific T-cell responses after allogeneic bone marrow transplant: correlation with CMV disease and effect of ganciclovir prophylaxis. Blood 1994; 83:1971–9.
32. Reusser P, Attenhofer R, Hebart H, Helg C, Chapuis B, Einsele H. Cytomegalovirus-specific T-cell immunity in recipients of autologous peripheral blood stem cell or bone marrow transplant. Blood 1997; 89:3873–3879.
33. Torok-Storb B, Boeckh M, Hoy C, Leisenring W, Myerson D, Gooley T. Association of specific cytomegalovirus genotypes with death from myelosuppression after marrow transplantation. Blood 1997; 90:2097–2102.
34. Shepp DH, Match ME, Ashraf AB, Lipson SM, Millan C, Pergolizzi R. Cytomegalovirus glycoprotein B groups associated with retinitis in AIDS. Journal of Infectious Diseases 1996; 174:184–7.
35. Cray C, Levy RB. CD8+ and CD4+ T cells contribute to the exacerbation of class I MHC disparate graft-vs-host reaction by concurrent murine cytomegalovirus infection. Clinical Immunology & Immunopathology 1993; 67:84–90.
36. Via CS, Shanley JD, Shearer GM. Synergistic effect of murine cytomegalovirus on the induction of acute graft-vs-host disease involving MHC class I differences only. Analysis of *in vitro* T cell function. Journal of Immunology 1990; 145:3283–9.
37. Jones M, Cray C, Levy RB. Concurrent MCMV infection augments donor antihost-specific activity and alters clinical outcome following experimental allogenic bone marrow transplantation. Transplantation 1996; 61:856–61.
38. Wang FZ, Dahl H, Linde A, Brytting M, Ehrnst A, Ljungman P. Lymphotropic herpesviruses in allogeneic bone marrow transplantation. Blood 1996; 88:3615–20.
39. Boström L, Ringden O, Sundberg B, Ljungman P, Linde A, Nilsson B. Pretransplant herpes virus serology and chronic graft-versus-host disease. Bone Marrow Transplantation 1989; 4:547–52.
40. Lönnqvist B, Ringden O, Wahren B, Gahrton G, Lundgren G. Cytomegalovirus infection associated with and preceding chronic graft-versus-host disease. Transplantation 1984; 38:465–8.
41. Jacobsen N, Andersen HK, Skinhoj P, *et al.* Correlation between donor cytomegalovirus immunity and chronic graft-versus-host disease after allogeneic bone marrow transplantation. Scandinavian Journal of Haematology 1986; 36:499–506.
42. Söderberg C, Larsson S, Rozell BL, Sumitran-Karuppan S, Ljungman P, Möller E. Cytomegalovirus-induced CD13-specific autoimmunity--a possible cause of chronic graft-vs-host disease. Transplantation 1996; 61:600–9.
43. Söderberg C, Sumitran-Karuppan S, Ljungman P, Möller E. CD13-specific autoimmunity in cytomegalovirus-infected immunocompromised patients. Transplantation 1996; 61:594–600.
44. Boeckh M, Bowden R. Cytomegalovirus infection in marrow transplantation. Cancer Treatment & Research 1995; 76:97–136.
45. Boeckh M, Gooley TA, Myerson D, Cunningham T, Schoch G, Bowden RA. Cytomegalovirus pp65 antigenemia-guided early treatment with ganciclovir versus ganciclovir at engraftment after allogeneic marrow transplantation: a randomized double-blind study. Blood 1996; 88:4063–71.
46. Krause H, Hebart H, Jahn G, Müller CA, Einsele H. Screening for CMV-specific T cell proliferation to identify patients at risk of developing late onset CMV disease. Bone Marrow Transplantation 1997; 19:1111–1116.
47. Bowden RA, Digel J, Reed EC, Meyers JD. Immunosuppressive effects of ganciclovir on *in vitro* lymphocyte responses. Journal of Infectious Diseases 1987; 156:899–903.
48. Ljungman P, Aschan J, Azinge JN, *et al.* Cytomegalovirus viraemia and specific T-helper cell responses as predictors of disease after allogeneic marrow transplantation. British Journal of Haematology 1993; 83:118–24.
49. Boeckh M, Gooley T, Bowden RA. CMV infection and viral load after peripheral blood stem cell transplant versus marrow transplant, 6th International Cytomegalovirus Workshop, Perdido Beach, AL, March 5–9, 1997.
50. Boeckh M. Unpublished results.
51. Gluckman E, Rocha V, Boyer-Chammard A, *et al.* Outcome of cord-blood transplantation from related and unrelated donors. New England Journal of Medicine 1997; 337:373–381.
52. Wagner JE, DeFor T, Rubinstein P, Kurtzberg J. Transplantation of unrelated umbilical cord blood: outcome and analysis of risk factors. Blood 1997; 90 (Supplement 1):398.

53. Wingard JR, Chen DY, Burns WH, *et al.* Cytomegalovirus infection after autologous bone marrow transplantation with comparison to infection after allogeneic bone marrow transplantation. Blood 1988; 71:1432–7.
54. Reusser P, Fisher LD, Buckner CD, Thomas ED, Meyers JD. Cytomegalovirus infection after autologous bone marrow transplantation: occurrence of cytomegalovirus disease and effect on engraftment. Blood 1990; 75:1888–94.
55. Ljungman P, Biron P, Bosi A, *et al.* Cytomegalovirus interstitial pneumonia in autologous bone marrow transplant recipients. Infectious Disease Working Party of the European Group for Bone Marrow Transplantation. Bone Marrow Transplantation 1994; 13:209–12.
56. Boeckh M, Gooley TA, Reusser P, Buckner CD, Bowden RA. Failure of high-dose aciclovir to prevent cytomegalovirus disease after autologous marrow transplantation. Journal of Infectious Diseases 1995; 172:939–43.
57. Holland HK, Dix SP, Geller RB, *et al.* Minimal toxicity and mortality in high-risk breast cancer patients receiving high-dose cyclophosphamide, thiotepa, and carboplatin plus autologous marrow/stem-cell transplantation and comprehensive supportive care. Journal of Clinical Oncology 1996; 14:1156–64.
58. Boeckh M, Stevens-Ayers T, Bowden RA. Cytomegalovirus pp65 antigenemia after autologous marrow and peripheral blood stem cell transplantation. Journal of Infectious Diseases 1996; 174:907–12.
59. Hebart H, Schröder A, Löffler J, *et al.* Cytomegalovirus monitoring by polymerase chain reaction of whole blood samples from patients undergoing autologous bone marrow or peripheral blood progenitor cell transplantation. Journal of Infectious Diseases 1997; 175:1490–1493.
60. Boeckh M, Bowden RA, Goodrich JM, Pettinger M, Meyers JD. Cytomegalovirus antigen detection in peripheral blood leukocytes after allogeneic marrow transplantation. Blood 1992; 80:1358–64.
61. Einsele H, Ehninger G, Steidle M, *et al.* Polymerase chain reaction to evaluate antiviral therapy for cytomegalovirus disease. Lancet 1991; 338:1170–2.
62. Brytting M, Mousavi-Jazi M, Bostrom L, *et al.* Cytomegalovirus DNA in peripheral blood leukocytes and plasma from bone marrow transplant recipients. Transplantation 1995; 60:961–5.
63. Gerna G, Furione M, Baldanti F, Percivalle E, Comoli P, Locatelli F. Quantitation of human cytomegalovirus DNA in bone marrow transplant recipients. British Journal of Haematology 1995; 91:674–83.
64. Zaia JA, Gallez-Hawkins GM, Tegtmeier BR, *et al.* Late Cytomegalovirus Disease in Marrow Transplantation is Predicted by Virus Load in Plasma. Journal of Infectious Diseases 1997; 176:782–785.
65. Boeckh M, Boivin G. Quantitation of cytomegalovirus: methodological aspects and clinical applications. Clinical Microbiology Reviews 1998:*in press.*
66. Ljungman P, Plotkin SA. Workshop on CMV disease: definitions, clinical severity scores, and new syndromes. Scandinavian Journal of Infectious Diseases 1995; 99:S87-S88.
67. Crawford SW, Bowden RA, Hackman RC, Gleaves CA, Meyers JD, Clark JG. Rapid detection of cytomegalovirus pulmonary infection by bronchoalveolar lavage and centrifugation culture. Annals of Internal Medicine 1988; 108:180–5.
68. Ruutu P, Ruutu T, Volin L, Tukiainen P, Ukkonen P, Hovi T. Cytomegalovirus is frequently isolated in bronchoalveolar lavage fluid of bone marrow transplant recipients without pneumonia. Annals of Internal Medicine 1990; 112:913–6.
69. Cathomas G, Morris P, Pekle K, Cunningham I, Emanuel D. Rapid diagnosis of cytomegalovirus pneumonia in marrow transplant recipients by bronchoalveolar lavage using the polymerase chain reaction, virus culture, and the direct immunostaining of alveolar cells. Blood 1993; 81:1909–14.
70. Slavin MA, Gleaves CA, Schoch HG, Bowden RA. Quantification of cytomegalovirus in bronchoalveolar lavage fluid after allogeneic marrow transplantation by centrifugation culture. Journal of Clinical Microbiology 1992; 30:2776–9.
71. Boivin G, Olson CA, Quirk MR, Kringstad B, Hertz MI, Jordan MC. Quantitation of cytomegalovirus DNA and characterization of viral gene expression in bronchoalveolar cells of infected patients with and without pneumonitis. Journal of Infectious Diseases 1996; 173:1304–1312.
72. Hackman RC, Wolford JL, Gleaves CA, *et al.* Recognition and rapid diagnosis of upper gastrointestinal cytomegalovirus infection in marrow transplant recipients. A comparison of seven virologic methods. Transplantation 1994; 57:231–7.
73. Ljungman P, Griffith P. Definitions of cytomegalovirus infection and disease, multidiscipinary approach to understanding cytomeglaovirus disease, 1993. Exerpta Medica, Amsterdam.
74. Cinque P, Vago L, Dahl H, *et al.* Polymerase chain reaction on cerebrospinal fluid for diagnosis of virus-associated opportunistic diseases of the central nervous system in HIV-infected patients. AIDS. 1996; 10:951–8.
75. Torok-Storb B, Simmons P, Khaira D, Stachel D, Myerson D. Cytomegalovirus and marrow function. Annals of Hematology 1992; 64:A128–31.

76. Boeckh M, Hoy C, Torok-Storb B. Occult cytomegalovirus infection of marrow stroma. Clinical Infectious Diseases 1998; 26:209–10.
77. Meyers JD, Reed EC, Shepp DH, *et al.* Aciclovir for prevention of cytomegalovirus infection and disease after allogeneic marrow transplantation. New England Journal of Medicine 1988; 318:70–5.
78. Prentice HG, Gluckman E, Powles RL, *et al.* Impact of long-term aciclovir on cytomegalovirus infection and survival after allogeneic bone marrow transplantation. European Aciclovir for CMV Prophylaxis Study Group. Lancet 1994; 343:749–53.
79. Boeckh M, Gooley T, Bowden R. Effect of high-dose aciclovir on survival in patients who receive ganciclovir prophylaxis at engraftment or for pp65 antigenemia after allogeneic marrow transplantation. Journal of Infectious Diseases, *in press.*
80. Goodrich JM, Bowden RA, Fisher L, Keller C, Schoch G, Meyers JD. Ganciclovir prophylaxis to prevent cytomegalovirus disease after allogeneic marrow transplant. Annals of Internal Medicine 1993; 118:173–8.
81. Winston DJ, Ho WG, Bartoni K, *et al.* Ganciclovir prophylaxis of cytomegalovirus infection and disease in allogeneic bone marrow transplant recipients. Results of a placebo-controlled, double-blind trial. Annals of Internal Medicine 1993; 118:179–84.
82. Salzberger B, Bowden RA, Hackman R, Davis C, Boeckh M. Neutropenia in allogeneic marrow transplant recipients receiving ganciclovir for prevention of CMV disease: risk factors and outcome. Blood 1997; 90:2502–2508.
83. Atkinson K, Downs K, Golenia M, *et al.* Prophylactic use of ganciclovir in allogeneic bone marrow transplantation: absence of clinical cytomegalovirus infection. British Journal of Haematology 1991; 79:57–62.
84. Atkinson K, Arthur C, Bradstock K, *et al.* Prophylactic ganciclovir is more effective in HLA-identical family member marrow transplant recipients than in more heavily immune-suppressed HLA-identical unrelated donor marrow transplant recipients. Australasian Bone Marrow Transplant Study Group. Bone Marrow Transplantation 1995; 16:401–5.
85. Przepiorka D, Ippoliti C, Panina A, *et al.* Ganciclovir three times per week is not adequate to prevent cytomegalovirus reactivation after T cell-depleted marrow transplantation. Bone Marrow Transplantation 1994; 13:461–4.
86. von Bueltzingsloewen A, Bordigoni P, Witz F, *et al.* Prophylactic use of ganciclovir for allogeneic bone marrow transplant recipients. Bone Marrow Transplantation 1993; 12:197–202.
87. Yau JC, Dimopoulos MA, Huan SD, *et al.* Prophylaxis of cytomegalovirus infection with ganciclovir in allogeneic marrow transplantation. European Journal of Haematology 1991; 47:371–6.
88. Reusser P, Gambertoglio JG, Lilleby K, Meyers JD. Phase I-II trial of foscarnet for prevention of cytomegalovirus infection in autologous and allogeneic marrow transplant recipients. Journal of Infectious Diseases 1992; 166:473–9.
89. Bacigalupo A, Tedone E, Van Lint MT, *et al.* CMV prophylaxis with foscarnet in allogeneic bone marrow transplant recipients at high risk of developing CMV infections. Bone Marrow Transplantation 1994; 13:783–8.
90. Goodrich JM, Mori M, Gleaves CA, *et al.* Early treatment with ganciclovir to prevent cytomegalovirus disease after allogeneic bone marrow transplantation. New England Journal of Medicine 1991; 325:1601–7.
91. Einsele H, Ehninger G, Hebart H, *et al.* Polymerase chain reaction monitoring reduces the incidence of cytomegalovirus disease and the duration and side effects of antiviral therapy after bone marrow transplantation. Blood 1995; 86:2815–20.
92. Boivin G, Handfield J, Murray G, *et al.* Quantitation of cytomegalovirus (CMV) DNA in leukocytes of human immunodeficiency virus-infected subjects with and without CMV disease by using PCR and the sharp signal detection system. Journal of Clinical Microbiology 1997; 35:525–526.
93. Bitsch A, Kirchner H, Dupke R, Bein G. Cytomegalovirus transcripts in peripheral blood leukocytes of actively infected transplant patients detected by reverse transcription-polymerase chain reaction. Journal of Infectious Diseases 1993; 167:740–3.
94. van der Bij W, Torensma R, van Son WJ, *et al.* Rapid immunodiagnosis of active cytomegalovirus infection by monoclonal antibody staining of blood leucocytes. Journal of Medical Virology 1988; 25:179–88.
95. van der Bij W, van Son WJ, van der Berg AP, Tegzess AM, Torensma R, The TH. Cytomegalovirus (CMV) antigenemia: rapid diagnosis and relationship with CMV-associated clinical syndromes in renal allograft recipients. Transplantation Proceedings 1989; 21:2061–4.
96. Verdonck LF, Dekker AW, Rozenbergarska M, Vandenhoek MR. A risk-adapted approach with a short course of ganciclovir to prevent cytomegalovirus (CMV) pneumonia in CMV-seropositive recipients of allogeneic bone marrow transplants. Clinical Infectious Diseases 1997; 24:901–907.
97. Ljungman P, Lore K, Aschan J, *et al.* Use of a semi-quantitative PCR for cytomegalovirus DNA as a basis for pre-emptive antiviral therapy in allogeneic bone marrow transplant patients. Bone Marrow Transplantation 1996; 17:583–7.

98. Bacigalupo A, van Lint MT, Tedone E, *et al.* Early treatment of CMV infections in allogeneic bone marrow transplant recipients with foscarnet or ganciclovir. Bone Marrow Transplantation 1994; 13:753–8.
99. Ljungman P, Oberg G, Aschan J, *et al.* Foscarnet for pre-emptive therapy of CMV infection detected by a leukocyte-based nested PCR in allogeneic bone marrow transplant patients. Bone Marrow Transplantation 1996; 18:565–8.
100. Prentice HG, Gluckman E, Powles RL, *et al.* Long-term survival in allogeneic bone marrow transplant recipients following aciclovir prophylaxis for CMV infection. Bone Marrow Transplantation 1997; 19:129–133.
101. Meyers JD, Leszczynski J, Zaia JA, *et al.* Prevention of cytomegalovirus infection by cytomegalovirus immune globulin after marrow transplantation. Annals of Internal Medicine 1983; 98:442–6.
102. Winston DJ, Ho WG, Lin CH, Budinger MD, Champlin RE, Gale RP. Intravenous immunoglobulin for modification of cytomegalovirus infections associated with bone marrow transplantation. Preliminary results of a controlled trial. American Journal of Medicine 1984; 76:128–33.
103. Winston DJ, Ho WG, Lin CH, *et al.* Intravenous immune globulin for prevention of cytomegalovirus infection and interstitial pneumonia after bone marrow transplantation. Annals of Internal Medicine 1987; 106:12–8.
104. Ringden O, Pihlstedt P, Volin L, *et al.* Failure to prevent cytomegalovirus infection by cytomegalovirus hyperimmune plasma: a randomized trial by the Nordic Bone Marrow Transplantation Group. Bone Marrow Transplantation 1987; 2:299–305.
105. Bowden RA, Fisher LD, Rogers K, Cays M, Meyers JD. Cytomegalovirus (CMV)-specific intravenous immunoglobulin for the prevention of primary CMV infection and disease after marrow transplant. Journal of Infectious Diseases 1991; 164:483–7.
106. Sullivan KM, Kopecky KJ, Jocom J, *et al.* Immunomodulatory and antimicrobial efficacy of intravenous immunoglobulin in bone marrow transplantation. New England Journal of Medicine 1990; 323:705–12.
107. Ruutu T, Ljungman P, Brinch L, *et al.* No prevention of cytomegalovirus infection by anti-cytomegalovirus hyperimmune globulin in seronegative bone marrow transplant recipients. The Nordic BMT Group. Bone Marrow Transplantation 1997; 19:233–6.
108. Bass EB, Powe NR, Goodman SN, *et al.* Efficacy of immune globulin in preventing complications of bone marrow transplantation: a meta-analysis. Bone Marrow Transplantation 1993; 12:273–82.
109. Guglielmo BJ, Wong-Beringer A, Linker CA. Immune globulin therapy in allogeneic bone marrow transplant: a critical review . Bone Marrow Transplantation 1994; 13:499–510.
110. Messori A, Rampazzo R, Scroccaro G, Martini N. Efficacy of hyperimmune anti-cytomegalovirus immunoglobulins for the prevention of cytomegalovirus infection in recipients of allogeneic bone marrow transplantation: a meta-analysis. Bone Marrow Transplantation 1994; 13:163–7.
111. Wolff SN, Fay JW, Herzig RH, *et al.* High-dose weekly intravenous immunoglobulin to prevent infections in patients undergoing autologous bone marrow transplantation or severe myelosuppressive therapy. A study of the American Bone Marrow Transplant Group. Annals of Internal Medicine 1993; 118:937–42.
112. Boeckh M, Bowden RA, Chao NJ, *et al.* Randomized, placebo controlled, double-blind study of a CMV glycoprotein H-specific monoclonal antibody (MSL-109) for prevention of CMV infection after allogeneic hematopoietic stem cell transplant, 24th Annual Meeting European Group for Blood and Marrow Transplantation, Courmayeur, Italy, 1998.
113. Bowden RA, Slichter SJ, Sayers M, *et al.* A comparison of filtered leukocyte-reduced and cytomegalovirus (CMV) seronegative blood products for the prevention of transfusion-associated CMV infection after marrow transplant. Blood 1995; 86:3598–603.
114. Boeckh M, Riddell SR, Woogerd P, Cunningham T, White K, Bowden RA. Primary CMV infection via marrow: incidence, response to early treatment, CMV-specific immune response, and risk of late CMV disease, 9th International Symposium on Infections in the Immunocompromised Host, Assisi, Italy, 1996.
115. Reed EC, Bowden RA, Dandliker PS, Lilleby KE, Meyers JD. Treatment of cytomegalovirus pneumonia with ganciclovir and intravenous cytomegalovirus immunoglobulin in patients with bone marrow transplants. Annals of Internal Medicine 1988; 109:783–8.
116. Schmidt GM, Kovacs A, Zaia JA, *et al.* Ganciclovir/immunoglobulin combination therapy for the treatment of human cytomegalovirus-associated interstitial pneumonia in bone marrow allograft recipients. Transplantation 1988; 46:905–7.
117. Emanuel D, Cunningham I, Jules-Elysee K, *et al.* Cytomegalovirus pneumonia after bone marrow transplantation successfully treated with the combination of ganciclovir and high-dose intravenous immune globulin. Annals of Internal Medicine 1988; 109:777–82.
118. Ljungman P, Engelhard D, Link H, *et al.* Treatment of interstitial pneumonitis due to cytomegalovirus with ganciclovir and intravenous immune globulin: experience of European Bone Marrow Transplant Group. Clinical Infectious Diseases 1992; 14:831–5.

119. Anonymous. Mortality in patients with the acquired immunodeficiency syndrome treated with either foscarnet or ganciclovir for cytomegalovirus retinitis. Studies of Ocular Complications of AIDS Research Group, in collaboration with the AIDS Clinical Trials Group. New England Journal of Medicine 1992; 326:213–20.
120. Reed EC, Wolford JL, Kopecky KJ, *et al.* Ganciclovir for the treatment of cytomegalovirus gastroenteritis in bone marrow transplant patients. A randomized, placebo-controlled trial. Annals of Internal Medicine 1990; 112:505–10.
121. Ljungman P, Cordonnier C, Einsele H, *et al.* Use of intravenous immune globulin in addition to antiviral therapy in the treatment of CMV gastrointestinal disease in allogeneic bone marrow transplant patients; A report from the European group for blood and marrow transplantation EBMT). Bone Marrow Transplantation 1998;21:473–476.
122. Aschan J, Ringden O, Ljungman P, Lonnqvist B, Ohlman S. Foscarnet for treatment of cytomegalovirus infections in bone marrow transplant recipients. Scandinavian Journal of Infectious Diseases 1992; 24:143–50.
123. Drew WL, Miner RC, Busch DF, *et al.* Prevalence of resistance in patients receiving ganciclovir for serious cytomegalovirus infection. Journal of Infectious Diseases 1991; 163:716–9.
124. Slavin MA, Bindra RR, Gleaves CA, Pettinger MB, Bowden RA. Ganciclovir sensitivity of cytomegalovirus at diagnosis and during treatment of cytomegalovirus pneumonia in marrow transplant recipients. Antimicrobial Agents & Chemotherapy 1993; 37:1360–3.
125. Reusser P, Cordonnier C, Einsele H, *et al.* European survey of herpesvirus resistance to antiviral drugs in bone marrow transplant recipients. Infectious Diseases Working Party of the European Group for Blood and Marrow Transplantation (EBMT). Bone Marrow Transplantation 1996; 17:813–7.
126. Boeckh M, Zaia J, Skettino S, *et al.* Oral ganciclovir after allogeneic marrow transplantation: a phase I/II study., 6th International Cytomegalovirus Workshop, Perdido Beach, AL, March 5–9, 1997.
127. Lewis RA, Carr LM, Doyle K, *et al.* Parenteral cidofovir for cytomegalovirus retinitis in patients with AIDS - the HPMPC Peripheral Cytomegalovirus Retinitis Trial - a Randomized, Controlled Trial. Annals of Internal Medicine 1997; 126:264 ff.
128. Bacigalupo A, Bregante S, Tedone E, *et al.* Combined foscarnet-ganciclovir treatment for cytomegalovirus infections after allogeneic hemopoietic stem cell transplantation. Transplantation 1996; 62:376–380.
129. Walter EA, Greenberg PD, Gilbert MJ, *et al.* Reconstitution of cellular immunity against cytomegalovirus in recipients of allogeneic bone marrow by transfer of T-cell clones from the donor. New England Journal of Medicine 1995; 333:1038–44.
130. McLaughlin-Taylor E, Pande H, Forman SJ, *et al.* Identification of the major late human cytomegalovirus matrix protein pp65 as a target antigen for CD8+ virus-specific cytotoxic T lymphocytes. Journal of Medical Virology 1994; 43:103–10.

10

CIDOFOVIR

Review of Current and Potential Clinical Uses

Sharon Safrin, Julie Cherrington, and Howard S. Jaffe

Gilead Sciences Incorporated
333 Lakeside Drive
Forest City, California 94404

1. MECHANISM OF ACTION AND *IN VITRO* ACTIVITY

Cidofovir (1-[(*S*)-3-hydroxy-2-(phosphonomethoxy)-propyl]cytosine dihydrate) is a cytidine nucleotide analogue with *in vitro* and *in vivo* activity against a broad spectrum of herpesviruses, including CMV, HSV-1 and -2, VZV, EBV, HHV-6, HHV-7, and KSHV (HHV-8) [Table 1], as well as against adenovirus, human papillomavirus(HPV), polyomaviruses, and human poxviruses. [1–9] Intravenous (IV) cidofovir has been approved in the United States and in Europe as a therapy for CMV retinitis in patients with AIDS.

Cellular uptake studies have shown that cidofovir enters cells by fluid-phase endocytosis.[10] Once internalized, cidofovir is phosphorylated first to its monophosphate metabolite and then to cidofovir diphosphate (CDVpp). Since cidofovir is itself a phosphonate, CDVpp is functionally analogous to a nucleoside triphosphate. [2,11,12] In contrast to the metabolism of ganciclovir, the phosphorylation of cidofovir is neither dependent on nor facilitated by viral enzymes. [2,11–13] Mono- and diphosphorylation are catalyzed by cellular rather than virus-encoded enzymes, suggesting a reduced potential for the development of resistance through mutation of viral activation genes. The active intracellular metabolite, CDVpp, does not readily exit through the cell membrane and thus has a prolonged half-life of 17 to 65 hours.[2,11–13] A separate metabolite, cidofovir phosphate-choline, has a half-life of at least 87 hours and may serve as an intracellular reservoir for active cidofovir diphosphate.[2,11,13,14]

CDVpp suppresses herpesvirus replication by selective inhibition of viral DNA synthesis, acting as both a potent inhibitor of and an alternate substrate for viral DNA polymerase, in competition with the natural substrate deoxycytidine triphosphate (dCTP).[2,15,16] Incorporation of cidofovir into the growing viral DNA chain dramatically reduces the rate of viral DNA synthesis; incorporation of two consecutive CDVpp molecules abrogates viral DNA synthesis.[17]

Antiviral Chemotherapy 5, edited by Mills *et al.*
Kluwer Academic / Plenum Publishers, New York, 1999.

Table 1. *In vitro* sensitivity of DNA viruses to cidofovir

Virus	$IC_{50\,(mM)}$	Reference
HCMV	0.3	20,61
HSV-1	14	61
TK^- HSV-1	7.2	61
HSV-2	36	61
VZV	0.6	5,61
EBV	0.03	61
HHV-6	14	61
HHV-7	21	4
HHV-8 (KSHV)	0.5	6
Adenovirus (human Ad5 & Ad8)	0.06-0.54	49,61
Vaccinia virus	14	61
Human polyomavirus	17	46,61

* IC50 = the concentration of drug to inhibit viral replication by ≥ 50%.

Cidofovir derives its specificity in part from a higher affinity for viral DNA polymerases than for host-cell polymerases, having a selectivity index of 1000-fold in cell culture.[15] Recent reports indicate that the binding affinity of CDVpp for human CMV (HCMV) DNA polymerase, as represented by the inhibition constant (Ki) is approximately 8 to 80 times greater than that for the human DNA polymerases.[2,16,18] The inhibition constants of cidofovir for other herpesviruses have also been determined; the Ki values against HSV-1 DNA polymerase and HSV-2 polymerase are 0.86 μM and 1.4 μM, respectively, providing a selective binding affinity of up to 600-fold for the viral enzymes.[2]

Cidofovir in combination with either ganciclovir or foscarnet has demonstrated synergistic inhibition of human CMV replication *in vitro*; ganciclovir and foscarnet showed additive inhibition rather than synergy.[19] None of the combinations showed increased cytotoxicity in comparison with each drug alone. Snoeck *et al.* have demonstrated that cidofovir in combination with ganciclovir, AZT, foscarnet, or aciclovir showed additive to synergistic inhibition of HCMV replication *in vitro*.[20]

2. ANIMAL STUDIES

Cidofovir is active against systemic HSV-1 and HSV-2 infections in mice and guinea pigs, murine and rat CMV infection, simian varicella virus (SVV) infection in African Green monkeys, and in the cottontail rabbit papillomavirus (CRPV) model.[21] The success of single-dose therapy of HSV-2 infection in guinea pigs underscores the prolonged antiviral effects of cidofovir.[22] Prophylactic efficacy of cidofovir administered prior to virus inoculation has also been shown in mouse, rat, and guinea pig models.

Toxicologic evaluation has been performed *in vitro* and *in vivo* in mice, rats, guinea pigs, rabbits, and monkeys after intravenous cidofovir administration. The major dose-limiting toxicity *in vivo* is route- and species-dependent nephrotoxicity (characterized by degeneration, necrosis, and/or regeneration of the proximal convoluted tubule cell), which can be alleviated by concomitant, high-dose probenecid administration.[23]

Dose-related increases in mammary adenocarcinomas were observed in female Sprague-Dawley rats receiving subcutaneous and intravenous injections of cidofovir.[24] In contrast, no tumors were observed in intravenous toxicological studies of intravenous cidofovir in cynomolgus monkeys.[24] To date, no excess frequency of malignancies has been observed in long-term follow-up of patients who have received cidofovir.[24]

3. CLINICAL STUDIES

3.1. CMV Retinitis

Three Phase II/III studies (Studies 105, 106, and 107) have been performed to investigate the safety and efficacy of cidofovir injection for the treatment of CMV retinitis in patients with AIDS. Study 106 was a randomized, controlled, multicenter study conducted at 7 sites in the United States and 1 site in Great Britain in AIDS patients with peripheral (i.e. non-sight threatening), previously untreated CMV retinitis.[25] Twenty-five patients were randomized to receive deferred therapy, while 23 patients began therapy with cidofovir immediately (5 mg/kg once weekly for 2 weeks [induction], then once every other week [maintenance]), along with concomitant saline hydration and oral probenecid (2 g 3 hours before infusion, and 1g 2 and 8 hours after each cidofovir infusion). The median time to progression of retinitis, evaluated by masked assessment of bilateral full-field retinal photographs, was 22 days in the deferred group and 120 days in the immediate treatment group ($p < .001$).[25] Seventy-three percent of cidofovir-treated patients had decreased or no retinitis activity during the study, compared with 15% of patients on deferred therapy. In the 16 patients who crossed over from deferred to active cidofovir therapy at the time of progression, the median time to the next progression of retinitis was 169 days.

Study 105 was a randomized, controlled, multicenter study conducted by the Studies of the Ocular Complications of AIDS (SOCA) and AIDS Clinical Trials (ACTG) collaborative groups.[26] In the first phase of the study, 29 patients with newly diagnosed, peripheral CMV retinitis were randomized to receive either deferred therapy or immediate intravenous cidofovir (5 mg/kg/dose for 2 consecutive once weekly doses (induction), then 3 mg/kg/dose once every other week (maintenance), along with concomitant saline hydration and oral probenecid). Following an interim analysis, 35 additional patients were randomized to either deferred therapy or one of two immediate intravenous cidofovir dose groups (5 mg/kg/dose for 2 consecutive once weekly doses (induction) followed by either 5 mg/kg/dose or 3 mg/kg/dose once every other week as maintenance). The median times to progression of retinitis, assessed by masked reading of retinal photographs, were 64 days for the 3 mg/kg maintenance group compared with 21 days for the deferred group ($p=.05$). The median time to progression was not reached for the 5 mg/kg maintenance group, compared with 20 days for the deferred group ($p=.004$).[28] Rates of increase in the retinal area of CMV involved were 1.81% vs. 0.8% in the deferred and lower (i.e. 3 mg/kg maintenance dosing) groups, respectively ($p=.01$), and were 2.2% vs. 0.34% in the deferred versus higher (i.e. 5 mg/kg maintenance dosing) groups, respectively ($p=.008$). There were no differences in the rate of loss of visual acuity in the 3 groups.

Study 107 was a randomized, open-label, multicenter, dose-ranging study conducted at 12 sites in the United States and 2 sites in Great Britain to determine the safety and efficacy of cidofovir injection for the treatment of active, relapsing CMV retinitis despite therapy with, or intolerance to, ganciclovir and/or foscarnet.[27] In this trial, 150 patients were randomized to receive induction therapy with 5 mg/kg cidofovir once weekly for 2 consecutive weeks, followed by maintenance cidofovir doses of either 5 mg/kg (Group A) or 3 mg/kg (Group B) once weekly, along with concomitant saline hydration and oral probenecid. In the first 100 patients, progression of retinitis was assessed by full-field, bilateral retinal photographs evaluated at a central reading center by an ophthalmologist unaware of randomization assignment. The median time to CMV retinitis progression was not reached in Group A [95% confidence interval (CI): 115 days to upper limit not yet reached] vs. 49 days in Group B (95% CI: 35 to 52 days) [p = 0.0006; log-rank test;]. Median time to progression, as assessed by ophthalmologic examination in the last 50 patients randomized, was not reached in Group A (95% CI: 215 days to upper limit not yet reached) vs. 105 days in Group B (95% CI: 43 to 170 days) [p = 0.2; log-rank test].[27] The proportion of patients with decreased or no retinitis activity during the study was 77% vs. 50% in the 5 mg/kg and 3 mg/kg dosing groups, respectively (p=.009).

These 3 studies can be pooled for an analysis of the safety of intravenous cidofovir. Dose-limiting toxicities of the drug are nephrotoxicity and ocular toxicity. In the 247 patients receiving intravenous cidofovir in studies 105, 106, or 107, serum creatinine ≥ 2.0 mg/dl occurred in 9%, and ≥ 2+ proteinuria in 44%. Ocular hypotony, defined as a decrease in intraocular pressure of ≥ 50%, occurred in 12% of patients in which serial testing was performed, and anterior uveitis or iritis was reported in 7% of cidofovir-treated patients. Of note, the frequency of both renal and ocular toxicities were greater in patients receiving 5 mg/kg biweekly maintenance dosing than those receiving 3 mg/kg dosing. Neutropenia (absolute neutrophil count ≤ 500/mm^3) occurred in 25%; no dose-related difference in rates of neutropenia was noted. Probenecid related side effects were reported in 40% of patients in these controlled clinical trials, and these were dose-limiting in 6%of cases.

The median times to retinitis progression are comparable in patients on deferred therapy in the above studies to those observed in prior studies of treatment with ganciclovir or foscarnet for AIDS patients with CMV retinitis.[28,29] Thus, despite the absence of controlled comparison of cidofovir to either ganciclovir or foscarnet for this indication, the lack of retinitis progression in 50% of patients receiving the 5 mg/kg biweekly maintenance dose of cidofovir in both studies 105 and 107 are suggestive of potent anti-CMV activity of this compound in patients with either previously untreated or relapsing retinitis. In addition to demonstrating efficacy, these studies have been important in defining the optimal dosing regimen for cidofovir in the treatment of CMV retinitis. Cidofovir is administered as an intravenous infusion in doses of 5 mg/kg once weekly for 2 consecutive weeks, followed by 5 mg/kg every other week thereafter. Utilization of maintenance doses of 3 mg/kg every other week or 5 mg/kg at intervals of longer than every 2 weeks may be considered on an individual basis, according to the activity of retinitis, concomitant use of highly active antiretroviral therapy, and overall condition of the patient. Oral probenecid is administered 2 hours before the infusion (2 grams), 3 hours after the infusion (1 gram), and 8 hours after the infusion (1 gram). Intravenous hydration of 2 or more liters of saline before and during infusion of cidofovir appears to be important in the prevention of nephrotoxicity. Cidofovir therapy is contraindicated in patients with serum creatinine > 1.5 mg/dl, calculated creatinine clearance ≤ 55 ml/min, or in the presence of ≥ 2+ proteinuria. Administration of agents with potential nephrotoxicity, such as aminoglycosides, amphotericin B, foscarnet, pentamidine, non-steroidal anti-inflammatory drugs, or contrast dye, should be avoided during and for 7 days preceding the initiation of cidofovir therapy. Maintenance dose reduction from 5 mg/kg to 3 mg/kg every other week (or from 3 mg/kg to 1.5 mg/kg every other week) should be instituted for rises in serum creatinine of 0.3–0.4 mg/dl; if serum creatinine rises by > 0.5 mg/dl or if ≥ 3+ proteinuria develops, cidofovir therapy should be discontinued. Probenecid intolerance can generally be treated pre-emptively, with antihistamines or acetaminophen as needed. Avoidance of probenecid ingestion on an empty stomach seems also to improve gastrointestinal tolerance. Finally, serial monitoring of intraocular pressure and slit lamp examination should be performed during cidofovir therapy, for the early detection and treatment of ocular hypotony and uveitis.

3.2. Resistance to CMV

Since cidofovir is a nucleotide analog, it does not require phosphorylation by the UL97 gene product.[30] Ganciclovir-resistant viruses carrying a mutation in UL97 are thus susceptible to cidofovir, while those carrying DNA polymerase mutations are cross-resistant to cidofovir *in vitro*.[31–34]

Cherrington and coworkers described the *in vitro* selection of human CMV variants that had reduced susceptibility (12-fold) to cidofovir following exposure to increasing concentrations of cidofovir.[35] Cidofovir-resistant viruses were cross-resistant to ganciclovir (6-fold) but remained susceptible to foscarnet: a single point mutation (K513N) in a conserved region of the CMV DNA polymerase gene was demonstrated in the cidofovir-selected viruses. *In vitro*-selected ganciclovir-resistant mutants from the same study were 35-fold resistant to ganciclovir, cross-resistant to cidofovir (15-fold) but not to foscarnet, and carried 2 mutations (K513N and V812L) in conserved regions of the polymerase gene, as well as a single point mutation in UL97 (M460V).[35] Snoeck *et al.* selected cidofovir-resistant viruses *in vitro* which were cross-resistant to ganciclovir but susceptible to foscarnet.[36] In studies of five laboratory-generated foscarnet-resistant mutants, cross-resistance to cidofovir was not observed.[36, 37]

Recent work has investigated the genotypic basis for cross-resistance to cidofovir among 28 ganciclovir-resistant isolates.[34] All of the isolates were obtained from patients who had received ganciclovir or both ganciclovir and foscarnet; none of the patients had received cidofovir. Isolates exhibiting high-level resistance to ganciclovir (IC50 > 30 μM; n = 11) were associated with mutations in both UL97 and DNA polymerase genes and were cross-resistant to cidofovir *in vitro*. In contrast, low-level ganciclovir resistance (IC50 8.5–30 μM; n = 17) was associated with mutations in the UL97 gene only; such isolates remained sensitive to cidofovir. Interestingly, this study also demonstrated a correlation between the length of ganciclovir therapy and level of ganciclovir resistance. Among the ganciclovir-resistant isolates, 19% showed high-level resistance when receiving up to 9 months of therapy; after 9 months of ganciclovir therapy, 64% had acquired high-level ganciclovir resistance, a statistically significant difference.[34] This study confirmed previous reports that cross-resistance to cidofovir *in vitro* requires at least one point mutation in the DNA polymerase gene.[31–33] To date, all clinical isolates that have exhibited resistance to ganciclovir due to mutations in the CMV DNA polymerase gene have been cross-resistant to cidofovir *in vitro*.[31–34]

Resistance to cidofovir as a result of cidofovir administration has not been documented to date in any clinical isolate obtained from patients receiving cidofovir. Cherrington *et al.* studied 29 pairs of isolates recovered from semen, before and after cidofovir exposure, from 22 AIDS patients with asymptomatic CMV shedding in semen or urine, and no prior receipt of anti-CMV therapy.[38] They received a median 30 mg/kg (range, 3 to 67) cumulative cidofovir dose for a median of 8 weeks (range, 2 to 38). Nine patients received cidofovir for >12 weeks; seven patients received two courses of cidofovir, and their total cidofovir exposure was 12 to 48 weeks. The IC50 values for pre-exposure isolates ranged from less than 0.5 μM to 1.9 μM, and those for post-exposure isolates ranged from less than 0.5 μM to 2 μM,[38] demonstrating that resistance to cidofovir had not developed.

Isolates from patients enrolled in cidofovir CMV retinitis trials have been assayed for their drug susceptibilities.[39] Blood culture isolates from patients receiving cidofovir for newly diagnosed retinitis (first-line) or relapsing retinitis (second-line) therapy were obtained for *in vitro* susceptibility analysis. Twelve of 24 patients tested had at least 1 positive blood culture during cidofovir first-line therapy. Isolates from 6 of these patients obtained at a mean time of 14.3 weeks on cidofovir were tested; all had complete susceptibility to cidofovir, ganciclovir and foscarnet *in vitro*. Twenty-one of 52 patients tested had at least 1 positive blood culture during second-line cidofovir therapy. These patients had received prior ganciclovir or foscarnet therapy for mean durations of 49 weeks and 17 weeks, respectively. Isolates from 20 of these patients obtained at a mean time of 17.3 weeks on cidofovir were tested; 4 showed complete susceptibility to all 3 drugs listed above, 7 showed moderately reduced susceptibility (≤ 4-fold) to cidofovir and high level resistance (8–23-fold) to ganciclovir, and 1 showed complete susceptibility to cidofovir but low level resistance to ganciclovir *in vitro*. Four of the 7 isolates with moderately reduced susceptibility to cidofovir and high-level ganciclovir resistance also showed reduced susceptibility (4–10-fold) to foscarnet *in vitro*. Based on prior work determining the genetic basis of cross-resistance to cidofovir among ganciclovir-resistant isolates,[34] these data suggest that *in vitro* reduced susceptibility to cidofovir may have been selected for by prior ganciclovir therapy rather than cidofovir therapy. There was no difference in time to retinitis progression on cidofovir when comparing patients with isolates that did or did not show reduced susceptibility to cidofovir *in vitro*.

3.3. Other Potential Clinical Applications

Administration of cidofovir by the intraocular route has been evaluated in a preliminary manner for the treatment of patients with CMV retinitis. Several open-label, uncontrolled reports from investigators at the University of California, San Diego have suggested the safety and efficacy of 10–20 μg of cidofovir injected intravitreally for this indication.[40–43] Potential complications include iritis and ocular hypotony, apparently due to ciliary body damage and/or atrophy.[44] In a multicenter, randomized, double-blind clinical trial of intraocular cidofovir for relapsing cytomegalovirus retinitis in patients with AIDS (Study 501), patients with active CMV retinitis despite at least 4 weeks of treatment with either ganciclovir or foscarnet were randomized to receive a single intraocular injection of cidofovir in doses of 5, 10, or 15 μg.[45] Following CMV retinitis progression, as assessed by masked reading of full-field, bilateral retinal photographs, responding patients received multiple intraocular cidofovir injections at their initial dose given at least once every 4 weeks, and nonresponding patients received a single dose of 15 μg. The study was terminated prematurely after enrollment of 31 of 90 planned patients by the recommendation of the Data Safety and Monitoring Board, due to suboptimal efficacy and ocular toxicity. Median times to progression *by patient* after a single intraocular dose of cidofovir (analyzing data from all treated eyes) were 29, 29, and 43 days for the 5, 10, and 15 μg dose groups, respectively; median times to progression *by eye* were 34, 45, and 43 days, respectively, for these 3 dose groups. After the initial masked single dose, iritis was observed in 24 of 31 (78%) patients. For all phases of the study combined, a total of 79 episodes of iritis or vitritis occurred in 28 patients (90%) after 97 injections of cidofovir (81%) into 87 eyes. Ocular hypotony, defined as a decrease in intraocular pressure by 50%, ≥ occurred in 4 of 31 (13%) patients after a single dose of cidofovir. For all phases of the study combined, 11 patients (36%) developed ocular hypotony, to total 20 events following 97 cidofovir injections (21%) into 87 eyes. Of these, 7 eyes in 6 patients had an associated decline in visual acuity of 3 or more lines on the standardized eye chart, and 4 events in 3 patients met the definition for severe hypotony (i.e., intraocular pressures of 0–1 mm Hg). Thus the therapy was minimally effective and poorly tolerated. Due to suggestions of improved safety and efficacy when oral probenecid is used concomitantly with intraocular cidofovir,[43] a multicenter, open-label, dose-escalation study was designed and is currently underway.

The broad *in vitro* spectrum of cidofovir, as described above, has stimulated investigation of a number of potential clinical applications. The *in vitro* activity of cidofovir against HHV-8,[6] for example, has triggered the initiation of an open-label trial of IV cidofovir for the treatment of cutaneous Kaposis sarcoma at the National Cancer Institute of the National Institutes of Health. An ongoing randomized, placebo-controlled multicenter trial is evaluating the utility of cidofovir gel for the treatment of cutaneous molluscum contagiosum in patients with AIDS, given the *in vitro* potency of cidofovir against poxviruses[9] and encouraging results in a case series of AIDS patients with molluscum contagiosum lesions who were treated with cidofovir.[45a] The *in vitro* activity of cidofovir against polyomaviruses,[46] combined with anecdotal reports of improvement in patients with progressive multifocal leukoencephalopathy (PML) receiving IV cidofovir,[47,48] has spurred the development of 2 clinical trials evaluating IV cidofovir for the therapy of AIDS patients with PML in the U.S. and in Europe. Based upon its activity against both herpesviruses and adenovirus,[1,8,49] a topical ophthalmic solution of cidofovir is being studied in controlled trials for the treatment of viral keratoconjunctivitis.

Two studies of cidofovir gel, in one applied as a single dose for treatment of recurrent genital herpes,[50] and in the other applied on five consecutive days to treat aciclovir-resistant HSV infection in patients with AIDS,[51] have demonstrated anti-HSV activity. In study 305, 96 immunocompetent patients with recurrent genital herpes were randomized to receive a single application of placebo, 1%, 3%, or 5% cidofovir gel in a double-blinded manner.[50] Median time to cessation of virus shedding was decreased by 63% in cidofovir-treated patients compared with placebo ($p < 0.001$). Treatment-related trends in decreasing time to complete healing were evident; however the study was underpowered to demonstrate statistical significance. Application site reactions were dose-related, occurring in 3% of patients receiving placebo gel, 5% at 1% cidofovir gel strength, 19% at 3% cidofovir gel, and 22% at 5% cidofovir gel. Study 301 was a Phase I/II, randomized, double-blinded, placebo-controlled parallel comparison of placebo, 0.3% cidofovir gel, and 1% cidofovir gel for the treatment of mucocutaneous HSV infection in patients with AIDS.[51] Compared with placebo, treatment with cidofovir gel provided statistically significant benefits in evoking complete healing (30% vs. 0%; $p = 0.03$), complete or good response (i.e. > 50% decrease in lesion size; 50% vs. 0%; $p = 0.008$), cessation of virus shedding (87% vs. 0%; $p < 0.001$), and reduction in pain ($p = 0.04$). Application site reactions occurred in 25% of cidofovir recipients and in 20% of placebo recipients; there was no evidence of significant systemic toxicity.

Cidofovir has anti-HPV activity in both the cottontail rabbit papilloma virus and HPV-11 athymic mice animal models.[52] Additionally, cidofovir has been shown to induce apoptosis in the human papilloma virus-containing cell lines CK1, SiHA, CaSki, and HeLa[53] and to have an antiproliferative effect on cells with an elevated proliferation rate.[54] Preliminary clinical studies suggest the potential utility of cidofovir against HPV-associated diseases. Snoeck *et al.* described successful resolution and prolonged remission in 3 AIDS patients with severe, relapsing anogenital condylomata in association with topical application of cidofovir.[55] Study 302, an open-label, sequential, dose-escalation study of cidofovir gel in HIV-infected patients with persistent anogenital condylomata despite at least 4 weeks of standard therapy revealed response rates of approximately 65%, as well as recurrence of warts in only 1 of 10 patients followed for 6 months.[56] A double-blind, placebo-controlled study of cidofovir gel is close to completion in immunocompetent patients with HPV-associated anogenital warts. A recent report described successful therapy of 2 patients with verruca vulgaris using topical cidofovir.[57]

Additionally, Snoeck *et al.* described complete resolution of HPV-associated cervical intraepithelial neoplasia (CIN) grade III lesions in 5 of 12 women following three topical applications of 1% cidofovir gel.[58]

Intralesional injection of cidofovir has shown promise in early studies for the treatment of respiratory papillomatosis. Van Cutsem *et al.* reported the prolonged resolution of a squamous papilloma of the hypopharynx-esophagus after intralesional injections of cidofovir,[59] while Snoeck *et al.* reported complete resolution in 14 of 17 patients with severe respiratory papillomatosis following intralesional injections of cidofovir.[60] A placebo-controlled, dose-escalation multicenter trial is soon to begin to evaluate this therapy in children with recurrent respiratory papillomatosis, under the auspices of the Collaborative Antiviral Study Group of the National Institutes for Allergy and Infectious Diseases.

In conclusion, the broad *in vitro* spectrum of cidofovir has sparked multiple clinical investigations as to its potential uses in intravenous, topical and intraocular formulations. Although currently licensed as a parenteral treatment for AIDS patients with CMV retinitis, other possible indications await results from ongoing studies.

REFERENCES

1. De Clercq E, Holy A, Rosenberg I, *et al.* A novel selective broad-spectrum anti-DNA virus agent. Nature 1986; 323:464–467.
2. Ho H-T, Woods KL, Bronson JJ, *et al*, Intracellular metabolism of the antiherpes agent (S)-1-[3-hydroxy-2-(phosphonylmethoxy)propyl]cytosine. Mol Pharmacol 1992; 41:197–202.
3. Bronson JJ, Ghazzouli I, Hitchcock MJM, *et al.* Synthesis and antiviral activity of the nucleotide analogue (S)-1-[3-hydroxy-2-(phosphonylmethoxy)propyl]cytosine. J Med Chem 1989; 32:1457–1463.
4. Takahashi K, Suzuki M, Iwata Y, *et al.* Selective activity of various nucleoside and nucleotide analogues against human herpesvirus 6 and 7. Antiviral Chem Chemother 1997; 8:24–31
5. Andrei G, Snoeck R, Reymen D, *et al.* Comparative activity of selected antiviral compounds against clinical isolates of varicella-zoster virus. Eur J Clin Microbiol Infect Dis 1995; 14:318–328
6. Kedes DH, Ganem D. Sensitivity of Kaposis sarcoma-associated herpesvirus to antiherpesvirus drugs: Implications for potential therapy. J Clin Invest 1997;99:2082–2086.
7. Hitchcock MJM, Jaffe HS, Martin JC, *et al.* Cidofovir, a new agent with potent anti-herpesvirus activity. Antiviral Chem Chemother 1996; 7:115–127.
8. de Oliveira CBR, Stevenson D, LaBree L, *et al.* Evaluation of cidofovir (HPMPC, GS-504) against adenovirus type 5 injection *in vitro* and in a New Zealand rabbit ocular model. Antiviral Res 1996; 31:165–172.
9. Huggins J, Robertson M, Kefauver D, *et al.* Potential antiviral therapeutics for smallpox and other poxvirus infections. Presented at the 11th International Meetings on Poxvirus and Iridoviruses. Toledo Spain 1996.
10. Connelly MC, Robbins BL, Fridland A. Mechanism of uptake of the phosphonate analog (S)-1-(3-hydroxy-2-phosphonylmethoxypropyl) cytosine (HPMPC) in Vero cells. Biochem Pharmacol 1993; 46:1053–1057.
11. Cihlar T, Votruba I, Horsk K, *et al.* Metabolism of 1-(S)-(3-hydroxy-2-phosphonomethoxypropyl) cytosine (HPMPC) in human embryonic lung cells. Collect Czech Chem Commun 1992; 57:661–672.
12. Cihlr T, Chen MS. Identification of enzymes catalyzing two-step phosphorylation of cidofovir and the effect of cytomegalovirus infection on their activities in host cells. Mol Pharmacol 1996; 50:1502–1510
13. Aduma P, Connelly MC, Srinivas RV, *et al.* Metabolic diversity and antiviral activities of acyclic nucleoside phosphonates. Mol Pharmacol 1995; 47:816–822.
14. De Clercq E. Therapeutic potential of HPMPC as an antiviral drug. Rev Med Virol 1993; 3:85–96.
15. Neyts J, Snoeck R, Schols D, *et al.* Selective inhibition of human cytomegalovirus DNA synthesis by (S)-1-(3-hydroxy-2-phosphonylmethoxypropyl)cytosine [(S)-HPMPC] and 9-(1,3-dihydroxy-2-propoxymethyl)guanine (DHPG). Virology 1990;179:41Ð50.
16. Xiong X, Smith JL, Kim C, *et al.* Kinetic analysis of the interaction of cidofovir diphosphate with human cytomegalovirus DNA polymerase on DNA elongation. Biochem Pharmacol 1996; 51:1563–1567
17. Xiong XF, Smith JL, Chen MS. Effect of incorporation of cidofovir into DNA by human cytomegalovirus DNA polymerase on DNA elongation. Antimicrob Agents Chemother 1997; 41:594–599.

18. Cherrington JM, Allen SJ, McKee BH, *et al.* Kinetic analysis of the interaction between the diphosphate of (S)-1-(3-hydroxy-2-phosphonylmethoxypropyl)cytosine, ddCTP, AZTTP, and FIAUTP with human DNA polymerases β and Υ. Biochem Pharmacol 1994; 48:1986–1988.
19. Mulato MS, Cherrington JM, and Chen MS. Anti-HCMV activity of cidofovir in combination with antiviral compounds and immunosuppressive agents: in-vitro analyses. Antiviral Chemistry and Chemotherapy 1996; 7:203–208
20. Snoeck R, Andrei G, Schols BJ, *et al.* Activity of different antiviral drug combinations against human cytomegalovirus replication *in vitro*. Eur J Clin Microbiol Infect Dis 1992; 11:1144–1155
21. Kern ER. Value of animal models to evaluate agents with potential activity again human cytomegalovirus. Tranplant Proc 1991; 23 (Suppl 3):152–155
22. De Clercq E and Holy A. Efficacy of (S)-1-(3-hydroxy-2-phosphonylmethoxypropyl)cytosine in various models of herpes simplex virus infection in mice. Antimicrob Agents Chemother 1991; 34:701–706
23. Stals FS, De Clercq E, and Bruggeman CA. Comparative activity of (S)-1-(3-Hydroxy-2-Phosphonylmethoxypropyl)cytosine (HPMPC) and 9-(1,3-Dihydroxy-2-Propoxymethyl)guanine (DHPG) against rat cytomegalovirus infection *in vitro* and *in vivo*. Antimicrob Agents Chemother 1991: 35:2262–2266.
24. Gilead Sciences Inc., data on file.
25. Lalezari JP, Stagg RJ, Kuppermann BD, *et al.* Intravenous cidofovir for peripheral cytomegalovirus retinitis in patients with AIDS. A randomized, controlled trial. Ann Intern Med 1997; 126:257–263.
26. The Studies of Ocular Complications of AIDS Reseach Group in collaboration with the AIDS Clinical Trials Group. Cidofovir (HPMPC) for the treatment of cytomegalovirus retinitis in patients with AIDS: the HPMPC peripheral cytomegalovirus retinitis trial. Ann Intern Med 1997; 126:264–274.
27. Kupperman B, Lalezari JP, Holland GN *et al.* A randomized, controlled study of intravenous cidofovir (CDV) for treatment of relapsing CMV retinitis: final results. Abstract 4279, ARVO meeting, Fort Lauderdale, 1997.
28. Spector SA, Weingeist T, Pollard RB *et al.* A randomized, controlled study of intravenous ganciclovir therapy for cytomegalovirus retinitis in patients with AIDS. J Infect Dis 1993; 168:557–563.
29. Palestine AG, Polis MA, De Smet MD *et al.* A randomized, controlled trial of foscarnet in the treatment of cytomegalovirus retinitis in patients with AIDS. Ann Intern Med 1991; 115:665–673.
30. Biron KK, Fyfe JA, Stanat SC, *et al.* A human cytomegalovirus mutant resistant to the nucleoside analog 9-{[2-hydroxy-1-(hydroxymethyl) ethoxy]methyl}guanine (BW B759U) induces reduced levels of BW B759U triphosphate. Proc Natl Acad Sci U S A 1986; 83:8769–8773.
31. Lurain NS, Thompson KD, Holmes EW, *et al.* Point mutations in the DNA polymerase gene of human cytomegalovirus that result in resistance to antiviral agents. J Virol 1992; 66:7146–7152.
32. Sullivan V, Biron KK, Talarico C, *et al.* A point mutation in the human cytomegalovirus DNA polymerase gene confers resistance to ganciclovir and phosphonylmethoxyalkyl derivatives. Antimicrob Agents Chemother 1993; 37:19–25.
33. Tatarowicz WA, Lurain NS, and Thompson KD. A ganciclovir-resistant clinical isolate of human cytomegalovirus exhibiting cross-resistance to other DNA polymerase inhibitors. J Infect Dis 1992; 166:904–907.
34. Smith IL, Cherrington JM, Jiles RE, *et al.* High-level resistance of cytomegalovirus to ganciclovir is associated with alterations in both the UL97 and DNA polymerase genes. J Infect Dis 1997; 176:69–77.
35. Cherrington JM, Mulato AS, Fuller M, *et al.* *In vitro* selection of a human cytomegalovirus (HCMV) that is resistant to cidofovir. Abstract H117, the 35th Interscience Conference on Antiviral Chem Chemother. San Francisco, CA.
36. Snoeck R, Andrei G, De Clercq E. Patterns of resistance and sensitivity to antiviral compounds of drug resistant strains of human cytomegalovirus selected *in vitro*. Eur J Clin Microbiol Infect Dis 1996; 15:574–579.
37. Sullivan V, and Coen DM. Isolation of foscarnet-resistant human cytomegalovirus patterns of resistance and sensitivity to other antiviral drugs. J Infect Dis 1991; 164:781–784.
38. Cherrington JM, Miner R, Hitchcock MJM, *et al.* Susceptibility of human cytomegalovirus (HCMV) to cidofovir is unchanged after limited *in vivo* exposure to various clinical regimens of drug. J Infect Dis 1996; 173:987–992.
39. Cherrington JM, Fuller MF, Lalezari JP, *et al.* *In vitro* susceptibilities of isolates from CMV retinitis patients receiving first or second line cidofovir therapy: relationship to clinical outcome. Abstract 304, the 4th Conference on Retroviruses and Opportunistic Infections, 1997, Washington, D.C.
40. Kirsch L, Arevalo F, De Clercq E, *et al.* Phase I/II study of intravitreal cidofovir for the treatment of cytomegalovirus retinitis in patients with the acquired immunodeficiency syndrome. Am J Ophthalmology 1995; 119:466–476.

41. Kirsch L, Arevalo F, Chavez de la Paz E, *et al*, Intravitreal cidofovir (HPMPC) treatment of cytomegalovirus retinitis in patients with acquired immune deficiency syndrome. Ophthalmology 1995; 102:533–543.
42. Rahhal F, Arevalo J, Chavez de la Paz E, *et al.* Treatment of cytomegalovirus retinitis with intravitreous cidofovir in patients with AIDS - A preliminary report. Ann Intern Med 1996; 125:98–103.
43. Chavez-de la Paz E, Arevalo JF, Kirsch LS, *et al.* Anterioir nongranulomatous uveitis after intravitreal HPMPC (cidofovir) for the treatment of cytomegalovirus retinitis. Analysis and preventon. Ophthalmology 1997; 104:539–544.
44. Banker AS, Arevalo JF, Munguia D *et al.* Intraocular pressure and aqueous humor dynamics in patients with AIDS treated with intravitreal cidofovir (HPMPC) for cytomegalovirus retinitis. Am J Ophthalmology 1997; 124:168–180.
45. Kuppermann B, Wolitz R, Stagg R, *et al.* A phase II randomized, double-masked study of intraocular cidofovir for relapsing cytomegalovirus retinitis in patients with AIDS. Presented at the Vitreous Society Annual Meeting, 1996, Cancun Mexico
45a. Meadows KP, Tyring SK, Pavia AT, *et al.* Resolution of recalcitrant molluscum contagiosum virus lesions in Human Immunodeficiency Virus-infected patients treated with cidofovir. Arch Dermatol 1997; 133:987–990.
46. Andrei G, Snoeck R, Vandepuite M, *et al.* Activity of various compounds against murine and primate polyomaviruses. Antimicrob Agents Chemother 1997; 41: 587–593.
47. Snoeck R, De Wit S, Ross C, *et al.* Treatment of progressive multifocal leukoencephalopathy with cidofovir in an AIDS patients. (Abstract 113). Antiviral Res 1996; 30:A51.
48. Brosgart C, Roberts T, Safrin S, *et al.* Cidofovir (CDV) therapy for progressive multifocal leukoencephalopathy (PML) in two AIDS patients. Abstract I-5, the 37th Interscience Conference on Antimicrobial Agents and Chemotherapy. Toronto Canada, September 1997.
49. Gordon YJ, Romanowski EG, and Araullo-Cruz T. Topical HPMPC inhibits adenovirus type 5 in the New Zealand rabbit ocular replication model. Invest Ophthal Vis Sci, 1994; 35:4135–4143.
50. Sacks S. Cidofovir topical gel in recurrent genital herpes. Presented as a latebreaker abstract at the 9th International Society for Antiviral Research, 1996, Japan.
51. Lalezari J, Jaffe HS, Schacker T, *et al.* A randomized, double- blinded, placebo-controlled study of cidofovir topical gel for aciclovir-resistant herpes simplex virus infections in patients with AIDS. Abstract 403, 3rd National Conference on Retroviral Infections, 1996, Washington D.C.
52. Kurtzman G, Pickel M, Christensen, *et al.* Phosphonate nucleoside analogs are potent anti-papillomavirus agents in animal models. Abstract 1593, the 33rd Interscience Conference on Antimicrobial Agents and Chemotherapy, 1993.
53. Andrei G, Snoeck R, Piette J, *et al.* Cidofovir-induction of apoptosis in human papilloma virus(HPV)-containing cell lines. Abstract 107, Tenth International Conference on Antiviral Research, Atlanta GA 1997.
54. Snoeck R, Andrei G, Piette J, *et al.* Antiproliferative potential of acyclic nucleoside phosphonates on human papillomavirus (HPV)-harboring cell lines compared with various HPV-negative cell lines. Abstract H-159, the 37th Interscience Conference on Antimicrobial Agents and Chemotherapy. Toronto Canada 1997.
55. Snoeck R, Van Ranst M, Andrei G, *et al.* Treatment of anogenital papillomavirus infections with an acyclic nucleoside phosphonate analogue (letter). N Engl J Med 1995; 333:943–944.
56. Tyring S, Douglas J, Schacker T, *et al.* A phase I/II study of cidofovir topical gel for refractory condyloma acuminatum in patients with HIV infection. Abstract 334, 4th Conference on Retroviruses and Opportunistic Infections, 1997, Washington D.C.
57. Zabawski EJ, Sands B, Goetz D, *et al.* Treatment of verruca vulgaris with topical cidofovir (letter). JAMA 1997; 278:1236
58. Snoeck R, Arens M, Muller C, *et al.* Preliminary results on the activity of cidofovir for the treatment of cervical intraepithelial neoplasia (CIN) grade III. Abstract 108, Tenth International Conference on Antiviral Research, Atlanta GA 1997.
59. Van Cutsem R, Snoeck R, Van Ranst M, *et al.* Successful treatment of a squamous papilloma of the hypopharyn-esophagus by local injections of (S)-1-(3-hydroxy-2-phosphonylmethoxypropyl)cytosine. J Med Virol 1995; 45:230–235
60. Snoeck R, Wellens W, Desloovere C, *et al.* Treatment of severe recurrent laryngeal papillomatosis by local injections of (S)-1-(3-hydroxy-2-phosphonylmethoxypropyl)cytosine (cidofovir). (Abstract 24). Antiviral Res 1996; 30:A25.
61. Naesens L, Snoeck R, Andrei G, *et al.* HPMPC (cidofovir), PMEA (adefovir) and related acyclic nucleoside phosphonate analogues: a review of their pharmacology and clinical potential in the treatment of viral infections. Antiviral Chem Chemother 1997; 8:1–23.

11

ORAL GANCICLOVIR

Stephen A. Spector

University of California, San Diego
Department of Pediatrics/Infectious Diseases
9500 Gilman Drive
La Jolla, California 92093-0672

Currently, three drugs are licensed for the treatment of cytomegalovirus (CMV) retinitis in persons with AIDS: ganciclovir, foscarnet and cidofovir. Of the three antivirals, only ganciclovir is available in an oral formulation. Studies that have examined the pharmacokinetics of oral ganciclovir indicate that although the peak levels of oral drug when administered at 1000 mg three times daily are considerably less than the standard intravenous dose of 5 mg/kg, the AUC for the oral drug approximates that of the intravenous formulation (Figure 1).[1] Oral ganciclovir has demonstrated efficacy as maintenance therapy for CMV retinitis, prevention of contralateral retinitis and extraocular disease and prevention of CMV disease when used either for "true" prophylaxis or as preemptive treatment.

The Roche 2226 study helped to clarify some of the differences seen in the studies that had previously examined oral ganciclovir for maintenance treatment of CMV retinitis.[5] As shown in Figure 2A, when patients with CMV retinitis had received ganciclovir treatment for <100 days prior to receiving oral maintenance therapy, there was no significant difference between the use of oral ganciclovir at daily doses of 3g, 4.5g and 6g compared to IV maintenance. However, for patients previously treated for >100 days, the 3g daily dose was inferior to the other regimens (Figure 2B). These data would suggest that a 4.5g dose (1.5g three times daily) is the dose of preference for maintenance when the patients are able to tolerate the higher dose.

Another recent study has examined the use of oral ganciclovir as systemic therapy for patients receiving intravitreal ganciclovir for treatment of CMV retinitis.[6] Patients with AIDS were randomized to one of three arms: oral ganciclovir 4.5g/day plus ganciclovir implant (n = 123); oral placebo plus ganciclovir implant (n = 122); or intravenous ganciclovir alone (n = 132). Patients randomized to receive oral ganciclovir plus implant had fewer cases of extraocular disease and contralateral retinitis than did patients randomized to implant plus placebo (Table 2). In this study oral ganciclovir was shown to reduce the incidence and delay the onset of new CMV disease.

Antiviral Chemotherapy 5, edited by Mills *et al.*
Kluwer Academic / Plenum Publishers, New York, 1999.

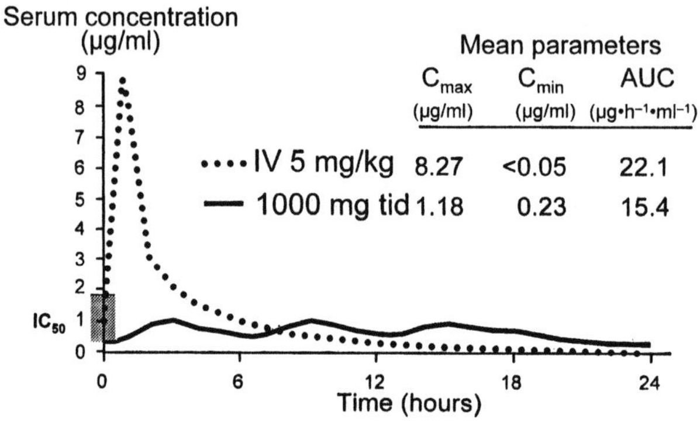

Figure 1. Intravenous and oral ganciclovir: Comparison of pharmacokinetic profiles (adapted from reference 1). Several studies have evaluated the use of oral ganciclovir for CMV maintenance. The 1653 study examined newly diagnosed patients with CMV retinitis, stable after IV ganciclovir therapy;[2] 1774 examined previously treated CMV retinitis stable after IV ganciclovir;[3] and 034 examined newly diagnosed or previously treated patients, stable after IV ganciclovir induction.[4] As shown in Table 1, although not always significantly superior, IV maintenance was consistently better than oral maintenance.

The frequency and severity of CMV disease in persons with AIDS strongly support the development and institution of strategies designed to prevent CMV-related diseases. Recent studies have demonstrated the feasibility of prophylaxis for CMV disease.

The Roche 1654 study was a randomized, masked clinical trial that evaluated oral ganciclovir for its ability to prevent CMV disease in persons with AIDS.[7] Study participants were randomized in a 2:1 fashion to receive either oral ganciclovir 1000 mg every eight hours or a matched placebo. Participants were evaluated at baseline with clinical and laboratory monitoring. Additionally, prior to study entry, dilated ophthalmologic examinations were performed by an experienced ophthalmologist and were repeated every 2 months for the duration of the study. Of the 725 persons enrolled into the study (ganciclovir - 486, placebo - 239), the median $CD4^+$ lymphocyte count was $22/mm^3$, with 88% of study participants having baseline $CD4^+$ lymphocytes below $50/mm^3$. The incidence of CMV disease at 18-months using Kaplan-Meier estimates was 39% in the placebo group versus 20% in the ganciclovir group ($P = 0.0001$ with a percent risk reduction of 49%) (Figure 3). CMV retinitis was most dramatically reduced, but other CMV events also were decreased in the ganciclovir prophylaxis group. Although there was a trend of overall survival benefit (relative risk ganciclovir versus placebo - 0.80), the difference was not statistically significant ($P = 0.1$).

Further virologic and clinical findings in the 1654 study support the efficacy of ganciclovir for prophylaxis. The prevalence of CMV positive urine cultures at baseline was 42% and 44% for the ganciclovir group and the placebo group, respectively. However, af-

Table 1. Intravenous versus oral ganciclovir mean time from start of maintenance to progression of CMV retinitis (days) by photographic assessment

Study	IV Ganciclovir	Oral Ganciclovir	Difference
1653	62	57	-5
1774	66	54	-12
034	62	51	-11

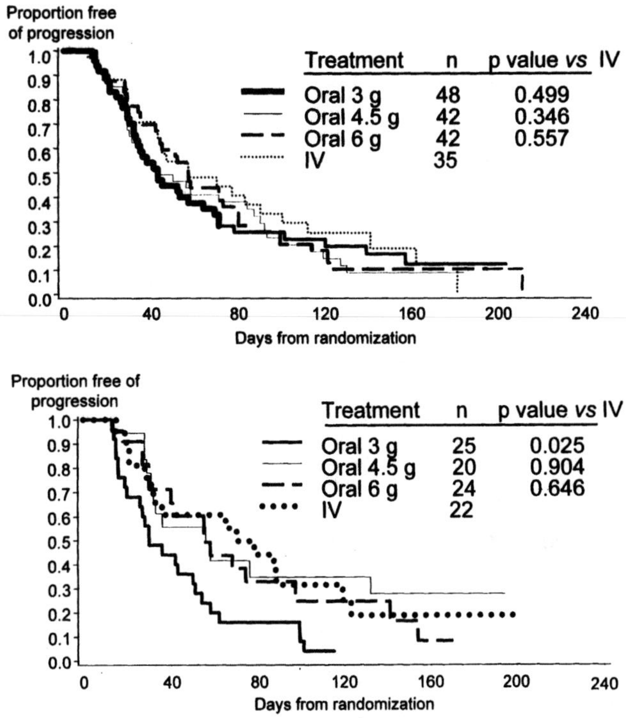

Figure 2. Time to progression of CMV retinitis: A. CMV duration <100 days at entry (n = 167); B. CMV duration >100 days at entry (n = 91, median = 8 months) (from reference 5).

ter 2 months on study, the ganciclovir group decreased to 12% with positive cultures compared to a constant 44% in the placebo recipients (P < 0.0001). These differences remained constant throughout the course of the study. In addition to the observed virologic effect of ganciclovir, the frequency of immediately sight-threatening (zone 1) retinitis was also decreased in those patients randomized to ganciclovir. Whereas 17% of ganciclovir recipients experienced zone 1 lesions at the time of diagnosis of CMV retinitis, 39% of persons in the placebo group experienced immediately sight-threatening retinitis (P = 0.02). Thus, study participants randomized to receive ganciclovir appeared to have smaller retinal lesions at the time of diagnosis than those randomized to placebo.

Oral ganciclovir used as prophylaxis was generally well-tolerated when compared to placebo. Gastrointestinal adverse events including diarrhea, nausea, and pancreatitis did not differ significantly between the two groups. Neuropathy and paresthesias also did not differ significantly. Although neutropenia did not differ significantly between the two groups (10% in the ganciclovir group and 6% in the placebo group, P = 0.1), granulocyte-colony stimulating factor (G-CSF) was used significantly more often in the ganciclovir group (24%) than the placebo group (9%) (P < 0.00001). The duration of G-CSF use per

Table 2. New CMV disease incidence at 6-months. Contralateral retinitis by photos or biopsy-proven extraocular disease (Kaplan-Meier estimates)

	IV Alone	Placebo + Implant	P-value
New CMV Disease	17.1%	37.8%	0.001
Extraocular Disease	3.3%	9.7%	0.081
Contralat. Retinitis	19.3%	36.8%	0.013

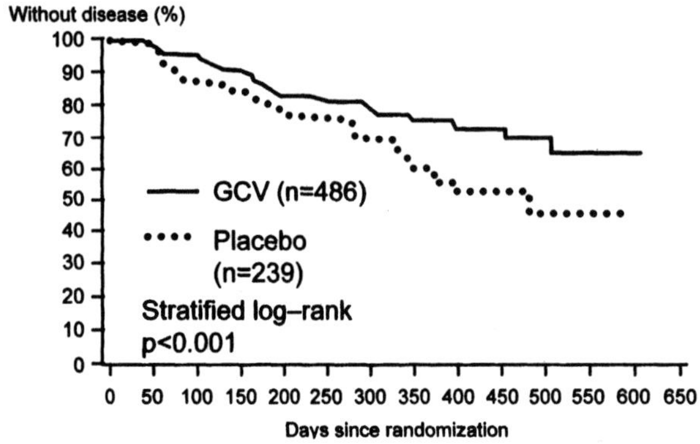

Figure 3. Kaplan-Meier event rates from baseline to development of CMV disease for all randomized patients (from reference 7).

person/year was 47 days for the ganciclovir group and 16 days for placebo recipients. Similarly, although the incidence of anemia was similar among the two groups (5% for ganciclovir recipients and 4% for placebo recipients), overall 14% of the ganciclovir group and 6% of the placebo group used erythropoietin ($P < 0.01$). A slight increase in maximum creatinine clearance was also observed in the ganciclovir group, but dose adjustment was not required because of increases in creatinine clearance.

A number of potential complications have been suggested with the use of oral ganciclovir prophylaxis. Of particular importance is the concern that resistant virus would develop, and thus decrease the efficacy of ganciclovir treatment. Additionally, the development of CMV disease at remote sites such as the CNS or adrenal glands might occur more frequently in persons receiving prophylaxis. To date, most of these complications have not been observed. As would be expected, CMV retinitis followed by gastrointestinal disease were the most common clinical end points observed in persons receiving ganciclovir prophylaxis, with other sites being uncommonly associated with disease. Additionally, in preliminary studies performed by the laboratory of Dr. W.L. Drew, only 3 of 40 CMV strains isolated from persons receiving ganciclovir prophylaxis were resistant to ganciclovir.[8] Because only 11% of persons receiving ganciclovir were culture positive during the study period, it is estimated that ≤1% of persons receiving ganciclovir prophylaxis, to date, have developed resistant virus.

We have also completed studies that have examined the ability of CMV DNA PCR of plasma at baseline to identify those persons at highest risk for development of CMV disease. Plasma samples from 619 study participants at baseline were amplified by qualitative and quantitative PCR.[9, 10] Our findings indicate that 45% of samples were plasma PCR positive at baseline. The 12 month Kaplan-Meier event rates for the PCR negative participants were 14% for the placebo group compared to 1% for ganciclovir recipients ($P < 0.0001$). For participants' qualitative plasma PCR at baseline, there was a Kaplan Meier 12 month event rate of 43% in the placebo group compared to 26% in the ganciclovir group ($P = .017$).The development of CMV disease significantly correlated with the quantity of CMV DNA present in plasma for both the placebo and ganciclovir groups. For each increase in CMV DNA categorized at study entry, there was a corresponding increase in the percentage of study participants who developed CMV disease (Figure 4A and 4B). There was also a correlation between the quantity of CMV DNA in plasma and survival. Study participants randomized to placebo who were PCR negative at baseline had an overall mortality at 12-months of 6.1% compared to 37.7% for those PCR positive; mortality increased progressively with increased baseline CMV DNA copy number (Figure 5A). Similarly, ganciclovir recipients who were CMV plasma PCR negative at baseline had a

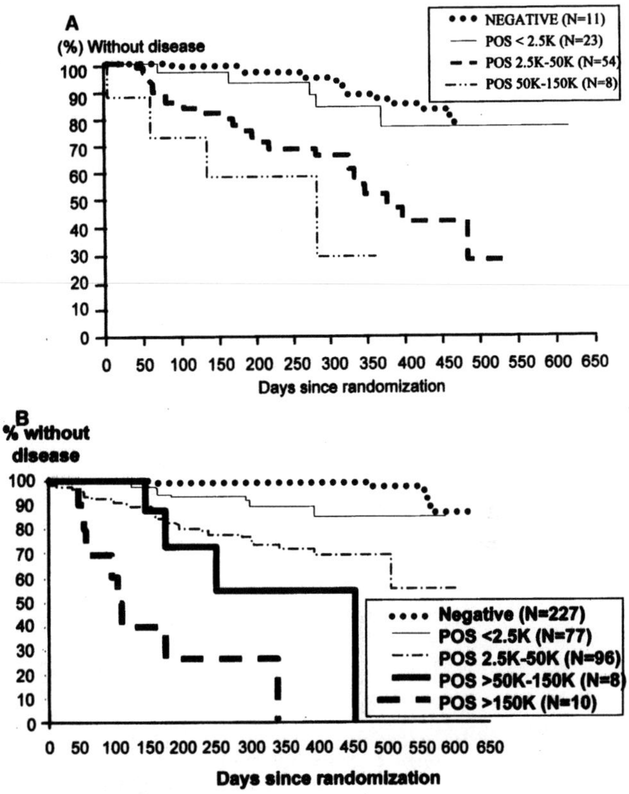

Figure 4. Kaplan-Meier curves from baseline to development of CMV disease for study participants based on CMV DNA status and quantity of DNA in plasma for subjects randomized to (A) placebo and (B) ganciclovir (from references 9 and 10).

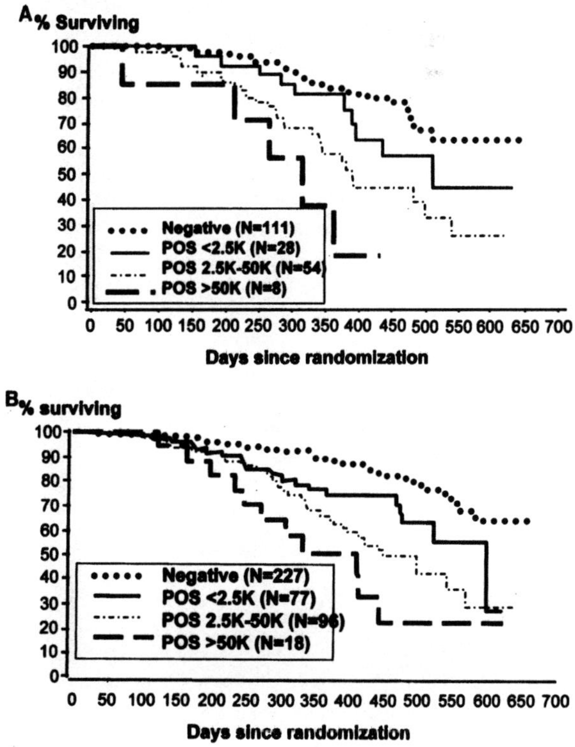

Figure 5. Kaplan-Meier curves from baseline to death for study participants based on CMV DNA status and quantity of DNA in plasma for subjects randomized to (A) placebo and (B) ganciclovir (from references 9 and 10).

12-month mortality rate of 11.4% compared to 32.5% who were PCR positive. Overall mortality progressively increased as the baseline CMV plasma DNA copy number increased (Figure 5B).

CMV PCR positivity was a better predictor of the risk of developing CMV disease than was CD4+ lymphocyte count in this group of patients with advanced AIDS. As assessed using a Cox regression model including both variables applied to patients who received placebo, baseline CMV PCR positivity was associated with a 3.4-fold increase in the risk of developing CMV disease ($P < 0.0001$). CD4+ lymphocyte count showed no significant association with CMV disease risk ($P = 0.47$). Quantified CMV load was also highly effective in predicting the risk of CMV disease and provided additional predictive value relative to qualitative PCR positivity alone. Among patients who were CMV PCR positive at baseline and received placebo, each $\log_{10}$ increase in baseline viral load was associated with a 3.1-fold increase in the risk of developing CMV disease ($P<0.0001$).

CMV PCR positivity also correlated strongly with survival time in a Cox regression analysis. Among patients receiving placebo, the relative risk of death associated with a positive PCR at baseline was 2.5 ($P = 0.0006$). CD4+ lymphocyte counts had no statistically significant association with survival time among these patients ($P = 0.28$). Quantified CMV load also correlated with survival time over and above the correlation noted for CMV positivity. Among patients who were CMV PCR positive at baseline and received placebo, each $\log_{10}$ increase in baseline viral load was associated with a 2.2-fold increase in the risk of death ($P<0.0001$).

The finding that quantification of CMV DNA is predictive of CMV disease and survival is strikingly similar to recent studies in HIV-1 infected individuals demonstrating that single determinations of the quantity of HIV-1 RNA in plasma are predictive of the rate of disease progression in subsequent years. Important criteria for demonstrating the adequacy of a marker for CMV disease have been met: (i) The presence or absence of CMV DNA in plasma is highly predictive of the subsequent development of CMV disease and mortality; (ii) There is a strong correlation between the quantity of CMV DNA present in plasma with the development of CMV disease and survival; and (iii) An active anti-CMV agent, ganciclovir, decreases the incidence of CMV disease in persons positive for plasma CMV DNA.

In summary, oral ganciclovir appears to be useful for maintenance treatment of CMV retinitis and for prevention of CMV disease. The recent development of a valine form of ganciclovir (valganciclovir) that gives equivalent serum levels when taken orally as the 5 mg/kg intravenous dose suggests that ganciclovir when administered orally will have greater efficacy than the current formulation for treatment and prevention of CMV disease.

ACKNOWLEDGMENTS

This research was supported in part by NIH AI-28270, AI-36214 [UCSD Center for AIDS Research] and MH-45294, and Roche Global Development, Palo Alto, CA.

REFERENCES

1. Spector SA, Busch DF, Follansbee S *et al*, Pharmacokinetic, safety, and antiviral profiles of oral ganciclovir in persons infected with human immunodeficiency virus: a phase I/II study. J Infect Dis 1995; 171:1431–7.

2. Drew WL, Ives D, Lalezari JP *et al*, Oral ganciclovir as maintenance treatment for cytomegalovirus retinitis in patients with AIDS. N Engl J Med 1995; 333:615–620.
3. Squires K, Oral ganciclovir for cytomegalovirus retinitis in patients with AIDS: results of two randomized studies. AIDS 1996; 10(suppl4):S13-S18.
4. The Oral Ganciclovir European and Australian Cooperative Study Group, Intravenous versus oral ganciclovir: European/Australian comparative study of efficacy and safety in the prevention of cytomegalovirus retinitis recurrence in patients with AIDS. AIDS 1995; 9:471–477.
5. Lalezari J, Drew L, Jordan C *et al*, *In vivo* anti-CMV activity and safety of oral lobucavir in HIV-infected patients. Abstracts of the 4th Conference on Retroviruses and Opportunistic Infections 1997;Session 31:301.
6. Martin D, Kuppermann B, Wolitz R *et al*, Combined oral ganciclovir (GCV) and intravitreal GCV implant for treatment of patients with cytomegalovirus (CMV) retinitis: a randomized, controlled study. Abstract. 37th Interscience Conference on Antimicrobial Agents and Chemotherapy, Toronto, Canada, September 28 - October 1, 1997.
7. Spector SA, McKinley GF, Lalezari JP *et al*, Oral ganciclovir for the prevention of cytomegalovirus disease in persons with AIDS. N Engl J Med 1996; 334:1491–1497.
8. Drew WL, Miner RC, Crager M *et al*, Prevalence of ganciclovir cytomegalovirus during oral ganciclovir prophylaxis. Abstract. Third Conference of Retroviruses and Opportunistic Infections, January 1996.
9. Spector SA, Wong R, Hsia K *et al*, Plasma Cytomegalovirus (CMV) DNA load predicts CMV disease and survival in AIDS patients. J Clin Invest 1998; 101:497–502.
10. Spector SA, Detection and quantification of human cytomegalovirus (CMV) as a marker for development of CMV disease and survival in patients with AIDS. Antiviral Research 1997; 2:200–205.

12

DEVELOPMENT OF NOVEL BENZIMIDAZOLE RIBOSIDE COMPOUNDS FOR TREATMENT OF CYTOMEGALOVIRUS DISEASE

Jeffrey Chulay,[1] Karen Biron,[1] Laurene Wang,[1] Mark Underwood,[1] Stanley Chamberlain,[1] Lloyd Frick,[1] Steven Good,[1] Michele Davis,[1] Robert Harvey,[1] Leroy Townsend,[2] John Drach,[2] and George Koszalka[1]

[1]GlaxoWellcome Incorporated
Five Moore Drive
Research Triangle Park, North Carolina 27709
[2]The University of Michigan
College of Pharmacy
Ann Arbor, Michigan 48109-1065

Three drugs are currently approved for treatment of cytomegalovirus (CMV) disease: ganciclovir, foscarnet and cidofovir1. Each has reasonably good antiviral activity, but significant toxicities limit their safety and effectiveness, and poor oral bioavailability further limits their clinical utility. In addition, drug resistance can develop during therapy and, because the mechanism of action of each of these drugs depends ultimately on inhibition of CMV DNA polymerase, cross resistance among these drugs can occur. There is a need for safe and effective drugs that are orally bioavailable and directed against novel viral targets.

1. IDENTIFICATION OF BDCRB

Townsend *et al.* synthesized 2-bromo-5,6-dichloro-1-b-D-ribofuranosyl benzimidazole (BDCRB; Figure 1) and its 2-chloro analogue (TCRB) and demonstrated that they are active against CMV at non-cytotoxic concentrations.[2] Both are derivatives of 5,6-dichloro-1-b-D-ribofuranosyl benzimidazole (DRB), a compound whose antiviral activity is likely mediated by inhibition of cellular RNA polymerase II.[3] However, BDCRB does not inhibit CMV RNA synthesis, nor does it inhibit DNA or protein synthesis in CMV-infected cells,[4] indicating that it has a mechanism of action that is different from that of previously identified anti-CMV drugs (Table 1).

Antiviral Chemotherapy 5, edited by Mills *et al.*
Kluwer Academic / Plenum Publishers, New York, 1999.

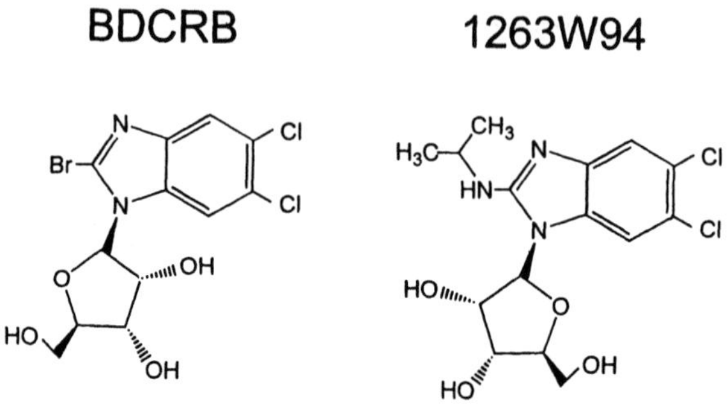

Figure 1. Structures of the novel benzimidazoles, BDCRB and 1263W94.

To investigate the mechanism of action of BDCRB, viral DNA was characterized by pulsed field gel electrophoresis. In the absence of drug, polygenomic concatemeric CMV DNA, which is detectable by 24 hours after infection, is then processed to genome-length DNA up through 120 hours after infection. In the presence of 20 mM BDCRB the processing of concatemeric CMV was strongly inhibited.[4] To discover the locus of action, BDCRB-resistant mutants of CMV were generated by selective drug pressure *in vitro*. Genetic mapping experiments with five mutant strains demonstrated that low level resistance mapped to one amino acid mutation (D344E) and high level resistance mapped to two amino acid mutations (D344E plus A355T) in the UL89 gene.[4] The CMV UL89 gene has sequence similarities to a known ATP-dependent endonuclease (terminase) from bacteriophage T4. These findings suggest that BDCRB prevents viral DNA maturation by interacting principally with the UL89 gene product, and that the UL89 gene encodes an endonucleolytic subunit of the putative CMV terminase. The homologue of UL89 in herpes simplex virus (HSV) has been shown to be an essential gene for viral replication.[5] Studies of a UL89 mutant selected for higher levels of TCRB resistance showed a point mutation in a second target gene, UL56.[6] The UL56 gene product is a DNA binding protein that binds specifically to "pac" sequences in the cleavage site for DNA maturation.[6]

Table 1. Characteristics of Anti-CMV Compounds

	Ganciclovir	Foscarnet	Cidofovir	BDCRB	1263W94
Inhibits DNA synthesis?	Yes	Yes	Yes	No	Yes
Inhibits DNA polymerase?	Yes•	Yes	Yes•	No	No
Inhibits DNA processing?	No	No	No	Yes	No
Site(s) of resistance mutations	UL97, DNA pol	DNA pol	DNA pol	UL89, UL56	Not yet known
Antiviral spectrum	all herpes viruses	all herpes viruses	all herpes viruses	CMV	CMV, EBV
Anti-CMV activity *in vitro*†					
IC_{50} (μM): strain AD169	0.4	19	0.2	0.5	0.1
IC_{50} (μM):clinical isolates	1-7.3	77-170	1.1-1.8	ND§	0.12-0.56

•DNA polymerase is inhibited by phosphorylated metabolites of ganciclovir and cidofovir.

† IC_{50} measured by DNA hybridization assay in MRC-5 cells for strain AD169 and by plaque reduction assay for clinical isolates. Ranges for ganciclovir and 1263W94 are for a set of 10 representative clinical isolates. Ranges for foscarnet and cidofovir are from Smith et al.[15]

§ND = not done.

Table 2. Characteristics of 1263W94

- Potent antiviral activity specific against CMV and EBV
- Novel mechanism of action
- Good oral bioavailability
- Favorable preclinical and early clinical safety profile
- Linear pharmacokinetics over a wide dose range
- Extensive plasma protein binding

Since host cell DNA replication does not involve a DNA maturation step, compounds which inhibit viral DNA maturation should be selective and safe.

BDCRB was highly selective for CMV and had little cytotoxicity *in vitro*. When administered to animals, however, BDCRB was rapidly metabolized to the aglycone (2-bromo-5,6-dichloro benzimidazole), which is inactive against CMV. Thus, because of a poor pharmacokinetic profile, BDCRB did not possess the characteristics of a potentially useful medicine.

2. IDENTIFICATION OF 1263W94

Of the large number of benzimidazole analogues that have been synthesized, 5,6-dichloro-2-isopropylamino-1-b-L-ribofuranosyl-1*H*-benzimidazole (1263W94; Figure 1) has emerged as a promising clinical candidate[7] (Table 2). 1263W94 is a potent antiviral agent *in vitro*, with an IC_{50} approximately 4- to 10-fold lower than ganciclovir[8] (Table 1). It is highly selective, with potent activity against human CMV and Epstein-Barr virus (EBV)[9] but not against murine CMV or several other herpes viruses. 1263W94 showed no extraneous pharmacologic activity in preclinical testing, has good oral bioavailability (absolute oral bioavailability >90% in rats and ~50% in monkeys), and has limited metabolism and an adequately long elimination half-life in laboratory animals.[10] Unlike other anti-CMV drugs, 1263W94 is eliminated primarily by biliary excretion. 1263W94 is extensively bound to human plasma proteins (98.5% binding, primarily to albumin), although a high dissociation constant indicates that the binding is readily reversible. There is limited but measurable penetration of 1263W94 into the brain and vitreous humor of laboratory animals. 1263W94 has a favorable preclinical safety profile in studies conducted to date, with very little cytotoxicity against bone marrow precursor cells *in vitro*; limited toxicity in animals (mild reversible anemia, reversible changes in the gastrointestinal tract); and lack of mutagenicity in a variety of preclinical assays.

Somewhat surprisingly, the mechanism of action of 1263W94 is distinct from both BDCRB and other characterized anti-CMV drugs.[8] Unlike BDCRB, 1263W94 does not inhibit the processing of CMV DNA and is active against BDCRB-resistant virus. Although 1263W94 inhibits DNA synthesis, this inhibition differs from that of ganciclovir, foscarnet and cidofovir, as it does not appear to involve DNA polymerase. 1263W94 is not phosphorylated in CMV-infected (or uninfected) cells; and neither 1263W94 nor the chemically synthesized mono- or tri-phosphates of 1263W94 inhibit purified viral (or host) DNA polymerase *in vitro*.[8] In addition, 1263W94 is active against viral mutants that are resistant to ganciclovir, foscarnet or cidofovir. The mechanism of action of 1263W94 is viral specific, since viral mutants resistant to 1263W94 can be generated by selective drug pressure

Table 3. Pharmacokinetic Parameters of 1263W94

Dose (mg)	Subjects*	N	Mean values C_{max} (μg/mL)†	t_{max} (h)	$AUC_{0-\infty}$ (h•μg/ml)	$t_{1/2}$ (h)
100	NHV	10	3.3	2.3	16	3.0
	HIV	12	3.4	1.6	16	4.0
400	NHV	10	17	1.8	98	3.9
	HIV	13	18	1.4	83	4.1
1600	NHV	10	49	1.9	437	4.8
	HIV	12	48	2.8	335	4.8

* NHV = normal healthy volunteers; HIV = HIV-infected volunteers.
† To convert μg/mL to μM, multiply by 2.66.

in vitro.[8] Experiments are in progress to define the locus of resistance and clarify the mechanism of action.

3. CLINICAL EVALUATION OF 1263W94

Because of its favorable preclinical safety profile, including lack of genotoxic potential in preclinical assays, 1263W94 was initially evaluated in a Phase I study in normal healthy volunteers.[11] Escalating single doses of 1263W94, from 50 mg to 1600 mg, were administered at weekly intervals to a total of 13 subjects. At each dose, ten subjects were randomly assigned to receive active drug and two to receive placebo. The drug was well tolerated and, except for one subject who withdrew for personal reasons and was replaced, all subjects completed the full dose escalation. The most common adverse event (altered taste) was reported with dose-dependent frequency and intensity, but was not dose-limiting. Other adverse events were uncommon and did not appear to be dose-dependent. After oral administration, 1263W94 was rapidly and well absorbed, with plasma total drug concentrations well above the *in vitro* IC_{50}. Pharmacokinetic analysis demonstrated nearly linear increases in C_{max} and AUC with increasing doses to 1600 mg (Table 3).

A subsequent Phase I study was conducted in HIV-infected volunteers.[12] Escalating single doses of 1263W94 (100, 400, 800 and 1600 mg) were administered at weekly intervals to a total of 17 subjects. At each dose, 12 subjects were randomly assigned to receive active drug and four to receive placebo. The safety and tolerability of 1263W94 in this second Phase I study were the same as in the first study (see above). Pharmacokinetic parameters in HIV-infected subjects were similar to those in healthy volunteers (Table 3), indicating that neither HIV-infection nor the concomitant medications taken by these HIV-infected subjects altered the pharmacokinetics of 1263W94.

Based on these encouraging results, a repeated dosing study is being conducted in HIV-infected volunteers with asymptomatic excretion of CMV in semen and urine. 1263W94 will be administered for 28 days (as divided doses two or three times daily) to cohorts of volunteers, with successive cohorts receiving increasing total daily doses of active drug or placebo.

4. FUTURE STUDIES

Control of CMV retinitis depends on adequate delivery of anti-CMV drugs to the retina, which can be reflected by drug concentrations in vitreous humor.[13] Because

1263W94 is extensively bound to plasma proteins, and penetration of 1263W94 into the vitreous humor of animals is limited, it will be important to determine if this drug can penetrate into the human eye. The availability of patients who choose to be treated with ganciclovir ocular implants,[14] and the ability to obtain vitreous samples safely from such patients, has enabled a study to measure vitreous drug concentrations in volunteers treated with 1263W94 for 7 days prior to surgical insertion of an ocular implant. Results of this study should help determine directions for further investigation of 1263W94.

5. SUMMARY

Benzimidazole ribosides are a new class of compounds with novel mechanisms of action against CMV. One compound in this series, BDCRB, inhibits CMV DNA processing by the UL89 gene product (putative terminase), but rapid metabolism to an inactive compound makes it unsuitable for development as a medicine. Another benzimidazole analogue, 1263W94, has many characteristics that make it an attractive candidate for development, including high potency *in vitro*, selectivity, good oral bioavailability, and lower toxicity than therapies currently available for treatment of CMV disease. Initial clinical trials have provided encouraging results, including good tolerability and linear pharmacokinetics over a wide dose range. Ongoing and planned clinical trials that will study the safety and tolerability of repeated dosing and evaluate the *in vivo* antiviral activity and ocular penetration of 1263W94, will help to determine the potential of this drug as an improved therapy for CMV disease.

REFERENCES

1. Jacobson MA. Treatment of cytomegalovirus retinitis in patients with the acquired immunodeficiency syndrome. New England Journal of Medicine 1997; 337:105–14.
2. Townsend LB, Devivar RV, Turk SR, Nassiri MR, Drach JC. Design, synthesis, and antiviral activity of certain 2,5,6-trihalo-1-(beta-D-ribofuranosyl)benzimidazoles. Journal of Medicinal Chemistry 1995; 38:4098–105.
3. Yankulov K, Yamashita K, Roy R, Egly JM, Bentley DL. The transcriptional elongation inhibitor 5,6-dichloro-1-beta-D-ribofuranosylbenzimidazole inhibits transcription factor IIH-associated protein kinase. Journal of Biological Chemistry 1995; 270:23922–5.
4. Underwood MR, Harvey RJ, Stanat SC, *et al.* Inhibition of HCMV DNA maturation by a benzimidazole ribonucleoside is mediated through the UL89 gene product. Journal of Virology 1998; 72:717–25.
5. Poon AP, Roizman B. Characterization of a temperature-sensitive mutant of the UL15 open reading frame of herpes simplex virus. Journal of Virology 1993; 67:4497–503.
6. Krosky PM, Feng KW, Townsend LB, Drach JD. Resistance of HCMV to benzimidazole ribonucleosides results from mutations in both UL89 and UL56, 22nd International Herpesvirus Workshop, San Diego, 2–8 Aug 1997.
7. Koszalka GW, Chamberlain SD, Harvey RJ, *et al.* Benzimidazoles for the treatment of human cytomegalovirus, 9th International Conference on Antiviral Research, Fukushima, Japan, 19–24 May 1996.
8. Biron KK, Davis MG, Stanat SC, *et al.* Antiviral activity and mechanism of action of 1263W94, a benzimidazole riboside inhibitor of human cytomegalovirus, 36th Interscience Conference on Antimicrobial Agents and Chemotherapy, New Orleans, 15–18 Sep 1996.
9. Zacny VL, Davis MG, Chamberlain SD, Townsend LB, Biron KK, Pagano JS. Activity against Epstein-Barr virus of benzimidazole riboside 1263W94 (5,6-dichloro-2(isopropylamino)-1-b-L-ribofuranosyl-1-*H*-benzimidazole), 37th Interscience Conference on Antimicrobial Agents and Chemotherapy, Toronto, 28 Sep-1 Oct 1997.

10. Frick LW, Hsieh AY, Good SS. Pharmacokinetics and metabolism in rats and monkeys of 1263W94, a potent and selective inhibitor of human cytomegalovirus, 36th Interscience Conference on Antimicrobial Agents and Chemotherapy, New Orleans, 15–18 Sep 1996.
11. Wang LH, Yogendran L, Weller S, Sallabank P, Wiggs R, Bye A. A phase I trial evaluating the tolerability and pharmacokinetics of 1263W94 following single oral administration of escalating doses in normal healthy volunteers, 36th Interscience Conference on Antimicrobial Agents and Chemotherapy, New Orleans, 15–18 Sep 1996.
12. Wang LH, Peck R, Chan PQ, Youle M, Eaves J, Bye A. A phase I tolerability and pharmacokinetics trial of 1263W94, a novel anti-HCMV agent, in HIV-infected volunteers, 4th Conference on Retroviruses and Opportunistic Infections, Washington, DC, 22–26 Jan 1997.
13. Arevalo JF, Gonzalez C, Capparelli EV, *et al.* Intravitreous and plasma concentrations of ganciclovir and foscarnet after intravenous therapy in patients with AIDS and cytomegalovirus retinitis. Journal of Infectious Diseases 1995; 172:951–956.
14. Musch DC, Martin DF, Gordon JF, Davis MD, Kuppermann BD. Treatment of cytomegalovirus retinitis with a sustained-release ganciclovir implant. The Ganciclovir Implant Study Group. New England Journal of Medicine 1997; 337:83–90.
15. Smith IL, Cherrington JM, Jiles RE, Fuller MD, Freeman WR, Spector SA. High-level resistance of cytomegalovirus to ganciclovir is associated with alterations in both the UL97 and DNA polymerase genes. Journal of Infectious Diseases 1997; 176:69–77.

13

FAMCICLOVIR/PENCICLOVIR

Stephen L. Sacks and Bruce Wilson

Department of Pharmacology and Therapeutics
Faculty of Medicine
The University of British Columbia and
Viridae Clinical Sciences Incorporated
Vancouver, British Columbia, Canada

Since the early 1980's, the gold standard for treatment for both varicella zoster (VZV) and herpes simplex virus (HSV-1, HSV-2) infections has been acyclovir. In general, the clinical significance of acyclovir's impact on clinical outcomes has been directly proportional to the severity of the natural history of disease. Unfortunately, acyclovir's limitations are governed by poor oral absorption with a bioavailability of about 15% along with maintenance of plasma antiviral activity in accordance with a short plasma T1/2 of 2.5 hours. Accordingly, novel approaches to address these limitations were required. Famciclovir is the oral prodrug of the antiviral nucleoside deoxyguanosine analog, penciclovir. Famciclovir is characterized by its high rate of absorption and rapid conversion to the active compound, penciclovir. Penciclovir triphosphate, the active intracellular nucleotide has a prolonged intracellular T1/2. Studies performed to date have shown these agents to be both effective and safe and in most cases, more conveniently dosed compared with acyclovir. Famciclovir has shown significant benefits in both active treatment and suppression of genital herpes infections in normal and immunocompromised hosts. Moreover, in the US, the currently approved dosing regimen of famciclovir for the treatment of herpes zoster (shingles) is 500 mg tid for 7 days, while for episodic treatment of recurrent genital herpes, it is 125 mg bid for 5 days.[1] Famciclovir is also effective for treatment of first episode genital herpes at a dose of 250 mg tid for 5 or 10 days and is used in chronic suppression of recurrences at a dose of 250 mg bid. Penciclovir is being developed in its own right as both an intravenous and topical formulation. Parenteral penciclovir (5 mg per kg intravenously [IV] every 8 hours for 5 days) is effective against recurrent genital herpes, although this is an impractical regimen for this indication.[2] More importantly, penciclovir IV (5 mg/kg) every 12 hours was also equally effective compared with penciclovir IV 8 hourly or acyclovir IV (5 mg/kg) every 8 hours in the treatment of herpes simplex infections in immunocompromised hosts, offering a significant dosing reduction in the hospital setting.[3] Penciclovir 1% cream is also effective in reducing the duration and severity of herpes labialis when used early and frequently (every 2 hours) at the onset of either symp-

Antiviral Chemotherapy 5, edited by Mills *et al.*
Kluwer Academic / Plenum Publishers, New York, 1999.

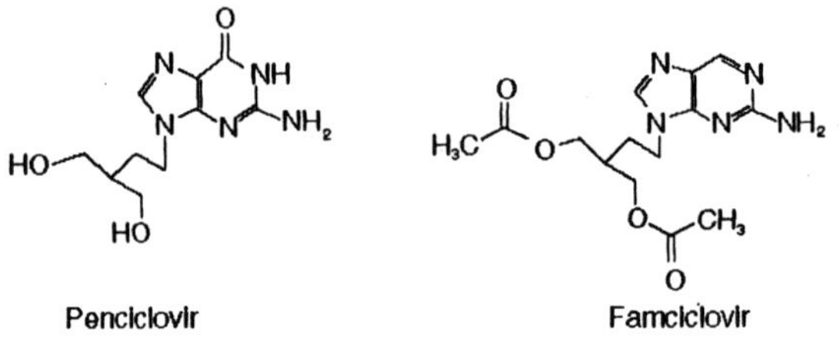

Figure 1. Chemical structures of the nucleoside analogue, penciclovir, and its oral prodrug, famciclovir (2-[2-(2-amino-9H-purin-9-yl)ethyl]-1,3-propanedioldiacetate).

toms or signs of an active episode,[4] and is the treatment of choice for herpes labialis according to the Medical Letter. In fact, it is the only antiviral which has clearly shown benefits in this clinical setting in large, placebo-controlled trials and is the only product licensed for this indication in the US.

1. STRUCTURE AND MECHANISM OF ACTION

Famciclovir is the diacetyl ester of 6-deoxy penciclovir (a synthetic acyclic guanine derivative). Like acyclovir and ganciclovir, penciclovir is an acyclic guanosine analog which inhibits herpes virus DNA synthesis (Figure 1).[5,6] Although active intravenously, penciclovir itself is very poorly absorbed orally. It was found in animal studies that its dipropionyl and diacetyl 6-deoxy derivatives (BRL 43599 and famciclovir, respectively) gave much improved blood levels of penciclovir.[7–9]

There are some qualitative differences in the mechanism of action between penciclovir and acyclovir in terms of rates of phosphorylation, stability and concentration of the triphosphate derivatives, and affinity for viral DNA polymerase. Penciclovir freely enters both virus-infected and uninfected cells. However, antiherpesviral selectivity relies on rapid phosphorylation to the active form, penciclovir triphosphate, in virus-infected cells only. Viral thymidine kinase is responsible for the rate-limiting initial phosphorylation to penciclovir-monophosphate and conversion to the triphosphate is via other cellular enzymes.[5, 6, 10] Acyclovir is converted to acyclovir triphosphate in a similar manner and has similar selectivity, but there are differences in both the rate of phosphorylation and the intracellular stability of nucleoside triphosphate between the two compounds.[5] Compared with acyclovir, penciclovir is preferentially phosphorylated in HSV- and VZV-infected cells. Further, penciclovir triphosphate enjoys a very prolonged intracellular half-life in infected cells (7 to 20 hours versus 0.7 to 1 hour for acyclovir triphosphate). However, more penciclovir triphosphate is required for inhibition of HSV DNA polymerase compared with acyclovir triphosphate.[6,10]

2. PHARMACOKINETICS AND SAFETY

After oral administration, famciclovir undergoes rapid and extensive absorption in the upper intestine and is quickly metabolized in the intestinal wall and liver to the active compound penciclovir.[11,12] Little or no famciclovir is detected in plasma or urine. Inactive metabolites formed include 6-deoxy penciclovir, monoacetylated penciclovir, and 6-deoxy monoacetylated penciclovir.[12] Acyclovir has a 10–20% bioavailability,[13] while oral penciclovir has a bioavailability of only about 5%. However, the bioavailability of penciclovir after oral administration of famciclovir is about 77%.[7,8] Compared with acyclovir, penciclovir is far more extensively phosphorylated by HSV- and VZV-infected cells.[5,6] Furthermore, penciclovir triphosphate has a very prolonged intracellular half-life in infected cells

(7 to 20 hours versus 0.7 to 1 hour for acyclovir triphosphate). Penciclovir is not metabolized, but is eliminated unchanged in the urine.[9] Plasma half-lives are essentially identical for both penciclovir and acyclovir (2.5 hours).

The pharmacokinetics of penciclovir remain largely unaltered in the elderly, making it unnecessary to adjust the dosage in these patients.[14,15] In a study of patients with renal impairment, urinary excretion of penciclovir decreased in proportion to the decrease in estimated creatinine clearance leading to a recommendation that the dosage interval of famciclovir should be prolonged from 8 hours to 12 hours for patients with moderate renal impairment, and to 24 hours for patients with severe renal impairment.[16] There is no evidence for the need to alter dosage levels of famciclovir in those with well-compensated hepatic impairment.[17] The pharmacokinetics of famciclovir have not been studied in patients with severe uncompensated hepatic impairment.

In three studies that investigated the effects of food on the pharmacokinetics and bioavailability of famciclovir,[18–20] no clinically significant differences were observed in healthy male volunteers who received famciclovir (250 and 500 mg single doses) 30 minutes after food compared to those who received famciclovir 2 hours before food. When famciclovir 250 mg was administered 30 minutes after food, Cmax was reduced by only 20%.

Famciclovir has a good safety profile. Studies in healthy male volunteers showed no evidence of a clinically significant pharmacokinetic interaction between famciclovir and cimetidine, theophylline, allopurinol or digoxin.[21] Safety data from all clinical studies, to date, consistently demonstrate that famciclovir has been well tolerated by patients with hepatitis B, herpes zoster or genital herpes infection with a safety profile comparable to placebo, unrelated to the degree of underlying immunocompetence.

3. SPECTRUM OF ANTIVIRAL ACTIVITY

Penciclovir is similar to acyclovir in terms of spectrum of antiherpesviral activity with 50% inhibition of viral replication (EC50) concentrations similar to those found for acyclovir in both laboratory and wild-type strains of HSV and VZV.[10,11] Penciclovir has limited activity against cytomegalovirus (CMV), but retains good activity against Epstein Barr virus.[11] When a range of different types of human cells were grown in the presence of either penciclovir or acyclovir, both are essentially non-toxic at clinical concentrations.

The vast majority of acyclovir-resistant HSV and VZV clinical isolates will be clinically and virologically cross-resistant to penciclovir. However, cross-resistance is not universal and examples of thymidine kinase (TK) - altered and DNA polymerase acyclovir-resistant strains have been described which are fully susceptible to penciclovir.[22] When the activities of penciclovir and acyclovir were tested against thymidine kinase-deficient strains, more were susceptible to penciclovir than to acyclovir although only at high EC50 values.[23]

4. CLINICAL APPLICATIONS IN THE IMMUNOCOMPETENT HOST

4.1. Herpes Zoster

The efficacy and safety of famciclovir in the treatment of varicella zoster (VZV) infections and the management of HSV infections have been studied in a number of multicenter clinical trials worldwide. In an early multicenter, randomized, double-blind clinical trial of famciclovir against VZV, 545 patients were treated with 250, 500 and 750 mg tid or acyclovir 800 mg five times daily. Each famciclovir dose was at least as effective as acyclovir for cutaneous healing but accelerated resolution of pain. With early treatment, there was approximately 1.5–1.8 fold acceleration in loss of pain over acyclovir, confirming the benefit of early treatment.[24] In a placebo-controlled trial of famciclovir in immunocompetent patients with VZV, Tyring et al showed acceleration of the events of cutaneous healing.[25] Patients were assessed both for acute phase healing, acute phase pain, for post-herpetic neuralgia (PHN; pain at or after healing in this study). Both doses of famciclovir were equally effective at reducing the durations of viral shedding and times to healing, as well as acute phase pain. Moreover, the duration of post-herpetic neuralgia was reduced from 128 days in the placebo group to 62 and 55 days in the 500 and 750 mg famciclovir groups. Older patients, at increased risk of developing post-herpetic neuralgia, experienced an even greater treatment response. However, recent recommendations for analyzing pain in trials of antiviral agents in this setting, suggest that specific timepoints be identified for analysis rather than adjusting the onset of PHN duration based on rash healing.[26] The study was analysed again according to those recommendations and demonstrated that the famciclovir benefit against PHN increases with time following healing, suggesting that the benefit is extended most to those patients with prolonged PHN.[27] Famciclovir is approved worldwide for the treatment of herpes zoster infection, at a dose of 250 mg tid (Europe) or 500 mg tid (US/Canada). A single daily dose of 750 mg has also been approved in the UK. PHN benefits have been studied most at the 500 mg tid dose. Treatment of herpes zoster with famciclovir has been shown to be cost-effective.[28] The patients most likely to benefit from therapy are those above the age of 50, those who present early in the course of the rash (<72 hours but <48 hours is even better), and probably those who present with severe pain or severe rash, regardless of age.[27]

4.2. Genital Herpes

The efficacy of famciclovir in the treatment of recurrent genital herpes has been reported in two double-blind, placebo-controlled clinical trials.[29,30] In one trial, treatment with famciclovir was initiated in the clinic and, in the second trial, famciclovir administration was initiated by the patient. In each study, famciclovir 125 mg, 250 mg, and 500 mg twice daily for 5 days were compared with placebo. In both studies, treatment was initiated within 6 hours of lesion/prodrome or signs/symptoms onset. In the clinic-initiated study,[29] all doses of famciclovir significantly reduced the times to cessation of viral shedding (p=.0001 for all dose groups based on estimated hazard ratios). Median times to lesion healing were 4.4, 4.0, and 3.9 days for the 125 mg, 250 mg, and 500 mg famciclovir groups, respectively, compared with 5.0 days for the placebo group (p<.001 for all dose groups based on estimated hazard ratios.[29] In addition, treatment with famciclovir at all three doses significantly reduced time to loss of lesion tenderness, pain, itching, burning, and tingling compared with placebo (p<.001).

In the patient-initiated study, times to cessation of viral shedding (p=.00001 for all dose groups based on estimated hazard ratios) were significantly reduced for each dosage group compared with placebo.[30] Similarly, times to loss of all symptoms were significantly reduced for the 125 mg, 250 mg, and 500 mg famciclovir groups (p<.05 based on estimated hazard ratios) compared with the placebo group (Figure 2).

In both trials, there were few differences noted among the three dosage groups. Treatment benefits against the viral shedding of episodic recurrences were shown in each

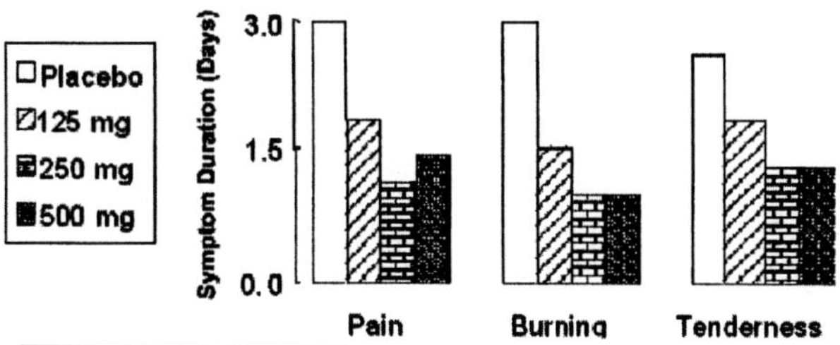

Figure 2. Duration (days) of moderate-to-severe lesion-associated symptoms during episodic (patient-initiated) treatment with famciclovir or placebo.[30]

study. Treatment also interrupted the natural history of lesion evolution by hastening times to lesion healing. Moreover, famciclovir reduced the duration of genital herpes-associated symptoms. A meta-analysis of these two trials including 775 patients was also performed and reconfirmed the virological and clinical benefits of famciclovir as shown in the individual trials.[31] Adverse events most commonly reported by both the famciclovir and placebo recipients included headache and nausea. These occurred at a similar incidence for both groups. In addition, no clinically important laboratory abnormalities were associated with administration of famciclovir.

Following 15 years' experience with the safe use of chronic antiherpesviral therapy, and in the setting of increasing awareness of asymptomatic shedding and its likely role in transmission, clinicians are in general moving to more widespread use of chronic suppressive therapy. The latter route has the advantage of reducing the frequency of both symptomatic and asymptomatic episodes. With acute, episodic therapy, the patient generally sustains a symptomatic episode with reduction of duration of lesions and concomitant discomfort. However, by definition this approach fails to address the problem of asymptomatic shedding. Clinical decision-making now requires specifically tailoring the approach to therapy. The physician should consider disease severity and frequency, the patient's lifestyle and relationship(s), their ability to understand the requirements of proper dosing and compliance, and perhaps most importantly, their own preference for management of their infection. For example, a patient in a herpes concordant monogamous relationship will be unconcerned about asymptomatic shedding, while a person with low to moderate outbreak frequency may wish to be on suppression in a herpes discordant relationship. The old formulas requiring 6 to 10 episodes per year as a minimum for commencing suppressive therapy no longer apply.

Famciclovir has been very effective in the management and control of genital herpes recurrences through suppressive therapy. Mertz, et al.[32] evaluated the efficacy and safety of oral famciclovir in a US placebo-controlled, dose-ranging trial (Figure 3). In this study, 375 women 18 years of age or older with a history of 6 or more episodes of genital herpes during 12 of the last 24 months (absent suppressive therapy) were treated for 4 months with 125 mg qd or bid, 250 mg qd or bid, 500 mg qd, or placebo. The median time to first recurrence was 82 days in placebo recipients, compared with 114 days in women receiving famciclovir, 125 mg qd, and more than 120 days in the other treatment groups. The time to the first clinical recurrence was significantly prolonged in women treated with famciclovir, 125 mg bid (hazard ratio, 1.8; 95% confidence interval, 1.0–3.0; $P = .03$), and in those treated with famciclovir, 250 mg twice daily (hazard ratio, 3.6; 95% confidence interval, 1.9–6.9; $P < .001$). Famciclovir was very well tolerated. The dose of 250 mg was chosen as the optimal suppressive regimen.

This study was extended to 12 months' suppression in both genders by Diaz-Mitoma et al.[33] who studied long-term suppression with famciclovir (Figure 4). In this double-

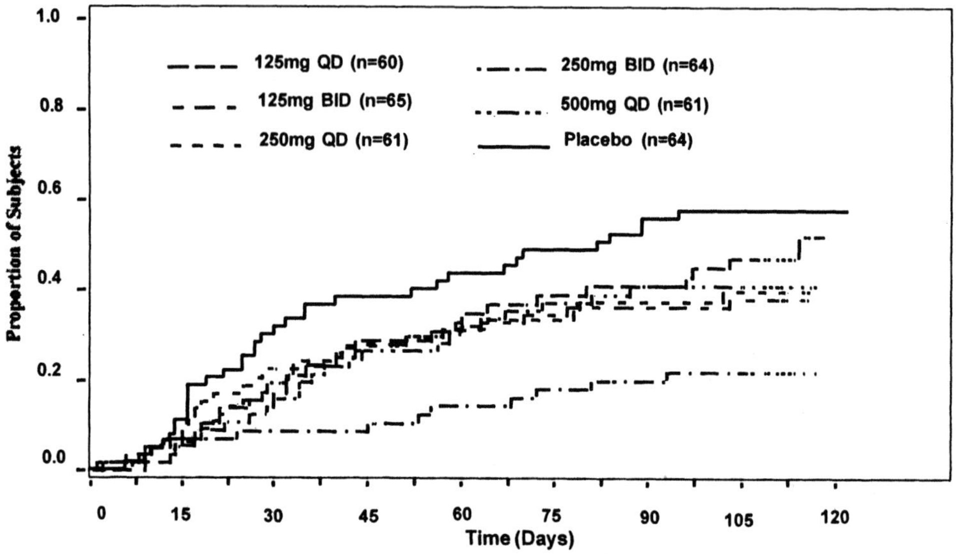

Figure 3. Time to first clinically confirmed recurrence of genital herpes on varying doses of famciclovir and placebo. Twice daily famciclovir 250 is the optimal dosing regimen.[32]

blind, randomized, placebo-controlled, multinational trial, 455 patients with frequently recurring genital herpes (6 episodes per year) were treated with famciclovir (125 mg tid, 250 bid, 250 tid) or placebo. Results showed that 61% of the patients experienced >12 lesional episodes of genital herpes per year, a modestly severe group. Of the patients on famciclovir (all doses), 77% were free from HSV lesional recurrences at 6 months compared to fewer than 30% on placebo. At 12 months, more than 70% of patients on famciclovir were recurrence-free compared with 21% on placebo. These differences were highly statistically significant. Famciclovir 250 mg bid was as effective as both tid regimens and adverse effects were, again, comparable to placebo.

Famciclovir has also been shown to reduce both symptomatic and asymptomatic viral shedding from the external genitalia/perineal/perianal and from internal cervicovaginal areas in women. Sacks, et al.[34] studied women with moderately frequent genital recurrences of type 2 herpes for 112 days using tid famciclovir 125 mg, 250 mg, or placebo in connection with daily self-cultures and symptom diaries. Of 178 women randomized, 177 took medication and collected swabs. Asymptomatic shedding was found at some point

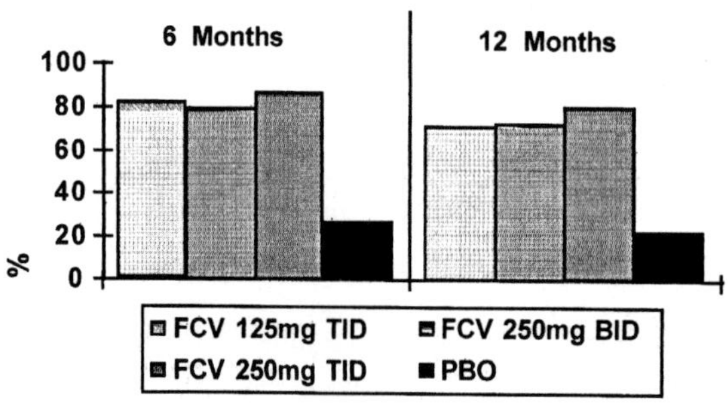

Figure 4. The proportion of subjects remaining free of genital herpes recurrences at each dose of famciclovir or placebo. All comparisons are statistically significant (P<0.001) (Reference: 33).

from any site in 19/60 (32%) women taking 125 mg tid, 16/59 (27%) taking 250 mg tid, and 38/58 (66%) taking placebo. Rates of asymptomatic shedding from any site were reduced from 3.1% of days in placebo recipients to 0.52%, 0.41%, in 125 mg and 250 mg tid recipients, respectively ($P < 0.0001$). Symptomatic viral shedding was similarly reduced from 5.5% to 0.72% and 0.19%, respectively ($P < 0.0001$). Median times to first positive culture (both symptomatic and asymptomatic were only 19 days in placebo recipients compared with 105, and more than 112 days in the 125 mg and 250 mg tid recipients, respectively ($P < 0.0001$). The different dosage formats were not statistically different, overall, although the higher dose was somewhat more effective for symptomatic shedding. The discrepancy of dose effect between symptomatic and asymptomatic shedding suppression is unexplained. Again, famciclovir was well tolerated.

Asymptomatic shedding was also studied in a Vancouver cohort of 16 men receiving famciclovir 250 mg bid compared with placebo[35] in a subset of patients undergoing year-long suppression to determine safety and effects on semenology parameters of this treatment. In the asymptomatic shedding study, the ability of famciclovir to reduce asymptomatic shedding reduced from 1.09% of days to 0.08% of days ($P<0.05$ compared with placebo). Although the cohort of men was smaller and the background shedding rates for men lower, this study demonstrated that men also receive the shedding benefit, and further, that the approved suppression dose of 250 mg bid is also effective in suppressing asymptomatic shedding. Semenology parameters were unaffected by famciclovir compared with placebo in this one year suppression study.[36]

First episode disease has also been studied extensively with famciclovir. It has been compared with acyclovir in three double-blind studies.[37–39] Famciclovir in three dosages (125 mg, 250 mg, 500 mg) tid for 10 days as well as famciclovir 250, 500, or 750 mg tid for 5 days was compared with acyclovir 200 mg 5 times daily. In each instance, famciclovir was equivalent to acyclovir. No apparent differences could be identified between treatment for 5 days and treatment for 10 days, although these study durations were not directly compared. No differences in times to cessation of viral shedding, complete healing, or loss of any or all symptoms were identified. The recommended dose for first episode genital herpes is famciclovir 250 mg tid. Famciclovir has been shown to be superior to valaciclovir in several mouse model studies from Hugh Field's group from Cambridge.[40–43] He has demonstrated a significant reduction in quantitative viral latency burden using famciclovir (but not valaciclovir) within 5 days of murine infection. It is not clear whether these findings have any clinical relevance, however. Stemming from these observations, new clinical studies are underway in first episodes to determine whether high dose famciclovir used early in first episode type 2 genital herpes can modify the subsequent recurrence rates in patients. No data are yet available from these trials and no therapeutic decisions would be appropriate based on these studies until clinical data become available.

4.3. Herpes Labialis

Despite many attempts at treatment of active Herpes Labialis infection, prior to studies with penciclovir cream, no therapy had resulted in consistent benefits in clinical trials enough to warrant routine episodic use.[45] Spruance et al.[46] have reviewed the factors which led to lack of efficacy in past trials, including late onset of treatment, lack of objective measurements of disease efficacy parameters, lack of placebo controls, failure to penetrate skin, and small patient numbers.

Penciclovir, the active antiviral nucleoside catabolite of famciclovir was studied as a 1% topical cream preparation in herpes labialis in two large multicenter trials. The cream preparation is well-tolerated by patients.[46] This cream was applied for 4 days in two randomized, double-blind, placebo-controlled, patient-initiated trials. The patients selected were all immune competent and had no other health problems. Patients were instructed to begin treatment within 1 hour of the first sign or symptom of a new herpes labialis episode by applying the cream every 2 hours while awake (6–9 times daily).

In the US multicenter study (31 centers; 1573 treated patients), Spruance et al[47] demonstrated that penciclovir-treated patients experienced a significantly shorter time to healing of classical lesions, with a 0.7 day difference (4.8 versus 5.5 days; $P<0.001$). In addition, symptoms were reduced, as demonstrated by reduction of duration of lesion pain (3.5 versus 4.1 days; $P<0.001$), and the cessation of viral shedding ($P<0.003$). As lesions were not aggressively probed for culture in this study, the viral shedding data are not as robust as they might have been. It is interesting to note that the difference in the days to loss of classical lesions between penciclovir treated patients and placebo treated patients increased as more of the lesions reached the end point (loss of crust). These data suggest that penciclovir may have been truncating the severity of lesions thereby preventing more severe attacks. However, more data are needed to sustain this hypothesis. Surprisingly, patients who were treated early in their episode (prodrome, erythema stages) received the same benefit in all three measures as those treated late (papule, vesicle stages). This was not predicted by previous studies. The safety and adverse effects profile for penciclovir cream was comparable to placebo.

A parallel multinational (Euro-Canadian) trial reported by Raborn et al.[48] was performed with the same protocol design (1484 treated patients). In this study, penciclovir-treated individuals lost their classical lesions 29% faster than placebo recipients (based on hazard ratios; $P=0.0001$) and resolved pain 32% faster ($P=0.0001$), while cessation of viral shedding was 47% faster ($P<0.0002$). A meta-analysis confirmed the virtual identity of the trials and further strengthened the results.

Based upon these trials, penciclovir cream became the first treatment available for episodic use in herpes labialis in the US. The Medical Letter also chose penciclovir cream as the treatment of choice for recurrences of herpes labialis, although patients who suffer from frequent recurrences of oral-labial herpes infections, or who for cosmetic or occupational reasons wish to suppress their outbreaks, or who know their trigger in advance of exposure (e.g., sun exposure on skiing or sunbathing), will benefit from chronic suppression and use of sun-blocking agents (49,50).

4.4. Famciclovir/Penciclovir in Immunocompromised Patients

Famciclovir has been shown to be effective in managing HIV in immunocompromised patients such as those infected with HIV, those undergoing bone marrow or solid organ transplants, and those undergoing chemotherapy or radiotherapy for the treatment of cancer or leukemia. This patient population is uniquely susceptible to developing resistance to antiviral therapy. In these conditions, famciclovir is equivalent in efficacy with oral acyclovir, albeit far more conveniently dosed.

Famciclovir has been used as acute treatment for HIV in the HIV-infected patients at a dose of 500 mg bid for 7 days.[51] In this double-blind, double-dummy, multicenter, multinational, randomized, clinic-initiated trial, famciclovir 500 mg bid was initiated within 48 hours of lesion onset. Approximately two-thirds of the patients had anogenital infection, while the remainder were suffering from orofacial infection. New lesion formation was reduced by both treatments. Famciclovir was found to be equivalent to acyclovir 400 mg five times daily at reducing time to healing, cessation of symptoms, and cessation of viral shedding, regardless of CD4+ count or lesion location.

Famciclovir studies in HIV-positive patients were further extended in genital herpes to suppression and observations of asymptomatic shedding by Schacker, et al.[52] This study was double-blind, placebo-controlled, with a crossover design. Specifically, famciclovir 500mg bid was used for 8 weeks followed by a washout period of one week, followed by placebo bid for 8 weeks. Patients were randomized to receive either placebo or active drug first. They were followed with daily home cultures (oral, genital, perirectal), and diary card assessments (symptoms, genital herpes lesions). Total herpes simplex virus type 2 shedding was reduced from 9.7% of days to 1.3% of days by active therapy (P=0.001), while asymptomatic shedding was reduced from 5.1% of asymptomatic days to 1.2% of asymptomatic days (P=0.02). It is important to note that famciclovir's clean adverse event profile was maintained throughout all of the active episode and suppression trials. No cases of hemolytic uremic syndrome (HUS) or thrombotic thrombocytopenic purpura (TTP) were observed.

Acute intravenous therapy of herpes simplex virus disease in immunocompromised patients was also studied with penciclovir, the nucleoside metabolite of famciclovir.[33] In this study, approximately one-fourth of the subjects were in the post-transplantation period, while the remainder suffered from hematological malignancies (leukemia, lymphoma, multiple myeloma, myelodysplastic syndrome, aplastic anemia and bone marrow). Penciclovir 5mg/kg intravenously was compared with acyclovir 5mg/kg intravenously for 7 days therapy in a multicenter, randomized, double-blind controlled trial. Four infusions were given per day to maintain the blind to the clinicians. The pharmacist was unblinded. Specifically, these studies sought to determine whether penciclovir's prolonged intracellular half-life could be capitalized upon to reduce the frequency of intravenous dosing. Penciclovir was tested both as 12 hourly and 8 hourly infusions compared with acyclovir standard dose of 8 hourly. Using the efficacy parameters of prevention of new lesion formation, time to healing, time to loss of viral shedding and loss of pain/symptoms, all treatment groups experienced identical outcomes. Accordingly, penciclovir 5 mg per kg every 12 hours was identical to 8 hourly dosing of either drug and penciclovir offers reduced frequency of dosing compared to acyclovir every 8 hours. All treatments were well-tolerated.

Herpes zoster infections have also been studied in immunocompromised hosts. Famciclovir 500 mg tid was compared with acyclovir 800 mg five times daily for 10 days in the treatment of herpes zoster in oncology and transplant patients.[53] In this study, time to complete healing was comparable in both the famciclovir and acyclovir groups and there was no significant difference in withdrawals due to lack of efficacy as assessed by dissemination (3% famciclovir vs. 8% acyclovir) or new lesion formation after day 10 (4% famciclovir vs. 1% acyclovir). Both treatments were well tolerated. Famciclovir offered a significant convenience advantage in dosing.

5. SUMMARY OF SAFETY FINDINGS

To date, 3996 patients have been treated for acute herpes simplex or herpes zoster with famciclovir in completed studies using doses of up to 750mg tid administered for up to 10 days. The mean age of treated patients has been 47 years, with a range from 15 to 102 years. Of the patients, 26% have been older than 65 years. Approximately 88% patients have been white, 5% black, and 7% from other races. In all cases, famciclovir has been well tolerated with an adverse event profile comparable to placebo. The most frequently reported adverse events have been headache and nausea, however, these are reported with similar frequencies amongst placebo patients. Famciclovir and placebo recipients have always shown a comparable low incidence of laboratory abnormalities.

Suppression studies have been similarly clean. The longest duration treatment trials of one year, again showed headache to be the most common adverse event (9% famciclovir; 10% placebo). Other low incidence adverse events (seen with essentially equal rates in placebo recipients) include abdominal pain, increased serum lipase, nausea, dyspepsia, dizziness, and hyperbilirubinemia.

6. CONCLUSION

Parenteral and topical penciclovir and its oral formulation, famciclovir, have proven to be beneficial antiviral agents in the treatment of both herpes simplex and herpes (varicella) zoster virus infections. In part because of the long intracellular half-life of the nucleoside triphosphate, famciclovir dosing intervals in active treatment are more convenient in both zoster (tid vs. five times daily) and simplex (bid/tid vs. five times daily) than acyclovir. Intravenous penciclovir in the compromised host can be used twice daily compared with tid intravenous acyclovir. Suppression of genital herpes requires twice daily dosing but is effective in suppressing both clinical and subclinical reactivations. Efficacy equivalence with acyclovir has been demonstrated in the compromised host and in normal hosts with first genital episodes. Comparative trials were otherwise not performed against another active drug in recurrent genital herpes, although the symptomatic relief and reduction of time to lesion resolution was quite robust in famciclovir's placebo-controlled trials, despite relatively modest patient population sizes in the two pivotal trials. Comparative trials with acyclovir against zoster demonstrated superiority in specific subpopulations, although these were the predicted subpopulations where effects on PHN might have been expected in the first place (early therapy, patients older than age 50). Against placebo, however, famciclovir 500 mg tid for 7 days has been the first treatment to significantly reduce the duration of PHN. Although some controversy has arisen over the definition of PHN as pain following rash healing, subsequent re-analyses of these data utlizing a variety of PHN definitions as suggested by expert panels have revealed that the strength of significance of PHN duration remains robust and may actually increase with time over the 6 months following the resolution of acute disease. Safety of this drug has been excellent, to date, and is not associated with any significant adverse effects of concern in either normal or immunocompromised hosts. Famciclovir tablets and penciclovir intravenous and topical formulations have taken a significant place in the treatment and prevention of herpes simplex virus and herpes zoster virus infections.

REFERENCES

1. Famvir Product Brochure. SmithKline Beecham Pharmaceuticals.
2. Sacks SL, Varner T, Macintosh M. Double-blind, randomised, placebo-controlled, lesional, clinic-initiated treatment of culture positive recurrent genital herpes with intravenous penciclovir: clinical benefits of rapid-onset antiviral activity. 3rd Congress of the European Academy of Dermatology and Venereology. Copenhagen, 1993: Abstract p.135.
3. Lazarus H, Belanger R, Candon A, *et al*, PCV IC study group. Efficacy and safety of penciclovir (PCV) for the treatment of HSV infections in immunocompromised (IC) patients [abstract H72]. 1997; 37th Interscience Conference on Antimicrobial Agents and Chemotherapy, Toronto, Ontario. p. 226.

4. Spruance SL, Rea TL, Thoming C, *et al* and the Topical Penciclovir Collaborative Study Group. Penciclovir cream for the treatment of herpes simplex labialis. A randomized, multicenter, double-blind, placebo-controlled trial. JAMA 1997; 277:1374–1379.
5. Vere Hodge RA, Cheng Y-C. The mode of action of penciclovir. Antiviral Chem Chemother 1993, 4 (suppl 1):13–24.
6. Vere Hodge RA. Famciclovir and penciclovir: the mode of action of famciclovir including its conversion to penciclovir. Antiviral Chem Chemother 1993; 4:67–84.
7. Vere Hodge RA, Sutton D, Boyd MR, *et al.* Selection of an Oral Prodrug (BRL 42810; Famciclovir) for the Antiherpesvirus Agent BRL 39123 [9-(4-Hydroxy-3-Hydroxymethylbut-1-yl)Guanine; Penciclovir]. Antimicrob Agents Chemother 1989: 33(10):1765–1773.
8. Pue MA, Benet LZ. Pharmacokinetics of famciclovir in man. Antiviral Chem Chemother 1993: 4(suppl 1):47–55.
9. Filer CW, Allen GD, Brown TA, *et al.* Metabolic and pharmacokinetic studies following oral administration of 14C-famciclovir to healthy subjects. Xenobiotica 1994; 24:357–68.
10. Boyd MR, Bacon TH, Sutton D, *et al.* Antiherpesvirus activity of 9-(4-hydroxy-3-hydroxymethylbut-1-yl)guanine (BRL†39123) in cell culture. Antimicrob Agents Chemother 1987; 31:1238–42.
11. Gnann Jr JW. New antivirals with activity against varicella-zoster virus. Ann Neurol. 1994; 34:569–72.
12. Winton CF, Fowles SE, Vere Hodge RA, *et al.* Assay of famciclovir and its metabolites, including the antiherpes agent penciclovir, in plasma and urine of rats, dog, and man. In Reid E, Wilson ID, editors. Analysis of drugs and metabolites. Cambridge: Royal Soc Chemistry, 1990:163–71.
13. Weller S, Blum R, Doucette M, *et al.* Pharmacokinetics of the acyclovir pro-drug valacyclovir after escalating single and multiple-dose administration to normal volunteers. Clin Pharmacol Ther 1993; 54:595–605.
14. Fowles SE, Pue MA, Pierce D, *et al.* Pharmacokinetics of penciclovir in healthy elderly subjects following a single oral administration of 750 mg famciclovir [abstract]. Br J Clin Pharmacol 1992; 34:450P.
15. Cirelli R, Herne K, McCrary M, *et al.* Famciclovir: review of clinical efficacy and safety. Antiviral Res 1996; 29:141–151.
16. Boike SC, Pue MA, Freed MI, *et al.* Pharmacokinetics of famciclovir in subjects with varying degrees of renal impairment. Clin Pharmacol Ther 1994; 55:418–26.
17. Boike SC, Pue M, Audet PR, *et al.* Pharmacokinetics of famciclovir in subjects with chronic hepatic disease, J Clin Pharmacol 1994; 34(12):199–207.
18. Fowles SE, Fairless AJ, Pierce DM, *et al.* A further study of the effect of food on the bioavailability and pharmacokinetics of penciclovir after oral administration of famciclovir [abstract]. Br J Clin Pharmacol 1991; 32:657P.
19. Fowles SE, Pierce DM, Prince WT, *et al.* Effect of food on bioavailability and pharmacokinetics of penciclovir, a novel antiherpes agent, following oral adminstration of the pro-drug, famciclovir [abstract]. Br J Clin Pharmacol 1990; 29:620P-1P.
20. Pratt SK, Standring-Cox R, Writer D, *et al.* Penciclovir pharmacokinetics in fed and fasted subjects following oral famciclovir in relation to in-vitro antiviral activity [abstract]. Proc 6th International Congress for Infectious Diseases (ICID). 1994 April 26–30.
21. Daniels S, Schentag JJ. Drug interaction studies and safety of famciclovir in healthy volunteers. Antiviral Chem Chemother. 1993; 4(suppl. 1):57 -64.
22. Chatis, PA, Crumpacker C.S. Resistance of herpesviruses to antiviral drugs. Antimicrob Agents Chemother 1992; 36:1589–1595.
23. Boyd MR, Safrin S, Kern ER. Penciclovir: a review of its spectrum of activity, selectivity and cross-resistance pattern. Antivir Chem Chemother. 1993; 4(suppl 1):3–11.
24. Degreef H. Famciclovir, a new oral antiherpes drug: results of the first controlled clinical study demonstrating its efficacy and safety in the treatment of uncomplicated herpes zoster in immune competent patients. Int J Antimicrob Agents. 1994; 4:241–246.
25. Tyring S, Barbarash RA, Nahlik JE, *et al.* Famciclovir for the treatment of acute herpes zoster: effects on acute disease and postherpetic neuralgia. A randomized, double-blind, placebo-controlled trial. Collaborative Famciclovir Herpes Zoster Study Group. Ann Intern Med. 1995; 123:89–96.
26. Dworkin RH, Carrington D, Cunningham A, *et al.* Assessment of pain in herpes zoster: lessons learned from antiviral trials. Antiviral Res 1997; 33:73–85.
27. Dworkin RH, Boon RJ, Griffin DRG, *et al.* Postherpetic Neuralgia: Impact of Famciclovir, Age, Rash Severity, and Acute Pain in Herpes Zoster Patients. J Infect Dis, 1998, In Press.
28. Gruger J, Backhouse ME. Economic evaluation of antiviral therapy for the treatment of herpes zoster in immunocompetent adults. Pharmacoeconomics 1997; 11:262–73.

29. Sacks SL, Martel A, Aoki F, *et al.* Early, Clinic-Initiated Treatment of Recurent Genital Herpes Using Famciclovir: Results of a Canadian, Multicenter Study. Clinical Research Meeting, American Federation of Clinical Research, Baltimore, Maryland, 1994.
30. Sacks SL, Aoki FY, Diaz-Mitoma F, *et al.* Canadian Cooperative Study Group. Patient-initiated, twice daily oral famciclovir for early recurrent genital herpes reduces the onset and duration of viral shedding, the duration of lesions and their components and symptoms: a Canadian, multicenter, randomized, placebo-controlled, dose-ranging trial. JAMA; 1996, 276:44–49.
31. Sacks SL. Canadian Cooperative Study Group: Episodic clinic- and patient-initiated famciclovir (FCV) for recurrent genital herpes (GHSV): A combined analysis. 5th Congress of European Academy of Dermatology and Venereology,:S75, 1995.
32. Mertz GJ, Loveless MO, Levin MJ, *et al.* Oral famciclovir for suppression of recurrent genital herpes simplex virus infection in women. A multicenter, double-blind, placebo-controlled trial. Collaborative Famciclovir Genital Herpes Research Group. Arch Intern Med 1997; 157:343–9.
33. Diaz-Mitoma F, Sibbald RG, Shafran SD, *et al* for the Collaborative Famciclovir Genital Herpes Research Group. Oral famciclovir for the suppression of recurrent genital herpes. JAMA. 1998; 280:887–892.
34. Sacks SL, Hughes A, Rennie B, Boon R. Famciclovir for suppression of asymptomatic and symptomatic recurrent genital herpes shedding: a randomized, double-blind double-dummy, parallel-group, placebo-controlled trial. 37th Interscience Conference on Antimicrobial Agents and Chemotherapy, Toronto, 1997.
35. Sacks SL, Shafran SD. Famciclovir suppression of Asymptomatic HSV Shedding in Recurrent Genital Herpes. 38th Interscience Conference on Antimicrobial Agents and Chemotherapy, San Diego, 1998.
36. Sacks SL, Sasadeusz J, Shafran S. Effect of long-term famciclovir treatment on sperm parameters in patients with recurrent genital herpes. International Conference on Infectious Diseases (ICID), Boston, 1998.
37. Murphy SM, Ruck F, Kitchin VS, *et al.* Oral famciclovir (FCV) a new antiherpes agent: comparative study with acyclovir in clinic initiated treatment of first episode genital herpes (FGH) [abstract] EADV/Traena Congress, Athens, 1991.
38. Loveless M, Harris W, Sacks S. Treatment of first episode genital herpes with famciclovir. 35th Interscience Conference on Antimicrobial Agents and Chemotherapy, 17–20 September, 1995; San Francisco, California.
39. Loveless M, Sacks SL, Harris JRW. Famciclovir in the management of first-episode genital herpes. Infectous Disease in Clinical Practice. 1997; 6(suppl. 1): S12-S16.
40. Thackray AM. Field HJ. Differential effects of famciclovir and valaciclovir on the pathogenesis of herpes simplex virus in a murine infection model including reactivation from latency. J Infect Dis 1996; 173:291–9
41. Thackray AM. Field HJ. Famciclovir and valaciclovir differ in the prevention of herpes simplex virus type 1 latency in mice: a quantitative study. Antimicrob Agents Chemother 1998; 42:1555–62.
42. Thackray AM, Field HJ. Comparison of effects of famciclovir and valaciclovir on pathogenesis of herpes simplex virus type 2 in a murine infection model. Antimicrob Agents Chemother 1996; 40:846–51.
43. Field HJ, Tewari D, Sutton D, *et al.* Comparison of efficacies of famciclovir and valaciclovir against herpes simplex virus type 1 in a murine immunosuppression model. Antimicrob Agents Chemother 1995; 39:1114–9.
44. Spruance SL. Current status of herpes simplex labialis. Infect Dis Newsletter 1988 7:81–8.
45. Spruance SL. Treatment of herpes simplex labialis. In: DeClerq E, ed. Clinical use of antiviral drugs. Boston: Martinu Nijhoff, 1988; 67–86.
46. Lavender EA, Laroche JP, Gill CR, *et al.* Dermal irritancy potential of penciclovir cream compared with acyclovir 5% cream in health volunteers. 3rd Congress of the European Academy of Dermatology and Venereology. Copenhagen, 1993: Abstract B130.
47. Spruance SK, Rea TL, Thoming C, *et al* for the Topical Penciclovir Collaborative Study Group. Penciclovir cream for the treatment of herpes simplex labialis. A randomized, multicenter, double-blind, placebo-controlled trial. JAMA 1997; 277:1374–1379.
48. Raborn GW for the Penciclovir Topical Collaborative Study Group. Penciclovir cream for recurrent herpes simplex labialis. 36th Interscience Conference on Antimicrobial Agents and Chemotherapy, New Orleans, 1996.
49. Spruance SL, Hamill ML, Hoge WS, *et al.* Acyclovir prevents reactivation of herpes simplex labialis in skiers. JAMA 1988; 260:1597–9.
50. Rooney JF, Straus SE, Mannix ML, *et al.* Oral acyclovir to suppress frequently recurrent herpes labialis. A double-blind, placebo-controlled trial. Ann Intern Med 1993; 118:268–72.
51. Frechette G, Romanowski B. Famciclovir Study Group. Efficacy and safety of famciclovir for the treatment of HSV infection in HIV+ patients [abstract 301]. 6th Canadian Conference on HIV/AIDS Research. Can J Infect Dis 1997;8 (Suppl A): 44A.

52. Schacker T, Hui-lin H, Koelle DM, *et al.* Famciclovir for the suppression of symptomatic and asymptomatic herpes simplex virus reactivation in HIV-infected persons. Ann Intern Med 1998;128:21–28.
53. Tyring S, Belanger R, Bezwoda W, *et al.* and the Famciclovir Immuncompromised Study Group. The efficacy and safety of famciclovir (FCV) for the treatment of herpes zoster (HZ) in immunocompromised (IC) patients. Presented as a poster at the 20th International Congress of Chemotherapy (ICC), 29 June-3 July 1997; Sydney, Australia.

14

VALACICLOVIR UPDATE

Andrew R. Bell

Glaxo Wellcome Research and Development
Greenford Road
Greenford, Middlesex, United Kingdom UB6 OHE

Valaciclovir (Valtrex™, Zelitrex™, VACV), the L-valine ester of aciclovir, increases oral aciclovir bioavailability by 3–5 fold. After oral administration, valaciclovir is rapidly absorbed and almost completely converted to aciclovir and the essential amino acid L-valine (Figure 1). The conversion to aciclovir is mediated by an enzyme referred to as valaciclovir hydrolase, which has recently been purified from rat liver mitochondria and characterised.[1] Bioavailability in a wide range of patient populations including the elderly, transplant recipients and those with advanced HIV disease or liver impairment is similar to that in healthy volunteers.[2–6] Valaciclovir has been studied for treatment of herpes zoster, treatment and suppression of herpes simplex virus infections and prophylaxis of CMV infection.[7] This paper reviews valaciclovir data generated since the previous publication in this series.[8]

1. HERPES ZOSTER

Valaciclovir was compared with aciclovir in the largest controlled study to date in herpes zoster.[9] This study included 1141 otherwise healthy patients of at least 50 years of age, who were randomised to receive valaciclovir 1g 3 times daily for 7 or 14 days or aciclovir 800mg 5 times daily for 7 days. Valaciclovir accelerated the resolution of pain (measured as zoster-associated pain[10,11] or post-herpetic neuralgia (PHN)) by approximately one-third compared with aciclovir (Figure 2). There was no additional benefit of 14 days' treatment compared with 7 days. A recently published cost consequence analysis,[12] using the data from this study, demonstrated that valaciclovir (compared with aciclovir) reduced direct medical costs per patient by 17% ($60) on average, and indirect costs by 25% ($46).

Two further studies have been completed more recently. One included 121 patients with ophthalmic zoster. This is a particularly serious form of herpes zoster, in which aciclovir has previously been shown to be effective in reducing the severity and persistence

Antiviral Chemotherapy 5, edited by Mills *et al.*
Kluwer Academic / Plenum Publishers, New York, 1999.

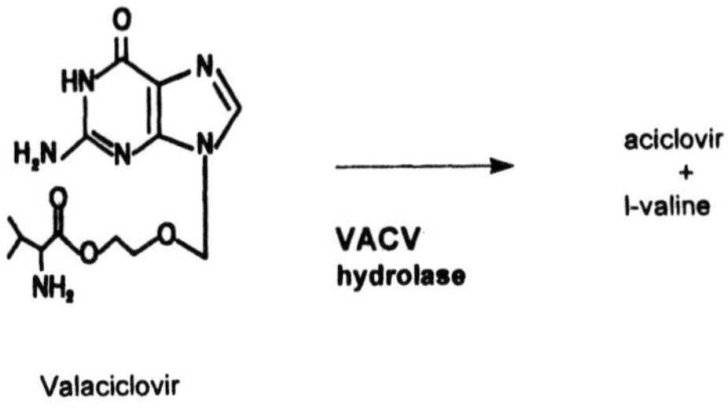

Figure 1. Conversion reaction valaciclovir.

of ocular complications.[13] In this study patients were randomised to receive valaciclovir 1g 3 times daily or aciclovir 800mg 5 times daily for 7 days.[14] Clinical endpoints included ocular complications as well as zoster-associated pain and lesion healing. The results (Table 1) indicated that valaciclovir achieved similar efficacy to aciclovir on the skin lesions and ocular complications of herpes zoster ophthalmicus, and a similar safety profile, but with the advantage of a convenient 3 times daily dosing regimen.

A large observational study (VIZA: Valaciclovir International Zoster Assessment) included 1897 patients with herpes zoster, from 24 countries, treated with valaciclovir 1g 3 times daily for 7 days, and followed up for 6 months. The objectives of the study were to gain experience of valaciclovir in a wide range of primary care and hospital settings, and to determine differences in the response to valaciclovir between pre-defined subgroups:

- Age ≤ 60 vs > 60 years
- Ophthalmic vs non-ophthalmic zoster

Interim results[15] show that, as expected, the duration of pain is longer in older patients, and suggest that it may also be marginally longer in ophthalmic than non-ophthalmic zoster. Severe prodromal or acute pain was shown to be an independent predictor of prolonged pain. The findings of this study confirm what has been known for 40 years,[16,17]

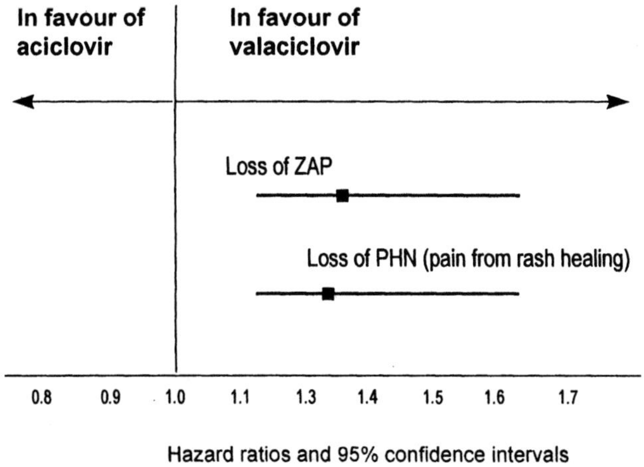

Figure 2. Valaciclovir for treatment of herpes zoster.

Table 1. Valaciclovir for treatment of herpes zoster ophthalmicus

	Valaciclovir		Aciclovir	
	No	%	No	%
Ocular complications				
Keratitis	12	21	12	22
Stromal keratitis	8	14	7	13
Uveitiis	7	13	9	17
Episcleritis	4	7	1	2
Time to healing*				
Patients healed at day 14	43	83	41	87
Pain				
Patients with pain at day 10	22	39	27	50
Patients with pain at week 24	3	5	3	6

* Defined as 100% healing or scabbing of lesions

but using rigorous modern analytical methods, and in the presence of antiviral therapy. Valaciclovir was well tolerated in this large study. Only 4 patients out of 1897 experienced serious adverse events during treatment which were considered possibly attributable to study drug.

2. GENITAL HERPES

Genital herpes is the most common cause of genital ulceration in the western world. It is usually caused by herpes simplex virus type-2 (HSV-2), although HSV-1 can also cause genital herpes.[18] Despite all attempts to promote safer sex, the incidence of genital herpes continues to increase. Approximately 22% of adults in the USA are seropositive for HSV-2; the seroprevalence increased by 30% between the late 1970s and the early 1990s.[19] Only a small proportion of those who are seropositive are clinically diagnosed, and only a small proportion of those are adequately treated.

Valaciclovir has been extensively studied for both episodic treatment and chronic suppression (prevention of episodes) of genital herpes.

2.1. Acute or Episodic Treatment of Genital Herpes

Over 3600 patients have been included in clinical studies of valaciclovir for treatment of initial and recurrent genital herpes (Table 2). Valaciclovir (1000mg or 500mg twice daily) was compared with aciclovir and/or placebo.

In a large scale international study in first-episode genital herpes, 643 patients were randomised to receive valaciclovir 1000mg twice daily or aciclovir 200mg 5 times daily for 10 days.[20] The results are summarised in Table 3, and as an example, the Kaplan - Meier curve for time to healing is shown in Figure 3. The hazard ratios comparing valaciclovir and aciclovir are very close to 1.0, with narrow 95% confidence intervals (95% CI), allowing the conclusion that the two dose regimens studied were clinically equivalent. Pharmacokinetically valaciclovir 500mg twice daily provides aciclovir exposure equivalent to that of aciclovir 200mg 5 times daily, so it can be concluded that 500mg twice daily would provide efficacy equivalent to aciclovir. Valaciclovir therefore provides effi-

Table 2. Studies of valaciclovir for treatment of genital herpes

	Number of patients treated				
	Valaciclovir 1000mg 2x/day	Valaciclovir 500mg 2x/day	Aciclovir 200mg 5x/day	Placebo	Total
A (first episode)*	323	-	320	-	643
B (recurrent)†	512	-	506	182	1200
C (recurrent)†	-	378	361	-	739
D (recurrent)†	368	360	-	259	987
Total	1203	738	1187	441	3569

* Therapy continued for 10 days
† Therapy continued for 5 days

cacy equal to that established for aciclovir in first episode genital herpes, but with the advantage of twice daily dosing.

In recurrent genital herpes,[21–24] valaciclovir 500mg twice daily or 1000mg twice daily was significantly superior to placebo in accelerating resolution of signs and symptoms of genital herpes (Figure 4). Valaciclovir also accelerated lesion healing and pain resolution and reduced the duration of viral shedding. Importantly, valaciclovir also significantly increased the proportion of patients whose lesions failed to develop to the vesicle stage or beyond[24] (see Figure 5 and Table 4). This had not previously been conclusively demonstrated for aciclovir in prospective controlled trials. Valaciclovir provides an effective twice daily treatment for recurrent genital herpes.

2.2. Suppression of Genital Herpes

Two large-scale placebo controlled studies have investigated the efficacy of valaciclovir for suppression (prevention of recurrences) of genital herpes.[25,26]

In both studies, patients with frequent recurrences were enrolled, so that without treatment most patients would have experienced at least one recurrence during the study. The principal endpoint was the time to first recurrence.

In one study,[25] 382 patients were randomised to receive valaciclovir 500 mg once daily or placebo in the ratio 3:1. To be eligible for this study patients had to have a history of at least 8 recurrences per year when not receiving suppressive therapy. Double-blind treatment was given for 16 weeks or until the first recurrence, whichever was earlier. Patients could then receive open-label valaciclovir treatment until 1 year after initial enrollment into the study. Valaciclovir 500mg once daily significantly prolonged the time to first recurrence; the hazard ratio for valaciclovir vs placebo was 0.15 (95% CI 0.11, 0.21; $p < 0.0001$) indicating that valaciclovir treatment prevented or delayed 85% of genital herpes

Table 3. Valaciclovir for first episode genital herpes

	Summary of key results (median days)			
	Valaciclovir (n=319)	Aciclovir (n=320)	Hazard Ratio (95% CI)	p - value
Virus shedding	3	4	1.0 (0.9,1.2)	0.74
Healing time	9	9	1.1 (0.9,1.3)	0.31
Pain duration	5	5	1.0 (0.9,1.2)	0.95
Complete resolution	13	13	1.1 (0.9,1.3)	0.43

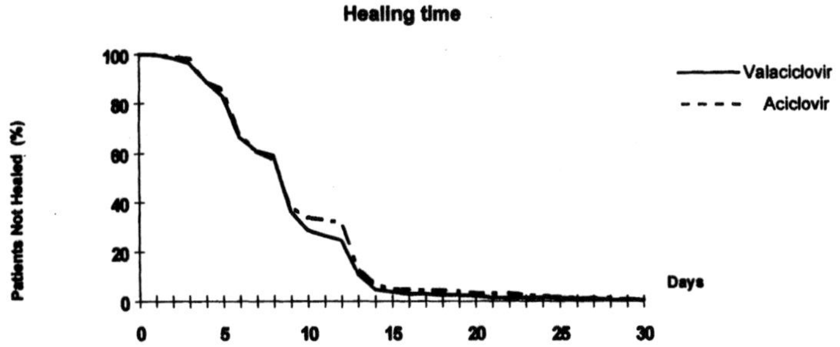

Figure 3. Valaciclovir for first episode genital herpes.

recurrences. By the end of the 16 week double-blind treatment period 69% of valaciclovir recipients were recurrence-free, compared with 9% of placebo patients.

A larger, dose ranging study has been recently reported.[26] In this study, 1479 patients with a history of at least 6 recurrences per year were randomised (in a 2:2:2:2:2:1 ratio) to receive valaciclovir (1000 mg, 500 mg or 250 mg once daily, or 250 mg twice daily), aciclovir 400 mg twice daily or placebo. The duration of treatment was 1 year. To protect against type-1 error, statistical analyses were performed on individual dose comparisons in a pre-determined, conditional, step-down sequence. Intent-to-treat analysis showed that all once daily valaciclovir regimens were highly effective compared with placebo ($p<0.0001$), although 250mg once daily was clearly sub-optimal. The hazard ratio (95% CI) for time to first recurrence for valaciclovir 500mg once daily compared with placebo was 0.29 (0.22, 0.37) $p<0.0001$.

A further randomised, controlled study[27] evaluated valaciclovir for HSV suppression in 1062 HIV infected subjects ($CD4^+$ >100/μl). The subjects were randomised to receive valaciclovir 500mg twice daily, valaciclovir 1000mg once daily or aciclovir 400mg twice daily. Treatment continued for one year, and the principal endpoint was the time to first recurrence. There were no significant differences between either of the valaciclovir doses and aciclovir, but the results suggested that valaciclovir 500mg twice daily was the best treatment regimen of those studied for this patient population. Hazard ratios and 95% CIs are given in Table 5.

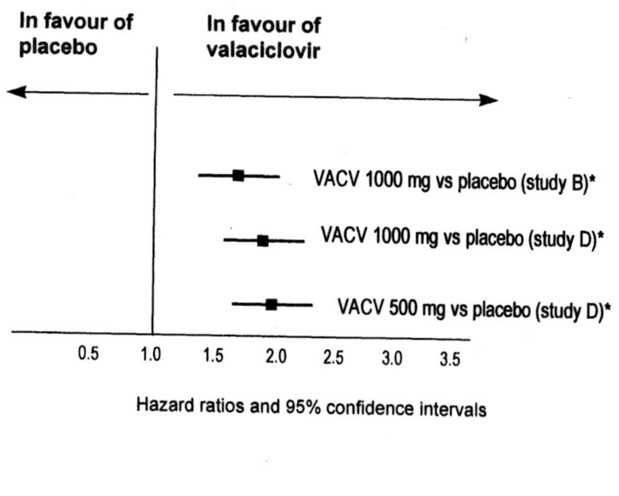

Figure 4. Valaciclovir for treatment of recurrent genital herpes: time to resolution of all signs and symptoms.

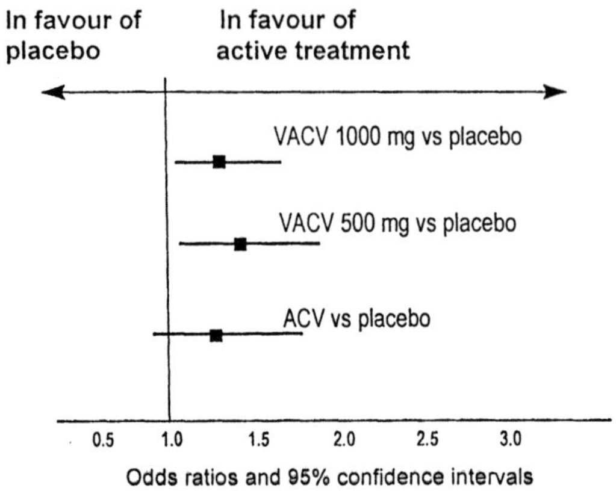

Figure 5. Valaciclovir for treatment of recurrent genital herpes: prevention of lesion development to vesicle stage.

Long term administration of valaciclovir was well tolerated in all of these studies in both immunocompetent and immunocompromised individuals. The commonest adverse event in both studies in immunocompetent individuals, and across all treatment groups, was headache (reported by 34–38% of patients in each treatment group including aciclovir and placebo in the dose ranging study). Nausea, flu-like syndrome, infection, pharyngitis and rhinitis were amongst the next most frequently reported. These, and other events reported less frequently, were generally mild, and reported in similar proportions of patients receiving valaciclovir, aciclovir or placebo. In immunocompromised patients the most common adverse events were diarrhoea, headache and infection. These were generally mild or moderate, and were reported in a similar proportion of patients in each treatment group.

3. CMV PROPHYLAXIS

CMV disease is an important cause of morbidity and mortality in severely immunocompromised patients. Once established, it is difficult to treat and available therapeutic options are associated with significant side-effects. Thus an effective, well tolerated, oral therapy for reducing the incidence and severity of CMV disease, that can be administered long-term on an out-patient basis, would be of substantial clinical benefit.

A programme of studies investigating the potential of valaciclovir for prevention of CMV disease in immunocompromised patients has been undertaken. The studies included individuals with advanced HIV disease, bone marrow and renal transplant recipients.

Table 4. Prevention of lesion development in episodic treatment of genital herpes

	Percentage of patients in whom vesicular lesions were prevented			
	VACV 1000mg bid	VACV 500mg bid	Aciclovir 400mg 5X daily	Placebo
Study B*	25.9	–	24.8	19.8
Study C*	–	22.0	20.8	–
Study D*	28.0	31.3	–	21.2

Table 5. Valaciclovir for HSV suppression in HIV infected individuals

	Hazard Ratio	95% CI	p-value
VACV 500mg twice daily vs ACV	0.73	0.50, 1.06	0.10
VACV 1000mg once daily vs ACV	1.31	0.94, 1.82	0.11

The first study to be completed evaluated valaciclovir (2g, 4 times daily) and aciclovir (800mg 4 times daily or 400mg twice daily) in 1227 patients with HIV infection and $CD4^+$ counts <100/μl.[28] Final analysis showed a significant protective effect of valaciclovir, with CMV disease being reduced by one-third, from 18% in aciclovir recipients to 12% in valaciclovir recipients (p=0.03). However, there was a trend towards higher mortality in the valaciclovir group, and valaciclovir at this dose was not well tolerated by these ill patients. A few cases of thrombotic microangiopathy occurred in this study, with an excess in the valaciclovir group.[29] There have been no such reports in >3000 patients, including HIV-infected subjects, in long-term valaciclovir studies for HSV suppression.[30]

Preliminary results in renal transplant recipients have been presented recently.[31] In this study patients received valaciclovir (2g 4 times daily adjusted according to renal function) or placebo for 90 days, and were evaluated for 6 months. The primary endpoint was CMV disease, as defined in the protocol and evaluated by an endpoints committee. The study enrolled 208 CMV seronegative recipients of a graft from a seropositive donor (D^+R^-, high risk) and 408 CMV seropositive recipients (R^+, moderate risk).

In the high risk patients, valaciclovir prevented or delayed 80% of CMV disease compared with placebo (hazard ratio 0.20; 95% CI 0.11, 0.38; p<0.0001) with 14/102 (14%) of valaciclovir and 41/106 (39%) of placebo recipients developing disease. Only 3 (3%) valaciclovir vs 41 (39%) placebo recipients developed CMV disease during the treatment period. In the moderate risk patients, as expected, disease was less common. Two out of 204 (1%) valaciclovir and 11/204 (5%) placebo recipients developed CMV disease. Valaciclovir prevented or delayed 81% of CMV disease (hazard ratio 0.19; 95% CI 0.042, 0.854; p=0.03). No case of CMV disease occurred in a valaciclovir recipient during the 90 day treatment period, whereas all cases in the placebo group occurred during this period. Preliminary data show that, with the expected main exception of neurological events, serious adverse events occurred in similar proportions of valaciclovir and placebo patients.

4. CONCLUSIONS

In treatment of herpes zoster, valaciclovir has demonstrated consistent clinical and statistical superiority over aciclovir in speeding resolution of zoster-associated pain and PHN. A recent study has confirmed the utility of valaciclovir in ophthalmic zoster, and a large scale study has confirmed that valaciclovir is well tolerated.

Twice daily valaciclovir has proven at least as effective and safe as aciclovir (500 mg twice daily) for treatment of both first and recurrent episodes of genital herpes.

For suppression of genital herpes valaciclovir provides efficacy with a patient-friendly once daily regimen. For HIV-infected individuals 500mg twice daily appears to be the appropriate dose regimen. In both immunocompetent and immunocompromised individuals, long-term administration of valaciclovir is well tolerated.

The potential for valaciclovir in CMV prophylaxis is now established, although further refinement of dosage regimens may be required.

Overall, valaciclovir offers advantages over aciclovir not only in more acceptable dose regimens but also in superior efficacy in management of HSV, VZV and CMV infections. These benefits are achieved while maintaining the established safety profile of aciclovir.

REFERENCES

1. Burnette TC, Harrington JA, Reardon JE *et al.* Purification and characterisation of a rat liver enzyme that hydrolyses valaciclovir, the L-valyl ester prodrug of aciclovir. J Biol Chem 1995; 270:15827–31
2. Weller S, Blum MR, Doucette, M *et al.* Pharmacokinetics of the aciclovir prodrug valaciclovir after escalating single- and multiple-dose administration to normal volunteers. Clin Pharmacol Ther 1993; 54:595–605
3. Wang LH, Schultz M, Weller S, *et al.* Pharmacokinetics and safety of multiple-dose valaciclovir in geriatric volunteers with and without concomitant diuretic therapy. Antimicrob Agents Chemother 1996; 40:80–5
4. First MR, Blum MR, Brennan P, *et al.* Pharmacokinetics of 256U87, an aciclovir prodrug, in renal transplant patients. Clin Pharmacol Ther 1993; 53:238
5. Jacobson MA, Gallant J, Wang LH, *et al.* Phase I trial of valaciclovir, the L-valyl ester of aciclovir, in patients with advanced human immunodeficiency virus disease. Antimicrob Agents Chemother 1995; 38:1534–40
6. Posner J, Soul-Lawton J, Weller S, *et al.*Valtrex (valaciclovir) - bioavailability in different patient populations. Can J Infec Dis 1995; 6(supplC):287C
7. Patel R. Valaciclovir: development, clinical utility and potential. Exp Opin Invest Drugs 1997; 6:1–17
8. Smiley ML, Murray AB, de Miranda P. Valaciclovir HCl (Valtrex): An aciclovir prodrug with improved pharmacokinetics and better efficacy for treatment of zoster. In: Mills J *et al*, eds. Antiviral Chemotherapy 4. Plenum, New York, 1996. 33–9.
9. Beutner KR, Friedman DJ, Forszpaniak C, *et al.* Valaciclovir compared with aciclovir for improved therapy for herpes zoster in immunocompetent adults. Antimicrob Agents Chemother 1995; 39:1546–53
10. Wood MJ, Kroon S (eds). Management strategies in herpes: reducing the burden of zoster-associated pain - update. PPS Europe Ltd 1995, on behalf of the IHMF.
11. Wood MJ, Balfour H, Beutner K, *et al.* How should zoster trials be conducted? J Antimicrob Chemother 1995; 36:1089–10
12. Grant DM, Mauskopf JA, Bell L, Austin R. Comparison of valaciclovir and aciclovir for the treatment of herpes zoster in immunocompetent patients over 50 years of age: a cost-consequence model. Pharmacotherapy 1997; 17:333–41
13. Harding SP, Porter SM. Oral aciclovir in herpes zoster ophthalmicus. Current Eye Research 1991; 10 Suppl:177–82.
14. Colin J, Hoang-Xuan T, Rolland B, *et al.* A double blind randomised trial to compare the safety and efficacy of valaciclovir and aciclovir for treatment of herpes zoster ophthalmicus. 3rd International Conference on the Varicella Zoster Virus. March 1997, Florida USA.
15. Gho CL, Crooks RJ, Shukla S. Valaciclovir International Zoster Assessment Group. Valaciclovir for the treatment of herpes zoster - influence of patient age and dermatome. 7th International Congress for Infectious Diseases, Hong Kong 1996. Abstract 95.010
16. Hope-Simpson RE. The nature of herpes zoster: a long-term study and a new hypothesis. Proc Roy Soc Med. 1965; 58:1–12
17. Burgoon CF, Burgoon JS, Baldridge GD. The natural history of herpes zoster. J Am Med Assoc. 1957; 164:265–9
18. Corey L. The current trend in genital herpes. Progress in prevention. Sex Transm Dis 1994; 21:s38–44
19. Fleming DT, McQuillan GM, Johnson RE, *et al.* Herpes simplex virus type 2 in the United States, 1976 to 1994. New England J Med 1997; 337:1105–11
20. Fife KH, Barbarash RA, Rudolph T *et al.* Valaciclovir versus aciclovir in the treatment of first-episode genital herpes infection. Sexually transmitted diseases 1997; 24:481–6
21. Spruance SL, Tyring SK, DeGregorio B *et al.* A large-scale, placebo-controlled, dose-ranging trial of peroral valaciclovir for episodic treatment of recurrent herpes genitalis. Arch Int Med 1996; 156:1729–35
22. Bodsworth NJ, Crooks RJ, Borelli S *et al.* Valaciclovir versus aciclovir in patient initiated treatment of recurrent genital herpes: a randomised, double blind clinical trial. Genitourinary Medicine. 1997; 73:110–6

23. Tyring SK, Douglas, JM, Corey, L *et al.* A randomised, placebo-controlled comparison of oral valaciclovir and aciclovir in immunocompetent patients with recurrent genital herpes infections. Arch Dermatol. 1998, In press.
24. Kinghorn GR, Fiddian AP, International valaciclovir HSV study group. Prevention of lesion development in recurrent genital herpes with oral valaciclovir. J European Acad Dermatol Venereol. 1995;5(Suppl 1):Abstract S162
25. Patel R, Bodsworth NJ, Woolley P *et al.* Valaciclovir for the suppression of recurrent genital HSV infection: a placebo controlled study of once daily therapy. Genitourinary Medicine. 1997; 73:105–9
26. Reitano M, Tyring S, Lang W, *et al.* A comparative trial of valaciclovir, aciclovir and placebo for the suppression of genital herpes infection in immunocompetent patients. J Inf Dis 1998 (Accepted for publication).
27. Lawrence AG, Bell AR. Valaciclovir International HSV Study Group. Valaciclovir for prevention of recurrent herpes simplex virus infection in HIV infected individuals - a double-blind controlled trial. J Clin Microbiol and Infection 1997; 3:P1422 (Abstract)
28. Feinberg J, Hurwitz S, Cooper D, *et al.* A randomised, double-blind trial of valaciclovir prophylaxis for cyomegalovirus diseasae in patients with advanced human immunodeficiency virus infection. J Inf Dis 1998; 117:48–56.
29. Bell WR, Chulay JD, Feinberg JE. Manifestations resembling thrombotic microangiopathy in patients with advanced human immunodeficiency virus disease in a cytomegalovirus prophylaxis trial (ACTG 204). Medicine 1997; 76:369–380
30. Chulay JD, Bell AR, International Valaciclovir HSV study Group. Long-term safety of valaciclovir for suppression of herpes simplex virus infections. Clin Inf Dis 1996; 23:879.
31. Neumayer H-H, Kreis H, Squifflet J-P, *et al.* Efficacy and safety of valaciclovir for suppression of cytomegalovirus disease in recipients of a cadaveric renal allograft. 37th Interscience Conference on Antimicrobial Agents and Chemotherapy, Toronto 1997. Late Breaker Abstract.

15

THERAPEUTIC APPROACHES TO THE MANAGEMENT OF HERPES ZOSTER

Richard J. Whitley and John W. Gnann, Jr.

Department of Pediatrics, Microbiology and Medicine
University of Alabama at Birmingham
616 Children's Hospital
1600 Seventh Avenue
South Birmingham, Alabama 35233-0011

Over the past five years herpes zoster has been one of the most intensely studied viral infections of the immunocompetent host, due mainly to the high incidence of disease and the resultant morbidity in elderly individuals. The introduction of the prodrugs valaciclovir and famciclovir for the treatment of herpes zoster has provided significant advances in clinical outcome. In the evaluation of thousands of immunocompetent individuals with herpes zoster, these drugs provide both pharmacokinetic advantages and enhanced clinical benefits over the standard of therapy, aciclovir. Both valaciclovir and famciclovir accelerate cutaneous healing and decrease acute neuritis and zoster-associated pain as compared to placebo. Additionally, the value of adjunctive therapy with corticosteriods has been rigorously assessed. While concomitant corticosteriod therapy does not have an effect on postherpetic neuralgia, it improves quality of life parameters such as return to usual activity, total cessation of analgesic use, and ability to sleep uninterrupted at night. In studies performed by the National Institute of Allergy and Infectious Diseases (NIAID) Collaborative Study Group (CASG), adjunctive corticosteriod therapy was reserved for older individuals who did not have contraindications for this management approach. While these recent advances provide improved outcomes for elderly individuals afflicted with herpes zoster, the persistence of pain and its attendant morbid complications remains a viable therapeutic target for antiviral medications and/or adjunctive therapies. This review will stress the advances achieved in the treatment of herpes zoster with valaciclovir and famciclovir; aciclovir studies provide an essential foundation.

1. NATURAL HISTORY OF HERPES ZOSTER

Herpes zoster results from the reactivation of varicella-zoster virus (VZV) from the dorsal root ganglia. Disease usually occurs in the elderly with an estimated annual inci-

Antiviral Chemotherapy 5, edited by Mills *et al.*
Kluwer Academic / Plenum Publishers, New York, 1999.

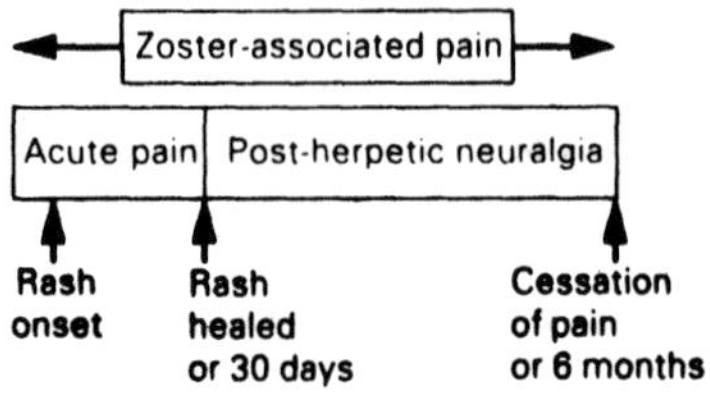

Figure 1. Definitions used in clinical trials to evaluate pain attributed to herpes zoster as a function of time. Pain that persists over the first month is defined as acute neuritis. Pain present at one month after disease onset is defined as post-herpetic neuralgia (irrespective of total healing of skin). Zoster-associated pain is defined as the continuum of pain from clinical presentation with disease to the last clinical evaluation (usually 6–12 months after disease onset).

dence of 300,000 to 500,000 cases in the United States.[1] Of note, the frequency of herpes zoster increases by decade, such that by the age of 80, the incidence is over 10 cases per 1000 population annually.[2] There is no preference for gender.

The hallmark of herpes zoster is a dermatomal rash, which is usually preceded by a prodrome of localized pain and malaise of 24 to 48 hours duration on average.[2] The clinical confirmation of disease is the appearance of erythema and a vesicular rash with a unilateral dermatomal distribution. In the immunocompetent host, the total number of lesions averages approximately 30 but can exceed 500, virtually filling in the entire dermatome. New lesions form over a period of five to seven days and evolve into pustules, which rapidly crust over a period of approximately ten days from the onset of lesions. Total healing, often difficult to evaluate because of altered pigmentation and scarring, requires three weeks or longer.

The most troublesome symptom of herpes zoster is the attendant pain.[3] This has been the subject of intensive evaluation and spirited debate. The intensive evaluation stems from the recognition that lessening pain is, indeed, the most important clinical endpoint for any clinical trial of new therapy for herpes zoster. Debate, on the other hand, focuses on the statistical methodology applied to pain evaluation of new therapy for herpes zoster. At the time of onset of disease, pain—acute neuritis—is a nearly uniform finding in all patients. Over the first month after the onset of disease, in conjunction with antiviral therapy, pain will resolve in the majority of patients. Pain that persists beyond one month, defined as post-herpetic neuralgia, can persist in a minority of patients for prolonged periods of time. To some investigators, a more accurate description of the pain response is "zoster-associated pain" which considers pain as a continuum from onset to resolution. Notably, advancing age and severity of pain at presentation predict the persistence of pain regardless of its descriptor (i.e., post-herpetic neuralgia or zoster-associated pain). Figure 1 illustrates the three definitions of pain associated with herpes zoster. Of note, contemporary clinical trials have, in general, considered acute neuritis and zoster-associated pain as the primary endpoints of therapeutic evaluations.

2. ANTIVIRAL THERAPY STUDIES

2.1. Studies with Aciclovir

Aciclovir was the first orally bioavailable antiviral medication which was proven efficacious in the therapy of herpes zoster in the immunocompetent host. The mechanism of

Table 1. Pain outcome post-aciclovir therapy

Time to Event (days)	Placebo	Aciclovir	P value
First cessation of pain	60.5	49.8	0.09
Total cessation of pain (mean)	85.6	49.1	<0.001
Estimated incidence of Post-herpetic neuralgia	19%	11%	<0.05

action of aciclovir involves monophosphorylation by VZV-encoded thymidine kinase with subsequent di- and tri-phosphorylation by cellular kinases. The triphosphate derivative is both a competitive inhibitor of viral DNA polymerase as well as a chain terminator. Five placebo-controlled clinical trials have established the efficacy of aciclovir.[4–8] In these studies, aciclovir was administered at 800 mg five times daily for seven to 21 days. Each of the clinical trials had different sample sizes and included patients with varied manifestations of herpes zoster (i.e., zoster ophthalmicus). In the early evaluations of herpes zoster, the primary endpoints focused more on cessation of new lesion formation and acute healing with less emphasis on acute pain and post-herpetic neuralgia. As a consequence, the rigor of evaluation was much less than in the more contemporary trials of valaciclovir and famciclovir. Regardless, aciclovir both accelerated cutaneous healing and resolution of acute neuritis. Retrospective analyses of the older data sets also indicated clinical benefit for the more recently defined parameter of zoster-associated pain.

The benefit of aciclovir therapy of herpes zoster has been re-examined in a recent meta-analysis[9] of four placebo controlled trials.[4–7] Table 1 summarizes pain outcome, indicating a consistent beneficial effect for each of the pain parameters that was evaluated.

These data are more accurately depicted by the assessment of hazard ratios with 95% confidence limits. Figure 2 provides the estimates of the hazard ratios according to the various pain milestones. The data are adapted from reference 9 with permission. Taken together these data define the value of aciclovir therapy on the events of pain resolution. Not indicated in these data, however, is the impact of aciclovir therapy on zoster-associated pain. These data are depicted in the Kaplan Meier survival plot, as illustrated in Figure 3. Differences between the treated and the placebo recipients were significant at $p<0.001$.

The clinical benefit of aciclovir therapy was achieved without significant adverse effects. However, this regimen required a five-times-daily dosing schedule because of the relatively poor oral bioavailability of aciclovir. Lower doses of aciclovir or less frequent administration did not appear as efficacious in Phase II studies. In fact, the plasma levels achieved with five times daily dosing were not considered optimal for the management of herpes zoster.

2.2. Studies with Valaciclovir

Valaciclovir is the L-valine ester of aciclovir which, when administered orally, provides significantly greater plasma levels of aciclovir than can be achieved with the parent compound. Obviously, the mechanism of valaciclovir is the same as aciclovir. Plasma levels of aciclovir following one gram of valaciclovir administered three times daily approximate that achieved with 5 mg/kg administered every eight hours intravenously.[10] Additionally, total exposure, as defined by area under the curve determinations, exceeds that achieved with doses of aciclovir normally employed to treat herpes zoster. Theoreti-

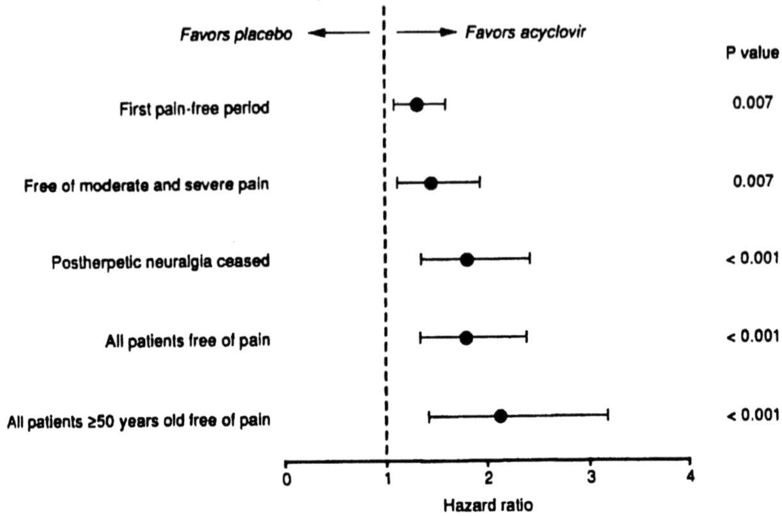

Figure 2. Hazard ratios.

cally, this dosing regimen should provide both improved convenience and greater drug exposure, offering a potential therapeutic advantage.

Valaciclovir has been evaluated in a multinational, double blind, controlled trial in volunteers greater than 50 years of age whose lesions were present less than 72 hours.[11] Over 1100 patients were randomized to one of three treatment regimens: valaciclovir at 1000 mg three times daily for either 7 or 14 days or aciclovir 800 mg five times daily for 7 days. Utilizing an intent-to-treat analysis, the 7 day valaciclovir regimen accelerated the resolution of zoster-associated pain, as indicated by the hazard ratios shown in Table 2. Additionally, the point prevalence of pain six months after the onset of therapy was decreased from 26% in aciclovir recipients to 19.9% in the counterpart valaciclovir recipients. Valaciclovir was well tolerated and adverse events were infrequent in all treatment

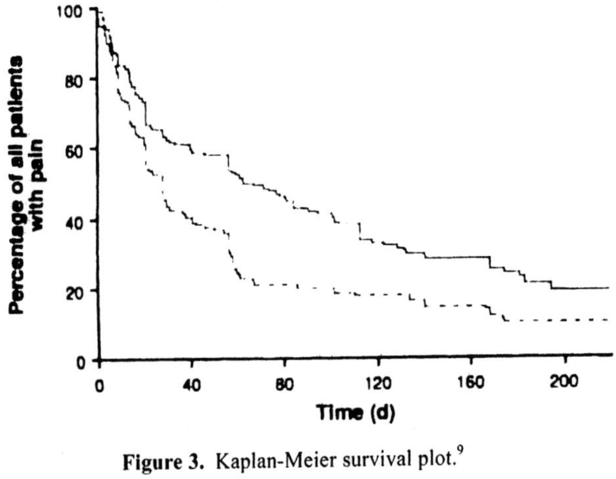

Figure 3. Kaplan-Meier survival plot.[9]

Table 2. Relative risks for the duration of zoster associated pain in patients according to randomization group (valaciclovir [VACV] versus aciclovir [ACV])[11]

Treatment group	Relative risk (95% CI)	P value
VACV-7 days-vs ACV	1.34 (1.12, 1.60)	0.001
VACV-14 days-vs ACV	1.22 (1.03, 1.46)	0.03
VACV-14 vs VACV-7	1.10 (0.92, 1.30)	ns

groups. Valaciclovir is licensed to treat herpes zoster at a dose of 1000 mg three times daily for seven days worldwide.

2.3. Studies with Famciclovir

Famciclovir is the diacetyl ester prodrug of penciclovir, a guanine nucleoside analogue. Famciclovir is converted to penciclovir by hepatic metabolism, providing an oral bioavailability of penciclovir of 77% in humans. In infected cells, penciclovir triphosphate is a competitive inhibitor of viral DNA polymerase but not a chain terminator. The intracellular half-life of the triphosphate derivative is much longer that the aciclovir equivalent.[12]

Famciclovir has been evaluated for treatment of herpes zoster in two large controlled studies. In an aciclovir-controlled study, over 500 patients (18 years or older) with cutaneous lesions less than 72 hours were randomized to famciclovir 250 mg, 500 mg, or 750 mg three times daily or to aciclovir at standard doses. For the intent-to-treat population, cutaneous healing and resolution of pain were judged equivalent for each of the treatment groups.[13] However, in a placebo controlled trial, famciclovir at either 500 mg or 750 mg three times daily for seven days demonstrated significant acceleration of both cutaneous healing and resolution of zoster-associated pain. These results were achieved without significant adverse events. In the United States, famciclovir is licensed for the treatment of herpes zoster at 500 mg three times daily for 7 days. In the United Kingdom, the approved doses are 250 mg three times daily for 7 days or 750 mg once daily, although controlled data to support the latter dose are non-existent.

2.4. Adjunctive Therapy of Herpes Zoster with Prednisone

The merits of concomitant therapy of herpes zoster with an antiviral and corticosteroids has been hotly debated for some time. Recently, two large clinical trials evaluated the effects of concomitant therapy. One study, which included individuals greater than 18 years of age and randomized volunteers to prednisone plus aciclovir therapy for either 7 or 21 days versus aciclovir alone for 7 days failed to demonstrate significant benefit on either time to first cessation or to complete cessation of pain. In addition, adverse effects were more frequent in prednisone recipients.[15] Since there were no placebo recipients in this trial, it is difficult to judge the relative lack of benefit. Nevertheless, the authors concluded that concomitant corticosteriod therapy did not provide significant added benefit and did not recommend its routine use.

More recently, the NIAID CASG completed a placebo-controlled study utilizing a two by two factorial design.[8] In contrast to the aforementioned trial, placebo recipients allowed a more accurate assessment of outcome events. In this study, outcome events not only included resolution of acute and chronic pain, but also quality of life endpoints such

Table 3. Disease resolution according to Cox regression models

Dependent Variable	Relative Risk
Cutaneous healing at one month	
Crusting	2.27 (1.46, 3.55)
Total Healing	2.07 (1.26, 3.38)
Quality of life evaluations at one month	
Cessation of actue neuritis	3.02 (1.42, 6.41)
Time to uninterrupted sleep	2.12 (1.25, 3.58)
Time to 100% usual activity	3.22 (1.92, 5.40)
Time to cessation of analgesic use	3.15 (1.69, 5.89)
Six month evaluation of pain	
Zoster associated pain	1.56 (0.92, 2.66)

as return to usual activity, ability to sleep uninterrupted, and total cessation of analgesic use. The clinical trial was designed such that patients received prednisone (60 mg daily for 7 days, followed by 30 mg daily for 7 days, then 15 mg daily for the final 7 days), aciclovir at 800 mg three times daily for 21 days, both drugs or placebo. Importantly, individuals for whom high dose corticosteriod therapy was contraindicated were excluded from enrollment.

This trial demonstrated an unequivocal benefit for patients who received the combination therapy. As illustrated in Table 3, the relative risks for patients who received both therapeutics varied from 2.12 to 3.15 with the respective p values <0.001.[8] Significance for an effect on long term zoster-associated pain was not achieved. These results were achieved without significant toxicity, prompting the authors to conclude that older individuals do benefit from concomitant therapy with prednisone and an antiviral agent.

We have interpreted these data as demonstrating direct benefit of aciclovir/prednisone combination therapy for individuals who are at risk for complications of herpes zoster. The more rapid return to a normal life style is of value for these patients.

3. SUMMARY

The past several years have provided exciting advances in the management of herpes zoster in the normal host. In spite of these advances, a significant portion of zoster patients still have persistent complications from this disease. Persistent pain is the most debilitating sequela and it occurs in at least 15% of individuals over 50 years of age. Future research efforts must embrace alternative approaches for pain control.

Studies performed by the authors and reported in this article are funded in whole or in part with Federal funds from the National Institutes of Health, National Institute of Allergy and Infectious Diseases (NO-AI-65306), grants from the National Cancer Institute (NCI) CA RO-1-13148, the Division of Research Resources (DRR) RR-032, and the state of Alabama and an unrestricted grant in Infectious Diseases from Bristol Myers Squibb.

REFERENCES

1. Gershon AA, Silverstein, SJ: Varicella-zoster virus. In: Richman D, Whitley R, and Hayden, F (Eds.) Clinical Virology. Churchill-Livingstone (New York), 1997, pp 421–444.

2. Hope-Simpson RE. The nature of herpes zoster: A long-term study and a new hypothesis. Proc R Soc Med. 1965; 58:9–20.
3. Kost R and Straus S. Postherpetic neuralgia—pathogenesis, treatment and prevention. N Engl J Med. 1996; 335:33–42.
4. Huff JC, Bean B, Balfour HH, *et al.* Therapy of herpes zoster with oral aciclovir. Am. J. Med. 1988; 85:84–89.
5. Harding SP, Porter SM. Oral aciclovir in herpes zoster ophthalmicus. Curr Eye Res. 1991; 10 (supp):177–182.
6. Morton P, Thomason AN. Oral aciclovir in the treatment of herpes zoster in general practices. N Z Med J. 1989; 102:93–95.
7. Wood MJ, Ogan PH, McKendrick MW, *et al.*, Efficacy of oral aciclovir in the treatment of acute herpes zoster. Am J Med. 1988; 85 (supp 2A):79–83.
8. Whitley RJ, Weiss H, Gnann JW, *et al.* Aciclovir with and without prednisone for the treatment of herpes zoster. Ann Intern Med. 1996; 125:376–383.
9. Wood MJ, Kay R, Dworkin, *et al.* Oral aciclovir therapy accelerates pain resolution in patients with herpes zoster: a meta-analysis of placebo-controlled trials. Clin Infect Dis. 1996; 22:341–347.
10. Weller S., Blum MR, Doucette M, *et al.* Pharmacokinetics of the aciclovir pro-drug valaciclovir after escalating single- and multiple-dose administration to normal volunteers. Clin Pharmacol Ther. 1993; 54:595–605.
11. Beutner KR, Friedman DJ, Fordszpainiak C, *et al.* Valaciclovir compared with aciclovir for improved therapy for herpes zoster in immunocompetent patients: . Antimicrob Agents Chemother. 1995;39:1546–1553.
12. Vere Hodge RA. Review: famciclovir and penciclovir. The mode of action of famciclovir including its conversion to penciclovir. Antiviral Chemistry and Chemotherapy. 1993; 4(2):67–84.
13. Degreef H and the Famciclovir Herpes Zoster Clinical Study Group. Famciclovir, a new oral antiherpes drug: results of the controlled clinical study demonstrating its efficacy and safety in the treatment of uncomplicated herpes zoster in immunocompetent patients. Int J Antimicrob Agents. 1994; 4(4):241–246.
14. Trying S, Barbarash R, Nahlik J, *et al.* Famciclovir for the treatment of acute herpes zoster: effects on acute disease and post-herpetic neuralgia. A randomized, double-blind, placebo-controlled trial. Collaborative Famciclovir Herpes Zoster Study Group. Ann Intern Med. 1995; 123:89–96.
15. Wood MJ, Johnson RW, McKendrick M, *et al.* A randomized trial of aciclovir for 7 days or 21 days with and without prednisolone for the treatment of acute herpes zoster. N Engl J Med. 1994; 330:896–900.

16

MANAGEMENT OF VARICELLA-ZOSTER VIRUS INFECTIONS IN CHILDREN

Ann M. Arvin

Infectious Diseases Division
Department of Pediatrics
Stanford University School of Medicine
Stanford, California 94305

Most disease caused by varicella-zoster virus (VZV) in children results from primary infection with the virus, which is manifest clinically as varicella or chickenpox.[1–3] Varicella is a very common illness in early childhood. VZV maintains a high prevalence because cases of herpes zoster, caused by reactivation of latent VZV in adults, provide a source of infectious VZV for spread to susceptible contacts. Varicella occurs in a worldwide geographic distribution and causes annual epidemics in temperate climates. These epidemics are less common in tropical areas; as a result, more adults remain susceptible to varicella.

VZV differs from other human herpesviruses in that most primary infections are associated with clinical symptoms. Whereas herpes simplex virus, cytomegalovirus, HHV-6, HHV-7 and Epstein-Barr virus infections are also transmitted to many children during the first few years of life, these infections are usually silent. VZV is also different from these viruses in its capacity to spread by respiratory transmission. It is highly contagious, resulting in attack rates of 80–90% among susceptible close contacts of an index case. No distinct subtypes of VZV have been identified although epidemiologically unrelated viruses can be distinguished by restriction enzyme digest patterns.

Herpes zoster is quite rare in children with the exception of those whose immunity is compromised by treatment for malignancy, bone marrow or organ transplantation or steroid therapy or those who have HIV infection. Intrauterine exposure or varicella during the first year of life increases the risk of herpes zoster in childhood.

1. PATHOGENESIS

Events in the pathogenesis of primary VZV infection include inoculation of mucous membrane sites, probable spread to regional lymph nodes and primary and secondary vire-

Antiviral Chemotherapy 5, edited by Mills *et al.*
Kluwer Academic / Plenum Publishers, New York, 1999.

mic phases.[1] The secondary viremia occurs four or five days before onset of symptoms and persists for 1–2 days after. Recent observations demonstrate that VZV is lymphotropic for CD4+ as well as CD8+ T-lymphocytes.[4] In the course of viremia, infected T-cells transport the virus to skin, where it replicates in dermal and epidermal cells and produces the characteristic vesicular rash. The virus has the pathogenic potential to cause disseminated infection involving lungs, liver, central nervous system and other organs if the T-cell viremia is not terminated by the host response. VZV establishes latent infection in dorsal root ganglia. Reactivation from latency causes the clinical syndrome of herpes zoster. Symptomatic reactivation of VZV is rare in children. Although the virologic mechanisms involved in latency are not known, protection from herpes zoster is associated with preservation of cell-mediated immunity to the virus.

2. CLINICAL DISEASE

Understanding the usual clinical course as well as the risks associated with varicella and herpes zoster in healthy and immunocompromised children is relevant for decisions about the use of antiviral therapy.

2.1. Varicella

The diagnosis of varicella is usually made clinically, based on its characteristic pattern of clinical signs and symptoms.[2] In many cases, especially among older children, varicella is associated with a prodrome of fever, malaise, headache and abdominal pain, occurring 24–48 hours before rash. The typical vesicular rash usually appears first on the face, scalp, or trunk. The lesions are pruritic, erythematous macules, that become fluid-filled vesicles within a few hours. Many patients have ulcerative lesions involving mucous membranes . In untreated cases of varicella, new lesions appear for 1–7 days. The extent of the exanthem ranges from fewer than 10 to more than 1500 lesions. The severity of lesions varies with age and whether the patient has acquired the disease from a household contact (Table 1). The average number of lesions in children ages 2–12 years is about 350 whereas adolescents and adults usually have many more (Table 1). The rash is also more severe in secondary household cases and patients with chronic skin disorders, such as eczema, or sunburn. Laboratory abnormalities observed in otherwise healthy patients with varicella include lymphopenia and granulocytopenia and slightly elevated liver function tests are common.

Table 1. Correlations with varicella severity in untreated children*

Cohort	Endpoint	
Secondary case	Increase in total lesions (432 vs.351)	{p = .03
	New lesions > 3 days	{p = .03
	Fever in first 24 hours	{NS
Age 5 - 12 years	Increase in total lesions (460 vs. 310)	{p < .001
	New lesions > 3 days	{NS
	Fever in first 24 hours (59% vs. 43%)	{p = .005

*Study population: healthy children age 2-12 years

Secondary bacterial infections caused by *Staphylococcus aureus* or *Streptococcus pyogenes* (group A beta-hemolytic streptococcus) are the most common complication of varicella.

Varicella pneumonia is a serious complication of primary VZV infection in adults, affecting as many as 15% of patients. The common neurologic complications of varicella are meningoencephalitis and cerebellar ataxia. Encephalitis is usually transient, resolving within 24–72 hours and permanent sequelae are rare; cerebellar ataxia often persists for days or weeks but also usually resolves completely. Children given aspirin during primary VZV infection are at risk for Reye's syndrome. Hepatitis can progress to acute liver failure. Hemorrhagic complications result from thrombocytopenia and disseminated intravascular coagulopathy.

Resolution of the symptoms of varicella is mediated by the induction of virus-specific immunity, particularly with cell-mediated immune responses.[1] As a consequence of diminished host responses, the clinical course of varicella is more severe in children with lymphoproliferative malignancies or solid tumors. Immunosuppressive therapy given during the incubation period and an absolute lymphocyte count of less than 500 cells at the onset of rash are associated with an increased risk of disseminated varicella. Patients who are immunosuppressed for organ or bone marrow tranplantation, patients receiving high dose, or chronic steroid therapy and patients with congenital cellular immunodeficiency disorders are also at high risk for progressive varicella. Although varicella is usually not life-threatening in children with HIV infection, unusual hyperkaratotic lesions and chronic rash occur.[3] The course of untreated varicella in high risk populations is characterized by prolonged new lesion formation, increased numbers of lesions, and the risk of visceral dissemination. Manifestations of disseminated disease include pneumonia, hepatitis, encephalitis and disseminated intravascular coagulopathy. Most fatal cases of varicella in healthy as well as immunocompromised patients result from viral pneumonia. Fulminant hepatitis and disseminated intravascular coagulopathy are also associated with high mortality rates.

If maternal varicella occurs during the last few days of gestation, the infant is at risk for neonatal varicella, with an attack rate of approximately 20%.[5] Infants who are born within 4 days after or 2 days before the onset of maternal varicella are likely to be born lacking transplacentally acquired antibodies to the virus and may develop progressive varicella with a significant risk of fatal infection without therapy. Congenital varicella syndromes a rare complication of maternal varicella during the first 20 weeks of gestation, affecting the fetus in less than 2% of cases.

2.2. Herpes Zoster

Herpes zoster is characterized by vesicular lesions clustered unilaterally in the dermatomal distribution of one or more adjacent sensory nerves.[1–3] When herpes zoster occurs in children, the cutaneous rash is usually limited and resolves rapidly, within a few days or a week. Healthy children are not at risk for postherpetic neuralgia. Herpes zoster in immunocompromised children is characterized by a more severe local dermatomal infection and a higher risk of cell-associated viremia and visceral dissemination. Viremia secondary to VZV reactivaiton can cause pneumonia, hepatitis, encephalitis, and disseminated intravascular coagulopathy. Children with HIV infection or prolonged immunosuppression for other reasons may have unusual, chronic or relapsing VZV reactivations, causing cutaneous disease or retinitis and central nervous system infections.

3. LABORATORY DIAGNOSIS

Although laboratory diagnosis is not important in healthy patients, rapid diagnosis may be important to guide decisions about antiviral therapy in immunocompromised children. Early varicella lesions can be difficult to identify. In these circumstances, the etiology of the lesions can be established by direct detection, in which cells from the base of a cutaneous vesicle are stained with antibody reagents that bind to VZV proteins or vesicular fluid is tested for viral antigens with enzyme immunoassay methods.[6] VZV replicates in tissue culture cells but it requires 3–7 days for cytopathic effect to appear. Viral culture can be used to confirm diagnoses made with rapid antigen detection methods. Although VZV can also be detected by in situ hybridization or polymerase chain reaction, these methods are used primarily in research settings. Serologic tests for VZV IgG and IgM antibodies are not useful for rapid diagnosis. Antibodies to VZV are present at the time of onset of the rash in some individuals, are detected in almost all patients by three days and continue to rise during convalescence. Testing for VZV IgG antibodies can be used to establish immune status and help to identify susceptible individuals. Enzyme immunoassay methods have a relatively high incidence of false negatives but a positive result is usually reliable; latex agglutination tests for VZV IgG compare well with research laboratory methods.

4. ANTIVIRAL THERAPY

Aciclovir, valaciclovir and famciclovir are licensed for the treatment of VZV infections in adults, but aciclovir is the only drug approved for use in children. Although vidarbine and interferon-α modified the severity of varicella in immunocompromised patients in early clinical trials, aciclovir is the drug of choice because it has less toxicity and was more effective against recurrent VZV infection than vidarabine.[3] Sorivudine modified the course of varicella in healthy adults but it is not approved in the United States because of its adverse interactions with 5-fluoruracil. Foscarnet is the only drug now available for the treatment of infections due to aciclovir-resistant VZV but the emergence of resistance is rare.

4.1. Varicella in Healthy Children

Oral aciclovir has been shown to be safe and to provide measurable effects on the clinical course of varicella in placebo-controlled studies. The clinical symptoms are reduced in otherwise healthy children, adolescents and adults when aciclovir is given within 24 hours after the appearance of the initial cutaneous lesions.[7, 8, 9] Treatment reduces fever, pruritis, the number of days of new lesion formation and the total number of cutaneous lesions compared with placebo (Table 2).[7] Children treated within 24 hours after onset can

Table 2. Benefits of aciclovir: treatment of varicella in healthy children

Reduction of fever from first day of treatment
• Aciclovir > placebo {p = 0.002
Reduction in maximum number of lesions
• Aciclovir > placebo {p < 0.001

Table 3. Dosing recommendations for aciclovir therapy of VZV infections

Administration	Dosage for VZV	Peak plasma concentrations
Intravenous	15mg/kg or 500mg/m^2 every 8 hours	15-25ul/ml = ~5-15 fold above VZV inhibitory concentration
Oral (about 20% absorbed)	Adult: 400 mg per dose 5 times/day Child: 20 mg/kg per dose 4 times/day	1.0-2.0 ug/ml = at or below VZV inhibitory concentration

Aciclovir is eliminated primarily (≥85%) by renal clearance. Give adequate hydration (oral or intravenous) to avoid crystalluria, and reduce the dose for impaired renal function.

be expected to have fever for less than 24–48 hr, new lesion formation for less than 72 hr, and fewer than 300 total lesions. The dose of oral aciclovir for children, 2–12 years old, is 20 mg/kg four times a day for five days (maximum 800 mg/dose) (Table 3). The clinical benefit is considered limited for otherwise healthy children in this age group. Therefore, antiviral therapy is optional; it may be useful in somewhat older children, in the 5–12 year old age range, in secondary household contacts or in children with eczema or other chronic skin disorders. Because varicella complications are rare, it is not known whether aciclovir therapy reduces their frequency in healthy children. The American Academy of Pediatrics recommends that aciclovir be considered for patients with varicella who have chronic cutaneous or pulmonary diseases, diseases requiring chronic salicylate therapy or who are being treated with short or intermittent courses of steroids or aerosolized corticosteroids.[10]

Adolescents are predisposed to more severe varicella. Oral aciclovir therapy initiated within 24 hours provides significant clinical benefit. In these patients, oral aciclovir, given at 800 mg/dose, four times a day for five days, reduces the duration of fever, new lesion formation and total number of lesions. Although clinical trials were not large enough to demonstrate the effect, the pathogenesis of primary VZV infection is such that early antiviral therapy is likely to reduce the risk of varicella pneumonia.

Oral aciclovir is well-tolerated in healthy children. Antiviral therapy initiated within 24 hours after the appearance of the rash does not interfere with the development of immunity to VZV.

Intravenous aciclovir therapy is required for patients who develop varicella pneumonia, hepatitis, thrombocytopenia or encephalitis. The recommended regimen is 500 mg/m^2/dose every 8 hours, for 7 days.

4.2. Varicella in Immunocompromised Children

Aciclovir therapy stops VZV viremia and virus replication in skin, compensating for the delayed acquisition of virus specific immunity in high risk patients (Table 4). Intravenous aciclovir is indicated to prevent progressive varicella in high risk patients.[11] Intravenous aciclovir therapy should be initiated within the first 24–72 hours after the onset of the rash because visceral dissemination occurs within this time interval. Patients at risk for life-threatening varicella include those receiving immunosuppressive therapy for malignancy, organ transplantation, bone marrow transplantation, autoimmune diseases or other chronic disorders. Progressive varicella is also observed in newborn infants born within 4 days before to 2 days after the onset of maternal varicella (Table 5). These infants have the onset of varicella symptoms after 5 days of age and are at risk for varicella hepatitis, pneumonia and other manifestations of disseminated infection.[5] Management of progressive

Table 4. Aciclovir: threapy of varicella in immunocompromised patients

Rationale: Compensate for host immunodeficiency
•Stop VZV cell-associated viremia
•Stop VZV repliction in skin lesions
•Stop VZV spread to lungs, liver, other organs
Intravenous aciclovir
•15 mg/kg or 500 mg/m^2/dose, one hour infusion every 8 hours
Early initiation
•Dissemination occurs in first 2-5 days

varicella in high risk patients requires intensive care support for pulmonary, hemorrhagic and neurologic complications.

Varicella mortality is reduced by antiviral treatment of immunocompromised patients for the most part by preventing varicella pneumonia. The dosage of intravenous aciclovir for varicella in high-risk patients is 500 mg/m2/dose every 8 hours. The drug is given for 7 days, or until no new lesions have appeared for 48 hours. Patients who have evidence of disseminated infection at the time of diagnosis also benefit from intravenous aciclovir therapy although fatal cases may occur.

Some immunocompromised children develop a second phase of viral replication when aciclovir is discontinued. These episodes do not indicate the development of VZV resistance to aciclovir but reflect impaired host responses. A second course of antiviral therapy may be needed in these cases.

4.3. Herpes Zoster in Immunocompromised Children

Aciclovir is used for treatment of herpers zoster in immunocompromised children based on its efficacy in placebo-controlled trials in adults.[3] Severely immunocompromised patients who are considered to be at risk for VZV viremia and visceral dissemination, should receive intravenous aciclovir (500 mg /m2 or 10 mg/kg given every 8 hours). The duration of therapy is 7 days or for 2 days after no new cutaneous lesions are observed. The response to treatment is optimal when aciclovir is given within 72 hours after the onset of symptoms but clinical benefit is observed with later treatment. Oral aciclovir may be given to immunocompromised children with herpes zoster who have lymphocyte counts within the normal range and stable underlying disease.

Table 5. Aciclovir: therapy of varicella in the newborn

Indications
•Infants with varicella born from 4 days after to 2 days before maternal varicella
•Onset of varicella in infant at 5-10 days of age
•High risk infant with progression in spite of VZIG
Not indicated
•Infants with lesions at birth or < 5 days of age
Dose
•Intravenous: 10-15 mg/kg/dose, one hour infusion every 8 hours

5. PROPHYLAXIS

5.1. Varicella-Zoster Immune Globulin

Passive antibody prophylaxis with varicella-zoster immune globulin (VZIG) is recommended for high risk patients who have been exposed to varicella in a close contact.[10] VZIG prophylaxis is recommended for immunocompromised children, seronegative pregnant women and newborn infants exposed to maternal varicella. VZIG should be given to susceptible high risk patients within 96 hours, and if possible within 48 hours, after exposure to an individual with varicella or herpes zoster. It has no antiviral effect when given after the onset of varicella. High dose intravenous immune globulin (100–400 mg/kg) also provides high titers of VZV IgG antibodies. Since high risk patients may develop varicella despite VZIG prophlaxis, exposed children should be monitored for symptoms and treated with antiviral therapy.

5.2. Aciclovir

Oral aciclovir has been given as prophylaxis to healthy children after exposure to varicella. Use of aciclovir in these circumstances is not recommended because information about its safety and efficacy as prophylaxis is limited.[10] Treatment before the appearance of the cutaneous lesions could interfere with acquisition of immunity. When exposure is known to have occurred, high risk patients can be observed for symptoms and managed effecctively by immediate initiation of aciclovir therapy. Aciclovir prophylaxis is not recommended for suppression of VZV reactivation in high risk patients because recurrences develop with the same frequency when suppressive therapy is stopped.

5.3. Varicella Vaccine

The live attenuated varicella vaccine, made from the Oka strain of VZV is approved for administration to healthy children in several countries.[12] Seroconversion rates are more than 95% with a single dose of vaccine in children 12 years old or younger; adolescents and adults require two doses to achieve equivalent levels of protection.[13] It is recommended for children 12–15 months of age in the United States and for susceptible adolescents and adults.[10] The vaccine is safe and well-tolerated by healthy vaccinees. Although it is attenuated, the vaccine virus can be transmitted to susceptible contacts if it causes rash; nevertheless, the risk of transmission is low because few vaccinees have vaccine-associated rash. The protective efficacy of the vaccine is high for healthy children with household exposures.[12] Some vaccinees develop mild breakthrough varicella despite immunization.[13,14] Breakthrough infection is usually associated with fewer than 50 cutaneous lesions and no fever. Persistence of immunity with current vaccine preparations has been documented for more than six years. Varicella vaccine has also been evaluated in children with acute leukemia in remission.[12] The attack rate after household exposure was reduced to 13%, but the vaccine is not approved for administration to high risk patients because the vaccine virus can cause viremia in immunocompromised children. Reactivation of the vaccine virus, causing herpes zoster, has occurred but it appears to be less common than reactivation of wild type VZV.

Since aciclovir inhibits the replication of the vaccine virus, serious varicella-like illness caused by the vaccine virus can be treated with antiviral therapy.

6. SUMMARY

The introduction of varicella vaccine for immunization of healthy children is expected to have a gradual impact on the incidence of VZV infections in the population but antiviral therapy remains an important intervention in clinical practice. The efficacy of aciclovir for treatment of primary and recurrent VZV infections in children has reduced the morbidity and mortality of these illnesses in immunocompromised children dramatically. Oral aciclovir is an effective and useful for the management of varicella in healthy children and adolescents.

REFERENCES

1. Arvin A. Varicella-zoster Virus. In: Virology, 3rd Edition, Fields BN, Knipe DM, Chanock R, Hirsch M, Melnick J, Monath T, Roizman B (eds). Virology. Lippincott-Raven Press. 1995:2547–86.
2. Arvin AM, Varicella-Zoster Virus, in Principles and Practices of Pediatric Infectious Diseas es. Long S, Prober C, Pickering L. eds. Lippincott-Raven, Phildelphia/New York, 1997.
3. Whitley RJ, Varicella Zoster Virus infections. In: Galasso G, Whitley R, Merigan T. eds. Antiviral Agents and Viral Diseases of Man. New York: Lippincott-Raven Press. 1997:279–304.
4. Moffat JF, Stein MD, Kaneshima H, Arvin AM. Tropism of varicella-zoster virus for human CD4+ and CD8+ T-lymphocytes and epiderman cells in SCID-hu mice. J Virol. 1995; 69:5236–42.
5. Prober CG, Gershon AA, Grose C, McCracken GH Jr, Nelson JD. Consensus: varicella-zoster infections in pregnancy and the perinatal period. Ped Infect Dis. J 1990; 9(12):865–9.
6. Gershon AA, Steinberg SP, Schmidt NJ. Varicella-zoster virus in Manual of Clinical Microbiology, eds. Balows A, Hauseler WJ, Herrman KL, Isenberg HD, Shadomy HJ. American Society for Microbiology, Washington D.C. 1991:838–52.
7. Dunkle LM, Arvin AM, Whitley RJ, *et al.* A controlled trial of aciclovir for chickenpox in normal children. N Engl J Med. 1991; 325:1539–44.
8. Balfour HH Jr, Rotbart HA, Feldman S, *et al.* Aciclovir treatment of varicella in otherwise healthy adolescents. The Collaborative Aciclovir Varicella Study Group. J Peds. 1992; 120:627–33.
9. Wallace MR; Bowler WA; Murray NB; Brodine SK; Oldfield EC 3d. Treatment of adult varicella with oral aciclovir. A randomized, placebo-controlled trial. Ann Intern Med. 1992;117:358–63.
10. Report of the Committee on Infectious Diseases, American Academy of Pediatrics, 24th Edition, Varicella-zoster Infections. 1997:573–585.
11. Prober CG, Kirk LE, Keeney RE. Aciclovir therapy of chickenpox in immunosuppressed children: a collaborative study. J Pediatr. 1982; 101:622–625.
12. Arvin AM, Gershon AA. Live attenuated varicella vaccine. Ornston LN, Balows A, Greenberg EP, eds. Annual Review of Microbiology. Annual Reviews, Inc. 1996; 50:59–100.
13. White C Jo. Clinical Trials of Varicella Vaccine in Healthy Children. Ellis R, White C Jo, Moellering RC Jr. eds. The Varicella Vaccine, Infectious Disease Clinics of North America, WB.Saunders, Inc. 1996; 10:595–608.
14. Krause P, Klinman DM. Efficacy, immunogenicity, safety and use of live attenuated chickenpox vaccine. J Pediatr. 1995; 127:518–525.

17

FAILURE OF HIV-1 PROTEASE INHIBITORS TO FULLY SUPPRESS VIRAL REPLICATION

Implications for Salvage Therapy

Steven G. Deeks

AIDS Program
University of California San Francisco
995 Potrero Road
San Francisco, California 94110

HIV-1 protease inhibitors have had a dramatic impact on the management of HIV disease.[1] As part of an aggressive three-drug combination, protease inhibitor therapy results in durable, potent suppression of viral replication. Recent data, however, indicates that a significant proportion of patients may not achieve long-term viral suppression.[2,3,4] The prognostic implications of protease inhibitor drug failure, and its clinical management, remain unknown. This chapter will review the clinical characteristics and genotypic resistance patterns associated with each of the protease inhibitors, and then discuss the limited clinical data addressing drug failure and salvage therapy.

1. VIRAL DYNAMICS: IMPLICATIONS FOR THERAPY

HIV infection is a dynamic process, characterized by a high rate of viral replication.[5] Considering the error prone nature of HIV reverse transcriptase, this high level of viral turnover results in as many as 10^4 to 10^5 mutations at each site in the HIV genome each day. Over time, countless variants, or "quasi-species", emerge. In the presence of selective pressure from an antiretroviral agent, mutations that confer a decreased sensitivity to the drug are rapidly selected, leading to drug resistance. Since viral DNA containing resistance mutations becomes permanently integrated in the host chromosome, drug resistance may be permanent. Treatment strategies are designed primarily to prevent the emergence of drug resistance by minimizing viral replication as much as possible.[6,7]

Antiviral Chemotherapy 5, edited by Mills *et al.*
Kluwer Academic / Plenum Publishers, New York, 1999.

Table 1. HIV-1 protease inhibitors

Protease inhibitor	Dosing	Side effects
Saquinavir (Invirase™)	600 mg tid with meals	Limited side effects
Ritonavir (Norvir™)	300 mg bid; increase by 100 mg/day over 10-14 days to 600 mg bid (with meals)	Malaise, fatigue, abdominal symptoms, fevers, circumoral parasthesias
Indinavir (Crixivan™)	800 mg every 8 hrs (fasting)	Nephrolithiasis, abdominal symptoms
Nelfinavir (Viracept™)	750 mg tid (with meals)	Diarrhea
Saquinavir (Fortovase™)	1200 mg tid (with meals)	Nausea, abdominal symptoms
Ritonavir/Saquinavir (combination)	Ritonavir: 400 mg bid; Saquinavir: 400 mg bid	Gastrointestinal symptoms, liver function abnormalities

2. HIV-1 PROTEASE INHIBITORS: SUMMARY OF SAFETY AND EFFICACY

There are five protease inhibitors currently approved for use by the U.S. Food and Drug Administration: saquinavir-hard gel capsule (Invirase), saquinavir-soft gel capsules (Fortovase), ritonavir (Norvir), indinavir (Crixivan) and nelfinavir (Viracept) [see Table 1]. Each has a unique safety and efficacy profile.[1]

Saquinavir-hard gel capsule (saquinavir-hgc), approved in December 1995, is safe and well tolerated. Saquinavir is extensively metabolized by intestinal and hepatic cytochrome P450 CYP 3A, resulting in limited oral bioavailability (approximately 4%). In early phase II trials, saquinavir (dosed at 600 mg three times daily) had only a modest effect on reducing plasma viral load.[8] To enhance the oral bioavailability and activity of saquinavir, Hoffman-La Roche developed a soft-gel formulation (saquinaivr-sgc or Fortovase). Preliminary data indicates that saquinavir exposure is increased ten fold with soft-gel saquinavir (dosed at 1200 mg three times daily) when compared to hard-gel saquinavir (dosed at 600 mg three times daily). In a limited number of treatment-naive patients, saquinavir-sgc, when used in combination with zidovudine and lamivudine, markedly inhibited HIV replication.[9]

Ritonavir, approved in March 1996, is a potent inhibitor of HIV replication, but side effects and a formidable list of drug interactions have limited its widespread use. Ritonavir also inhibits P450 CYP 3A activity, a characteristic which allows it to be used in combination with other protease inhibitors (see below). In phase I/II clinical trials, ritonavir had a marked effect on plasma viral load and CD4 T cell counts.[10]

Indinavir, approved in March 1997, is also a highly effective antiretroviral. In general, indinavir is well tolerated. Its main adverse event has been nephrolithiasis, which occurs in approximately 5% of patients. Due to its rapid metabolism *in vivo*, indinavir must be given every 8 hours in a fasting state. This rigid dosing schedule may be inconvenient for some patients. Twice daily dosing of indinavir is currently being investigated. When used in combination with zidovudine and lamivudine in minimally pre-treated patients, indinavir results in a potent and durable anti-HIV effect.[11]

Nelfinavir, approved in March 1997, is also an effective agent. It is generally well tolerated; moderate diarrhea is the most common side effect. As with indinavir and ritonavir, nelfinavir provides strong, lasting suppression of viral replication when used in combination with two nucleoside reverse transcriptase inhibitors in minimally pre-treated patients.[12]

Table 2. Possible dual protease inhibitor combinations

Combination	Dose	Status
Ritonavir/Saquinavir-hgc	400/400 mg bid	Virologic data available
Nelfinavir/Saquinavir-sgc	750/800 mg tid	Virologic data available
Ritonavir/Indinavir	400/400 bid	Pharmacokinetic data available
Ritonavir/Nelfinavir	500/750 mg bid	Pharmacokinetic data available
Indinavir/Nelfinavir	1000/1250 mg bid	Pharmacokinetic data available

[1]Ritonavir/saquinavir is a recommended first line option for therapy[6]

[2]Limited clinical data are available to support the use of nelfinavir and saquinavir-sgc; other combinations require further study

[3]Preliminary *in vitro* data suggests that there may be an antagonistic interaction between saquinavir and indinavir; further investigation is necessary

Dual protease inhibitor therapy is becoming increasingly popular. To varying degrees, each protease inhibitor is both an inhibitor and a substrate for cytochrome P450 CYP 3A, leading to favorable drug-drug interactions. For example, ritonavir, an inhibitor of CYP 3A, increases the oral bioavailability of saquinavir approximately 20 fold [unpublished data, Abbott Laboratories]. The combination of ritonavir (400 mg bid) and saquinavir-hgc (400 mg bid), both used at reduced levels, provides effective and long term suppression of viral replication.[13] Nelfinavir (750 mg tid) and saquinavir-sgc (800 mg tid) also appears to be an effective combination. Other combinations, including ritonavir/nelfinavir, ritonavir/indinavir, nelfinavir/saquinavir and nelfinavir/indinavir are undergoing clinical evaluation (see Table 2).

3. PROTEASE INHIBITOR RESISTANCE: *IN VITRO* STUDIES

In vitro, each protease inhibitor selects for high level resistance. Genotypic patterns of resistance are unique for each drug (see Table 3). In clinical trials, each drug typically selects for mutations consistent with those seen *in vitro*. The clinical significance of these mutations, and their ability to confer cross-resistance to other protease inhibitors, remains an area of active clinical investigation.

Saquinavir therapy selects for the L90M and G48V mutations (the latter mutation may be more common with the soft-gel formulation of saquinavir).[14] Each alone is associated with a modest increase in phenotypic resistance to saquinavir. *In vitro*, either mutation confers limited or no cross-resistance to other protease inhibitors, suggesting that sequential therapy with a second protease inhibitor may be effective in patients previously treated with saquinavir.

Table 3. Common mutations associated with protease inhibitor therapy

Drug	Critical mutations	Other mutations*
Saquinavir (Invirase™)	G48V, L90M	L10I, I54V, A71V, V82A, I84V, others
Ritonavir (Norvir™)	V82A/F, I84V	L10I, K20R, M36I, M46I, I54V/L, A71V/L, L90M, others
Indinavir (Crixivan™)	V82A/F, M46I/L	L10I, K20M, I54V/T, A71T/V, I84V, L90M, others
Nelfinavir (Viracept™)	D30N	M36I, A71V, V77I, N88D, L90M, others
Amprenavir	I50V	M46I, I47V, I84V
ABT378	I84V	L10F, M46I, I47V

*Partial list

Indinavir and ritonavir select for similar mutation patterns.[15,16] Typically, several mutations are seen before high level phenotypic resistance emerges. Many investigators believe that an initial mutation occurs in the active site of the protease enzyme (i.e., the V82A mutation), leading to reduced activity of the inhibitor. The resulting virion is "less fit" and unable to replicate efficiently. Subsequent "compensatory" mutations then occur, resulting in a replication-competent resistant virus. Indinavir and ritonavir resistance is commonly associated with high-level cross-resistance among all studied protease inhibitors, suggesting that salvage therapy with a protease inhibitor-containing regimen after failing one of these agents may be difficult.

Nelfinavir selects predominantly for the D30N mutation. After prolonged exposure *in vitro*, and possibly *in vivo*, the L90M mutation may also emerge. A wide variety of other mutations occur during nelfinavir therapy. Since many of these mutations occur in the absence of protease inhibitor therapy (so-called "natural polymorphisms"), the clinical significance of these mutations remains unknown.[17] As with saquinavir, *in vitro* data suggests that nelfinavir resistance does not confer high level resistance to other protease inhibitors.

4. EXPERIMENTAL PROTEASE INHIBITORS

Information about the next generation of protease inhibitors is still fragmentary. The Glaxo-Wellcome agent (amprenavir, formally 141W94), dosed at 1200 mg twice daily, has shown so far to be well-tolerated and safe. Like indinavir, it is a moderate inhibitor of P450 CYP3A. *In vitro* and *in vivo*, amprenavir selects primarily for a unique mutation at position 50 of the protease gene (I50V). Mutations at positions 46 and 47 also occur. Preliminary *in vitro* data indicates that the V82A mutation, common after failing indinavir or ritonavir, confers some cross-resistance to amprenavir.

Abbott Laboratories has a new protease inhibitor in development, ABT 378. In the prescence of human serum, ABT 378 is 10 fold more potent than ritonavir (IC_{50} 0.1 M). Like saquinavir, ABT 378 is heavily metabolized by P450, and therapeutic levels are therefore difficult to achieve. Preliminary data indicates that high levels of ABT 378 will be readily obtainable in patients treated concurrently with low dose ritonavir, a potent inhibitor of cytochrome P450 CYP3A. *In vitro*, ABT 378 selects L10F, M46I, I47V and I84V. Although variants containing the V82A mutation have reduced susceptibility to ABT 378, it may be possible to use increased doses of ABT 378 to suppress the replication of resistant variants.

5. PROTEASE INHIBITOR FAILURE: INCIDENCE AND PREDICTORS

Antiretroviral treatment failure is now generally defined based on plasma viral load.[6,7] Although no standard definition for failure exists, clinical guidelines state that the goal of therapy is to decrease viral load to levels below the limit of quantitation of currently available assays (400 to 500 copies/ml). HIV RNA concentrations should become undetectable after no more than 4 to 6 months of therapy.[6] Failure to achieve durable viral suppression with antiretroviral regimens suggests the emergence of drug resistance.[6,7]

Using this definition of drug failure, only 10 to 20% of patients failed protease inhibitor therapy in phase II/III studies.[11,12] These studies generally enrolled healthy, minimally pre-treated patients. When initiated late in the natural history of HIV infection, failure rates increase significantly.[2,3,4]

Table 4. Possible therapeutic strategies after failure of initial protease inhibitor therapy to suppress viral replication[1]

First protease inhibitor	May be effective[1]	Unlikely to be effective
Saquinavir	Indinavir + NNRTI[2]	Nelfinavir[3]
	Nelfinavir + NNRTI[2]	Indinavir or Ritonavir[4,5]
Indinavir or Ritonavir	Nelfinavir + Saquinavir + NNRTI[2]	Saquinavir
	Refer to clinical study	Nelfinavir
		Ritonavir + Saquinavir[6]
Nelfinavir	Ritonavir + Saquinavir[7]	Saquinavir
	Indinavir +/- NNRTI[2]	
Multi-protease inhibitor failure	Continue best tolerated regimen	
	Refer to a clinical trial	
	Recycle prior therapies	

[1]All regimens should include two nucleoside analogues that the patient is unlikely to have resistance to at the time of the switch
[2]NNRTI = non-nucleoside reverse transcriptase inhibitor (ie: nevirapine, delavirdine). Protease inhibitor doses may need to be modified in the presence of an NNRTI
[3]Lawrence et al., International Workshop on HIV Resistance, St. Petersburg FL 1997
[4]Para et al., 35th IDSA, San Francisco CA 1997
[5]Dulioust et al., International Workshop on HIV Resistance, St. Petersburg FL 1997
[6]Deeks et al., 37th ICAAC, Toronto, Canada 1997
[7]Henry et al., 37th ICAAC, Toronto, Canada 1997

Early retrospective data suggests that the failure rate of protease inhibitor therapy may be significantly higher in the primary care setting than that seen in controlled clinical trials. At the San Francisco General Hospital AIDS Clinic, a large public health clinic serving an urban, predominantly medically indigent patient population, the failure rate of indinavir or ritonavir in 1996 was approximately 50%. Predictors of a poor response to therapy included advanced immunodeficiency (CD4 T cell count less than 200 cells/mm^2), high baseline viral load, prior therapy with nucleoside analogues or saquinavir, and evidence of non-adherence. Adding a protease inhibitor to a pre-existing, failing regimen was strongly associated with drug failure. The virologic success rate was much higher in minimally pre-treated patients who modified their nucleoside analogues at the time indinavir or ritonavir was initiated.[4]

6. SEQUENTIAL PROTEASE INHIBITOR THERAPY: CLINICAL STUDIES

The data from clinical studies addressing sequential, or salvage, protease inhibitor therapy are limited. Only one prospective, randomized study has been reported (AIDS Clinical Trials Group 333). Due to the limited nature of available data, definitive conclusions regarding sequential protease inhibitor therapy are not possible. Regimens to consider are outlined in Table 4.

6.1. Salvage Therapy after Saquinavir Failure

Several reports indicate that saquinavir exposure compromises the activity of subsequent protease inhibitor therapy. In AIDS Clinical Trials Group Study 333, the only prospective randomized study evaluating sequential protease inhibitor therapy reported to date, 72 patients with prior saquinavir-hard gel capsule experience were randomized to receive: (1) continued saquinavir-hard gel capsule, (2) saquinavir-soft gel capsule or (3) indinavir. Nucleoside reverse transcriptase inhibitors were not modified at the time of the switch. In subjects switching to indinavir, a modest 0.58 log decline in plasma viral load was observed (at week 8). This trend persisted over 24 weeks of treatment. Although the interpretation of this data is limited by the fact that nucleoside reverse transcriptase inhibitors were not modified at the time of the switch, the modest virologic effect of indinavir in these patients suggests that saquinavir confers cross-resistance to indinavir.[18] In a prospective study performed in France, 22 patients failing saquinavir therapy were switched to indinavir. Twelve of these 22 patients subsequently failed indinavir therapy. Genotypic resistance assays were performed on samples obtained from patients after indinavir failure occurred. Notably, saquinavir-related mutations (L90M, G48V) were common in this group failing indinavir. Typical indinavir-related mutations such as V82A, were rare, even after prolonged exposure to indinavir.[19] This data indicates that resistance to saquinavir may confer cross-resistance to indinavir *in vivo*, despite limited cross-resistance *in vitro*.

6.2. Salvage Therapy after Indinavir or Ritonavir Failure

There are limited data regarding the use of sequential protease inhibitor therapy after failing indinavir or ritonavir. *In vitro* data predicts that salvage therapy with other protease inhibitor containing regimens will have limited activity in this setting. At the San Francisco General Hospital (SFGH) AIDS Program, 18 patients who failed indinavir therapy were subsequently switched to the combination of ritonavir plus saquinavir (12 of 18 patients modified their background nucleoside analogues at the time of the switch). After 4 weeks of follow-up, a 1.6 log decline in plasma viral load was noted. Unfortunately, this effect was not durable. By week 16 to 24 of follow-up, plasma viral load in 15 of the 18 patients had returned to within 0.5 log of baseline.[20] A more durable response to ritonavir/saquinavir was reported in 16 patients failing indinavir therapy at Johns Hopkins University. Twelve of 16 patients had an undetectable viral load after a median follow-up duration of 28 weeks. The success of ritonavir/saquinavir and 2 nucleoside reverse transcriptase inhibitors in this group was attributed to the strategy of switching early after indinavir began to fail, prior to the emergence of high-level indinavir resistance.[21] The combination of nelfinavir, saquinavir-soft gel capsule, abacavir and nevirapine, but not nelfinavir, saquinavir-soft gel capsule and abacavir, appeared promising in one ongoing study of indinavir failures.[22] This suggests that non-nucleoside reverse transcriptase inhibitors may play an important role in salvage therapy for patients who have failed an initial regimen containing indinavir and 2 nucleoside reverse transcriptase inhibitors. Salvage therapy after indinavir failure is undergoing evaluation in several clinical trials.

6.3. Salvage Therapy after Nelfinavir Failure

As with saquinavir, *in vitro* data indicates that nelfinavir resistance does not confer cross-resistance to indinavir or ritonavir. In a prospective study, 19 patients who failed nelfinavir-containing regimens were switched to the combination of ritonavir, saquinavir, stavudine and lamivudine. After 16 weeks of follow-up, 13 of 19 evaluable patients had an undetectable viral load.[23]

7. SUMMARY

When used as initial therapy in combination with two nucleoside reverse transcriptase inhibitors, indinavir, ritonavir, nelfinavir and possible saquinavir-soft gel capsule (saquinavir-sgc) are highly effective agents. Patients who have been extensively pretreated, have advanced immunodeficiency or are unable to adhere with therapy are at high risk of failing protease inhibitor therapy.

Although relevant prospective, randomized controlled clinical trails have not been reported, failure of a first protease inhibitor appears to compromise future therapeutic options. After resistance to indinavir or ritonavir emerges, salvage therapy with a protease inhibitor-containing regimen may be difficult. Preliminary data indicate that salvage therapy after resistance to saquinavir or nelfinavir emerges may be possible; however, this requires further investigation. The activity of newer agents, such as amprenavir or ABT 378, in salvage regimens is unknown. Until these issues are addressed in large prospective studies, clinicians should assume that cross-resistance will be common among all protease inhibitors.

REFERENCES

1. Deeks S, Smith M, Holodniy M, Kahn J. HIV-1 protease inhibitors: a review for clinicians. JAMA 1997; 277:145–53.
2. Hirsch, M for the protocol 039 (indinavir) study group. Indinavir (IDV) in combination with zidovudine (ZDV) and lamivudine (3TC) in ZDV-experienced patients with CD4 cell counts <=50 cells/ mm^3. In: Program and abstracts of the IV Conference on Retroviruses and Opportunistic Infections, Washington, DC, January 22–26, abstract LB7.
3. Hammer S, Demeter LM, Fischl MA, Squires KE, Bosch RJ, Hughes MD for the ACTG 320 Study and Virology Teams. Clinical, immunological and virological outcomes in ACTG 320, a randomized, placebo controlled trial of indinavir in combination with two nucleosides in HIV-1 infected persons with CD4+ cell/counts < 200 cells/mm^3. In: Program and abstracts of the International Workshop on HIV Resistance, Treatment Strategies, and Eradication; St. Petersburgh FL, USA June 25–28, 1997 [abstract 67].
4. Deeks S, Loftus R, Cohen P, Chin S, Grant RM. Incidence and predictors of virologic failure to indinavir (IDV) or/and ritonavir (RTV) in an urban health clinic. In: Programs and abstracts of the 37th International Conference on Antimicrobial Agents and Chemotherapy. Toronto, Canada 1997 [Abstract LB-2].
5. Havlir DV, Richman DD. Viral dynamics of HIV: implications for drug development and therapeutic strategies. Ann Intern Med 1996;124:984–94.
6. Department of Health and Human Services, Panel on Clinical Practices for Treatment of HIV Infection. Guidelines for the use of antiretroviral agents in HIV-1 infected adults and adolescents. November 5, 1997 document.
7. Carpenter CC, Fischl M, Hammer SM, Hirsch MS, Jacobsen DM, Katzenstein DA, *et al.* Antiretroviral therapy for HIV infection in 1997: Updated recommendations of the International AIDS Society-USA Panel. JAMA 1997; 277:1962–69.
8. Collier AC, Coombs RW, Schoenfeld DA, *et al.* Treatment of human immunodeficiency virus infection with saquinavir, zidovudine and zalcitabine. N Eng J Med 1996; 334:1011–17.
9. Farthing C, Sension M, Churchill H, Chang H, Siemon-Hryczyk P, Pilson R, Fischer L. Soft gel capsule saquinavir (sgcSQV) in combination with AZT and 3TC in antiretroviral-naive HIV-1 infected patients. Program and abstracts of the Infectious Diseases Society of America 35th Annual Meeting, San Francisco, September 13–16, 1997. [Abstract 220].
10. Markowitz M, Saag M, Powderly WG, *et al.* A preliminary study of ritonavir, and inhibitor of HIV-1 protease, to treat HIV-1 infection. N Eng J Med 1995; 333:1534–9.
11. Gulick R, Mellowr, J, Havlir D, *et al.* Indinavir (IDV), zidovudine (ZDV) and lamivudine (3TC): concurrent or sequential therapy in ZDV-experienced patients. In: Programs and abstracts of the 37th International Conference on Antimicrobial Agents and Chemotherapy. Toronto, Canada 1997. [Abstract 1–89].
12. Saag M, Knowles M, Chang Y, *et al.* Combination use of ritonavir and saquinavir in HIV-infected patients: preliminary safety and activity data. In: Program and abstracts of the XI International Conference on AIDS, July 7–12, 1996; Vancouver, Canada. [Abstract I-101].

13. Cameron W, Sun E, Markowitz M, *et al.* Combination use of ritonavir and saquinavir in HIV-infected patients: preliminary safety and activity data. In: Program and abstracts of the XI International Conference on AIDS, July 7–12, 1996: Vancouver, Canada. [Abstract TH.B.934].
14. Jacobsen H, Hanggi M, Ott M, Duncan IB, Owen S, Andreoni M, Vella S, Mous J. *In vivo* resistance to a human immunodeficiency virus type 1 protease inhibitor: mutations, kinetics, and frequencies. J Inf Dis 1996; 173:1379–87.
15. Condra JH, Holder DJ, Schlief WA, *et al.* Genetic correlates of *in vivo* viral resistance to indinavir, a human immunodeficiency virus type 1 protease inhibitor. J Virology 1996; 70:8270–76.
16. Molla A, Korneyeva M, Gao Q, *et al.* Ordered accumulation of mutations to an HIV-1 protease confers resistance to ritonavir. Nature Med 1996; 2:760–766.
17. Patick AK, Kuritzkes D, Johnson VA, *et al.* Genotypic and phenotypic analyses of HIV-1 variants isolated from patients treated with nelfinavir and other HIV-1 protease inhibitors. In: Program and abstracts of the International Workshop on HIV Resistance, Treatment Strategies, and Eradication; St. Petersburgh FL, USA June 25–28, 1997. [Abstract 18].
18. Para MF, Collier A, Coombs R, *et al* for the ACTG 333 Study Team. ACTG 333: antiviral effects of switching from saquinavir hard capsule (SQVhc) to saquinavir soft gelatin capsule (SQVsgc) vs. switching to indinavir (IDV) after prior saquinavir. In: Program and Abstracts of the Infectious Diseases Society of America 35th Annual Meeting, San Francisco, September 13–16, 1997. [Abstract 21].
19. Dulioust A, Paulous S, Guillemot L, Boue F, Galanaud P, Clavel F. Selection of saquinavir-resistant mutants following a switch from saquinavir. In: Program and abstracts of the International Workshop on HIV Resistance, Treatment Strategies, and Eradication; St. Petersburgh FL, USA June 25–28, 1997. [Abstract 16].
20. Deeks S, Grant RM, Horton C, Simmonds N, Follansbee S, Eastman S. Virologic effect on ritonavir (RTV) plus saquinavir (SQV) in subjects who have failed indinavir (IDV). In: Programs and abstracts of the 37th International Conference on Antimicrobial Agents and Chemotherapy. Toronto, Canada 1997. [Abstract 1205].
21. Gallant JE, Hall C, Barnett S, Raines C. Ritonavir/saquinavir (RTV/SQV) as Salvage Therapy after Failure of Initial Protease Inhibitor (PI) Regimen. In: Program Abstracts of the 5th Conference on Retroviruses and Opportunistic Infections, Chicago, Illinois, February 1–5, 1998 (abstract 427).
22. Deeks SG, Grant RM, Symmonds W, Becker M, Hellmann NS, Chesney M, Volberding PA. Pilot study of nelfinavir, saquinavir-soft gel capsule, abacavir with and without nevirapine in patients who have failed an indinavir-containing regimen. In: Program and abstracts of the 2nd International Workshop on HIV Drug Resistance and Treatment Strategies; Lake Maggiore, Italy, June 24–27, 1998.
23. Henry K, Kane E, Melrome H, Simpson J, Patick A, Winslow D. Experience with ritonavir/saquinavir based regimen for the treatment of HIV-infection in subjects developing increased viral loads while receiving nelfinavir. In: Programs and abstracts of the 37th International Conference on Antimicrobial Agents and Chemotherapy. Toronto, Canada 1997. [Abstract I-204].

18

NEW REVERSE TRANSCRIPTASE INHIBITORS

Suzanne Crowe

Macfarlane Burnet Centre for Medical Research
Yarra Bend Road
Fairfield, Victoria 3078 Australia

Reverse transcriptase inhibitors used in the treatment of HIV-1 infection include the nucleoside analogues zidovudine, didanosine, zalcitabine, lamivudine and stavudine. More recently a number of other classes of reverse transcriptase inhibitors have been discovered, and are in various phases of clinical trial. This chapter will focus on the nucleoside analogue abacavir or 1592U89, the non-nucleoside analogues MKC-442, nevirapine, delavirdine, loviride, efavirenz or DMP 266, and the nucleotide analogues adefovir dipivoxil or bis-pom PMEA, and bis-poc PMPA (Table 1).

1. NEW NUCLEOSIDE ANALOGUES THAT INHIBIT HIV-1 REVERSE TRANSCRIPTASE

1.1. Abacavir (1592U89)

Marketed by Glaxo-Wellcome, abacavir is a carbocyclic guanosine analogue, with the chemical name of: (1*R*,4*S*,)-4-[2-amino-6-(cyclopropyl amino)-9H-purin-9-yl]-2-cyclo-pentene-1-methanol. It is activated to the triphosphate by a novel intracellular pathway,[1] initially being phosphorylated to the monophosphate derivative by adenosine phosphotransferase.[2] Abacavir monophosphate is then converted to carbovir mono- di- and triphosphates.[3] The intracellular half-life of carbovir triphosphate is approximately 3.3 hours, similar to that of zidovudine triphosphate.[4] Abacavir triphosphate is not present in cells incubated with abacavir. Less than 2% of abacavir is converted directly to carbovir[1] (see Figure 1).

Abacavir triphosphate inhibits the reverse transcriptase (RT) of both HIV-1 and HIV-2. Although the triphosphate of abacavir is a poor inhibitor of HIV RT, carbovir- triphosphate potently inhibits HIV RT by competitively inhibiting dGTP incorporation into DNA and also by acting as a DNA chain terminator. Abacavir has similar potency against HIV-1 clinical isolates to zidovudine (IC_{50} of 0.26 and 0.23 uM respectively).The drug is syner-

Antiviral Chemotherapy 5, edited by Mills *et al.*
Kluwer Academic / Plenum Publishers, New York, 1999.

Table 1. New reverse transcriptase inhibitors

Drug	Class	Inhibits	Rapid emergence of resistance with monotherapy
abacavir (1592U89)	nucleoside analogue	HIV-1 and HIV-2	minority of pts within 12 weeks; codons 65,74 and 184
MKC-442	non-nucleoside analogue (HEPT)	HIV-1	codon 103 in few pts within 1 month of monotherapy
nevirapine	non-nucleoside analogue	HIV-1	codon 181 +/- 98, 100, 103, 106, 108, 188, 190
delavirdine	non-nucleoside analogue	HIV-1	codon 181, 103, +/- 236
efavirenz (DMP 266)	non-nucleoside analogue	HIV-1	monotherapy results in 11/21 pts developing R by week 16: major mutation in codon 103
adefovir dipivoxil (bis-pom PMEA)	acyclic nucleotide analogue	HIV-1, HIV-2, SIV, FIV, hepatitis B, HSV1 and 2, CMV, EBV, HHV6	mutations associated with clinical resistance do not arise readily
PMPA and prodrug	acyclic nucleotide analogue	HIV-1, HIV-2, SIV, FIV, hepatitis B, less effective against herpesviruses	no data

gistic with zidovudine, nevirapine and the protease inhibitor 141W94, and shows additive to synergistic interactions with lamivudine, zalcitabine, didanosine and stavudine.[5,6,7] Abacavir has activity against hepatitis B virus, with an IC_{50} of 5 to 7 uM. There is no activity against herpes viruses or influenza virus in concentrations of up to 100 uM (Data on file).

Abacavir-resistant strains of HIV-1 have been generated by continued passage of the virus in increasing concentrations of the drug *in vitro*. Selection for resistance *in vitro* does not occur rapidly. Nucleic acid sequence analysis of RT has been performed to determine genotypic changes in resistant strains of HIV-1. Mutations at residues 65 (lys-->arg), 74 (leu-->val), 184 (met-->val) and 115 (tyr-->phe) have been described in association with

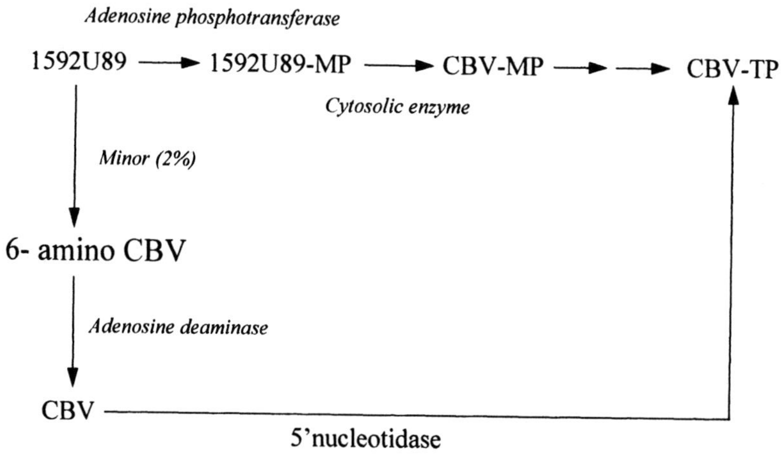

[abbreviations - CBV: carbovir; MP: monophosphate; TP: triphosphate]

Figure 1. The intracellular metabolism of abacavir or 1592U89.

the development of resistance. The mutation at 184 is the first to appear, but only confers a marginal reduction in susceptibility. More than 1 mutation is required to produce significant resistance to abacavir. Individual mutations conferred only 2–4 fold resistance; double and triple mutations conferred 7–11 fold resistance.[8,9,5]

The issue of cross-resistance is of obvious clinical relevance. Abacavir-resistant HIV remains sensitive to zidovudine, and stavudine.[5,9] The development of the mutation within codon 184 has been shown to partially revert zidovudine resistance. Some abacavir-resistant mutants are cross-resistant to didanosine, zalcitabine and lamivudine,[9] however, the development of the mutation within codon 184 resulting from lamivudine therapy may not preclude therapy with abacavir as alone it only confers a low-grade change in susceptibility *in vitro*.

Abacavir has good oral bioavailability (>75% in rat) and a plasma $T_{1/2}$ of approximately 1 hour.[5] Protein binding is approximately 50% in human plasma, with food lowering the plasma concentration by 5% (using the AUC value). The drug crosses the blood-brain barrier reasonably well with a CSF:plasma ratio of 0.19 when measured 1 hour post-dose.[10] Metabolism is not via cytochrome p450. Although there are few studies regarding drug interactions with abacavir, it has been shown that there is no clinically significant interaction between abacavir and the protease inhibitor 141W94.[11] The standard dose of abacavir is 300 mg bid.

In a phase II study (CNAA2001) to determine safety, pharmacokinetics and preliminary antiviral efficacy, HIV-1 infected adults (n = 80) with a CD4 count of between 200–500 cells per ul on entry, no AIDS defining illness, and less than 12 weeks of prior zidovudine therapy, were sequentially enrolled to receive 4 weeks of abacavir monotherapy at a dose of 200 mg tid, 300 mg bid, 400 mg tid, or 600 mg tid. At the end of the fourth week the participants were randomized to add zidovudine or placebo and continued on either abacavir as monotherapy or in combination with zidovudine from weeks 4 to 12, with a 4 week post-treatment observation period.

The baseline demographics were similar in all groups, with a median CD4 count of 355–396 cells per ul and a median plasma HIV RNA of 4.50–5.11 $\log_{10}$ copies per ml. At week 12, plasma HIV-1 RNA was less than 200 copies per ml in 48% and 20% of patients receiving combination therapy (combined cohorts) versus monotherapy, respectively, and a 1.7–2.1 $\log_{10}$ median decrease in plasma HIV RNA in the monotherapy and combination arms. The median CD4 count increased by 90 to 145 CD4 cells per ul in monotherapy and combination arms respectively. There were very few adverse events reported, with one case each of neutropenia, thrombocytopenia, elevated liver transaminases, and hypoglycemia. Other adverse events included nausea, headache, asthenia, diarrhea, insomnia, dizziness, vomiting, and abdominal pain. At week 12 mutations were found in codons 65, 74 or 184 in a minority of patients; combination therapy with zidovudine delayed development of these mutations. Cessation of therapy was associated with a rapid return of plasma HIV RNA to baseline values and reversion of mutations to predominantly wild-type virus.[8]

The safety and antiviral efficacy of abacavir in combination with the protease inhibitor 141W94 has been examined in a small study of 9 patients who had no prior exposure to protease inhibitors. They were treated with abacavir (300 mg bid) in combination with 141W94 (900 mg bid) for 4 weeks. At baseline, the median plasma HIV RNA was 4.17 $\log_{10}$ copies per ml, and the median CD4 count was 223 cells per ul. Preliminary data from 7 patients showed a median maximum decrease in plasma HIV RNA of 1.97 $\log_{10}$ from baseline and a median maximum increase in CD4 count of 79 cells per ul.[12]

Abacavir has also been evaluated in a small study of 9 patients who were previously enrolled in CNAA2001 and who had stopped abacavir for varying periods of time. These

Table 2. Virologic, clinical, and pharmacokinetic features of abacavir

- Nucleoside analogue reverse transcriptase inhibitor
- Similar *in vitro* potency to zidovudine
- Inhibits HIV-1 and HIV-2
- Requires intracellular phosphorylation for antiretroviral activity
- Resistance via development of a number of mutations
  single mutation in codon 184 results in minor decrease in sensitivity
- Reasonable CNS penetration
- Metabolism not via cytochrome p450
- Oral dose of 300 mg bid
- Limited clinical efficacy data (surrogate markers) of abacavir in combination with other nucleoside analogues and protease inhibitors
- Well tolerated side effects include rash, nausea, headache, and rarely neutropenia, raised liver transaminases

patients received abacavir in addition to treatments containing zidovudine or lamivudine plus indinavir or ritonavir. Preliminary data from patients who completed 24 weeks of therapy showed a mean 1.9 $\log_{10}$ decline in plasma HIV RNA and a mean increase of 174 CD4 cells per ul, compared to baseline values.[13]

2. NON-NUCLEOSIDE ANALOGUE INHIBITORS OF HIV-1 REVERSE TRANSCRIPTASE

2.1. Efavirenz (DMP 266, L-743,726,)

Marketed under the trade name of Sustiva_{TM} by DuPont Merck Pharmaceutical Company, it is a novel non-nucleoside reverse transcriptase inhibitor (NNRTI) that binds to an allosteric site on RT. It is a member of the 1,4-dihydro-2H-3,1-benzoxazin-2-one class of NNRTIs.

There is good *in vitro* inhibition of HIV-1 by efavirenz with a 90% inhibitory concentration of 2.5nM against HIV-1_{RF}.[14] The drug is also active against zidovudine-resistant and certain other NNRTI-resistant strains of HIV-1.[15,16] Resistance to the drug develops with *in vitro* passage of HIV-1 in MT-2 cells in the presence of increasing concentrations of efavirenz, with mutations in codons 179 (Val-->Asp), 100 (Leu-->Ileu), 108 (Val -->Ileu), 181 (Tyr-->Cys) which are associated with 1000-fold reduced susceptibility.[14] The mutation within codon 100 is apparently sufficient to reduce susceptibility by 28-fold.[14]

Resistance has also been observed to develop *in vivo*. In patients receiving a daily dose of 200 mg as monotherapy for 2 weeks followed by combination with indinavir, approximately half (11/21) developed a rebound in plasma HIV RNA by 16 weeks, with one or more mutations. The most common was within codon 103 (Lys-->Asn); changes were also noted in codons 106, 188, 190 and 100, but not in 101 or 181.[17]

A major advantage of efavirenz over most other antiretroviral drugs is the fact that it can be administered once daily due to its long plasma half-life (greater than 40 hr).[18] The drug is 99% protein bound[16] and in rats has an oral and absolute bioavailability of 16% and 42% respectively.[15] Efavirenz is a modest inducer of the cytochrome p450 isoenzyme CYP3A and a substrate for the CYP2B6 isoform. As a result it interacts with other compounds that are metabolized by these cytochrome p450 isoenzymes including protease inhibitors. Efavirenz decreases plasma indinavir and increases plasma nelfinavir

Table 3. Virologic, clinical and pharmacokinetic features of efavirenz

- Non-nucleoside reverse transcriptase inhibitor
- Inhibits only HIV-1
- Does not require intracellular metabolism for anti-HIV-1 activity
- Active against zidovudine-resistant and certain NNRTI-resistant strains of HIV-1
- High level resistance develops *in vitro,*, and *in vivo* with monotherapy
- Metabolized by cytochrome p450
- There are a number of potential drug interactions
  Eg. efavirenz reduces plasma indinavir levels by 30%
  Eg. efavirenz reduces plasma clarithromycin levels by over 40%, and the combination is poorly tolerated
- Highly protein bound
- Current dose is 600 mg per day
- Limited clinical efficacy data (surrogate markers) demonstrated for efavirenz in combination with indinavir (1000 mg tid) and stavudine
- Well tolerated (main side-effects are rash and dizziness).

concentrations by about 30% and 15% respectively; indinavir does not alter plasma efavirenz levels.[19] There is no significant interaction between efavirenz and fluconazole.[20] The area under the curve of clarithromycin is decreased by 44% and the combination is poorly tolerated. Unbound drug enters the CSF and CSF concentrations exceed the IC_{95} for most wild type and clinical strains of HIV-1 at a daily dose of 200mg.[21]

A double blind pilot study was performed to compare long-term (50 weeks) efficacy and tolerability (24 weeks) of efavirenz in combination with indinavir.[18,22] The study enrolled patients with a CD4 count of between 100 and 500 cells per ul, and a plasma HIV RNA of greater than 20,000 copies per ml. The patients were protease inhibitor and NNRTI naive with baseline demographics including a mean age of 36 years, of whom 73% had received prior nucleoside therapy, with a mean CD4 count of 270 cells per ul, and a mean plasma HIV RNA of 4.97 $\log_{10}$ copies per ml; greater than 80% were male Caucasians. The study design was for 2 weeks monotherapy with indinavir, followed by continued indinavir monotherapy (n=9) or combination therapy with indinavir and efavirenz at a dose of 200 mg per day (n=21) for the next 40 weeks. There were a number of protocol changes during the study. As indinavir levels were found to decrease by 30% in the presence of efavirenz, the dose of indinavir was increased to 1000 mg tid at week 12. After week 24, participants randomized to the indinavir monotherapy arm were given the option of adding stavudine and efavirenz. After week 36 the dose of efavirenz was increased to 600 mg per day.

At 50 weeks 73% of those receiving combination therapy with efavirenz and indinavir had plasma HIV RNA levels below detection (less than 400 copies per ml) with a mean decrease of 2.07 $\log_{10}$ HIV RNA copies per ml and an increase in CD4 cells of 127 cells per ul. Two thirds of those receiving efavirenz, stavudine and indinavir had viral load levels below detection (less than 400 copies per ml) with a mean decrease of 1.89 $\log_{10}$ HIV RNA copies per ml and an increase in CD4 cells of 197 cells per ul.[22]

Efavirenz is well tolerated with the main adverse effect being a rash that develops in 19% and 11% in combination and monotherapy arms respectively; only 2 patients receiving efavirenz discontinued treatment due to rash, and there were no grade 4 rashes.[18] Central nervous side effects such as dizziness have also been reported.

Table 4. Virologic, clinical, and pharmacokinetic features of MKC-442

- Nucleoside analogue, but functions as a non-nucleoside inhibitor of reverse transcriptase (HEPT derivative)
- Inhibits only HIV-1
- Does not require intracellular metabolism for anti-HIV-1 activity
- Metabolized by cytochrome p450
- Dose-escalation ongoing to define optimal dose
- Clinical trials in progress
- Appears to be well tolerated (main side-effects are rash, gastrointestinal complaints, headache, and elevated liver function tests)

2.2. MKC 442

Marketed by Triangle Pharmaceuticals, it is a HEPT derivative (1-[(2-hydroxyethoxy)methyl]-6-(phenylthio)thymine with the chemical name of 6-benzyl-1-ethoxymethyl-5-isopropyluracil. It is an acyclic nucleoside derivative, but phosphorylation of the hydroxyl group of the side chain at N-1 is not necessary for anti-HIV-1 activity, thus the drug is considered similar to non-nucleoside RT inhibitors.

The drug has good *in vitro* antiretroviral activity with an IC_{50} and IC_{90} of 3.8 and 12 nM respectively in PBMCs and 15 and 98 nM respectively in MT-4 cells[23]. It is active against zidovudine-resistant[23] and lamivudine-resistant[24] strains of HIV-1 and has additive to synergistic interactions with zidovudine, didanosine, zalcitabine, lamivudine, nevirapine, and saquinavir.[23,25,26,27]

In vitro culture of HIV-1 in the presence of MKC-442 results in the development of resistance, with a mutation at position 181 of RT (Tyr--> Cys).[23,24] A second mutation has also been described within codon 188(Tyr --> Cys), resulting in 32-fold decrease in sensitivity to MKC-442[27]. MKC-442-resistant HIV is cross-resistant to nevirapine.[23] *In vivo*, resistance has been reported with development of a mutation within codon 103 (Lys-->Asn) in one patient at day 15, and in 4 patients at day 22.

MKC-442 has equal distribution in the plasma and brain of rats.[28] The drug has a plasma $T_{1/2}$ of approximately 7 hours[28] and is metabolized by cytochrome P450 (isoenzymes CYP3A4 and 3A5, and to a lesser extent by CYP1A2).[29] Food does not significantly alter the pharmacokinetics of the drug, resulting in an increase in AUC of 21% and a delay in T_{max}.[28]

A phase 1B study has commenced to define safety, tolerability and preliminary efficacy using escalating multiple repeated oral doses.[29] The design was a double blind, randomized, placebo-controlled study, with repeated oral doses of MKC-442 for 2 months of up to 500 mg bid or one month of 750 mg bid with one month of follow-up following therapy. The study enrolled HIV-infected, asymptomatic, antiretroviral naive males and females, with CD4 counts of greater than 100 cells per ul, and plasma HIV RNA levels of greater than 10,000 copies per ml. At baseline, the mean CD4 count was 355 cells per ul and mean HIV RNA was 5.05 $\log_{10}$ copies per ml.

Preliminary results indicted a median maximum decrease in viral load at 1 week of 1.19 $\log_{10}$ copies per ml in those receiving 500 mg bid (n = 6); and a decrease of approximately 1.4 $\log_{10}$ copies per ml in those receiving 750 mg bid. The proportion of patients with greater than a 1.0 $\log_{10}$ reduction in HIV RNA at week 1 was more than 95% of those in the 750 mg bid arm, and approximately 80% of those in the 500 mg bid arm.

Table 5. Virologic, clinical, and pharmacokinetic features of nevirapine

- Non-nucleoside reverse transcriptase inhibitor (dipyridodiazepinone derivative)
- Inhibits only HIV-1
- Does not require intracellular metabolism for anti-HIV-1 activity
- *In vitro* and *in vivo* resistance with the development of mutations within codon 181, 188, and 190.
- Good CSF penetration in a small study of 6 persons (45% of plasma level)
- The oral dose is 200 mg daily for the first 14 days then 200 mg bid (due to metabolic autoinduction of hepatic enzymes that decrease plasma levels of nevirapine)
- Metabolism is via cytochrome p450
- There are a number of potential drug interactions
  Eg. ketoconazole plasma levels decreased by 60% thus ketoconazole should not be co-administered with nevirapine
- Clinical efficacy (surrogate markers) has been demonstrated for nevirapine in combination with nucleoside analogues
- Well tolerated — most common side-effects include rash, elevated liver enzymes, headache

MKC-442 is well tolerated, with the most common side-effects being nausea, vomiting or diarrhea, and headache. Rash and elevated liver function tests (gamma glutamyl transferase) have also been reported.

2.3. Nevirapine

Marketed by Boehringer Ingelheim under the trade name of Viramune. It is a non-nucleoside reverse transcriptase inhibitor. The details of the *in vitro* activity of nevirapine, its pharmacokinetics and results from early clinical trials are summarized by Cheeseman 1993,[30] by Kilby[31] and Saag 1996, and by Crowe 1997.[32]

ACTG 241 was a randomized, double-blind, placebo-controlled trial assessing the efficacy of triple therapy with nevirapine, zidovudine and didanosine versus double therapy with zidovudine and didanosine over a 48 week period in 398 patients with less than 350 CD4 cells per ul and extensive prior antiretroviral therapy. At baseline, the mean CD4 count was 153 cells per ul, with the mean plasma HIV RNA of 4.59 $\log_{10}$ copies per ml. Participants had a mean of 25 months of prior antiretroviral treatment with one third of patients having received greater than 3 years of prior therapy. At the end of the 48 weeks the improvement in CD4 count and viral load in those receiving the triple combination was greater than those randomized to double therapy (Data on file).

BI-1046 was an international, randomized, double-blind, placebo-controlled trial assessing triple therapy with nevirapine, zidovudine and didanosine versus double therapy with either zidovudine and didanosine or zidovudine and nevirapine in 151 participants with CD4 counts of betwen 200 and 600 cells per ul, who were treatment-naive at entry. The study was of 12 months duration. At entry the mean CD4 count was 375 cells per ul, and mean plasma HIV RNA was 4.41 $\log_{10}$ copies per ml. In those randomized to either triple therapy or zidovudine plus didanosine, there was a significant increase in CD4 count above baseline of 150 cells per ul sustained for 1.5 years. The decline in viral load was significantly greater in the patients randomized to the triple therapy arm, compared to those receiving zidovudine and didanosine. Those randomized to receive zidovudine and nevirapine had a short-lived improvement in viral load and CD4 count (Data on file).

The VAN study was an open label, uncontrolled study of 21 patients with advanced disease (less than 50 CD4 cells per ul) who were heavily pretreated with antiretroviral drugs, but naive to indinavir and nevirapine. They received lamivudine, indinavir and nevirapine for 20 weeks. At baseline, the mean viral load was 5.13 $\log_{10}$ copies per ml. There was a reduction in plasma HIV RNA of 3.12 $\log_{10}$ and a median increase in the CD4 count of 75 cells per ul. Viral load was below detection (500 copies per ml) in 56% of patients (Data on file).

2.4. Delavirdine

Marketed under the trade name Rescriptor_{TM} by Pharmacia and Upjohn. It is a BHAP (bisheteroaryl piperazine) derivative that is only active against HIV-1. It binds to HIV-1 RT at a different site to nucleoside analogues and inhibits activity of the enzyme allosterically. Its inhibition is non-competitive with the dNTP substrate.

In vitro, delavirdine inhibits HIV-1 with IC_{50} of 0.066–0.26uM,[33] and has synergistic activity against HIV-1 with zidovudine, didanosine, zalcitabine, saquinavir and combinations of nucleoside analogue reverse transcriptase inhibitors.[34,35] It is also active against strains of HIV-1 that are resistant to zidovudine, didanosine, zalcitabine and lamivudine.

In vitro resistance (10–100-fold) develops within 3–5 passages of virus in peripheral blood lymphocytes in the presence of drug.[36] Mutations found within codons 236 (Pro-->Leu), 228 (Leu-->Phe), 273 (Gly-->Arg) that appear to be responsible for resistance.[36] No mutations have been detected within codons 103 and 181. Delavirdine-resistant strains of HIV-1 remain sensitive to zidovudine.

In patients treated with delavirdine monotherapy, mutations develop in codons 181, 103 +/- 236, resulting in reduced sensitivity to delavirdine. The 236 mutation is thought to delay the emergence of zidovudine resistance. Mutations in codon 181 with or without 103 develop in 50% of patients within 8 weeks of commencing monotherapy.[37] In patients receiving delavirdine and zidovudine combination therapy, there is protection against developing the mutation in codon 181, although mutations develop within codons 236 and 101[38] (Freimuth, personal communication). In 6 of 7 nucleoside-experienced patients receiving delavirdine and didanosine combination, mutations in codons 103 and/or 181 developed within 4–10 weeks of therapy.[39]

Delavirdine has non-linear pharmacokinetics with reduced oral clearance and increases in plasma $T_{1/2}$ with increasing dose. The drug may be taken without regard to timing of meals and the recommended dose is 400 mg tid. The major metabolic pathway is via cytochrome p450 (CYP3A4 isoform), and it both inhibits, and is a substrate for CYP3A4.[40] There are many drug interactions due to the metabolism of delavirdine by cytochrome p450. eg Rifampicin is a potent inducer of CYP3A4 and decreases plasma delavirdine levels 50-fold; it is contraindicated in patients receiving delavirdine.[41]

For information regarding phase I/II clinical trials, please refer to Freimuth 1996.[42] A number of more recent trials are described below.

Protocol 0021[42] (data on file) enrolled HIV-infected persons with CD4 counts between 200–500 cells per ul. Participants were randomized to 1 of 4 arms (n=180 per arm): delavirdine placebo plus zidovudine; delavirdine 200 mg tid plus zidovudine; delavirdine 300 mg tid plus zidovudine; delavirdine 400 mg tid plus zidovudine. At baseline, 60% of participants were zidovudine naive and 40% had received zidovudine for less than 6 months; the mean CD4 count in the arms ranged from 325–340 cells per ul. Patients randomized to receive delavirdine and zidovudine combination therapy had better responses (sustained reduction in HIV RNA of about 0.5 $\log_{10}$ copies per ml for up to 52 weeks, and a sustained increase in CD4 count for up to 60 weeks) than those randomized to zidovudine monotherapy.

Table 6. Virologic, clinical, and pharmacokinetic features of delavirdine

- Non-nucleoside reverse transcriptase inhibitor (BHAP derivative)
- Inhibits only HIV-1
- Does not require intracellular metabolism for anti-HIV-1 activity
- *In vitro* resistance with the development of mutations within codons 236, 228, 273
- *In vivo* resistance is associated with mutations in codons 181, 103 +/- 236
- Very little CSF penetration
- The major metabolic pathway is via cytochrome p450.
- There are a number of potential drug interactions
  Eg. Rifampicin is a potent inducer of CYP3A4 and decreases plasma delavirdine levels 50-fold; it is contraindicated in patients receiving delavirdine.
- The recommended oral dose is 400 mg tid
- Clinical efficacy (surrogate markers) has been demonstrated for delavirdine in combination with nucleoside analogues

In Protocol 0021, part II (data on file), newly-enrolled patients with CD4 counts of between 200 and 500 cells per ul, less than 6 months prior exposure to zidovudine, and who were lamivudine and delavirdine naive were eligible to enrol into a study to evaluate triple therapy using delavirdine, zidovudine and lamivudine, with 2 dual therapy regimens (delavirdine plus zidovudine and delavirdine plus lamivudine). In an interim analysis: triple therapy was associated with a significantly greater mean reduction in HIV RNA than zidovudine plus lamivudine at weeks 8–32.

The number of persons with viral load below detection (less than 400 copies per ml) was above 60% in the triple therapy group from weeks 8–32, 30% for zidovudine plus lamivudine, and less than 10% for those receiving delavirdine plus zidovudine.

In Protocol 0017, persons with advanced HIV infection with CD4 counts of 0–300 cells per ul were randomized to receive delavirdine placebo plus didanosine, or delavirdine 400 mg tid plus didanosine. About 595 patients were enrolled into each arm. At baseline all were zidovudine experienced (mean 1 year), and 25% had previously received didanosine (less than 4 months). The mean CD4 count was 135 cells per ul, and mean plasma HIV RNA about 5.8 $\log_{10}$ copies per ml. Following an interim analysis, the DSMB recommended terminating the trial because the goal of showing significant clinical difference between the 2 arms would not likely be achieved. In a clinical endpoint analysis (intent to treat) there was no difference between the 2 arms in terms of clinical progression (data on file).

The most prominent adverse event associated with delavirdine use is the development of a rash which occurs in 12–18% more treated patients than controls; serious rashes are rare (0.4%).

3. NUCLEOTIDE ANALOGUE INHIBITORS OF RT

3.1. Adefovir Dipivoxil (Bis-Pom PMEA)

Marketed by Gilead Sciences, Inc., adefovir dipivoxil is an acyclic nucleotide analogue, with the chemical name of 9-[2-[bis[(pivaloyloxy) methyl]phosphinyl]methoxy]ethyl]adenine. It is the orally bioavailable pro-drug of nucleotide analog PMEA (adefovir). Adefovir dipivoxil is converted to adefovir in cells and then phosphorylated by mitochondrial adenylate kinase[43] to the diphosphate derivative.

Table 7. *In vitro* activity of adefovir and adefovir dipivoxil against HIV, hepatitis B, and herpes simplex virus type 2

Vir	Compound	IC_{50} uM	CC_{50} uM	Selectivity index
HSV-2	adefovir	30 +/- 19	2666 +/- 471	88
HSV-2	adefovir dipivoxil	0.5 +/- 0.2	135	255
HIV	adefovir	16 +/- 6	160 +/- 10	10
HIV	adefovir dipivoxil	0.5 +/- 0.2	2 +/- 0.3	4
HBV	adefovir	0.2	106	530

(Data from Foli *et al.*, 1996[49,50], data on file). Abbreviations. IC_{50}: 50% of inhibitory concentration; CC_{50}: 50% cytotoxic concentration. Values are given +/- standard deviation of 3 experiments for all except HBV).

The drug has broad spectrum of activity *in vitro* against a wide range of viruses, including HIV and other retroviruses, hepatitis B virus, herpes simplex virus types 1 and 2, Epstein Barr virus, cytomegalovirus, and human herpes virus type 6.

Adefovir is less potent than zidovudine against HIV-1 in peripheral blood lymphocyte cultures but is active against HIV replication in macrophages.[44] Adefovir dipivoxil is 32-fold more potent than adefovir against $HIV\text{-}1_{IIIB}$ but has a lower selectivity index (data on file). Strains of HIV-1 from patients receiving zalcitabine or didanosine with a mutation in codon 65 (Lys --> Arg) of RT are 12–15 times less sensitive to adefovir dipovoxil than wild type HIV-1.[45,46] Zidovudine- resistant HIV-1 and lamivudine-resistant mutants with M184V mutation remain sensitive to adefovir.[47] When administered subcutaneously to macaques on days 1 and 2 prior to inoculation with SIV and continued for 30 days, PMEA (20 mg/kg) did not protect from infection but resulted in lower viral load and sustained health, compared to untreated animals.[48]

In vitro selection for mutations in codons 65 and/or 70 (Lys --> Arg, and Lys --> Glu respectively) occurs when infected cells are cultured in the presence of adefovir, resulting in approximately a 10-fold decrease in susceptibility to adefovir. However, mutations associated with clinical resistance do not arise readily during 12 months of adefovir therapy.[51,52]

Adefovir is the only metabolite detected in rats after oral administration of adefovir dipivoxil: no pro-drug is detected.[53] There is low oral bioavailability of adefovir in humans, due to limited intestinal permeability of the phosphonate (ionized at physiological pH)[53]. Adefovir dipivoxil has better permeability and a reported oral bioavailability of <12%-40% in humans.[54,55] The half-life of adefovir is long, allowing once-daily administration of the drug.

Five phase I or I/II trials of adefovir dipivoxil for HIV infection have been completed. Adefovir dipovoxil is currently in phase III clinical trials in HIV-infected individuals (and in phase II trials for chronic hepatitis B). In general, adefovir dipivoxil causes a modest decrease in plasma HIV RNA (approx 0.5 $\log_{10}$ copies per ml after 14 days of treatment).[56]

In a multicentre, randomized double-blind placebo controlled trial (GS-94–403) to evaluate safety and efficacy of adefovir dipivoxil monotherapy in HIV-infected adults.[57] HIV-infected persons with a CD4 count of greater than 200 cells per ul, and plasma HIV RNA levels of greater than 10,000 copies per ml were randomized to receive one of two dose levels of adefovir dipivoxil (125 versus 250 mg/kg/day) or placebo, for a period of 6 weeks, after which all participants were offered active drug. At baseline, the mean CD4 count was 326–348 cells per ul, and mean plasma HIV RNA was 4.9–5.0 $\log_{10}$ copies per

Table 8. Virologic, clinical, and pharmacokinetic features of adefovir dipivoxil

- Acyclic nucleotide analogue
- Orally bioavailable pro-drug of nucleotide analog PMEA (adefovir)
- Broad spectrum of antiviral activity against retroviruses, and herpes viruses
- *In vitro* selection for mutations in codons 65 and/or 70. *In vivo*, mutations conferring resistance are rare
- Long plasma half-life, thus once daily dosing
- Modest decline in plasma HIV RNA in clinical trials (about 0.5 log10 copies per ml)
- Reduces serum carnitine levels
- Well tolerated: main side-effects are nausea, diarrhea, and aesthenia.

ml. Between 71–83% (higher in placebo group) of participants had received prior antiretroviral treatment, with the mean duration of prior treatment being 2.3–3.5 years (higher in placebo group).

There was a mean change in CD4 count in placebo, 125 and 250 mg groups of -41, +46 and + 15 cells per ul, respectively, at week 6. There was no mean change in plasma HIV RNA from baseline in the placebo arm, but in the 125 and 250 mg arms there was a significant decline of 0.5 and 0.4 $\log_{10}$ copies per ml, respectively ($p = 0.003$, and 0.01 versus placebo respectively). Preliminary evidence of anti cytomegalovirus activity was found in a nested pilot study of 13 persons with detectable CMV DNA in semen at baseline who had a semen sample collected at week 6. Patients receiving placebo ($n = 3$) had a median increase in semen CMV levels of 0.3 log luminometer units; those receiving 125 mg ($n = 5$) and 250 mg ($n = 5$) had median decreases of 0.7 and 0.3 units respectively.[57]

The drug was well tolerated with the most commonly reported adverse events being nausea, diarrhea and asthenia. Moderately raised transaminases and neutropenia that were judged to be possibly drug related were infrequently observed. Serum carnitine levels decreased by 42% and 62% by week 12 in the 125 and 250 mg groups respectively. Mutations in codons 65 and 70 were not seen; a mutation was observed in codon 184 (M-->V) but this patient had a sustained decline in HIV RNA of 1.95 $\log_{10}$ copies per ml, and therefore the significance of this mutation remains unclear.

3.2. PMPA Prodrug (GS-4331-05)

Another closely related acyclic nucleotide analogue (adenosine nucleoside monophosphate) under development by Gilead Sciences, Inc. It has the chemical name of 9-[2-(R)-[[Bis[[isopropoxy-carbonyl) oxy]methoxy] phosphinoyl]methoxy]propyl]adenine fumarate. PMPA is metabolized to the diphosphate derivative which inhibits HIV replication by direct inhibition of RT and premature DNA chain termination.

PMPA prodrug is 15–230-fold more active than PMPA against HIV-1. PMPA is active against HIV-1, HIV-2, SIV, FIV and is more potent in inhibiting HIV-1, HIV-2 and FIV than PMEA *in vitro*.[58] It has an IC_{50} of 0.0004–0.003 uM with a high selectivity in-

Table 9. *In vitro* activity of PMPA and its prodrug against HIV

Compound	IC_{50} uM*	CC_{50} uM	Selectivity Index
PMPA	0/095	1200	12600
PMPA prodrug	0.0004	30	75000

* in PBMCs using a clinical isolate of HIV-1[51].

dex.[59] The drug also potently inhibits HIV replication in macrophages with an IC_{50} of 0.025uM.[60] There is additive to synergistic activity against HIV-1 with lamivudine, stavudine, didanosine, zalcitabine, nelfinavir, ritonavir, saquinavir, indinavir, and strong synergy with zidovudine (Data on file). The drug potently inhibits hepatitis B virus with an IC_{90} of 10–19uM (Data on file).

The data from animal studies regarding PMPA or its prodrug are very promising, with 100% efficacy of topical PMPA in preventing vaginal transmission of SIV,[59] and 100% efficacy of once daily PMPA administered subcutaneously 48 hours before, 4 hours after or 24 hours after intravenous inoculation of SIV in macaques and continued for 4 weeks.[61]

In vitro passage of HIV-1_{IIIB} in MT2 cells selects for virus with mutations within codon 65 (Lys -->Arg) *in vitro*[62]. Zidovudine- resistant, lamivudine-resistant (M184V or M184I), multidrug resistant (Q151M), and PMEA-resistant (codon 70, Lys --> Glu) strains of HIV remain susceptible to PMPA.[63]

PMPA has a prolonged intracellular plasma $T_{1/2}$ of 4.5 -7.2 hours. The active intracellular metabolite is PMPA diphosphate whose plasma $T_{1/2}$ exceeds 10 hours.[63] There is significant transfer of PMPA across the placenta with a maternal/fetal concentration ratio of 17% (Data on file). The drug is excreted primarily via the kidney, with active tubular secretion (Data on file).

In a randomized double-blind placebo-controlled dose-escalation study to assess the pharmacokinetics, safety and efficacy of PMPA,[63] a single intravenous dose of PMPA (n = 16) or placebo (n = 4) was administered to HIV-infected persons with CD4 counts of greater than 200 cells per ul and greater than 10 000 copies per ml of plasma HIV RNA. This single dose of PMPA monotherapy was followed by 7 day wash-out period and then 7 consecutive days of once-daily intravenous therapy, with 2 sequential doses of 1 and 3 mg per kg per day. At baseline, the median CD4 count was 239, 496 and 804 cells per ul in arms receiving 3 mg per kg per day, 1 mg per kg per day and placebo respectively, and median HIV RNA of 5.1, 4.7 and 4.3 $\log_{10}$ copies per ml respectively. After completion of 7 days treatment, the change in plasma HIV RNA from baseline in those receiving 3 mg per kg per day, 1 mg per kg per day, or placebo was a decline of 1.1, a decline of 0.6 and and increase of 0.1 $\log_{10}$ copies per ml respectively. Adverse events were not greater or different to those observed in the placebo group, apart from fatigue. There were no grade 3 or 4 adverse events.

REFERENCES

1. Faletto MB, Miller WH, Garvey EP, *et al.* Unique intracellular activation of the potent anti-human immunodeficiency virus agent 1592U89. Antimicrob Ag Chemother. 1997; 41:1099–1107.
2. Garvey EP, Krenitsky TA. A novel human phosphotransferase highly specific for adenosine. Arch Biochem Biophys. 1992; 296:161–9.
3. Miller WH, Daluge SM, Garvey EP, *et al.* Phosphorylation of carbovir enantiomers by cellular enzymes determines the stereoselectivity of antiviral activity. J Biol Chem. 1992; 267:21220–4.
4. Parker WB, Shaddix SC, Bowdon BJ, *et al.* Metabolism of carbovir, a potent inhibitor of human immunodeficiency virus type 1, and its effects on cellular metabolism. Antimicrob Ag Chemother. 1993; 37:1004–9.
5. Daluge SM, Good SS, Faletto MB, *et al.* 1592U89, a novel carbocyclic nucleoside analog with potent, selective anti-human immunodeficiency virus activity. Antimicrob Ag Chemother. 1997; 41:1082–1093.
6. Bilello JA, Bilello PA, Symonds W, *et al.* (1997). 1592U89 A novel carbocyclic nucleoside analog with potent anti-HIV activity, is synergistic in combination with 141W94, on HIV protease inhibitor. Abstract. 4th Conference on Retroviruses and Opportunistic Infections, Washington.

7. St. Clair MH, Millard J, Rooney J, *et al.* *In vitro* antiviral activity of 141W94 (VX-478) in combination with other antiretroviral agents. Antiviral Res, 1996; 29:53–56.
8. Harrigan R, Stone C, Griffin P, *et al.* Antiretroviral activity and resistance profile of the carbocyclic nucleoside HIV reverse transcriptase inhibitor 1592U89. 1997.Abstract. 4th Conference on Retroviruses and Opportunistic Infection, Washington.
9. Tisdale M, Alnadaf T, Cousens D. Combination of mutations in human immunodeficiency virus type 1 reverse transcriptase required for resistance to the carbocyclic nucleoside 1592U89. Antimicrob Ag Chemother, 1997; 41:1094–1098.
10. LaFon S. 1592U89 - a second generation nucleoside analogue. Abstract Proceedings from the 3rd International Congress on Drug Therapy in HIV Infection. November 4th, 1996. Glaxo Wellcome Symposium.
11. McDowell JA, Sadler BM, Millard J, *et al.* Evaluation of potential pharmacokinetic (PK) drug interaction between 141W94 and 1592U89 in HIV+ patients. Abstract A-62. 1997.37th Interscience Conference on Antimicrobial Agents & Chemotherapy, Toronto.
12. Schooley RT and the 141W94 International Study Group. Preliminary data from a phase I/II study on the safety and antiviral efficacy of the combination of 141W94 plus 1592U89 in HIV-infected patients with 150 to 400 CD4+ cells/mm. Abstract. 1997. 4th Conference on Retroviruses and Opportunistic Infections.Washington.
13. Sonnerborg A, Lancaster D, Torres R, *et al.* The safety and antiviral effect of 1592U89, alone and in combination with zidovudine in HIV-1 infected patients with CD4 counts 200–500 cells/mm^3. Abstract OP4.1. 1997. Proceedings from the 3rd International Congress on Drug Therapy in HIV Infection.
14. Winslow DL, Garber S, Reid C, *et al.* Selection conditions affect the evolution of specific mutations in the reverse transcriptase gene associated with resistance to DMP 266. AIDS, 1996; 10:1205–1209.
15. Young SD, Britcher SF, Tran LO, *et al.* L-743–726 (DMP-266); a novel, highly potent nonnucleoside inhibitor of the human immunodeficiency virus type 1 reverse transcriptase. Antimicrob Ag Chemother, 1995; 39:2602–5.
16. Bacheler LT, Anton E, Baker D, *et al.* Impact of mutation, plasma protein binding and pharmacokinetics on clinical efficacy of the HIV-1 non nucleoside reverse transcriptase inhibitor, DMP 266. Abstract I-115. 1997. 37th Interscience Conference on Antimicrobial Agents and Chemotherapy, Toronto.
17. Bacheler LT, George H, Ambremski K, *et al.* Mutations associated with viral load rebound in patients treated with the HIV-1 non nucleoside reverse transcriptase inhibitor DMP 266 in combination with the HIV-1 protease inhibitor crixivan®. Abstract I-172. 1997. 37th Interscience Conference on Antimicrobial Agents and Chemother, Toronto.
18. Wagner K, Kahn J, Mayers D, *et al.* Long-term anti-HIV-1 activity and tolerability of DMP 266 in combination with indinavir. Abstract. 1997. 8th Conference on Clinical Microbiology and Infectious Diseases, Lausanne.
19. Fiske WD, Benedek IH, White SJ, *et al.* Pharmacokinetic interaction between DMP 266 and nelfinavir mesylate (NFV) in healthy volunteers. Abstract I-174.1997. 37th Interscience Conference on Antimicrobial Agents and Chemotherapy, Toronto.
20. Benedek IH, Fiske WD, White SJ, *et al.* Plasma of fluconazole (FL) are not altered by coadministration of DMP 266 in healthy volunteers. Abstract A-4. 1997. 37th Interscience Conference on Antimicrobial Agents and Chemotherapy, Toronto.
21. Fiske WD, Nibbelink DW, Brennan JM, *et al.* DMP 266 cerebrospinal fluid concentrations (CSF) after oral administration. Abstract A-12. 1997 37th Interscience Conference on Antimicrobial Agents and Chemotherapy, Toronto.
22. Kahn J., Havlir D., Mayers D., *et al.* Long term (50 weeks) anti-HIV-1 activity and tolerability (24 weeks) of DMP 266 in combination with indinavir (IDV) (DMP 266–003, cohort III). Abstract. 1997. IDSA 35th Annual Meeting, San Francisco.
23. Baba M, Shigeta S, Yuasa S, *et al.* Preclinical evaluation of MKC-442, a highly potent and specific inhibitor of human immunodeficiency virus type 1 *in vitro*. Antimicrob Ag Chemother, 1994; 38:688–692.
24. Balzarini J, Pelemans H, Jesus M, *et al.* Marked inhibitory activity of non-nucleoside reverse transcriptase inhibitors against human immunodeficiency virus type 1 combined with (-)2',3'-dideoxy-3'-thiacytidine. Molecular Pharmacol; 1996; 49:882–890.
25. Brennan TM, Taylor DL, Bridges G, *et al.*, The inhibition of human immunodeficiency virus type 1 *in vitro* by a non-nucleoside reverse transcriptase inhibitor MKC-442, alone and in combination with other anti-HIV compounds. Antiviral Res. 1995; 26:173–187.
26. Okamoto M, Makino M, Yamada K, *et al.* Complete inhibition of viral breakthrough by combination of MKC-442 with AZT during a long term culture of HIV-1-infected cells. Antiviral Res, 1996; 31:69–77.
27. Piras G, Nakade K, Yuasa S, *et al.* Three-drug combination of MKC-442, lamivudine and zidovudine *in vitro*: potential approach towards effective chemotherapy against HIV-1. AIDS; 1997; 13:469–475.

28. Moxham CP, Szczech GM, Blum MR, *et al.* Pharmacokinetics (PK) and safety of single escalating doses of MKC-442 in HIV-infected volunteers. Abstract.1997. 8th Conference on Clinical Microbiology and Infectious Diseases, Lausanne.
29. Moxham CP, Borroto-Esoda K, Noel D, *et al.* Preliminary efficacy and safety of MKC-442 in HIV infected patients. Abstract. 1997. 37th Interscience Conference on Antimicrobial Agents and Chemotherapy, Toronto.
30. Cheeseman SH. Nevirapine. Antiviral Chemotherapy, New Directions for Clinical Application and Research. Eds: J Mills, L Corey. Pub: PTR Prentice Hall. 1993; 3:345–349.
31. Kilby JM, Saag MS. Clinical experience with non nucleoside reverse transcriptase inhibitors L-697,661 and nevirapine. Antiviral Chemotherapy, New Directions for Clinical Application and Research. Eds: J Mills, P Volberding, L Corey. Publisher Plenum Press. 1996;4;291–299.
32. Crowe S. Antiviral Drugs. The Use of Antibiotics, A Clinical Review of Antibacterial, Antifungal and Antiviral Drugs. Eds: Kucers A, Crowe S, Grayson L, Hoy J. Pub: Butterworth Heinemann. 1997; 5:1509–1855.
33. Dueweke TJ, Pope SM, Romero DL, *et al.* U-09152 a potent inhibitor of human immunodeficiency virus type 1 replication. Antimicrob Ag Chemother; 1993; 37:1127–31.
34. Chong KT, Pagnon PJ, Hinshaw RR. Bisheteroarylpiperazine reverse transcriptase inhibitor in combination with 3'-azudi-3'-deoxythymidine or 2',3'-dideoxytytidine synergistically inhibits HIV-1 replication *in vitro*. Antimicrob Ag Chemother; 1994; 38:288–93.
35. Pagano PJ, Chong KT. *In vitro* inhibition of human immunodeficiency virus type 1 by a combination of delavirdine (U-90152) with protease inhibitor U-75875 or interferon. J Infect Dis; 1995; 171:61–7.
36. Dueweke TJ, Pushkarskaya T, Poppe SM *et al.* A mutation in reverse transcriptase of bis(heteraryl)piperazine-resistant human immunodeficiency virus type 1 that confers increased sensitivity to other nonnucleoside inhibitors. Proc Natl Acad Sci USA.1993;90:4713–7.
37. Demeter LM, Nawaz T, Morse G, *et al.* Development of zidovudine resistance mutations in patients receiving prolonged didanosine monotherapy. J Infect Dis; 1995; 172:1480–5.
38. Been-Tiktak AMM, Joly V, Sibon G, *et al.* Combination therapy with delavirdine mesylate (Delavirdine) and AZT; virology data from an Europe phase II trial. J Acquir Immune Defic Syndr Hum Retrovir; 1995; 10:S23–4.
39. Demeter LM, Meehan PM, Morse G, *et al.* HIV-1 drug susceptibilities and reverse transcriptase mutations in patients receiving combination therapy with didanosine and delavirdine. J AIDS Hum Retrovirol; 1997; 14:136–144.
40. Cheng CL, Smith DE, Carver PL, *et al.* Steady state pharmacokinetics of delavirdine in HIV-positive patients: Effect on erythromycin breath test. Clin Pharmacol Ther; 1997; 61:531–43.
41. Borin MT, Chambers JH, Carel BJ, *et al.* Pharmacokinetic study of the interaction between rifampin and delavirdine mesylate. Clin Pharmacol Therapeutics. 1997; 61:544–553.
42. Freimuth WW. Delavirdine mesylate, a potent non-nucleoside HIV-1 reverse transcriptase inhibitor. Antiviral Chemotherapy: New Directions for Clinical Application and Research.. Edited by J Mills, P Volberding, L Corey., Plenum Press, New York. 1996; 4:279–289
43. Robbins BL, Greenhaw J, Connelly MC, *et al.* Metabolic pathways for activation of the antiviral agent 9-(2-phosphonylmethoxyethyl)adenine in human lymphoid cells. Antimicrob Ag Chemother. 1995; 39:2304–2308.
44. Balzarini J, Perno CF, Schols D, *et al.* Activity of acyclic nucleoside phosphonate analogues against human immunodeficiency virus in monocyte/macrophages and peripheral blood lymphocytes. Biochem & Biophysical Res Communications. 1991; 178:1025–1030.
45. Gu Z, Gao Q, Fang H, *et al.* Identification of a mutation of codon 65 in the IKKC motif of reverse transcriptase that encodes HIV resistance to 2',3'-dideoxycytidine and 2',3'-dideoxy-3'-thiacytidine. Antimicrob Ag Chemother. 1994; 38:275–281.
46. Zhang D, Caliendo A, Eron J, *et al.* Resistance to 2',3'-dideoxycytidine conferred by a mutation in codon 65 of the HIV-1 reverse transcriptase. Antimicrob Ag Chemother. 1994; 38:282–287.
47. Gu Z, Salomon H, Cherrington J, *et al.* The K65R mutation of human immunodeficiency virus type 1 reverse transcriptase encodes cross resistance to 9-(2-phosphonylmethoxyethyl)adenine. Antimicrob Ag Chemother. 1995; 39:1888–1891.
48. Joag SV, Zhuang LI, Foresman L, *et al.* Early treatment with 9-(2-phosphonylmethoxyethyl)adenine reduced virus burdens for a prolonged period in SIV-infected rhesus macaques. AIDS Res Hum Retrovir, 1997; 13:241–246..
49. Foli A, Sogocio K, Anderson B, *et al.* *In vitro* selection and molecular characterisation of human immunodeficiency virus type 1 with reduced sensitivity to 9-(2-phosphonylmethoxyethyl)adenine. Antiviral Res. 1996; 32;91–8.

50. Yokota T, Mochizuki S, Konno K, *et al.* Inhibitory effects of selected antiviral compounds on human hepatitis B virus DNA synthesis. Nucleosides and Nucleotides, 1991; 14:567–570.
51. Cherrington J, Mulato A, Fuller M, *et al.* A novel mutation (K70E) in human immunodeficiency virus type 1 reverse transcriptase confers decreased susceptibility to 9-[2-(phosphonomethoxy)ethyl]adenine (PMEA) *in vitro*. Antiviral Chem Chemother. 1996; 40:2212–2216.
52. Mulato AS, Lamy PL , Li W, *et al.* Genotypic characterisation of HIV-1 variants isolated from AIDS patients after prolonged therapy with adefovir dipivoxil (bis-POM-PMEA). Abstract I-114. 1997 37th Interscience Conference on Antimicrobial Agents and Chemotherapy, Toronto.
53. Shaw JP, Louie MS, Krishnamurthy VV, *et al.* Pharmacokinetics and metabolism of selected prodrugs of PMEA in rats. Drug Metabolism Disp. 1996; 25:362–366.
54. Barditch-Crovo PA, Cundy KC, Wachsman M, *et al.* Pharmacokinetic profile of 9-[2-(bis-pivaloyloxymethyl) phosphonylmethoxyethyl]adenine (bis-POM PMEA), an orally bioavailable prodrug of the antiviral nucleotide, PMEA. Abstract #488. 1995 The Second National Conference on Human Retroviruses. Washington.
55. Cundy KC, Barditch-Crovo P, Walker RE, *et al.* Clinical pharmacokinetics of adefovir in human immunodeficiency virus type 1 infected patients. Antimicrob Ag Chemother. 1995; 39:2401–2405.
56. Barditch-Crovo P, Toole J, Hendrix CW, *et al.* Anti-human immunodeficiency virus (HIV) activity, safety and pharmacokinetics of adefovir dipivoxil (9-[2-(bis-pivaloyloxymethyl)-phosphonylmethoxyethyl]adenine) in HIV-infected patients. J Infect Dis. 1997; 176:406–13
57. Deeks SG, Collier A, Lalezari J, *et al.* The safety and efficacy of adefovir dipivoxil a novel anti-HIV therapy, in HIV infected adults. J Infect Dis 1997.(in press).
58. Balzarini J, Vahlenkamp T, Egberink H, *et al.* Antiretroviral activities of acyclic nucleoside phosphonates [9-(2-phosphonylmethoxyethyl)adenine, 9-(2-phosphonylmethoxyethyl) guanine, (R)-9-(2-phosphonylmethoxypropyl)adenine, and MDL 74,968] in cell cultures and murine sarcoma virus-infected newborn NMRI mice. Antimicrob Ag Chemother. 1997; 41:611–616.
59. Bischofberger N, Naesens L, de Clercq E, *et al.* Bis(POC)PMPA an orally bioavailable prodrug of the antiretroviral agent PMPA. Abstract 1997. 4th Retrovirus Conference, Washington.
60. Perno CF, Balestra E, Aquaro S, *et al.* Potent inhibition of human immunodeficiency virus and herpes simplex virus type 1 by 9-(2-phosphonylmethoxyethy)adenine in primary macrophages is determined by drug metabolism, nucleotide pools and cytokines. Molecular Pharmacol. 1996; 50:359–366.
61. Tsai CC, Follis KE, Sabo A, *et al.* Prevention of SIV infection in macaques by (R)-9-(2-phosphonylmethoxypropyl)adenine. Science. 1995; 270:1197–1199.
62. Cherrington J, Chandok R, Mulato AS, *et al.* *In vitro* selection and characterisation of HIV-1 variants with reduced susceptibility to PMPA. Abstract I-113.1997. 37th Interscience Conference on Antimicrobial Agents and Chemotherapy, Toronto
63. Barditch-Crovo P, Deeks S, Kahn J, *et al.* PMPA safety, pharmacokinetics and antiretroviral activity when administered as a single dose and for seven consecutive days to patients with HIV-infection. Abstract. 1997. International Conference on Antiviral Research, Atlanta.

19

HIV VIRAL LOAD MONITORING

Douglas Richman,[1] Suzanne Crowe,[2] and Katya Harvey[2]

[1]Departments of Pathology and Medicine
San Diego Veterans Affairs Medical Center and
University of California, San Diego
9500 Gilman Drive, La Jolla California 92093-0679
[2]Macfarlane Burnet Centre for Medical Research
Yarra Bend Road
Fairfield, Victoria 3078 Australia

Markers of HIV infection are used in the clinical setting to predict prognosis in HIV-infected persons, measure disease progression and monitor responses to therapy. In the past, the most commonly used tests were CD4 cell counts and serum p24 antigen levels, although serum beta-2 microglobulin, serum neopterin, serum IgA levels and CD8 cell numbers were used as adjunctive measures by some clinicians. Since the early to mid 1990s new technologies for the detection of HIV RNA in human plasma have allowed relatively accurate and precise measurement of plasma HIV RNA concentrations, which directly relate to the production of HIV-1 in tissues, particularly the lymphoid tissues. The available assays to quantify plasma HIV RNA, in combination with CD4 counts, enable clinicians to monitor responses to antiretroviral therapy and to predict the risk for disease progression.[1,2,3,4]

1. THE QUANTITATIVE ASSAYS FOR HIV RNA

There are currently 3 commercial assays available for the quantification of HIV RNA levels in plasma. These are the Amplicor HIV Monitor 1.0 assay (RT-PCR; Roche Molecular Systems), which has a lower limit of detection of 200 to 400 copies/ml; Quantiplex HIV RNA 2.0 assay (bDNA; Chiron Corporation) which has a stated lower limit of detection of 500 copies/ml, and the NucliSens HIV-1 QT assay (NASBA; Organon Teknika) (see Table 1).

To measure viremia below the threshold of the above assays, more sensitive assays have been developed based upon either using larger volumes of plasma and incorporating a centrifugation step prior to RNA extraction (RT-PCR) or by improving signal amplifica-

Antiviral Chemotherapy 5, edited by Mills *et al.*
Kluwer Academic / Plenum Publishers, New York, 1999.

Table 1. Comparison of the 3 assays

	AMPLICOR RT-PCR (Roche)	QUANTIPLEX bDNA (Chiron)	Nuclisens NASBA (Organon)
Technology	Reverse transcription of RNA with amplification of cDNA. Internal quantification standard. Pair of gag primers used.	Target sequences of viral RNA are anchored to a solid phase. Target probes mediate capture and bind pre-amplifier. These are hybridized to bDNA amplifiers, which are labelled and detected by chemiluminescence. 49 specific probes to the pol region.	RNA isolation using guanidine thyocyanate lysis. Silicon dioxide solid phase. Three enzymes: AMV-RT, RNase H and T7 polymerase, produce multiple copies of target RNA at one temperature. Three internal quantification standards.
Samples	Single 200 ul aliquot plasma collected into EDTA or ACD. Heparin inhibits polymerase activity. Plasma should be separated and used or frozen at -70 within 6 hours of collection.	Duplicate 1ml aliquots of plasma collected into preferably EDTA (or ACD). Samples and controls run in duplicate. Plasma should be separated and used or frozen at -70 within 6 hours of collection.	Variable volume from 100ul-2000ul collected into EDTA (or ACD or heparin). Plasma should be separated within 4 hours of collection. Plasma can be added to lysis buffer and stored until extraction is possible.
Limit of detection	Amplicor HIV Monitor 1.0 assay 500 copies/ml Amplicor HIV Monitor 2.0 assay 50 copies/ml (dependent on input)	Quantiplex HIV RNA 2.0 assay 500 copies/ml stated; 1500 actual Quantiplex HIV RNA 3.0 assay 50 copies/ml stated; actual probably higher	40 copies/ml (dependent on input)
Reproducibility	<5-fold inter-assay variability	<3-fold inter-assay variability	<2-fold inter-assay variability
Dynamic range	Linear over 3 log concentration	Linear over 3 log concentration	Linear over 5 log concentration
HIV-1 subtypes detected (Alaeus et al., 1997; Dunne et al., 1997).	Amplicor 1.0 assay able to detect subtypes B,C,D. Amplicor 2.0 assay able to detect A-F subtypes	able to detect A-F subtypes	able to detect A-F subtypes (see text for details)

(Adapted from table prepared by Katya Harvey and Amanda Dunne, Clinical Research Laboratory, Macfarlane Burnet Centre, and Elizabeth Dax and Sue Best, National Reference Laboratory, Melbourne).

tion (bDNA); these modifications increase the ability to detect plasma HIV-1 about 10-fold. The lower limits of detection range from 20 to 50 copies/ml. The Amplicor HIV Monitor 2.0 assay, and the Quantiplex HIV RNA 3.0 assay, are available in certain research settings. Additional assays are in development.

1.1. Amplicor HIV Monitor 1.0 Assay (RT-PCR)

The Amplicor HIV Monitor Assay is a target amplification assay, in which HIV-1 RNA is converted to DNA by reverse transcription (RT); the DNA is then amplified by polymerase chain reaction (PCR). This two-step reaction is achieved with a single enzyme, by using the recombinant Thermus thermophilus DNA polymerase (rTth pol) enzyme which has both RT and PCR activity. The rTth enzyme thus catalyzes both the conversion

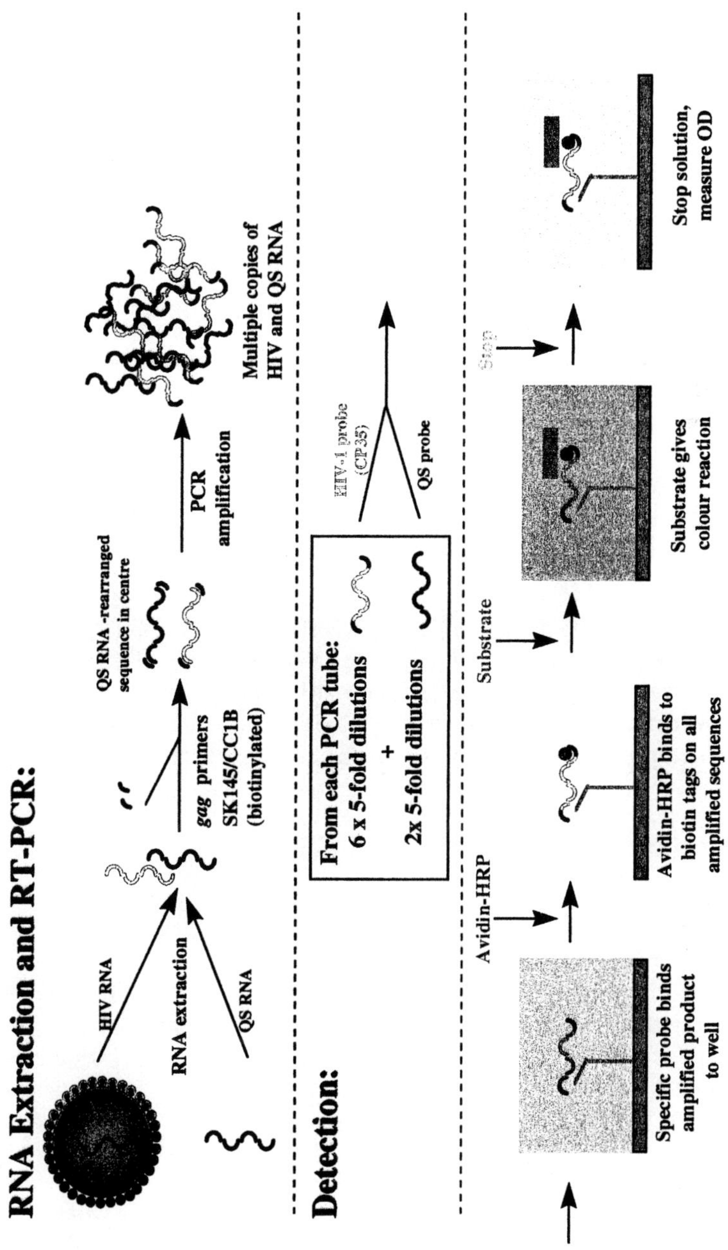
RNA Extraction and RT-PCR:
HIV RNA
RNA extraction
QS RNA
gag primers
SK145/CC1B
(biotinylated)
QS RNA -rearranged
sequence in centre
PCR
amplification
Multiple copies of
HIV and QS RNA
Detection:
From each PCR tube:
6 x 5-fold dilutions
+
2x 5-fold dilutions
HIV-1 probe
(CP35)
QS probe
Avidin-HRP
Substrate
Stop
Specific probe binds
amplified product
to well
Avidin-HRP binds to
biotin tags on all
amplified sequences
Substrate gives
colour reaction
Stop solution,
measure OD

Figure 1

of RNA into cDNA and the subsequent PCR amplification of the cDNA. The assay is performed in a tube, with the incorporation of uracil-N-glycosylase and dUTP (rather than dTTP) in the PCR to prevent amplification of contaminating DNA.[5]

Biotinylated primers, SK462 and SK431 are used to amplify a 142 base target sequence within a highly conserved region of the HIV-1 gag gene (based on the sequence of the subtype B HIV-1 ARV2 strain).[6] An internal quantitation standard (QS) which is a synthetic RNA transcript (219 base) containing identical primer binding sites to those on the HIV-1 target and a unique probe binding region distinguishing the QS from the HIV-1 amplicon, is incorporated into each reaction mixture at known copy number, in order to normalize for differences in amplification efficiency.[7,8,9] The biotinylated HIV-1 and QS amplicons are then detected in separate wells of a microwell plate coated with HIV-1 specific and QS specific probes. The bound amplicons are quantified with an avidin-horseradish peroxidase (Av-HRP) conjugate and by a colorimetric reaction to HRP.

Five-fold serial dilutions of the amplified product allow quantification of the HIV-1 amplicons over a dynamic range of greater than 3 $\log_{10}$. The HIV-1 copy number, expressed as HIV-1 RNA copies/ml of plasma, is determined by comparing the HIV signal to the QS signal for each specimen[10] The lower detection limit of the assay is 400 HIV-1 RNA copies/ml of plasma. Quantification is linear from this lower limit to 750,000 copies/ml.

1.2. Amplicor HIV Monitor 2.0 Assay

The modified assay includes a high speed centrifugation step to pellet the virus from a larger volume of plasma than that used in the earlier assay, providing an increased sensitivity, with a lower detection limit of 50 copies of viral RNA/ml.[11] In addition, new primers have been included (SK145, SKCC1B) which amplify a 155 base pair amplicon and improve the performance of the assay for quantification of subtypes A, E and F.[12]

1.3. Quantiplex HIV RNA 2.0 Assay (bDNA)

The branched DNA assay quantifies HIV-1 in human plasma by amplifying the signal from the target HIV RNA.[13,14,15]

Virus is concentrated from duplicate 1ml plasma specimens by centrifugation and subsequently lysed to liberate HIV-1 genomic RNA from virions. The assay uses a total of 49 different specific probes to the pol region to recognize the HIV-1 sequence. A specific set of 10 oligonucleotide target probes complementary to the HIV-1 pol gene captures the RNA, hybridizing it to a microwell plate during a 16–18hr incubation. A second set of 39 target probes hybridizes the captured RNA to the pre-amplifier probes, which are subsequently hybridized to the bDNA amplifiers. This immobilized complex is further hybridized to multiple copies of an alkaline phosphatase labeled probe which amplifies the signal. Detection is achieved by incubating the complex with a chemiluminescent substrate triggered by the alkaline phosphatase resulting in a chemiluminescent signal which is directly proportional to the amount of HIV-1 RNA present in each sample.[16,17] The stated lower detection limit of the assay is 500 HIV-1 RNA copies/ml; however, with external standards this value represents approximately 1500 copies with the other assays.[18] Quantification is linear from 500 to 1.2 million copies/ml.

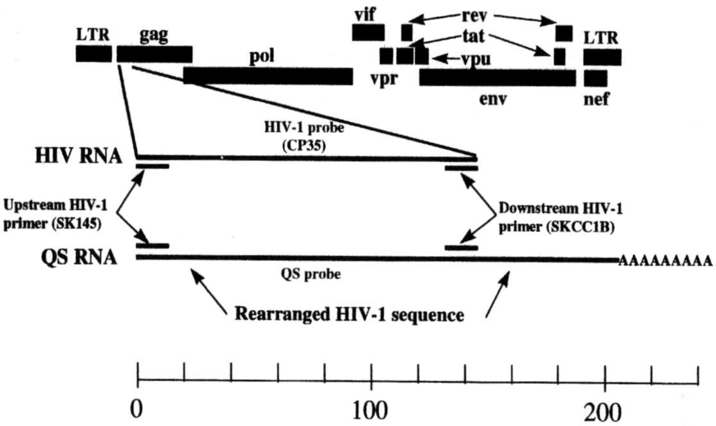

Figure 2. Location of specific primers and probes for RT-PCR assay.

1.4. Quantiplex HIV RNA 3.0 Assay

The modified Quantiplex assay achieves a stated lower detection limit of approximately 50–100 copies/ml by the inclusion of novel nucleotides in the amplification sequences that prevent non-specific hybridization, thus permitting increased signal amplification. The number of specific oligonucleotide probes within the pol region has also been increased in this assay from 49 to 74.[19]

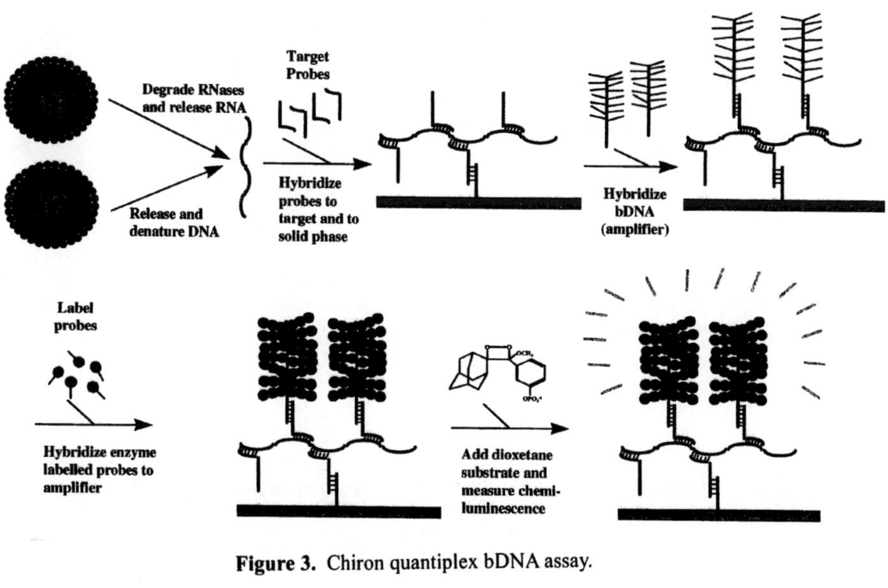

Figure 3. Chiron quantiplex bDNA assay.

1.5. NucliSens HIV-1 QT Assay (Nasba)

Nucleic Acid Sequence Based Amplification (NASBA) technology involves isothermal nucleic acid amplification of target RNA, achieved by concomitant activity of avian myeloblastosis virus reverse transcriptase (AMV-RT), Rnase H and T7 RNA polymerase.[20,21] Quantification of an internal standard (calibrator) RNA ensures equal efficiency of amplification within the assay. The NASBA can detect and amplify RNA in a DNA background and therefore quantification of HIV RNA will not be obscured by amplification of proviral DNA.[10]

A lysis buffer is used to inactivate the viral activity of the HIV-1 plasma sample and release and stabilize the nucleic acid. Silicon dioxide is used as the solid phase to attract released nucleic acid, which is subsequently washed and the RNA eluted. Primer 1 anneals to the upstream, or sense RNA, and from this AMV-RT transcribes the cDNA. The original RNA is then degraded by RNase H which allows the annealing of primer 2 to the newly synthesized cDNA. The AMV-RT promotes elongation of primer 2 thereby synthesising a double stranded cDNA copy of the original RNA. T7 RNA polymerase then transcribes antisense RNA corresponding to the original RNA target, which serves as a template for additional amplification. The single stranded RNA products are hybrid captured to paramagnetic beads via a streptavidin-biotin labeled complex, and reacted with ruthenium-labeled detector probes which are detected by electrochemiluminescence (ECL).[12] Following amplification, the single reaction mixture is divided into separate detection tubes for probe hybridisation and ECL quantification of viral and calibrator RNAs.[20] The lower detection limit of the assay is 40 HIV-1 RNA copies/ml.

1.6. Correlation between bDNA, RT-PCR, and NASBA

Whilst there are some differences between the current commercially-available assays, the disparities are not large and all are suitable for monitoring HIV-infected patients. However, because of the differences, individual patients should be monitored consistently

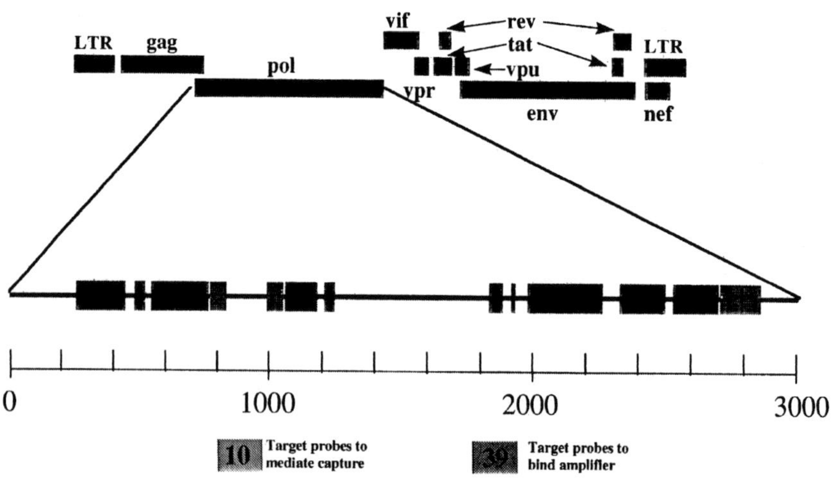

Figure 4. Location of specific probes for bDNA assay.

using one assay. For example, the Amplicor RT-PCR values are generally a little higher than Quantiplex bDNA levels, while NASBA results are about 0.4 to 0.5 log10 copies/ml higher than those by either Quantiplex bDNA or Amplicor RT-PCR assays, when tested using a panel of reconstructed plasma.[23]

Early studies showed that the inter-assay variability for the Quantiplex bDNA assay when tested over 20 assay runs had a coefficient of variation of 19%; this translates to a 2.8 fold change in viral load being significant.[24] Most comparisons indicate that the Quantiplex bDNA assay is more reproducible than the Amplicor RT-PCR assay, in part attributable to the fact that Amplicor RT-PCR uses a single evaluation whereas Quantiplex bDNA is performed in duplicate. Recent data show that the Amplicor RT-PCR assay performs well in experienced hands, with a median absolute difference between duplicate samples of 0.2 $\log_{10}$ copies/ml, and the absolute difference between duplicate samples being larger in samples with low HIV RNA levels (less than 500 copies/ml).[25] These data support the findings of Iozepet *et al*[26] who found in reproducibility experiments designed to measure inter-assay (mean 0.09 $\log_{10}$ copies/ml) and inter-laboratory (mean 0.14 $\log_{10}$ copies/ml) variation that the Amplicor assay can detect between 3-fold (inter-assay) and 5-fold (inter-laboratory) differences with 95% confidence. The correlates of intra-assay and inter-assay variation for the RT-PCR 1.0 and 2.0 assay are similar.[27]

There have been several studies to compare assays for quantification of plasma HIV RNA. In a comparative study of NASBA, Amplicor RT-PCR and Quantiplex bDNA assays (defining deviations of less than 0.5 from the mean $\log_{10}$ copies/ml as not significant) there was a correlation of 83% between the 3 assays[10]. In a second study, the Quantiplex bDNA assay was found to have the smallest mean standard deviation (less than 0.10 $\log_{10}$ units), with the NASBA and Amplicor RT-PCR assays having slightly higher mean standard deviations (less than 0.20 $\log_{10}$ units) using a reconstructed plasma panel.[23] A third study found that the frequency of concordant results (defining differences of less than 0.5 $\log_{10}$ copies/ml as not significant) was 77.5% between Amplicor RT-PCR and Quantiplex bDNA, 80.4% between RT-PCR and NASBA and 58.6% between bDNA and NASBA,

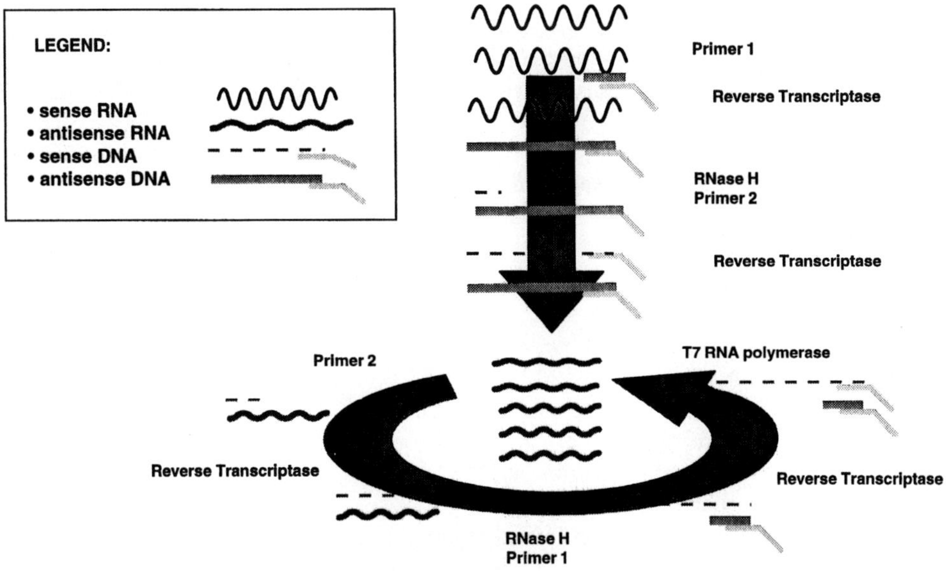

Figure 5. NASBA assay.

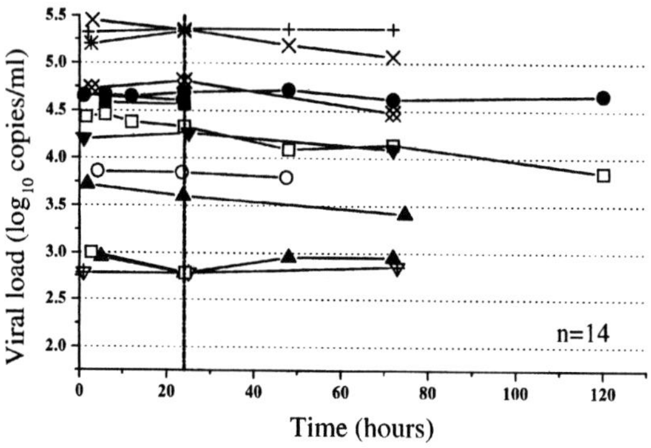

Figure 6. Stability of HIV RNA in whole blood.

with coefficients of variation from 11 to 40% for RT-PCR, 6 to 35% for bDNA and 13 to 62% for NASBA.[28]

It is important to appreciate that with all these assays the coefficients of variation increase at the extremes of the dynamic ranges. Thus at the lowest levels, the precision of stated values diminishes. In addition at the lower end of the dynamic range, for example 50 copies/ml, not all specimens containing 50 copies/ml will generate a positive value. As a result at the lower values one should consider separately the reliability and limits of quantitation and of detection.

1.7. Comparison between Viral Load and Other Surrogate Markers

HIV RNA assays are more sensitive than commercial p24 assays[24,29,10] and newer p24 assays, in which diluted serum is boiled rather than acidified in order to destroy immune complexes, are more sensitive than the standard p24 assay.[30] In combination with heat denaturation, a signal-amplification-boosted p24 antigen assay has been developed which, in a recent study, was found to be as sensitive as PCR technology for measuring viral load.[31] Although earlier studies showed that quantification of HIV RNA in plasma correlated significantly with other viral markers and CD4 counts.[32,33] More recent data have shown that the correlation with CD4 counts is weak[1] and that plasma HIV RNA predicts the subsequent rate of decline in the CD4 cell count, but does not correlate closely with CD4 numbers.[4]

1.8. Diurnal and Day to Day Variation; Plasma versus Serum; Sample Preparation

HIV RNA within the plasma of patients with clinically stable HIV-1 infection has a variance of approximately 0.4 $\log_{10}$ copies/ml when assessed frequently over a 48 hour period, without evidence of any diurnal pattern.[17] Monthly assessments of HIV RNA have also been found it to be stable within the plasma of treated and untreated individuals.[34] These data support earlier studies.[35]

HIV RNA remains stable for at least 6 months when stored in lysis buffer at -70°C.[36] EDTA or ACD are preferred to heparin as an anticoagulant, due to interference by heparin with the Amplicor RT-PCR assay, resulting in inhibitory effects on gene amplification.[6] HIV

RNA within EDTA-containing whole blood is stable for 72 hours at 25°C, but declines rapidly if stored in EDTA or heparin for 24 hours at 4°C,[36] (Crowe *et al*, unpublished data. Figure 6). These data differ from earlier findings, where storage at room temperature prior to centrifugation resulted in a significant decline in HIV RNA.[37] Ideally, plasma should be separated from whole blood and frozen at -70°C within 6 hours of collection, and whole blood should remain at room temperature until plasma is separated. When stored at -70 to 80°C, plasma HIV RNA is stable for up to 12 months,[38] and for at least 3 freeze-thaw cycles.[14] Thawing should be performed at controlled temperatures, not using a 37° water bath. Plasma HIV RNA levels are higher than serum levels by approximately 0.4 to 0.5 $\log_{10}$ copies/ml and serum is generally not used for HIV RNA determination.[36,39]

1.9. Detection of HIV-1 RNA in Diverse Subtypes

In view of the genetic diversity of HIV-1 the performance of the various kits in detecting HIV-1 of different subtypes is of importance. There are significantly lower levels of HIV-1 RNA of subtypes A, E and F, measured by the Amplicor RT-PCR 1.0 assay than the Amplicor bDNA 2.0 assay,[40] and of subtypes A, B and F when compared to the NASBA assay.[41,42] NASBA detects lower levels of subtype H than the Amplicor RT-PCR 1.0 assay,[42] but also appears to be unable to accurately measure HIV RNA in a number of subtype A samples.[42] These differences are likely to be based upon differences in the primers used by the Amplicor RT-PCR and NASBA assays, and primer mis-match with the HIV sequence. As the Quantiplex bDNA assay uses 49 probes within pol, lack of recognition is less likely than with the Amplicor RT-PCR 1.0 which uses a single set of primers within the gag region. The Amplicor RT-PCR 2.0 assay includes additional primers in the gag region for detection of non-B subtypes, and has been shown to perform well in quantifying subtypes A to F.[12]

1.10. Measurement of HIV RNA in Other Body Fluids

Although the assays are designed to quantify HIV RNA in plasma, their utility in quantifying HIV RNA in CSF[43] and semen[44,45] has been studied. Amplification of HIV RNA in cell-free seminal plasma has been found to be frequently inhibited using the Amplicor RT-PCR assay, but NASBA has been reported to be reliable for quantification of seminal plasma HIV RNA.[44] The Boom extraction method of the NASBA assay for extraction of HIV RNA from semen renders all assays applicable to quantitate semen RNA. In a study using the NASBA assay to quantify HIV RNA in semen (both whole semen and seminal plasma) and blood-derived plasma, viral load was consistently higher in the latter than in semen.[45]

2. THE BIOLOGICAL BASIS OF PLASMA HIV RNA AND IMPLICATIONS OF VARIOUS VALUES

2.1. Prognosis

With a baseline CD4 cell count and measurement of plasma HIV RNA one can predict the time to disease with a high degree of confidence.[46] In fact the level of plasma HIV RNA correlates well with the rate of decline of CD4 cells in the peripheral blood. A reduction in plasma of HIV RNA correlates well with the clinical benefit from antiretroviral treatment.[47–50]

2.2. Dynamics of Infection

The administration of potent antiretroviral therapy permits calculations to estimate the half-life of plasma virions (at most 6 hours.[51]) Because the first phase decay constant appears to be independent of baseline CD4 or HIV RNA levels,[51,52] the steady state level of plasma HIV RNA levels must be determined by the level of production. Haase *et al* have determined that the number of acutely infected CD4 lymphocytes in the lymphoid tissue correlates well with steady state levels of plasma HIV RNA.[53] These observations are consistent with the argument that the acutely infected activated CD4 lymphocytes in the lymphoid tissue are the source of approximately 99% of the plasma HIV RNA and the level of plasma HIV RNA reflects the rate at which such cells are destroyed by viral replication. By extrapolation the benefit of antiretroviral therapy largely reflects the interruption of this process.

2.3. The Impact of Potent Combination Chemotherapy

The decay of plasma HIV RNA with potent regimens of antiretroviral drugs that can reduce HIV RNA levels more than 100 fold permits the recognition of the second phase of HIV RNA decay, which displays a half-life of one to four weeks.[52] This phase was proposed to be attributable to the half-life of infected antigen presenting cells,[51] however, the decay of extracellular RNA trapped in the germinal centers of lymphoid tissue also appears to have a similar half-life.[54]

Several observations of plasma HIV RNA in patients receiving such potent regimens now permit additional insights.

a. The lower the nadir of plasma HIV RNA the longer it takes for the rebound from suppression which is associated with HIV drug resistance.[55]
b. The level of suppression of HIV RNA in the plasma is correlated with the degree of reduction of HIV RNA in the lymph nodes.[56]
c. Even patients who have levels of plasma HIV RNA in the hundreds (copies/ml) have detectable ongoing evolution of HIV RNA in the lymph node as determined by sequence changes (even before development of drug resistance mutations).[57] The presence of multiply spliced HIV RNA which indicates ongoing transcription can also be detected in such patients. In contrast patients with less than 50 copies of HIV RNA per ml of plasma have no evidence of viral evolution or transcription in their lymph nodes.[57]
d. Enhanced virus isolation techniques in patients with less than 50 copies HIV RNA per ml of plasma have revealed a cellular reservoir of replication competent HIV- the latently infected CD4 RO45 (memory) CD4 lymphocyte.[58,59]

3. HOW TO USE ASSAYS OF HIV RNA IN CLINICAL PRACTICE

From the observations summarized above, guidelines[60] can be proposed for the use of these assays in clinical management.

1. All patients should have a baseline plasma HIV RNA assessment to determine steady state level and consequent risk for progression. This level can then be used to help decide whether to initiate antiretroviral treatment.[61]

2. Following the initiation of an antiretroviral regimen, therapeutic response should be monitored closely (perhaps every 1–2 months) to assess success in achieving suppression below the levels of detection.
3. Failure to achieve notable suppression indicates therapeutic failure (whether due to poor adherence, pharmacologic problems or resistance). The sooner the response to this failure by changing the treatment regimen, the fewer the number of resistance mutations that will have accumulated.
4. Assays with detection limits above 50 copies/ml are suboptimal and are not detecting those patients with a reduced but continuing risk for failure. Any detectable HIV RNA reflects ongoing replication.
5. Adherent patients with undetectable levels of plasma HIV RNA should be monitored for levels of HIV RNA and CD4 at intervals of every 3 months.

ACKNOWLEDGMENTS

This work was supported by grants AI 27670, AI 38858, UCSD Center for AIDS Research (AI 36214) , AI 29164, from the National Institutes of Health, and the Research Center for AIDS and HIV Infection of the San Diego Department of Veterans Affairs. Blood was collected from 14 HIV-infected persons into multiple EDTA-containing tubes, and kept at ambient temperature for varying periods of time until plasma was separated (within 2 hours, after 6, 12, 24, 48, 72, and 120 hours) and stored at -70°C for batch testing to quantify HIV RNA by Quantiplex bDNA assay. Not all participants had plasma separated at all times (depending on the volume of blood available).

REFERENCES

1. Mellors JW, Rinaldo Jr CR, Gupta P, *et al.* Prognosis in HIV-1 infection predicted by the quantity of virus in plasma. Science. 1996; 272:1167–1170.
2. Coombs RW, Welles SL, Hooper C, *et al.* Association of plasma human immunodeficiency virus type 1 RA level with risk of clinical progression in patients with advanced infection. J Infect Dis. 1996; 174: 704–712.
3. Saag MS, Holondniy M, Kuritzkes DR, *et al.* HIV viral load markers in clinical practice. Nature Medicine. 1996; 2:625–629.
4. Mellors JW, Munoz A, Giorgi JV, *et al.* Plasma viral load and CD4+ lymphocytes as prognostic markers of HV-1 infection. Ann Intern Med. 1997; 126:946–54.
5. Longo MC, Berininger MS, Hartley JL. Use of uracil DNA glycosylase to control carry-over contamination in polymerase chain reactions. Gene. 1990; 93:125–128.
6. Holodiny M, Katzenstein DA, Israelski DM, *et al.* Reduction in plasma human immunodeficiency virus ribonucleic acid following dideoxynucleoside therapy as determined by the polymerase chain reaction. J Clin Invest. 1991;88:1755–9.
7. Steiger M, Demolliere C, Ahlborn-Laake L, *et al.* Competitive polymerase chain reaction assay for quantitation of HIV-1 DNA and RNA. J Virol Methods. 1991; 34:149–160.
8. Sninsky JJ, Kwok S (1993). The application of quantitative polymerase chain reaction to therapeutic monitoring. AIDS. 7: S29.
9. Mulder J, McKinney N, Christopherson C, *et al.* Rapid and simple PCR assay for quantitation of human immunodeficiency virus type 1 RNA in plasma; application to acute retroviral infection. J Clin Microbiol. 1994; 32:292–300.
10. Revets H, Marissens D, De Wit S, *et al.* Comparitive Evaluation of NASBA HIV-1 RNA QT, AMPLICOR-HIV Monitor, and QUANTIPLEX HIV RNA Assay, Three methods for quantification of human Immunodeficiency virus type 1 RNA in plasma. J Clin Microbiol. 1996; 34:1058–1064.

11. Mulder J, McKinney N, Kwok S (1996). A rapid and simple sample preparation method dramatically increases sensitivity of HIV-1 plasma RNA assay. Abstract.Th.B.914. XI International Conference on AIDS; Vancouver.
12. Michael N, Robb M, Birx D, *et al.*, (1997). Performance of the amplicor HIV-1 monitorTM test and a modified HIV-1 monitor test on HIV-1 subtypes A to F. Abstract. 4th Conference on Retroviruses and Opportunistic Infections, Washington DC.
13. Urdea MS, Wilber JC, Yeghiazarian T, *et al.* Direct and quantitative detection of HIV-1 RNA in human plasma with branched DNA signal amplification assay. AIDS. 1993; 7:S11-S14.
14. Todd J, Pachl C, White R, *et al.* Performance characteristics for the quantitation of plasma HIV-1 RNA using branched DNA signal amplification technology. J Acquir Immune Defic Syndr Hum Retroviral. 1995; 10:S35.
15. Kern D, Collins M, Fultz T, *et al.* An enhanced sensitivity branched DNA (ES bDNA) assay for the quantification of HIV-1 RNA in plasma. J Clin Micro. 1996;34:3196–3202.
16. Dunne A, Janky S, Crowe S. Signal Amplification: A Direct Approach to Viral Quantitation. Today's Life Science. 1995; 7:48–53.
17. Deeks SG, Coleman RL, White R, *et al.* Variance of plasma human immunodeficiency virus type 1 RNA levels measured by branched DNA within and between days. J Infect Dis. 1997; 176:514–517.
18. Mellors JW, Munoz A, Giorgi JV, *et al.* Plasma viral load and CD4+ lymphocytes as prognostic markers of HIV-1 infection. Ann Intern Med 1997; 126:946–954.
19. Collins ML, Irvine B, Tyner D, *et al.* A branched DNA signal amplification assay for quantification of nucleic acid targets below 100 molecules/ml. Nucleic Acids Res. 1997; 25:2979–2984.
20. Compton J. Nucleic acid sequence based amplification. Nature (London). 1991; 350:91–92.
21. Romano JW, Williams KG, Shurtliff RN, *et al.* NASBA technology: isothermal RNA amplification in qualitative and quantitative diagnostics. Immunol Invest. 1997; 26:15–28.
22. van Gemen B, van Beuningen R, Nabbe A. A one tube quantitative HIV-1 RNA NASBA nucleic acid amplification assay using electrochemiluminescent (ECL) labelled probes. J Virologic Methods. 1994; 49:157–168.
23. Schuurman R, Descamps D, Weverling GJ, *et al.* Multicenter comparison of three commercial methods for quantification of human immunodeficiency virus type 1 RNA in plasma. J Clin Microbiol. 1996; 34:3016–3022.
24. Dewar R, Highbarger H, Sarmiento MD, *et al.* Application of branched DNA signal amplification to monitor human immunodeficiency virus type 1 burden in human plasma. J Infect Dis. 1994; 170:1172–9.
25. Raboud JM, Montaner JSG, Conway B, *et al.* Variation in plasma RNA levels, CD4 cell counts, and p24 antigen levels in clinically stable men with human immunodeficiency virus infection. J Infect Dis. 1996; 174:191–4.
26. Izopet J, Poggi C, Dussaix E, *et al.* Assessment of a standardized reverse transcriptase PCR assay for quantifying HIV-1 RNA in plasma and serum. J Virologic Methods. 1996; 60:119–129.
27. Schockmel GA, Yerly S, Perrin L. Detection of low HIV-1 RNA levels in plasma. J Acq Immune Defic Syndr Hum Retrovir. 1997; 14:179–183.
28. Coste J, Montes B, Reynes J, *et al.* Comparative evaluation of three assays for the quantitation of human immunodeficiency virus type 1 RNA in plasma. J Med Virol. 1996; 50:293–302.
29. Ercoli L, Sarmati L, Es-Sawaf G, *et al.* Plasma viremia titration and RNA quantitation in ICD-p24 negative HIV type-1 infected patients. AIDS Res Hum Retrovir. 1995; 11:1203–1207.
30. Schupbach J, Boni J, Tomasik Z, *et al.* Sensitive detection and early prognostic significance of p24 antigen in heat denatured plasma of human immunodeficiency virus type 1 infected infants. J Infect Dis. 1994; 170:318–24.
31. Boni J, Opravil M, Tomasik Z, *et al.* Simple monitoring of antiretroviral therapy with a signal amplification boosted HIV-1 p24 antigen assay with heat denatured plasma. AIDS. 1997; 11:47.
32. Cao Y, Ho DD, Todd J, *et al.* Clinical evaluation of branched DNA signal amplification for quantifying HIV type 1 in human plasma. AIDS Res Hum Retrovir. 1995; 11:353–361.
33. Pachl C, Todd JA, Kern DG, *et al.* Rapid and precise quantification of HIV-1 RNA in plasma with use of a branched DNA signal amplification assay. J AIDS. 1995; 5: 446–454.
34. Paxton WB, Coombs RW, McElrath MJ, *et al.* Longitudinal analysis of quantitative virologic measures in human immunodeficiency virus-infected subjects with ≥ 400 CD4 lymphocytes; implications for applying measurements to individual patients. National Institute of Allergy & Infect Diseases AIDS Vaccine Evaluation Group. J Infect Dis. 1997; 175:247–54.
35. Holodniy M, Mole L, Winters M, *et al.* Diurnal and short term stability of HIV virus load as measured by gene amplification. J Acq Imm Def Syndr. 1994; 7:363–368.

36. Bruisten SM, Oudshoorn P, van Swieten P, *et al* (1997). Stability of HIV-1 RNA in blood during specimen handling and storage prior to amplification by NASBA-QT. J Virol Methods. 67: 199–207.
37. Holodniy M, Mole L, Yen-Lieberman B, *et al.* Comparative stabilities of quantitative human immunodeficiency virus RNA in plasma from samples collected in VACUTAINER CPT, VACUTAINER PPT, and standard VACUTAINER tubes. J Clin Microbiol. 1995; 33:15620–1566.
38. Winters M, Tan L, Katzenstein D, *et al.* Biological variation and quality control of plasma human immunodeficiency virus type 1 RNA quantitation by reverse transcriptase polymerase chain reaction. J Clin Microbiol. 1993; 31:2960–2966.
39. Ioannidis JPA, Cappelleri JC, Lau J, *et al.* Predictive value of viral load measurements in asymptomatic untreated HIV-1 infection; a mathematical model. AIDS. 1996; 10:255–62.
40. Dunne A, and Crowe S. Comparison of branched DNA and reverse transcriptase polymerase chain reaction for quantifying six different HIV-1 subtypes in plasma. AIDS. 1997; 11:126–127.
41. Gobbers E, Fransen K, Oosterlaken T, *et al.* Reactivity and amplification efficiency of the NASBA HIV-1 RNA amplification system with regard to different HIV-1 subtypes. J Virologic Methods. 1997; 66:293–301.
42. Alaeus A, Lidman K, Sonnerborg A, *et al.* Subtype-specific problems with quantification of plasma HIV-1 RNA. AIDS. 1997; 11:859–865.
43. Brew B, Pemberton L, Cunningham P (1996). Abstract WeB 3287. Cerebrospinal fluid HIV-1 RNA levels correlate with AIDS dementia complex. XI International Conference on AIDS, Vancouver.
44. Dyer JR, Gilliam BL, Eron JJ, *et al.* Quantitation of human immunodeficiency virus type 1 RNA in cell free seminal plasma; comparison of NASBA with Amplicor reverse transcription PCR amplification and correlation with quantitative culture. J Virol Methods. 1996; 60:161–70.
45. Gupta P, Mellors J, Kingsley L, *et al.* High viral load in semen of human immunodeficiency virus type 1 infected men at all stages of disease and its reduction by therapy with protease and nonnucleoside reverse transcriptase inhibitors. J Virol. 1997; 71: 6271–5.
46. Mellors JW, Munoz A, Giorgi JV, *et al.* Plasma viral load and $CD4^+$ lymphocytes as prognostic markers of HIV-1 infection. Ann Intern Med 1997; 126:946–954.
47. Coombs RW, Welles SL, Hooper C, *et al.* Association of plasma human immunodeficiency virus type-1 RNA level with risk of clinical progression in patients with advanced infection. J Infect Dis 1996; 174:704–712.
48. Welles SL, Jackson JB, Yen-Lieberman B, *et al.* Prognostic value of plasma Human Immunodeficiency Virus Type 1 (HIV-1) RNA levels in patients with advanced HIV-1 disease and with little or no prior zidovudine therapy. J Infect Dis 1996; 174:696–703.
49. Katzenstein DA, Hammer SM, Hughes MD, *et al* and AIDS Clinical Trials Group Study 175 Virology Study Team. The relation of virologic and immunologic markers to clinical outcomes after nucleoside therapy in HIV-infected adults with 200 to 500 CD4 cells per cubic millimeter. N Engl J Med 1996; 335:1091–1098.
50. Marschner IC, Collier AC, Coombs RW, *et al.* Use of changes in plasma levels of human immunodeficiency virus type 1 RNA to assess the clinical benefit of antiretroviral therapy. J Infect Dis. 1998; 177:40–47.
51. Perelson AS, Neumann AU, Markowitz M, *et al.* HIV-1 dynamics *in vivo*: virion clearance rate, infected cell lifetime, and viral generation time. Science. 1996; 271:1582–1586.
52. Perelson AS, Essunger P, Cao Y, *et al.* Decay characteristics of HIV-1-infected compartments during combination therapy. Nature. 1997; 387:188–191.
53. Haase AT, Henry K, Zupancic M, *et al.* Quantitative image analysis of HIV-1 infection in lymphoid tissue. Science. 1996; 274:985–989.
54. Cavert W, Notermans DW, Staskus K, *et al.* Kinetics of response in lymphoid tissues to antiretroviral therapy of HIV-1 infection. Science. 1997; 276:960–964.
55. Kempf D, Rode R, Xu Y, *et al.* The durability of response to protease inhibitor therapy is predicted by viral load, International Workshop on HIV Drug Resistance, Treatment Strategies and Eradication St. Petersburg, FL:June 25–28 1997 (Abstract).
56. Wong JK, Gunthard H, Havlir DV, *et al.* Reduction of HIV-1 in blood and lymph nodes following potent anti-retroviral therapy and the virologic correlates of treatment failure, Proc Natl Acad Sci USA 1997;94:12574–12579.
57. Günthard HF, Wong JK, Ignacio CC, *et al.* HIV replication and genotypic resistance in blood and lymph nodes after one year of potent antiretroviral therapy. J Virol 1998; 72:2422–2428.
58. Finzi D, Hermankova M, Pierson T, *et al.* Identification of a reservoir for HIV-1 in patients on highly active antiretroviral therapy. Science. 1997; 278:1295–1300.

59. Wong JK, Hezareh M, Günthard H, *et al.* Recovery of replication-competent HIV despite prolonged suppression of plasma viremia. Science. 1997; 278:1291–1294.
60. Saag MS, Holodniy M, Kuritzkes DR, *et al.* HIV viral load markers in clinical practice. Nature Medicine 1996; 2:625–629.
61. Carpenter CCJ, Fischl MA, Hammer SM, *et al.* Antiretroviral therapy for HIV infection in 1997: updated recommendations of the International AIDS Society-USA panel. JAMA 1997; 277:1962–1968.

20

POST-EXPOSURE PROPHYLAXIS FOR HIV

Julie Louise Gerberding[1] and Mitchell H. Katz[2]

[1]Departments of Medicine (Infectious Diseases) and Epidemiology and Biostatistics
University of California, San Francisco,and San Francisco General Hospital
Building 100, Room 301, San Francisco General Hospital
1001 Potrero Avenue, San Francisco, California 94110
[2]San Francisco Department of Public Health and University of California, San Francisco
San Francisco General Hospital
1001 Potrero Avenue, San Francisco, California 94110

In 1996, the Centers for Disease Control and Prevention (CDC) published a case-control study demonstrating that post-exposure treatment with zidovudine was associated with a 81% reduction in the odds of occupational HIV transmission.[1,2] Since publication of this report, provision of antiretroviral treatment after parenteral exposures to HIV among health care workers has become a standard of care in the United States.[3–5] In the past year, the potential merits of offering chemoprophylactic treatment to persons sustaining sexual and injection drug use exposures to HIV have been debated. Although not endorsed by CDC as a public health prevention program, provision of chemoprophylaxis in individual cases of high risk exposure is gaining acceptance in clinical practice.[3,6,7] Moreover, offering antiretroviral therapy to victims of sexual assault is now recommended by CDC.[8,9]

The rationale for post-exposure prophylaxis is based on the premise that immediate antiretroviral treatment will prevent or at least attenuate initial cellular infection and HIV propagation and allow the host's immune defenses to clear the virus inoculum. Cutaneous dendritic cells (epidermal Langerhan's cells and/or dermal dendritic cells) are probably essential to this process.[10] These cells take up HIV and present viral antigens to T-lymphocytes, NK cells, and other immune effector cells. Antigen-specific cytotoxic T-lymphocyte responses have also been observed in exposed but non-infected persons, although this has not been proven to be a protective response.[11,12]

Sexual exposure to HIV through mucosal surfaces is not completely analogous to transcutaneous exposure, but may involve similar host responses. In the simian model of intravaginal HIV infection, simian immunodeficiency virus (SIV) is initially taken up by Langerhan's cells and macrophages in the lamina propria underlying the epithelium, but

Antiviral Chemotherapy 5, edited by Mills *et al.*
Kluwer Academic / Plenum Publishers, New York, 1999.

local replication is not conspicuous.[13–15] Within two days, infected cells migrate via the afferent lymphatics to the T-cell dependent sections of the regional lymph nodes where active virus replication occurs. Infected T cells and cell-free virus can be detected in the peripheral blood within five days after exposure.[13] The pathogenesis of infection through urethral and rectal mucosa may be similar.[16]

Although zidovudine and other antiretroviral drugs do not prevent the entry of HIV into target cells, local propagation or dissemination could be prevented if there is a sufficient window of opportunity for drug to be administered and distributed to target cells in its active form before irreversible infection occurs. In theory, treatment could also favorably alter the natural history of HIV infection by attenuating initial viremia, even if it is not effective in preventing infection.

Post-exposure zidovudine treatment has been efficacious in animal models of feline and murine retrovirus infection if administered in the first few hours after exposure.[17–19] Zidovudine was shown to suppress HIV replication in mice with severe combined immunodeficiency reconstituted with human lymphoid tissue (SCID-hu mice).[20,21] Early studies of post-exposure zidovudine treatment in rhesus macaques intravenously inoculated with high titers of SIV failed to demonstrate protection. However, more recent primate studies have demonstrated that treatment efficacy is dependent on at least 3 variables: a) the inoculum size; b) interval between exposure and treatment, and c) duration of treatment.[22] Extrapolation from animal data to humans is hindered by differences in the biological properties of retroviruses and experimental conditions employed. For most occupational exposures, the HIV inoculum is very small compared to that used in animal studies, treatment is initiated within a few hours of exposure, and is continued for four weeks. Hence, the current guidelines for managing occupational exposures are designed to maximize the potential for post-exposure therapy to be efficacious.

1. PROPHYLAXIS AFTER OCCUPATIONAL EXPOSURES TO HIV

1.1. HIV Transmission Risk

HIV is transmitted to 0.2–0.4% of health care workers sustaining needle punctures or similar percutaneous injuries involving HIV-infected blood.[23–26] This risk estimate is based almost entirely on exposures to blood from patients with advanced HIV infection in the pre-treatment era, who are apt to have circulating virus titers that are several logs higher than those who are asymptomatic.[27] The risks associated with contamination of mucous membranes or non-intact skin are not zero, but are too low to be quantified in the prospective studies performed to date.[26] Many factors can affect the probability of transmission in individual cases, including volume of blood involved in the exposure, titer of virus in the source fluid or tissue, characteristics of the virus strain, and the integrity of the host response.

1.2. Prophylactic Treatment Efficacy

The outcome of two studies prompted development of new treatment recommendations for post-exposure prophylaxis after occupational exposures. First, data from the AIDS Clinical Trial Group (ACTG) protocol 076 demonstrated that zidovudine treatment could reduce the risk of perinatal HIV transmission by 67%.[28,29] This placebo-controlled study evaluated treatment of infected mothers (who had no prior zidovudine therapy) during the last two trimesters and of their newborns for six weeks. The benefit of treatment could not be explained solely by a reduction in maternal virus titer. Although this experience cannot be generalized to transcutaneous exposures experienced by health care providers, it does demonstrate the prophylactic efficacy of antiretroviral treatment under some clinical circumstances.

The second study, the CDC's case control study of occupational HIV infections, is directly relevant to health care providers.[1,2] In this retrospective study, those who received zidovudine treatment after needle exposure were 81% less likely (OR = .19) to acquire infection than those who did not get treated, independent of the other variables affecting the probability of transmission (deep punctures, visibly bloody needle, needle used for venous or arterial access, source patient with pre-terminal disease). The study did not evaluate dosing regimens or the duration of zidovudine treatment. Based on this study, the USPHS now recommends provision of antiretroviral drugs to health care workers who sustain risky exposures to HIV.[4,30]

Failure of zidovudine prophylaxis has been documented in at least eleven health care providers (5 in the U.S.) who were treated after occupational exposures.[31] These cases indicate that zidovudine treatment is not always successful, but in the absence of a controlled trial, the efficacy of post-exposure treatment cannot be established. The reasons for treatment failure are not known. Pharmacologic issues (low dose, short duration, delayed onset) may have contributed to failure in some cases. Zidovudine resistance may also play a role, but in at least one case, treatment was started within one hour of the exposure, the virus was not resistant to zidovudine, and infection still occurred. In these reports, treatment was not associated with a delay in seroconversion or other adverse effects on the natural history of infection.

1.3. Prophylactic Treatment Regimen

The frequency of zidovudine resistance among source patients undergoing treatment increases with the duration of therapy. In addition, transmission of zidovudine-resistant virus has been documented, and may account for more than 10% of new infections in some locales. Zidovudine prophylaxis is unlikely to be effective when the exposure involves resistant virus. Hence, the current treatment guidelines recommend combinations of two or more antiviral drugs.[30] Treatment with zidovudine (200 mg tid or 300 mg bid) plus lamivudine (150mg bid) is currently advised for most HIV exposures conferring a risk of transmission. A protease inhibitor (e.g. nelfinavir or indinavir) or a non-nucleoside reverse transcriptase inhibitor is added when the exposure is especially high risk and/or the source patient has recently taken both zidovudine and lamivudine.

The protocol in current use at San Francisco General Hospital is based on the USPHS recommendations (Table 1). When the source patient is currently taking (i.e. in the past 30 days) zidovudine and lamivudine, an alternate regimen may be indicated, especially if the source has a high viral load (suggestive of treatment failure and drug resistance). One reasonable treatment option in this case is to use DDI plus D4T, since cross-resistance is less likely. Some experts use D4T plus 3TC as the initial regimen because these drugs are well-tolerated by most patients. However, use of a regimen that does not contain AZT is controversial, since the only efficacy data are derived from studies where AZT was the primary drug used to prevent infection. Most experts believe that prophylactic treatment should be started as soon as possible (within hours) after the exposure to maximize the chance of efficacy.[26, 30] Initiating treatment more than 72 hours after exposure is not advised, but may be considered for high risk exposures. The optimal dose and

Table 1. Current post-exposure prophylaxis (PEP) protocol for health care workers exposed to HIV at San Francisco General Hospital

Basic regimen	
Zidoudine	200 mg PO tid or 300 mg PO bid
Lamivudine	150 mg PO bid
Alternate regimen I*	
Didanosine (DDI)	200 mg bid (125 mg bid if < 60 kg)
Stavudine (D4T)	40 mg bid (30 mg bid if < 60 kg)
Alternate regimen II**	
Stavudine (D4T)	40 mg bid (30 mg bid if < 60 kg)
Lamivudine (3TC)	150 mg bid
Protease inhibitor***	
Indinavir (IND)	800 mg q8h
or Nelfinavir (NEL)	750 mg tid
Initiation	ASAP after known or potential exposure to HIV
Duration	4 weeks

* use when source patient has taken zidovudine and lamivudine in the past month
** use when source patient has taken zidovudine but not lamivudine in the past month
***add to 2-drug combination when nucleoside analogue resistance is suspected and/or the exposure is especially risky

duration of treatment has not been established, but a 4 week regimen is currently recommended.

Most (99.7%) parenteral occupational exposures to HIV do not transmit HIV infection. Antiretroviral chemoprophylaxis is a promising strategy that appears to be protective in select cases. Data on the safety of post-exposure zidovudine treatment among health care workers indicates that serious objective toxicity is uncommon, but one third of subjects stop therapy because of subjective drug intolerance, such as nausea, fatigue, insomnia, and headache.[23,26,34] Neither teratogenic or oncogenic effects of antiretroviral treatment have been observed, but long-term toxicity remains unknown. While it is hoped that the incidence of serious adverse events associated with combination treatment will be low, experience with the new agents alone or in combination with zidovudine is limited. When treatment is used, close monitoring for known and potential adverse effects must be included in the follow-up protocol. Health care workers electing treatment should carefully consider the potential risks as well as benefits associated with treatment and provide written consent (or perhaps declension) to document their decision about treatment. Adverse reactions should be reported to the Food and Drug Administration and to the drug manufacturer(s).

Treatment of pregnant women or persons not practicing effective contraception is complicated by lack of data about the safety of antiretroviral drugs in pregnancy. As a general rule, the decision to treat should be based on the same criteria used for managing other exposures.[30] Zidovudine plus lamivudine is the most common regimen employed at the present time. Indinavir use is complicated by the risk of kidney stones and hyperbilirubinemia, two problems which may be more serious among pregnant women. Nelfinavir is the only protease inhibitor that is approved for pediatric use, but there is little experience with use during pregnancy.

Periodic testing for HIV antibody is advised for six months after exposure. Seroconversion beyond that point may occur but is rare.[35] Viral load tests are not recommended for routine follow-up, but should be done if clinical symptoms suggestive of acute retrovirus infection occur. Even though most health care providers can be reassured that HIV transmission is unlikely, referral to mental health professionals for stress management or crisis intervention is sometimes required during follow-up. Avoidance of unsafe sexual behaviors and other practices that could pose a transmission risk to others is recommended, but adherence often creates additional problems that necessitate additional counselling for exposed individuals and their sexual partners.

The pharmacy acquisition cost of a four week course of zidovudine and lamivudine is about $300.00. Laboratory monitoring and office visits would add $100.00-$200.00. A formal analysis will be needed to determine the cost-effectiveness of prophylaxis, but since the lifetime medical costs of HIV/AIDS are very high (approximately $100,000) the cost-benefit ratio will probably be consistent with those of other recommended preventive medical therapies.[36,37]

2. PROPHYLAXIS AFTER SEXUAL AND INJECTING DRUG USE EXPOSURES TO HIV

The USPHS guidelines for managing occupational exposures advise that prophylaxis should be offered after mucosal as well as percutaneous exposures to HIV.[4,30] These recommendations have led to advocacy for access to treatment after sexual (mucosal) and injection drug use (percutaneous) exposures to HI.[3,6,7] Although not yet a standard of care, acceptance of this potentially useful prevention intervention is gaining momentum in some areas.

2.1. HIV Transmission Risk Estimates

Inadvertant exposures to HIV are not rare, even among couples committed to safer sex. For example, condom breakage occurs in 1%-7.3% of episodes of intercourse, slippage occurs in 1%-6.5% of episodes, and reported total failure rates are 2.7%-14.6%.[38–40] The absolute risks associated with a single episode of sexual exposure to HIV are not completely defined. Even when the average per-contact transmission risk can be calculated, the estimate cannot be extrapolated to a specific exposure event because many variables (e.g. partner infectivity, host characteristics, the nature of the sexual act, and characteristics of the virus strain) influence the probability of transmission in individual cases. However, the estimated transmission risks associated with unprotected receptive anal and vaginal exposures are similar in magnitude to that associated with an HIV-contaminated needle puncture.[41–46] There are no published estimates of transmission risks associated with oral sex, although there are numerous documented cases of transmission through this route. Intuitively, the probability of HIV infection following use of contaminated injection drug equipment would be expected to be similar to that of parenteral occupational exposures. In one study, the measured risk was slightly higher,[47] which may reflect the greater volume of blood transferred in sharing of needles.[48]

2.2. Prophylactic Treatment Advice

Care for all exposed individuals should include a comprehensive HIV risk assessment and on-going risk reduction counselling. Those exposed through injecting drug use should be referred to drug treatment, needle exchange, or peer-led risk reduction programs.[6,7,49–51] We recommend antiretroviral treatment as one component of a comprehensive post-exposure prevention program, but only after those exposures posing the highest

Table 2. Factors that may facilitate sexual transmission of HIV

HIV-infected partner
High viral load (early and late infection)
High viral load in semen/vagina
Menses at the time of exposure
Inflammatory genital disease (e.g., urethritis, cervicitis, vaginitis)
Lack of circumcision
Uninfected host
Genetic susceptibility
Genital ulcer disease
Inflammatory genital disease (e.g., urethritis, cervicitis, vaginitis)
Lack of circumcision
Cervical ectopy
Characteristics of the sexual act
Trauma (including activities such as fisting)
Douching prior to anal or vaginal intercourse
Increased duration or intensity of intercourse
Characteristics of the virus
Macrophage tropism
Nonsyncytium-inducing phenotype

risk of HIV transmission. In San Francisco, eligible exposures include parenteral blood contact, rectal and vaginal sexual exposure, and oral receptive sex.[6,7] For activities less likely to transmit HIV, the risks of treatment may outweigh the benefits, and treatment is not routinely offered. However, the presence of factors facilitating transmission (e.g. traumatic intercourse, presence of genital mucosal lesions) could increase the infection risk and the potential benefit of treatment (Table 2). The decision about treatment should also reflect the patient's willingness to adhere to a drug regimen and avoid unsafe practices in the future. In our view, post-exposure prophylaxis should be provided only with the informed consent of the patient.

The HIV status of the partner may not be known at the time the exposure is reported. If the partner is willing to be tested, then treatment can be started and stopped if a negative HIV antibody test result is obtained. Access to rapid HIV testing can facilitate this evaluation, since a negative rapid test reliably excludes infection. In cases of partners who are unavailable or unwilling to be tested, decisions about treatment should be based both on the exposure and the likelihood that the partner is HIV positive. As with occupational exposures, factors to consider include the partner's risk factors for HIV and the prevalence of HIV/AIDS in the community. Post-exposure drug treatment is probably not the best option for persons who are repeatedly exposed to HIV. Enhanced prevention counselling and case management are more likely to decrease the risk of transmission among these individuals, without increasing the risk of infection with drug resistant virus strains.

The treatment regimen used in San Francisco is based on the same approach advised for occupational exposures (above) (Table 3). Treatment with both zidovudine and lamivudine for 4 weeks is appropriate for most eligible exposures, and the regimen is adjusted in individual cases, depending on the probability of drug resistance and exposure risk. Recommendations for prophylaxis of health care personnel stress the importance of initiating treatment as soon as possible, preferably within a few hours after the exposure. Whether this will prove to be feasible for non-occupational exposure management is an important question, but preliminary experience in San Francisco suggests that many exposed persons will seek care in time.

Table 3. Guidelines for post-exposure prophylaxis (PEP) for sexual and drug exposures[1]

Consider post-exposure prophylaxis if conditions 1, 2, 3, 4, and 5 are met:

1. High risk exposure (in descending order of risk):
   - Unprotected[2] receptive anal intercourse
   - Sharing needles or drug paraphernalia
   - Unprotected receptive vaginal intercourse
   - Unprotected insertive vaginal intercourse
   - Unprotected insertive anal intercourse
   - Unprotected receptive fellatio with ejaculation
2. Partner is known to be HIV-infected, or is in an HIV risk group (gay or bisexual men, injection drug user, sex worker), or patient was raped.
3. Exposure is isolated event (eg broken condom) or patient expresses willingness to avoid HIV exposures in the future through safer sex and clean needle use.
4. Patient presents for care within 72 hours of exposure.
5. Patient desires treatment and agrees to adhere to the treatment regimen.

If decision is made to treat patient:

1. Basic treatment regimen[3]
   - Zidovudine (ZDV) 200 mg po tid (or 300 mg po bid)
   - Lamivudine (3TC) 150 mg po bid
   - Treat for 28 days
   - Note: Consider adding a protease inhibitor (indinavir 800mg po tid or nelfinavir 750mg po tid) if source partner known to have a high viral load (e.g. > 50,000) or other factors potentiating transmission are present (mucosal bleeding or inflammation) or when antiretroviral dug resistance is suspected.

   Alternative treatment regimen (eg partner is taking both ZDV/3TC) and resistance suspected).
   - Stavudine (D4T) 40 mg po bid (30 mg po bid if < 60 kg)
   - Didanosine (DDI) 200 mg po bid (125 mg po bid if < 60 kg)

   Alternative treatment regimen (eg partner is taking ZDV and resistance suspected)
   - Stavudine (D4T) 40 mg po bid (30 mg po bid if < 60 kg)
   - Lamivudine (3TC) 150 mg po bid
2. Recommended baseline laboratories for exposed individual:
   - HIV antibody
   - Hepatitis B virus antibody
   - Hepatitis C virus antibody
   - Gonorrhea, syphilis, chlamydia (if appropriate)
   - Pregnancy testing (if appropriate)
   - Complete blood count
   - Liver function tests
   - Renal function tests
3. Partner evaluation:
   - If partner's current HIV status is unknown
     - HIV risk behaviors of partner (eg injection drug use)
     - Prior HIV-antibody test results
     - Perform partner HIV testing and stop PEP if partner is found to be negative
   - If partner known to be HIV infected
     - Clinical stage of HIV infection
     - Quantitative HIV RNA titer (viral load)
     - Antiretroviral treatment history (past 30 days is most relevant)
4. All patients should receive:
   - HIV risk reduction counselling. Major components are:
     - Review modes of HIV transmission and the relative risks of transmission associated with specific behaviors (eg anal intercourse is riskier than oral sex)
     - Discuss specific HIV risk behaviors relevant to the individual patient, including circumstances contributing to recent exposure (eg substance abuse)
     - Assess the patient's readiness to change HIV risk behaviors
     - Strengthen the patient's skills to reduce HIV risk (eg proper use of condoms, communication with partners about condom use)

(continued)

Table 3. (*Continued*)

Negotiate a realistic and incremental risk reduction plan that focuses on the steps the patient will take to reduce their HIV risk
Refer the patiet to ongoing prevention programs to help them avoid risky behavior

5. Follow-up evaluation:
   If patient is found to be HIV-positive on baseline evaluation, change from PEP to optimal treatment for infected person depending on stage of infection (ie primary HIV infection, long standing HIV infection) and viral load
   Assess side effects of antiretroviral treatment at two and four weeks (if post-exposure prophylaxis is prescribed)
   1) complete blood count
   2) liver function tests
   3) amylase (if indinavir is used)
   4) urinalysis (if indinavir is used)
   Assess HIV risk behavior during interval since prior visit
   Reinforce prevention messages
   Assess adherence with medication
   Test for HIV-antibody at six weeks, twelve weeks, and six months
   Assess symptoms and signs of primary HIV infection

[1] Department of Public Health (DPH)/San Francisco General Hospital (SFGH)
[2] Unprotected includes broken condoms
[3] Basic Treatment Regimen is the regimen recommended by the PHS for occupational exposures and is useful for most exposures (default). Alternative regimens reflect current practices for special cases at SFGH, but experience is limited and too small to predict efficacy and safety. Recommendation will likely change as new drugs are approved and/or PHS guidelines change.

Although post-exposure prophylaxis will hopefully benefit some individuals, the outcome of treatment programs on HIV incidence in the community is difficult to predict. Persons who experience an HIV exposure are by definition at increased risk for HIV infection and other sexually transmitted diseases, and comprise a priority population for prevention services. Marketing exposures as an opportunity to initiate needed prevention services could improve access to HIV risk assessment and testing, identify persons who do not realize they are already infected, and promote primary prevention interventions.[6,7] In this context, chemoprophylaxis is just one component of a comprehensive prevention program that stresses safer behavior and avoidance of future risks, and is used only as a back-up when other prevention efforts fail. However, awareness that such "morning after" treatment is available could increase unsafe behavior in the community and lead to a net increase in HIV incidence. Carefully constructed public health messages that place prophylaxis in the overall context of health promotion should help prevent this perverse outcome. The message should be "got exposed - get help!", not "got exposed - get prophylaxis".

3. SUMMARY

The advent of preventive treatment for HIV highlights the urgent need for basic, clinical, and epidemiologic research targeting the pathogenesis and prevention of cutaneous and mucosal infection. In addition, the impact of HIV prophylaxis on the frequency of risk behaviors and antiretroviral drug resistance, especially in cities with high HIV prevalence, must be evaluated. In our view, the available data (albeit not definitive) are strong enough to support the provision of post-exposure prophylaxis in select cases of sexual and injection drug use exposure in addition to occupational exposures. However, post-exposure treatment should be used judiciously and only in the context of a comprehensive prevention program. Ultimately, primary exposure prevention, whether in health care facilities or in the community, is the most important strategy to prevent AIDS.

REFERENCES

1. CDC. Case-control study of HIV seroconversion in health-care workers after percutaneous exposure to HIV-infected blood-France, United Kingdom, and United States, January 1988-August 1994. MMWR. 1995; 44:929–933.
2. Cardo DM, Culver DH, Ciesielski CA, et al. A case-control study of HIV serovonversion in health care workers after percutaneous exposure. New Engl J Med 1997; 337(21):1485–90.
3. Carpenter CCJ, Fischl MA, Hammer SM, et al. Antiretroviral therapy for HIV infection in 1996: recommendations of an international panel. JAMA. 1996; 276:146–154.
4. CDC. Update: provisional public health service recommendations for chemoprophylaxis after occupational exposure to HIV. MMWR. 1996; 45:468–472.
5. Gerberding JL. Post-exposure prophylaxis for human immunodeficiency virus at San Franisco General Hospital. Am J Med 1997; 102:85–89.
6. Katz MH, Gerberding JL. Post-exposure treatment of people exposed to the Human immunodeficiency virus through sexual contact or injection-drug use. New Engl J Med 1997; 336(15):1097–1100.
7. Katz MH, Gerberding JL. Managing recent sexual exposures to HIV. Ann Intern Med 1998; 128:306–12.
8. CDC. Guidelines for Treatment of Sexually Transmitted Diseases. MMRW 1998; 47(RR-1):109–111.
9. Gostin LO, Lazzarini Z, Alexander D, et al. HIV testing, counselling, and prophylaxis after sexual assault. JAMA. 1994; 271:1436–1444.
10. Blauvelt A, Katz SI. The skin as target, vector, and target effector organ in human immunodeficiency disease. J Invest Dermatol. 1995; 105:122S-126S.
11. Clerici M, Levin JM, Kessler HA, et al. HIV-specific T-helper activity in seronegative health care workers exposed to contaminated blood. JAMA 1994; 271:42–6.
12. Pinto LA, Landay AL, Berzofsky JA, et al. Immune response to human immunodeficiency virus (HIV) in health care workers occupationally exposed to HIV-contaminated blood. Am J Med 1997; 102:21–24.
13. Spira AI, Marx PA, Patterson BK, et al. Cellular targets of infection and route of viral dissemination after an intravaginal inoculation of simian immunodeficiency virus into rhesus macaques. J Exp. Med. 1996; 183:215–225.
14. Lu Y, Brosio P, Lafaile M, et al. Vaginal transmission of chimeric simian/human immunodeficiency viruses in rhesus macaques. J Virology. 1996; 70:3045–3050.
15. Milman G, Sharma O. Mechanisms of HIV/SIV mucosal transmission. AIDS Research and Human Retroviruses. 1994; 10:1305–1312.
16. Miller CJ. Mucosal transmission of simian immunodeficiency virus. Current Topics in Microbiology and Immunology. 1994; 188:107–122.
17. Tavares L, et al. 3'-azido-3'deoxythymidine in feline leukemic virus-infected cats: a model for therapy and prophylaxis of AIDS. Cancer Research 1987; 47:3190–3194.
18. Mathes LE, Polas PJ, Hayes KA, et al. Pre and post-exposure chemoprophylaxis: evidence that 3'-azido-3'-dideoxythymidine inhibits feline leukemia virus disease by a drug-induced vaccine response. Antimicrob Agents Chemother 1992; 36:2715–21.
19. Ruprecht RM, et al. Suppression of mouse viraemia and retroviral disease by 3'-azido-3'-deoxythymidine. Nature 1986; 323:467–469.
20. McCune JM, Namikawa R, Shih CC, Rabin L, Kaneshima H: 3'azido-3'-deoxythymidine suppresses HIV infection in SCID-hu mouse. Science 1990; 247:564–6.
21. Shih CC, Kaneshima H, Rabin L, et al. Post-exposure prophylaxis with zidovudine suppresses human immunodeficiency virus type 1 infection in SCID-hu mice in a time-dependent manner. J Infect Dis 1991; 163: 625–7.
22. Black R. Animal studies of post-exposure prophylaxis. Am J Med 1997; 102:39–44.
23. Tokars JI, Marcus R, Culver, DH, et al. Surveillance of human immunodeficiency virus (HIV) infection and zidovudine use among health care workers with occupational exposure to HIV-infected blood. Ann Intern Med 1993;

24. Henderson DK, Fahey BJ, Willy M, et al. Risk for occupational transmission of human immunodeficiency virus type-1 (HIV-1) associated with clinical exposures. A prospective evaluation. Ann Intern Med 1990; 113:47–6.
25. Gerberding JL. Incidence and prevalence of human immunodeficiency virus, hepatitis B virus, hepatitis C virus, and cytomegalovirus infection among health care personnel at risk for blood exposure: final report from a 10 year longitudinal study. J Infect Dis 1994; 70:1410–7.
26. Gerberding JL. Management of occupational exposures to blood-borne viruses. N Engl J Med. 1995; 332:444–451. CDC.
27. Ho DD, Moudgil T, Alam M. Quantitation of human immunodeficiency virus type 1 in the blood of infected persons. New Engl J Med 1989; 321:1622–25.
28. Connor EM, Sperling RS, Gelber R, et al. Reduction of maternal-infant transmission of human immunodeficiency virus type 1 with zidovudine treatment. N Engl J Med. 1994; 331:1173–1180.
29. Sperling RS, Shapiro DE, Coombs RW, et al. Maternal viral load, zidovudine treatment, and the risk of transmission of human immunodeficiency virus type 1 from mother to infant. New Engl J Med 1996; 225:1621–29.
30. Management of health care worker exposures to HIV and recommendations for post-exposure prophylaxis. MMWR. 1998; 47:1–39.
31. Jochimsen EM. Failures of zidovudine post-exposure prophylaxis. Am J Med 1997; 102:52–55.
32. Gerberding JL. Prophylaxis for occupational exposure to HIV. Ann Intern Med. 1996; 125:497–501.
33. Beekmann SE, Fahrner R, Henderson DK, et al. Zidovudine safety and tolerance among uninfected health care workers: a brief update. Am J Med 1997; 102:63–4.
34. Puro V, Ippolito G, Guzzanti E, et al. Zidovudine prophylaxis after accidental exposure to HIV: the Italian experience. The Italian Study Group on Occupational Risk of HIV Infection. AIDS. 1992; 6:963–9.
35. Ciesielski CA, Metler RP. Duration of time between exposure and serroconversion in healthcare workers with occupationally acquired infection with human immunodeficiency virus. Am J Med 1997; 102:115–116.
36. Hurley SF, Kaldor JM, Gardiner S, et al. Lifetime cost of human immunodeficiency virus-related health care. J Acquir Immune Defic Syndr. 1996; 12:371–378.
37. Gable CB, Tierce JC, Simison D, et al. Costs of HIV+/AIDS at CD4+ counts disease stages based on treatment protocols. J Acquir Immune Defic Syndr. 1996; 12:413–420.
38. Sparrow MJ, Lavill K. Breakage and slippage of condoms in family planning clients. Contraception. 1994; 50:117–129.
39. Steiner M, Piedrahita C, Glover L, et al. Can condom users likely to experience condom failure be identified? Family Planning Perspectives. 1993; 25:220–223 & 226.
40. Thompson JLP, Yager TJ, Martin JL. Estimated condom failure and frequency of condom use among gay men. Am J Public Health. 1993; 83:1409–1413.
41. DeGruttola V, Seage GR III, Mayer KH, et al. Infectiousness of HIV between male homosexual partners. J Clin Epidemiol. 1989; 42:849–856.
42. Downs AM, De Vincenzi. Probability of heterosexual transmission of HIV: relationship to the number of unprotected sexual contacts. J Acquir Immune Defic Syndr. 1996; 11:388–395.
43. Mastro TM, Satten GA, Nopkesorn T, et al. Probability of female-to-male transmission of HIV-1 in Thailand. Lancet. 1994; 343:204–207.
44. Padian NS, Shiboski SC, Jewell NP. Female-to-male transmission of human immunodeficiency virus. JAMA. 1991; 226:1664–1667.
45. Peterman TA, Stoneburner RL, Allen JR, et al. Risk of human immunodeficiency virus transmission from heterosexual adults with transfusion-associated infections. JAMA. 1988; 259:55–58.
46. Wiley JA, Herchkorn SJ, Padian NS. Heterogeneity in the probability of HIV transmission per sexual contact: the case of male-to-female transmission in penile-vaginal intercourse. Statistics in Medicine. 1989; 8:93–102.
47. Kaplan EH, Heimer R. A model-based estimate of HIV infectivity via needle sharing. J Acquir Immune Defic Syndr. 1992; 5:1116–1118.
48. Gaughwin MD, Gowans E, Ali R, et al. Bloody needles: the volumes of blood transferred in simulations of needle-stick injuries and shared use of syringes for injection of intravenous drugs. AIDS. 1991; 5:1025–7.
49. CDC technical guidance on HIV counseling. MMWR 1993; 42(RR-2)8–17.
50. Rietmeijer CA, Kane MS, Simons PZ, et al. Increasing the use of bleach and condoms among injecting drug users in Denver: outcomes of a targeted, community-level HIV prevention program. AIDS. 1996; 10:291–298.
51. Choi K-H, Coates TJ. Prevention of HIV infection. AIDS 1994; 8:1371–89.

21

ANTIRETROVIRAL THERAPY IN PRIMARY HIV

Lawrence Corey and M. Michelle Berrey

Fred Hutchinson Cancer Research Center
Program in Infectious Diseases
University of Washington
Virology Division, 1100 Fairview Ave. North
D3-100, P.O. Box 19024, Seattle, Washington 98109-1024

Increasingly, evidence is accumulating that combination antiretroviral therapy can markedly prolong the time to development of AIDS-defining opportunistic infections and death.[1–2] As such, a natural extension of this concept is the use of antiviral therapy very early in the course of infection, especially at the time of HIV-1 seroconversion. Several recent articles have shown that the initial days and weeks after exposure and establishment of HIV are critical in defining host/virus interactions and subsequent disease course. Within 4 months after infection a viral inflection point is reached; after this point the level of mRNA in plasma is predictive of clinical progression and time to death.[3–5] In addition, host immune responses, especially cytotoxic T-lymphocyte (CTL) responses, correlate with plasma virus levels, and therefore with subsequent progression of disease.[6] Diversity of host cell response is also predictive: individuals with a less restricted immune response to HIV appear to have a lower rate of disease progression.[7] These data suggest that early intervention affords the best opportunity to maximize the care of the HIV-infected patient.

The initiation of therapy before the advent of significant clinical immunodeficiency has potential to preserve diverse host CD4+ T cell responses. This in turn could reduce the need for prophylaxis against, and improve host control of microorganisms associated with opportunistic infections, subsequently reducing morbidity, mortality and cost of care. The early control of HIV replication could theoretically reduce the quantitative viral reservoir of HIV-1. A decrease in the number of replication cycles would, in turn, decrease the frequency of spontaneous mutations of drug resistant isolates.[8] This could result in prolonged effectiveness of antiretroviral therapies. Balanced against these potential benefits are both the cost and difficulty of maintaining the administration of several medications. Whether very early antiretroviral therapy can suppress initial plasma HIV-RNA and preserve the immunologic repertoire, and therefore accomplish the same level of disease non-progression as that seen in patients with innate immunity is not yet established.

Antiviral Chemotherapy 5, edited by Mills *et al.*
Kluwer Academic / Plenum Publishers, New York, 1999.

Immune suppression begins within days of infection with HIV, as evidenced by frequent diagnoses of oral and genital candidiasis, and reported Pneumocystis pneumonia,[9] and cytomegalovirus encephalitis[10] during primary HIV infection. CD4+ T cells fall to low levels during primary infection, rebound to near-normal levels within the first few weeks, and then decrease at a rate of 50 to 130 cells per ml per year.[3,11] Cytotoxic T-lymphocyte proliferation targeted at HIV-envelope epitopes are associated with reduced HIV RNA plasma levels, but may be exhausted within the first few months of infection. Higher cytotoxic T-lymphocyte response to HIV envelope antigen in primary HIV patients correlates with decreased plasma HIV RNA within the first 18 months of HIV infection.[6] Viral escape mutations may also result from HIV-1 specific CTL pressures detected during primary infection.[12]

The degree to which antiretroviral therapy is effective correlates directly with the magnitude and duration of viral suppression, presumably by reducing the antigen-driven immunologic dysfunction. Aggressive antiretroviral therapy instituted late in the disease course may provide an increase in the absolute number of CD4+ cells and more normal CD4:CD8 ratios, but the predominant cells regained are memory, and not naïve cells. Beginning antiretroviral therapy very early in HIV infection may prevent clonal deletion and subsequent immune dysregulation. Destruction of memory cell populations to a wide variety of opportunistic pathogens may be reduced, suggesting long-lasting effects. Recent evidence indicates that therapy during primary infection is not only associated with the anticipated decrease in plasma viremia, but also with the generation of a p24-specific proliferative response.[13] This HIV-specific response was also detected in persons who, in the absence of antiretroviral therapy, were able to control HIV plasma virus loads, so-called slow-progressors.[13] Preservation of such T-cell help may improve host responses to HIV per se, in essence converting treating individuals to the immunologic phenotype of long-term non-progressors. Full suppression of viral replication may serve to avoid emergence of resistant strains and more hostile phenotypes, such as syncytium-inducing strains.

1. MONOTHERAPY FOR ACUTE HIV INFECTION

Several studies have used monotherapy with either AZT or D4T for the treatment of acute primary HIV infection. These studies in general have shown that nucleoside monotherapy may boost the CD4+ lymphocyte count and restore normal CD4:CD8 ratios, but the degree of clinical benefit and reduction in overall viral replication appears modest.

An early study[14] in 11 subjects with symptomatic seroconversion showed zidovudine, 1000mg/d, was well-tolerated in primary infection patients, with only one of the eleven patients discontinuing therapy due to nausea. However, there was no improvement in T-cell counts when compared with historical controls, and no clear clinical benefit in terms of decreased duration of symptoms associated with seroconversion.

A placebo-controlled double-blinded trial in Australia and Europe involving 77 patients evaluated the use of zidovudine monotherapy in primary HIV infection. At the end of the six month trial, the group receiving zidovudine had a mean increase of 8.9 CD4+ cells per month. The group receiving placebo continued to sustain CD4+ cell loss at an average of 12.0 cells per ml per month.[15] A reduction of HIV plasma RNA in both the placebo and treatment groups was observed, with a trend toward decreased viral load in the zidovudine group. Some reduction in minor opportunistic infections was seen, however, the study design was such that zidovudine treatment was discontinued and long-term evaluation of the effects is thus not available.

2. COMBINATION ANTIRETROVIRAL THERAPY FOR ACUTE HIV

Combination therapy with zidovudine 500mg/d and a non-nucleoside reverse transcriptase inhibitor, L-697,661 was reported in a pilot study of four primary infection patients.[16] Plasma viremia in three patients fell below the level of detection (<200 copies/mL), but was sustained in only two. Three of the 4 patients maintained normal CD4+ counts during the 6 months of therapy. The quantity of proviral DNA in PBMC decreased progressively during the treatment phase. Upon discontinuation of treatment, both plasma and proviral DNA increased. One patient in four studied had a persistently high plasma viremia, and a decrease in CD4+ counts with abnormal CD4:CD8 ratio. Mutations associated with low-level resistance to L-697,661 emerged during therapy; no AZT-resistance mutations were detected.

A more potent antiviral effect was observed in 10 symptomatic primary HIV infection patients given zidovudine, didanosine, and lamivudine.[17] The plasma RNA in all patients fell to below the level of detection (<200 copies/mL) after a mean of 108 (+/- 32) days. A more rapid decline in plasma viremia was noted in patient treated before day 15 of symptoms.

The above studies suggest that if the goal of treatment of acute primary infection is rapid viral suppression, a regimen of at least three initial antiretroviral agents is preferable. As such, several groups have initiated pilot studies of triple therapy consisting of a combination of two nucleosides and a protease inhibitor. Some authorities have suggested that such early regimens should consist of multiple drugs focused on the same viral target, as they may provide complete suppression of replication at a single step.[18] Preliminary data from these studies has shown suppression of plasma virus to below the level of detection within 90 days of initiating such therapy. This approach also has the advantage of following the concept that a 3-drug regimen provides longer lasting antiviral effects than an initial 2-drug regimen followed by the later addition of a third drug.

Alteration of initial therapy based on the antiretroviral regimen of the presumed source partner, or perhaps on community virus genotype, is also an important consideration in selecting a therapeutic regimen for acute HIV. Transmission of HIV resistant to nevirapine and zidovudine is well-described[19] and the resulting resistant mutants are fully pathogenic. To date, the prevalence of high-grade ZDV resistance (215 mutation) remains less than 20%, even in heavily treated areas. The frequency of ZDV resistance in our Seattle primary HIV cohort appears to be less than 10%. Five patients infected with HIV with the altered genotype were recently followed by the Sydney Primary HIV Infection Study Group and did not have clinical differences at one year of infection.[20]

2.1. Evaluation of Efficacy

Development of long term outcome studies comparing treatment of acute HIV versus initiation at later time periods will be a formidable task. Clinical progression rates with the initiation of early antiretroviral therapy are low. The issue to be answered is whether there is significant benefit for initiating therapy early in infection versus excellent suppression at later time periods. Would one achieve better immunologic function with treatment of acute infection? Will long-term therapy require fewer drugs later on? All of these questions remain unanswered at present. The benefits of treating acute infection is largely being assessed with short-term studies of immunologic restoration and viral load measured in selected compartments such as CNS, lymph node, and genitourinary tract. Larger clinical studies are required to answer these issues.

2.2. Adherence

Beginning antiretroviral therapy during primary infection requires rapid diagnostic screening, adequate pre- and post-test counseling, and personnel available to spend hours with a newly diagnosed individual discussing the theoretical benefits of early intervention. Screening patients for alcohol and other drug abuses, which may impact their ability to adhere to a complex drug regimen, is an important step in aggressive early intervention plans. Clearly, beginning therapy in non-adherent patient populations will risk the future treatment success of individuals as well as the community. Twice-daily dosing may enhance adherence to a complicated drug regimen.

Our experience with initiation of a triple-therapy (AZT, 3TC, indinavir) regimen in primary infection patients has confirmed the difficulty of assessing prospectively patients who will remain adherent. Patients recently diagnosed with HIV infection must face the reality of a lifelong infection. Weeks into therapy the side effects of nausea, headache, and fatigue can be intrusive on a young working person's career, and the benefits of long-term immune preservation are less tangible. Use of recreational drugs and alcohol in our patients substantially impacted the ability to adhere to a strict medication regimen, particularly the midday dose and dietary restrictions. Of 20 patients we enrolled into an early treatment protocol utilizing ZDV 300 BID, 3TC 150 BID, and indinavir 800 mg TID, six were switched from AZT to D4T (provided for in the protocol) secondary to persistent nausea (n=3), or rash and acute anemia (n=3). One patient discontinued indinavir secondary to recurrent indinavir crystalluria, hematuria, and renal colic and 2 patients stopped therapy due to problems associated with recreational drug use. Thus, at 6 months, 11 of 20 were on the same starting regimen and 17 of 20 on another 3-drug regimen. This attrition rate reflects the need for additional exploration for well-tolerated potent drug regimens.

2.3. Will Treatment Affect Latency?

Early studies suggest that both persistent and latent pools of cells infected by HIV are established early on post-infection. These observations help account for the prompt release of virus in plasma upon discontinuing therapy. Recent studies have shown that prolonged therapy, even up to 30 months, is associated with no discernible change in the level of latently infected cells.[21] Whether extremely early therapy will lead to reduction in this reservoir of infection remains unclear. Moreover, the effects of such a reduced reservoir on the subsequent clinical course of disease are unknown.

3. SUMMARY

Initiation of antiretroviral therapy during primary HIV-1 infection may well redefine the course of the disease by controlling the burst of viral replication and subsequent immune dysfunction. At present, treatment of acute primary HIV-1 should remain largely in the domain of clinical investigation. There are significant problems in providing constant vigilance to medication and subsequent viral suppression. Only clinical trials can define the relative benefit of initiating antiretrovirals during this stage of infection.

REFERENCES

1. Lewis JS, Terriff CM, Coulston DR, *et al.* Protease inhibitors: a therapeutic breakthrough for the treatment of patients with human immunodeficiency virus. Clin Ther 1997; 19(2):187–214.
2. Denning D, Michelet C, Jouan M, *et al.* Impact of highly active antiretroviral therapy (HAART) on opportunisitic infections (abstract) 37th ICAAC, Toronto.
3. Schacker T, Hughes JP, Shea T, Coombs RW, Corey L. Biological and virologic characteristics of primary HIV infection. Ann Intern Med (In press, 1998; 128 (8)
4. Craib KJP, Strathdee SA, Hogg RS, *et al.* Serum levels of human immunodeficiency virus type 1 (HIV-1) RNA after seroconversion: A predictor of long-term mortality in HIV infection. J Infect Dis 1997: 176:798–800.
5. Mellors JW, Rinaldo CR, Gupta P, *et al.* Prognosis in HIV-1 infection predicted by the quantity of virus in plasma. Science 1996; 272:1167–70.
6. Musey L, Hughes J, Schacker T, *et al.* Cytotoxic T cell responses, viral load, and disease progression in early human immunodeficiency virus type 1 infection. N Engl J Med 1997; 337:1267–74.
7. Pantaleo G, Soudeyno H, Demarest JF, *et al.* Evidence for rapid disappearance of initially expanded HIV-specific CD8+ T-cell clones during primary HIV-1 infection. Proc Natl Acad Sci USA. 1997; 94(18):9848–53.
8. Havlir DV, Richman DD. The role of viral dynamics in the pathogenesis of HIV disease and implications for antiviral therapy. Springer Semin Immunopath 1997; 18(3):267–83.
9. Vento S, Di Perri G, Garofano T, *et al.* Pneumocystis carinii pneumonia during primary HIV-1 infection. Lancet 1993; 342:24–5.
10. Berger DS, Bucher G, Nowak JA, Gomatos PJ. Acute primary human immunodeficiency virus type 1 infection in a patient with concomitant cytomegalovirus encephalitis. Clin Infect Dis 1996; 23:66–70.
11. Perrin L, Yerly S. Acute HIV infection. J Biol Reg Homeostatic Agents 1995;9:95–9.
12. Borrow P, Lewicki H, Wei X, *et al.* Antiviral pressure exerted by human immunodeficiency virus type-1 specific cytotoxic T-lymphocytes (CTL) during primary infection demonstrated by rapid selection of CTL escape virus. Nat Med 1997; 3:205–11.
13. Rosenberg ES, Billingsley JM, Caliendo AM, *et al.* Vigorous HIV-1 specific CD4+ T cell responses associated with control of viremia. Science 1997; 278:1447–50.
14. Tindall B, Gaines H, Imrie A, *et al.* Zidovudine in the management of primary HIV-1 infection. AIDS 1991;5:477–84.
15. Kinloch-de Loes S, Hirschel BJ, Hoen B, *et al.* Zidovudine versus placebo in primary HIV infection: a randomized study. N Engl J Med 1995; 333:408–13.
16. Perrin L, Rakik A, Yerly S, *et al.* Combined therapy with zidovudine and L-697,661 in primary HIV infection. AIDS 1996;10:1233–7.
17. Lafeuillade A, Poggi C, Tamalet C, *et al.* Effects of a combination of zidovudine, didanosine, and lamivudine on primary human immunodeficiency virus type 1 infection. J Infect Dis 1997; 175:1051–5.
18. Perrin L, Kinloch-de Loes S. Therapeutic interventions in primary HIV infection. J AIDS Hum Retrovir 1995; 10(S1):S69–76
19. Imrie A, Beveridge A, Genn W, Vizzard J, Cooper D. Transmission of human immunodeficiency virus type I resistant to nevirapine and zidovudine. J Infect Dis 1997; 175:1502–6.
20. Imrie A, Carr A, Duncombe C, *et al.* Primary infection with zidovudine-resistant human immunodeficiency virus type 1 does not adversely affect outcome at 1 year. J Infect Dis 1996; 174:195–8.
21. Chun T-W, Stuyver L, Mizell S, *et al.* Presence of an inducible HIV-1 latent reservoir during highly active antiretroviral therapy. Proc Natl Acad Sci USA 1997; 94(24):13193–7.

22

INITIAL TREATMENT FOR HIV INFECTION

When, Why, and with What?

James O. Kahn

UCSF and
AIDS Program, San Francisco General Hospital
995 Potrero Street
San Francisco, California 94110

There are no clinical trials that define the optimal time to begin treatment for HIV-1 infection, nor is there data defining the best initial treatment. The expected benefits of early effective treatment are a maximal reduction of viral replication leading to immunologic restoration, slowed disease progression, longer survival and reduced disability. Current combination regimens of reverse transcriptase and protease inhibitors appear to be achieving many of these goals, at least in the short term. However, there are potential drawbacks to early therapy. Early therapy may limit the availability of effective treatments in the late stages of HIV infection when there is a high risk of disease progression. Early therapy may also lead to resistant strains of HIV; in addition to being a significant public health risk through transmission to others,[1] these strains vitiate the efficacy of antiretroviral drug regimens. Further, antiretroviral chemotherapy is expensive, and is associated with both recognized and as-yet-unrecognized adverse reactions, drawbacks which may become particularly significant with long-term therapy.

Three different consensus panels provide different treatment options, and opinions, regarding optimal time to initiate treatments and treatment options.[2, 3, 4] Despite these differences there is a general consensus that in most circumstances clinicians should not be in a rush to begin therapy; that patients should generally not begin therapy before they are confident that they can sustain full adherence to the proposed treatment regimen; and that once the decision to treat is made, the regimen must include effective doses of antiretrovirals in combinations which maximally suppress viral replication.

This chapter will evaluate the different treatment strategies regarding optimal initial antiretroviral medications. It represents a single clinician's opinion, and like much in clinical medicine that is incompletely studied but with ongoing real world care concerns, is subject to significant individual biases.

Antiviral Chemotherapy 5, edited by Mills *et al.*
Kluwer Academic / Plenum Publishers, New York, 1999.

1. RATIONALE FOR EARLY INITIATION OF HIV THERAPY

The justification for initiating treatment early in HIV-1 infection relies on interpretation of studies involving small numbers of patients and elegant laboratory based evaluations. These studies suggest that early therapy preserves the antiviral immune response, which augments antiretroviral suppression of viral replication; in turn, this limits ongoing destruction of the immune system and creates the potential for full immunologic recovery.[5] The general admonition to proceed cautiously with initiation of early treatment does not extend to primary HIV infection or to post-exposure prophylaxis (see Chapter 19 by Gerberding). In these two later cases, rapid initiation of treatment is probably essential.[6]

"Early therapy" of HIV infection has not been strictly defined, but in general it means therapy initiated before the patient has developed clinically-significant damage to the immune system (generally measured by CD4 lymphocyte numbers), and well before any clinical signs or symptoms of HIV infection have appeared. "Early therapy" therefore applies generally to treatment of asymptomatic patients with CD4 lymphocyte counts above 500/uL. This is contrasted with "late therapy", which indicates treatment of patients with the beginnings of significant damage to the immune system (e.g. a CD4 lymphocyte count below 350/uL), or with signs or symptoms of HIV-related disease.

Whether early antiretroviral therapy will prevent ongoing immune destruction and will reconstitute the immune response is a matter of considerable debate.[7] Although significant immune destruction occurs early in the course of HIV infection,[8] delayed initiation of combination antiretroviral therapy appears to restore previously lost (or at least undetected) immune responses.[9] It has been hypothesized that the maintenance of these lost immune responses will translate to reduced disease progression.[10] Prompt reduction in HIV replication has the additional benefit of maintaining completely normal CD4 lymphocyte numbers, an essential cell population for maintaining immunologic integrity.[11] Finally, it is undeniable that in developed countries with access to combination treatment regimens, mortality from HIV infection has been remarkably decreased and the reduction is clearly related to wide-spread and effective use of antiretroviral drugs.[12, 13] It is primarily the analogy to the obvious benefit conferred by effective combination therapy observed for those with advanced disease that leads clinicians to suggest early treatment for antiretroviral naive persons (Table 1).

Table 1. Rationale for treatment decisions

Possible reasons to initiate treatment early

- Prevent irreversible damage to the immune system
- Limit HIV dissemination
- Achieve long-term prevention of disease progression cure HIV infection altogether

Possible reasons to delay therapy

- Clinical benefit of combination antiretroviral chemotherapy has not been shown in patients with relatively intact immune systems
- Cumulative short-term toxicities that may be life-threatening
- Unknown long-term toxicities
- Substantial cost
- Increased risk of poor adherence with long term treatment leading to HIV strains resistant to antiretrovirals

2. RATIONALE FOR DELAYED INITIATION OF HIV THERAPY

The rational for delaying initial anti-HIV therapy until there is evidence of progressive damage to the immune system relates to the risk-benefit analysis referred to previously, and has been well summarized (Table 1).[14] There have been no clinical trials of currently accepted optimum HIV therapy (i.e. triple combination antiretroviral regimens) initiated shortly after infection showing delayed disease progression. The benefits thus far have been demonstrated solely among persons with advanced disease or among previously-treated patients. Clinicians acknowledge the benefit of antiretroviral therapy observed among those with advanced immune suppression and assume that similar benefits accrue any time effective combination treatment is initiated. This assumption has not been confirmed.

Ultimately the decision to begin antiretroviral therapy in a person with HIV infection is the result of a risk-benefit analysis. The benefits are listed above, but what are the risks (Table 1)? First, a person beginning therapy early in HIV infection may not have the same treatment option years, or even decades later when it is most needed, when the immune system has deteriorated. Treatment options may be lost because of the development of side effects precluding the use of one or more antiretrovirals, or because of the development of drug-resistant HIV strains. If resistance develops it is often class specific.[15] Thus, resistance which develops in response to a specific protease or non-nucleoside reverse transcriptase inhibitor will often confer resistance to other compounds in the same group. Therefore, when antiretrovirals fail due to resistance, the opportunity to switch to other antiretrovirals may be very limited or non-existent. Like other medications for chronic disease, antiretrovirals exact a toxicity price. The HIV protease inhibitors, which were thought to be relatively free of significant toxicity in short-term studies, are now, after only a few years, known to cause lipodystrophy, abnormal glucose metabolism and accelerated atherosclerosis.[16–19] These toxicities have been observed with a relatively short duration of therapy (a few years) in small numbers of patients. As antiretrovirals are disseminated more widely, and as patients are treated for decades rather than for a few years, new toxicities will undoubtedly be revealed.

Current antiretroviral regimens are relatively complex, and involve different dosing intervals (e.g. twice versus thrice daily), differing requirements for timing in relation to meals, and complex drug interactions. Undertaking combination antiretroviral therapy requires a strong—and long-lasting—commitment by the patient, and often forces changes in lifestyle. With some medications consumption of only 50–80% of intended doses may still result in a beneficial biological effect;[2] however, consumption of only 50% of the intended doses of an antiretroviral regimen might be close to useless, and even the benefit of 80% adherence is unknown and may not confer adequate viral suppression. It is difficult for asymptomatic HIV infected people to adhere fully to a medication schedule that compromises their life-style and induces toxicities without apparent improvement in their clinical situation (since they are still asymptomatic). Those clinicians who favor delaying therapy are merely suggesting waiting until the patient and the health care provides are concerned that irreversible harm may be done. A heightened sense of increased risk for disease progression on the patient's part is likely to coincide with renewed determination to adhere to medication, thus ensuring maximal benefit from therapy. Although laboratory tests can provide quantitative estimates of the future risk of HIV disease progression,[20] individuals vary tremendously in their subjective assessment of any particular degree of risk, and it is that subjective assessment which governs their enthusiasm for accepting the hazards and discomforts of treatments. The art of HIV therapy in its present form is to assess the risks and benefits fully at every time point, and to initiate therapy at the point of maximal benefit and minimal risk.

3. EVALUATION OF LARGE RANDOMIZED CLINICAL TRIALS AMONG PERSONS WHO INITIATE ANTIRETROVIRAL MEDICATION

The goal for the early treatment of HIV infection is clear: to slow disease progression. Early treatment with active combinations of anti-HIV medications has not been evaluated in randomized clinical trials. The key clinical trials which have been conducted in asymptomatic patients using clinical endpoints as outcome measures have revealed the following. If a patient has less than 500 CD4 lymphocytes/μl, zidovudine monotherapy will reduce disease progression by about 50% compared with placebo.[21] Combinations such as zidovudine plus didanosine or zidovudine plus zalcitabine reduce disease progression more than zidovudine alone.[22, 23] Zidovudine plus lamivudine has not been shown to be more effective than zidovudine alone in any single trial; however, this was shown in a meta-analysis.[24] Analysis of the laboratory markers embedded in these large trials has consistently shown that reductions in plasma HIV RNA concentrations and increases in CD4 lymphocyte numbers are associated with a decreased risk of disease progression assessed on the basis of purely clinical endpoints.[25, 26] Short-term studies not powered to determine clinical benefit among persons initiating therapy have used reductions in viral load and increases in CD4 lymphocyte numbers as surrogates for reduced disease progression.[22, 28, 29] Thus, many of the double and triple antiretroviral combinations have suppressed HIV replication (as assessed by viral load measurements) and increased CD4 lymphocyte numbers over periods of several years. Clinicians have made the assumption that the relatively short-term improvement in laboratory variables observed in these studies can be extrapolated to a durable clinical benefit persisting for decades.

Other clinical trials relevant to early therapy in asymptomatic patients have been conducted among patients with advanced HIV disease or those previously treated with antiretrovirals. Since such patients tend not to respond as well as previously-untreated patients with early disease, clinicians have again made the assumption that if a patient with advanced disease or with extensive pretreatment benefited from a specific regimen, then a healthier person given identical therapy would receive at least as great a benefit. The US AIDS Clinical Trials Group (ACTG) Study 320 showed that among persons with advanced HIV disease the combination of zidovudine, lamivudine and indinavir reduced disease progression compared with zidovudine plus lamivudine alone.[30] The Abbott Pharmaceutical Company Study 019 demonstrated that in persons with far advanced HIV-related immune dysfunction ritonavir added to a variety of treatment regimens also reduced disease progression.[31] Finally ACTG Study 193 demonstrated that among patients with less than 50 CD4 lymphocytes per μl the triple reverse transcriptase inhibitor combination of zidovudine, didanosine and nevirapine conferred some survival advantage compared with other double nucleoside reverse transcriptase inhibitor combinations.[32] Taken together, these studies show that in general treatment with three antiretrovirals reduces disease progression compared to regimens containing only two agents. This generalization is also supported by analysis of plasma in HIV RNA concentrations and CD4 lymphocyte numbers.

Table 2. Combinations of therapeutic agents for initial treatment of persons with HIV infection

A. Two compatible reverse transcriptase agents plus a non-nucleoside		
Zidovudine/lamivudine)		(Nevirapine 400 mg once a day
or	plus	or
Didanosine/stavudine)		(Efavirenz 600 mg once a day at night
or		
Stavudine/lamivudine)		
B. Two compatible reverse transcriptase agents plus a protease inhibitor		
Zidovudine/lamivudine)		(Indinavir at 800 mg every 8 hours
or		or
Didanosine/stavudine)	plus	(Nelfinavir at 750 mg three times a day
or		or
Stavudine/lamivudine)		(Saquinavir 400 mg twice a day with
		(Ritonavir 400 mg twice a day
C. A non-nucleoside reverse transcriptase inhibitors plus a protease inhibitor		
Efavirenz)	plus	(Indinavir 1000 mg every 8 hours
		or
		(Nelfinavir 750 mg three times a day

4. TREATMENT GUIDELINES RECOMMENDED BY THE CONSENSUS PANELS

The International AIDS Society-USA (IAS-USA) panel "recommend[s] antiretroviral therapy for any patient with established HIV infection and confirmed plasma HIV-1 RNA levels greater than 5,000 to 10,000 copies/ml who is committed to the complex, long-term therapy."[2] The IAS-USA also states that asymptomatic subjects, those with HIV RNA levels below the 5,000–10,000 copies and with CD4 lymphocyte numbers above 350/μL may consider deferring therapy. This panel of experts recommended initiating therapy for anyone with detectable HIV-1 RNA and with CD4 lymphocyte numbers below 500/μL (and especially if below 350/μL). The IAS-USA recommended a 3-drug regimen consisting of a potent protease inhibitor and 2 nucleoside reverse transcriptase inhibitors as first-line therapy. For this group of experts deferring a protease inhibitor-containing regimen was not a favored option; in cases where protease inhibitors were unwanted, a non-nucleoside reverse transcriptase inhibitor (usually nevirapine or efavirenz) would replace the protease inhibitor in a 3-drug regimen. The Department of Health and Human Services and Henry J Kaiser Family Foundation panel of experts also recommended initiating therapy based upon plasma HIV-1 RNA concentration, suggesting that therapy should be initiated in any patient with more than 10,000 to 20,000 HIV RNA copies/mL plasma and with a CD4 lymphocyte number count less than 500/μL.[4] The treatment choices of this panel were similar to those of the IAS-USA (and there was considerable overlap between the two in panel membership).

The British HIV Association's (BHA) guidelines differ from those of the IAS-USA.[3] The BHA recommends individualization of therapy and deferral of therapy as long as CD4 lymphocyte numbers are above 350/μL. The interpretation of viral load measurements is left wholly to the discretion of clinicians, although they acknowledge that, "patients with high viral loads before treatment are significantly more likely to develop clinical illness over defined periods of follow-up than are CD4 matched [sic] people with low viral loads." The BHA guidelines include an emphasis on Treatment option C (see Table 2) if longer term data support this combination. However, the BHA does use viral load as a guide to specific antiretroviral combinations. They suggest that for patients with less than 50,000 HIV RNA copies/mL, the first-choice combination would be a non-nucleoside reverse transcriptase inhibitor and two nucleoside reverse transcriptase inhibitors (Treatment option B, Table 2). For patients with viral loads above 50,000 HIV RNA copies/mL they recommend that a protease inhibitor and two nucleosides be the initial combination of choice (Treatment option A, Table 2).

All three consensus panels agreed that antiretroviral therapy needed to be individualized, that all patients with CD4 lymphocyte counts below 350/μL should receive triple combination therapy, and that the goal of combination therapy was, if possible, to reduce plasma HIV RNA concentrations below the limit of detection (<400–500 copies/ml with conventional assays, but preferably < 50 copies/ml as determined by the ultrasensitive assays). Another recommendation common to all three panels was the belief that patients with high levels of plasma HIV RNA require therapy with two nucleoside reverse transcriptase inhibitors and at least one (and perhaps two) protease inhibitors. Finally the three panels agree that the initial treatment choice is a critical decision, since the maximum and most durable suppression of HIV replication, and the greatest immunologic benefit, usually results from the first treatment regimen. Not explicitly stated but also common to the three panels condemnation of any strategy of "induction and maintenance," in which patients are begun on triple combination antiretroviral regimens, but the number of agents is reduced after some period of time during which HIV RNA has been undetectable in plasma. The limited experience with the "induction and maintenance" strategy so far suggest that it cannot sustain the level of suppression of HIV replication achieved with continued triple combination antiretroviral chemotherapy.[33]

5. RATIONALE TREATMENT COMBINATION FOR EARLY TREATMENT

The present goal of early treatment is to reduce HIV replication (as assessed by viral load measurements) to below the limits of quantitation. This goal can only be accomplished consistently with combinations of antiretrovirals, usually as three agents together. The most potent and clinically tolerable combinations are listed in Table 2. Other choices are possible but may lack the potency of the medications listed, have a higher frequency of adverse reactions, or have other characteristics that limit their clinical utility.

When choosing among various antiretroviral combinations, the potency, durability, convenience, ease of adherence, short and long-term toxicities, patterns of resistant mutations and subsequent rescue treatment options should all be considered. There have been no direct comparisons among the different treatment options (Table 2). The listed regimens simply represent a rational synthesis of available treatment options. If a patient desires treatment with the regimen considered most active, then I would suggest option B, based on the apparent extraordinary potency of the protease inhibitors. If the patient wants the most convenient regimen then I would choose option A, in which patients may consume as few as three pills a day. If the patient wants the initial treatment option that will allow for the most effective treatment when the initial choice fails to continue suppressing HIV replication, then option A again would be the first choice. However, this point is controversial and represents perhaps the most critically unknown piece of data for patients and their clinicians.

Table 3. Key concerns when initiating combination antiretroviral therapy

1. Establish a reliable baseline for CD4 lymphocyte counts and plasma HIV-1 RNA concentrations.
2. Begin therapy only when the patient is ready for and committed to the regimen.
3. Always and only use maximally suppressive therapy (see Table 2).
4. Prepare your patient for toxicities and then monitor appropriately.
5. Never initiate therapy and plan to reduce its intensity (induction-maintenance).
6. Never initiate therapy by adding one drug at a time.
7. Never simply discontinue one drug at a time (as in managing an adverse reaction).

6. PRACTICAL RECOMMENDATIONS

The success of an HIV-infected patient's first exposure to antiretroviral chemotherapy will be increased by some simple practical guidelines (Table 3).[34,35] To maximize adherence, do not begin treatment until the patient is fully committed to a long term treatment strategy. Consider using medication trays or vitamin pills to mimic the complexity of the treatment regimen so that the patient can realistically appraise the "cost" of initiating treatment. Clinicians should anticipate toxicities, informing their patients of likely side effects, and performing the appropriate laboratory testing at regular intervals. Alerting patients to the adverse effects of their regimen enhances the likeliness that patients will be able to recognize common toxicities and initiate the proper response to mitigate their severity.

Management of adverse reactions in patients in combination chemotherapy is a complex area where there is little or no data to guide evidence-based practice; it is also somewhat beyond the scope of this chapter. If an adverse reaction occurs which is not immediately life-threatening and which can be fairly-clearly ascribed to only one agent (eg anemia in a patient on a zidovudine + lamivudine + nevirpine combination), and the patient has a stable virologic response (e.g. plasma HIV RNA continues to be undetectable), then discontinuing the offending agent (in this instance, of course, zidovudine) and immediately substituting a similar agent generally lacking such adverse reactions (eg stavudine or didanosine) is a reasonable strategy. If the source of the adverse reaction cannot be pinpointed, then discontinuing the regimen altogether is the preferred option, as it is also in patients with life-threatening adverse reactions.

7. CONCLUSIONS

Initiating effective antiretroviral therapy represents the most important intervention for persons with HIV infection. There are no data bearing on the optimal time to begin treatment nor is there sufficient evidence to allow one to select the best combination. Consensus exists among international experts that antiretroviral chemotherapy must be individualized and that only maximally suppressive combination therapy should be given (admitting that data bearing on this are sorely lacking). All HIV-1-infected patients with less than 350 CD4 lymphocyte numbers/μL should receive effective combination therapy, and persons with high viral loads (eg >10,000 HIV RNA copies/ml) should also consider initial therapy even at CD4 lymphocyte counts well above 350/μl. However, experts disagree as to what level of plasma HIV RNA should invariably serve as a trigger for initiating antiretroviral chemotherapy, regardless of CD4 lymphocyte count. Protease inhibitor and dual nucleoside therapy may represent the most active combination, but dual nucleoside plus a non-nucleoside is also acceptable therapy and may have fewer toxicities, a smaller pill requirement and a less complex dosing regimen.

REFERENCES

1. Hecht, FM; Grant, RM; Petropoulos, CJ, *et al.* Sexual transmission of an HIV-1 variant resistant to multiple reverse-transcriptase and protease inhibitors N Engl J Med, 1998; 339:307–11.
2. Carpenter CCJ, Fischl MA, Hammer SM, *et al.* Antiretroviral therapy for HIV infection 1998. JAMA 1998; 280:78–86.
3. Gazzard G, Moyle G on behalf of the BHIVA Guidelines Writing committee. 1998 revision to the British HIV Association guidelines for antiretroviral treatment of HIV seropositive individuals. Lancet 1998; 352:314–16.
4. US Department of Health and Human Services and the Henry J. Kaiser Family Foundation Guidelines for the use of antiretroviral agents in HIV-1 infected adults and adolescents. MMWR Morb. Mortl. Wkly Rep. 1998; 47(RR-05); 43–82.
5. Walker BD, Basgoz N. Treat HIV-1 infection like other infections-Treat it. JAMA 1998; 280:91–3.
6. Kahn, JO, Walker, BD. Acute human immunodeficiency virus type 1 infection. N Engl J Med. 1998; 339:33–9.
7. Powderly WG, Landy A, Lederman MM. Recover of the immune system with antiretroviral therapy. JAMA 1998; 280:72–77.
8. Pantaleo G, Cohen OJ, Schacker T, *et al.* Evolutionary pattern of human immunodeficiency virus (HIV) replication and distribution in lymph nodes following primary infections implications for antiviral therapy. Nat Med. 1998; 4:341–5.
9. Gorochov G, Neumann AU, Kereveur A, *et al.* Perturbation of CD4+ and CD8+ T-cell repertoires during progression to AIDS and regulation of the CD4+ repertoire during antiviral therapy. Nat Med 1998; 4:215–221.
10. Cohen OJ, Fauci AS. HIV/AIDS in 1998-Gaining the upper hand? JAMA 1998; 280:87–8.
11. Rosenberg ES, Billingsley JM, Caliendo AM, *et al.* Vigorous HIV-1 specific CD4+ T-cell responses associated with control of viremia. Science 1997; 278:1447–50.
12. Palella FJ, Delaney KM, Moorman AC *et al.* Declining morbidity and mortality among patients with advanced human immunodeficiency virus infection N Engl J Med 1998; 338:853–60.
13. Hogg RS, Heath KV, Yip B *et al.* Improved survival among HIV-infected individuals following initiation of antiretroviral therapy. JAMA 1998; 279:450–4.
14. Burman WJ, Randall RR, Con DL. The case for conservative management of early HIV disease. JAMA 1998; 280:93–5.
15. Deeks SG, Smith M, Holodniy M, *et al.* HIV-1 protease inhibitors: A review for clinicians. JAMA 1997; 277:145–53.
16. Lo JC, Mulligan K, Tai VW, *et al.* "Buffalo hump" in men with HIV-1 infection Lancet 1998; 351:867–70.
17. Miller KD, James E. Yyanovski JA, *et al.* Visceral abdominal-fat accumulates associated with use of indinavir. Lancet 1998; 351:871–75.
18. Carr A, Samaras K, Burston S, *et al.* A syndrome of peripheral lipodystrophy, hyperlipidemia and insulin resistance due to HIV protease inhibitors. Presented at The 5th Conference on Retroviruses and Opportunistic Infections, American Society for Microbiology, February 1–5; 1998; Chicago, IL. Abstract 410.
19. Henry K, Melroe H, Heubsch J, *et al.* Severe premature coronary artery disease with protease inhibitors. Lancet 1998; 351:1328.
20. Mellors JM , Munoz AM, Giorgi JV, *et al.* Plasma viral load and CD4+ lymphocytes are prognostic markers of HIV-1 infection. Ann Intern Med 1997; 126:946–54.
21. Volberding PA, Lagakos SW, Koch MA, *et al.* Zidovudine in asymptomatic human immunodeficiency virus infection: a controlled trial in persons with fewer than 500 CD4-positive cells per cubic millimeter. N Engl J Med 1998; 322:941–9.
22. Hammer S, Katzenstien DA, Hughes MD, *et al.* A trial comparing nucleoside monotherapy with combination therapy in HIV-infected adults with CD4 cells counts from 200 to 500 per cubic millimeter. N Engl J Med 1996; 335:1081–90.

23. Delta Coordinating Committee. Delta: a randomized double-blind controlled trial comparing combinations of zidovudine plus didanosine or zalcitabine with zidovudine alone in HIV-infected individuals. Lancet 1996; 348:283–91.
24. Staszewski, S; Hill, AM; Bartlett, et al. Reductions in HIV-1 disease progression for zidovudine/lamivudine relative to control treatments: a meta-analysis of controlled trials. AIDS, 1997; 11:477–83.
25. Katzenstein DA, Hammer SM, Hughes M, *et al.* Virologic and immunologic markers and clinical outcomes after nucleoside therapy in adults with 200 to 500 CD4 cells per cubic millimeter. NIAID sponsored AIDS Clinical Trials Groups Study 175, a virology substudy. N Engl J Med 1996; 335:1091–98.
26. Brun-Vizinet F, Boucher C, Loveday C, *et al.* HIV-1 viral load, phenotype, and resistance in a subset of drug-naïve participants from the Delta trial. Lancet 1998; 350:983–90.
27. Eron JE, Benoit SL, Jensek J, *et al.* Treatment with lamivudine, zidovudine or both in HIV-positive patients with 200–500 CD4+ cells per cubic millimeter. N Engl J Med 1996; 333:1662–69.
28. Katlama C, Ingrand D, Loveday C, *et al.* Safety and efficacy of lamivudine-zidovudine combination therapy in antiretroviral-naïve patients. JAMA 1996: 276:118–25.
29. Montaner JSG, Reiss P, Cooper D, *et al.* A randomized double-blind trial comparing combinations of nevirapine, didanosine and zidovudine for HIV-infected patients. JAMA 1998; 279:930–37.
30. Hammer SM, Hughes MD, Squires K, *et al.* A controlled trial of two nucleoside analogues plus indinavir in persons with human immunodeficiency viral load infection and CD4 cells counts of 200 per cubic millimeter or less. N Engl J Med. 1997; 337:725–33.
31. Cameron DW, Heath-Chiozzi M, Danner S, *et al.* Randomised placebo-controlled trial of ritonavir in advanced HIV-1 disease. The Advanced HIV Disease Ritonavir Study Group. Lancet. 1997; 351:543–549.
32. Henry WK, Erice A, Tierney C, *et al.* The AIDS Clinical Trial Group 193A. Study team JAIDS (in press).
33. Reijers MHE, Wevering GJ, Jurriaans S et al. Maintenance therapy after quadruple induction therapy in HIV-1 infected individuals: Amsterdam Duration of Antiretroviral Medication (ADAM) study. Lancet 1998; 352:185–190.]
34. Lerner BH, Gulick RM, Dubler NN. Rethinking nonadherence: historical perspectives on triple-drug therapy for HIV disease. Ann Intern Med 1998; 129:573–8.
35. Sackett DL, Haynes RB, Gibson ES, *et al.* Randomized clinical trials of strategies for improving medication compliance in primary hypertension. Lancet 1975; 1:1205–7.

23

STRATEGIES FOR SECOND-LINE ANTIRETROVIRAL THERAPY IN ADULTS WITH HIV INFECTION

Ann C. Collier and Margot A. Schwartz

University of Washington
Department of Medicine
Box 359929
325 9th Avenue, Seattle, Washington 98104

With the use of combination antiretroviral therapy for treatment of human immunodeficiency virus type one (HIV) infection, patient outcome has improved. Rates of deaths from acquired immunodeficiency syndrome (AIDS) and rates of new AIDS-defining illnesses have decreased in countries with resources to make treatment widely available to persons with HIV infection.[1,2] However, as the number of agents and tools available to treat and monitor HIV infection have increased, the complexity of management has also increased.[3,4] The management issues about patients on antiretroviral therapy who are having less than an ideal response are especially challenging, and will be discussed in this chapter.

Several sets of guidelines on the treatment of HIV infection have been published recently.[5–8] All of the guidelines emphasize that long-term clinical outcome data, upon which are based decisions about when to change therapy and which salvage regimens to use, are inadequate. The benefits and risks of continuing a regimen versus changing to an alternative regimen vary with each patient. In approaching theses issues, it is important to understand principles underlying use of antiretroviral agents, reasons why therapy may need to be changed, available efficacy information about antiretroviral regimens in patients with prior therapy, future potential options for such patients, and relevant practical issues about antiretroviral management

1. PRINCIPLES OF ANTIRETROVIRAL THERAPY

In 1997, a federal panel sponsored by the National Institutes of Health reviewed and summarized available data about the natural history, pathogenesis, and treatment of HIV,

Antiviral Chemotherapy 5, edited by Mills *et al.*
Kluwer Academic / Plenum Publishers, New York, 1999.

Table 1. NIH panel's principles of therapy for chronic HIV infection in adults*

1. HIV replication leads to immune system damage and AIDS.
2. Plasma HIV RNA indicates the amount of HIV replication and CD4+ cells indicate the extent of HIV-induced immune damage.
3. Treatment decisions should be individualized by level of risk of disease progression.
4. The goal of therapy should be maximum suppression of HIV replication.
5. The most effective therapy is simultaneous use of new, non-cross resistant anti-HIV drugs.
6. Each drug in a combination should be used in an optimal schedule and dose.
7. Antiretroviral drugs are limited (in number, mechanisms of action, by cross resistance); changes affect future potential options.
8. Women should receive optimal therapy regardless of pregnancy

*Excludes the three principles that relate to primary HIV, children, and transmission. Adapted from Report of NIH Panel to Define Principles of Therapy of HIV Infection.[6]

and suggested eleven principles for therapy of HIV infection.[6] Eight of these principles are relevant to the treatment of adults with chronic HIV infection. They are summarized in Table 1, and key aspects are briefly discussed below.

1.1. Natural History and Predictors of Disease Progression

The natural history of HIV suggests that HIV infection is progressive in most infected persons who do not receive anti-HIV therapy, and eventually results in severe immune impairment and increased susceptibility to opportunistic infections and malignancies. Data from several studies, most notably the Multicenter AIDS Cohort Study, have demonstrated that CD4+ T cell counts and plasma HIV RNA levels are useful predictors for the time to development of subsequent AIDS or death in HIV-infected men.[9,10] Data from a cohort study of injection drug users (ALIVE) suggest that patterns of CD4+ T cells and plasma HIV RNA are similar in women and men, although absolute values of HIV RNA were lower for women.[11] These data strongly support the principles that HIV infection is harmful, that CD4+ T cells and plasma HIV RNA measurements are useful prognostic markers, and that decisions about treatment should be individualized, based upon a patient's clinical condition, and these laboratory markers.

1.2. HIV Replication and Resistance

Several investigators have demonstrated that in the absence of treatment, persons with HIV infection have ongoing viral replication from the time of acquisition of HIV throughout the course of infection, even when they are asymptomatic.[12–14] An estimated 10 billion HIV virions are produced daily. The high rate of viral replication is accompanied by a high rate of turnover in some types of infected cells, such as CD4+ T cells, suggesting a very dynamic process for both the virus and the host.[15] This active process of HIV replication is accompanied by evolution in the genome of HIV.[16, 17] Persistence of HIV replication in the presence of antiretroviral agents is associated with selection of specific codon changes.[18–23] Table 2 lists the most common mutations associated with decreased susceptibility of HIV to individual antiretroviral agents. Available data suggest a correlation between genotypic changes and phenotypic assay results.[19, 24, 25] The rate at which viral resistance develops varies with different agents. Agents such as nevirapine and lamivudine, for which only a single codon change in the viral genome is required to alter susceptibility rapidly develop resistance during non-suppressive therapy with these agents

Table 2. Codon changes associated with decreased susceptibility of HIV-1*

Nucleoside Reverse Transcriptase Inhibitors						
Abacavir	Adefovir**	Didanosine	Lamivudine	Stavudine	Zalcitabine	Zidovudine
65,69,74,115, 184,151,214	41,70,215	65,74,75, 151, 184	65,151,184	41,50,69,70, 75,151	65,69,74, 151, 184	41,67,70, 151, 215,219
Nonnucleoside Reverse Transcriptase Inhibitors						
Delavirdine	Efavirenz	Nevirapine				
103,181,236	103,108,225	98,100,103,106, 108,181, 188,190				
Protease Inhibitors						
Amprenavir	Indinavir	Nelfinavir	Ritonavir	Saquinavir		
10,20,46,50,82, 84	10,20,24,46, 54,63, 64,71,82, 84,90	30,36,46,63, 71,77,88,90	10,20,36,46, 54,63,71, 82,84,90	10,36,48,54, 63,71,82,84,90		

*Codons that are underlined are the most closely associated with decreased susceptibility
**Adefovir is a nucleotide

over a time course of days to a few weeks.[26–29] For protease inhibitors and some of the nucleosides like zidovudine, where decreased susceptibility results from the accumulation of several mutations, resistance develops more slowly.[18,30,31] These data support the principle that the goal of antiretroviral therapy should be to suppress HIV to the greatest extent possible and to prevent the development of drug-resistant variants.[6] There may be selective circumstances where a less aggressive goal is chosen, such as when therapeutic options are very limited as a result of drug intolerances or multiple prior therapies, or when a patient is unwilling to take or unable to comply with a particular regimen.

The mutations that occur when a drug is used in combination regimens may not be the same as with monotherapy. Interactions between certain mutations have been described. For example, HIV isolates with high level genotypic resistance to zidovudine and a mutation at reverse transcriptase codon 184 associated with lamivudine are more sensitive to zidovudine than in the absence of the codon 184 change.[32,33] A similar observation has been described with adefovir and the 184 mutation.[34] However, there are some generalizations that can be made about the clinical effects of mutations in the HIV genome. In studies of nucleoside reverse transcriptase inhibitors, zidovudine resistance has been associated with a worse clinical outcome.[35–37] In retrospective studies of specimens from patients with prior nucleoside and protease inhibitor use, phenotypic resistance to an antiretroviral compound more accurately predicted failure to respond to that therapy than the lack of resistance predicted subsequent clinical success.[38,39] Cross-resistance is common among non-nucleoside reverse transcriptase inhibitors as well as between the protease inhibitors indinavir and ritonavir. The initial protease codon changes which occur in patients treated with protease inhibitors (saquinavir, ritonavir, indinavir, nelfinavir), in settings where HIV replication continues, may vary but isolates with high level resistance to a particular protease inhibitor frequently develop cross-resistance to other protease inhibitors.[40]

The most effective strategy to maximize viral suppression with the available agents is to use combinations of antiretroviral agents.[5–7] Most recommended regimens include three or four agents, although pilot data are available about salvage regimens using up to six different drugs.[41] Ideally, in order to minimize the potential for the development of resistance, drugs should be started simultaneously. Data from several studies have demonstrated that the sequential addition of agents in a regimen is substantially less effective than starting agents at the same time.[42,43]

1.3. Viral Latency

The complexity of HIV's pathogenesis is underscored by recent demonstrations that the virus can be latent in some CD4+ T cells, with estimates of the body burden of these latently infected cells varying from 10^4 to 10^6.[44–47] The existence of this latently infected pool and its slow decay in patients treated with potent antiretroviral therapy suggest that treatment will need to be continued for prolonged periods of time, if not lifelong, until new treatment strategies are developed.[46] New approaches to decrease this latent pool of infected cells are under investigation, but are not clinically relevant at present.

2. REASONS WHY THERAPY SHOULD BE CHANGED

Although there is substantial controversy about the optimal time to change therapy in those with detectable plasma HIV RNA, little controversy exists about the reasons that a change in therapy should be considered. Reasons for changing therapy include clinical failure, such as clinical deterioration or intolerance to one or more agents in a regimen, inadequate responses in virological or immunological measurements, and use of a suboptimal regimen.[5,7] Another, more specialized reason to consider changing therapy in women with HIV is pregnancy, because of potential teratogenicity or other adverse effects on the fetus.

2.1. Clinical Failure

The most common type of clinical deterioration in patients with HIV infection is the development of opportunistic infections or malignancies. Historically, AIDS-defining illnesses were considered a marker of antiretroviral therapy failure, although they actually reflect immunological deterioration and the loss of pathogen-specific immunity that may occur in the setting of prolonged HIV replication. In addition, some opportunistic infections may reflect a lack of use of appropriate infection-specific prophylaxis or a failure of prophylactic therapy.[48] Thus, decisions about whether a particular illness reflects failure of anti-HIV therapy or other factors should be individualized. For example, if *Pneumocystis carinii* pneumonia (PCP) occurs in a patient not taking or poorly compliant with PCP prophylaxis, ensuring that adequate PCP prophylaxis is used is a better strategy than changing antiretrovirals. Also, if a patient develops an opportunistic infection or malignancy soon after starting on an antiretroviral regimen, it is reasonable to continue the antiretroviral regimen. In contrast, if a patient on a stable, potent antiretroviral regimen, taking appropriate opportunistic prophylaxis develops a new opportunistic infection, a change in antiretrovirals should be considered.

For patients with HIV infection on an antiretroviral regimen who have persistent and progressive unexplained symptoms, such as significant weight loss or persistent fatigue, and have no evidence of a concurrent illness which explains the symptoms, consideration should be given to changing antiretroviral therapy. Evaluation of endocrine status, including thyroid function and testosterone levels should be part of this evaluation.

2.2. Drug Intolerance

Each antiretroviral agent is associated with characteristic potential toxicities, although the spectrum of common adverse events and their prevalence varies (reviewed in[5,49]). Table 3 lists the most common toxicities associated with several Food and Drug Administration (FDA)-approved and investigational antiretroviral agents.

In addition to known toxicities associated with specific protease inhibitors, other adverse events have been reported with use of these agents. An increase in spontaneous bleeding in patients with hemophilia A and B receiving protease inhibitor therapy has been described. Metabolic complications have been described to be more common in patients using protease inhibitors than in patients on no therapy or non-protease inhibitor regimens; these have been linked to each of the FDA approved protease inhibitors (Table 4).[50–55] An alert was issued by the FDA in 1997 describing a possible relationship between protease inhibitor use and the onset or worsening of diabetes mellitus and hyperglycemia.[56] Limited data suggest that the prevalence of overt diabetes mellitus in this setting is one to two percent, although insulin resistance is much more frequent. The prevalence of body image changes in patients on protease inhibitors was over 80% from one clinic in Australia; the prevalence of hyperlipidemia has been up to 50% in other case series.[57,58] The pathogenesis, clinical significance, and optimal management of these events, as well as the relationship to specific antiretrovirals are under investigation.[59]

2.3. Virological or Immunological Failure

A uniform definition of treatment failure using laboratory measures has not been established, but several definitions of inadequate responses have been proposed. The U.S Public Health Service Guidelines suggest six circumstances under which laboratory test responses to antiretroviral therapy should be considered inadequate.

Either plasma HIV RNA or CD4+ T cells trends over time may meet the criteria for changing therapy (a specific rate of fall of CD4+ T cell was not specified in these guidelines). The International AIDS Society USA guidelines suggest similar strategies based upon HIV RNA and CD4+ T cell trends, but emphasize that optimal management varies with the available options.[7] For research studies, definitions of "virological failure" have included failure of plasma HIV RNA to achieve values below the limits of quantification or detection of available assays, less than a decrease of 0.5 or 1.0 $\log_{10}$ copies/mL from baseline, HIV RNA results that become detectable after having values below the limit of detection, or a composite definition, which includes several of these concepts.

Data from several studies suggest that durability of viral suppression is associated with the nadir of response after initiating therapy.[60–62] Less data are available about the predictive factors for virological suppression after changing therapy, although they may be similar. The difference between the nadir and baseline HIV RNA values will vary with the initial viral load, and the true nadir may be masked by the limit of quantification or detection of an assay. The desire to maximize viral suppression and the potential importance of nadir values have increased interest in using the most sensitive HIV RNA assays available. With HIV RNA polymerase chain reaction assays, there is more variability at lower copy levels than with higher values. Transient increases in plasma HIV RNA may occur in the setting of antigenic stimulation such as immunizations and intercurrent infections, so repeating tests before changing therapy is advised to avoid premature discontinuation of a regimen.[63]

Table 3. Main toxicities associated with antiretroviral drugs

Nucleoside reverse transcriptase inhibitors						
Abacavir*	**Adefovir***	**Didanosine**	**Lamivudine**	**Stavudine**	**Zalcitabine**	**Zidovudine**
Diarrhea	Decreased carnitine	Peripheral neuropathy	Few distinct toxicities	Peripheral neuropathy	Peripheral neuropathy	Anemia
Rash	Nausea	Nausea			Stomatitis	Neutropenia
Hypersensitivity reaction	Proximal renal tubule dysfunction	Diarrhea				Headache
		Pancreatitis				Nausea
						Asthenia
						Insomnia
Nonnucleoside reverse transcriptase inhibitors						
Delavirdine	**Efavirenz**	**Nevirapine**				
Rash	Dizziness	Rash				
Headache	Rash	Hepatitis				
		Steven's Johnson Syndrome				
Protease inhibitors**						
Amprenavir*	**Indinavir**	**Nelfinavir**	**Ritonavir**	**Saquinavir**		
Nausea	Nausea	Diarrhea	Nausea	Nausea		
Diarrhea	Nephrolithiasis		Diarrhea	Diarrhea		
Rash	Hyperbilirubinemia		Vomiting	Headache		
Headache			Paresthesias	Increased transaminases		
			Asthenia			
			Taste change			
			Increased triglycerides			
			Increased transaminases			
			Increased uric acid			

* These agents are investigational
**Metabolic complications have been reported with all agents; see text and Table 4

Table 4. Metabolic complications associated with use of protease inhibitor-containing regimens

Peripheral lipodystrophy	Diabetes mellitus
Central adiposity	Hyperglycemia
Hypertriglyceridemia	Insulin resistance
Hypercholesterolemia	Ketoacidosis
Accelerated atherosclerosis	

2.4. Suboptimal Therapy

Since the recommendations for antiretroviral therapy and available treatment options have changed substantially during the past few years, patients who have been on the same antiretroviral regimen for more than two years may be on regimens which are now viewed as suboptimal. One such situation is patients on dual nucleoside reverse transcriptase inhibitors. It is important to review patients' antiretroviral regimens in light of new clinical data.

3. ANTIRETROVIRAL REGIMENS

Limited data from controlled studies are available about the long-term outcomes of HIV-infected persons who have "failed" a treatment regimen and switched to more potent therapies. Some data exist about patients with prior nucleoside reverse transcriptase inhibitor treatment who were subsequently treated with protease inhibitor-containing regimens. Data are also available from observational studies of clinic patients who have had inadequate virological responses to protease inhibitor-based regimens and who switched to other therapies. Most of these reports have had limited durations of follow-up.

3.1. Treatment Studies of Nucleoside-Experienced Patients

Clinical outcome data are available from several randomized, double-blind studies which enrolled persons who had previously received nucleoside reverse transcriptase ther-

Table 5. Criteria for changing antiretroviral therapy based upon laboratory tests

Test	Time after start of therapy	Value*
Public health service guidelines[5]		
HIV RNA	4 weeks	<-0.5 to -0.75**
	8 weeks	<-1.0**
	4-6 months	Detectable
	---	Repeatedly detectable after undetectable
	---	>3 fold increase from nadir
CD4+ cells	---	Persistent decline
International AIDS Society		
HIV RNA	---	Detectable HIV RNA but individualize level per available options
CD4+ cells	---	Declining or <100 cells/mm^3

*For HIV RNA $\log_{10}$ copies/mL of plasma
**Compared with baseline

apy. Switching to regimens which contained a protease inhibitor improved clinical outcome, compared to remaining on nucleoside-based regimens (Table 6). In separate studies, patients who were assigned to an indinavir, ritonavir, or saquinavir-containing regimen had delayed development of AIDS-defining illnesses and death, compared to those assigned to remain on nucleoside therapies, which was standard therapy during the era when these studies were conducted.[64–66] Survival rates and mean plasma HIV RNA and CD4+ T cell results were also significantly better in the enrollees assigned to protease inhibitor-containing regimens. These studies were important in establishing the clinical efficacy of protease inhibitor therapy in patients with prior antiretroviral therapy, although some of the strategies tested, such as adding a protease inhibitor, do not reflect current antiretroviral recommendations. The issue of clinical outcome over years or decades has not yet been addressed in studies of potent antiretroviral combinations.

A more recent randomized, open-label study compared the use of indinavir and ritonavir, combined with no, one, or two nucleosides in 1251 patients with advanced HIV, CD4+ T cell counts less than 50 cells/mm^3 and prior nucleoside reverse transcriptase inhibitor therapy.[67] The clinical outcomes of development of a new AIDS-defining illness or death were similar in the two treatment groups. However, treatment-limiting toxicities occurred much more commonly in the patients initially assigned to ritonavir.

Clinical studies in nucleoside-experienced patients have suggested that other regimens are associated with favorable short-term responses in plasma HIV RNA and CD4+ cell counts (Table 7), but it is important to remember that differences in study design, patient populations, virological assays, and analysis methods make it difficult to compare treatment responses between studies, and longer-term follow-up is needed to assess the durability of most of these regimens. Virological suppression appears better with regimens containing multiple new drugs, especially from different classes of agents. In observational studies conducted in clinical settings, less than half of nucleoside-experienced patients starting protease inhibitors have achieved suppression of HIV RNA to below the limits of detection 16 to 24 weeks later (Table 8). However, most of these series included patients who added a protease inhibitor to reverse transcriptase inhibitors, as well as those who changed the other antiretrovirals at the time of starting a protease inhibitor. Failure to change the background therapy has been identified in several studies as one of the predictors for lack of favorable virological responses.[69, 74, 78, 79] The Merck 035 study also emphasized that viral suppression was improved if at least one other new drug was started simultaneously with a protease inhibitor, rather than adding new agents sequentially[42] (Table 7). Similar findings have been reported with nelfinavir.[43] Other predictors of virological failure have included higher viral load at the time of therapy switch, lower CD4+ T cells, and non-adherence to therapy.[74, 76, 79, 80]

3.2. Treatment Studies of Protease Inhibitor-Experienced Patients

Available data about the effects of salvage regimens for patients who have suboptimal responses to protease inhibitors come mostly from observational studies. Table 9 summarizes selected studies in which at least 25 patients were observed for a median of 12 or more weeks after switching therapy. Widely varying proportions of patients who achieved plasma HIV RNA responses below 400–500 copies/ml have been reported in different studies. One observational study of 128 patients tried to correlate the initial protease inhibitor used with treatment outcome following a switch in therapy to one of four different protease inhibitors.[86] However, without randomized allocation of new treatments, it is difficult to conclude which regimen was better. In a randomized study of 20 patients "fail-

Table 6. Major clinical outcome studies of protease inhibitors in nucleoside-experienced patients

	Baseline characteristics						
Study name	Regimens	N per Arm	Rx history	Mean HIV RNA (Log_{10} Copies/ml)	Mean CD4 (cells/mm^3)	Mean F/U (mos)	% with disease progression
AIDS Clinical Trials Group 320[64]	IDV/ZDV/3TC*	577	Mean 21 mos prior ZDV	5.0**	87**	9.5**	6%
	ZDV/3TC*	579					11%
Abbott 247[65]	RTV + NRTIs†	543	Mean 25-26 mos prior ZDV	5.4**	20-30**	7**	22%
	NRTIs	547					38%
Hoffmann-LaRoche NV14256[66]	SQV††/ddC	308	Intolerant or failing ZDV, median 17 mos prior ZDV	5.0**	170**	17**	23%
	SQV	318					38%
	ddC	314					39%

*d4T could be substituted for ZDV

**For all enrollees

†Up to 2 RTIs were allowed; RTIs were stable for 6 weeks prior to and for first 16 weeks of the study

††This study used the original hard capsule formulation of saquinavir

N=number; Rx=treatment; F/U=follow-up; mos=months; NRTI=nucleoside reverse transcriptase inhibitor,; IDV=indinavir, ZDV=zidovudine, 3TC=lamivudine, RTV=ritonavir, SQV=saquinavir; ddC=zalcitabine; d4T=stavudine

Table 7. Selected studies of switching therapy in nucleoside-experienced patients

Study name or investigator	Baseline characteristics						Results		
	Regimens	N per arm	Rx history	Mean HIV RNA (Log_{10} copies/ml)	Mean CD4(cells/mm^3)	Mean F/U (weeks)	Change in HIV RNA (Log_{10} copies/ml)	% below LOQ*	Change in CD4 (cells/mm^3)
Hoffmann-LaRoche NV15182[68]	SQV-SGC+ 2NRTI	444	NRTIs in 96% Median 44 mos prior Rx	4.1	227	24	–	40%	+80
AIDS Clinical Trials Group 364[69]	1-2 new NRTIs +		Extensive NRTIs	3.9	388	16	–		79**
	NFV or	66						64%	
	EFZ or	65						69%	
	NFV/EFZ	64						81%	
Martinez[70]	d4T/3TC+ IDV or	49	NRTIs >6 mos with CD4 return to baseline	–	–	24	-1.7	50%	+100
	RTV or	49					-2.2	60%	+167
	SQV	49					-1.3	30%	+105
Merck 035[42,71]	IDV/ZDV/3TC	33	Mean 30 mos prior ZDV	4.6†	144†	100†	-2.0	78%	+209
	IDV, + ZDV/3TC later	31					-1.4	45%	+163
	ZDV/3TC + IDV later	33					-1.3	30%	+101
Harris[72]	NVP/3TC/IDV	22	CD4 <50 with intolerance, toxicity, or disease progression on NRTIs	5.2	30	24	-3.1	73%	+95

*HIV RNA assays lower limit of quantitation of 200-500 copies/mL

**CD4 cell trends were similar in all groups

†For all enrollees

Rx=treatment; LOQ=limit of quantification; F/U=follow-up; NRTI=nucleoside reverse transcriptase inhibitor; NFV=nelfinavir, EFZ=efavirenz, IDV=indinavir, N=number, NVP=nevirapine, RTV=ritonavir; SQV-SGC=saquinavir soft gel capsule

Table 8. Protease inhibitor use in clinical settings in nucleoside-experienced patients

Lead investigator	N	Baseline HIV RNA (Log_{10} Copies/mL)	Baseline CD4+ Cells (per mm^3)	Protease Inhibitor**	% Undetectable HIV RNA* or Change from Baseline ≥ 12 weeks	% Undetectable HIV RNA* or Change from Baseline 16-24 weeks
Clumeck[73]	224	5.1	46	IDV	–	−1.24†
	228	5.1	40	RTV		−0.78†
Deeks[74]	196	4.7	129	IDV or RTV	–	47%
Piketty[75]	177	4.8	70	IDV or RTV	68%	47%
Clough[76]	131	–	–	IDV or RTV	45%	–
Izopet††[77]	349	5.0	116	IDV, RTV or SQV		37%§

*Less than 200-500 copies/mL on Amplicor or bDNA assay
**Given with reverse transcriptase inhibitors
†Change from baseline
††14% were naïve to antivirals
§10% had HIV RNA <20 copies/mL
N=number; IDV=indinavir; RTV=ritonavir; SQV=saquinavir

ing" indinavir regimens, a four-drug regimen of nevirapine, abacavir, nelfinavir, and saquinavir was substantially better at week 20 in suppressing HIV RNA than a regimen of a nucleoside, abacavir, nelfinavir, and saquinavir.[39] Patients with HIV sensitive to 2 or 3 of the drugs had an HIV RNA decrease of 2.5 $\log_{10}$ copies/ml. No one regimen can be considered superior, although better responses have generally been seen with regimens that included multiple new agents (e.g. two protease inhibitors and two new reverse transcrip-

Table 9. Selected studies of salvage therapy in patients with protease inhibitor treatment failure

Lead investigator	N	Baseline HIV RNA: Entry criteria	Baseline HIV RNA: (Log_{10} Copies/mL)	New Regimens	Median F/U (Weeks)	HIV RNA criteria for response (Copies/mL)	HIV RNA response
Barbour[81]	32	HIV RNA >1000 copies on PI triple regimen	4.9	RTV/SQV + 2 RTIs	48	% <400	62% (87% in an on-treatment analysis)
Gallant[82]	46	Detectable HIV RNA on IDV or NFV regimens	3.9-4.2	RTV/SQV + 2 new RTIs	33	% <400	71%
Kaufmann[83]	56	HIV RNA >5000*	4.9	RTV/SQV + 2 RTIs	25	% <400	50% at week 12 49% at week 48
Lohmeyer[84]	25	Failure on PI triple regimen	4.3	NFV/SQV +2NRTIs	24	% <500 % <40	45% 35%
Walmsley[85]	56	HIV RNA >500 copies on PI	4.6	NFV + RTIs	12	% <500 copies/mL % >1 $\log_{10}$ decrease	36% 18%

*69% had prior PIs and 90% had prior NRTIs
N=number; PI=protease inhibitor; IDV=indinavir; NFV=nelfinavir; RTV=ritonavir; SQV=saquinavir; RTI=reverse transcriptase inhibitor; NRTI=nucleosideRTI

tase inhibitors). More complex regimens with up to six different antiretrovirals have been studied in small numbers of patients, although logistics and toxicity are issues for long-term use.[41]

One controlled study enrolled patients with extensive prior saquinavir (hard capsule) therapy, and compared switching to indinavir with switching to saquinavir soft gelatin capsules or the short-term continuation of saquinavir (hard capsules).[87, 88] In this study, concurrent nucleoside antiretrovirals remained stable at the time of the switch in protease inhibitors, limiting the applicability of this study's results to more recent treatment strategies where multiple antiretrovirals are changed simultaneously. Plasma HIV and CD4+ T cell responses were modest in both groups with a trend toward a better anti-HIV effect in the patients switched to indinavir. Several studies are currently taking place to compare the efficacy and toxicity of salvage treatment regimens for patients who have had poor responses to their initial protease inhibitor regimens, and will provide more useful data upon which to base treatment recommendations.

3.3. Hydroxyurea as a Treatment for HIV

Hydroxyurea is a drug that has attracted interest as a treatment for HIV infection. It is FDA-approved for treatment of other diseases and has been used for nearly three decades for treating malignancies.[89, 90] Hydroxyurea blocks ribonucleotide reductase, a cellular enzyme involved in the synthesis of deoxynucleoside triphosphates, thus decreasing levels of dATP. The *in vitro* inhibition of HIV by hydroxyurea is synergistic with nucleoside or nucleotide reverse transcriptase inhibitors, especially adenosine-based compounds such as didanosine, adefovir, and PMPA, whose active metabolites compete with dATP.[90–92] In two small, uncontrolled studies, hydroxyurea monotherapy did not show any benefit in treating HIV infection,[93, 94] but combinations of hydroxyurea and didanosine, have shown anti-HIV activity.[95–103]

Much of the data about the combination of hydroxyurea and didanosine are from studies of antiretroviral-naïve patients, and therefore not relevant for this discussion. However, Rossero et al used hydroxyurea, didanosine, and stavudine in HIV-infected persons with prior antiretroviral theray.[104] Mean HIV RNA titers decreased by 1.3 $\log_{10}$ copies/ml (range −0.1 to −2.4) in 14 patients on therapy at week 12, and 8 of these patients had HIV RNA levels less than 500 copies/ml. Hydroxyurea at a dose of 500 or 1000 mg daily was added to didanosine therapy in a randomized, open-label study.[99] After four weeks, those who received the higher does of hydroxyurea had a decrease in HIV RNA of 0.63 $\log_{10}$ copies/ml compared to 0.02 $\log_{10}$ in those who received the lower dose. The extension phase of a study which compared didanosine plus stavudine with and without hydroxyurea provided information about the addition of hydroxyurea to this dual nucleoside regimen.[105] The patients who received delayed hydroxyurea had similar antiviral responses to the patients started on the triple regimen initially. In an uncontrolled, observational study of 18 patients with extensive prior antiretroviral therapy who had been on a median of seven prior antiviral regimens, hydroxyurea, stavudine and lamivudine therapy resulted in a median HIV RNA decrease of 1.8 $\log_{10}$ copies/ml after 8 weeks of therapy. However, severe toxicity, especially myelosuppression, was common,[106] and most studies of hydroxyurea-containing regimens suggest that initial CD4+ T cell increases are not as high with hydroxyurea as in the treatment arms without hydroxyurea.

3.4. Regimen Simplification

Because of the difficulties of adhering to complex antiretroviral combinations, an important thrust of antiretroviral research has been to identify simpler regimens. Three studies have reported results about decreasing the number of drugs in a regimen, after a period of initial multi-drug therapy.[107–109] In these studies, patients were treated with potent three or four drug regimens containing one or two protease inhibitors, followed by a decrease in the number of drugs in patients who had sustained plasma HIV RNA suppression. Each of these studies demonstrated that rebound in plasma HIV titers above the limit of detection was more common in patients whose therapy was simplified than in those who continued the initial multi-drug regimen. These observations emphasize the importance of maintaining a potent antiretroviral regimen.

A more successful strategy of regimen simplification has been to simplify dosing schedules. Separate studies evaluating twice daily dosing of indinavir, nelfinavir and saquinavir (soft gelatin capsule), compared to the currently recommended thrice daily dosing of these protease inhibitors (in combination with reverse transcriptase inhibitors) are underway in patients with no or limited previous antiretroviral therapy.[110–112] The preliminary results from ongoing studies suggested that similar HIV RNA, CD4+ T cell counts, and toxicity profiles were seen with the twice daily dosing regimens than with more frequent dosing. If these results are confirmed with longer follow-up, decreasing the frequency of protease inhibitor administration will be an important strategy to improve adherence to therapy. However, recently released data from a large comparative study of indinavir-containing regimens demonstrated that a TID regimen was more effective than a BID regimen. An additional strategy to simplify regimens, which has been widely utilized, is to use two protease inhibitors which have favorable pharmacokinetic interactions that result in higher drug exposure with less frequent dosing, in conjunction with reverse transcriptase inhibitors. The combination of ritonavir and saquinavir has been the most widely used.[82,113–122] Several other dual protease regimens are under study including nelfinavir/saquinavir, nelfinavir/indinavir, nelfinavir/ritonavir, and amprenavir in combination with either indinavir, nelfinavir, or saquinavir.[84, 112, 123–128] In addition to protease inhibitors, reverse transcriptase inhibitors including didanosine and nevirapine are also being studied in regimens with less frequent drug administration, specifically once daily instead of twice daily.[129, 130] Use of newer more potent agents with longer half lives is another important potential strategy for simplifying therapy.

4. FUTURE OPTIONS

Newer therapies for the treatment of HIV, which have potential advantages over previously available agents, are under investigation or just entering widespread use (Table 10). Most are members of the same three classes of antiretrovirals that are in widespread use. The rationale for continued development of these agents is to identify agents with improved characteristics. These include increased anti-viral potency, especially against viral strains that are resistant to available agents; more favorable pharmacokinetic characteristics that will allow for less frequent dosing; less unfavorable pharmacokinetic interactions with other medications; and/or less toxicity. Several compounds with one or more of these potential advantages are in development. However, many compounds which appear promising in pre-clinical studies do not always have favorable results in human studies. ABT-378 and tipranavir have *in vitro* activity against

Table 10. Investigational antiretrovirals

Nucleoside/Nucleotide reverse transcriptase inhibitors
Abacavir
Adefovir dipivoxil*
Bis-POC PMEA*
DOTC (BCH10652)
FTC
Lodanosine (FddA)
MKC-442**
Non-nucleoside reverse transcriptase inhibitors
AG1549 (S-1153)
(+) Calanolide A
Carboxanilides (UC-781)
Efavirenz***
HPY 097
PNU 142721
Protease inhibitors
ABT-378
Amprenavir
DMP450
PD-178390
Tipranavir (PNU-140690)
Fusion inhibitor
T-20

* Nucleotide analogue
**Functions as a non-nucleoside
***FDA approved in Oct., 1998.

HIV strains resistant to the available protease inhibitors and are being studied in both protease inhibitor-naïve and experienced patients.[131,132] Compounds which have new modes of action are also under study, including a fusion inhibitor called T-20, which has been shown to have anti-HIV activity in a short-term study.[133] Additional data are needed to determine if any of these compounds will improve the outcome of patients who have failed other antiretroviral therapies. Table 10. Investigational antiretrovirals.

Information about four investigational compounds likely to become widely available in the near future is briefly summarized, as it relates to use in patients who have received other antiretroviral therapies. Efavirenz was approved by the U.S. FDA in October, 1998.

4.1. Abacavir

Abacavir, formerly called 1592U89, is a nucleoside analogue with potent activity against HIV *in vitro*[134] and *in vivo*.[135–137] High level resistance does not occur rapidly *in vitro*, and multiple mutations are required to confer a significant reduction in sensitivity.[138] Abacavir is given in a twice daily regimen. Over two thousand patients have received abacavir in controlled studies or through an expanded access program. Most available data about abacavir from controlled studies is from antiretroviral naïve patients, and potent antiviral activity has been seen with several different antiretroviral combinations containing abacavir in this setting.[137, 139, 140] Data from the abacavir expanded access program has suggested that phenotypic resistance testing may be useful in identifying patients more and less likely to respond to abacavir. Patients with HIV isolates resistant to either zidovudine or lamivudine had similar HIV RNA responses to abacavir as patients with wild-type isolates, but patients with HIV that was resistant to both zidovudine and lamivudine had decreased responses to abacavir.[38] Analysis of the safety data from this program has shown that a hypersensitivity reaction to abacavir, which consists of fever, malaise, rash, and/or nausea, occurred in about three percent of patients.[141] Patients who have experienced this reaction should not be rechallenged with abacavir. Ongoing studies will provide more data about the use of abacavir as a second-line therapy. In the meantime, abacavir is most likely to be useful as part of a regimen in patients who have not had extensive prior treatment with zidovudine and lamivudine.

4.2. Adefovir

Adefovir [9-(2-phosphonomethoxyethyl)adenine], formerly called PMEA, is a nucleotide analogue which has anti-HIV activity *in vitro* and *in vivo*.[142–145] Early studies showed that adefovir had low oral bioavailability.[146] To improve bioavailability, two pivalic acid moieties were attached to adefovir, resulting in a prodrug, adefovir dipivoxil, that has 40% oral bioavailability and complete conversion to adefovir.[147] When used with other oral prodrugs, pivalic acid derivatives have been safe, but associated with reduced serum carnitine concentrations.[148] Adefovir has a pharmacokinetic profile which allows for once daily administration, along with a daily carnitine supplement. A randomized, double-blind, placebo-controlled study in 72 patients, most of whom had received prior antiretrovirals, showed that adefovir had anti-HIV activity with a median decrease in HIV RNA of 0.5 $\log_{10}$ copies/mL, and that a dose of 125 mg daily was safe and well tolerated for 12 weeks.[149] A subset of patients continued on adefovir, usually in conjunction with other antiretrovirals, and plasma HIV RNA suppression and the absence of adefovir-associated mutations was seen for up to 6 to 12 months.[150] Data from a randomized, controlled study of adefovir versus placebo added to participants' current antiretrovirals confirmed the antiviral activity of adefovir in 442 patients in a 24 week study.[151] Unfortunately, recent information from adefovir dipivoxil studies has suggested that proximal renal tubular dysfunction occurred in over 30% of patients treated with 120 mg of adefovir for more than six months. This syndrome is characterized by elevated creatinine, hypophosphatemia, proteinuria, glycosuria, and/or low serum bicarbonate. These changes are usually gradual in onset, and reversible if the drug is stopped. The efficacy and toxicity profile of lower doses of adefovir are under investigation. Adefovir may be a reasonable addition to a regimen for patients with limited options for new antiretrovirals, but such patients must be carefully monitored for proximal renal tubular dysfunction.

4.3. Efavirenz

Efavirenz, formerly called DMP-266, is a non-nucleoside reverse transcriptase inhibitor that has potent *in vitro* activity against a variety of HIV isolates.[152] It has anti-HIV activity against HIV variants with some mutations associated with resistance to other non-nucleoside reverse transcriptase inhibitors, and *in vitro* data suggests emergence of highly resistant virus develops only after multiple passages in tissue culture.[152,153] Several hundred HIV-infected persons have received efavirenz in studies, and additional patients have received the drug through an expanded access program. Although most available data are from studies of patients without prior antiretroviral therapy, preliminary results in patients who had previously received other antiretrovirals are also available. One study compared a regimen of efavirenz and indinavir (N=59) to indinavir alone with addition of efavirenz and stavudine at week 12 (N=42).[154] Patients assigned to the delayed triple regimen did not have as good an anti-HIV response as the patients receiving combination therapy from the start of the study. However, 55% of the patients who received the delayed triple regimen had less than 400 copies/ml of HIV RNA at 72 weeks, and CD4+ T cells increased by more than 200 per mm^3 in this group. Efavirenz, indinavir, and two nucleosides were compared to indinavir plus two nucleosides in a randomized, double blind study which enrolled 184 nucleoside-experienced patients (who were naïve to protease inhibitors and non-nucleoside reverse transcriptase inhibitors).[155] Enrollees had HIV RNA greater than 4.0 log_{10}copies/ml and CD4+ T cells over 50/mm^3. At week 24, a greater proportion (54%) of patients on the four-drug regimen had plasma HIV RNA less than 50 copies/ml than on the three drug regimen (37%). Addition of efavirenz to ongoing therapy with zidovudine and lamivudine in patients with plasma HIV RNA levels greater than 3.4 log_{10}copies/ml was associated with high rates of treatment failure.[156] Preliminary reports of a study which administered efavirenz and nelfinavir to nucleoside-experienced and naïve patients suggested good anti-HIV activity and tolerability after 16 weeks.[157] Efavirenz is likely to be a useful therapy when combined with other agents for patients with limited prior non-nucleoside antiretrovirals.

4.4. Amprenavir

Amprenavir, formerly VX-478, and 141W94, is a protease inhibitor with potent *in vitro* activity against HIV. It appears to be synergistic with zidovudine, didanosine, abacavir, and saquinavir, and additive with indinavir and ritonavir in inhibition of HIV.[158] The IC_{90} for HIV is increased by only 1.5 fold in the presence of 45% human plasma, which is a minor change compared with some FDA-approved HIV protease inhibitors.[159] Development of resistance by HIV to amprenavir appears to be associated with an initial mutation which is unique among the currently available proteases, suggesting that there may be less potential for cross-resistance.[138, 160] In a pharmacokinetic study, the average trough concentration of amprenavir 1200 mg at 12 hours was greater than 10 times the IC_{50} for HIV[158,161]. Recent amprenavir studies have utilized twice daily dose regimens.

Most available data from studies of amprenavir are from antiretroviral-naïve patients, and have shown anti-HIV activity in several combination regimens. In 60 patients who had not had prior protease inhibitors or lamivudine three different doses of amprenavir were studied in combination with zidovudine and lamivudine.[162] Preliminary 24 week results showed a median decrease in HIV RNA of 1.9 to 2.4 log_{10} copies/ml, and CD4+ T cell increases of 99 to 135 cells/mm^3. In a double-blind randomized study which compared amprenavir alone (1200 mg BID) to a triple combination of amprenavir, zidovudine and lamivudine, half of the patients had received prior antivirals (but not lamivudine or protease inhibitors).[163] After a median of 88 days of therapy, the study was terminated as a result of an interim review, which showed that the triple therapy arm had a much better response than the amprenavir monotherapy arm. Only one of 46 patients in the triple therapy arm met the study definition of virological failure versus 13 of 46 in the monotherapy arm. Combination therapy studies of amprenavir with abacavir suggest that this combination is associated with potent antiretroviral activity over 16 weeks in antiretroviral naïve patients.[164, 165] Additional studies using amprenavir in other combinations of antiretrovirals are underway, including studies of patients who have failed other protease inhibitor-containing combinations.

Table 11. Potential regimens for patients who have failed treatment*

Prior therapy	New regimens
2 NRTIs plus	2 new NRTIs plus
IDV	RTV/SQV, NFV/NNRTI, or NFV/SQV
NFV	IDV, RTV, RTV/SQV, NNRTI/IDV, or NNRTI/RTV
RTV	RTV/SQV, NFV/SQV, or NFV/NNRTI
SQV	RTV/SQV, NFV/SQV, or NFV/NNRTI
2 NRTIs/NNRTI	2 new NRTIs/1 or 2 new PIs
1 or 2 NRTIs/2 PIs	2 new NRTIs/1 or 2 new PIs/NNRTI
NRTI/NNRTI/PI	2 new NRTIs/2 new PIs
+2 NRTIs	2 new NRTIs/1 or 2 PIs, or 1 new NRTI/NNRTI/PI

*Adapted from Guidelines for the Use of Antiretroviral Agents in HIV-infected Adults[5] and Antiretroviral Therapy for HIV infection in 1998[7]
NRTI=nucleoside reverse transcriptase inhibitor. NNRTI=nonnucleoside reverse transcriptase inhibitor, PI=protease inhibitor, NFV=nelfinavir, RTV=ritonavir, SQV=saquinavir

5. PRACTICAL MANAGEMENT ISSUES

Issues to consider in the choice of second-line regimens include the previous regimen(s), reasons for failure of previous regimen(s), the efficacy and toxicities of the available options, and the impact of the proposed regimen on future therapeutic choices. A difference in approach exists with dose-limiting toxicity, where a single drug in a regimen can be changed. In contrast, in other circumstances, multiple agents should be changed. The available guidelines suggest potential regimens or approaches for designing potential regimens for patients who have failed antiretroviral therapy. Examples are shown in Table 11. These regimens are recommended based upon theoretical considerations, since limited clinical data exist. The major theme of the recommendations is to change as many drugs as possible in patients failing treatment and not use agents cross resistant to previously used drugs. Similar to tuberculosis therapy, because of the potential for the development of viral resistance and based upon the available data, a single drug should not be added or substituted in a failing regimen. Adding potent agents from new classes is logical, when possible. However, despite the association between genotypic and phenotypic resistance and outcome, data from several studies has shown that a substantial proportion of patients with poor virological responses to combination regimens still have wild-type virus (without evidence of resistance to the therapies being used). The clinical role for resistance testing remains to be defined, although it is hoped that use of such tests will improve outcome.[4]

5.1. Drug Toxicity

Management of adverse events varies by drug class, and specific agents. Dose decreases for some nucleoside reverse transcriptase inhibitors (zidovudine, stavudine, and didanosine) can be made for dose-limiting side-effects. However, in the desire to maximize efficacy, substitution of a different antiretroviral is often done when dose-limiting toxicity occurs with these agents. The most common substitution is stavudine for zidovudine. For toxicities associated with non-nucleoside reverse transcriptase inhibitors and for protease inhibitors, total daily doses are not usually altered. Symptomatic remedies are used for mild or moderate toxicities. For severe intolerance, substitution of another agent or agents is preferred. Since the frequency of side-effects with ritonavir increases with dose, one approach to side-effects with ritonavir given as a sole protease inhibitor (600 mg BID) is to use it concurrently with another protease inhibitor, such as saquinavir, since this allows for use of a lower dose of ritonavir (because of the pharmacokinetic interaction between ritonavir and saquinavir).[113, 115]

Available data are conflicting about whether the frequency of hyperlipidemia and body composition changes vary with different protease inhibitors, and only anecdotal data are available about the effects of switching protease inhibitors in the setting of severe metabolic abnormalities. Until additional data are available, it is appropriate to measure serum glucose, triglycerides, and cholesterol periodically in patients on potent antiretroviral therapy. Approaches to these abnormalities depend on the severity and whether they are associated with symptoms. Management of diabetes is similar to patients without HIV; it is unclear if discontinuation of the protease inhibitor alters its clinical course. Dietary counseling, weight loss for obese patients, and increased exercise should be recommended for patients with metabolic abnormalities. Severe hyperlipidemia should be treated with antilipidemic drugs.

5.2. Treatment Failure

Obtaining a repeat plasma specimen for HIV RNA and/or CD4+ T cells prior to making therapeutic decisions is recommended if a change in therapy is based upon deterioration in laboratory test results.[5, 7] Data from several observational studies suggest that virological responses to subsequent regimens may be better when patients had lower levels of plasma HIV RNA at the time of switching therapy. However, insufficient data are available to specify an exact optimal level of plasma HIV RNA which should dictate a switch in therapy. As reflected in the treatment guidelines, the threshold for switching therapy is controversial, and may range from just above the detection limit of HIV RNA to 10^4 copies/ml or more. For patients with extensive prior antiretroviral therapy and few new treatment options, delaying a switch in order to take advantage of future treatments may be reasonable if patients are clinically and immunologically stable. Evaluating patients for potential explanations for treatment failure is important before designing a new regimen, since the best strategy to avoid failure of a new regimen may vary depending upon the underlying reasons for poor response to the current regimen. Viral resistance (as described above), pharmacokinetic, and behavioral reasons are the underlying reasons which appear to be most common as explanations for poor responses to antiretroviral treatment.

5.3. Pharmacokinetic Issues

The major pharmacokinetic reason for treatment failure is inadequate drug exposure. Reasons for inadequate exposure include dosing errors, interference with drug absorption by food or other medications, and pharmacokinetic interactions between drugs which enhance the metabolism of antiretrovirals. Several clinically relevant pharmacokinetic interactions have been identified which can decrease plasma concentrations of protease inhibitors, and may be potential causes of poor viral suppression (Table 12) (reviewed in[49]).

Two-way drug interaction data are available for interactions between many of the antiretroviral agents, and between antiretroviral agents and some agents which are affected or affect the hepatic cytochrome p450 system.[128, 166–168] However, few data are available about more complex situations, such as with three agents that each have different cytochrome p450 effects. Data supporting and refuting a relationship between plasma con-

Table 12. Pharmacokinetic interactions with potential to decrease plasma levels of antiretrovirals

Agent	Antiretroviral	Effect on antiretroviral*
Didanosine	Indinavir	Interferes with absorption when taken simultaneously**
	Ritonavir	Interferes with absorption when taken simultaneously**
Efavirenz	Indinavir	35% decrease
Grapefruit juice	Indinavir	26% decrease
Nevirapine	Indinavir	28% decrease†
	Nelfinavir	None to 46% decrease††
	Saquinavir	25% decrease
Rifabutin‡	Indinavir	32% decrease
	Nelfinavir	32% decrease
	Ritonavir	35% decrease
	Saquinavir	40% decrease
Rifampin§	Amprenavir	82% decrease
	Indinavir	92% decrease
	Nelfinavir	82% decrease
	Ritonavir	35% decrease
	Saquinavir	80% decrease

*Clinical significance of most of these interactions is unknown
**Doses should be separated by ≥ 2 hours
†Antiviral effect has been seen with standard doses of indinavir
††Data are conflicting. Antiviral effect has been seen with standard doses of nelfinvir
‡Not recommended for use with ritonavir or saquinavir
§Not recommended for use with amprenavir, indinavir, nelfinavir, or saquinavir

centrations of certain protease inhibitors and virological response are available.[118,169–174] There is substantial interperson variability in drug concentrations with indinavir, nelfinavir, and saquinavir, which argues for the potential usefulness of therapeutic drug monitoring.[175] At present, monitoring of drug concentrations is generally limited to research settings. Reviewing explicit details of when patients took which medications in relationship to each other and food may reveal potential misunderstandings about timing of medications or timing in relationship to food. Reviewing all medications, in addition to anti-HIV medications, should be done in the setting of treatment failure.

5.4. Behavioral Issues

Several behavioral issues have the potential to impact on use of antiretroviral medications, including medication taking behavior and concurrent use of mind-altering drugs such as alcohol and illicit drugs. Adherence to complex regimens is difficult. Although demographic characteristics do not predict adherence to antiretroviral therapy, other factors, including patient attitudes, beliefs, physical and mental health, relationships between patients and their health care providers, and regimen complexity (numbers of drugs and number of doses per day) do impact on adherence.[176–182] Self-reported adherence to antiretrovirals has been reported to be significantly associated with better viral suppression.[80] If inability to adhere to a complex regimen contributes to treatment failure, it is important to try to develop a regimen with simpler logistics, but without decreasing its potency.

5.5. Suboptimal Therapy

Management plans for these patients should be individualized. It is likely that the majority of such patients should have their therapy changed to a more potent regimen, to prevent further resistance mutations from developing. The new therapy should include new reverse transcriptase inhibitors that do not have the same resistance profile as their prior therapy as well as other agents, such as protease inhibitors or non-nucleoside reverse transcriptase inhibitors. If patients have undetectable or stable very low levels of plasma HIV RNA and adequate, stable CD4+ T cells with a well tolerated, nucleoside analog regimen, continuation of the therapies may be reasonable. However, this is likely to be the exceptional situation, and such patients should be counseled that their therapy is not in accordance with current treatment guidelines.

6. CONCLUSION

Although knowledge about HIV and its treatment are increasing at a rapid rate, the treatment options for persons with HIV in many developed countries far exceed the available information. The numbers of combinations of antiretroviral drugs is extremely large; 11 approved antiretrovirals, 83 triple and 260 four-drug protease inhibitor-containing regimens exist. Controlled clinical studies will not be conducted for all of these regimens, thus understanding the principles underlying therapy and important drug interactions is important, as is knowing what data are available about antiretroviral drugs and regimens. Trying to understand the reasons underlying poor responses to antiretrovirals, and ensuring that new regimens include multiple new agents which are not cross-resistant to prior therapies, are key elements in maximizing the success of second-line therapy.

REFERENCES

1. Palella FJ, Delaney KM, Moorman AC. Declining morbidity and mortality among patients with advanced human immunodeficiency virus infection. N Engl J Med. 1998; 338:853–60.
2. Mouton Y, Alfandari S, Valette M, *et al.* Impact of protease inhibitors on AIDS-defining events and hospitalizations in 10 French AIDS reference centres. AIDS 1997; 11:F101–5.
3. Saag MS, Holodniy M, Kuritzkes DR, *et al.* HIV viral load markers in clinical practice. Nature Medicine. 1996; 2:625–9.
4. Hirsch MS, Conway B, D'Aquila R, *et al.* Antiretroviral drug resistance testing in adults with HIV infection: implications for clinical management. JAMA 1998; 279:1984–91.
5. Centers for Disease Control and Prevention. Guidelines for the use of antiretroviral agents in HIV-infected adults and adolescents. Morbidity and Mortality Weekly Report 1998; 47(RR-5):43–82.
6. Centers for Disease Control. Report of the NIH Panel to define principles of therapy of HIV infection and guidelines for the use of antiretroviral agents in HIV-infected adults and adolescents. Morbidity and Mortality Weekly Report 1998; 47 (RR-5):1–39.
7. Carpenter CCJ, Fischl MA, Hammer SM, *et al.* Antiretroviral therapy for HIV infection in 1997: updated recommendations of the International AIDS Society - USA Panel. JAMA 1998; 280:78–86.
8. Gazzard BG, Moyle GJ, Weber J, *et al.*, for the BHIVA Guidelines Co-ordinating Committee. British HIV Association guidelines for antiretroviral treatment of HIV seropositive individuals. Lancet. 1997; 349:1086–92.
9. Mellors JW, Rinaldo CR, Gupta P, *et al.* Prognosis in HIV-1 infection predicted by the quantity of virus in plasma. Science. 1996; 272:1167–70.
10. Mellors JW, Muñoz A, Giorgi JV, *et al.* Plasma viral load and CD4+ lymphocytes as prognostic markers of HIV-1 infection. Ann Intern Med. 1997; 126:946–54.
11. Farzadegan HY, Hoover DR, Astemborski J, *et al.* Gender differences in the relationship between HIV-1 and progression to AIDS. 12th World AIDS Conference. Abstract 13384. Geneva, Switzerland, June 28 - July 3, 1998.
12. Piatak M, Saag MS, Yang LC, *et al.* High levels of HIV-1 in plasma during all stages of infection determined by competitive PCR. Science. 1993; 259:1749–54.

13. Wei X, Ghosh SK, Taylor ME, *et al.* Viral dynamics in human immunodeficiency virus type 1 infection. Nature. 1995; 373:117–22.
14. Ho DD, Neumann AU, Perelson AS, *et al.* Rapid turnover of plasma virions and CD4 lymphocytes in HIV-1 infection. Nature. 1995; 373:123–6.
15. Perelson AS, Neumann AU, Markowitz M, *et al.* HIV-1 dynamics *in vivo*: virion clearance rate, infected cell life-span, and viral generation time. Science. 1996; 271:1582–6.
16. Coffin JM. HIV population dynamics *in vivo*: implications for genetic variation, pathogenesis, and therapy. Science. 1995; 267:483–9.
17. Havlir DV, Richman DD. Viral dynamics of HIV-implications for drug development and therapeutic strategies. Ann Intern Med. 1996; 124:984–994.
18. Larder BA, Kemp SD. Multiple mutations in HIV-1 reverse transcriptase confer high-level resistance to zidovudine (AZT). Science 1989: 246:1115–58.
19. Larder BA. Ineractions between drug-resistance mutations in human immunodeficiency virus type 1 reverse transcriptase. J Gen Virol. 1994; 75:951–7.
20. Condra JH, Schleif WA, Blahy OM, *et al.* *In vivo* emergence of HIV-1 variants resistance to multiple protease inhibitors. Nature. 1995; 374:569–71.
21. Markowitz M, Mo H, Kempf DJ, *et al.* Selection and analysis of human immunodeficiency virus type 1 variants with increased resistance to ABT-538, a novel protease inhibitor. J Virol. 1995; 69:701–6.
22. Condra JH, Holder DJ, Schleif WA, *et al.* Genetic correlates of *in vivo* viral resistance to indinavir, a human immunodeficiency virus type 1 protease inhibitor. J Virol. 1996; 70:8270–6.
23. Roberts NA. Drug-resistance patterns of saquinavir and other HIV proteinase inhibitors. AIDS 1995; 9:S27–32.
24. Harrigan PR, Kinghorn I, Bloor S, *et al.* Signifiance of amino acid variation at human immunodeficiency virus type 1 reverse transcriptase residue 210 for zidovudine susceptibility. J Virol. 1996; 70:5930–4.
25. Hooker DJ, Tachedjian G, Solomon AE, *et al.* An *in vivo* mutation from leucine to tryptophan at position 210 in human immunodeficiency virus type 1 reverse transcriptase contributes to high-level resistance to 3'-azido-3'-deoxythymidine. J Virol. 1996; 70:8010–8.
26. Havlir D, McLaughlin MM, Richman DD. A pilot study to evaluate the development of resistance to nevirapine in asymptomatic human immunodeficiency virus-infected patients with CD4 cell counts of >500/mm^3: AIDS Clinical Trials Group Protocol 208. J Infect Dis. 1995; 172:1379–83.
27. Havlir D, Cheeseman SH, McLaughlin M, *et al.* High-dose nevirapine: safety, pharmacokinetics, and antiviral effect in patients with human immunodeficiency virus infection. J Infect Dis. 1995; 171:537–45.
28. Schuurman R, Nijhuis M, van Leeuwen R, *et al.* Rapid changes in human immunodeficiency virus type 1 RNA load and appearance of drug-resistant virus populations in persons treated with lamivudine (3TC) J Infect Dis. 1995; 171:1411–9.
29. Wainberg MA, Salomon G, Gu Z, *et al.* Development of HIV-1 resistance to (-)2'-deoxy-3'-thiacytidine in patients with AIDS or advanced AIDS-related complex. AIDS 1995; 9:351–7.
30. Larder BA, Darby G, Richman DD. HIV with reduced sensitivity to zidovudine (AZT) isolated during prolonged therapy. Science. 1989; 243:1731–4.
31. Jacobsen H, Hänggi, Ott M, *et al.* *In vivo* resistance to a human immunodeficiency virus type 1 proteinase inhibitor: mutations, kinetics, and frequencies. J Infect Dis. 1996; 173:1379–81.
32. Tisdale M, Demp SD, Parry NR, *et al.* Rapid *in vitro* selection of human immunodeficiency virus type 1 resistant to 3'-thiacytidine inhibitors due to a mutation in the YMDD region of reverse transcriptase. Proc Natl Acac Sci USA 1993; 90:5653–6.
33. Larder BA, Kemp SD, Harrigan PR. Potential mechanism for sustained antiretroviral efficacy of AZT-3TC combination therapy. Science. 1995; 269:696–9.
34. Miller M, Anton KE, Mulato AS, *et al.* HIV-1 expressing the 3TC-associated M184V mutation in reverse transcriptase (RT) shows increased sensitivity to adefovir and PMPA as well as decreased replication capacity *in vitro*. 12th World AIDS Conference. Abstract 41214. Geneva, Switzerland, June 28 - July 3, 1998.
35. D'Aquila RT, Johnson VA, Welles SL, *et al.* Zidovudine resistance and HIV-1 disease progression during antiretroviral therapy. Ann Intern Med 1995;122:401–8.
36. Japour AJ, Welles S, D'Aquila RT, *et al.* Prevalence and clinical significance of zidovudine resistance mutations in human immunodeficiency virus isolated from patients after long-term zidovudine treatment. J Infect Dis. 1995; 171:1172–9.
37. Miller V, Phillips A, Rottmann C, *et al.* Dual resistance to zidovudine and lamivudine in patients treated with zidovudine-lamivudine combination therapy: association with therapy failure. J Infect Dis. 1998; 177:1521–32.

38. Lanier ER, Smiley ML, St. Clair MH, *et al.* Phenotypic HIV resistance *in vitro* correlates with viral load response to abacavir (1592, ABC) *in vivo.* 12th World AIDS Conference. Abstract 231/32289. Geneva, Switzerland, June 28 - July 3, 1998.
39. Deeks, S. Twenty-week HIV RNA response and correlation with baseline phenotypic drug susceptibility during treatment with a novel quadruple salvage regimen after failure of indinavir combination therapy. 12th World AIDS Conference. Abstract 22490. Geneva, Switzerland, June 28 - July 3, 1998.
40. Hertogs K, Larder B, Mellors J, *et al.* Patterns of phenotypic and genotypic cross-resistance among protease inhibitors in over 1,000 clinical HIV-1 isolates. 12th World AIDS Conference. Abstract 42195. Geneva, Switzerland, June 28 - July 3, 1998.
41. Workman C, Mussen R, Sullivan J. Salvage therapy using six drugs in heavily pretreated patients. 5th Conference on Retroviruses and Opportunistic Infections. Abstract 426. Chicago, IL, February 1–5, 1998.
42. Gulick RM, Mellors, JW, Havlir D, *et al.* Simultaneous vs sequential initiation of therapy with indinavir, zidovudine, and lamivudine for HIV-1 infection: 100-week follow-up. JAMA 1998; 280:35–41.
43. Saag M, Knowles M, Chang Y, *et al* for the Viracept Cooperative Study Group. Durable effect of Viracept (nelfinavir mesylate, NFV) in triple combination therapy. 37th Interscience Conference on Antimicrobial Agents and Chemotherapy. Poster I-101. Toronto, Canada, September 28-October 1, 1997.
44. Wong JK, Hezareh M, Günthard HF, *et al.* Recovery of replication-competent HIV despite prolonged suppression of plasma viremia. Science. 1997; 278:1291–5.
45. Finzi D, Gallant J, Chadwick K, *et al.* Latent viral reservoirs in patients on highly active antiretroviral therapy: implications for virus eradication. 12th World AIDS Conference. Abstract 273/11150. Geneva, Switzerland, June 28 - July 3, 1998.
46. Finzi D, Hermankova M, Pierson T, *et al.* Identification of a reservoir for HIV-1 in patients on highly active antiretroviral therapy. Science. 1997; 278:1295–300.
47. Chun T-W, Stuyver L, Mizell SB, *et al.* Presence of an inducible HIV-1 latent reservoir during highly active antiretroviral therapy. Proc Natl Acad Sci USA 1997; 94:13193–7.
48. Centers for Disease Control. 1997 USPHS/IDSA Guidelines for the prevention of opportunistic infections in persons infected with human immunodeficiency virus. Morbidity and Mortality Weekly Report 1997; 46 (RR-12):1–46.
49. Flexner C. HIV-protease inhibitors. N Engl J Med. 1998; 338:1281–92.
50. Carr A, Samaras K, Burton S, *et al.* A syndrome of peripheral lipodystrophy (LD) hyperlipidemia and insulin resistance due to HIV protease inhibitors (PI's). 5th Conference on Retroviruses and Opportunistic Infections. Abstract 410. Chicago, IL, February 1–5, 1998.
51. Walli RK, Herfort O, Michl GM, *et al.* Treatment with protease inhibitors associated with peripheral insulin resistance and impaired oral glucose tolerance in HIV-1 infected patients. AIDS 1998; 12:F167–173.
52. Hengel RL, Watts NB, Lennox J. Benign symmetric lipomatosis associated with protease inhibitors. Lancet. 1997; 350:1596.
53. Henry K, Melroe H, Huebsch J, et al. Severe premature coronary artery disease with protease inhibitors. Lancet. 1998; 351:1328.
54. Lo JC, Mulligan K, Tai VW, *et al.* "Buffalo hump" in men with HIV-1 infection. Lancet 1998; 351:867–70.
55. Miller KD, Jones E, Yanovski JA, *et al.* Visceral abdominal-fat accumulation associated with use of indinavir. Lancet. 1998; 351:871–5.
56. Nightingale SL. From the Food and Drug Administration. JAMA 1997:278:379.
57. Carr A, Cooper D, Thorisdottir A, *et al.* Prevalence and severity of protease inhibitor (pI) induced lipodystrophy (LD) and insulin resistance. 12th World AIDS Conference. Abstract LB 8. Geneva, Switzerland, June 28 - July 3, 1998.
58. Bernasconi E, Carota A, Magenta L, *et al.* Metabolic changes in HIV-infected patients treated with protease inhibitors. 12th World AIDS Conference. Abstract 178/12375. Geneva, Switzerland, June 28 - July 3, 1998.
59. Carr A, Samaras K, Chisholm DJ, *et al.* Pathogenesis of HIV-1 protease inhibitor-associated peripheral lipodystrophy, hyperlipidaemia, and insulin resistance. Lancet. 1998; 351:1881–3.
60. Kempf DJ, Rode RA, Xu Y, *et al.* The duration of viral suppression during protease inhibitor therapy for HIV-1 infection is predicted by plasma HIV-1 RNA at the nadir. AIDS 1998; 12:F9–14.
61. Montaner JSF, *et al.* A randomized double-blind trial comparing combinations of nevirapine, didanosine, and zidovudine for HIV-infected patients: The INCAS Trial. JAMA 1998; 279:930–7.
62. Raboud JM, Montaner JSG, Conway B, *et al.* Suppression of plasma viral load below 20 copies/ml is required to achieve a long-term response to therapy. AIDS 1998; 12:1619–24.
63. Stanley SK, Ostrowski MA, Justement JS, *et al.* Effect of immunization with a common recall antigen on viral expression in patients infected with human immunodeficiency virus type 1. N Engl J Med. 1996; 334:1222–30.

64. Hammer SM, Squires KE, Hughes MD, *et al,* for the AIDS Clinical Trials Group 320 Team. A controlled trial of two nucleoside analogues plus indinavir in persons with human immunodeficiency virus infection and CD4 cell counts of 200 per cubic millimeter or less. N Engl J Med. 1997; 337:725–33.
65. Cameron DW, Heath-Chiozzi M, Danner S, *et al,* for the Advanced HIV Disease Ritonavir Study Group. Randomized placebo-controlled trial of ritonavir in advanced HIV-1 disease. Lancet. 1998; 351:543–9.
66. Haubrich R, Lalezari J, Follansbee SE, *et al.* Improved survival and reduced clinical progression in HIV-infected patients with advanced disease treated with saquinavir plus zalcitabine. Antiviral Therapy 1998; 3:33–42.
67. Tomino C, Vella S, Fragola FV, *et al.* A multicenter randomised study comparing ritonavir and indinavir in 1251 previously treated patients with CD4 below 50/mm^3 (Italian ISS-IP1 Study Group). 12th World AIDS Conference. Abstract 132/12237. Geneva, Switzerland, June 28 - July 3, 1998.
68. Gill MJ, Beall G, Beattie D, *et al,* for the NV15182 Study Team. Safety of saquinavir soft gelatin capsule (SQV-SGC) in combination with other antiretroviral agents: multicenter study NV15182: 24 week analysis. 37th Interscience Conference on Antimicrobial Agents and Chemotherapy. Abstract I-90. Toronto, Canada, September 28-October 1, 1997.
69. Albrecht M, Katzenstein D, Bosch RJ, Liou SH, Hammer SM. ACTG 364: virologic efficacy of nelfinavir (NFV) and/or efavirenz (EFV) in combination with new nucleoside analogs in nucleoside experienced subjects. 12th World AIDS Conference. Abstract 12203. Geneva, Switzerland, June 28 - July 3, 1998.
70. Martinez E, Gatell J, Buira E, *et al.* An open randomized trial comparing the effect of a triple combination therapy including d4T, 3TC, and a protease inhibitor (saquinavir, ritonavir, or indinavir) in adult HIV-1 infected patients previously treated with nucleoside reverse transcriptase inhibitors. 5th Conference on Retroviruses and Opportunistic Infections. Abstract 370. Chicago, IL, February 1–5, 1998.
71. Gulick RM, Mellors JW, Havlir D, *et al.* Treatment with indinavir, zidovudine, and lamivudine in adults with human immunodeficiency virus infection and prior antiretroviral therapy. N Engl J Med. 1997; 337:734–9.
72. Harris M, Durakovic C, Rae S, *et al.* A pilot study of nevirapine, indinavir, and lamivudine among patients with advanced human immunodeficiency virus disease who have had failure of combination nucleoside therapy. J Infect Dis. 1998; 177:1514–20.
73. Clumeck N, Colebunders B, Vandercam B, *et al.* Randomized comparative outcome trial of indinavir (I) and ritonavir (R) in protease inhibitors (PI) naïve HIV patients (p) with CD4 below 100 cells/µl. 5th Conference on Retroviruses and Opportunistic Infections. Abstract 386. Chicago, IL, February 1–5, 1998.
74. Deeks S, Loftus R, Cohen P, *et al.* Incidence and predictors of virologic failure of indinavir and/or ritonavir in an urban health clinic. 37th Interscience Conference on Antimicrobial Agents and Chemotherapy. Abstract. Toronto, Canada, September 28-October 1, 1997.
75. Piketty C, Castiel P, Gilquin J, *et al.* Prospective follow-up of 177 HIV-infected patients treated with indinavir in combination with nucleoside analogues. 37th Interscience Conference on Antimicrobial Agents and Chemotherapy. Poster I-100. Toronto, Canada, September 28-October 1, 1997.
76. Clough LA, Haas DW, Raffanti S. Predictor and prevalence of incomplete virologic response to protease inhibitor-based highly active antiretroviral therapy (HAART) for HIV infection. 35th Annual Meeting of the Infectious Diseases Society of America. Abstracts 223. San Francisco, CA, September 13–16, 1997.
77. Izopet J, Salama G, Pasquier C, *et al.* Ultra sensitive detection of plasma HIV-1 RNA for predicting the durability of 3-drug antiretroviral therapy. 5th Conference on Retroviruses and Opportunistic Infections. Abstract 326. Chicago, IL, February 1–5, 1998.
78. Fessel WJ, Hurley LB. Outcomes of triple therapy that included a protease inhibitor (PI) among 2139 patients. 5th Conference on Retroviruses and Opportunistic Infections. Abstract 145. Chicago, IL, February 1–5, 1998.
79. Staszewski S, Phillips A, Miller V, *et al.* Virological response to protease inhibitors in an HIV clinic cohort. 12th World AIDS Conference. Abstract 12294. Geneva, Switzerland, June 28 - July 3, 1998.
80. Hecht FM, Colfax G, Swanson M, *et al.* Adherence and effectiveness of protease inhibitors in clinical practice. 5th Conference on Retroviruses and Opportunistic Infections. Abstract 151. Chicago, IL, February 1–5, 1998.
81. Barbour C. Long term viral suppression using quadruple drug combination therapy in treatment experienced HIV/AIDS patients, following initial protease failure. 12th World AIDS Conference. Abstract 32295. Geneva, Switzerland, June 28 - July 3, 1998.
82. Gallant JE, Barnett S, Raines BC, *et al.* Efficacy and durability of ritonavir/saquinavir (RTV/SQV) as salvage therapy after failure of initial protease inhibitor (PI) regimen. 12th World AIDS Conference. Abstract 12330. Geneva, Switzerland, June 28 - July 3, 1998.

83. Kaufmann GR, Duncombe C, Cunningham P, *et al.* Treatment response and durability of a double protease inhibitor therapy with saquinavir and ritonavir in an observational cohort of HIV-1 infected individuals. AIDS 1998; 12:1625–30.
84. Lohmeyer J, Selhorst J, Friese G, *et al.* Combined nelfinavir/saquinavir protease inhibitor treatment in a BID regimen as salvage therapy in advanced HIV disease. 12th World AIDS Conference. Abstract 12303. Geneva, Switzerland, June 28 - July 3, 1998.
85. Walmsley S, Walach C, Moses A, *et al.* Salvage therapy with a combination including nelfinavir in patients failing treatment including a protease inhibitor (PI). 12th World AIDS Conference. Abstract 12285. Geneva, Switzerland, June 28 - July 3, 1998.
86. Cavorus A, Thompson CE, Salvato PD. Salvage therapy for antiretroviral failure in HIV+ patients. 12th World AIDS Conference. Abstract 22402. Geneva, Switzerland, June 28 - July 3, 1998.
87. Para MF, Collier A, Coombs R, *et al*: ACTG 333. Antiviral effects of switching from saquinavir hard capsule (SQVhc) to saquinavir soft gelatin capsule (SQVsgc) vs. switching to indinavir (IDV) after prior saquinavir. Abstract 21. Thirty-fifty Annual Meeting of the Infectious Diseases Society of America (IDSA), San Francisco, California, September 14–16, 1997.
88. Para MF, Coombs R, Collier A, *et al.* Relationship of baseline genotype to RNA response in ACTG 333 after switching from long term saquinavir (SQVhc) to indinavir (IDV) or saquinavir soft gelatin capsule (SQVsgc). 5th Conference on Retroviruses and Opportunistic Infections. Abstract 511. Chicago, IL, February 1–5, 1998.
89. Lori F, Malykh A, Cara A, *et al.* Hydroxyurea as an inhibitor of human immunodeficiency virus-type 1 replication. Science. 1994; 266:801–5.
90. Gao, W-Y, Johns DG, Mitsuya H. Anti-human immunodeficiency virus type 1 activity of hydroxyurea in combination with 2', 3'-dideoxynucleosides. Molecular Pharmacology 1994; 46:767–72.
91. Malley SD, Grange JM, Hamedi-Sangsari F, *et al.* Synergistic anti-human immunodeficiency virus type 1 effect of hydroxamate compounds with 2',3'-dideoxyinosine in infected resting human lymphocytes. Proc Natl Acad Sci. 1994; 91:11017–21.
92. Gao W-Y, Johns DG, Chokekijchai S, *et al.* Disparate actions of hydroxyurea in potentiation of purine and pyrimidine 2',3'-dideoxynucleoside activities against replication of human immunodeficiency virus. Proc Natl Acad Sci. 1995; 92:8333–7.
93. Giacca M, Zanussi S, Comar M, *et al.* Treatment of human immunodeficiency virus infection with hydroxyurea: virologic and clinical evaluation. J Infect Dis. 1996; 174:204–9.
94. Mole L, Holodniy M. A pilot study of the tolerance and potential antiviral activity of hydroxyurea alone and in combination with ddI in HIV-infected subjects. Thirty-fifth Interscience Conference on Antimicrobial Agents and Chemotherapy (ICAAC), San Francisco, California, September 17–20, 1995, Abstract I110.
95. Lori F, Jessen H, Foli A, *et al.* Long-term suppression of HIV-1 by hydroxyurea and didanosine. JAMA 1997; 277:1437–8.
96. Biron F, Lucht F, Peyramond D, *et al.* Anti-HIV activity of the combination of didanosine and hydroxyurea in HIV-1-infected individuals. J Acquired Immune Deficiency Syndromes. 1995; 10:36–40.
97. Biron F, Lucht, F, Peyramond D, *et al.* Pilot clinical trial of the combination of hydroxyurea and didanosine in HIV-1 infected individuals. Antiviral Res. 1996; 29:111–13.
98. Vila J, Biron F, Nugier F, Vallet T, Peyramond D. 1-year follow-up of the use of hydroxycarbamide and didanosine in HIV infection. Lancet. 1996; 348:203–4.
99. Montaner JSG, Zala C, Conway B, *et al.* A pilot study of hydroxyurea among patients with advanced human immunodeficiency virus (HIV) disease receiving chronic didanosine therapy: Canadian HIV Trials Network Protocol 080. J Infect Dis. 1997; 175:801–6.
100. Rusnak J, Bery A, Stinnette B, *et al.* DDI once daily vs twice daily alone and also with hydroxyurea once daily. 12th World AIDS Conference. Abstract 12352. Geneva Switzerland, June 18 – July 3, 1998.
101. Seminari E, Maserati R, Lori F, *et al.* Long term suppression of HIV with a simple, non toxic, inexpensive combiantion of didanosine (ddI) plus hydroxyurea (HU). 12th World AIDS Conference. Abstract 41180. Geneva Switzerland, June 18 – July 3, 1998,
102. Federici ME, Lupo S, Cahn I, *et al.* Hydroxyurea in combination regimens for the treatment of antiretroviral naïve, HIV-infected adults. 12th World AIDS Conference, Geneva Switzerland, June 18 – July 3, 1998, Abstract 12235.
103. Hellinger J, Torres R, Schrader S, *et al.* Safety and antiretroviral activity of hydroxyurea with ddI in HIV-infected individuals. Thirty-seventh Interscience Conference on Antimicrobial Agents and Chemotherapy (ICAAC)., Abstract I-27. Toronto, Ontario, Canada, September 28-October 1, 1997.

104. Rossero R, McKinsey D, Green S, *et al.* Open label combination therapy with stavudine, didanosine, and hydroxyurea in nucleoside experienced HIV-1 infected patients. 5th Conference on Retroviruses and Opportunistic Infections, Chicago, IL. Abstract 653. February 1–8, 1998.
105. Rutschmann OT, Opravil M, Iten A, *et al.* A placebo-controlled trial of didanosine plus stravudine with and without hydroxyurea for HIV infection. The Swiss cohort study. AIDS 1998; 12:F71–77.
106. Miles S, Winters RE, Ruane P. Salvage of multi-drug resistant HIV infection with d4T/3TC/hydroxyurea. 12th World AIDS Conference. Abstract 12205. Geneva, Switzerland, June 28 - July 3, 1998.
107. Havlir DV, Marschner IC, Hirsch MS, *et al.* Maintenance antiretroviral therapy in HIV infected subjects with undetectable plasma HIV RNA after triple-drug therapy. N Engl J Med 1998; 18:1261–1268.
108. Pialoux G, Raffi F, Brun-Vezinet F, *et al.* A randomized trial of three maintenance regimens given after three months of induction therapy with zidovudine, lamivudine and indinavir in previously untreated HIV-1 infected patients. N Eng J Med 1998; 18:1269–1276.
109. Reijers MHE, Weverling GJ, Jurriaans S, *et al.* Maintenance therapy after quadruple induction therapy in HIV-1 infected individuals: Amsterdam Duration of Antiretroviral Medication (ADAM) study. Lancet. 1998; 352:185–90.
110. Nguyen B-Y, Haas DW, Ramirez-Ronda C, *et al.* A pilot, multicenter, open-label, randomized study to compare the safety and activity of indinavir sulfate (IDV) administered every 8 hours (h) versus every 12 h in combination with zidovudine (ZDV) and lamivudine (3TC). 37th Interscience Conference on Antimicrobial Agents and Chemotherapy. Abstract I91. Toronto, Canada, September 28-October 1, 1997.
111. Petersen A, Johnson M. Longterm comparison of BID and TID dosing Viracept (nelfinavir) in combination with stavudine (d4T) and lamivudine (3TC) in HIV patients. 12th World AIDS Conference. Poster 12224. Geneva, Switzerland, June 28 - July 3, 1998.
112. Cohen C, Siemon-Hryczk P, Pilson R, *et al.* Potent and convenient Fortovase™ (SQV) SGC BID regimens in combination with 2 nucleosides or nelfinavir (NFV) plus 1 nucleoside in HIV-1 infected patients. 12th World AIDS Conference. Abstract 12314. Geneva, Switzerland, June 28 - July 3, 1998.
113. Kempf DJ, Marsh KC, Kumar G, *et al.* Pharmacokinetic enhancement of inhibitors of the human immunodeficiency virus protease by coadministration with ritonavir. Antimicrobial Agents and Chemotherapy. 1997; 41:654–60.
114. Cameron DW, Japour A, Mellors J, *et al.* Antiretroviral safety and durability of ritonavir (RIT)-saquinavir (SQV) in protease inhibitor-naïve patients in year two of follow-up. 5th Conference on Retroviruses and Opportunistic Infections. Abstract 388. Chicago, IL, February 1–5, 1998.
115. Merry C, Barry MG, Mulcahy F, *et al.* Saquinavir pharmacokinetics alone and in combination with ritonavir in HIV-infected patients. AIDS 1997;11:F29–33.
116. Cassano P, Hermans P, Sommereijns B, *et al.* Combined quadruple therapy with ritonavir-saquinavir (RTV-SQV) + nucleosides in patients (p) who failed in triple therapy with RTV, SQV or indinavir (IDV). 5th Conference on Retroviruses and Opportunistic Infections. Abstract 423. Chicago, IL, February 1–5, 1998.
117. de Truchis P, Force G, Zucman D, *et al.* Effects of "salvage" combination therapy with ritonavir + saquinavir in HIV-infected patients previously treated with protease-inhibitors (PI). 5th Conference on Retroviruses and Opportunistic Infections. Abstract 425. Chicago, IL, February 1–5, 1998.
118. Fatkenheuer G, Hunn N, Jutte A, *et al.* Efficacy and drug levels of ritonavir/saquinavir combination salvage therapy. 12th World AIDS Conference. Abstract 12316. Geneva, Switzerland, June 28 - July 3, 1998.
119. Puig T, Bonjoch A, Ruiz L, *et al.* Usefulness of ritonavir and saquinavir combination therapy for HIV-advanced patients failing on indinavir. 37th Interscience Conference on Antimicrobial Agents and Chemotherapy. Abstract I-201. Toronto, Canada, September 28-October 1, 1997.
120. Rodriguez-Rosado R, Soriano V, Jimenez I, *et al.* Efficacy and safety of the combination of ritonavir plus saquinavir, with two nucleoside analogues. 12th World AIDS Conference. Abstract 12336. Geneva, Switzerland, June 28 - July 3, 1998.
121. Ruane PJ, Tam JT, Libraty DH, *et al.* Salvage therapy using ritonavir/saquinavir with a non-nucleoside reverse transcriptase inhibitor after prolonged failure with indinavir or ritonavir. 12th World AIDS Conference. Abstract 32308. Geneva, Switzerland, June 28 - July 3, 1998.
122. Tebas P, Kane E, Klebert M, *et al.* Virologic responses to a ritonavir/saquinavir-containing regimen in patients who have previously failed nelfinavir. 5th Conference on Retroviruses and Opportunistic Infections. Abstract 510. Chicago, IL, February 1–5, 1998.
123. Gallant J, Heath-Chiozzi M, Anderson R, *et al.* Phase II study of ritonavir-nelfinavir combination therapy: an update. 12th World AIDS Conference. Abstract 12207. Geneva, Switzerland, June 28 - July 3, 1998.
124. Havlir D, Riddler S, Squires K, *et al.* Co-administration of indinavir (IDV) and nelfinavir (NFV) in a twice daily regimen: preliminary safety, pharmacokinetic and anti-viral activity results. 5th Conference on Retroviruses and Opportunistic Infections. Abstract 393. Chicago, IL, February 1–5, 1998.

125. Moyle G. Study of protease inhibitor combination in Europe (SPICE): saquinavir soft gelatin capsule (SQV-SGC) plus nelfinavir (NFV) in HIV-infected individuals. 12th World AIDS Conference. Poster 12222. Geneva, Switzerland, June 28 - July 3, 1998.
126. Eron J, Haubrich R, Richman D, *et al.* Preliminary assessment of 141W94 in combination with other protease inhibitors. 5th Conference on Retroviruses and Opportunistic Infections. Abstract 6. Chicago, IL, February 1–5, 1998.
127. Saah A, Riddler S, Havlir DV, *et al.* Co-administration of indinavir and nelfinavir: pharmacokinetics, tolerability, anti-viral activity, and preliminary viral resistance. 12th World AIDS Conference. Abstract 22352. Geneva, Switzerland, June 28 - July 3, 1998.
128. Merry C, Barry MG, Mulcahy F, *et al.* Saquinavir pharmacokinetics alone and in combination with nelfinavir in HIV-infected patients. AIDS 1997; 11:F117–20.
129. Dusek A, Hall D, Lamson M, *et al.* Once-daily dosing of nevirapine: a retrospective, cross-study analysis. 12th World AIDS Conference. Poster 12360. Geneva, Switzerland, June 28 - July 3, 1998.
130. Reynes J, Denisi R, Bicart-see A, *et al.* STADI: sustained efficacy of once daily administration of didanosine in combination with stavudine. 12th World AIDS Conference. Abstract 22352. Geneva, Switzerland, June 28 - July 3, 1998.
131. Japour A, Murphy R, Hicks C, *et al.* Safety and efficacy of ABT-378/ritonavir in antiretroviral naïve patients: preliminary phase II results. 12th World AIDS Conference. Abstract LB4/12460. Geneva, Switzerland, June 28 - July 3, 1998.
132. Wang Y, Tutton CM, Borin MT, *et al.* The safety, tolerance, pharmacokinetics, and efficacy of PNU-140690, a new non-peptidic HIV protease inhibitor, in a phase I/II study. 12th World AIDS Conference. Abstract 41176. Geneva, Switzerland, June 28 - July 3, 1998.
133. Kilby JM, Hopkins S, Venetta TM, *et al.* Potent suppression of HIV-1 replication in humans by T-20, a peptide inhibitor of gp41-medicated virus entry. Nature 1998; 4:1302.
134. Daluge SM, Good SS, Faletto MB, *et al.* 1592U89, a novel carbocyclic nucleoside analog with potent, selective anti-human immunodeficiency virus activity. Antimicrob Agents Chemother 1997; 41:1082–93.
135. Sonnerborg A, Lancaster D, Torres R, *et al.* The safety and antiviral effect of abacavir, alone and in combination with zidovudine in HIV-1 infected patients with CD4+ counts 200–500 cells/mm^3. AIDS 1996, volume 10, supplement 2, abstract OP4.1
136. Torres R, Saag M, Lancaster D, *et al.* Antiviral effects of abacavir (1592) following 36 weeks of therapy. 5th Conference on Retroviruses and Opportunistic Infections. Abstract 659. Chicago, IL, February 1–5, 1998.
137. Fischl M, Greenberg S, Clumeck N, *et al.* Safety and activity of abacavir (1592, ABC) with 3TC/ZDV in antiretroviral naïve subjects. 12th World AIDS Conference. Abstract 12230. Geneva, Switzerland, June 28 - July 3, 1998.
138. Tisdale M, Myers RE, Snowden W. Genotypic and phenotypic analysis of HIV from patients on ZDV/3TC/amprenavir combination therapy. 12th World AIDS Conference. Abstract 32312. Geneva, Switzerland, June 28 - July 3, 1998.
139. Mellors J, Lederman M, Haas D, *et al.* Antiretroviral effects of therapy combining abacavir with HIV protease inhibitors. 5th Conference on Retroviruses and Opportunistic Infections. Abstract 4. Chicago, IL, February 1–5, 1998.
140. Staszewski S, Morales-Ramirez J, Flanigan T, *et al.* A phase II, multicenter, randomized, open-label study to compare the antiretroviral activity and tolerability of efavirenz (EFV) + indinavir (IDV), versus EFV + zidovudine (ZDV) + lamivudine (3TC), versus IDV + ZDV + 3TC at 24 weeks [DMP266–006]. 12th World AIDS Conference. Abstract 22336, Geneva, Switzerland, June 28-July 3, 1998.
141. Hetherington S, Steel HM, Lafon S, *et al.* Safety and tolerance of abacavir (1592, ABC) alone and in combination therapy of HIV infection. 12th World AIDS Conference. Abstract 12353. Geneva, Switzerland, June 28 - July 3, 1998.
142. Balzarini J, Zhang H, Herdewijn P, *et al.* Intracellular metabolism and mechanism of antiretrovirus action of 9-(2-phosphonylmethoxy)adenine, a potent anti-human immunodeficiency virus compound. Proc Natl Acad Sci USA 1991; 88:1499–1503.
143. Perno CF, Balestra E, Aquaro S, *et al.* Potent inhibition of human immunodeficiency virus and herpes simplex virus type 1 by 9- (2-phosphonylmethoxyethyl) adenine in primary macrophages is determined by drug metabolism, nucleotide pools, and cytokines. Mol Pharmacol 1996; 50:359–66.
144. Gong YF, Marshall DR, Srinivas RV, *et al.* Susceptibilities of zidovudine-resistant variants of human immunodeficiency virus type 1 to inhibition by acyclic nucleoside phosphonates. Antimicrob Agents Chemother 1994; 38:1683–7.
145. Arends S, van Halteren E, Kamp W, *et al.* Safety of 9- (2-phosphonylmethoxyethyl) adenine (PMEA) in patients with human immunodeficiency virus infection: a pilot study. Pharm World Sci. 1996; 18:30–4.

146. Cundy KC, Barditch-Crovo P, Walter RE, *et al.* Clinical pharmacokinetics of adefovir in human immunodeficiency virus type 1-infected patients. Antimicrobial Agents and Chemotherapy. 1995; 39:2401–5.
147. Barditch-Crovo PA, Toole J, Hendrix CW, *et al.* Anti-human immunodeficiency virus (HIV) activity, safety, and pharmacokinetics of adefovir dipivoxil (9-[2-(bis-pivaloyloxymethyl)-phosphonylmethoxyethyl]adenine) in HIV-infected patients. J Infect Dis. 1997; 176:406–13.
148. Holme E, Greter J, Jacobson CE, *et al.* Carnitine deficiency induced by pivampicillin and pivmecillinam therapy. Lancet. 1989; ii:469–73.
149. Deeks SG, Collier A, Lalezari J, *et al.* The safety and efficacy of adefovir dipivoxil, a novel anti-HIV therapy, in HIV infected adults: a randomized, double-blind, placebo controlled trial. J Infect Dis. 1997; 176:1517–23.
150. Mulato AS, Lamy PD, Miller MD, *et al.* Genotypic and phenotypic characterization of human immunodeficiency virus type 1 variants isolated from AIDS patients after prolonged adefovir dipivoxil therapy. Antimicrob Agents Chemother. 1998; 42:1620–28.
151. Kahn J, Lagakos S, Weng D *et al.* A multi-center, randomized, double-blind placebo controlled study of the efficacy and safety of adefovir dipivoxil (ADV) when added to standard antiretroviral therapy (ART). 38th Interscience Conference on Antimicrobial Agents and Chemotherapy Annual Meeting, San Diego, CA, September 24–27, 1998, Abstract I-108.
152. Yount SD, Britcher SF, Tran LO, *et al.* L-743,726 (DMP-266): a novel, highly potent nonnucleoside inhibitor of the human immunodeficiency virus type 1 reverse transcriptase. Antimicrob Agents Chemother 1995; 39:2602–5.
153. Winslow DL, Garber S, Reid C, *et al.* Selective conditions affect the evolution of specific mutations in the reverse transcriptase gene associated with resistance to DMP-266. AIDS 1996; 10:1205–9.
154. Riddler S, Kahn J, Hicks C, *et al.* Durable clinical anti-HIV-1 activity (72 weeks) and tolerability for efavirenz (DMP266) in combination with indinavir (IDV) [DMP266–003, Cohort IV]. 12th World AIDS Conference. Abstract 12359. Geneva, Switzerland, June 28-July 3, 1998.
155. Fessel WJ, Haas DW, Delapenha RA, *et al.* A phase III, double-blind, placebo-controlled, multicenter study to determine the effectiveness and tolerability of the combination of efavirenz (EFV, SUSTIVA™, DMP 266) and indinavir (IDV) versus indinavir in HIV-1 infected patients receiving nucleoside analogue (NRTI) therapy at 24 weeks [DMP266–020]. 12th World AIDS Conference. Abstract 22343 Geneva, Switzerland, June 28-July 3, 1998.
156. Mayers D, Jemsek J, Eyster E, *et al.* A double-blind, placebo-controlled study to assess the safety, tolerability, and antiretroviral activity of efavirenz (EFV, SUSTIVA ™, DMP in combination with open-label zidovudine (ZDV) and lamivudine (3TC) in HIV-1 infected patients [DMP 266–004]. 12th World AIDS Conference. Abstract 22340.Geneva, Switzerland, June 28-July 3, 1998.
157. Mildvan D, Martin G, Eyster M, *et al.* Initial effectiveness and tolerability of nelfinavir (NFV) in combination with efavirenz (EFV, SUSTIVA™, DMP 266) in antiretroviral therapy naïve or nucleoside analogue experienced HIV-1 infected patients: Characterization in a phase II, open-label, multicenter study at 16 weeks [DMP 266–024]. 12th World AIDS Conference. Abstract 22386 Geneva, Switzerland, June 28-July 3, 1998.
158. St. Clair MH, Millard J, Rooney J, *et al.* *In vitro* antiviral activity of 141W94 (VX-478) in combination with other antiretrovirals. Antiviral Res. 1996; 29:53–6.
159. Livingston DJ, Pazhanisamy S, Porter DJ, *et al.* Weak binding of VS-478 to human plasma proteins and implications for anti-human immunodeficiency virus therapy. J Infect Dis 1995; 172:1238–45.
160. Partaledis JA, Yamaguchi K, Tisdale M, *et al.* *In vitro* selection and characterization of human immunodeficiency virus type 1 (HIV-1) isolates with reduced sensitivity to hydroxyethylamino sulfonamide inhibitors of HIV-1 aspartyl protease. J Virol. 1995; 69:5228–35.
161. Piscitelli S, Vogel S, Sadler B, *et al.* Effect of efavirenz (DMP 266) on the pharmacokinetics of 141W94 in HIV-infected patients. 5th Conference on Retroviruses and Opportunistic Infections. Abstract 346. Chicago, IL, February 1–5, 1998.
162. Haubrich R. Phase 2 study of amprenavir, a novel protease inhibitor, in combination with zidovudine/3TC. 12th World AIDS Conference. Abstract 12321. Geneva, Switzerland, June 28 - July 3, 1998.
163. Murphy R, DeGruttola V, Gulick R, *et al.* 141W94 with or without zidovudine/3TC in patients with no prior protease inhibitor or 3TC therapy—ACTG 347. 5th Conference on Retroviruses and Opportunistic Infections. Abstract 512. Chicago, IL, February 1–5, 1998.
164. Mellors J, Lederman M, Haas D, *et al.* Antiretroviral effects of therapy combining abacavir with HIV protease inhibitors. 5th Conference on Retroviruses and Opportunistic Infections. Abstract 4. Chicago, IL, February 1–5, 1998.

165. Bart PA, Rizzardi GP, Gallant S, *et al.* Combination abacavir (1592, ABC)/amprenavir (141W94) therapy in HIV-1 infected antiretroviral naïve subjects with CD4 counts >400 cells/mL and viral load >5000 copies/mL. 12th World AIDS Conference. Abstract 12204. Geneva, Switzerland, June 28 - July 3, 1998.
166. Isaacs R, Havlir D, Pottage J, *et al.* Sixteen week follow-up of indinavir sulfate (IDV) administered q 8 hours (q8H) versus q12H in combination with efavirenz (EFV). 12th World AIDS Conference. Abstract 12290. Geneva, Switzerland, June 28 - July 3, 1998.
167. Skowron G, Leoung G, Dusek A, *et al.* Stavudine (d4T), nelfinavir (NFV), and nevirapine (NVP): preliminary safety, activity and pharmacokinetic (PK) interactions. 5th Conference on Retroviruses and Opportunistic Infections. Abstract 350. Chicago, IL, February 1–5, 1998.
168. Sahai J, Stewart F, Swick L, *et al.* Rifabutin reduces saquinavir plasma levels in HIV-infected patients. 36th Interscience Conference on Antimicrobial Agents and Chemotherapy. Abstract A27. New Orleans, LA, September 15–18, 1996.
169. Acosta EP, Henry K, Weller D, *et al.* Indinavir pharmacokinetics and relationships between exposure and antiviral effect. 37th Interscience Conference on Antimicrobial Agents and Chemotherapy. Abstract A-15. Toronto, Canada, September 28-October 1, 1997.
170. Hoetelmans RMW, Heeswijk RPG, Meenhorst PL, *et al.* Plasma concentrations of saquinavir (SQV) determine HIV-1 RNA response over a 48-week period. 12th World AIDS Conference. Abstract 511/42261. Geneva, Switzerland, June 28 - July 3, 1998.
171. Hoetelmans RMW, Reijers MHE, Weberling GF, *et al.* The rate of decline of HIV-1 RNA in plasma correlates with nelfinavir concentrations in plasma. 12th World AIDS Conference. Abstract 510/42259. Geneva, Switzerland, June 28 - July 3, 1998.
172. Chodakewitz J, Deutsch P, Leavitt R, *et al.* Relationship between indinavir (IDV) pharmacokinetics and antiviral activity in plase I/II trials. 12th World AIDS Conference. Abstract 42266. Geneva, Switzerland, June 28 - July 3, 1998.
173. Burger DA, Hoetelmans RMW, Mulder JW, *et al.* Low plasma levels of indinavir (IDV) are highly predictive of virological treatment failure in patients using IDV-containing triple therapy. 12th World AIDS Conference. Abstract 42275. Geneva, Switzerland, June 28 - July 3, 1998.
174. Kerr B, Pithavala Y, Zhang M, *et al.* Virologic response-plasma drug concentration relationship in Phase III study of nelfinavir mesylate (Viracept®). 12th World AIDS Conference. Abstract 12304. Geneva, Switzerland, June 28 - July 3, 1998.
175. Vanhove GF, Kastrissios H, Gries J-M, *et al.* Pharmacokinetics of saquinavir, zidovudine, and zalcitabine in combination therapy. Antimicrob Agents and Chemo 1997;41:2428–32.
176. Ickovics JR, Meisler AW. Adherence in AIDS clinical trials: a framework for clinical research and clinical care. J Clin Epid. 1997; 5:385–91.
177. Broers B, Morabia A, Hirschel B. A cohort study drug users' compliance with zidovudine treatment. Arch Inter Med. 1994; 154:1121–7.
178. Eraker SA, Kirscht JP, Becker MH. Understanding and improving patient compliance. Ann Intern Med. 1984; 100:258–68.
179. Eisen SA, Miller DK, Woodward RS, *et al* The effect of prescribed daily dose frequency on patient medication compliance. Arch Intern Med. 1990; 150:1881–4.
180. Besch CL. Compliance in clinical trials. AIDS 1995; 9:1–10.
181. Morse EV, Simon PM, Walter J. Issues of recruitment, retention, and compliance in community-based clinical trials with traditionally underserved populations. App Nurs Res. 1995; 8:8–14.
182. Klaus BD, Grodesky MJ. Assessing and enhancing compliance with antiretroviral therapy. Nurse Pract. 1997; 22:211–2, 215, 219.

INDEX

Culture, Ecology and Economy of Fire Management in North Australian Savannas

REKINDLING THE *WURRK* TRADITION

Jeremy Russell-Smith • Peter Whitehead • Peter Cooke

CSIRO
PUBLISHING

National Library of Australia Cataloguing-in-Publication entry

Culture, ecology and economy of savanna fire management in Northern Australia : in the tradition of Wurrk / editors: Jeremy Russell-Smith, Peter Whitehead, Peter Cooke.

9780643094024 (pbk.)

Includes index.
Bibliography.

Savanna ecology – Australia, Northern.
Prescribed burning – Australia, Northern.
Burning of land – Australia, Northern.
Wilderness areas – Fire management – Australia, Northern.

Russell-Smith, Jeremy.
Whitehead, Peter J.
Cooke, Peter.

333.74099429

Published by
CSIRO PUBLISHING
150 Oxford Street (PO Box 1139)
Collingwood VIC 3066
Australia

Telephone:	+61 3 9662 7666
Local call:	1300 788 000 (Australia only)
Fax:	+61 3 9662 7555
Email:	publishing.sales@csiro.au
Web site:	www.publish.csiro.au

Front cover (clockwise from top left): planning meeting of traditional owners and others (photo: Peter M Cooke); on-ground patch burning (photo: Peter M Cooke); line of strategically placed fires ignited from aircraft (photo: Peter M Cooke); MODIS satellite image showing active areas of fire and scars of previous fires (http://modis.gsfc.nasa.gov/); early dry season preventative burn lit in sparse Spinifex on the Arnhem Land Plateau (photo: Peter M Cooke)

Back cover: aerial prescribed burning ignitions along the mid reaches of the Mann River (photo: Peter M Cooke)

Set in Adobe Minion 10/12 and Vectora
Edited by Peter Storer Editorial Services
Cover and text design by James Kelly
Typeset by Desktop Concepts Pty Ltd, Melbourne
Index by Russell Brooks
Printed in China by Bookbuilders

CSIRO PUBLISHING publishes and distributes scientific, technical and health science books, magazines and journals from Australia to a worldwide audience and conducts these activities autonomously from the research activities of the Commonwealth Scientific and Industrial Research Organisation (CSIRO).

CONTENTS

ACKNOWLEDGEMENTS

The work described in this volume reflects an ongoing partnership comprising a large number of individuals and organisations. The Tropical Savannas Cooperative Research Centre is thanked foremost for providing critical resources, including for the publication of this book, but even more importantly, for the cultural and intellectual space that enabled and supported us collectively to embark on the 10-plus year journey described herein. Other critical funding and support has been provided through: Australian Government sources including the Community Development Employment Program, Natural Heritage Trust (now Caring for Our Country), Australian Greenhouse Office (now the Department of Climate Change), Land and Water Australia, and the Natural Disaster Mitigation Program; Indigenous (Aboriginal) communities and their representative organisations, including Adjumarlarl Rangers (Kunbalanya), Djelk Rangers and the Bawinanga Aboriginal Corporation (Maningrida), Jawoyn Association (Katherine), Manwurrk Rangers and the Wardekken Land Management Limited (Kabulwarnamyo), Mimal Rangers (Bulman), the Caring for Country Unit at the Northern Land Council, and the North Australia Indigenous Land and Sea Management Alliance (NAILSMA); the Northern Territory Government's Department of Natural Resources, Environment, the Arts and Sport, and especially its rural fire management agency Bushfires NT; two CSIRO Divisions – Sustainable Ecosystems, and Atmospheric Research (now Marine and Atmospheric Research); the Remote Sensing Section of the Western Australia Department of Land Information (Landgate); universities – Charles Darwin University, Australian National University, University of Queensland; and ConocoPhillips Company. Senator Robert Hill, as Federal Minister for Environment, gave early support to the initiatives on which this program built.

Expert reviews of chapters or sections of them were provided by Jon Altman, Sally Brockwell, Garry Cook, Nicholas Evans, Rolf Gerritsen, Lindsay Hutley, Lynda Prior, David Ritchie, Steve Sutton and Dick Williams, as well as the editors. Ian Thynne provided advice and support in regard to governance issues. The *International Journal of Wildland Fire* and its publisher, CSIRO Publishing, are thanked for permission to reproduce in full two papers (Chapters 9 and 13) independently reviewed and recently published in that journal.

Cameron Yates and Felicity Watts from Bushfires NT generated the maps. Photographs were mostly provided by the authors, but we are particularly grateful for permission to use historical images held by the Australian Institute of Aboriginal and Torres Strait Islander Studies (Chapter 3). Thanks are due to all individuals who gave permission for their (contemporary) images to be reproduced in this book. Otto Campion authorised use of an image including his deceased father.

Finally, we acknowledge the critical influence of key individuals from both Indigenous and western scientific knowledge traditions who have actively helped inform and promote cross-cultural perspectives concerning *wurrk* and landscape fire management, respectively. Based on our own experience, for western Arnhem Land those ideas have been substantially influenced by Nipper Kapirigi, Billy Gunbunuka, Rhys Jones, Mick Kubarkku, and George Manyita (all deceased); Lofty Bardayal Nadjamerrek AO, Chris Haynes, Dean Yibarbuk, Jack Djandjomerr, Mary Kolkkiwarra, Bill Birriyabirriya, Ruby Bilindja, Jimmy Kalarriya, Peter Biless, Otto Bulmaniya Campion, Matthew Ryan, and George Djanggawanga. We trust the contributions in this volume add usefully to that continuing discourse.

Note: Some images include Indigenous people who are now deceased. Please respect family and regional protocols in relation to displaying images of deceased people when using and sharing this book.

The editors

LIST OF CONTRIBUTORS

Jon C Altman
Centre for Aboriginal Economic Policy Research
The Australian National University, Canberra ACT 0200, Australia

Alan N Andersen
Tropical Savannas Management Cooperative Research Centre, and
CSIRO Sustainable Ecosystems, PMB 44, Winnellie, Northern Territory 0822, Australia

Damian Barrett
Sustainable Minerals Institute
University of Queensland, Brisbane, Queensland 4072, Australia

Peter Biless
Ngarridj subsection, Djordi clan
c/- Warddeken Land Management Limited
PO Box 785, Nightcliff, Northern Territory 0814, Australia

Ruby Bilindja
Ngarridjdjan subsection, Wurrbbarn clan
c/- Warddeken Land Management Limited
PO Box 785, Nightcliff, Northern Territory 0814, Australia

Bill Birriyabirriya
Bulanj subsection, Kardbam clan
Marrkolidjban Outstation
c/- Bawinanga Aboriginal Corporation
PMB 102, Winnellie, Northern Territory 0821, Australia

Kym Brennan
Tropical Savannas Management Cooperative Research Centre, and
Department of Natural Resources, Environment, the Arts and Sport, PO Box 496, Palmerston, Northern Territory 0831, Australia

Peter Brocklehurst
Tropical Savannas Management Cooperative Research Centre, and
Department of Natural Resources, Environment, the Arts and Sport, PO Box 496, Palmerston, Northern Territory 0831, Australia

Garry D Cook
Tropical Savannas Management Cooperative Research Centre, and
CSIRO Sustainable Ecosystems, PMB 44, Winnellie, Northern Territory 0822, Australia

Peter M Cooke
Tropical Savannas Management Cooperative Research Centre, and
Warddeken Land Management Limited
PO Box 785, Nightcliff, Northern Territory 0814, Australia

Jack Djandjomerr
Kamarrang subsection, Bolmo clan
Marlkawo Outstation
c/- Warddeken Land Management Limited
PO Box 785, Nightcliff, Northern Territory 0814, Australia

Andrew C Edwards
Tropical Savannas Management Cooperative Research Centre, and
Bushfires Cooperative Research Centre, and
Bushfires NT, Department of Natural Resources, Environment, the Arts and Sport, PO Box 37346, Winnellie, Northern Territory 0821, Australia

Murray Garde
School of Languages and Linguistics
The University of Melbourne, Melbourne, Victoria 3010, Australia

A Malcolm Gill
Fenner School of Environment and Society
Australian National University, Canberra, Australian Capital Territory 0200, Australia

Lindsay Hutley
Tropical Savannas Management Cooperative Research Centre, and
School of Environmental and Life Sciences, Charles Darwin University, Darwin, Northern Territory 0909, Australia

Jimmy Kalarriya
Kodjok subsection, Wurrbbarn clan
c/- Warddeken Land Management Limited
PO Box 785, Nightcliff, Northern Territory 0814, Australia

Sarah Kerin
Parks Australia North
PO Box 71, Jabiru, Northern Territory 0886, Australia

Mary Kolkkiwarra
Kodjdjan subsection, Wakmarranj clan
Kabulawarnamyo Outstation
c/- Warddeken Land Management Limited
PO Box 785, Nightcliff, Northern Territory 0814, Australia

Mick Kubarkku
Balang subsection, Kulmarru clan
formerly of Yikarrakkal Outstation
Western Arnhem Land, Northern Territory, Australia

Robert Levitus
Anthropos Pty Ltd
Canberra ACT 2600, Australia

Adam Liedloff
Tropical Savannas Management Cooperative Research Centre, and
CSIRO Sustainable Ecosystems, PMB 44, Winnellie, Northern Territory 0822, Australia

Stefan Maier
Satellite Remote Sensing Services, Western Australian Land Information Authority, and
Charles Darwin University, Darwin, Northern Territory 0909, Australia

John McCartney
Tropical Savannas Management Cooperative Research Centre and
Parks Division, Department and of Natural Resources, Environment, the Arts and Sport, PO Box 496, Palmerston, Northern Territory 0831, Australia

Mick CP Meyer
Bushfires Cooperative Research Centre, and
CSIRO Marine and Atmospheric Research, PMB 1, Aspendale, Victoria 3195, Australia

Brett P Murphy
Tropical Savannas Management Cooperative Research Centre, University of Tasmania, and Bushfires NT, Department of Natural Resources, Environment, the Arts and Sport, PO Box 37346, Winnellie, Northern Territory 0821, and School of Plant Science, Private Bag 55, Hobart, Tasmania 8001, Australia

Bronwyn A Myers
School of Environmental and Life Sciences
Charles Darwin University, Darwin, Northern Territory 0909, Australia

Bardayal Lofty Nadjamerrek AO
Wamud subsection, Mok clan
Kabulawarnamyo Outstation
c/- Warddeken Land Management Limited
PO Box 785, Nightcliff, Northern Territory 0814, Australia

Paul Purdon
Environment, Heritage and the Arts
Department of Natural Resources, Environment, the Arts and Sport
PO Box 496, Palmerston, Northern Territory 0831, Australia

David Ritchie
Department of Local Government and Housing
GPO Box 4621, Darwin, Northern Territory 0801, Australia

Jeremy Russell-Smith
Tropical Savannas Management Cooperative Research Centre, North Australia Indigenous Land and Sea Management Alliance, School of Environmental Research, Charles Darwin University, and Bushfires NT, Department of Natural Resources, Environment, the Arts and Sport, PO Box 37346, Winnellie, Northern Territory 0821, Australia

Jon Schatz
Tropical Savannas Management Cooperative Research Centre and
CSIRO Sustainable Ecosystems, PMB 44, Winnellie, Northern Territory 0822, Australia

Stephen Sutton
Bushfires NT, Department of Natural Resources, Environment, the Arts and Sport, PO Box 37346, Winnellie, Northern Territory 0821, Australia

Felicity A Watt
Tropical Savannas Management Cooperative Research Centre, and
Bushfires NT, Department of Natural Resources, Environment, the Arts and Sport, PO Box 37346, Winnellie, Northern Territory 0821, Australia

Peter J Whitehead
Tropical Savannas Management Cooperative Research Centre,
Charles Darwin University, Darwin, Northern Territory 0909 Australia

Richard J Williams
Tropical Savannas Management Cooperative Research Centre, and
CSIRO Sustainable Ecosystems, PMB 44, Winnellie, Northern Territory 0822, Australia

Steve Winderlich
Parks Australia North
PO Box 71, Jabiru, Northern Territory 0886, Australia

John CZ Woinarski
Tropical Savannas Management Cooperative Research Centre, and
Department of Natural Resources, Environment, the Arts and Sport, and School of Environmental Research, Charles Darwin University, PO Box 496, Palmerston, Northern Territory 0831, Australia

Cameron P Yates
Tropical Savannas Management Cooperative Research Centre, and
Bushfires NT, Department of Natural Resources, Environment, the Arts and Sport, PO Box 37346, Winnellie, Northern Territory 0821, Australia

1

Challenges and opportunities for fire management in fire-prone northern Australia

Jeremy Russell-Smith, Peter J Whitehead, Peter M Cooke and Cameron P Yates

INTRODUCTION

In the southern summer of late 2002 through to early 2003, large bushfires ravaged somewhere between 20 000–30 000 km^2 of mostly forested terrain in south-eastern Australia. The south-east corner of Australia is the most densely populated portion of a mostly arid, flammable and very sparsely settled continent. Terrifying images of huge fires sweeping onto the smoke-enshrouded outlying suburbs of Sydney – and of the fast moving inferno that engulfed over 500 homes in the western suburbs of the nation's capital, Canberra – played out across the global media. At the time, much of southern Australia was in the grip of a prolonged and savage drought and, for an already nervous populace, these images presaged the enormity of an emerging climate-induced increase in numbers of days of extreme temperature and associated wildfire activity and diminishing regional water availability in the decades to come (CSIRO and Bureau of Meteorology 2007; Garnaut 2008). The even greater loss of life and property in the 2009 fires in rural Victoria will undoubtedly drive intense re-examination of present approaches to fire prevention and response.

An important political response to the 2002/3 southern Australian fire crisis was establishment of the Council of Australian Government's (COAG) *National Inquiry on Bushfire Mitigation and Management* (Ellis *et al*. 2004). This was the first ostensibly national bushfires inquiry in a long history of such inquiries following fire 'disasters' in various southern Australian settings – commencing with Justice Stretton's formative report into the Black Friday fires of 1939.

As part of its terms of reference, the COAG inquiry undertook an assessment of 'the facts in relation to major bushfires in the 2002–03 season'. The report observed (p. 6) that whereas 'community and media interest during the 2002–03 fire season focused on fires that affected about 3 million hectares in south-eastern Australia … in the same fire season, however, around 38 million hectares was affected by fire in northern and central Australia'. In fact, this figure represents only the Northern Territory, where, between April 2002 and March 2003, fires affected 28.6% of the jurisdiction. Although these figures include an exceptionally large area of fire in arid central Australia in 2002 following the most significant rainfall for decades in 2001, and the subsequent build-up of flammable grassy fuels, the inquiry also noted (p. 16) that 210 000 km^2 of the Northern Territory's tropical savannas region is affected by fire in an average burning year.

Three weeks before the report's release on 31 March 2004, two of the authors of this chapter (Peter Cooke and Jeremy Russell-Smith) – invited as representatives of northern Australian Indigenous and research interests, respectively – participated in a small 'invited specialists forum' to review the then close-to-complete draft document. The draft presented to us at the time reflected simply a deep lack of cultural understanding and appreciation for the vast landscape scales and resourcing issues confronting fire management in fire-prone regional and northern Australia. Among the various significant issues we documented are:

- a lack of understanding of the continental scale of, and the need to provide support for, the use of fire as a management tool for production, conservation and cultural management goals – as opposed to being concerned singularly about risk-management and emergency-response issues – especially in southern Australia
- a general lack of recognition that the largest areas of annual fires (and wildfires) occur predominantly in northern Australia – and periodically in the arid centre – and that current fire regimes are exerting significant impacts on regional biodiversity, environmental, greenhouse gas emissions, human health, social and community values. In fact, in the (subsequently amended) summary on page 1, the draft report found that, despite over a quarter of the Northern Territory being burnt by wildfires in 2002, there was 'little detrimental impact' in that year!
- a limited appreciation of the necessity for landscape-scale remote sensing applications for informing regional real-time fire management operations, and regional and national policy agendas
- a lack of comprehension of the national and global significance, challenges and opportunities surrounding greenhouse gas emissions from savanna burning
- a total lack of understanding of contemporary and emerging social demographic issues in regional and northern Australia: particularly opportunities and requirements for engaging Indigenous (Aboriginal) skills and knowledge in developing economically viable fire-management enterprises and partnerships.

While acknowledging that the final report did in fact begin to address several of our seemingly 'parochial' concerns as expressed to the meeting – and subsequently to the authors in frank correspondence – this recent national inquiry usefully illustrates a contemporary stereotypical national policy focus on bushfire management in Australia. We acknowledge that risk-management and emergency-response issues in densely concentrated southern Australian electorates are legitimate political prerogatives that need to be tackled. In response to this generic national policy myopia, however, we recognise that key challenges for the fire-prone north include: setting out a cogent and compelling case that savanna fire-management issues do warrant national attention; that we are in the process of developing novel, collaborative regional approaches for delivering environmentally, culturally and economically sustainable solutions; and that such solutions provide useful examples for fire management in various other regional, national and international savanna contexts. The obligation to communicate the elements of this case provides a first strong motivation for compiling this book.

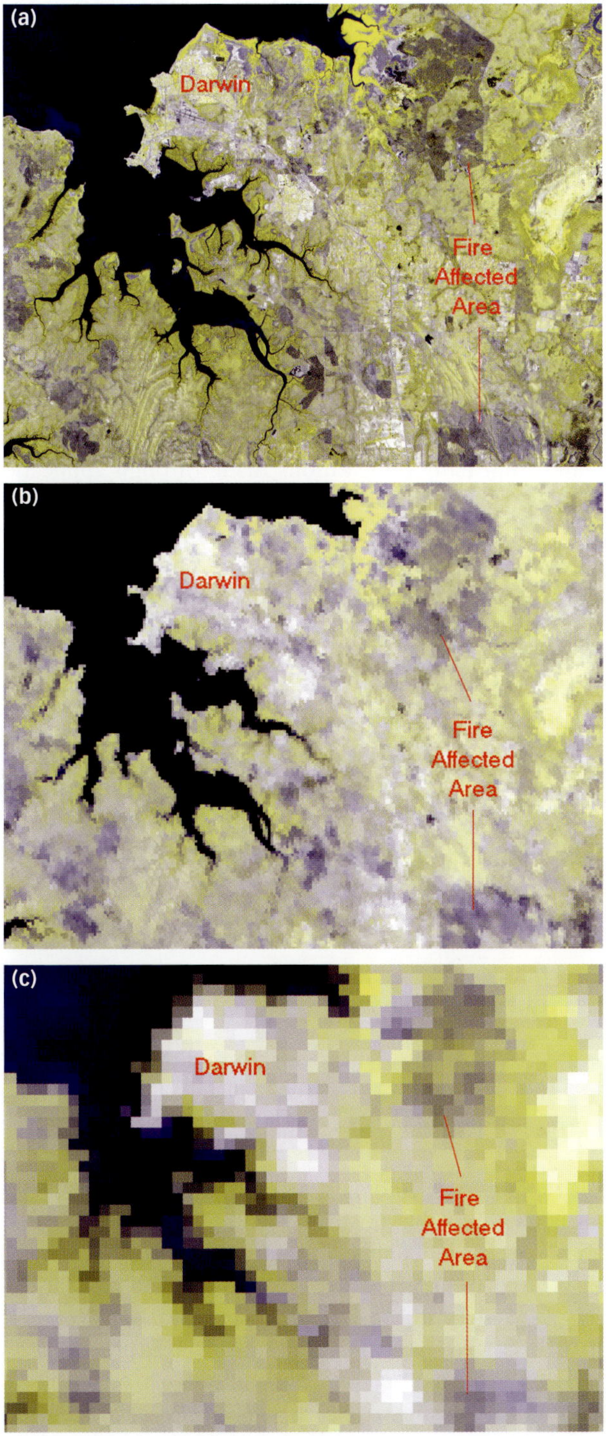

Figure 1.1 Images for the Darwin region, Northern Territory, 8 June 2008, illustrating fire-affected areas at different imagery scales: (a) Landsat, (b) MODIS, (c) AVHRR. All images display red and near-infrared bands.

FIRE MAPPING AND CONTEMPORARY BURNING PATTERNS

Our contemporary understanding of fire patterning in northern Australian landscapes has been informed largely through the application of satellite imagery at three scales of resolution. Dating back to the early 1980s, relatively fine-resolution LANDSAT imagery (pixel sizes ~0.1–0.5 ha, depending on the sensor; Figure 1.1a) has been used in a large number of regional studies – including assessments included in this book (Chapters 9 and 10) – for characterising savanna fire regimes (e.g. fire extent, seasonality, frequency, interval-between-fires and patchiness) and addressing attendant ecological, greenhouse and land-management issues. With individual scenes with a dimension of 180 × 180 km, combined with relatively high cost (at least to date) and relatively infrequent sampling (every 16 days), this imagery has been applied typically in spatially and temporally limited regional snapshots or, more rarely, in multi-scene, decadal studies of regional fire regimes such as for Kakadu, Litchfield and Nitmiluk National Parks (Chapter 10).

Since the late 1990s, continental-wide understanding of the occurrence of fires has been developed substantially from daily observations made through application of the relatively coarse resolution (~1.1 × 1.1 km pixels at orbital nadir) Advanced Very High Resolution Radiometer (AVHRR) instrument on the United States' National Oceanic and Atmospheric Administration (NOAA) series of satellites (Figure 1.1c). Currently, assembled fire observation data derived from AVHRR are available from 1997 for the whole of the continent, and from 1990 for Western Australia and the Northern Territory (Craig *et al.* 2002; Meyer 2004).

More recently, a second continental-scale sensor – the Moderate Resolution Imaging Spectroradiometer (MODIS; 250 × 250 m pixels), with purpose-built daily fire detection and mapping capabilities – has become available (Figure 1.1b). Both AVHRR and MODIS have a sample swath exceeding 2000 km in any one overpass, and both sensors enable automatic detection of active fire 'hotspots', and post-fire assessments and mapping of burned or, more appropriately, fire-affected areas. Hotspot data derived from both satellite sensors are routinely made available within hours of satellite overpasses on a number of Australian fire management websites:

- http://www.firewatch.dli.wa.gov.au
- http:// www.firenorth.org.au
- http://sentinel2.ga.gov.au/acres/sentinel.

Fire-mapping products are also available from the first two listed sites. The 'firenorth' – or North Australia Fire Information (NAFI) website developed by the Tropical Savannas Cooperative Research Centre – also provides fire-mapping products derived from MODIS imagery (Figure 1.2).

Collectively, these remotely sensed data sources and website portals have started to transform widely held misperceptions that bushfire events – and their social, environmental and economic implications – are a particularly southern Australian phenomenon. On the basis of continental mapping of large fire-affected areas (> ~2–4 km^2) for the period 1997–2004 derived from AVHRR imagery, Russell-Smith *et al.* (2007) observed that 76% of total mean fire affected area (508 000 km^2 p.a.) occurred in the northern savannas. Expressed as a proportion of continental land area defined by rainfall classes: a mean of 0.6% of southern Australia (53% of continental land area) was affected by fire each year; 5% of central Australia (25% of continent); and 23% of northern Australia (22% of continent). These general patterns are reflected also in updated (1997–2007) continental mapping of the frequency of large fires derived from AVHRR imagery (Figure 1.3).

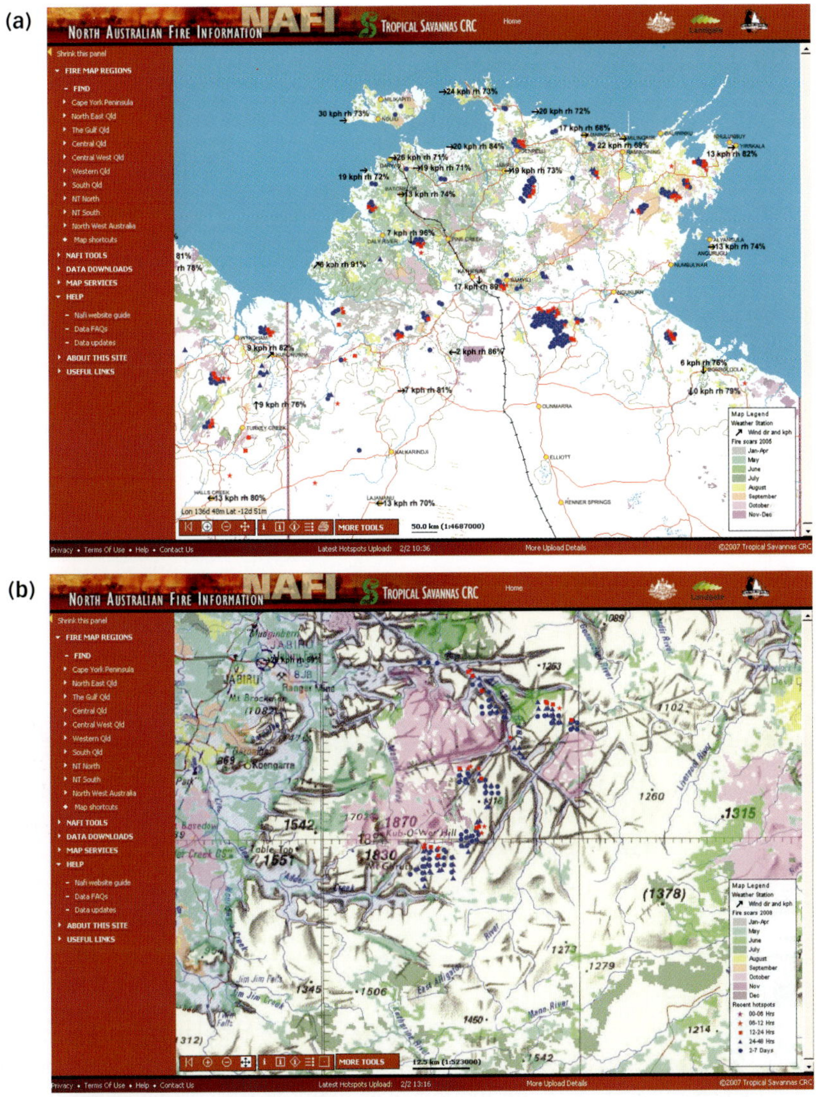

Figure 1.2 Examples of fire mapping from the North Australia Fire Information website (www.firenorth.org.au): (a) an overview of the Top End region, (b) a subset of Kakadu and western Arnhem Land. Pink and red dots indicate fires in the previous 6 and 24 hours, respectively, with blue dots indicating fire locations over the preceding week. Coloured areas represent fire-affected areas mapped by month (see legend in the bottom right-hand corner of each map).

At a continental scale, fire extent is substantially explained by rainfall seasonality; that is the ratio of the amount of rainfall received in the wettest to driest quarter (Russell-Smith *et al.* 2007). In the monsoonal north, intense bursts of high summer rainfall, separated by an extended annual dry season (southern winter and spring) 'drought' – during which next to no rain may fall for 6 months or more – drive an annual cycle of elevated fire risk, alternating with periods of little or no fire risk. In the arid centre, erratically variable periods of above average rainfall and subsequent high rates of plant growth may alternate with dry periods extending over several years to decades. Susceptibility of landscapes to fire therefore changes over longer

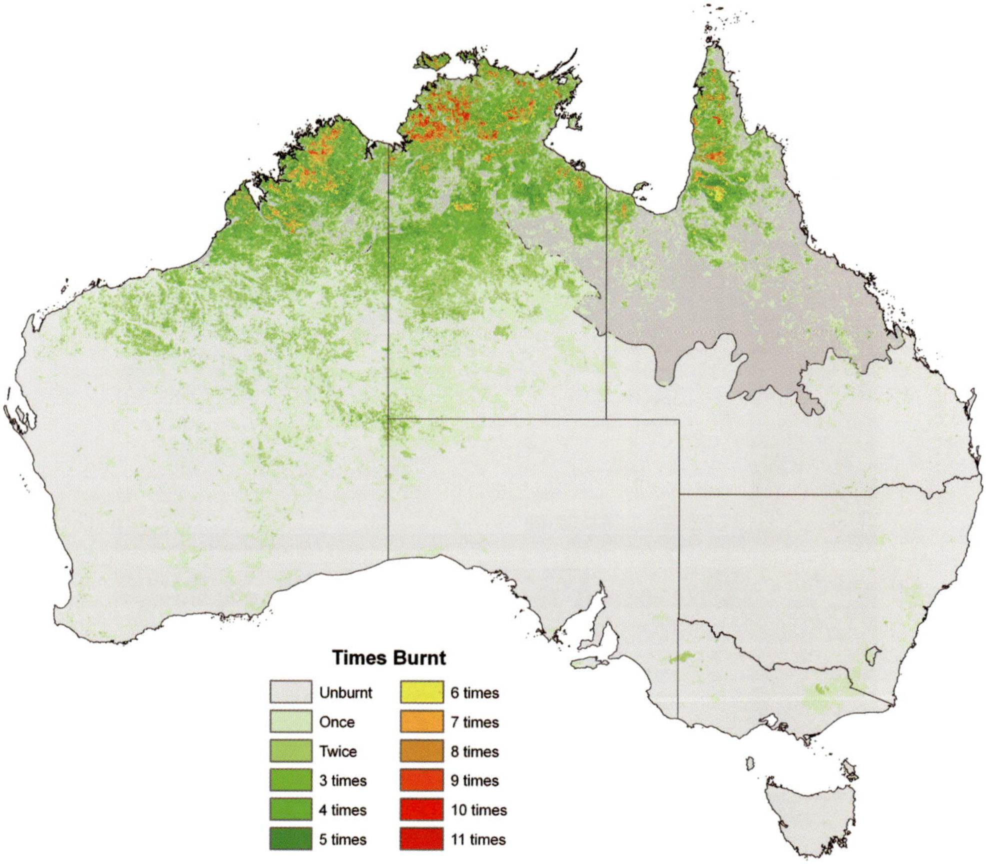

Figure 1.3 Frequency of large fire-affected areas (>~2–4 km^2) derived from AVHRR imagery, 1997–2007. North of the line indicates the tropical savannas region as defined by the Tropical Savannas Cooperative Research Centre.

than annual cycles, but very large wildfires occur with some regularity: fuelled mostly by grassy fuels such as Spinifex. In parts of mesic southern Australia (especially forested regions), vulnerabilities increase with fuel accumulation over many years, so that when fires occur, mostly in summer, they can be extraordinarily intense.

For the tropical savannas (Figure 1.3), mean annual rainfall declines rapidly from more than 2000 mm in parts of the far north to about 500 mm in southern regions (Figure 1.4). The annual mean extent of large fires follows this general trend (Figure 1.5). Significantly, modelling of fire extent with a range of biophysical variables (e.g. antecedent rainfall, or fire, in previous year(s); or land use) suggests that rainfall is sufficiently reliable in the north to support annually recurrent burning (Russell-Smith *et al.* 2007). With declining, less-reliable annual rainfall, fire propagation relies on cumulative antecedent rainfall for the development of adequate fuel loads and fuel continuity (Allan and Southgate 2002; Meyer 2004).

A critical salient feature of contemporary savanna fire regimes concerns the predominance of fires occurring in the latter part of the dry season: typically under severe fire weather conditions (periodically strong south-easterly winds, high temperatures, low humidities and fully

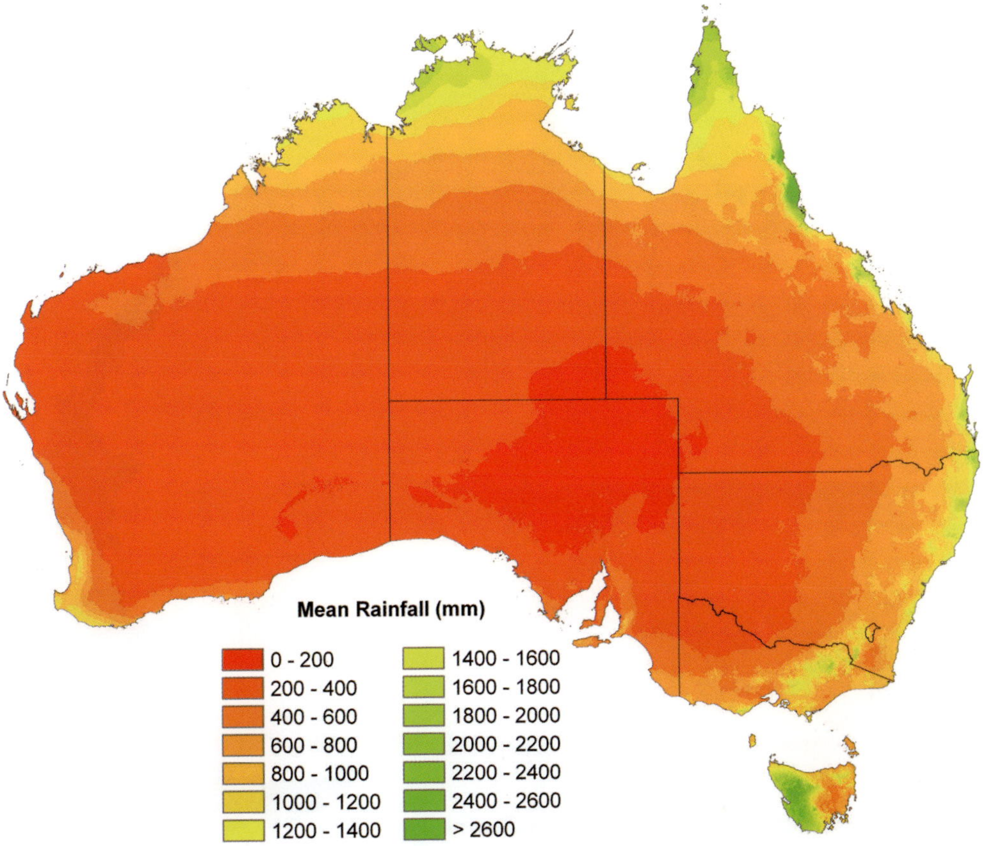

Figure 1.4 Mean annual rainfall distribution, 1976–2006 (Bureau of Meteorology).

cured fuels). For example, of the annual mean 351 000 km^2 of the tropical savannas region affected by large fires over the period 1997–2007 (Figure 1.3), 67% occurred in the late dry season months of August–November. Fire regimes dominated by frequent, large, late dry season fires are commonplace in many regions of northern Australia, especially the Kimberley region in the north-west, the Top End of the Northern Territory, and western Cape York Peninsula (Figure 1.3). Contemporary north Australian fire regimes are having significant impacts on regional biodiversity values, and have significant implications for greenhouse gas emission estimates and related carbon dynamics (Chapters 8, 9, 12 and 13).

Of particular relevance here, 'prescribed burning of savannas' is listed as an accountable activity under the provisions of the Kyoto Protocol, and although only recently ratified by Australia, Australia's National Greenhouse Gas Inventory (NGGI) reports annually on savanna burning emissions. Typically, accountable greenhouse gas emissions annually contribute between 1 and 3% of Australia's NGGI (AGO 2007). However, following international convention, Australia's NGGI considers only emissions of the accountable greenhouse gases methane (CH_4) and nitrous oxide (N_2O) – CO_2 itself is not accounted for because it is assumed (often incorrectly, see Cook *et al.* 2005; Chapter 12) that CO_2 emissions in one burning season are negated by vegetation growth in subsequent growing seasons (IPCC 1997). Nevertheless, CO_2, together with more reactive species (e.g. the ozone precursors, comprising CO, volatile organic

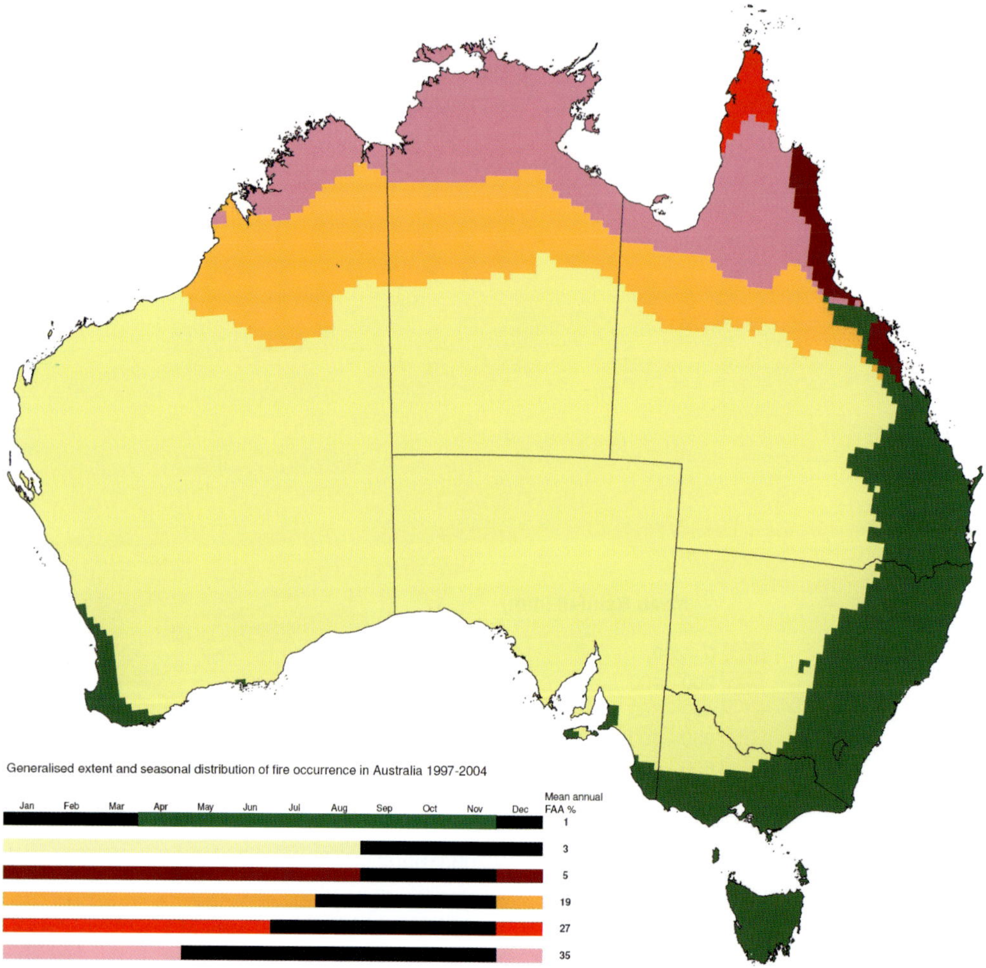

Figure 1.5 Contemporary Australian fire seasonality and extent, summarised in classes defined from Australian rainfall records 1969–2004, and fire mapping derived from AVHRR imagery 1997–2004; refer to Russell-Smith *et al.* (2007) for details.

compounds and oxides of nitrogen) that are released into the atmosphere over a typically long burning season are likely to have a substantial impact on regional atmospheric composition, and its inter-annual variability.

Achieving significant greenhouse gas emissions abatement, and more effective savanna landscape fire management generally, are thus components of the same integral problem – how do we practically and economically reduce the incidence and extent of contemporary late dry season wildfires?

THE ECOLOGICAL AND CULTURAL LANDSCAPES

Why has it proved so difficult to manage Australia's savanna landscapes to avert unwelcome change, despite the absence of intense population pressures?

Ecology

Savanna environments are relatively intact structurally and many areas are recognised for their important biodiversity values, including major centres where there are many endemic plants and animals that occur nowhere else (Woinarski *et al.* 2007a). However, there have been losses of ecological function – chiefly evidenced in biodiversity declines – at a number of levels and across large areas (Woinarski *et al.* 2001; Franklin *et al.* 2005). Scattered richer patches of the landscape are asked to maintain natural production, but at the same time support economic (agricultural and pastoral) production. More resilient and reliably productive environments (such as wetlands) are valued highly by many different interests or sectors (Jonauskus 1996). Despite increasing demand for water, our rivers are mostly unregulated. Many rivers carry high wet season flows, but most cease flowing during parts of the annual dry season. As a consequence, primary industries (agriculture and mining) rely heavily on groundwater extraction: even in the high rainfall areas.

Temperatures are uniformly high, and often very high. Rainfall is intensely seasonal (wet–dry tropics) with no equivalent to the more equable wet tropics of the east coast of Australia. In the seasonal tropics, the timing of the onset and cessation of wet conditions is highly unpredictable. Evaporation rates greatly exceed precipitation for most of the year. Hence the length of the growing season is highly unpredictable, but often short. Extreme weather events (storms and cyclones) are common and recur on a range of spatial and temporal scales. Landscapes are old, frequently reworked and highly weathered. Soils are often of low fertility or low water-holding capacity and, in many settings, highly erodible and so limiting for agriculture (Woinarksi and Dawson 2002).

Many areas of northern Australia contain exploitable concentrations of minerals and fossil energy, some of which are already in production and many others under development. Mineral extraction often involves relatively low-grade deposits requiring movement of large quantities of overlying or intervening rock, which build long-term problems of acid formation during the subsequent oxidation of waste and drainage from it (e.g. Harries 1997). On-site processing can add to pressures on water availability and also compromise water quality.

Society

It is more difficult to present a comprehensive statement of the social and economic character of the savannas. Most routinely produced demographic and socioeconomic data are presented to jurisdictional or regional boundaries that do not align with biophysical domains. The outline we present here is therefore based on a number of individual studies and one-off estimates made over the last few years.

The human population of the savannas is relatively small (495 000 in 2006), widely dispersed and 18.7% Indigenous. Average population density of 0.29 persons.km^{-2} is low by Australian (2.5 persons.km^{-2}) and global standards, and lower still away from major centres. Excluding the larger centres of Darwin and Townsville, there is 1 person for each 700 ha (0.14 persons.km^{-2}). By any standard, the tropical savannas of Australia are sparsely inhabited (Taylor *et al.* 2006) and the scope to marshal human resources for intensive land management correspondingly weak.

The proportion of the population that is Indigenous is much greater outside the major centres. In the Kimberley, nearly half of the population is Indigenous; in the Northern Territory savannas outside Darwin, it is more than 70%; and in very remote regions generally, more than 90%. Nationally, 45.4% of the population living in areas classified by the Australian

Bureau of Statistics as very remote is Indigenous (Taylor 2006). In the Northern Territory, most indigenous people (70%) live on lands held under Aboriginal communal title (Taylor 2003). As highlighted by Taylor (2006), this means that Indigenous people and their institutions predominate over most of the Australian land mass, and the northern savannas in particular. The closeness and durability of association between people and place – and connected obligations to land and dependence on it – are recognised over much of north Australia in use of terms such as 'on country and 'caring for country' to indicate much more than location.

The savanna population is growing, with an increase of 8% between the 1996 and 2001 Censuses. The rate of increase is higher among the Indigenous population, which grew by 14.9% compared with 6.5% in the non-Indigenous community. Non-Indigenous population growth is confined largely to the major centres. Projections to 2021 see a continuation of higher rates of growth in the Indigenous population (25.7%) over this period than the non-Indigenous population (14.8%) (Taylor *et al.* 2006).

At regional scales, rates of population growth are highly variable, both through time and by location: being strongly influenced by shifts in immigration and emigration of non-Indigenous people tracking employment and other economic opportunities. Sites of high local Indigenous population growth are often poorly matched to areas of likely job growth (Taylor 2003).

Morbidity and mortality rates are unacceptably high among the Indigenous population, and proving resistant to simple correction (Burgess *et al.* 2005; SCRGSP 2003). Educational systems have broken down in many remote areas, so that many Indigenous people suffer from poor literacy and numeracy and experience difficulty in taking advantage of any mainstream employment opportunities that may be available in the regions (Collins 1999).

Compounding this educational deficit, the savanna is subject to other knowledge deficits. The formal scientific understanding of regional environments and natural resources is weaker than in more densely settled jurisdictions because of historical and contemporary limitations on technical capacity and research investment. Our understanding of regional human demography is limited by dependence on Census data aggregated to inappropriate boundaries. Regional economies are weakly characterised and their dynamics poorly understood.

In contrast to these formal knowledge deficits, many Indigenous people have very detailed knowledge of the landscapes and resources for which they are responsible, but mechanisms for applying this and other local knowledge are poorly developed and often contested (see e.g. Brook and McLachlan 2005; Gilchrist *et al.* 2005).

Economy

Like its ecology, a key defining attribute of the mainstream savanna economy is its geographic patchiness and variability through time. Volatility is associated with low enterprise diversity, dependence on variable commodity prices and extractive industries, and relatively small size (e.g. NTG 2006). A recent (2006) estimate of gross incomes from mineral extraction in the savannas (excluding coal) is about $6 billion p.a. Pastoralism – with a total herd of nearly 5 million cattle – generates incomes of about $1.4 billion annually.

Tourism – which is highly dependent on access to natural environments and related activities such as recreational fishing – generates expenditure exceeding $2 billion p.a. and is an important source of employment, although in some locations costs of servicing tourists may be high compared with benefits captured locally (Greiner *et al.* 2004). Approximately 8% of the savanna population is engaged in tourism or hospitality industries, but employees are mostly non-Indigenous.

The formal savanna economy is very open, which is reflected in a very high ratio of imports and exports to gross product, with little local production of manufactured, or otherwise highly processed, items and limited value-adding to exports. Correlation between unemployment and employment rates is weak: reflecting gaps in local and regional labour markets and consequent importation of skilled and semi-skilled labour during economic upswings. Many savanna residents enjoy well-above-average incomes, but a mostly unemployed and growing regional Indigenous population is on chronically low to very low incomes (e.g. Taylor 2003).

It follows that, at the regional scale, multipliers (measures of the extent to which money circulates and stimulates additional activity in the economy) from activities such as pastoralism and mining are low, because local communities are often poorly placed to provide the goods or services sought by these industries, including labour. Multipliers from public sector investments in services such as health are substantially greater (Stoeckl *et al.* 2007).

Defence facilities and their personnel make a large contribution to economic activity. There is a strong general dependence on public expenditure, which is well above national per capita averages. A low capacity to raise revenue locally, combined with the costs imposed by remoteness, mean that maintaining services, especially in the regions, depends on national fiscal equalisation policies implemented through the Commonwealth Grants Commission (Morris 2003). Total subventions are based on the size of the population requiring services. It follows that funds needed to manage issues influenced more strongly by land area and biophysical pressures are often inadequate (e.g. Altman *et al.* 2007).

Operating mostly separately from the mainstream economy is an Indigenous customary economy based on direct non-commercial exploitation of natural resources (Altman 1987). That economy is stronger in the remoter areas, which there intersects with the mainstream economy chiefly through production of arts and crafts, which often draws on continued associations with country and the use of natural products (Altman 2003). In many parts of the savannas, Indigenous society has been substantially supported by welfare payments from the Australian Government. Some critical parts of that support are being withdrawn or restructured, so that need for productive engagement with the mainstream economy is increased.

Land use

The great majority of the northern savanna is used ostensibly for pastoral production, especially the grazing of cattle (*Bos taurus* and *Bos indicus*), and also sheep in parts of western Queensland (Figure 1.6). Most of this land is leasehold; that is, leased from state and Northern Territory governments. Despite the very small area used for mining purposes, such land use constitutes by far the greatest economic return to the regional economy (see above).

The predominance of Indigenous people living outside the major towns and associated rapid population growth is reflected increasingly in changing patterns of legal ownership of, and interests in, land. Recent available data indicate that around 19% of the tropical savannas region is presently owned or managed by Indigenous people (Figure 1.7a): ranging from 35% of savannas in the Northern Territory, to 6% in Queensland. Additionally, Indigenous interests in land – as expressed through determinations of and applications made for Native Title under the Commonwealth of Australia's *Native Title Act 1993* – indicate that, as of May 2007: determinations of Native Title have been granted for a further 7%, predominantly in Western Australia (Figure 1.7b); and Registered or Scheduled Native Title applications (i.e. still to be determined) cover more than 50% of the tropical savannas region, ranging from 41% of Western Australian savannas to 59% in Queensland (Figure 1.7c).

Summary

These observations, although sketchy and incomplete, illustrate some important features of land and resource management in northern Australia. Low human population density and

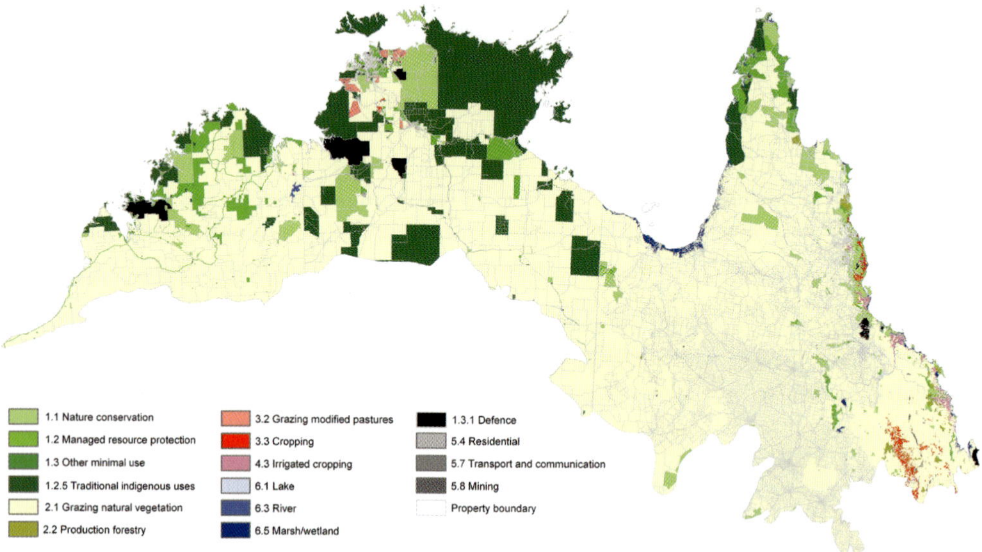

Figure 1.6 Generalised land use (Commonwealth of Australia 2007).

weak infrastructure compromise capacity to manage pervasive damaging processes such as wildfire that require active intervention to manage their impacts (Whitehead 1999; Whitehead *et al.* 2002). However, the north Australian jurisdictions have historically found it hard to fund interventions demanding coordinated action over large areas, because their funding sources and models for allocating funds are often based on population.

Mainstream economic developments are rarely capable of complementing or substituting for government intervention. Developments such as mines are patchy in space and time and local or regional interests often find it difficult to connect to them. Immigrants responding successfully to development opportunities may stay for relatively short periods and export benefits as savings. As a consequence, the economic and related social benefits and environmental costs of savanna development can be inequitably distributed, with many Indigenous people remaining severely disadvantaged, especially in the regions, and despite being 'land rich'.

Disadvantage for Indigenous people is compounded because distinct 'cultural' values may not be easily communicated and so receive inadequate consideration in areas subject to development. In most parts of the savannas, it is possible to identify people with particular ties to, and obligations to care for, sites subject to development. It is nearly always necessary to deal with complex cross-cultural issues and competing interests. Despite the sparseness of population, there is no *terra nullius*.

In part, this book is about new approaches to fill the gap between demonstrable need for active coordinated management of savanna environments, and the financial and human capacity to intervene effectively.

GOALS OF THIS BOOK

As we have seen, Australian regional perspectives on, and approaches to, the management of fire are strongly influenced by interplay among the frequency of fire, its intensity, the assets that are put at risk, and the resources that are available to assert control. In densely settled southern Australia, the dominant objective is to protect human life and property from the infrequent,

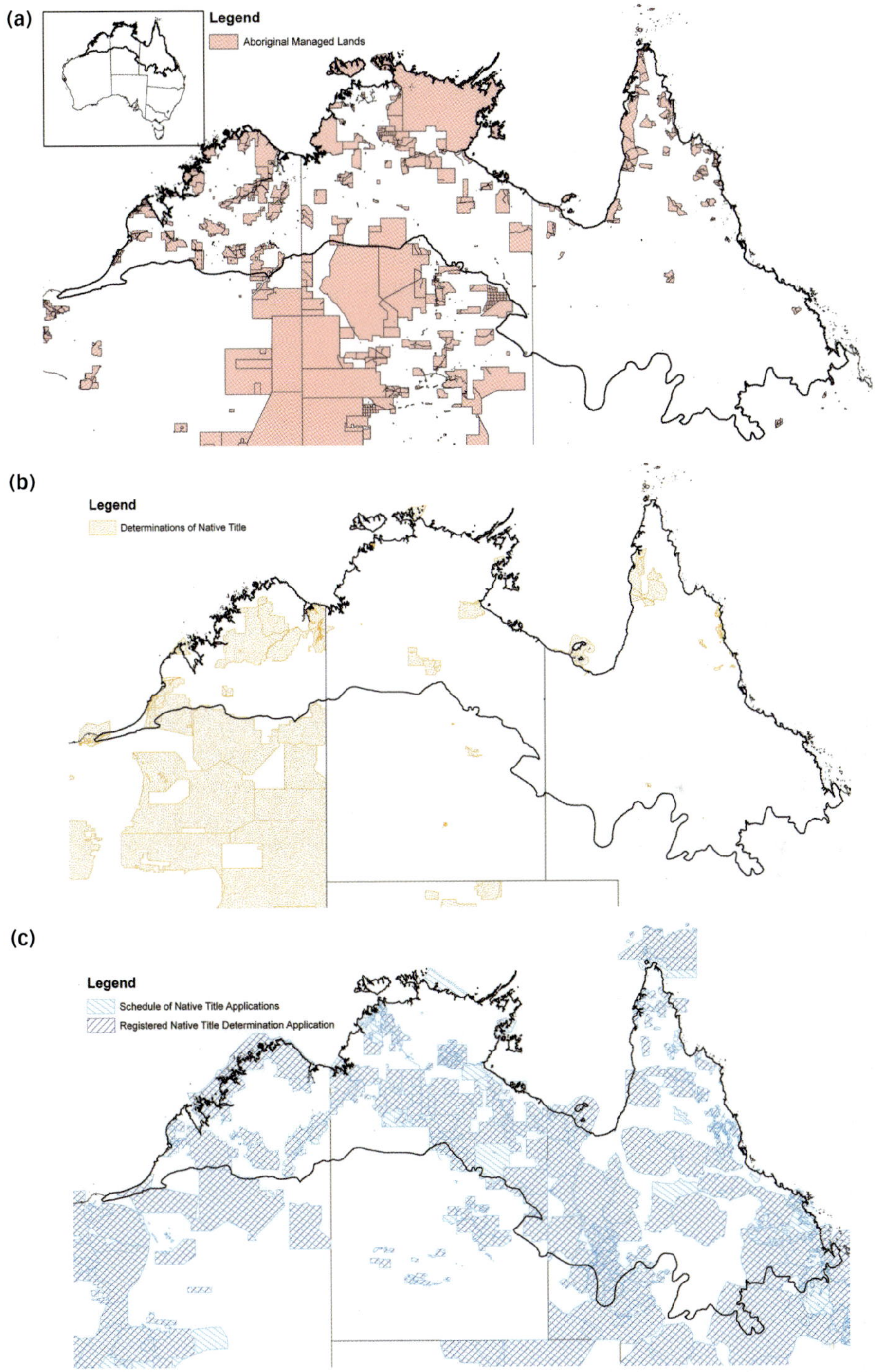

Figure 1.7 Current status of Aboriginal interest in land. (a) Aboriginal-owned or managed lands (Source: mapped from land parcel data provided by Indigenous Land Corporation 2008). (b) Determinations of Native Title (Commonwealth of Australia 2008a). (c) Native Title applications (Commonwealth of Australia 2008b).

but very intense, wildfires that can invade urban or peri-urban landscapes or smaller rural settlements. In central Australia, occasional huge fires sweep through the region: putting at risk built assets, pastures and the biota. In the last decade, the mainland extinction of an endangered macropod, the Mala (*Lagorchestes hirsutus*), was attributed to such a fire (Woinarski *et al.* 2007b). In the tropical savannas of northern Australia, which encompass about a quarter of the nation's land area (or more than 1.9 million km^2), fire is frequent, occurs over more than 6 months of each year, affects large areas and varies greatly in intensity. Instead of treating occurrence of fire as aberrant or as evidence of breakdown of flawed management systems, responses to fire risks should focus on the reasons for fire use and the ways in which fire's active application can produce and, indeed, may be essential for delivering conservation and social benefits of local, regional, and national significance.

Unfortunately, regional diversity is not always reflected in fire policy, and law imposed from mostly southern Australian capital cites: where approaches are dictated by an ongoing antagonism between threats to life and property presented by infrequent highly destructive fires, and demands to forgo regular fuel reduction burning in the interests of aesthetics, health effects on local human populations, and biodiversity conservation. Because fires are viewed as aberrations, too little attention is paid to the influence of fire management on carbon dynamics at the national scale and implications for global warming.

This book is about redressing this imbalance. We seek to articulate a distinctly northern Australian understanding of and approach to the management of fire, matched to the biophysical and social realities of the north. We examine in detail the biophysical and social determinants of fire regimes in one of the most fire-prone regions of a flammable continent. This book is also about the way the feedbacks from those fire patterns in turn influence the structure and function of the north's ecological and social systems. We particularly seek to develop understanding of how – despite the many obstacles to effective management associated with remoteness, sparse population and limited transport and other infrastructure – management of fire can be improved for positive ecological and social outcomes over a large part of the Australian land mass.

The rationale for this treatment of fire management challenges and options in northern Australia derives from a number of key propositions. They are:

1. The natural and cultural assets of north Australian landscapes are highly valued by the Australian and international public and there is both a legal and moral obligation to protect those assets from ill-considered or inadvertent change.
2. Fire has shaped, and continues to influence, the structure and composition of northern Australian vegetation and other components of the biota, and hence the ecosystem services available from the savannas.
3. Contemporary fire regimes differ from those experienced in recent history, and in prehistory, and so are likely to be important contributors to change. An important driver of shifts in fire-management behaviour has been the movement of many Indigenous people away from their lands and the consequent breakdown of skilled fire management over large parts of the landscape.
4. Contemporary fire regimes in many parts of the savannas are demonstrably detrimental to natural and cultural assets and, in some cases, the damage is severe.
5. Knowledge of the spatial and temporal features of fire regimes needed to redress damage is growing, but remains limited. Nonetheless, it is recognised that total exclusion of fire from all or large parts of the savannas is impracticable and probably undesirable.
6. Irrespective of decisions about the desirability of different regimes in different parts of the landscape, capacity to implement effective change by asserting management control is

presently limited, and this constraint is unlikely to be redressed in the short term by substantial increases in entirely public funding.
7. Improved understanding of the ways in which residents of remote northern Australia, and users of the regions resources, can make greater contributions by choosing and implementing more favourable fire regimes is therefore critical.
8. Indigenous people presently make up most of the non-urban population of north Australia: they own much of the land, have interests in most of the remainder, and their populations are growing rapidly. Indigenous residents have particular knowledge of, and cultural obligations for, the considered use of fire in land and resource management. Indigenous people will play a major role in determining and implementing improved fire management for north Australia.
9. The capacity of the north Australian population to participate in formal systems of fire management is compromised by low educational standards in many regions, and weakness of other critical institutions.
10. Effective management of fire for achieving an array of environmental, social and economic goals demands an unprecedented level of collaboration among private and public sectors, Indigenous and non-Indigenous people, which, in turn, will depend on building new institutions and organisations, as well as better use of existing institutions and structures.

We consider each of these propositions and use them, and connected arguments, to frame practical management models matched to the existing and projected circumstances of northern Australia and the people. We illustrate many aspects of our treatment by reference to the particular fire-management issues arising in a remote, culturally rich region of high biodiversity in western Arnhem Land in the Top End of the Northern Territory, and the creative methods that local people, researchers, government and private enterprises have developed jointly to deal with those issues.

A large part of our treatment is necessarily about biophysical aspects of fire: accumulation of fuel; sources of ignition; fire behaviour; the resultant spatial and temporal patterns of fire; impacts on landforms, vegetation and wildlife; effects on greenhouse gas emissions and carbon dynamics; the ways that those patterns might be modified by human intervention; benefits from interventions; and approaches to measuring all of these attributes and issues. We particularly focus on options for controlling greenhouse gas emissions on a range of timescales. But, in positing options for human intervention to improve outcomes from management of fire, it is essential also to consider the plausibility of the interventions proposed.

We approach the question of plausibility in several ways. First, we closely examine the social history of northern Australia (especially the Top End) to improve understanding of the way different groups interpreted their actions and influence on patterns of fire and their benefits. Second, informed by this historical analysis, we consider relevant aspects of contemporary northern Australia society to assess options to secure greater engagement of the present population in shared approaches to fire management. Third, and perhaps most importantly, we consider the relationship between the well-being of northern Australia's human population and its use of fire. As well as examining closely how choices about fire use have in the past, and may in the future, directly influence livelihoods through effects on use of land and availability of resources, we also explore new opportunities that fire management may offer for some presently disadvantaged segments of the community to participate in the mainstream economy, and so build new capacity for continued improvement in fire-management performance. We consider the extent to which systems developed for fire management may underpin a larger suite of options for people in remote regions – with few other present or projected employment options – to deliver ecosystem services to regional, national and international markets.

ORGANISATION OF THIS BOOK

In 15 multi-authored chapters, this book deals with the key challenges and novel options for addressing chronic landscape-scale fire management issues in north Australian savannas through development of (1) collaborative, cross-cultural 'two toolkit' approaches, and (2) commercially supported environmental services programs.

The ***first section*** sketches the biophysical and social context of burning across northern Australia:

- a sparsely populated region with few towns and limited infrastructure
- a rapidly growing, but economically marginalised, decentralised Indigenous population with very substantial interests in, and customary responsibilities for, land
- cattle production is by far the dominant extensive land use
- mining and nature-based tourism provide the major economic drivers
- as a consequence of this land use history, a vast relatively unmodified natural landscape, showing increasing indications that important components of the biota are suffering under prevailing management regimes.

We begin by showing that the Australian view of fire and its management has been almost exclusively shaped by relatively infrequent, but sometimes catastrophic, wildfires in rural and peri-urban areas of southern Australia – explaining why Australia has yet to come to terms with the realities of fire management in its remoter regions (**Chapter 1**). We provide ecological, demographic and socioeconomic sketches of the area of northern Australia known as the tropical savannas. We describe prevailing fire regimes and relate them to the biophysical and social features of the region that drive the present patterns.

Having established the broad context, we then narrow our focus: first to the Top End of the Northern Territory and ultimately to the area of western Arnhem Land, which is the principal subject of the book. A number of contributors set out what is known of the social history of the region and discuss how that history has shaped contemporary patterns of human settlement, and the distribution and behaviour of people in the landscape. David Ritchie (**Chapter 2**) describes how from the mid-1800s interactions between the settler and Indigenous populations led to quite rapid depopulation of large areas and interruption of long-established patterns of fire use. Introduced disease was probably the most potent influence on the size of the Indigenous population, with violence between settlers and Indigenous people and among Indigenous groups a secondary factor. Demands for access to resources held by settlers – particularly tobacco – strongly influenced the geographic distribution of the remaining Indigenous population, with many aggregating around mining settlements, pastoral stations and moving as groups with mobile parties hunting buffalo for their hides. An important conclusion from his work is reaffirmation of the centrality of fire use to Indigenous life, but a failure of settlers to respect, or even to recognise, its importance in meeting obligations to lands and so asserting rights over country.

Robert Levitus (**Chapter 3**) picks up elements of this history to describe in more detail changes in Indigenous life and fire use in the Kakadu and western Arnhem Land regions to the east of Darwin. He particularly emphasises the patchiness of resource availability and the influence this had on patterns of movement through the landscape. Freshwater wetlands developing west of the Arnhem Land Plateau more than 1000 years ago attracted people from many parts of the surrounding landscape, including the Arnhem Land stone country. It appears that groups moved from the stone country to the floodplains in the early dry season and returned to elevated sites during the wet. Post-settlement populations fringing the plateau were also much reduced by disease, but there is little evidence of the organised violence (massacres) that occurred in some other areas. He argues that the buffalo industry, which operated more or less continuously during the first half of the 20th century, created more benign interactions because the white hunters provided resources sought by Indigenous people, who brought skills and labour essential to the industry. His treatment concludes with description of the formal recognition and development of customarily informed Indigenous fire management practices in present-day Kakadu National Park and, in particular, ongoing tensions between the fire management aspirations of Indigenous landowners and the expectations and institutional requirements of management staff.

In **Chapter 4**, Peter Cooke explores the histories of the *Wardekken* – people of the rock country – who used, and still use, the western areas of the Arnhem Land escarpment as a centre of activity, even though they travelled outside the escarpment to exploit seasonal abundance of fish or other foods and resources. He examines post-settlement influences on occupation of the stone country: emphasising the importance of buffalo, mining (tin), tobacco (baki) and exotic religion (Jesus) on the 'creation of a modern wilderness', which is only now being re-occupied and the walking tracks and trade routes re-established. He describes the impacts of this emptiness on the country, motivations for return and the process of reoccupation from the 1970s to the present. He describes the central role that fire use and management has played in the process of re-occupation.

Murray Garde, Lofty Bardayal Nadjamerrek and other Indigenous collaborators (**Chapter 5**) then set down an Indigenous 'language of fire'; a description of the behaviour and effects of fire and the behaviour and intent of the people who applied it to the management of their country and its resources. This remarkable linguistic and historical summary of exchanges of information and opinion – recorded over a period of more than 10 years – reveals patterns of Indigenous life in the stone country that persisted despite the pervasive, and sometimes extreme, social dislocations described in earlier chapters. Critically for this volume, they demonstrate the associated survival of a great breadth and depth of knowledge of fire, its influence on the health of the landscape and the resources that sustained a human population in such a seemingly inhospitable place. Their work provides an indispensable base for improved communication with Indigenous fire managers over the application of the 'two toolkits' (western science and Indigenous knowledge) approach that has characterised the 'Caring for Country' movement built by Indigenous groups in northern Australia. Close understanding is essential for collaborations to define goals and achieve outcomes that are clearly understood and genuinely shared by Indigenous and non-Indigenous interests.

In **Chapter 6**, Jon Altman gives a personal description of a large, well-organised kangaroo hunt associated with a major Kunabibi ceremony, in which the use of fire was integral. Skilled use of fire was needed not just during the hunt but in months of prior burning to ensure that the site would be protected from wildfire until it was used for the hunt. The sorts of skills and high levels of cooperation shown here in asserting control over fire in a chosen part of the landscape are clearly essential for any effective form of fire management in undeveloped landscapes, whatever the motive for seeking control. Jon's chapter concludes the contributions that set the social context and reveal the sorts of knowledge and skills that Indigenous people bring to management of fire in some of the most fire-prone (and fire-sensitive) landscapes in the world.

In the ***second section***, we bring together summaries of Western scientific understanding of fire regimes in the region's savannas and implications for regional and national goals in conservation. Much of that understanding has been accrued through regional fire management case studies in the Top End of the Northern Territory: particularly current developments in the Aboriginal-owned western Arnhem Land region, and on adjoining conservation reserves (Kakadu and Nitmiluk National Parks – both under Joint Management arrangements between their Aboriginal owners and Australian and Northern Territory Governments, respectively), and large pastoral properties. This vast, little-populated, regional landscape is recognised internationally for its cultural and biodiversity values, which are epitomised in the World Heritage status of Kakadu National Park.

Chapter 7, by Dick Williams and colleagues summarises fire research in northern Australia and its relevance to delivery of production and environmental goals. Although much scientific work has focused on relationships among fire regimes, vegetation pattern and biological diversity, more recently interest has shifted to implications for management of greenhouse gases, and this chapter introduces some of the ideas that will be explored in more detail later in the volume. Issues in climate change add to the complexity of the range of options that Australian society will wish to consider in making decisions on use and control of fire to deliver multiple benefits sought by different sectors of society.

In **Chapter 8**, John Woinarski and others focus on the interaction between fire management and biological diversity. They catalogue the extraordinary biodiversity values – including high levels of endemism – of the Arnhem Land Plateau, but note that there is evidence of recent decline in those values. They identify adverse fire regimes as one of the drivers of detrimental change and note that fire is cited as a major threat for a higher proportion of threatened species of plateau flora and fauna than other parts of the savanna landscape. They suggest that a reduction in fire frequency is a 'no regrets' policy for the plateau that can be achieved in part by supporting Indigenous fire managers to assert greater control. However, they caution that it would be unwise to assume that all conservation goals can be met by restoration of customary fire management, because some species are threatened by factors other than fire, or may benefit from fire regimes that differ from both presently weakly managed fire or reimposed traditional regimes.

Chapter 9 by Edwards and Russell-Smith sets out an approach to management of fire-sensitive vegetation that recognises thresholds for tolerance of fire frequency. Based on detailed fire history and vegetation mapping derived principally from Landsat imagery, the analysis focuses in part on a distinction between species that can re-sprout from rootstock or epicormic buds following fire and others that are unable to do so. The latter 'seeder' species depend on fire-free intervals of several or more years to reach reproductive age and produce the next generation. 'Seeder' persistence in the landscape depends on prevention of frequent large fires that burn large areas uniformly. The fire history and vegetation mapping methodologies developed for this assessment underlie the savanna burning greenhouse gas emissions inventory methodology reported in Chapter 13.

Edwards and colleague's perspective is expanded in the next contribution (**Chapter 10**) by Russell-Smith and others. They provide a decade-long assessment and comparison – using remote sensing and plot-based studies – of contemporary fire regimes and their effects on vegetation and fauna in Kakadu, Nitmiluk and Litchfield National Parks. The program involves observations of fauna and flora, including assessments of relative abundance, and, for woody plants, measures of growth and mortality of tagged individuals. Plots in sandstone heath have not been established long enough to be part of this analysis, but the study demonstrates the potential to relate change in relative abundance, and hence composition and above ground biomass, of vegetation in plots to fire regimes. The full utility of the monitoring plots will not be established until the program has continued long enough for all plots to be surveyed on at least three separate occasions, but already it has shown promise as a management tool that requires resource managers to articulate their objectives, design programs to deliver specified outcomes and then to review effectiveness of their actions. Moreover, it has shown promise as a source of information on responses of woody biomass to variation in fire regimes: especially the incidence of severe late dry season fires. The significance of this issue is explored further later in the book.

The ***third section*** focuses on the link between northern fire regimes to emissions of greenhouse gases and hence to climate change policies. In **Chapter 11,** Peter Whitehead and colleagues discuss the evolution of the West Arnhem Land Fire Abatement (WALFA) project and its significance as a test of practice in reducing fire frequency and incidence of destructive late dry season fires. They identify the features of the project that have contributed to its success in reducing emissions over 3 years by about 40% from a baseline. Given this success, they identify the challenges remaining to create a robust Indigenous business model for delivering similar results in other parts of the savannas. They also consider the approach to national policy in regard to emissions trading and emissions accounting that will allow Indigenous fire managers to engage with markets in emissions credits (offsets).

Chapter 12 by Garry Cook and Mick Meyer, describes the biophysical and chemical processes generating greenhouse gas emissions from savannas and, in particular, the way in which shifts in fire regimes alter the relative importance of the potent greenhouse gases methane (CH_4) and nitrous oxide (N_2O). The significance of this contribution is in explaining some of the more arcane aspects of greenhouse gas accounting, and why reductions in net emissions (and the potentially tradeable credits associated with them) are not directly dependent on changes in the amount of carbon stored in plant biomass. In **Chapter 13**, Jeremy Russell-Smith and a large group of collaborators describe the present 'state of the art' in accounting for greenhouse gas emissions from fire, and identify some of the remaining challenges to refine estimates and simplify measurement by, for example, automating mapping of fire-affected areas in savannas. **Chapter 14** (Brett Murphy and others) returns to the three-parks study to extract – from a subset of frequently assessed plots – some estimates of change in above-ground woody biomass in trees and shrubs and relates those estimates to fire severity. They derive a simple model to predict rates of storage of carbon under reduced frequencies of severe fires of the sort that the WALFA project is designed to prevent. They conclude that the commercial potential is considerable, should such changes be formally recognised under national and international arrangements as contributing to removal of carbon dioxide from the atmosphere through increased biosequestration.

Finally, in **Chapter 15**, we seek to draw together the many issues raised by other authors and weave their analyses and questions into a proposal for empowering north Australians, and Indigenous people in particular, to make nationally significant contributions to greenhouse gas abatement and biodiversity conservation. The significance of the opportunities we identify extends well beyond the environmental. Delivery of environmental services provides one of the few pathways for Indigenous people in remote areas to connect with the mainstream Australian economy. The management of fire illustrates the potential of a larger array of opportunities. The federal government and north Australian state and territory jurisdictions have been presented with an opportunity to deliver some ground-breaking domestic policies through the adjustments made for a robust carbon-influenced economy: policies that, from the outset, build Indigenous participation into what will be become important mainstream institutions. The options we identify here differ fundamentally from many failed approaches to northern development that treat the region and its people as unique problems to be solved by extraordinary measures – often positioned outside mainstream Australian institutions. Instead, we seek resolution of issues of national significance and concern through products offered on well-regulated markets serviced by northern people inhabiting, and using, northern landscapes. Indigenous residents of remote and regional north Australia are treated not as problems but as valuable members of Australian society who equip the nation to deal with critical national challenges.

REFERENCES

AGO (Australian Greenhouse Office) (2007) *National Greenhouse Gas Inventory 2005.* AGO, Canberra.

Allan GE and Southgate IR (2002) Fire regimes in the spinifex landscapes of Australia. In *Flammable Australia: The Fire Regimes and Biodiversity of a Continent.* (Eds RA Bradstock, JE Williams and AM Gill) pp. 145–176. Cambridge University Press, Cambridge, UK.

Altman JC (1987) *Hunter-gatherers Today.* Australian Institute of Aboriginal and Torres Strait Islander Studies, Canberra.

Altman JC (2003) People on country, healthy landscapes and sustainable Indigenous economic futures: the Arnhem Land case. *The Drawing Board: An Australian Review of Public Affairs* **4(2)**, 65–82.

Altman JC, Buchanan GJ and Larsen L (2007) 'The environmental significance of the Indigenous estate: Natural resource management as economic development in remote Australia'. CAEPR Discussion Paper No. 286/2007. Centre for Aboriginal Economic Policy Research, Australian National University, Canberra.

Brook RK and McLachlan SM (2005) On using expert-based science to 'test' local ecological knowledge. *Ecology and Society* **10**(2), r3, <http://www.ecologyandsociety.org/vol10/iss2/resp3/>

Burgess C, Johnston F, Bowman D, Whitehead PJ and O'Dea K (2005) Healthy land: healthy people? Exploring the health benefits of indigenous land management. *Australian New Zealand Journal of Public Health* **29**, 117–122.

Collins R (1999) *Learning Lessons: An independent Review of Indigenous Education in the Northern Territory.* Northern Territory Department of Education, Darwin.

Commonwealth of Australia (2007) *Australian Land Use.* Bureau of Rural Sciences, Department of Agriculture, Fisheries and Forestry, Canberra, <http://adl.brs.gov.au/mapserv/landuse/index.cfm?fa= in July 2008>.

Commonwealth of Australia (2008a) *National Native Title Register (NNTR, Determinations of Native Title) – Boundaries and Core Attributes.* Geoscience Australia, Canberra, <https://www.ga.gov.au/products/servlet/controller?event=DEFINE_PRODUCTS>.

Commonwealth of Australia (2008b) *Register of Native Title Claims – Boundaries and Core Attributes about application.* Geoscience Australia, Canberra, <https://www.ga.gov.au/products/servlet/controller?event=DEFINE_PRODUCTS>.

Cook GD, Liedloff AC, Eager RW, Chen X, Williams RJ, O'Grady AP and Hutley LB (2005) The estimation of carbon budgets of frequently burnt tree stands in savannas of northern Australia using allometric analysis and isotopic discrimination. *Australian Journal of Botany* **53**, 621–630.

Craig R, Heath B, Raisbeck-Brown N, Steber M, Marsden J and Smith R (2002) The distribution, extent and seasonality of large fires in Australia, April 1998–March 2000, as mapped from NOAA–AVHRR imagery. In 'Australian fire regimes: contemporary patterns (April 1998–March 2000) and changes since European settlement'. (Eds J Russell-Smith, R Craig,

AM Gill, R Smith and J Williams). Australia State of the Environment Second Technical Paper Series (Biodiversity). Department of the Environment and Heritage, Canberra, <http://www.ea.gov.au/soe/techpapers/index.html>

CSIRO and Bureau of Meteorology (2007) 'Climate change in Australia: technical report 2007'. CSIRO, Australia, <http://www.climatechangeinaustralia.gov.au/>

Ellis S, Kanowski P and Whelan R (2004) 'National inquiry on bushfire mitigation and management'. Commonwealth of Australia, Canberra, <http://www.coag.gov.au/>

Franklin DC, Whitehead PJ, Pardon G, Matthews J, McMahon P and McIntyre D (2005) Geographic patterns and correlates of the decline of granivorous birds in northern Australia. *Wildlife Research* **32(5)**, 399–408.

Garnaut R (2008) 'The Garnaut climate change review: final report'. Cambridge University Press, Melbourne.

Gilchrist G, Mallory M and Merkel F (2005) Can local ecological knowledge contribute to wildlife management? Case studies of migratory birds. *Ecology and Society* **10(1)**, 20. <http://www.ecologyandsociety.org/vol10/iss1/art20/>

Greiner R, Mayocchi C, Larson S, Stoeckl N and Schweigert R (2004). 'Benefits and costs of tourism for remote communities: Case study: Carpentaria Shire, north-west Queensland'. Tropical Savannas Cooperative Research Centre, Darwin.

Harries J (1997) 'Acid mine drainage in Australia: its extent and potential future liability'. Supervising Scientist Report 125. Office of the Supervising Scientist, Canberra.

IPCC (1997) *Revised 1996 Intergovernmental Panel on Climate Change (IPCC) Guidelines for National Greenhouse Gas Inventories*. 3 vols. IPCC/OECD/IEA, Paris, France.

Jonauskas P (Ed.) (1996) *Making Multiple Use work. Proceedings of the Wetland Workshop, 6–7 December 1994*. Department of Lands, Planning and Environment, Darwin.

Meyer CP (2004) 'Establishing a consistent time-series of greenhouse gas emission estimates from savanna burning in Australia'. Report to the Australian Greenhouse Office, Canberra. CSIRO Publishing, Melbourne.

Morris A (2003) 'Powerhouse or mendicant? Is the Territory an engine of growth or a "drag" on the Federation?' Paper presented to the *Symposium Beyond the Frontier: Sustainable Futures for Northern Australia, 17–18 July 2003*. Charles Darwin University, Darwin.

NTG (2006) 'Economic development framework'. Northern Territory Government, Darwin, <http://www.nt.gov.au/business/documents/general/EDF_Vatskalis.pdf>.

Russell-Smith J, Yates CP, Whitehead P, Smith R, Craig R, Allan G, Thackway R, Frakes I, Cridland, Meyer CP and Gill AM (2007) Bushfires 'down under': patterns and implications of Australian landscape burning. *International Journal of Wildland Fire* **16**, 361–377.

SCRGSP (Steering Committee for the Review of Government Service Provision) (2007) 'Overcoming indigenous disadvantage: key indicators 2007'. Productivity Commission, Canberra.

Stoeckl N, Stanley O, Brown V and Stoeckl W (2007) 'Regional economic multipliers in Australia's tropical savannas'. Report to the Tropical Savannas CRC. James Cook University, Townsville, <http://savanna.cdu.edu.au/savanna_web/publications/downloads/Regional-Economic-Multipliers-in-the-Tropical-Savannas.pdf>.

Taylor J (2003) 'Indigenous economic futures in the Northern Territory: the demographic and socioeconomic background'. CAEPR Discussion Paper 246/2003. Centre for Aboriginal Economic Policy Research, Australian National University, Canberra.

Taylor J (2006) 'Population and diversity: policy implications of emerging Indigenous demographic trends'. CAEPR Discussion Paper No. 283/2006. Centre for Aboriginal Economic Policy Research, Australian National University, Canberra.

Taylor J, Brown D and Bell M (2006) 'Scoping population dynamics and demographic accounting in arid and savanna Australia: methods, issues and outcomes'. Report of a scoping study for the Desert Knowledge and Tropical Savannas Cooperative Research Centres. Desert Knowledge Cooperative Research Centre, Alice Springs.

Whitehead PJ (1999) Promoting conservation in landscapes subject to change: lessons from the Mary River. *Australian Biologist* **12(1)**, 50–62.

Whitehead PJ, Woinarski JCZ, Franklin D and Price O (2002) Landscape ecology, wildlife management and conservation in northern Australia: matching policy, practice and capability in regional planning. In *Landscape Ecology and Resource Management: Linking Theory and Practice.* (Eds J Bissonette and I Storch) pp. 227–259. Island Press, New York.

Woinarski JCZ and Dawson F (2002) Limitless lands and limited knowledge: coping with uncertainty and ignorance in northern Australia. In *Ecology, Uncertainty and Policy: Managing Ecosystems for Sustainability.* (Eds JW Handmer, TW Norton and SR Dovers) pp. 83–115. Prentice-Hall, New York.

Woinarski JCZ, Milne DJ and Wanganeen G (2001) Changes in mammal populations in relatively intact landscapes of Kakadu National Park, Northern Territory, Australia. *Austral Ecology* **26**, 360–370.

Woinarski JCZ, MacKey B, Nix H and Traill B (2007a) *The Nature of Northern Australia: Natural Values, Ecological Processes and Future Prospects.* Australian National University, Canberra.

Woinarski JCZ, Pavey C, Kerrigan R, Cowie I and Ward S (Eds) (2007b). *Lost from our Landscape: Threatened Species of the Northern Territory.* Northern Territory Department of Natural Resources, Environment and the Arts, Darwin.

2

Things fall apart:

the end of an era of systematic Indigenous fire management

David Ritchie

SUMMARY

In the 1860s, European settlers gained their first permanent foothold in the Northern Territory close to where the city of Darwin now lies on the coast of the north-west Top End. One observable effect of this was that the population of the region's Indigenous people was reduced to about a tenth of pre-settlement levels within two generations. Introduced disease was the major cause of this catastrophe, which had devastating consequences for Indigenous social structure over a much wider area. Indigenous people were drawn to the towns and centres of European activity by irresistible attractions and established substantial camps on the fringes. High death rates and the relocation to sedentary semi-permanent settlements disrupted the traditional relationships with land, including the use of fire. Where people remained on their lands, settlers often objected, sometimes violently, to the use of fire – further suppressing traditional practice. By the late 19th century or early 20th century, fire regimes in much of the north-west had fundamentally changed. Further east, from the 1880s until the 1960s, hunting of buffalo on the coastal plains attracted Indigenous people from within the region and beyond. This activity required application of Indigenous skills in fire management. Patterns of movement to follow herds and locate new buffalo populations for exploitation may have mimicked some aspects of customary use of landscapes, but the resultant fire regimes are unlikely to have closely matched pre-contact patterns. After the Second World War, the buffalo industry collapsed and associations of Indigenous people with pastoral enterprises also broke down. In Arnhem Land, the return to traditional lands was fostered by the outstation movement in the 1970s, but was not accompanied by full restoration of pre-contact patterns of fire use. The land claim process,[1] established in the late 1970s, focused on the practice of Aboriginal traditions. However, in the first claims in this region, most references to Indigenous burning were to the threat this practice posed to neighbouring landowners. Over the next 20 years the 'right to burn country' became recognised as a very significant aspect of a traditional owner's primary spiritual responsibility for their land. Ironically, as this right was being recognised in the last decades of the 20th century, increased use of motor vehicles meant that fewer parts of the landscape were visited, and for shorter periods, than when most travel off the major tracks was by foot or horseback. These relatively well-documented demographic processes of collapse in Indigenous populations, and movement to permanent or semi-permanent settlements, and subsequent influence of vehicle availability on patterns of landscape use have been repeated throughout the savannas, including the Arnhem Land Plateau. The relative influence of different drivers and their timing may vary, but there has been one constant: Indigenous people continue to treat the use of fire as an essential feature of their relationship with land – as an assertion of rights and an important way of meeting obligations.

Figure 2.1 Traditional owner with an improvised beater retreats from advancing fire he lit earlier in the day (1981).

INTRODUCTION

We were driving through the bush on a hot dry day in May 1981 throwing lighted matches out the window. We were in the hills separating the catchments of the Adelaide River and the Finniss River. As it happened, this is the same area where, in 1866 surveyors from Escape Cliffs first recorded the way that burning by the Indigenous inhabitants promoted new growth in the native sorghum (*Sarga intrans)* (Wells 2003). The trip was the first the traditional owners had made since the wet season and they were obligated to, in their words, 'clean-up the country'. Burning the drying grass had both a practical and a political purpose. The year before, hearings for the Finniss River Land Claim had been held and in a few months the Daly River (Malak Malak) claim would begin. Burning became a public assertion of a right under Aboriginal tradition that had not been recognised prior to Land Rights. The Indigenous practice of setting fire to the county was well known. In the Finniss River Land Claim, opponents of the Claim

Figure 2.2 Edwin McGregor and a younger countryman burning country in Upper Adelaide River (1981).

had argued the risk from uncontrolled bushfires would be increased if the claim areas became Aboriginal land.

The earliest written accounts of the Indigenous groups in the country between the Daly to the Mary Rivers, in the north-west Top End, record the prevalence of deliberately lit fires in the landscape (e.g. Hutchinson 1864; McKinlay 1866a,b). Research now suggests that regular burning in this region was once widespread and systematic. This is no longer the case: today only a relatively small area is burnt by the Indigenous traditional owners of the country. Fires now occur more frequently in the later months of the dry season and, for reasons outlined in subsequent chapters, these recent patterns of burning have major implications for the condition of landscapes.

This chapter outlines the changes to Indigenous communities in the north-west Top End since the first European settlement in the mid-1860s and, in particular, the changes that reduced the capacity of Indigenous groups to live within their traditional estates and carry out stewardship in accordance with their traditions. It provides an overview of dominant influences on Indigenous demography and interactions with settler society and economy, and so establishes context for less-thoroughly documented change in other regions, including the Arnhem Land Plateau.

THE REGION AND ITS INDIGENOUS PEOPLE

The North-West corner of the northern Top End is where the first concerted European settlement in the Northern Territory occurred. Beginning in the mid-1860s at Escape Cliffs on the mouth of the Adelaide River, the settlement was re-established in 1869 on Darwin Harbour:

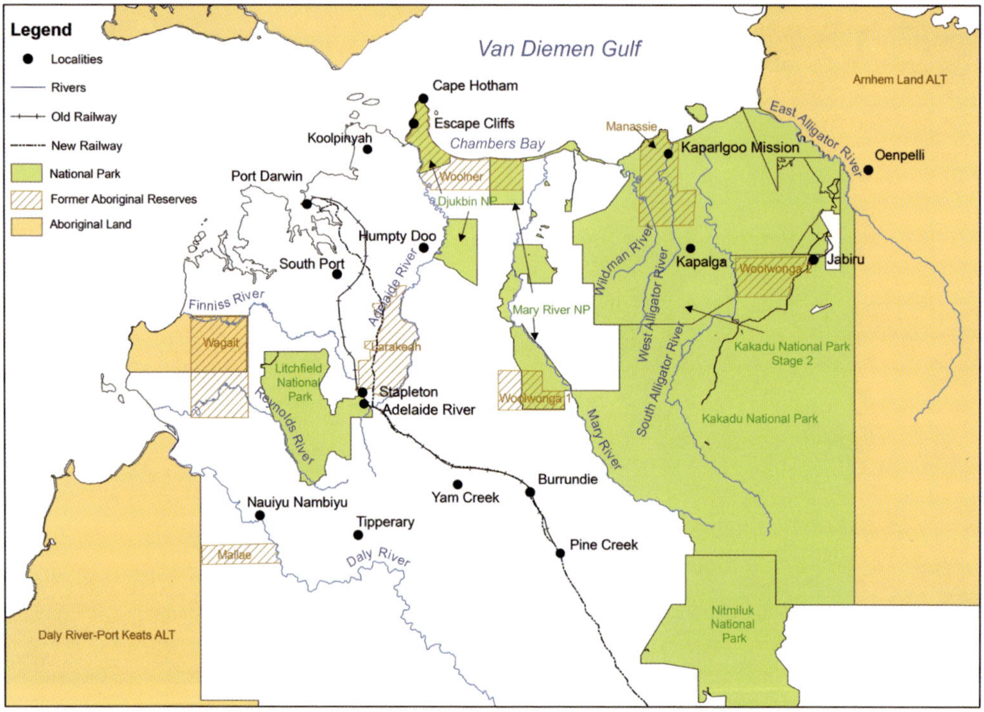

Figure 2.3 Major river systems and associated sites in the north-western Top End.

now a city with a population over 117 000.[2] In 1866, one of the first observations made about the Indigenous inhabitants of Port Darwin was that that they lit fires in the bush as they walked through the country (Young 1866).

Much of the north-west Top End comprises coastal wetlands associated with the river drainage basins of the Adelaide and Mary rivers and, further to the east, the Alligator Rivers, which all run north into the Van Diemen Gulf. To the west, the Finniss and Reynolds rivers run into the Timor Sea (Figure 2.3). The headwaters of these rivers drain from low hills and plateaus, which for half of the year are very dry. The area is subject to a monsoonal climate, with 90 per cent of the rainfall occurring in the wet season between November and March. During this time, the plains are inundated. The wetlands associated with the river plains are highly productive in both animal and vegetable food species and, according to 19th century records, supported large Indigenous populations, at least on a seasonal basis. These landscape features continue to dictate the extent of human activity in the region. The coastal plains of the major rivers have never again supported a population as large as the Indigenous presence at the time of first European contact.

The north-west Top End was the home of at least 12 distinct Indigenous languages. As elsewhere in the Northern Territory, the social group recognised by Indigenous people as having common rights in a contiguous area of land usually comprises a cluster of families notionally descended from a common ancestor – referred to in anthropological literature as 'the clan' (Harvey 2002b).[3] Such groupings may be recognised by the name given to the patrilineal totem inherited by its members – such as 'fire dreaming clan' – although this practice is not uniformly followed in the north-west Top End. Members of these groupings are characteristically linked to places or sites of traditional significance, which are regarded, among other things, as being the core of their country.[4]

There are several 'fire dreaming' sites and a 'smoke dreaming' located on the coastal plains (Keen 1980). In the Warray language of the upper Adelaide River, there are several different verbs for burning, including a specific verb for the act of deliberately setting fire to the bush. For instance, to burn is *no-y*, to light a camp fire is *dirrimbu-m*, while to light the grass or bush is *lukka-ngi* (Harvey 1986, p. 365). The existence of special language terms and both clan totems and sacred sites associated with fire attests to the centrality of traditions associated with fire and burning practices. The traditions governing use of fire in the period prior to the disruption caused by colonisation in the late 19th century can be inferred only partially from the surviving record. Indigenous groups in the north-west Top End were observed lighting fires frequently when they travelled through their country. When Ludwig Leichhardt travelled through the Alligator Rivers area in 1844 he concluded – because of the amount of smoke seen on 1 November – that the area was well populated (Keen 1980, p. 31). Leichhardt reported that:

> *natives burned the grass systematically along every watercourse and round every waterhole in order to have them surrounded with young grasses as soon as the rain sets in ... it is no doubt connected with the systematic management of their runs, to attract game to particular spots. (Ludwig Leichhardt 1845 quoted in Wells 2003, pp. 118–119).*

Burning by local Aboriginal groups was recorded by most surveying parties operating from Escape Cliffs and Port Darwin. Fires were apparently lit throughout the year. Most were recorded in the mid dry season, but this is also when most expeditions into the hinterland took place. Indigenous groups were skilled in use of fire as a hunting tool; in Wagiman there is a specific verb *gurrkkurrma* 'to hunt with fire' (Harvey *pers. comm.*). On several occasions, groups were recorded using fire as a weapon to harass the colonists (Wells 2003, pp. 61, 101). Some traditions relating to fire recalled to me by Indigenous elders in the 1980s included prohibitions on burning in vine-forests, burning in the vicinity of a platform burial for 5 years, and burning in association with certain sacred sites.[5]

THE EFFECTS OF COLONISATION ON THE INDIGENOUS PRESENCE

Face-to-face contact between Aborigines and Europeans in the north-west Top End occurred in 1864 when an expedition financed by the South Australian government, and led by Lieutenant Colonel B.T. Finniss as Government Resident, founded a settlement at the mouth of the Adelaide River. In the early dry season in May 1864, the survey schooner *Beatrice* anchored at a point approximately 32 nautical miles upstream from the Adelaide River mouth, adjacent to the area of coastal plains that now form Djukbinj National Park. Crew members recorded that the grass on either side of the river had been set alight and that they were visited by the men from a large Aboriginal camp located a kilometre from the river (Hutchinson 1864).

When Finniss and his party arrived at Escape Cliffs, the Indigenous population of the Adelaide and Mary River coastal wetlands probably numbered in the vicinity of 1500 people.[6] Once the settlement was established, there were regularly over 200 Aborigines camped in the vicinity (Young 1866). At the same time, exploring parties encountered bands of Aborigines over 50 strong (McKinlay 1866a, b). The Aboriginal practice of lighting fires as they travelled through the landscape was regularly mentioned in despatches.

In 1882, an exploration party travelled across the Adelaide and Mary River plains to the Wildman River and reported that 'natives are very numerous' and that the 'immense quantities of game' supported an even larger seasonal population evidenced by an extensive network of fish weirs and camping places (McMinn 1882). McMinn's account is the most detailed contemporary evidence that the coastal plains supported an Indigenous population of several thousand and that the large numbers of people used labour-intensive technologies, including fire.

McMinn discovered that the Aborigines had built extensive 'roads' through the Mary River wetlands using corduroyed logs. When asked to guide his expedition they carried 'tons of timber and earth' to make a crossing for his horses (McMinn 1882, 26 January). He recorded a large Aboriginal camp at Lake Finniss where Aborigines were using both dug-out canoes made from cotton-wood (*Bombax ceiba*) and huge rafts made from sheets of paperbark (*Melaleuca* spp.) bound with vines and capable of carrying between 10 and 12 people (McMinn 1882).

Similar large Indigenous populations were recorded on the coastal plains of the other rivers in the region in the early years of European settlement. The records of the Jesuit mission established on the lower Daly River suggest a pre-contact population in the area of over 1000. By 1899, the missionaries estimated that there were approximately 400 Aboriginal people on the Daly in the vicinity of the mission (O'Kelly 1967, p. 69). On the East Alligator River, Paddy Cahill started buffalo shooting at Oenpelli in the 1890s, and to the west, on the South Alligator River, Kapalga[7] mission operated between 1899 and 1903. Each of these settlements recorded local populations of several hundred Aboriginal people (Dewar 1989).

Indigenous groups living on the coastal plains and wetlands were described as having large healthy populations. In contrast, the groups of the hills and plateaus further inland were recorded as being small and unhealthy. The contrast is illustrated by the following accounts. In June 1864, the log of the *Beatrice* recorded that 16 men came on board from a much larger group camped on the lower Adelaide River in the area now known as Djukbinj National Park. All of these men were over 6 feet tall and 'very strong and muscular' (Hutchinson 1864, p. 3). Eighteen months later in the hills south of Mount Bundey, approximately 50 kilometres inland, McKinlay observed that in contrast to the health and vigour of the Aborigines of the coastal plains many of this inland group appeared 'poor, emaciated and half-starved … teeth literally worn level with the gum all round' (McKinlay 1866b, p. 11).[8]

The groups on the coastal plains controlled natural resources that were vastly richer than those of the hills and plateaus of the hinterland. It is likely that this control was continually tested by groups from the hinterland (Sutton 1990). In Chapter 3 of this book, Levitus describes 'an annual pulse of concentration and dispersal' across the Alligator Rivers whereby the Arnhem Land Plateau was seasonally depopulated during the mid to late dry season when Aboriginal groups moved towards the lowlands and the downstream swamps. There are many accounts from this period of inter-group hostility centred around contested resources. When the Europeans arrived at the mouth of the Adelaide River, and later Port Darwin, the dominant coastal groups rapidly established relationships with the settlers in order to obtain exotic commodities and defended their hegemony over this new resource. To do this, many relocated to the fringes of the growing settlements. Within a few decades, a succession of land-owning groups from the coastal plains south and east of Darwin relocated closer to the settlement. By the mid-1880s, 400 members of 'the Alligator River tribe' had arrived in Darwin (O'Kelly 1967, p. 18). Groups from the hills and plateaus of the hinterland moved onto the coastal plains to take their place. The anthropologist William Stanner concluded, based on his observations in the 1930s, that many groups:

> *had migrated, and large tracts had thus been emptied decades before the authorities or settlers were aware of it. Some of the small tribes … had ceased to exist. (Stanner 1979, p. 81)*

For the once-dominant coastal groups the move to townships and settlements meant that they were the first to be exposed to diseases introduced by the settlers and to suffer the debilitating effects of a poor diet, drug addiction and cultural dissonance.

Across the north-west Top End the Indigenous population declined dramatically as European settlement expanded. Epidemics continued to sweep through Indigenous groups well into the 20th century. One of the most significant effects was that, combined with a movement of the survivors to European settlements, large areas within the north-west Top End were abandoned and traditional fire-management practices were disrupted. As visits to country became less frequent, so did the opportunity to burn. In Chapter 4, Cooke argues that traditional burning practices ceased on the Arnhem Land Plateau because the 'critical mass of plateau managers' relocated to buffalo camps, mines, pastoral properties, government settlements and church missions. He records that the principal driving force for this move – recalled by today's senior generation of landowners – was a desire for *baki* or tobacco.

The impacts of disease

The most pervasive cause of mortality and declining birth-rate in the north-west Top End was introduced disease. Aborigines told officials from Darwin about the devastation of past epidemics among their people. These accounts, along with characteristic facial scarring, were interpreted as indications of smallpox introduced among coastal groups by Macassans (Foelsche 1881; Campbell 2002; Herbert nd).

In the wet season of 1884, smallpox broke out among the Wulna people camped on the Adelaide River and several deaths were officially recorded – but the extent of the epidemic was widespread (Northern Territory Times and Gazette 1884). Less than 10 years later, an influenza epidemic swept through the north-west Top End killing many people (O'Kelly 1967, p. 52). Early this century, the lessee of Koolpinyah estimated there were around 100 Aborigines living around the Adelaide River on the Koolpinyah, Humpty Doo and Marrakai pastoral leases (Sunter 1937). In 1910, he recorded that an epidemic – possibly influenza – killed half the members of this group. The mortality rate was particularly high among the young women 'between 12 and 20' (Sunter 1937, p. 200). Malaria outbreaks occurred periodically in the Daly River area from the 1890s (O'Kelly 1967, p. 52) and continued to be a threat until well into the 20th century. An oral tradition among the European residents of the Daly is that in the 1920s many Aborigines died in the Reynolds/Finniss area after a malaria outbreak.

After almost two decades of colonisation, the pathology of epidemic diseases including malaria, smallpox and influenza was exacerbated by the debilitating effects of narcotics, alcohol and poor diet on Indigenous groups living in close association with the mining settlements and other north-west Top End townships. Within two generations of European settlement, the population of Indigenous people in the north-west Top End had been reduced by over 90% (Stanner 1933; Harvey 1992; Harvey 2002a). It is likely that the massive drop in the Indigenous population had its causes in the fatal epidemics, outlined above, which swept through the Indigenous population in the last decades of the 19th century, as well as the longer term consequences of chronic low-level infections and sexually transmitted diseases – including lowered immune response and fertility (Reid 1996). These effects began to be manifest by the turn of the century and were documented by the first systematic medical survey of the north-west Top End conducted in 1911.

Indigenous use of colonial settlements

The surviving members of the depleted land-owning groups became drawn more and more into association with the colonial settlements. In some instances, they were physically displaced, as was the case in the mining regions around Pine Creek. Over time, this permanently changed the seasonal routines and traditional practices for many groups and large tracts of country were visited less frequently.

Most surviving Indigenous groups came at some time or another to the settlements along the Overland Telegraph, which was completed by the end of 1872. Gold had been found in the gullies around Pine Creek in 1871 and within 2 years a number of mines and crushing plants were operating at Pine Creek and Yam Creek (Bauer 1964). Within 5 years there were over 3500 Chinese working on the goldfields (South Australia 1881). Mining was predominantly alluvial, with the result that the surface of the landscape was altered – often beyond recognition – as billabongs and creeks were filled or diverted, the surface vegetation was stripped and the original surface buried beneath the tailings of hundreds of small workings (Daly 1887; McGrath and Coulthardt 1980). Indigenous groups whose traditional country was located in the hills where mining was concentrated found much of their country transformed beyond recognition and overrun with thousands of miners. The task of extracting mineral ore was labour intensive, monotonous and required skills that, unlike buffalo hunting, had no direct traditional parallels. The established mines soon became the hubs of interlocking walking paths that ran from the coastal plains and ultimately linked the people with the rail heads at Pine Creek, Adelaide River, Stapleton and Southport (Figure 2.2).

The Overland Telegraph, railway and mining towns located on the watershed between the two major river catchments of the north-west Top End displaced the traditional Indigenous landowners and simultaneously attracted Indigenous groups from further afield. This exacerbated the effect of the reduced population and permanently altered the way Indigenous groups moved through their traditional country; in turn, this changed burning practices.

At the same time that the north-west Top End was experiencing its first mining boom, there was a rush to establish pastoral and horticulture industries on the alluvial plains of the major rivers. In spite of this activity, the number of colonists living on the coastal plains by the end of the 19th century remained small. It became evident that the alluvial plains were third-rate cattle country owing to seasonal flooding, native grasses with poor nutritional value and high incidence of the cattle disease 'redwater' or tick fever. By 1890, the industry leader, the North Australian Territory Company, had removed most of its stock from the north-west Top End in favour of its extensive holdings in the Victoria River basin, hundreds of kilometres to the south-west.

Indigenous response to colonisation

The effect of colonisation on the Indigenous population of the north-west Top End was catastrophic. Aborigines from the coastal plains moved south to the goldfields where the disproportionate number of non-Aboriginal men created a steady demand for sexual services, while the general isolation ensured strong demand – particularly from the Chinese – for fresh game and bush foods.[9] Aborigines on the goldfields provided miners with these services. In addition to the established barter items, Aborigines were paid in tobacco, opium ash and alcoholic beverages (Price 1878; Northern Territory Times 1889). By 1900, missionaries at Kapalga on the South Alligator River recorded that they needed to relocate the mission further east into Arnhem Land to get away from the influence of miners on the goldfields after young women began deserting the mission in search of opium and to live as prostitutes (Freier 2008).

Groups from the Mary River established a seasonal camp on the Burrundie field near Pine Creek (Parkhouse 1895a, b). The goldfields attracted Aborigines from farther afield and local Indigenous groups violently contested this immigration. The rapidly expanding mining industry became the catalyst for serious conflicts between Indigenous groups in the area.

Although the first three European settlements in northern Australia between 1824 and 1849 were short lived (Bauer 1964), and over 300 kilometres east of the north-west Top End, they had one enduring consequence for the coastal plains of the north-west: they were the sites for introduction of Asian water-buffalo, which were firmly established in the Adelaide/Mary River system before the first European land exploration parties arrived in the area (Young 1866). While the pastoral stations north of Katherine were being destocked and abandoned, an industry developed on the coastal plains based on the seasonal hunting of feral buffalo for their hides and horns.[10]

This industry was made viable by the hunting skills and bushcraft of Aborigines. For the next 80 years, the work was carried out by teams consisting of one, or perhaps two, Europeans and a large entourage of Indigenous workers and their families. Buffalo hunting was seasonal and could only be undertaken in the dry season when the coastal wetlands dried out. The nature of the work meant that hunters could not remain working from a base camp, but had to move around the plains to follow the herds of buffalo, carrying only the most rudimentary provisions and equipment, and living off the land as they went. This nomadic style of operation apparently suited many Aborigines. Aspects of traditional burning were maintained in the buffalo country of the Adelaide and Mary Rivers because the practice aided the hunters.

On the East Alligator River, Paddy Cahill started buffalo shooting at Oenpelli in the 1890s. By 1909, shooting feral buffalo for their hides had become the main economic activity on the lower Adelaide River. Buffalo hunting was the preferred employment for the Aborigines of the coastal plains. They also provided labour for the small-time 'fossicking economy' of the region (Levitus 1982), including crocodile and brumby shooting, mining, horticulture and cattle raising. It was common for Aborigines to work unsupervised for long periods. This work environment allowed traditional fire management to continue – at least in the areas where the work camps were situated.

Between the 1880s and First World War, hundreds of Aborigines moved from the country on the plains to semi-permanent camps around the outskirts of Darwin and the various mining communities and depots along the north–south railway. In 1932, Stanner talked with Indigenous people he worked with on the Daly River about the reasons for their apparent voluntary movement from their traditional country a generation before his field work. He records that Aboriginal people told him:

> *that their appetites for tobacco and, to a lesser extent, for tea became so intense that neither man nor woman could bear to be without. Jealousy, ill will and violence arose over divisions of the small amounts, which came by gift and trade … individuals, families and parties of friends simply went away to places where the avidly desired things could be obtained. The movement had phases and fluctuations, but it was always a one way movement. (Stanner 1979, p. 47)*

He recorded that 'as late as the early 1930s I was able to see for myself the battles between the encroaching myalls and weakening, now sedentary groups who had monopolised European sources of supply and work' (Stanner 1979, p. 48). Cooke (Chapter 4) also records that the principal driving force for the move away from the Arnhem Land Plateau – recalled by today's senior generation of landowners – was a desire for *baki* or tobacco. By the 1930s, most of the traditional landowning groups of the north-west Top End were sedentary and were no longer moving through their traditional country and carrying out the burning regimes that had been observed in the mid-19th century.

SETTLER RESPONSES TO THE INDIGENOUS PRESENCE

The increasing number of Indigenous groups establishing themselves more or less permanently around Darwin and the other hinterland townships caused the Government Resident, J.L. Parsons, to recommend to the South Australian Government that 'reserves' be established in the hope that they would provide a place for the repatriation of the Aborigines already living in fringe camps (Parsons 1885).

The areas eventually gazetted as reserves were small and located on the basis that the land had no foreseeable value to the colony rather than on the basis of the lands' importance to the Indigenous groups of the area. Most survive as land scheduled as Aboriginal Land under the *Aboriginal Land Rights Act (NT) Act 1976*. They included the Larakeah Reserve, bordering the Adelaide River near Acacia Gap, and the Woolner Reserve, which included Cape Hotham in the Escape Cliffs area and black soil plains bordering Chambers Bay from the mouth of the Adelaide to the Mary River (now Djukbinj National Park and Mary River Wetlands National Park). The Woolwonga Reserve was established north of the goldfields area between the Adelaide and Mary Rivers. The Wangaits Reserve (now Wagait Land Trust), was established on the freshwater wetlands between the Finniss and Reynolds River. The reserves were proclaimed on 3 March 1892 and gazetted on 20 April (South Australia 1892). It was not until well after the Commonwealth takeover of the Northern Territory, in 1931, that nearby Arnhem Land Reserve around the mission lease at Oenpelli was established (Dewar 1992, pp. 1–3).

The creation of these reserves had very little effect on the trend of Indigenous people moving to Darwin and townships on the Telegraph Line or associating with the other small European settlements in the north-west Top End. Until the latter part of the 20th century, access to, or use of, these 'reserves' by the traditional owners of the country was unrecorded, despite some official incentive to do so. It is likely that visits were infrequent and traditional burning regimes were not able to be maintained, even on this reserved land.

The effect, then, of the first 50 years of European colonisation on the Indigenous population of the north-west Top End was devastating and irrevocable. Initially, the brunt of the changes brought about by colonisation had been borne by the groups on the Adelaide/Mary River area where the first settlement had been established in the mid-1860s. In contrast, the Reynolds/Finniss and Alligator Rivers wetlands remained inaccessible and comparatively distant from the centres of colonial activity, which concentrated around Port Darwin and along the corridor leading to the goldfields. While there were few intrusions into the Reynolds River and Alligator Rivers swamps by colonists during this period, a centre of colonial activity had been established on the Daly River, 50 kilometres south, which included mining, farming and the Jesuit mission. While on the Alligator Rivers, Paddy Cahill's buffalo hunting lease at Oenpelli was taken over as a Government Research Station that operated between 1916 and 1919, but in 1925 it was offered to the Anglicans (Church Missionary Society) for a new mission that operated for the next 50 years. Within a couple of years, the church-going population was between 90–100 people and the school taught children from five language groups (Dewar 1989).

Colonisation upset any equilibrium that existed between the populations on the coastal plains of the north-west Top End and those groups further inland. Centres of colonial activity, including the ports, the mines and railheads, were catalysts that accelerated the slowly pulsing movements that characterised the pre-contact period (Sutton 1990). Evidence suggests that movement of Indigenous people from the coastal plains in the years following colonisation was more complex than simply a drift towards the major township of Darwin and the various rail heads on the north–south line. From the turn of the 20th century onwards, non-Indigenous observers increasingly referred to mixed 'tribal identities' whereas, in the previous period, these were noted to be a rarity (Eylmann, 1908; Kelly, 1911; Parkhouse, 1895a). This suggests a social response to the high mortality and low birth rate that characterised the Indigenous population of the region at the beginning of the 20th century. By the time the Commonwealth Government assumed control of the Northern Territory in 1911, the disruption caused by European settlement was –for the Indigenous groups of the north-west Top End – inescapable.

In the 20th century, the surviving Indigenous communities developed new routines and practices, including ways of maintaining traditional burning practices. This included incorporating their knowledge of fire management into work on pastoral properties. Some managers of pastoral leases in the north-west Top End incorporated Indigenous knowledge of fire into their own management practices. The station diaries of Koolpinyah pastoral lease contain regular entries documenting Indigenous workers being directed to burn off in the Dry to promote new growth (Wells 2003, p. 119). The practice on Koolpinyah was not followed everywhere in the region. Other pastoralists actively tried to prevent traditional burning.

A traditional owner of the Reynolds River area recounted to me that, in the late 1930s, he and other kinsmen working for the Byrne brothers on Tipperary pastoral lease routinely burnt the country after the Wet. This practice was discouraged by the owners of the lease. One season, the fire he set burnt down an outstation hut and a mustering yard and he was caught and savagely beaten by the owners. He fled the property fearing for his life (Nugget Majar *pers. comm.* 9 May 1985).

The Indigenous groups that had remained on the coastal plains between 1910 and 1950 had retained a degree of independence not shared by those who had been drawn into the major townships. The enduring industry on the coastal plains – the hunting of feral buffalo – relied heavily on Indigenous labour. The seasonal nature of the industry discouraged formation of sedentary communities and, as such, was seen as undermining the Commonwealth Governments policy for Aborigines. There were several official attempts to stop Aborigines from gaining employment and forming attachments with 'undesirables' in industries such as feral buffalo shooting and alluvial prospecting.

The Second World War precipitated a number of changes. Army Aboriginal native settlements inadvertently promoted a vision of a more egalitarian future yet to come (Berndt and Berndt 1987). After the wartime Aboriginal native settlements were disbanded in 1946, many Indigenous families returned to the country where they had traditional affiliations. The effects of the massive reconstruction and policy shifts of the post-war period filtered through slowly to the north-west Top End. The concerns of the Indigenous population were periodically monitored by a bureaucracy in Darwin charged with implementing what became known as the 'assimilation policy'.

In the first half of the 20th century, surviving Indigenous groups had either moved off their traditional country more or less permanently or had become accommodated within the small pastoral, buffalo hunting or fossicking enterprises that allowed them to remain on country and continue to carry out traditional practices relating to their country. In the post-War period these arrangements started to break down. By the early 1960s, the Indigenous communities associated with several pastoral leases across the coastal plains were forced to disband as the properties were sold and broken up. Coinciding with this, the collapse of the buffalo hide industry meant that routines and work practices that had developed over half a century were no longer relevant or sustainable. On the Koolpinyah pastoral lease on the Adelaide River, the owners, who had adopted traditional fire management and lived in close association with the Indigenous people living on the station, died and the property passed into the hands of a company. On the Reynolds/Finniss area, the pastoralist family most closely associated with the area since the early 1920s had also sold out to a foreign company – displacing the Indigenous employees and their families. On most of the coastal plains pastoral leases, the lives of the lessees and those of the local Indigenous groups had become closely intertwined. What must have been seen by many Indigenous groups as long-term relationships with the European families who owned these properties were dissolved, creating a new level of uncertainty. By the 1970s, it was apparent that any expectations of continuing relationships based on reciprocity were unlikely to be realised (Hamilton 1972).

In the assimilation period, no value was placed on traditional affiliations with country and, in particular, the desire to stay close to ancestral lands and to live in close proximity to the other members of the group. Traditional burning was an example of the kind of collective practice that had to be abandoned in order for an individual to fulfil a 'productive role in society' (Wells 1995). As the official push to have the Indigenous groups of the north-west Top End become economically self-sufficient, a number of proposals for collectively owned economic enterprises, centring around small scale cattle raising, were promoted by the Welfare Branch and its successor, the Department of Aboriginal Affairs. The official emphasis in these projects was to subsume what was considered to be 'old tribal ways' beneath a new cooperative approach so that 'all members of the community benefited'. This, in turn, was a consequence of the assumption that underpinned the assimilation policy: that social change would be effected by giving individual Indigenous people the tools to become productive members of society (Wells 1995).

Recognition of indigenous land rights

From the failure of the assimilation policy grew an increasing official recognition of the centrality of Indigenous traditions to the identity and motivation of Indigenous people. The shift in Indigenous policy that marked the end of the assimilation period was given legislative expression in the *Aboriginal Land Rights (Northern Territory) Act* 1976. This Act recognised Aboriginal traditional cultural beliefs and, in particular, Northern Territory Aborigines' close affiliation with their land and sites of traditional significance.[11] Before the 1992 High Court ruling that Aboriginal Native title exists under Australian common law, Indigenous people's relation to the land was seen to be of a spiritual nature, manifested in complex ritual relationships.[12] These changes took some time to filter down to the Indigenous groups living on the north-west Top End. The early years of land rights were marked by a continuation of the community-based projects and enterprises that had been gestated in the dying days of the assimilation era. After 1972, the newly created Department of Aboriginal Affairs concentrated its efforts in the north-west Top End on planning acquisition of properties and the excision of living areas at Humpty Doo, Adelaide River and on the Finniss River. As Indigenous groups in the north-west Top End became aware of the potential benefits of the *Land Rights Act*, notions about group identity that had possibly never been articulated previously, were publicly stated for the first time. The immediate effect of land rights was to turn peoples' attention to the past in search of precedents and principles on which to authenticate their claims. One of the legal tests under the *Land Rights Act* for a group to be recognised as the traditional Aboriginal owners of an area of land is that they have a right to hunt and forage over the land. Surprisingly, in the first land claim hearings in the north-west Top End the continuity of traditional burning practices was given little attention as evidence of the claimants' exercise of these traditional rights (Commonwealth of Australia 1980).

The land claim process elicited an acknowledgement of the perseverance of traditional burning practices almost by default. In a reaction to the perceived threat imposed by Indigenous traditional burning, several pastoralists and miners submitted in the Finniss River Land Claim Hearings that the risk from uncontrolled bushfires would be increased if the claim areas became Aboriginal land. The mining company Uranerz Australia Pty Ltd suggested that Aboriginal activity in the area would necessarily increase the fire risk to neighbouring properties. The Land Commissioner in his findings stated:

> *There was no evidence that the risk of uncontrolled fire activity would be increased if the land was owned by Aboriginals nor was there any evidence that Aboriginal owners would be less co-operative in the control of fires which threatened adjacent properties. (Toohey 1981a, par 389)*

Similarly, there was no mention of traditional burning practices in evidence provided to the hearings of the Alligator Rivers Stage II Land Claim: an area that included most of Kakadu National Park.[13] Ironically, the only evidence about fire management was that the granting of the claim may disrupt scientific research on 'fire ecology' that had been ongoing in the area since 1971 (Keen 1980; Toohey 1981b). When the Jawoyn Land claim was heard a few years later, a scientific paper by the Australian National Parks and Wildlife Service on Aboriginal fire practices was tendered as an exhibit. The claimants gave evidence that their custodial responsibilities also included duties to clean sacred sites 'including burning the country'. The Land Commissioner Kearny J observed that 'there are obvious limitations on this practice nowadays' (Kearney 1988). As in the previous claims, attention focused on a submission (from the local volunteer brigade of the Northern Territory Bushfires Council) that there would be an increased risk of uncontrolled fires arising if the land was granted to the traditional owners (Kearney 1988).

Twenty years on from the first land claims, evidence about traditional burning had become more central. In the Kenbi (Cox Peninsula) claim, the Land Commissioner commented on evidence that the traditional owners had the right to burn the land claimed, including the right to decide when, where and to what extent, burning should take place. Gray noted in his report on the claim that 'the right to burn country is a very significant aspect of primacy in relation to responsibility for land' (Gray 2000).

The successful grant of land within the north-west Top End to the traditional owners, as well as the control over the former reserves and joint management of several national parks, has contributed to the revival of traditions relating to burning country. Increasingly, however, traditional owners and surrounding landholders alike have had to contend with the rapidly developing threat from highly flammable, large introduced grasses, especially Mission grass (*Pennisetum polystachyon*) and Gamba Grass (*Andropogon gayanus*).

The higher fuel loads combined with the increased number of small rural landholdings in the greater Darwin area has meant that the non-Indigenous population and authorities have a low tolerance for traditional owners lighting fires in situations where they may cross onto neighbours' property.

The Indigenous population of the north-west Top End is now increasing again after remaining stable since the 1950s at between 650 and 700 people. This statistic obscures the fact that the number of communities has been halved and the population of the coastal wetlands to the east of Darwin has contracted from 165 in 1956 to fewer than 50 today. However, the community at Acacia, located on the old Larrakeyah Reserve, has grown over the last two decades to a population of 72. West of Darwin, the population of Indigenous communities is 680. The community of Nauiyu Nambiyu has grown the most: in 1981 it had an officially recorded Indigenous population of 163 (Northern Territory Government 1981), which, in 2008, stood at 350. The increasing Indigenous presence is outstripped by growth in the non-Indigenous population and the intensification of horticulture and pastoral industries. Probably the only

Figure 2.4 Doris White surveys burnt landscape after a fire through introduced grasses and Mimosa on Adelaide River plains.

areas in the north-west Top End where traditional fire-management practices are either practiced systematically or encouraged are in the jointly managed National Parks. On the larger areas of Aboriginal Land, such as the Wagait, Malak Malak and Arnhem Land Trust, burning has been revived but, as Figure 2.4 shows, increased fuel loads resulting from weed infestations are resulting in major fire impacts on native vegetation communities.

The practice of setting fires by throwing burning matches from a vehicle while driving along bush tracks, described at the start of this chapter, is now the most usual way to 'clean-up the country' and tends to happen later in the season because this is when vehicles can be driven off-road without getting bogged. This work is no longer the exclusive domain of the traditional owners and is often done by visitors. Destructive fires that get out of control are not uncommon. Several years ago, a fire allegedly lit by visitors to the Wagait land trust burnt out of control into the peat fringing the wetlands in the late Dry destroying this valuable habitat and literally cooking alive thousands of aestivating tortoises: an important traditional bush tucker.

CONCLUSION

The effect, then, of colonisation on the Indigenous population of the north-west Top End was devastating and irrevocable. Firstly, disease reduced the population to a tenth of its pre-contact level within two generations. Secondly, the survivors' traditional routines and practices were disrupted by movements away from traditional country and establishment in sedentary communities. Thirdly, most land was alienated, and systems and laws discouraged maintenance of any traditions that offended the non-Indigenous community. Traditional burning practices were, with a few exceptions, discouraged. Regular burning by groups of traditional owners walking through their traditional territory ceased, or was greatly reduced, over much of the north-west Top End by the first decade of the 20th century. In the first land claim hearings, most references to Indigenous burning were to the threat this practice posed to neighbouring landowners. Over the next two decades, the 'right to burn country' became recognised as a very significant aspect of a traditional owner's primary spiritual responsibility for their land. Ironically, as this right was being recognised in the last decades of the 20th century, increased use of motor vehicles meant that fewer parts of the landscape were visited, and for shorter periods. While there has been a revival of interest in these practices over the last three decades, the practice is rarely carried out on foot. The right to burn country in accordance with tradition has become central to the political construction of Indigenous identity and there is a developing relationship between Indigenous groups, academic fire researchers and officials managing National Parks.

ENDNOTES

1 The Aboriginal Land Rights (NT) Act 1976 established Land Rights: a judicial process for identifying traditional Aboriginal owners on the basis of the strength of their traditional affiliation with sites on the land claimed.

2 Australian Bureau of Statistics accessed at <www.abs.gov.au/ausstats/abs@nsf/PrimaryMainFeatures/3222.0?OpenDocument> in September 2008.

3 For discussion of alternative models relevant to the Larrakia groups of Darwin and Cox Peninsula see Sutton (1996).

4 In anthropological literature, this core country has been referred to as the clan's 'estate'. The boundaries of such estates may be fluid, and members of such clan groupings have overlapping rights to adjoining estates.

5 For instance, in the Wagiman Language 'Wibin refers to areas where the grass has been burnt. Movement through wibin country is generally easy and visibility is good: allowing hunting and tracking to be undertaken with ease. Further, following the fires, there is usually some grass regrowth, and this green pick attracts grazing animals such as wallabies, which can then be easily hunted. Wibin is safe country to camp in as it cannot be razed by fire and approaching visitors or animals can be easily seen' Martin L (2006).

6 Harvey M (2002a) A Grammar of Gaagudju. Berlin/New York, Mouton de Gruyter. has surveyed the literature on estimating Aboriginal population in the coastal region between the Adelaide River and the Alligator Rivers region. A population density of one person per 4 square kilometres has been used to arrive at the above figure although as Harvey points out, evidence suggests population densities range from between one person for 8.5 square kilometres to one person per square kilometre.

7 In contemporary documents Kapalga is spelt 'Kaparlgoo'.

8 The contrast between the stature and physique of the coastal and inland people was remarked on by Parkhouse (1895b) 'the Aborigines of the Alligator and Mary River tribes, which possess similar characteristics ... are the finest in physique and apparently are the most numerous. They are feared by the other tribes, whom they hold in small esteem.'

9 Archaeological investigations of Chinese habitation areas on the goldfields reveal evidence of a high proportion of bush foods being consumed (Mitchell 1996).

10 The first buffalo shooters operated on Cobourg Peninsula and Melville Islands (Bauer 1964, p. 181). However, in 1884, hunters were operating on the Adelaide River (Hill 1951, p.

371) and by 1894 Paddy Cahill was operating in the Alligator Rivers area (Bradshaw 1905; Dahl 1926, p. 156).

11 These sites of traditional significance were termed 'sacred sites' in the legislation.

12 This was the legal situation reiterated in Milirrpum v Nabalco, otherwise known as the 1968 Gove Land Rights Case. In his judgement, Justice Blackburn. found that 'the relationship of the Rirratjingu and the Gumatj to particular areas of land cannot be shown to be some form of proprietary interest' and declared the connection between the plaintiff clans and their land to be primarily spiritual (Blackburn 1971).

13 Yarriburrik (fire) was recorded as one of the four 'Kunwinjku semi-moieties'.(Keen 1980, p. 94)

REFERENCES

Bauer FH (1964) 'Historical geography of white settlement in part of northern Australia part II'. CSIRO, Canberra.

Berndt CH and Berndt RM (1987) *End of an Era Aboriginal Labour in the Northern Territory.* Australian Institute of Aboriginal Studies, Canberra.

Blackburn J (1971) 'Reasons for judgment Milirrpum v Nabalco'. Northern Territory Supreme Court, Darwin.

Bradshaw J (1905) The north coast of Arnhem Land. *Victorian Geographical Journal* **23**, 1019.

Campbell J (2002) *Invisible Invaders: Smallpox and Other Diseases in Aboriginal Australia 1780–1880.* Melbourne University Press, Melbourne.

Commonwealth of Australia (1980) 'Transcript of proceedings Finniss River land claim'. Office of the Aboriginal Land Commissioner, Darwin.

Dahl K (1926) *In Savage Australia: An Account of a Hunting and Collecting Expedition in Arnhem Land.* Houghton Mifflin, Boston.

Daly HW (1887) *Digging Squatting and Pioneering Life in the Northern Territory of South Australia.* Sampson, London.

Dewar M (1989) Strange bedfellows: Europeans and Aborigines in Arnhem Land before World War II. History. MA (Hons) thesis. University of New England, Armidale.

Dewar M (1992) *The 'Black War' in Arnhem Land: Missionaries and the Yolngu 1908–1940.* North Australian Research Unit, ANU, Canberra.

Eylmann E (1908) *Die Eingeborenen de kolonie Sudaustralien.* Dietrich Reimer (Ernst Vohsen), Berlin.

Foelsche P (1881) Notes on the Aborigines of north Australia. *Transcript and Proceedings and Report of the Royal Society of South Australia* **5**, 1–18.

Freier P (2008) The Northern Territory Native Industrial Mission, Kaparlgoo, 1899 to 1903. *Journal of Northern Territory History* **19**, 17–32.

Gray J (2000) 'Report of the former Aboriginal Land Commissioner to the Minister for Aboriginal and Torres Strait Islander Affairs on the The Kenbi (Cox Peninsula) Land Claim No. 37'. Commonwealth of Australia, Canberra.

Hamilton A (1972) Blacks and whites: the relationships of change. *Arena* **30**, 34–48.

Harvey M (1986) Ngoni Waray Amungal-yang the Waray language from Adelaide River. MA thesis. Department of Linguistics, Australian National University, Canberra.

Harvey M (1992) The Gaagudju people and their language. Department of Linguistics. PhD thesis. University of Sydney, Sydney.

Harvey M (2002a) *A Grammar of Gaagudju.* Mouton de Gruyter, Berlin/New York.

Harvey M (2002b) Land tenure and naming systems in Aboriginal Australia. *The Australian Journal of Anthropology* **13**, 23–44.

Herbert SW (nd) *Reminiscences of Experiences in the Northern Territory*. Unpublished manuscript held in South Australian Archives (accession number 996), Adelaide.

Hill E (1951) *The Territory*. Angus & Robertson, Sydney.

Hutchinson J (1864) *Remark Book of Surveying Schooner Beatrice*. South Australian Parliament, Adelaide.

Kearney J (1988) 'Report to the Minister for Aboriginal Affairs on the Jawoyn (Katherine Area) Land Claim'. Commonwealth of Australia, Canberra.

Keen I (1980) 'Alligator Rivers stage II land claim – anthropologist's report'. Northern Land Council, Darwin.

Kelly JH and Beckett JT (1911) 'Report to the Chief Protector of Aborigines re Aborigines about Darwin and vicinity'. Department of Aboriginal Affairs, Adelaide.

Levitus R (1982) 'Everybody bin all day work: the social history of the Alligator Rivers region of the Northern Territory 1869–1973'. Australian National Parks and Wildlife Service, Canberra.

Martin L (2006) *Wagiman Plants and Animals: Aboriginal Flora and Fauna Knowledge from the mid Daly River Basin, Northern Australia*. Department of Infrastructure, Planning and Environment and Diwurruwurru Jaru, Katherine and Darwin.

McGrath A and Coulthardt L (1980) *Brief History of the Non-Aboriginal Activities in Vicinity of Finniss River Land Claim Area*. Northern Land Council, Darwin.

McKinlay J (1866a) *Journal and Report of Explorations 29 June 1866–25 August 1866*. South Australian Parliament, Adelaide.

McKinlay J (1866b) *Journal of Exploring Expedition 23 September 1865–29 June 1866*. South Australian Parliament, Adelaide.

McMinn GK (1882) 'Report on Exploration of the Mary River'. Australian Archives CRS A1640 Item 83/131.

Mitchell, S (1996). Archaeology of the Darwin hinterland: investigation at goldfields Chinatown. North Australian Research Unit, Australian National University, Darwin.

Northern Territory Government (1981) 'Index of Northern Territory Aboriginal communities, outstations and pastoral groups'. Department of Community Development, Darwin.

Northern Territory Times (1889) Letter re opium use. *Northern Territory Times*. Darwin.

Northern Territory Times and Gazette (1884) Committal proceedings against Jemmy, Tommy, Jacky, Daly and Long-Legged Charlie. *Northern Territory Times and Gazette*. Darwin.

O'Kelly GJ (1967) The Jesuit mission stations in the Northern Territory 1882–1899. BA (Hons) thesis. History Department, Monash University, Melbourne.

Parkhouse TA (1895a) 'Native tribes of Port Darwin and its neighbourhood'. *Report of the Australian Association for the Advancement of Science* **6**, 638–647.

Parkhouse TA (1895b) Remarks on the native tongues in the neighbourhood of Port Darwin. *Transcriptions of Royal Society of South Australia* **19**, 1–18.

Parsons J Langdon (1885) 'Aboriginal reserves'. Government Resident's Report, Darwin.

Price EM (1878) 'Report of Government Resident Palmerston, Northern Territory'. Office of Government Resident, Adelaide.

Reid B (1996) The health of the Aboriginal people of the Northern Territory during the South Australian government period of administration 1863–1910. *Northern Perspective* **19**(2), 26–31.

South Australia (1881) 'Report of the Mining Warden'. Parliamentary Papers, South Australian Parliament, Adelaide.

South Australia (1892) 'Proclamation of Aboriginal reserves in Northern Territory'. Adelaide, South Australian Government, Adelaide.

Stanner WEH (1933) The Daly River tribes: a report on field work in northern Australia. *Oceania* **3(4)**, 377–405.

Stanner WEH (1979) *White Man Got No Dreaming: Essays 1938–1973.* Australian National University, Canberra.

Sunter GH (1937) *Adventures of a Trepang Fisher: A Record without Romance Being a True Account of Trepang Fishing on the Coast of Northern Australia: and Adventures Met in the Course of the Same.* Hurst and Blackett, London.

Sutton P (1990) The pulsating heart: large scale cultural and demographic processes in Aboriginal Australia. In *Hunter-Gatherer Demography: Past and Present.* (Eds B Meehan and N White) pp. 71–80. Oceania Publications, Sydney.

Sutton P (1996) 'Familes of polity: post-classical Aboriginal society and native title'. National Native Title Tribunal, Perth.

Toohey J (1981a) 'Finniss River land claim report by the Aboriginal Land Commissioner to the Minister of Aboriginal Affairs and the Minister of the Northern Territory'. Office of the Aboriginal Land Commissioner, Canberra.

Toohey J (1981b) 'Report by the Aboriginal Land Commissioner to the Minister for Aboriginal Affairs on the Alligator Rivers stage II land claim'. Office of the Aboriginal Land Commissioner, Canberra.

Wells JT (1995) The long march: assimilation policy and practice in Darwin, the Northern Territory. PhD thesis. Department of History. University of Queensland, Brisbane.

Wells S (2003) Negotiating place in colonial Darwin. Interactions between Aborigines and Whites 1869–1911. PhD thesis. Faculty of Humanities and Social Sciences. University of Technology, Sydney.

Young C (1866) *Escape Cliffs Journal November 1865–August 1866.* South Australian Parliament, Adelaide.

3

Change and catastrophe:

adaptation, re-adaptation and fire in the Alligator Rivers region

Robert Levitus

SUMMARY

The archaeologically recent formation of lowland freshwater wetlands on the major rivers draining the Arnhem Land Plateau to the west introduced a major new component to the Aboriginal resource base in the region. It instigated a broad pattern of seasonal movement that integrated those swamps, lowland forests and sandstone outliers, escarpment valleys and large parts of the plateau into an expanded human ecological range. That seasonal adaptation acted as a conduit for the application of new patterns of fire to the Alligator Rivers landscape. The demographic underpinnings of that adaptation were removed by epidemic disease introduced by non-Aboriginal visitors and settlers. One can imagine the effects of catastrophic population decline not only as decimated numbers, but in a remnant society weakened by internal conflict and lost vitality. From the 1880s to the 1950s, the buffalo hides industry offered another form of bush living to many of the region's people, but one that involved a modified form of articulation with country, including in the continued use of fire on the lowland plains. The region was substantially depopulated of Aborigines by the mid-1970s, when policy changes opened the way for a substantial movement back into the newly declared Kakadu National Park. There, the acceptance of fire as an instrument of environmental management has created a new significance for traditional usage as a framework for debate among Park staff and Aborigines.

INTRODUCTION

The previous chapter by David Ritchie has provided an account of how the successive phases of colonisation in the Darwin hinterland and across the north-western Top End altered the demographic conditions underlying the incidence of fire in that environment. I want here to look at certain aspects of that history. Moving eastwards, my geographical focus is on the northern Alligator Rivers region (Figure 3.1). This is the area between the West and East Alligator River catchments and includes the western section of the Arnhem Land Plateau, and the

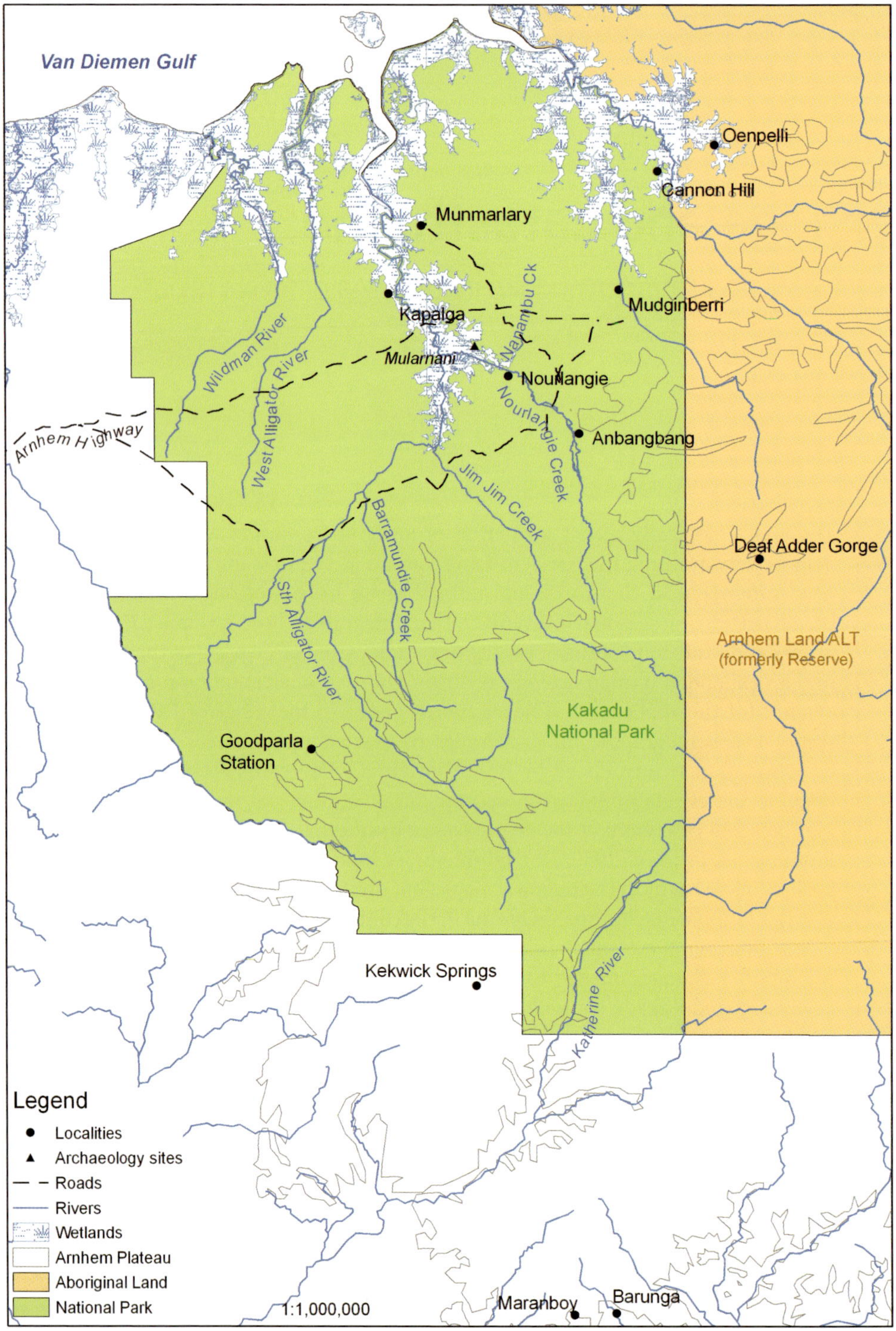

Figure 3.1 Map of the Alligator Rivers region showing the present boundaries of Kakadu National Park (green) in relation to the adjoining western portions of Arnhem Land. Place names mentioned in the text are shown.

northern part of what is now Kakadu National Park. It includes also the eastern half of that area known as the buffalo country, which consists of lowland forests, coastal floodplains and sub-coastal freshwater wetlands, over which herds of introduced feral water buffalo were hunted for their hides from the 1880s to the 1950s.

My timeframe is different also. I am interested in a broad historical succession in three phases. Much of the discussion makes no mention of fire, but seeks to identify successive themes in the cultural economy of the Alligator Rivers region that produced a social context for burning. The chapter begins with the archaeologically recent formation of lowland freshwater wetlands, and the way in which they influenced the late prehistoric human ecology of the region. Central to this ecology was a pattern of seasonal movement across physiographic zones from the plateau to the wetlands. Seasonal movement produced the traditional pattern of burning that was in place at the time of European arrival in the Top End. This recent landscape transformation and the human adaptation that it instigated are discussed in the first two sections.

Contact with European and Chinese colonists, and before them with Macassan trepang fishermen, produced a dramatic demographic collapse among Indigenous groups in the region. The cause of this collapse was introduced epidemic disease, and it marked a watershed between traditional and post-contact adaptations. In the third section, I indicate the scale of the decline, and suggest also the ways in which it might have affected the coherence and vitality of the society of the surviving groups. The next two sections discuss aspects of the new adaptation that developed from engagement with the colonisers in the buffalo-shooting industry. I identify the conditions that promoted this engagement, and the ways in which it allowed Aborigines to maintain a bush life, albeit one in which they knew and burned the bush in ways consistent with new production relationships.

The buffalo industry was the dominant arena of introduced economic activity in the Alligator Rivers region for several decades. When it finished, Aboriginal residence was for a time concentrated around other kinds of activities, but by the mid-1970s the region was substantially depopulated. The establishment of Kakadu National Park not only opened the way for Aborigines to return, but into an area in which environmental management was now the first priority. In the final section, I consider burning as a management issue. Both of the previous historical phases have had a bearing on this debate: the first as the source of an ideal model and the second as a cause for scepticism that modern Aborigines could speak authoritatively about that tradition. I argue that seasonality has supplied a framework that contains both contemporary Aboriginal and European management perspectives within a common discourse.

CHANGE AND ADAPTATION

Wetland formation and ecological range

The views of archaeologists concerning the duration of human occupation in the Top End vary between 40 000 and 60 000 years (Roberts *et al.* 1990; Hiscock and Kershaw 1992, pp. 49–50; Mulvaney and Kamminga 1999, pp. 139–42). Over such a timescale, the region has experienced major landscape changes contingent upon the end of the last major glaciation and accompanying changes in sea level (Hope *et al.* 1985; Clark *et al.* 1992; Woodroffe *et al.* 1993). Those changes have, in turn, established a succession of limiting environmental conditions for human ecology (Hiscock 1999; Brockwell 1989, pp. 224–28). However, for contemporary issues and patterns of fire management, the relevant periodicity starts only recently in archaeological time, with the latest major phase of landscape history and the establishment of a recognisably contemporary geography.

That occurred with the recent development of freshwater wetlands along a number of watercourses between their sources in the Arnhem Land Plateau and their downstream tidal reaches. Researchers have dated wetland formation with increased biomass on the South Alligator River and Magela Creek to between 1500 and 1000 years ago (Hope *et al.* 1985, p. 236; Clark *et al.* 1992, p. 108; Jones 1985, p. 293). A recent summary of landscape evolution (Brockwell *et al.* 2005, pp. 84–85) at the mouth of the Blyth River further east mirrors the sequence established in the Alligator Rivers region, leading, in this case, to the development of a small freshwater swamp behind the coastal sand ridges about 1400 years ago. The authors comment on the generality of this type of event in the late Holocene, when 'freshwater conditions became widespread on the northern floodplains, as a result of reduced tidal influence due to sedimentation and coastal progradation, and ponding of freshwater behind chenier ridges' (Brockwell *et al.* 2005, p. 84).

These dates from the South Alligator River and Magela Creek, and from the Blyth River, in addition to somewhat earlier dates for wetland formation on the Adelaide and Mary Rivers further west (Brockwell 2001b, p. 333), indicate that within the last 2000 years, a series of new freshwater swamp sites of varying size appeared on the lowlands bordering the Arnhem Land Plateau to its west and north. They vary in size from those, such as the Woolwonga wetlands on the South Alligator, that are large and complex landscapes in themselves to isolated swamp depressions of a few hectares.

The importance of this development for understanding burning regimes lies in the rich new resources available around these lowland freshwater swamps for Aboriginal exploitation (Jones 1985, p. 294). There, people found waterbirds attracted by feeding and breeding habitat, as well as abundant fish, turtles, file snakes, water lilies, rushes, and other animal and plant food species (see Brockwell 2001b, pp. 329–330 for the most important). Jones (1985, pp. 291–93) argues that the wetlands on the South Alligator River allowed a reorganisation of the Aboriginal economy and a major increase in regional population density. Meehan *et al.* (1985, pp. 135–36) estimate that the large open sites along the eastern side of the South Alligator wetlands south of Nourlangie Creek were probably host to a dry season population of between 600 and 900 people. In order to emphasise the focal significance of the wetlands for Aboriginal demography, Jones says:

> *that the population density in the Alligator River region, certainly around the wetlands, may have been as high as one or two people per square kilometre of country. This would be of the order of about 40–60 times the density of human occupation on the escarpment and in the dry inland country to the south and east of these estuarine plains. (Jones 1985, p. 291)*

Jones does not talk here about an increase in population size, only in density. Moreover, it is evident from other parts of Jones' discussion, and from the findings of his research team, that the model being proposed incorporated people from the drier eastern and southern zones into a seasonally migratory strategy. The density of occupation on the freshwater wetlands was thus increased by a migratory inflow to those resource zones.

Ethnographic data and the archaeology of the Anbangbang and other outlier sites on the Kakadu lowlands (Jones and Johnson 1985a, pp. 57, 63, 76) indicate that the formation of the South Alligator wetlands, after some intervening period, began attracting the dry season attention of groups whose own countries lay to the east and south-east, over the forested lowlands, plateau outliers and valleys, and the west Arnhem Land Plateau itself. Excavations in the plateau valley of Deaf Adder Gorge showed a peak period of site usage between 3500 and

Figure 3.2 Magpie geese and other waterbirds abundant on floodplains of the Alligator Rivers region.

1000 years ago, after which intensity of usage dropped by two-thirds at the same time as the large open wetland sites came into prominence on the eastern side of the South Alligator wetlands. For Jones and Johnson (1985b, p. 211), 'it was as if this riverine plain region was sucking people out of the plateau valleys at this time'. Hiscock (1999, p. 100) has also noted a decline in usage of sites around Darwin harbour at this time, raising the possibility that coastal people were also being attracted to upstream wetlands.

With people, came fire. The burning of Magela floodplain began about 700–800 years ago, some time after the establishment of productive freshwater swamp conditions (Clark *et al.* 1992, p. 86). The findings reported for the South Alligator are less clear. Hope *et al.* (1985) imply that the stabilisation of the modern wetlands with grasslands and lagoons, increased human usage and increased burning, were all contemporaneous. Their statement that charcoal deposits show that 'burning has been frequent for at least this time' (Hope *et al.* 1985, p. 236) seems from the context to refer to the last 1400 years. However, Meehan *et al.* (1985, p. 152) and Jones (1985, p. 293) argue that Aboriginal occupation of the South Alligator wetlands greatly increased within the last 1000 years, thus suggesting a gap between wetland formation and the human response, as in the Magela case.

Seasonal movement

This economic reorganisation involved a change of seasonal strategies. The people living around the eastern side of the South Alligator wetlands in one season included people who lived 'in the dry inland country to the south and east' in others. Seasonal migration integrated them into a single expanded resource range. This means that the formation of the freshwater wetlands on the riverine lowlands entailed a geographic extension of how the 'region' is to be understood in human ecological terms. One indicator of this is the distances over which stone artefacts must have travelled in order to be ultimately deposited at the wetland sites. The nearest sources of stone for these artefacts give a minimum range over which different ecological zones were economically integrated either by trade or by seasonal movement.

> *The presence of exotic raw materials (quartzite, chert, sandstone, haematite and ochre), and the fact that artefacts from these raw materials were not manufactured in situ, suggest that people using the wetland sites must have had contacts throughout Kakadu and perhaps beyond, or ranged over an extremely large area. Variation in distributions and colour of chert pieces indicate that occupants of sites north and south of Nourlangie Creek obtained this raw material from different sources. For the southern sites, the nearest known source was some 50 km to the south at Barramundie Creek and the upper South Alligator River. Mularnani, lying north of Nourlangie Creek, was close to the chert sources of the East Alligator River. Quartzite was available from the escarpment and the outliers some 30–50 km to the east of the South Alligator wetlands. (Brockwell 1989, p. 211)*

Brockwell *et al.* (2001) draw together the various sources of evidence for this adaptive pattern as it was enacted annually across a range that included sites in a plateau valley of the west Arnhem escarpment, outlier rock-shelters, around the edges of the wetlands and in the neighbouring woodlands.

On the South Alligator River, the wetland edges were occupied year-round. As the dry season advanced, the floodplains dried out and the water bodies retreated. People, their numbers swelled by incoming visitors, camped extensively around the swamps. With the arrival of the monsoonal storms (Figure 3.3) and replenishment of water bodies, the population dispersed. Some groups – presumably mostly local estate owners – could then have based

Figure 3.3 Storm approaching South Alligator River wetland at the end of the dry season.

themselves on elevated mounds located in the adjacent woodlands from which they hunted the fauna of the lowland forests during the main monsoonal storm period (Brockwell *et al.* 2001, p. 369). The floodwaters also offered a valuable fishing resource. These sites on the wetland margin were especially important for harvesting goose eggs from their swamp nests at the end of the wet season, and they remained useful during the early dry season until floodwaters began to contract (Brockwell 2001b, p. 336; 2006, pp. 53–54).

For the forested lowlands and plateau valleys, Clarke's analysis of plant remains at rock-shelter sites showed a range of usable local resources that would have allowed these zones to be occupied at almost any time of the year (Brockwell *et al.* 2001, p. 371). Permanent occupation was unlikely. This interpretation is supported by the frequency in archaeological deposits of bipolar cores. These artefacts indicate that a limited supply of stone on site has been exploited with a technique that allows continued production of flakes from smaller cores (Hiscock 1996, p. 152). The longer people stay in one place, the more they will resort to this technique. There is thus an inverse correlation between the frequency of bipolar cores and the level of mobility between sites. This is true independently of distance from stone sources. The relative abundance of these cores indicates the lowest level of mobility at the large open sites on the floodplain margins, much greater mobility in usage of outlier sites and of the woodland sites near the floodplain, and highest mobility around plateau valley sites (Hiscock 1996).

Although the study sites discussed above allow us to presume with some confidence a widespread traditional practice of seasonal movement around the western edges of the Arnhem Land Plateau, it is the plateau itself for which detailed analytical data has been most lacking. We have recent ethnohistory and the observations of explorers from which to infer that parts of the western plateau hinterland were included within the range of this seasonal adaptation. Chaloupka's (1981) reconstruction of a Badmardi band's annual movements, compiled from the testimony of a senior man in the 1970s, includes wetlands, lowland forests, escarpment and plateau in its route. In part, this account has the group travelling from south to north across the western plateau, following the courses of the upper Katherine and East Alligator Rivers (Figure 3.4), in mid dry season (1981, p. 168).

As the dry season advanced, however, there were some upland areas from which people appear to have moved away. Two explorers crossed sections of the plateau at the very end of the dry season. In November 1845, Leichhardt crossed the middle of the plateau from the Roper River to the South Alligator catchment. He descended from the escarpment into Jim Jim Creek valley on 21 November and met four Aborigines. By this time, his last encounter had been 18 days previously, and his last record of recent encampments on 6 November, about 70 km to the south-east (Leichhardt 1847, pp. 466–86). Lindsay (1884, pp. 16–19), recorded no signs of Aboriginal occupation during a 16-day north-to-south transect from the Liverpool to the Katherine Rivers in November 1883, though Cooke (Chapter 4) views a local rock art site as evidence that his passage was observed on the third day, above the Cuthbertson Falls. In the 1950s, Fred Pocock operated a sawmill at Kekwick Springs on the headwaters of Snake Creek, a tributary of the Katherine River. During one dry season, two groups of Aborigines travelled there from Maranboy to the south. A group that came in June had plenty of water, but two men who travelled the same route in November nearly perished.

Other parts of the plateau, however, offered living resources at any time of year. Garde (Chapter 5), while acknowledging a pattern of dry season movement from the plateau to the lowlands to the north and west, also documents the ecological knowledge and seasonal practices of people that preferred to maintain year-round residence in the rock country. There was overlap and exchange between all these neighbouring networks. However, once we begin to consider families whose identities and practices related them wholly to the central and northern

Figure 3.4 East Alligator River flowing through the Arnhem Land Plateau.

Figure 3.5 The western Arnhem Land escarpment, dividing the Arnhem Land Plateau from the Alligator Rivers lowlands.

plateau, we have crossed beyond the reach of the seasonal adaptation that integrated people of the plateau into the extended human ecological range of the west Arnhem escarpment and Alligator Rivers lowlands (Figure 3.5).

If we then treat the physiographic components of this adaptation in the plateau, plateau valleys and lowland outliers, and on the wetlands, as components of a regional system of land use, an annual dynamic becomes apparent. In the wet season, bands were dispersed across western Arnhem Land, whether camped on elevated mounds around the edges of lowland floodplains, living in rock-shelters further back in the lowland forests or along the escarpment, or within the stone country. In the dry season, there was a steady net movement towards the lowlands and the downstream swamps. By the end of the dry season, there were substantial concentrations of people on large open sites around the edges of those major wetland areas that were large enough to retain some water in an otherwise increasingly desiccated landscape. When the replenishing storms returned, travelling bands again followed habitual routes and pursued the cultural relationships and obligations of their own estates (Brockwell *et al.* 2001, p. 376). There was thus an annual pulse of concentration and dispersal across the Alligator Rivers region driven by the contraction and replenishment of productive bodies of freshwater, focussing at the driest time on the lowland riverine swamps.

Given, however, the size of the population and the dispersed and diverse ecological niches present across such a region, it is important to avoid any appearance of demographic absolutism in our efforts to model human ecology. Chaloupka's account of the Badmardi year provides a useful counterpoint here. Although this band did take advantage of the goose harvest on the shrunken swamps before the wet season set in, its late dry season hot weather time is said to have been spent hosting visiting groups for ceremony in its escarpment valley estate (Chaloupka

1981, p. 170). One particular technique of organised kangaroo fire drive (see Chapter 6) was also practised at this time. An island of grass left unburnt between creek anabranches was set alight at one end, sending sheltering wallabies towards hunters waiting at the other (Chaloupka and Giuliani 1984, pp. 72–3). These data remind us that seasonality in west Arnhem human ecology instigates net population movements, and their description can easily omit the particularities of small group spatial choices, preferences and obligations, especially those of short duration.

But Chaloupka's Badmardi story and Garde's linguistic ethnography lead to the further understanding that such particularities are not just anomalies outside the model. They suggest, rather, a different understanding of the drivers behind seasonal movement. I have written above of the increasingly desiccated landscape of the late dry season, and of the contraction and replenishment of freshwater sources. This is of course descriptively accurate, but it also implies that the availability of water was the cause of seasonal movement. This is also correct, but only partly so. It is true that as the dry season advanced, fewer living sites were available, and some large areas became unattractive.

However, along a series of creek and river courses, deep billabongs or perennial springs offered permanent fresh water in the forests and on the plateau. So the thing that attracted hundreds of people to the downstream swamps in October and November was not the absence of water elsewhere, but a massive and sustained protein harvest. There was seasonal drought in some places, at the same time as geese and turtles on the swamps exerted a special attraction. In between those forbidding and attracting places was a third class of campsite, those many places where smaller groups could make do until the rains returned, if they so wished. I have written elsewhere that seasonal movement 'needs to be understood as a strategy that selects from among the possibilities for material provisioning according to the correspondences that can be made between those possibilities and the wider concerns of life' (Brockwell *et al.* 2001, p. 376). Chaloupka and the contributions in this volume remind us that any large-scale model of seasonality in human ecology will tend to obscure the detail and diversity of those possibilities.

I have dwelt at length here on this seasonal dynamic simply because the pattern of burning followed the pattern of movement and residence. Fire travelled with people. Seasonal movement was a conduit for the human application of fire to the west Arnhem landscape. The importance of freshwater wetlands to this migratory impulse – though dramatically illustrated across the large-scale canvas of the Alligator Rivers and the west Arnhem escarpment country – varied in other environments. The sea coast of Arnhem Land around the Blyth River, east of Maningrida, provides probably the greatest contrast. There, the small freshwater swamp of Balpildja, which is located about 10 km south-west of the nearest coastal home base, offered resources that, at times, significantly supplemented a dietary regime that was otherwise dominated by seafood throughout the year (Meehan 1991, pp. 199–202).

CATASTROPHE AND RE-ADAPTATION

The collapse of numbers

For the generations of Aborigines who lived in the buffalo country, the one overwhelming impact of contact, even indirect, with foreigners was a massive population decline, during which a number of language groups appear to have entirely disappeared from the region. The exact extent of the decline in numbers is probably impossible to calculate. The worst of it seems to have happened quickly, before 1920, and largely outside the range of local genealogical memory. Keen (1980a, pp. 171–72; 1980b, pp. 37–40) used comparative ecological and demographic data, as well as historical reports, to estimate a population decline of 96% – from around 2000 to 80 – for the sub-coastal lands between the Adelaide and East Alligator Rivers, by the time he did his research in 1979. Keen was referring to traditional owners – and the actual number of people on the ground was partly restored by incoming people from other areas – but the cultural impact of this collapse deserves attention before I turn more specifically to the effects of the buffalo industry.

To a large extent what I have to say here must be speculative, because, as mentioned above, the most dramatic drop in Aboriginal numbers occurred in the decades immediately following the arrival of Europeans, and scanty documentation and the absence of a genuine Aboriginal oral tradition have left little in the way of contemporary observations. I think it is worth distinguishing between three possible causes of post-contact decline in an Aboriginal population. The first is large-scale violence and massacre. This is of little relevance in the buffalo country, for reasons I will discuss later. Of those massacres in the Northern Territory that have been documented, the closest occurred around Mount Bundey and the McKinlay River immediately to the south in 1884. A survey of the literature by Sutton and Palmer (1980, pp. 3–5) concludes that these punitive expeditions almost destroyed the 'Woolwonga' people – a group that Spencer (1914, p. 53) identified in 1912 as 'Warrai', who consisted of a 'remnant occupying the country between a place called Rum Jungle and Brock's Creek'.

One Aboriginal woman alive in the 1980s spoke of another massacre, but hers was an isolated recollection, and it referred to Goodparla station during its period of operation as a cattle station, when it was on the southern edge of the buffalo country. There is no shared memory and recounting of any 'killing time' as is found in other regions, no testimony to violent subjugation or random terrorism in the bush, and Keen (1980b, p. 36) concludes that the incidents that are known could not in total have made a significant contribution to population decline. The first possible cause, therefore, which might in other areas have generated Aboriginal resistance and produced in the end a subdued remnant, did not operate to that effect in the buffalo country.

The second possible cause is disease, which travels under its own impetus as a contagion and thus could have massive impacts upon Indigenous people even beyond the range of the colonisers themselves. For the people of the buffalo country, there were three main sources from which such contagion could have originated. Over a period of perhaps 200 years, Macassans occupied wet season camps on north Arnhem Land beaches while harvesting and processing trepang, and entered into trading relations with Aboriginal people. In the 1820s and the 1840s, the British established and then abandoned naval outposts at Raffles Bay and Port Essington on the north coast of Cobourg Peninsula. Finally, permanent non-Indigenous settlement in the Top End began in 1869 when South Australian colonists established Palmerston at Port Darwin. The small European population was enlarged greatly by Chinese workers imported for railway construction to the south-east, and both groups increased in the 1880s when a gold rush occurred between Adelaide River and Pine Creek.[1]

Ritchie (Chapter 2) has discussed the repeated impact of epidemics across the north-west Top End, beginning probably with the Macassans (Campbell 2002, pp. 72–82, 107–9, 170–74), though the first documented instance was of an influenza epidemic that spread from Port Essington on Cobourg Peninsula in 1847 (Spillett 1972, pp. 144–45). Both this, and other diseases, continued in the late 19th century to reduce the numbers and vitality of Aboriginal groups, particularly those around the mining areas whose resistance was impaired by drugs and malnutrition.

In the Aboriginal scheme, death by disease will likely appear as unexplained death and therefore be open to interpretation as the result of sorcery. The willingness of Gagudju speakers at Oenpelli to make sense of a death in those terms is portrayed in Spencer's (1914, pp. 246–47) description of the 'Kuderi' mourning ceremony in which the men and women, having brought their respective possessions, and the women a supply of foods, to the ceremonial ground, sit down and begin to wail.

> *Everyone is painted with red ochre and gradually they work themselves up into a state of great excitement and anger. The men ask each other 'who stole his korno?' That is, who killed him by evil magic. The question is repeated time after time.*

The evil magic involved here, though it might come from some distant source, would more likely be seen to have a local origin. Spencer (1914, pp. 257–62) documented four techniques of sorcery, and I was able to add three more. All depend on obtaining something connected with the victim – excrement, food or cloth – and therefore presume an evildoer within the victim's immediate social field. The word 'korno', or goonu, means faeces (Harvey 2002, p. 470). Any untimely death is appropriate for the perception of sorcery. A senior man stated the general principle: 'If somebody old like me dies', he said, 'well, that's just because we're old; but if a young person dies, there's something wrong'. One of Spencer's examples shows how, in such an environment of generalised suspicion, sorcery could be deployed as retrospective diagnosis for introduced disease.

> *Still another form of evil magic is associated with the mud that attaches itself to the foot of a native walking through a swamp. When he comes on to dry ground he naturally scrapes the mud off, generally using something such as a piece of paper bark to do so. If another man, who wishes to injure him, comes across his tracks, he gathers up some of the mud or paper bark to which it is attached. He wraps it in some more paper bark and ties it round with string. In his camp when it is quite dry, he pounds it up until he can roll it into a ball and then … places it in a hole that he makes in the base of an ant hill. By and by the victim's foot breaks out into sores which gradually spread all over it. The toes drop off, and the hands and feet decay. No medicine man can do anything to counteract this form of magic. It is a disease which is every now and then met with amongst the Kakadu natives and is, superficially at least, suggestive of leprosy. (Spencer 1914, pp. 260–61)*

Finally, it is in this context of socially divisive interpretations of frequent premature deaths that we can best understand the most common Aboriginal explanation for population decline; that is, that the old Alligator Rivers people constantly feuded and killed one another off, which was a factor also noted by Spencer (1914, p. 11). Ritchie (1998, p. 12) has recorded an equivalent statement about large-scale inter-tribal killing for the Adelaide and Mary Rivers to the west. In other words, it may be that the difference between European violence and an epidemic is that Aboriginal vengeance is directed externally in response to the former, but internally in response to the latter.

The third possible cause of population decline is a reduced birthrate. This is obviously related to the second in that disease may have been a major cause of infertility, but malnutrition and abortion are also relevant. Spencer (1914, p. 10) comments on it as 'a remarkable feature' of many of the groups in contact with Oenpelli in 1912, and the Census figures taken at annual blanket distributions from 1899 to 1905 also indicate a small proportion of children (Keen 1980b, p. 40). The cultural impact of this form of population decline is more difficult to document, but I think it deserves to be understood separately.

Deaths by violence or disease are events that Indigenous people have to deal with, emotionally, ritually and politically. Population decline by way of multiple deaths provides a series of such events that the survivors respond to and assimilate as life crises and as social transformation. In contrast, depopulation following a reduced birthrate occurs as an absence of events, by way of nothing happening, which denies the opportunity to respond. There is no individual trauma here, but there is a slow crisis. It stalls the progress of the generations and the reproduction of culture. In particular, elderly females never really achieve the status of 'old woman' because they remain childless, and the thinning ranks of adolescent males reduce the urgency, or even the opportunity, for holding initiation ceremonies, so the knowledge and authority of the senior men are denied their most effective and affecting forum for expression and reinforcement. Religious knowledge becomes academic, instead of reproductive and powerful, because its terror and mystery are denied their social object: the young.

Epidemic disease and a failure to have children appear on the evidence to be the principal causes of radical population decline across the buffalo country in the late 19th and early 20th centuries. I have suggested here that these phenomena not only thinned the numbers, but were also accompanied by secondary effects that tended to reduce social cohesion and cultural vitality among the surviving groups. The capacity of Aboriginal society to reproduce itself was being attacked in ways that went beyond a simple absence of people. Where the self-reproducing integrity of a social formation is so seriously compromised, its capacity to retain the allegiance of its members will be further reduced where alternatives are available to attract their participation.

The buffalo camps[2]

Given that knowledge and expertise are always differentially distributed within a population, the extent of the collapse in numbers was probably enough by itself to make the maintenance of many aspects of local pre-colonial culture impossible. The anthropologists Ron and Catherine Berndt visited two buffalo stations in 1945, by which time only a few representatives of the local Mary River language groups were left. The Berndts concluded that these were the only groups – compared with all the others they had surveyed on seven Territory pastoral stations and five settlements – for which the possibility of cultural revival had passed (Berndt and Berndt 1987, p. 212). They therefore took a critical view of the conditions of work in the buffalo industry. While it seems clear from the foregoing that it was the wider effects of contact that were primarily responsible for that cultural decline, the Berndts' general argument that the conditions of particular local regimes had their own repressive impact on Aboriginal culture leads us to now consider the specific local impact of the buffalo camps.

A convenient starting point is that made by the Berndts themselves (1987, p. 202): that the traditional socio-cultural system is 'much more vulnerable when work-patterns bear little or no relationship to traditional Aboriginal socioeconomic imperatives'. The buffalo industry provided a more benign theatre of contact than the cattle stations precisely because it had a less drastic impact on pre-colonial Aboriginal living patterns. This can be shown in two important respects.

Firstly, the establishment of the industry seems to have been a less violent process than that which accompanied pastoral settlement across north Australia. The explanation for this is to be found not in the character and values of the people who met in the buffalo country, because the Aborigines had shown their readiness to use violence against their own neighbours, and many of the Europeans came from pastoral areas with recent histories of violent pacification. At the level of interpersonal encounters and within particular camps, violence and coercion could figure in the range of interactions. The point here is that there was nothing in the nature of this new form of production that set Aborigines as a group in opposition to this group of colonisers. Rather, the material conditions of production in the buffalo industry, unlike those in the pastoral industry, were more conducive to engagement and cooperation than to competition and conflict between Aborigines and non-Aborigines.

Buffalo, unlike sheep and cattle, were not private property, but feral animals. After first introduction and release, they colonised the country independently, found food and water in most years and perished in drought years. There was no husbanding of the stock, and no point in claiming monopoly control, as pastoralists did, of waterholes and pasture. The buffalo shooter owned only the right to shoot buffalo found on his lease area. The industry harvested animals according to their natural availability. Aboriginal hunters were capable of killing buffalo, but not in numbers that would undermine a shooting operation even if they had wished to do so. In contrast to sheep, which could be slaughtered in large numbers, the buffalo was thick-skinned and dangerous. One buffalo shooter, Tom Cole (1988, pp. 308–9), was told by a group of Aborigines of their efforts to kill a buffalo, which required a number of wounds inflicted with shovel-nosed spears and several hours of tracking. In 1936, the anthropologist Donald Thomson (1983, p. 87) was told by his colleague from the Glyde River, Raiwalla, that 'although metal shovel-nosed spears were sometimes used for hunting buffalo, they preferred the stone-head spears because they made a greater wound and caused such a loss of blood'.

Moreover, the secondary effects of Aboriginal spearing of cattle – that it made the rest of the herd wild, more difficult to muster and less likely to fatten (McGrath 1987, p. 14) – were also irrelevant here. The shooters were interested in kills, not control, and whatever unsettling effect Aboriginal spearmen may have had was negligible compared to the shooting itself. Finally, there was no competition for the product. White shooters wanted hides, Aborigines wanted meat. Buffalo shooting for hides left lying in its wake large quantities of fresh meat. The most sensible approach from the Aboriginal point of view was not to resist the shooting camps, but to join with them, probably as hangers-on at first and then as workers.

There may indeed have been a further impetus towards joining. In numbers, buffalo, like sheep and cattle, have a destructive impact on the native environment and indigenous food sources. Buffalo can also directly threaten humans, and Aborigines are always wary, even terrified, of them. It is likely that where herds appeared in significant concentrations in the late 19th century, local Aborigines felt their access to major food sources around springs and billabongs to be under threat (Gillespie 1979). In the 1970s and 1980s, Meehan noted such impacts on the plains around the lower Blyth River, east of Maningrida. With increasing buffalo numbers, freshwater sources became polluted and dried up earlier, children were left at base camps for safety if foraging expeditions were to be carried out on foot, and foragers chose safe paths through wooded areas instead of fanning out over the plains to check the state of other resources while on the way to a swamp or other destination (Meehan 1988, pp. 18–20). Whereas on the grasslands of north Australia pastoralists were the source of the Aborigines' problems of competition with introduced herds, in the buffalo country the white shooters may well have appeared as a solution. As the pattern of pre-contact dry season movements seems to have been along the lowland waterways, the water buffalo would have presented a dry-season problem, and the shooting camps a dry-season alternative. The buffalo camps may have been the means by which Aborigines were able to re-occupy stretches of country being denied to them by an expanding buffalo population, or at least a means of more safely moving through the bush.

Secondly, this new adaptation showed certain continuities with pre-contact life. It did not interrupt the basic seasonality of Aboriginal life. It was a dry-season activity located on downstream lowlands and floodplains, which form part of the range across which Aborigines ordinarily camped at that stage of the year. It could not be done during the wet season because

Figure 3.6 Buffalo feeding on green pick after fire and early rains in September.

movement across country was difficult, buffalo were scattered and hides would rot. It could not be pursued even to the end of the dry season, because increasing humidity placed too much strain on the horses. So Aborigines were free to begin moving away from the floodplains and swamps towards their wet-season shelters around the time they ordinarily would have done. Production was by direct harvesting and involved only limited processing of the product, it required mobility throughout the season and introduced no mechanised infrastructure; that is to say the hides industry involved an extensive, rather than intensive, relationship to country and so shared important commonalities with central features of the pre-colonial Aboriginal economy.

Moreover, the male responsibility for dealing with larger game was preserved in the pursuit of buffalo, though here of course provision of meat was a secondary concern, and the technology and organisation of work was entirely different. The continuities were more pronounced for the women. Buffalo camps moved several times through a shooting season, which repeatedly provided access to fresh areas of bush tucker. The camps were often located on the boundary between floodplain and low woodland ridges, and always on bodies of still or flowing water that were necessary for washing the hides. These locations at the intersection of ecological zones were consistent with the pattern of Aboriginal dry-season camping around Magela Creek and the lower South Alligator River documented during the Alligator Rivers Stage II Land Claim (Layton 1981, p. 15). The camp work routine typically required women each morning to re-salt and re-stack all hides presently being processed, leaving some time free for going bush before fresh hides from the day's shooting came in to be washed. The buffalo industry, then, placed Aborigines in a central productive role, while allowing space for the exercise of individual Aboriginal independence and preserving some central principles of precontact economy.

Figure 3.7 Buffalo hunters dining on a carcass (photo: AIATSIS Audiovisual Archive Robins.R1.BW 611.33).

Once established, this relatively benign engagement was nevertheless attended by processes of Aboriginal cultural change. Although the buffalo industry involved direct extraction of the product, direct consumption was restricted to a by-product: the meat (Figure 3.7). To acquire other things, Aborigines exchanged their labour for commodities. At that point of exchange with their employers, Aborigines became the remote end-point of a series of linkages that drew buffalo hides overseas to markets in the Middle East and Europe, and forwarded provisions and capital equipment into the bush. However, although clothes, boots, swag covers, mosquito nets, sugar, flour and tea were desirable, the work was hard and sometimes dangerous, and bosses could be demanding and rough. Other places, such as the mission, the towns or the bush, had their own attractions, including a social life with relations.

Viewed in this light, the supply of addictive substances, especially tobacco, takes on particular significance (see Brady and Long 2003, and Chapter 4). By all accounts, tobacco was the most important single element in relationships between Aborigines and non-Aborigines in the buffalo country. It was the indispensable incentive to Aboriginal cooperation and service in European undertakings. When the tobacco ran out, the Aborigines withdrew. When a missionary at Kapalga on the South Alligator River pleaded for government support in 1901, the 'most meagre resources' provided in response included 'sufficient tobacco to retain the population at the mission' (Freier 2008, p. 20–21). Although alcohol, clothes or other commodities will often be mentioned alongside tobacco as attractive and important, it appears from the totality of evidence that availability of tobacco was the condition precedent to work. Thus, a white boss might be judged poorly by his Aboriginal workers if he distributed insufficient of other things, but a lack of tobacco was intolerable and justified abandonment of the camp. Other employers of Aborigines had to follow suit. The first missionary at Oenpelli on the eastern edge of the buffalo country, Alf Dyer, reported:

> *That gift box of tobacco has done well on the wages side. I hate the stuff and wish there was something else, but it is the only thing which the Aborigines seem to want. (Cole 1972, p. 58–59)*

But, while employment in the buffalo camps was seasonal, nicotine addiction was not, and it compromised the ability of Aborigines to remain in the bush – away from Europeans – for extended periods. When Paddy Cahill returned to his station at Oenpelli after a trip in May 1916, he found about 250 people from the east and north gathered there waiting, some 'nearly dead' for want of the tobacco that they used to receive from whites working along the coast (Mulvaney 2004, p. 112). One incident was well remembered in the Kakadu region because it was the occasion of a woman being taken by a crocodile while crossing the South Alligator River in the 1930s. After the work season had ended, she was one of four people who were making their way from their up-river bush camp to a white shooter's wet-season station to renew their tobacco supply. A more pointed tale is recounted by Dick Harris, another Oenpelli missionary. He recalls a middle-aged man, Naringi,

> *going 'walkabout' with others and having received a walkabout issue of tobacco. He returned to the Mission weeks short of the recognised time and when asked why he came back so quickly, although he was well equipped to live in the bush, he said 'Mission, tobacco'. (Harris 1981, p. 184)*

When the missionaries later imposed bans on the distribution of tobacco, many of their most experienced Aboriginal workers left for the buffalo camps. Over a period, tobacco, along with other commodities, made the more settled areas to the west – towns, mining camps and cattle stations – more attractive wet season destinations, but the incidents set out here suggest that even where people did retire to bush camps after the buffalo work season ended, they no longer moved as widely through their traditional range.

While there was this nagging element of compulsion behind the Aboriginal engagement with the buffalo industry, it should also be said that the work camps, purposeful and active

Figure 3.8 Buffalo camp storeroom at river landing (photo: AIATSIS Audiovisual Archive Robins.R1.BW 610.10).

places that they were, readily became a focus of Aboriginal commitment and aspirations. They drew the attention and energies predominantly of people in the younger age range, for whom work in a hides camp appears to have been a source of style and prestige. In the film *Something of the Times* (McKenzie 1985), one old veteran of the industry laughingly recalls how smart she and her contemporaries thought they were, working on buffalo hides. Such prestige attached most strongly to those few men who were horseback shooters. These are the people who, as old men in the 1980s, spoke with greatest pride of their ability and exploits. One, conscious of the glamour that surrounded his work, recalled how, as he pursued buffalo across an open plain, his uncle sat singing of his skill to the woman he planned to marry, as they watched from the camp.

Even more than the young adults, it seems the children were entirely seduced by the ethos of these camps. Noreen Freer recorded in her diary daily events in her husband's buffalo camp. 'All children learn to smoke very young', she wrote. 'One little girl of six smokes a pipe.' In another more romantic passage, she describes how the children joined in with the working men as they cleared and repaired a vehicle track. It is notable how the attire of the work camps, even ill-fitting or dilapidated, had become an object of fashion.

> *[T]he boy picanninnies [sic] ran catching up tops of shrubs and sticks then throwing them one side – they were laughing and loving it – each one looked quite a sketch, one's trousers, which had been men's shorts were no longer trousers the leg seams had come undone and the trouser seams also, which left him two independent side flaps which just hung round his waist, by some sort of belt – another's pants were intact, but how they kept on was a marvel, somehow it looked as if a button in front must be caught in the belt the rest suspended thereby, with the back part slipped well down below belt level – a smaller picaninny [sic] had only a handkerchief on, which now and then he pulled tighter and tucked the end in with great concern to the side part – another boy looked the typical stable boy – blue dungarees with a dirty torn shirt which hung loosely over the top of his pants – a cap pulled over his eyes at an angle of its own.*
>
> *When we came to a gutter, which needed a good deal of filling in the picaninies [sic] ran by with bits of anthill considering it great fun – the baby one just as anxious to carry – with his slightly sprained ankle as any – he would pick up lumps nearly the size of himself – throw it in to the gutter and brush the ants off himself – this finished they made a rush for the lorry, scrambled on board – and off. (Garry 1930)*

Country, work and knowledge

To the extent that the buffalo camps held the attention of younger Aborigines, they also mediated the Aborigines' daily relationships to the land and affected both what they learnt and the way they learnt it. This was obviously true for those whose traditional estates were located to the east of the buffalo country and who never returned to their own areas, where there was minimal European enterprise, and where non-Aborigines were ultimately not allowed after the Arnhem Land Aboriginal Reserve was declared in 1931. Among the senior people I worked with in the 1980s, one woman had left the Liverpool River area as a girl with her parents, and another man had been born in the buffalo country after his father had moved from another Liverpool River estate. When I suggested the possibility of visiting their traditional estates to the east, they laughingly admitted that they didn't know their own country, but they knew this country, where they had lived and worked. It was clear then that knowledge of country was a function of life history rather than traditional attachment. It later became evident that this could be true at a very detailed level.

I did a number of field trips into what had once been the Munmalary buffalo lease, which occupied the western half of the area bounded by the East and South Alligator Rivers, north of the Arnhem Highway. I was accompanied by two people: one the Liverpool River woman referred to before, and the other a local man whose traditional estate is located immediately south-east of Munmalary. Each had been associated for several consecutive years in the 1940s and 1950s with shooting operations on Munmalary and they remembered the country well. We were able to locate and map nine shooting camps on the lease area, each located on a billabong or spring, and I recorded names in the local Gagudju language for all of them, plus English names for some. The data about these places was clear and consistent and given freely.

Just as interesting, however, were other things that they didn't know. On two occasions during these trips I drove further east, to two springs and a billabong, all with plenty of water in them, even in November. The man thought he might once have camped on one of the springs, but neither knew anything about the other places. We had, they told me, left the area of Munmalary lease and crossed the unmarked boundary onto the Cannon Hill lease area, on which they had never worked and would not have ventured for fear of trouble from the white shooter there. Even on one of the springs that we had mapped within Munmalary, the man was unfamiliar with a swamp area downstream from the shooting camp because, as he explained, they had never shot over the swamp, but had gone west from the camp to shoot on the open plain. Knowledge of country, then, could be an exact reflection of experience at work.

Moreover, non-Aboriginal use of the buffalo country affected not only what country was learned, but how it was learned. Traditionally, place names were understood not to refer to an unrelated collection of points, but to occur in a certain geographical order or arrangement that had to be rehearsed and remembered, sometimes as songs or stories that recounted mythical or actual pathways. Because this corpus of knowledge was ordered according to patterns of movement, a new pattern of movement could make known country strange once again, and require a period of personal re-orientation during which the order of it had to be re-learned.

One modern example can be cited here. During the Alligator Rivers Stage II Land Claim, the anthropologist and an Aboriginal claimant were trying to locate Mosquito Dreaming. This man was used to approaching the site by travelling north-west up the old Munmalary track, and turning north onto another track along Namanbu Creek. The second track had become overgrown with vegetation after years of disuse and they missed the turn-off. They then tried again, approaching this time by travelling east along the Arnhem Highway, which had been built 5 years earlier The claimant suddenly recognised Mosquito Dreaming alongside the highway. Though he had travelled along the highway before, he had never tried to locate the dreaming site that way, and remembered it only by the old way (Keen pers. comm.).

It is reasonable to speculate that this kind of thing happened from time to time in the past. In the first instance, local people moving from the regime of the bush camp to that of the buffalo camp would have begun seeing country according to a different logic of movement that may well have been initially disorienting, especially when the camps crossed the forested ridge country from one watershed to another, intersecting with Aboriginal pathways that followed the drainage lines. As for migrants from distant areas, their introduction to new country was governed by a non-Aboriginal rationale. At a later stage, the same thing probably recurred when the principal mode of transport in the buffalo industry switched from horses to motor vehicles after the Second World War. Vehicle tracks tended to follow the ridge lines, running between drainage systems, with forks running off at intervals to billabongs or swamp pockets on either side. This was a different principle to that which governed Aboriginal movements. After the hides industry finished, this process continued. Not only did the pattern of moving through country change again with the introduction of many new tracks for buffalo catching and meat production, but environmental degradation changed the country itself. Starting with the buffalo shooters, then, a succession of non-Aboriginal agents interrupted the intimacy between Aborigines and the landscape.

Fire on the plains

Carl Warburton ventured out into the buffalo country in 1920. At that time there were still many Aborigines living in the bush. When he and his party reached Magela Creek and climbed a hill, he looked west to see numerous columns of smoke rising from the plains and open forest (Warburton 1934, pp. 71–72). One of his Aboriginal companions recommended that he burn the dry grass to provide a green pick for the horses. He later learned that:

> *on the coastal areas north of Katherine, it is the custom to burn the grass after the wet season – about March or April. Then the whole countryside is ablaze without any danger of bushfires as we know them in New South Wales or Victoria. (Warburton 1934, pp. 82–83)*

Tom Cole shot buffalo between the lower South Alligator and the Wildman Rivers in the 1930s. At the beginning of one season he observed:

> *It was going to be hard for a start; in the wet season the spear grass grows to a height of six or eight feet, and when the rain stops it falls over and mats up. When it dries out it has to be burnt off. The heat of the grass fire generates a growth of green grass referred to as 'burnt feed'; the horses love it. But right then it wouldn't burn; it was too damp underneath. It was a foot or so deep, hiding stumps and logs and it was too easy to have a bad fall. I didn't like it at all. … The mosquitos were bad, too, and would continue to be until we could get a decent burn. (Cole 1988, p. 288)*

Buffalo shooters burned, then, for ease and safety of movement, to produce new feed, and for general comfort, which were also some of the reasons for Aboriginal burning (see Chapter 5). In June 1936, Cole describes having to walk down the South Alligator River towards his camp at Kapalga after a vehicle breakdown:

> *By midday the first day we came to the head of the plains. Stretching away into the shimmering distance like a greyish-green sea was a dence [sic] cover of heavily matted grass. It was not going to be easy walking for me. I looked at it for a while, then tossed a match into it. With a strong southerly behind it, it roared quickly away, leaving only a thick layer of black ash. We camped for an hour, boiling the billy and waiting for the ground to cool before we made another start. (Cole 1988, p. 340)*

That fire burned through the afternoon, then flared up again the next morning and nearly caught Cole's party from behind.

Jim Doyle was one of the last generation of buffalo shooters in the industry's final decade after the Second World War. In his diaries for 1950 and 1951, he recorded burning around his Cannon Hill lease on the lower East Alligator River. In 1950, he was away in April, returning to Cannon Hill on 7 May:

> *May 8 I burnt grass around camp.*
> *May 29 I burnt country around Manning.*

June 25 Fred and myself went down to Mungarawar spring and burnt the country.
June 27 Myself and Tim went over to Mungerowie and down the spring to Ginean Plain. I burnt the grass on the plain.

In 1951 he was at Cannon Hill from mid-January:

March 30 I burnt spear grass.
April 26 I rode Opal ... and went up around the old yard camp and right around on ... the river. I burnt all the way.
May 3 Had a good burn on plain.
May 4 Fires at Callabarrie and Manning, also fire on ... to Mudgenbarrie.
May 17 I burnt all the bottom end.
June 11 I mustered horses and burnt plain.
June 12 I sent Harry to Calabarrie, to get spray and burn island.
August 13 Boys went up the river and burnt grass.

The record left by these three buffalo shooters shows no inclination to suppress fires that were not their own. Fires could be lit by the white shooter, by his employees under instruction, or independently by Aborigines. The fires on May 4 may have been Aboriginal fires. By Doyle's time, however, the widespread and numerous Aboriginal presence in the bush recorded by Warburton had been much reduced.

In summary, the burning recorded here was another aspect of the continuities that the new regime of the buffalo camps sustained with previous Aboriginal modes of livelihood on the lowlands. As with those other continuities, however, they were not replication. We do not know what subtleties of place were sacrificed in the shift from living in relation with country, to operating on country within new relationships of employment and mediated access to commodities. Perhaps the Aborigines involved did not know themselves. Most of the traditional owners of Cole's and Doyle's shooting leases were gone; their workers mostly came from further east and south-east. Adaptation via the relatively benign regime of the buffalo camps provided a cultural compromise that kept Aborigines in the bush, but living out the first phase of separation from its meanings.

MANAGEMENT: CONSERVATION AND ADAPTATION

That separation became greater in the following years. The buffalo hides industry collapsed in the mid-1950s when the Suez crisis in the Middle East cut off access to markets. The largest concentration of Aborigines in the region at that time, apart from Oenpelli mission, was located at the Arnhem Land Timber Company camp at Nourlangie. An official count listed nearly 60 people there (Northern Territory 1957), and visiting groups sometimes swelled these numbers. A local woman, Carla Ngalyorrun, remembered that many had come from Bamyili and north Arnhem Land, and some she had not seen in the area before. The attraction of European commodities at the timber camp was complemented by the bush resources of the nearby swamps and billabongs of the South Alligator wetlands. At the end of the 1950s, the timber camp too closed down (Stewart 1969, pp. 13–14), and was replaced by a safari camp.

People moved away to missions, settlements and towns. For a time in the late 1960s, another substantial community of people gathered around Mudginberri station and abattoir. By the mid-1970s, however, the Kakadu region's Aboriginal population had declined to a few dozen (Keen 1980b, pp. 63–64), consisting of workers and their families at two local abattoirs, a few safari camps and Conservation Commission camps, and one family living independently. This abandonment was reversed when the policy revolution of the late 1970s, which was initiated by land rights and the discovery of uranium, led to creation of Kakadu National Park under a Commonwealth management that accommodated Aboriginal residence at a number of locations. The Aboriginal population of the Park quickly rose to over 300.

In August 1983, again at Cannon Hill, the anthropologist Henry Lewis listened to the senior local Aboriginal owner, Bill Nayidji, teach the same lesson that Warburton had learned 60 years before. Earlier that year the Ash Wednesday fires had killed over seventy people in southern Australia.

> *While acknowledging the human tragedy, Bill Neidjie added that it was a 'crime', not simply a crime that a number of these fire storms had been set, but a crime in allowing a country to 'get so dirty'. People, he said, should 'take care of their country.' Such a fire would not happen in places that he was responsible for, country that he and his kinsmen regularly burn. What happened in South Australia and Victoria, he said, was a 'shame,' but people would not have been killed if the country had been 'kept clean'. (Lewis 1989, p. 940)*

The first stage of Kakadu National Park was declared in 1979, and further declarations would include most of the Alligator Rivers region within its boundaries. Lewis (1989, pp. 950–53) was recording the early moments of a debate within the Parks Service staff, and of disagreement and failed communication between European staff and Aborigines, over the use of fire. Aborigines argued that white rangers were too cautious about burning, that they did not learn from Aboriginal practice or respect Aboriginal knowledge, and tried to prevent some fires. The white staff thought Aboriginal burning was ad hoc, pragmatic and heedless of environmental consequences, and that cultural change meant that current Aboriginal practice could not claim the integrity of tradition. They wanted more scientific information before they burned, particularly around wetlands, and wanted burning to be part of systematic and professional management. Lewis (1989, p. 954) thus diagnosed the problem as a failure by European land managers to recognise the continued existence and value of traditional environmental knowledge, disguised as it was by the use of introduced technology by people whose way of life now appeared to owe little to tradition.

Later in the Park's history, I returned to the problem (Levitus 2005). By then, fire ecologists had accepted that traditional Aboriginal burning should be the model for Park burning (e.g. Press 1987, p. 247). This resolved the basic management dilemma. But still, Aborigines accused European rangers of too little burning, especially around wetlands, and rangers countered with complaints of delinquent fires lit outside the safe early dry season window or in areas needing protection. No doubt the historical transformations discussed in earlier parts of this chapter had emptied the corpus of Aboriginal knowledge of much of its fine grain with respect to the particularities of time and place in using fire. Discussion with Park staff in 1996 revealed some of the same scepticism that Lewis had found regarding the life history backgrounds of local people and the quality of the indigenous knowledge base available for management to draw on. It appeared to me, however, that official Parks acceptance of the traditional model (Australian National Parks and Wildlife Service and Kakadu National Park Board of Management 1991, pp. 43–44) meant that the perceived obstacle of incompatible knowledge systems had been removed, and that the remaining conflicts were of a different kind.

In an earlier section of this chapter, I discussed the traditional human ecology of seasonal movement and its carriage of fire across the landscape. Once this was accepted as a legitimate historical model for Parks fire management, it was then apparent that the contending views over current practice in Kakadu both originated from this source. The 1991 Plan of Management (Australian National Parks and Wildlife Service and Kakadu National Park

Board of Management 1991, pp. 42, 44) noted an existing management prescription whereby Parks staff burned 'as early as possible in the dry season (May–July) to achieve cool, low-intensity fires' and authorised the continuation of that practice. Even when the subsequent Plan stated that management burning should 'be done throughout the year as appropriate, but will be mainly done in the early dry season so as to restrict late fires and reduce their impact' (Kakadu Board of Management and Parks Australia 1998, p. 68), staff often took this to mean that the fire-management program needed to be delivered entirely within the early part of the year. By contrast, resident Aborigines confidently deployed fire as and when they thought the country needed it, or to facilitate their own purposes, such as hunting, on country, at any time they thought proper. At its most elemental, the stand-off was between staff adherence to a notion of burning when one should, and Aborigines' situational responses of burning when one could.[3]

The originating point of both attitudes, with their respective emphases on conservation and adaptation within a liveable environment, can be seen in the seasonal model. Parks staff considered the conservation ethos to be served by the modern equivalent of the many small and dispersed fires lit by bands beginning their dry season movements, needing to clean up country clothed in the previous wet season's vegetation. The Aboriginal adaptation ethos also required that level of burning, but later burning as well, to facilitate continued access and resource use as the country dries. When Leichhardt reached the South Alligator wetlands on 27 November 1845, after the first storms had come, he found Aborigines 'burning the grass on the plains'; further downstream on 1 December, '[n]umerous pillars of smoke were seen to the westward' (Leichhardt 1847, pp. 493, 501). Most of the disagreements, then, expressed contrasting understandings of the grounds for burning in the first place. They emerged as disagreements, however, when operational inadequacies prevented the seasonal model of burning from being fully realised each season.

Aboriginal liberality with fire presented a problem for the Parks Service's conservationist conservatism because there was still too much left to burn by the time the Parks burning season ended. To the extent that management burns approximate the coverage achieved traditionally by travelling Aboriginal bands during the first half of the dry season, fires lit later by contemporary Aboriginal users of the Park are not a cause for concern. The contest of values and purposes is a problem of coverage, not of knowledge systems. That is not to say that there are not other points about fire on which people will disagree and criticise, whether technical, operational, environmental or social. For example, there is an argument that aerial incendiaries should be dropped not along linear transects, but along the creek lines following the movement patterns of Aboriginal bands. In other words, the debate now tends to occur within a common frame of reference. So one family's claim for resourcing priority to be given to Indigenous burning was met with the question from some Park staff, 'why is your burning different to our burning?'

CONCLUSION

In this chapter I have jumped from a time scale of a millennium to one of decades in an effort to identify a succession of attractions and forces bearing upon Aboriginal society and people in the Alligator Rivers region, each of which caused them to re-think and re-focus the social and environmental adaptations that sustained them. There is, of course, a critical difference in the nature of these long-term prehistoric and short-term historic changes, in that the first transition was a human ecological response to an evolving physical environment, while the second was a social response to a radically transformed social environment and introduced adaptive pressures. From the point of view of modern environmental management, the first phase is an objective to approximate, the second a breakdown to repair.

In the years following the end of the buffalo hides industry in the mid-1950s, other agents of separation interposed themselves between Aborigines and their country (Chapters 2 and 4), and, until the 1970s, Aboriginal numbers in the bush continued to decline. Since then, as discussed in other chapters, a wave of recolonisation, research, and institutional innovation has reversed the demography and taken steps towards reinstating informed human control over fire. In Kakadu, that has occurred through the efforts of management to reinstate a pre-European fire regime. We are now witnessing in Arnhem Land efforts, from several directions, to restore Indigenous responsibility for land management. The ultimate environmental question will be how closely these new arrangements are able to achieve the same range and subtlety as the traditional system in the application of fire to western Arnhem Land. A more proximate question is whether the new impetus towards management will provide a context for a new form of Aboriginal adaptation that is both materially sustainable in itself and environmentally sustaining across that broad landscape scale.

ACKNOWLEDGEMENTS

My thanks to Jeremy Russell-Smith for inviting me to contribute to this volume and offering comments and support during its preparation. I have benefited from several discussions with Peter Cooke and Murray Garde. David Ritchie and Kathleen Oien commented helpfully on the draft, and Sally Brockwell gave advice on the archaeology sections. Thanks also to Felicity Watt for preparing the map. The Australian Institute of Aboriginal and Torres Strait Islander Studies provided electronic scans of the photographs from their collection.

ENDNOTES

1 For general histories of this period, see Powell (1996) and Donovan (1981).

2 For histories of the buffalo country and the buffalo industry, see Clinch (1988, 1990); Levitus (1995); Mulvaney (2004). For first-hand accounts see Cole (1988); Mulvaney (2004); Warburton (1934).

3 I would like to thank Monica Minnegal for expressing the difference in these terms.

REFERENCES

Australian National Parks and Wildlife Service and Kakadu National Park Board of Management (1991) *Kakadu National Park Plan of Management.* Commonwealth of Australia, Jabiru, Northern Territory.

Berndt RM and Berndt CH (1987) *End of an Era: Aboriginal Labour in the Northern Territory.* Australian Institute of Aboriginal Studies, Canberra.

Brady M and Long J (2003) Mutual exploitation? Aboriginal Australian encounters with Europeans, Southeast Asians, and tobacco. In *Drugs, Labor, and Colonial Expansion.* (Eds W Jankowiak and D Bradburd) pp. 31–58. University of Arizona Press, Tucson.

Brockwell CJ (1989) Archaeological Investigations of the Kakadu Wetlands, Northern Australia. MA Thesis. Australian National University, Canberra.

Brockwell CJ (2001a) Archaeological settlement patterns and mobility strategies on the Lower Adelaide River, northern Australia. PhD Thesis. Northern Territory University, Darwin.

Brockwell S (2001b) Wetlands archaeology in the Top End: models, mounds and mobility. In *Histories of Old Ages: Essays in Honour of Rhys Jones.* (Eds A Anderson, I Lilley and S O'Connor) pp. 327–340. Pandanus Books, Australian National University, Canberra.

Brockwell S (2006) Earth mounds in northern Australia: a review. *Australian Archaeology* **63**, 47–56.

Brockwell S, Clarke A and Levitus R (2001) Seasonal movement in the prehistoric human ecology of the Alligator Rivers region, north Australia. In *Histories of Old Ages: Essays in Honour of Rhys Jones*. (Eds A Anderson, I Lilley and O'Connor S) pp. 361–380. Pandanus Books, Australian National University, Canberra.

Brockwell S, Meehan B and Ngurrabangurraba B (2005) An-barra archaeological project: a progress report. *Australian Aboriginal Studies* **1**, 84–89.

Campbell J (2002) *Invisible Invaders: Smallpox and Other Diseases in Aboriginal Australia 1780–1880*. Melbourne University Press, Carlton.

Chaloupka G (1981) The traditional movement of a band of Aboriginals in Kakadu. In *Kakadu National Park: Education Resources*. (Ed. T Stokes) pp. 162–171. Australian National Parks and Wildlife Service, Canberra.

Chaloupka G and Giuliani P (1984) 'Gundulk abel gundalg: Mayali flora'. Unpublished report. Northern Territory Museum of Arts and Sciences, Darwin.

Clark RL, Guppy J, Mahon D, McBride P and Wasson RJ (1992) Late Quaternary evolution of the Magela plain. In 'Modern sedimentation and late Quaternary evolution of the Magela Creek plain'. (Ed. RJ Wasson) pp. 81–157. Supervising Scientist for the Alligator Rivers Region Research Report 6. Australian Government Publishing Service, Canberra.

Clinch M (1988) Home on the range: the role of the buffalo in the Northern Territory, 1824–1920. *Northern Perspective* **11**(2), 16–27.

Clinch M (1990) On the edge of the paperbark swamp: the role of the buffalo in the NT 1920–1956. *Northern Perspective* **13**(1), 14–24.

Cole K (1972) *Oenpelli Pioneer*. Church Missionary Historical Publications, Melbourne.

Cole T (1988) *Hell West and Crooked*. Angus and Robertson, Sydney.

Donovan P (1981) *A Land Full of Possibilities: A History of South Australia's Northern Territory*. University of Queensland, Brisbane.

Freier P (2008) The Northern Territory Native Industrial Mission, Kaparlgoo, 1899 to 1903. *Journal of Northern Territory History* **19**, 17–32.

Garry EN (1930) 'Papers relating to her stay in the Northern Territory, 1930–32, 1966. Items 1 and 2, ML MSS 3977'. Mitchell Library, Sydney.

Gillespie D (1979) Nganaburru gumwam (then the buffalo came): environmental, economic and social effects of the introduction of a feral species into the estates of hunters and gatherers in the Alligator Rivers region – hypotheses. Unpublished manuscript. University of New England, Armidale.

Harris GR (1981) *My story, notes on my life in North Australia 1928–1965*. Unpublished manuscript. (In author's possession)

Harvey M (2002) *A Grammar of Gaagudju*. Mouton de Gruyter, Berlin/New York.

Hiscock P (1996) Mobility and technology in the Kakadu coastal wetlands. *Indo-Pacific Prehistory Association Bulletin (Chiang Mai Papers Vol. 2)* **15**, 151–57.

Hiscock P (1999) Holocene coastal occupation of western Arnhem Land. In *Australian Coastal Archaeology*. (Eds J Hall and I McNiven) pp. 91–103. Research Papers in Archaeology and Natural History No. 31. ANH Publications, Australian National University, Canberra.

Hiscock P and Kershaw AP (1992) Palaeoenvironments and prehistory of Australia's tropical Top End. In *The Naïve Lands: Prehistory and Environmental Change in Australia and the South-west Pacific*. (Ed. J Dodson) pp. 43–75. Longman Cheshire, Melbourne.

Hope G, Hughes PJ and Russell-Smith J (1985) Geomorphological fieldwork and the evolution of the landscape of Kakadu National Park. In *Archaeological Research in Kakadu National*

Park. (Ed. Jones R) pp. 229–240. Special Publication 13. Australian National Parks and Wildlife Service, Canberra.

Jones R (1985) Archaeological conclusions. In *Archaeological Research in Kakadu National Park*. (Ed. Jones R) pp. 291–298. Special Publication 13. Australian National Parks and Wildlife Service, Canberra.

Jones R and Johnson I (1985a) Rockshelter excavations: Nourlangie and Mt Brockman massifs. In *Archaeological Research in Kakadu National Park*. (Ed. R Jones) pp. 39–76. Special Publication 13. Australian National Parks and Wildlife Service, Canberra.

Jones R and Johnson I (1985b) Deaf Adder Gorge: Lindner site, Nauwalabila 1. In *Archaeological Research in Kakadu National Park*. (Ed. R Jones) pp. 165–227. Special Publication 13. Australian National Parks and Wildlife Service, Canberra.

Kakadu Board of Management and Parks Australia (1998) *Kakadu National Park Plan of Management*. Commonwealth of Australia, Jabiru, Northern Territory.

Keen I (1980a) The Alligator Rivers Aborigines – retrospect and prospect. In *Northern Australia: Options and Implications*. (Ed. R Jones) pp. 171–186. Research School of Pacific Studies, Australian National University, Canberra.

Keen I (1980b) *Alligator Rivers Stage II Land Claim*. Northern Land Council, Darwin.

Layton R (1981) *Statement on the Alligator Rivers Stage Two Land Claim: (1) traditional foraging and camping patterns; (2) rock paintings*. Exhibit 66. Aboriginal Areas Protection Authority, Darwin.

Leichhardt L (1847) *Journal of an Overland Expedition in Australia, from Moreton Bay to Port Essington*. T and W Boone, London.

Levitus R (1995) Social history since colonisation. In *Kakadu: Natural and Cultural Heritage and Management*. (Eds A Press, D Lea, A Webb and A Graham) pp. 64–93. Australian Nature Conservation Agency and North Australia Research Unit, Australian National University, Darwin.

Levitus R (2005) Management and the model: burning Kakadu. In 'Sustainable environments, sustainable communities: potential dialogues between anthropologists, scientists and managers'. (Ed. M Minnegal) pp. 29–35. School of Anthropology, Geography and Environmental Studies Research Paper No. 21. University of Melbourne, Melbourne.

Lewis HT (1989) Ecological and technological knowledge of fire: Aborigines versus park rangers in northern Australia. *American Anthropologist* **91**, 940–61.

Lindsay D (1884) Mr. D. Lindsay's explorations through Arnheim's Land. *South Australia: Proceedings of Parliament and Papers 1883–84* **5**, 239.

McGrath A (1987) *'Born in the Cattle': Aborigines in Cattle Country*. Allen and Unwin, Sydney.

McKenzie K (1985) *Something of the times*. Documentary film. Australian Institute of Aboriginal Studies Film Unit, Canberra.

Meehan B (1988) Changes in Aboriginal exploitation of wetlands in northern Australia. In *Northern Australia: Progress and Prospects Volume 2: Floodplains Research*, Appendix 2. (Eds D Wade-Marshall and P Loveday) pp. 1–23. North Australia Research Unit, Australian National University, Darwin.

Meehan B (1991) Wetland hunters: some reflections. In *Monsoonal Australia: Landscape, Ecology and Man in the Northern Lowlands*. (Eds CD Haynes, MG Ridpath and MAJ Williams) pp. 197–206. A.A. Balkema, Rotterdam and Brookfield.

Meehan B, Brockwell S, Allen J and Jones R (1985) The wetlands sites. In *Archaeological Research in Kakadu National Park*. (Ed. Jones R) pp. 103–153. Special Publication 13. Australian National Parks and Wildlife Service, Canberra.

Mulvaney J (2004) *Paddy Cahill of Oenpelli*. Aboriginal Studies Press, Canberra.

Mulvaney J and Kamminga J (1999) *Prehistory of Australia*. Allen and Unwin, Sydney.
Northern Territory of Australia (1957) *Government Gazette: Schedule Welfare Ordinance Register of Wards*. Government Printer, Sydney.
Powell A (1996) *Far Country: A Short History of the Northern Territory*. Melbourne University Press, Carlton.
Press AJ (1987) Fire management in Kakadu National Park: the ecological basis for the active use of fire. *Search* **18(5)**, 244–48.
Ritchie D (1998) A painful wrench for a European mind: land claims and representations of history and Aboriginal tradition in the Darwin hinterland. PhD Thesis. Deakin University, Melbourne.
Roberts RG, Jones R and Smith MA (1990) Thermoluminescence dating of a 50,000-year-old human occupation site in northern Australia. *Nature* **345**, 153–56.
Spencer B (1914) *Native Tribes of the Northern Territory of Australia*. Macmillan, London.
Spillett PG (1972) *Forsaken Settlement*. Lansdowne, Melbourne.
Stewart A (1969) *The Green Eyes Are Buffaloes*. Lansdowne, Melbourne.
Sutton P and Palmer A (1980) *Daly River (Malak Malak) Land Claim*. Northern Land Council, Darwin.
Thomson D (1983) *Donald Thomson in Arnhem Land*. Compiled by N. Peterson. Currey O'Neil, Melbourne.
Warburton C (1934) *Buffaloes: Life and Adventure in Arnhem Land*. Angus and Robertson, Sydney.
Woodroffe CD, Mulrennan ME and Chappell J (1993) Estuarine infill and coastal progradation, southern van Diemen Gulf, northern Australia. *Sedimentary Geology* **83**, 257–75.

4

Buffalo and tin, baki and Jesus:

the creation of a modern wilderness

Peter M Cooke

SUMMARY

People have occupied the western Arnhem Land Plateau for at least 45 000 years, spanning periods of major shifts in climate, sea levels and landscapes. Continuity of occupation is shown by a rich chronicle of rock art and confirmed by accounts from a few older residents who describe themselves as 'stone country' people (*Warddeken*). But late in the 20th century, the people had all but gone and the country was under threat, because fire regimes had become chaotic and destructive. The thin historic record and accounts of long-term residents indicate that early post-settlement contacts in the mid-19th century were few and mostly inconsequential for either settlers or *Warddeken*. But by the late 19th and early 20th centuries, mines, missions and buffalo-hunting camps in the adjacent lowlands were important attractions, drawing people from their lands. Demand for tobacco was a principal driving force. The process of depopulation continued through the mid-20th century although slowed by the outstations movement during the 1970s, when some groups returned to ancestral lands. Return to outstations in itself failed to restore pre-settlement fire regimes to the wider landscape because outstations were relatively few and movement between them and major settlements was mostly along established roads or tracks by motor vehicle. Bushfire authorities also sought to prevent ignitions rather than foster restoration of traditional practice in fire use. Real progress in re-establishing human control over fire in the plateau began in 1997, when a loosely structured collaboration among Indigenous people, fire researchers, fire management authorities, and public and private funding agencies began to promote active restoration of customary methods. That effort has now evolved into the West Arnhem Land Fire Abatement (WALFA) project, which provides a strong focus for fire management based on an amalgam of traditional practice and use of modern technology.

THE OLDEST ILLUSTRATED HISTORY

On the Arnhem Land Plateau, the signs of enduring habitation can be found almost everywhere. The most striking of these signs are tens of thousands of ochred images left in caves, on cliffs and around sandstone outliers. These images provide an extensive and extraordinary record of cultural and biophysical changes embedded within the longest continuous art tradition on the planet (Chaloupka 1993). Now-extinct species are depicted by the early plateau artists. Among the animals depicted, identification of the marsupial tapir, *Palorchestes,* and marsupial lion, *Thylacaleo,* is controversial but some paintings of the Thylacine or Tasmanian Tiger, *Thylacinus,* and the long-beaked echidna, *Zaglossus,* are so clear and correctly anatomically detailed that their co-existence with people is undeniable.

Arguably, this pictorial chronicle of Australian life began soon after the arrival of humans in Australia from the north-west and their dispersal from a coastline that now lies beneath the Arafura Sea. Two sites, Malakunanja and Nauwilabila, on the western rim of the Arnhem plateau – in what is now Kakadu National Park – have been dated to a period between 45 000 and 60 000 years BP (Roberts *et al.* 1990, 1993). Although dating of the earliest phase of occupation remains controversial (Bird *et al.* 2002; O'Connell and Allen 2004), there is nevertheless broad acceptance that people have been in western Arnhem Land for at least 45 000 years (Johnson 2006).

During that time, both local and global climatic events brought landscape changes – large and small, welcome or unwelcome – for the first Australians to cope with, adapt to and take advantage of. For example, today's northern Australian landscapes show strong annual responses to a dependable annual wet season brought by the north-west monsoon. But during pre-historic times apparently this was not always so. From 30 000 to 12 000 years ago the monsoon appears to have shut down completely across northern Australia (Magee *et al.* 1995). Additionally, in low-lying areas, seas have retreated and advanced; 18 000 years ago – at the height of the Last Glacial Maximum when sea levels were about 130 m lower than at present – sites such as Malakanunja and Nauwalabila were some 400 km from the coast, and land stretched all the way to New Guinea (Chappell 1982; Lambeck and Chappell 2001). What is now the Arafura Sea floor was then most probably a predominantly treeless mixed shrubland and grassland with isolated forest stands including *Callitris, Casuarina* and various Myrtaceae (Yokohama *et al.* 2001).

From 18 000 years ago the seas began to rise again and, over time, the monsoons came back to create the annual wet–dry seasonal cycles that so much influence the lives of people, plants and animals in northern Australia today. The rock art records of flora and fauna reflect climate change, and coastlines came back closer to the sandstone hills: creating the lowland wetlands of Kakadu and providing resources for greater concentrations of populations in some areas and creating new drivers for seasonal movement. As Robert Levitus explained in Chapter 3, the high productivity of the very recently established lowland freshwater wetlands in Kakadu affected the size and density of some human populations and created new drivers for seasonal movement over the past 2000 years – a feature not just of the Alligator Rivers region but also for other wetlands eastwards across northern Arnhem Land.

But, while landscape changes in recent times increased the human carrying capacity of the lowlands, the uplands appear to have always been an area of scarce food resources. The venerable Bardayal Nadjamerrek AO, says of the traditional life on the plateau, 'we lived on kangaroo meat and wild honey'. Nevertheless, rock art from the high plateau shows us that habitats, tool and weapons technologies changed substantially over millennia. Boomerangs transformed from hunting and fighting tools suited to more arid and open early landscapes, to instruments of ritual music-making. Shields disappeared from the plateau, but continued to be used in savanna and the desert further inland.

In light of the visible evidence of long-term cultural continuity and change depicted on almost every suitable rock surface, it is impossible to regard the country of the Arnhem plateau as 'wilderness'. Yet when the Western Arnhem Land Fire Abatement (WALFA) project began in 1997, much of the northern and southern plateau had literally become a wilderness – an ancient

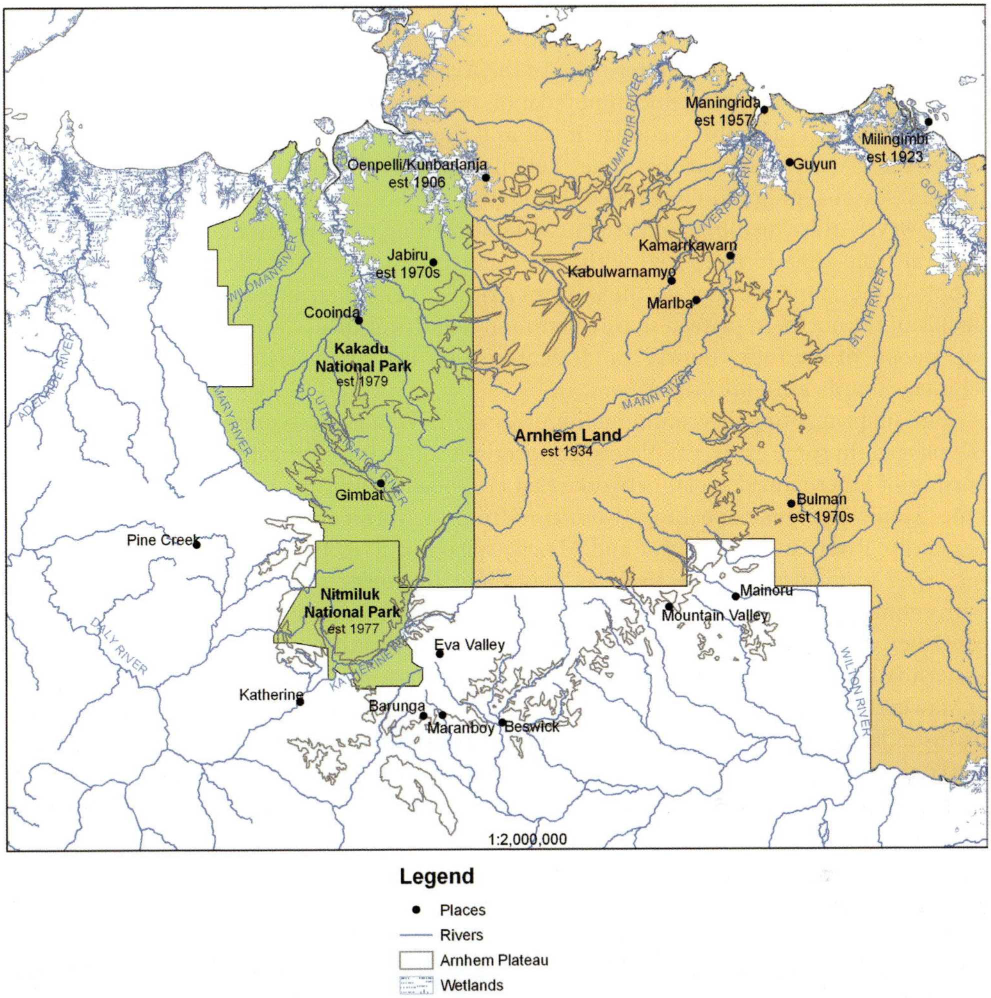

Figure 4.1 Map of the area of the Arnhem Land Plateau in relation to Kakadu National Park and place names of sites that played important roles in the social history of the region.

garden from which the gardeners had gone. Temporal and spatial patterns of annual deliberate burning that had been developed and applied over millennia were replaced by a chaotic, episodic and random fire regime in just a few short generations.

The breakdown of Aboriginal fire management through the departure of the critical mass of plateau managers over a period of a little more than 100 years was driven by particular historical circumstances in areas off the plateau; pulling people down from their rocky homelands in various directions. Many left and came back to the highlands. But, by the end of the 1950s, almost all had gone to the lowlands and into the communities of lowland clans: into buffalo camps, mines, pastoral properties, government settlements and church missions (Figure 4.1).

TRADE AND OTHER TIES THAT BIND

The formation of the lowland freshwater wetlands, and their associated rich resources, over the last few millennia doubtless resulted in new patterns of movement and social organisation for plateau people. However, whether these new patterns entailed seasonal or intermittent movements, or more permanent shifts, is unclear. Based on archaeological evidence from rock-shelter deposits fringing the western rim of the Arnhem plateau, Jones and Johnson (1985, p. 211) reflected on signs of decline in population numbers associated with plateau valleys over the past 1000 years and observed that 'it was as if this riverine plain region was sucking people out of the plateau valleys at this time'. Levitus (Chapter 3) says of the upland and lowland peoples 'seasonal migration integrated them into a single expanded resource range.'

However, older members of the plateau land-owning clans today claim a distinct identification from time immemorial with the relatively resource poor uplands and with other people of the 'stone country', as the western Arnhem Land Plateau is sometimes known. The people of the northern plateau share a common identification as *Warddeken*: a name signifying 'people of the rock country', and identify their regional variant of the *Bininj Kunwok* language family as *Kundedjnjenghmi*.

But, supplementing this commonality, other cultural and social connections extend beyond the plateau. In some cases, off-plateau connections following river catchments provide membership of other named social networks that cross-cut groupings by language and landscape type. *Kundedjnjenghmi*-speaking *Warddeken* from the rock country at the headwaters of the Gumardir River, and *Kunwinjku* and *Mawng*-speaking clans from the river's lower reaches, may refer to themselves collectively as *Manginburru* people: a term that includes all people with country in the Gumardir River Catchment – from rocky headwaters to coastal mudflats.

Other connections were forged by resource sharing with near neighbours – the north-eastern *Warddeken* came down from time to time to share in the early dry season plenty of fish at Mawurrk or Bulkay on the Tomkinson River plains. In the later dry season, people from the plains walked to the plateau to join plateau clans in hunting kangaroo with fire.

Trading partnerships for exchange of resources, such as the rare pigment huntite, from the Upper Gumardir, and stone spear points from the quarry at Marlba, linked the *Kundedjnjenghmi* with non-*Warddeken* neighbours such as the *Mbukarla* at Cooinda – the nearest source of bamboo spear shafts for the plateau people.

Figure 4.2 Recently re-established walking track used until the 1950s for trade among Aboriginal groups.

Trade – both for material objects but also concerning rights to perform ceremonies – created patterns of movement along known and regularly travelled bininj man-bolh or 'black-fella road' walking tracks. As described by Garde *et al.* in Chapter 5, these walking tracks were also instrumental in creating the pattern of fire in the landscape, and helped control the incidence of large late dry season wildfires.

WALKING INTO THE NEW WILDERNESS

In the year 2000, 12 Aboriginal people and two Balanda (non-Aboriginal people) followed one of these walking tracks westward from Kamarrkawarn outstation on the Mann River to the East Alligator River (Figure 4.2) – a distance of about 120 km, but still about 60 km from the South Alligator Wetlands in Kakadu National Park where the exchange of highland stone spear points for lowland bamboo spear shafts continued well into the 20th century. Bardayal Nadjamerrek says he walked this track perhaps a dozen or more times for trade: probably in the 1940s and 1950s. But the fires lit on these, and other, routes that were followed for trade or ceremony beyond this region, and from band to band on the plateau, seem to have finally gone out across most of the plateau by the 1960s. When our party walked through this newly emergent 'wilderness' in 2000, we found signs everywhere of that falling away of intensive management. In the words of today's elders, the country had become orphaned and it cried for its absent carers. Stands of *an-larrh* (the *Kundedjnjenghmi* name for cypress pine, *Callitris intratropica*) stood stark and dead as we passed by. After existing in some apparent viable population stability over long periods of Indigenous fire management, the rapidly escalating decline of this species in the absence of customary management was apparent both to western and indigenous observers alike. There was little evidence of recruitment in surviving *an-larrh* groves. The thinned out *an-larrh* patches – having lost many individual trees in all age classes and with survivors carrying few lower branches – could no longer put down the thick blankets of fire-suppressing needles that typify healthy groves.

Sheltering in deep gorges where they are protected from severe wildfires, the ancient and stately endemic tree, *an-binik* (*Allosyncarpia ternata*) continued as the dominant species of a mostly closed canopy habitat. But, in isolated *an-binik* patches on open sand plains among expanses of shrubby heath and woodland, the walkers likewise found *an-binik* suffering badly from the prolonged absence of customary fire management. Large older trees had been burnt out, exposing regenerating *an-binik* to increased competition from fire-carrying grasses and the intermittent ravages of further wildfires. Most of these same *an-binik* patches also exhibited significant trampling damage from feral buffalo. Bardayal Nadamerrek recalls seeing buffalo for the first time on the high plateau in the mid-20th century.

A THIN HISTORICAL RECORD

What happened to the plateau and its people over the past 150 years is poorly documented. In that time, only a couple of parties of Europeans crossed the high plateau. They saw very little of the *Warddeken* and so had little to say about occupation or management practice. And although the impacts of exotic diseases on upland clans who went westward to the buffalo shooters camps of the late 19th and early 20th centuries are understood in a general sense (Chapters 2 and 3), there is little in the way of an official record from European sources of who went where, when and what happened to them.

We do, however, have the testimony of today's *Warddeken* elders – pre-eminent among them Bardayal Nadjamerrek, who was born in 1926 and universally revered among the plateau people and its diaspora as their foremost authority. We also have some oral histories collected among those *Rembarrnga* people who moved south of the plateau from its eastern edge and into pastoral country down to Katherine after early bad experiences with 19th century pastoralism around the Arafura Swamp, in north-central Arnhem Land.

We have both European historical records and Aboriginal oral history of the impact of mining: in particular from the Maranboy tin mines. We have observations of movements westwards, from the northern rim of the plateau to Oenpelli and onto Darwin, and then the drawing in of people on the plateau's eastern side to the government settlement of Maningrida, which was established in 1957.

On some rare occasions, we have intersections of European written history and Indigenous history: sometimes oral and sometimes in rock art imagery. For example, the explorer Ludwig Leichhardt and his party passed to the south of *Kundedjnjenghmi* country, through Jawoyn territory and into the Alligator River catchment in 1846; but it was not until 1883 when Europeans travelled southwards through the catchments of the Mann and Liverpool Rivers that the first European presence was noted in *Kundedjnjenghmi* country.

Towards the end of June 1883, the surveyor David Lindsay set out from Glencoe Station on the Adelaide River with four white men, a Larrakia 'boy' and 27 horses. They travelled nearly 2000 miles – eastwards to the Gulf of Carpentaria around the south of the plateau, and back to Katherine in early November via a route between the Mann and Liverpool Rivers across the back of the plateau from north-east to south-west (Lindsay 1884).

Lindsay was interested in grass – as pasture – and the account of his trip published in 1884 makes many references to grass, often in the context of indigenous burning practice. From the beginning of his journey in June, he makes regular reference to country burning, just burned, or with young grass just showing: usually in the context of finding feed for his horses.

Although Lindsay notes in his journal and subsequent report scarcity of feed from day to day, at other times he speaks highly of 'splendid' and 'grand grass'. He makes this note to clarify the issue in his entry for 2 September 1883:

> *Although I am describing the country as well and splendidly grassed etc., it is to be understood that the country produces grasses as described; but during my journey the whole of the country is recently burnt or now burning, leaving but scanty feed for the horses.*

He had many encounters with native people, some friendly some not. At the end of September he experienced a tense stand-off with a group of about 100 Aboriginal men near the resource-rich Arafura Swamp and some subsequent incidents of stealthy stealing from camp as he travelled westwards.

But his account of travel into *Kundedjnjenghmi* country is without direct interaction with the *Warddeken*. Although he saw some native smoke rising as his party struggled up the Liverpool River into the 'stone country' in the oppressively hot and humid season known to the *Warddeken* as Kunumeleng, he saw neither people nor occupied encampments.

He named the Cuthbertson Falls on the Liverpool River after one of his travelling companions. His horses were finding it hard going and over the next few days he abandoned gear – three pack saddles, a fishing net, rifle and pistol ammunition and unneeded clothing. There is no mention of natives or fires as the travellers focused on making passage through the rocks and coping with the increasingly oppressive temperatures and thunderstorms. Lindsay's report no longer praises fine country, but relates stories of horses and men stumbling and falling among the rocks and spinifex. 'I lay for a time wondering whether life was worth living,' he writes later.

Figure 4.3 Images of Europeans captured in rock art in the Liverpool River region.

Figure 4.4 Images at the same site of what can be interpreted as a pack-saddle and rolled up fishing net, items of equipment abandoned in the region by the surveyor Lindsay.

But although he saw no Aborigines, it seems the *Warddeken* observed him quite closely. At a rock art site first photographed only 2 years ago, and within a few kilometres of the place where he abandoned surplus gear, is arguably a record of his passing. A number of pipe-smoking Europeans are depicted carrying rifles, sighting along the barrels and crouching as if stalking (Figure 4.3). A horse's head and neck, and parts of a rider can also be discerned. But most exciting are three unusual shapes that look very much like pack-saddle bags and a rolled up net (Figure 4.4). Unlike many other places where palimpsests of rock art end with images of contact, a large white kangaroo overlays and partly obscures the images of white men and their guns at this site.

BUFFALO AND TIN, BAKI AND JESUS

While the *Warddeken* were left to pursue their traditional lifestyles, events were happening to the east and west that suggest that these last few decades of the 19th century mark a crucial demographic turning point that led to the breakdown of the large landscape-scale customary burning by native land managers on the plateau. For example, in the last decade of the 19th century, European attempts at pastoralism in north central Arnhem Land brought widespread violence that disrupted customary land management for the *Warddeken*'s eastern neighbours: the *Rembarrnga*, the *Djinba* and others.

After some centuries of foreign contact occurring only on the beaches of Arnhem Land, it accelerated inland from the 1880s. A wagon track ran north-east from around the site of Bulman today up to the Arafura Swamp. In 1910, there was a short-lived mining 'bubble' at 'Boolman'. About the same time, tin was discovered at Maranboy on Jawoyn country and, although this was always a small-scale mining field, it continued until after the Second World War and was important in attracting both Jawoyn and other people from further afield. Mainoru Station was created in 1918 and was also to play a big role in drawing people off country and away from their responsibilities for fire management and other customary duties.

Establishment of Christian missions at Milingimbi and the Goulburn Islands early in the 20th century gathered in some groups and also introduced new sources of trade goods. Buffalo camps drew many *Warddeken* down to the Alligator Rivers region: partly for trade goods but also, as George Chaloupka (*pers. comm.*) points out, for the general excitement and social licence that attended life in the buffalo camps.

But, when today's senior generation of landowners from the western Arnhem Land Plateau and further east into Central Arnhem Land talk of the factors that led their parents or grandparents to leave their clan estates for buffalo camps, missions, mines and pastoral stations, the principal driving force most often identified is a desire – perhaps even a passion or a craving – for baki (tobacco).

Peter Biless, born in 1933, is a *Warddeken* elder from a generation that grew up at the time when the high plateau population was hitting a low point. In the north-east, only small groups of people were still on the plateau: mostly along the mid-reaches of rivers such as the Liverpool and the Mann.

'We missed learning about the names of many places', he says. 'Everyone was leaving the plateau for Maranboy, Katherine and so on. They were going for tobacco. By the time I was growing up nobody was living on the central plateau. I am of that first generation not to live up here but at the mission at Kunbarlanjnja'.

It is not known when the Aboriginal people of Arnhem Land first tasted *Nicotiana tabacum*, but it was certainly important in shaping and maintaining relationships between Aborigines and the Macassan fishermen who had been visiting the coast of Arnhem Land since the late/mid 1600s. In north-east Arnhem Land, the Aboriginal use of Macassan tobacco and long Macassan-style smoking pipes were elevated into an association with Indigenous sacred ceremonial life. The smoking pipes of some powerful men were painted or engraved with religious symbols so sacred that the pipe stems had to be wrapped in paperbark or cloth so they could not be seen by the uninitiated (Thomson 1939). Goodwill gifts or payments of tobacco are documented in records of early contacts with Europeans on the coast, and later with inland exploration.

The discovery of gold at Pine Creek began a rush of mostly Chinese miners: creating an inland source of valued alien trade goods, which began to draw in Aboriginal people westwards from the East Alligator regions and what is now western Arnhem Land. Not only tobacco, but also the dross of opium pipes, was available on the mining fields in these early days. A further decade on, the construction of the north–south railway line further strengthened the demographic pull to the west.

Tobacco certainly played a major part in the demographic chain reaction that emptied the plateau and put out the fires of the early dry season. Speaking in particular of people from Gimbat – now part of Kakadu National Park – Keen and Merlan (1990) remarked that many Aborigines were drawn towards the mines and towns, towards rations, tobacco 'and also to follow their fellows'. This double-action demographic dynamic continues even to the present time. For various reasons, some people move and then, for family reasons, others follow.

In Macassan times, tobacco was available only at a relatively small number of locations, only on the coast, and only during the wet season. During the Wet, travel on foot was inhibited by flooded rivers and there was a preference for staying in one place and seeing out these leaner subsistence times of the year. As European colonisation gathered pace, tobacco and other alien trade goods became available at many locations – often all year round.

As any smoker and medical research can attest, nicotine is a powerful drug – one for which it is easy to acquire an addiction, but from which withdrawal or supply interruption is difficult and painful, even today when smokers understand the personal health cost of maintaining the habit. For Aborigines, it was a pleasant drug to use, whether smoked or chewed: a trade 'good' of high value in an Indigenous social context and one whose use seemed to bring no negative consequences – as long as one could maintain supply. But when supplies ran out, it made good its reputation still widely understood by Aboriginal and non-Aboriginal addicts alike – tobacco is boss.

In a number of published and unpublished commentaries, the power of tobacco is attested by Aboriginal people who recall its effect on their lives as children and the lives of their parents in the first half of the 20th century. These commentaries tell us much about the timing and direction of demographic drift from western and central Arnhem Land.

Smiler Martin, a *Rembarrnga* man whose country lies not far south of Ramingining recalls:

> *I was a little fella at Mainoru (pastoral property). I used to play about in the camp. There were Rembarrnga, Jinba, Rittharngu people. Those were three tribes from Arnhem Land. We all travelled from Arnhem Land. As soon as they started tasting a bit of tobacco, they used to travel. … Our father carried us all the way because they had tasted tobacco and tea and sugar and tucker. We grew up in this area then. We forgot about country. We went back sometimes to look at our country, a couple of times. But no vehicles at that time, just foot walking. (Martin et al. 1995, p. 1)*

The missions, too, were a source of tobacco and other stores. Smiler Martin was among those who went north to the coast to trade at Milingimbi.

> *We used to go down Goyder River way and follow that river right along to Murruwangi. Me and Willie took alligator hides. We carried them down to Milingimbi one week,*

> *walking and sold them to the missionary, Mr Wells. He didn't give us any money but he gave us tea, sugar and tobacco. … We were having a good time in the bush. (but) As soon as people tasted smoke we said 'we'll have to walk this way now'. So everyone came to this place (Mainoru) then. (Martin et al. 1995, p. 1)*

The following commentary by Ronnie Buck hints at the furtive relationships between 'station blacks and bush blacks' during the mid-20th century when there were still groups of people living semi-independent lives in the bush in the south of western Arnhem Land, but when these 'bush people' were to some extent still fearful and hiding from white people.

> *I was born just before my mother saw the first white people … probably 1938 or 1940, or later. My parents moved to Mountain Valley … my father used to travel to Maranboy and Mataranka with the ceremony and then he'd come back to Mountain Valley where other family people were working. The first thing people saw was that black tobacco. They found something that tasted very good. 'Who did that special thing, nikki nikki (plug tobacco), come from?' they asked. Not only that, but sugar. … A lot of the people keep telling one another, 'where did you get this from?' They used to say, 'well, we got this from a place called Maranboy, or sawmill or Eva Valley'. So one, or many people, or a whole tribe, got a taste of that thing, tucker, or damper. All travelled, some as far as this. There were hardly any stations round here yet. First, people used to go to Mainoru or Mountain Valley station even though they were told there was a white bloke there, different from us, and a lot of the people were frightened.*
>
> *What usually happened was, if people ran out of tobacco and wanted to get some more tobacco, they used to wait till late at night. They used to sneak up and see the (Aboriginal) family that used to feed them, who'd give them tobacco, or even flour, and then sneak off. During the day they used to stay out where they thought no white people were going, or mainly just wandering around where there's lots of scrub.*
>
> *If ever they wanted anything they used to tell the other family, 'count the days. Give me four days, and I'll come back sometime in the night after three days and get some more'. … We were between Mainoru and Mountain Valley, just hiding, just going around the places such as Mainoru and Mountain Valley. We were there for some time, until I was, say, one and half years old. (Buck 1995, pp. 2–3)*

Maranboy tin mines and Eva Valley station appear in many Aboriginal accounts of early and important alien contact for the inland people of central and western Arnhem Land. One of those who went to Maranboy from Central Arnhem Land was Mandark, who was born about 1915 on a tributary of the Cadell River and who saw a white man for the first time at the Maranboy tin mines when he was in his late 20s. Mandark was one of the few who travelled to Maranboy and its mines, gardens and timber mill but who came home to stay and continue a traditional life with his five wives: raising their 25 children on the Upper Cadell River. Mandark and his family came to look after some eight estates on behalf of the bulk of the population who left permanently just after the Second World War.

Some of these families went south and west to Katherine, others to Oenpelli mission. In the 1950s, missionary Helen Wilson recorded over 600 names of Aboriginal people who visited Oenpelli over 4 years – but at any time there were rarely more than 200 people at the mission. On the coast to the east, Maningrida was established as a government settlement in 1957 and, by 1963, the first vehicle track was 'bush-bashed' into Maningrida from Oenpelli by a joint party of staff from Welfare and Forestry branches of the Australian Government's Northern

Territory Administration. This was the beginning of the end for long distance travel by foot. Within 10 years, some of the Aboriginal groups who had been gathered in to Maningrida were returning to the bush with tractors to establish outstations. The change to road travel was a major influence on the decline of customary burning. The missionary and ethnographer Gowan Armstrong said that by 1963 there were no people permanently living in the bush in the area of the Blyth and Liverpool Rivers but that 'large numbers of the Gunwinggu people continued to spend most of their dry seasons in their home territory' (Armstrong, undated).

South of the plateau, many *Rembarrnga* and *Jawoyn* had been brought in and 'settled down' by the government at new communities of Beswick Station and Bamyili: the former with a view to 'training Aboriginal men for employment on pastoral properties' (Welfare Branch 1967). Both the Beswick pastoral property and the Maningrida Forestry Project had one important thing in common – non-Aboriginal employees running both projects regarded Aboriginal traditional burning as behaviour to be extinguished. As Ronnie Buck says: 'You weren't allowed to burn any grass around Beswick and Bamyili because of cattle and horses'.

At Maningrida, Forestry put in extensive networks of roads and firebreaks and imposed fire exclusion on its management area. Chris Haynes, then a forester in charge of the Maningrida project, told a Senate enquiry in 1977 how the *Rembarrnga* outstation of Guyun and surrounding country 'had been declared a fire-free zone to encourage the development of cypress regeneration. Fires were persistently lit by Aborigines. I had my orders: the hunters lit up, and the forestry section put out. We did not have much understanding of each others' purpose.'

Although Aborigines continued to light up, the introduction of roads and the loss of critical mass for landscape-scale burning meant that – other than in areas close to outstations where people continued to walk and hunt – the large landscape effect of many mobile groups lighting many fires from early in the dry season had effectively stopped. As Haynes asserts (Haynes 1991), the customary creation of mosaic burning patterns that could control late dry season fires:

> *required extensive walking within the relatively few weeks of the cold weather. Depopulation then could be critical, especially if the fit and active people who lit the fires were the most common absentees. Progressively more fires would have been later in the year and injured the more sensitive species, the most obvious of which is cypress pine. This is indeed what happened.*

Forestry project managers were unable to see the management value of customary burning for protecting cypress. Forestry branch workers (and I was one of them briefly) routinely climbed specially constructed watchtowers to spot Aboriginal fires and provide locations to fire-fighting teams back at Maningrida. These exercises would often end with a fire fighting tanker truck slowly working its way through the bush extinguishing a low trickling fire until eventually it caught up with a sometimes bemused, but sometimes angry Aboriginal landowner with matches and gun or spear going about his customary business of hunting. Anger increased when Forestry's aggressive fire suppression led to incidents in which a bulldozer desecrated a sacred site and a grader was pushed through the sacred ground of the region's most strict and access-restricted religious ceremony, while a ceremony was underway (Haynes 1991).

With a change of federal government in 1972 and an expression of support for Aboriginal self-determination as Government policy, Aboriginal people were empowered to end the ill-conceived and harshly executed forestry project and also begin the return to traditional estates, which has become known as the outstation movement. After the demise of the Maningrida forestry project the former project manager Chris Haynes returned to work for Aboriginal people at Maningrida and to explore Aboriginal perceptions of why European pyrophobic forestry practice was rejected so clearly by Aboriginal land managers. Aboriginal land managers returning to their traditional estates pointed to the onset of patterns of large late dry season fires and consequent death of fire-sensitive species (such as *Callitris*) that had resulted from Government Forestry's failure to deal effectively with annually accumulating fuel loads.

No longer trying to stop Aboriginal fires, Haynes observed and recorded Aboriginal burning practice in word and deed. His work with the Gunei (Kune) land managers of the Cadell River provides remarkable insights into the sophisticated customary way in which fire is used to manage fire in Arnhem Land. His seminal papers (Haynes 1978, 1985, 1991) document how a wide range of factors inform the decision of Aboriginal experts as to where and when to burn, and to what effect. The work of Haynes, Kalakala, Midikuria, Nawilil and many others is today recognised as the first serious collaboration between indigenous expertise and western science on fire management in this region. It is a source of knowledge that has maintains its relevance into the new century, and today still informs contemporary collaborations on fire management.

BACK TO THE BUSH, BY ROAD NOT FOOT

By 1976 there were more than 12 'outstations' re-established around Maningrida and a formal organisation providing logistic and administrative support to enable year-round occupation. Many Aboriginal families who had moved west to the mission at Oenpelli (now Kunbarllanjnja) began to move back eastwards: some seeking to escape the damage being done by alcohol following the granting of a liquor license to the Border Store, which was outside Arnhem Land but only some 15 km from the Mission. Others just wanted to go home and be on country for which they had both rights and responsibilities of management. Escalating mining exploration from the late 1960s added a sense of urgency as people sought to assert their land ownership by their physical presence and also to protect sacred sites.

The missionary and aviator Alf Wilson provided support for the outstation pioneers moving eastwards from Kunbarllanja (Oenpelli) (Wilson 2005). Ten small family-based communities, each consisting of a cluster of bush timber and tin huts beside a bush airstrip, came into being during the 1970s and 1980s. Most were in the lowlands in country easy to access, but by the 1980s the *Warddeken* were on the way home with three emerging outstations on the upper Mann River and its tributaries at the northern fringe of the plateau. To the south of the plateau, the outstation movement was slower and more limited in extent, with only a small number of mostly seasonal outstations around Bulman.

But, although many people were now back on country, the creation of largely sedentary settlements near airstrips, acquisition of tractors and 4WD light trucks, and the creation of roads and tracks marked the end of the kind of landscape-scale social movement that underpinned the mechanics of customary fire regimes. Small, patchy, low-intensity, early dry season fires were typical around outstations, but at a larger scale – across most of the remote, rugged and difficult to resettle high plateau – patterns of customary fire management disappeared.

To the south-east of the plateau, from where the strong winds of the dry season blow across the plateau to the north-west, customary fire management had been very seriously weakened by non-Aboriginal negative attitudes to Aboriginal burning. Customary burning was often referred to as 'pyromania' by neighbouring pastoralists. In time, protected area managers at Kakadu and Nitmiluk National parks were making similar remarks. Reflecting pastoral attitudes, the Bushfires Council of the Northern Territory erected signs on the road from Katherine to Bulman warning of fines and gaol for fire lighting.

The major breakdown of burning to the south-east of the plateau was disastrous. Late dry season wildfires became the norm as indigenous land managers were intimidated into ceasing early burning. Every few years mega-fires swept through the thick accumulated fuels of the plateau: often penetrating Kakadu National Park and perpetuating the myth that Arnhem Land fire managers were ill-informed and irresponsible pyromaniacs.

It was in this context that in the mid-1990s the WALFA project was born. The creation of a Caring for Country Unit at the Northern Land Council in 1995 provided an overarching representation of Aboriginal interest in land management at a suitable scale to address the wildfire problem. The Caring for Country unit became a broker involving larger Aboriginal corporations in the region in the fire issue: linking the Jawoyn Association based in Katherine, Bawinanga Aboriginal Corporation (Maningrida), Demed Aboriginal Corporation (Kunbarllanja) and unincorporated groups from Bulman and other sites of the plateau diaspora.

Collaboration between science and indigenous knowledge was developed to address fire-management issues of the plateau, and was funded through a series of Australian Government Natural Heritage Trust grants from 1997. Younger people became involved in the emerging Aboriginal community ranger movement and Aboriginal elders travelled to meetings at locations around the plateau. Old social links were reforged and new links made across cultural and knowledge systems in the cause of 'fixing' fire management. Communication between experts such as Bardayal and Jeremy Russell-Smith were enhanced by the skills of linguist Murray Garde and indigenous fire ecologists Dean Yibarbuk and Otto Bulmaniya Campion.

Because the worst effects of the changed fire regimes were happening in the most remote and 'empty' parts of the region, the first step was to create an understanding of what was happening 'out of sight' at this large landscape scale. Science provided the tool to do this. Remotely-sensed fire-scar mapping derived from satellite imagery showed, by colour coding, the extent and sequence of fires at different times of the year. Annual fire scar maps provided a time series that younger people were quickly able to interpret and engage with. Within a short time, elders too could see what was happening as younger people explained the fire scar maps and asked them whether it had been like this when they were young.

Having established the nature and extent of the problem and that the new 'feral fire' regimes were 'killing country', the task for indigenous managers and scientific collaborators was to understand how Aboriginal customary burning had managed fire in a sustainable manner and then to develop innovative ways to emulate the best aspects of customary burning.

The involvement of elders was critical to success. The regional status of leaders such as Bardayal, Peter Biless, George Djanggawanga, Brian Nyinawanga, Jimmy Kalarriya, Mary Kolkkiwarra and Jack Djandjomerr enabled them to provide an endorsement and patronage of the project that provided a platform for younger leaders such as Dean Yibarbuk and Otto Campion to build a program of fire awareness and fire management.

Given that the only way to emulate traditional effects of walking and burning at the right season in this vast and rugged area was to use helicopters to drop incendiary capsules, gaining community approval for such high-tech 'whitefella' technology was essential to project success. In the past, Bushfires Council, and Government Forestry before them, had delivered incendiary runs by light plane across much of the area and this had been seen by Aboriginal landowners as something done without their involvement or, to any significant extent, permission.

But this time Aboriginal people, both younger and older, were involved in drawing up flight lines for aerial prescribed burning. Indigenous landowners were able to see that 'fire bombing' did not involve explosions and that incendiary capsules were not much more than very strong matches or traditional 'djalk', or firesticks. Walks into remote areas, such as the previously described walk across the plateau in 2000, involved elders flown in by helicopters to discuss the state of country now and in the past. Within a few years, Aboriginal community rangers were gaining endorsements in aerial prescribed burning and despatching incendiaries themselves.

In recent times, the WALFA project has received significant media attention and official accolades. Although winning the 2007 Eureka prize for 'innovative solutions to climate change' can be rightly seen as official recognition of the project's achievements to date, the greater achievement has been local. The principles of Indigenous customary fire management have been demonstrated to be sound, effective and wise. Non-indigenous science has provided remarkably useful tools such as the North Australian Fire Information fire-mapping website (www.firenorth.org.au). Aboriginal and non-Aboriginal staff deliver early burning from helicopters in the most remote areas, but often with elders aboard to navigate and place fires. Annual on-country, early burning walks provide opportunities to learn fire behaviour up close and to acknowledge the wisdom of the 'old people' while following in their footsteps. Employment and other opportunities associated with WALFA and other projects leveraged by WALFA's success are helping the plateau diaspora return home and return the wilderness of orphaned country to health again. Perhaps this time – to turn around Haynes' (1977) comments to a Senate inquiry – Aboriginal and non-Aboriginal people working to manage fire in western and central Arnhem Land at last do have some understanding of each others' purpose.

REFERENCES

Armstrong G (undated) *A Brief History of Maningrida*. Unpublished typescript.

Bird MI, Turney CSM, Fifield LK, Jones R, Ayliff LK, Palmer A, Cresswell R and Robertson S (2002) Radiocarbon analysis of the early archaeological site of Nauwalabila I, Arnhem Land, Australia: implications for sample stability and stratigraphic integrity. *Quaternary Science Reviews* **21**, 1061–75.

Buck R (1995) Ronnie Buck. *Bulman Oral History Series*. Barunga Press, Katherine.

Chaloupka G (1993) *Journey in Time*. Reed Books, Sydney.

Chappell J (1982) Sea levels and sediments: some features of the context of coastal archaeological sites in the tropics. *Archaeology in Oceania* **17**, 69–78.

Haynes CD (1977) Joint select committee on Aboriginal land rights in the Northern Territory. In *Hansard, Canberra 14 April 1977*, pp. 767–780. Commonwealth of Australia, Canberra.

Haynes CD (1978) Gunret, gundulk dja Bining. *Commonwealth Forestry Review* **57**, 99–106.

Haynes CD (1985) The pattern and ecology of *munwag*: traditional Aboriginal fire regimes in north-central Arnhem Land. *Proceedings of the Ecological Society of Australia* **13**, 203–214.

Haynes CD (1991) Use and impact of fire. In, *Monsoonal Australia: Landscape, Ecology and Man in the Northern Lowlands*. (Eds CD Haynes, MG Ridpath and MAJ Williams) pp. 61–71. Balkema, Amsterdam.

Johnson C (2006) *Australia's Mammal Extinctions: A 50,000 Year History*. Cambridge University Press, New York.

Jones R and Johnson I (1985) Deaf Adder Gorge: Lindner site, Nauwalabila I. In *Archaeological research in Kakadu National Park*. (Ed. R Jones) pp. 165–227. Australian National Parks and Wildlife Service, Canberra.

Keen I and Merlan F (1990) 'The significance of the Conservation Zone to Aboriginal people'. Report to the Resource Assessment Commission. Kakadu Conservation Zone Inquiry. AGPS, Canberra.

Lambeck K and Chappell J (2001) Sea level change through the last glacial cycle. *Science* **292**, 679–685.

Lindsay D (1884) *Mr D. Lindsay's Explorations through Arnheim's Land*. Government Printer, Adelaide.

Magee JW, Bowler JM, Miller GH and Williams DLG (1995) Stratigraphy, sedimentology, chronology and palaeohydrology of Quaternary lacustrine deposits at Madigan Gulf, Lake Eyre, South Australia. *Palaeogeography, Palaeoclimatology, Palaeoecology* **113**, 3–42.

Martin S, Martin R, Forbes J and Forbes R (1995) Smiler, Ronnie, June and Ricky. *Bulman Oral History Series.* Barunga Press, Katherine.

O'Connell JF and Allen J (2004) Dating the colonization of Sahul (Pleistocene Australia – New Guinea): a review of recent research. *Journal of Archaeological Science* **31**, 835–853.

Roberts RG, Jones R and Smith MA (1990) Thermoluminescence dating of a 50,000-year-old human occupation site in northern Australia. *Nature* **345**, 153–156.

Roberts RG, Jones R and Smith MA (1993) Optical dating at Deaf Adder Gorge, Northern Territory, indicating human occupation between 53,000 and 60,000 years ago. *Australian Archaeology* **37**, 58–59.

Thomson D (1939) Notes on the smoking pipes of North Queensland and the Northern Territory. *Man* **39**, 81–91.

Welfare Branch (1967) '1966–67 report'. Welfare Branch, Northern Territory Administration, Darwin.

Wilson A (2005) *The Changing Scene.* Acacia Printing, Humpty Doo, Northern Territory.

Yokoyama Y, Purcell A, Lambeck K and Johnston P (2001) Shore-line reconstruction around Australia during the Last Glacial Maximum and Late Glacial Stage. *Quaternary International* **83–85**, 9–18.

5

The language of fire:

seasonality, resources and landscape burning on the Arnhem Land Plateau

Murray Garde in collaboration with Bardayal Lofty Nadjamerrek, Mary Kolkkiwarra, Jimmy Kalarriya, Jack Djandjomerr, Bill Birriyabirriya, Ruby Bilindja, † Mick Kubarkku and Peter Biless

SUMMARY

There are few published data that detail the relationship between Aboriginal people and the environment of the Arnhem Land Plateau and even rarer in these studies are the voices of Aboriginal people themselves. What knowledge we do have has been collected from the fringes of this region from well-known archaeological studies in Kakadu National Park on the western margin, and for coastal groups in central-north Arnhem Land to the east. Although the process of depopulation on the Arnhem Land Plateau commenced at the end of the 19th century, small bands of Aboriginal people continued to occupy the plateau: living in the myriad of sandstone shelters and harvesting and managing the resources of the region up until the early 1970s. The material in this chapter represents an intersection of history, ethnography and human ecology recorded in the Bininj Kunwok language from those few remaining Aboriginal people – ecological experts in their own right – who resisted permanent settlement at regional towns and missions and continued their traditional occupation of the plateau until recent times. Accounts of the seasonally available plant and animal resources, the traditional patterns of movement and the specialist language used to talk about fire and its variety of uses are important in establishing baseline ethnographic data and an historical context in which contemporary ecological and cultural changes can be understood.

INTRODUCTION

The study of Australian Aboriginal burning practices has been of interest to a variety of academic disciplines across both the natural and social sciences. The literature is now extensive and deals with a diverse range of research questions, many of which relate to landscape and vegetation history (Bowman 1998) and more recently the role of fire in contemporary land-management policy development (Dyer *et al.* 2002; Whitehead *et al.* 2003). One focus to date has been investigations of traditional Aboriginal fire regimes in the context of the biological condition of the Australian landscape and the quest to understand the factors that have shaped it. The objective of this chapter, however, is to remedy a rather serious lacuna in the study of Aboriginal landscape burning: namely the absence of Aboriginal voices and especially Australian language texts as primary data. This is not to suggest that the slate in this area is completely blank. Important Aboriginal contributions or co-authored works include Langton (1998, 2000), Yibarbuk (1998), Yibarbuk and Cooke (2001), Yibarbuk *et al.* (2001) and the Aboriginal language texts in Bowman *et al.* (2001) and Hill and Nowakowski (2003). In addition, many long-term collaborations between Indigenous people and ecologists have resulted in important descriptions of Aboriginal landscape burning. For the area with which we are concerned in this chapter – the Top End of the Northern Territory and in particular the Arnhem Land Plateau – important studies include Haynes (1985, 1991), Russell-Smith (1985a, 1985b, 2002), Russell-Smith *et al.* (1997, 1998, 2002, 2003) and Yibarbuk *et al.* (2001).

The purpose of presenting extended texts on fire and landscape in an Aboriginal language here is not to argue for a particular point of view in relation to the history of Australian landscapes, nor as a demonstration of the nature of classical pre-contact Aboriginal fire regimes. The transcripts and translations of the texts that follow will certainly be of value for a number of other reasons. Firstly, recent research on biocultural diversity has highlighted the increasing threats to the viability of traditional ecological knowledge (Ellen *et al.* 2000; Florey 2001a, 2001b; Zent 2001; Stepp *et al.* 2002; Maffi 2005). Access to the ecological knowledge of Indigenous peoples is mostly available in the fullest detail to members of the speech community in question. It is here that collaboration among linguists, Indigenous people and natural science specialists can assist in the documentation and maintenance of threatened traditional ecological knowledge. For linguists, there is no reason why the documentation of a previously undescribed minority language should not include the detailing of cultural knowledge, and especially ecological knowledge, as part of the language description. In Australian linguistics, there has been such a dramatic loss of linguistic diversity since European settlement that, in many cases, opportunities to record unique types of Aboriginal ecological knowledge have been lost.

Secondly, the promotion and maintenance of traditional Aboriginal ecological knowledge is of relevance to the current discourse on the links between linguistic and biological diversity (Harmon 1996, 2002; Maffi 2001a, 2001b). It can be argued that the diversity and complexity of both the topographic and biological diversity of the Arnhem Land Plateau has engendered a diversity of adaptational ideas for the Aboriginal groups who have lived in the region. The plateau is a major centre for biodiversity and 'supports an unusually diverse biota, including very many relictual and endemic plant and animal species' (Commonwealth of Australia 2005). The correlating linguistic diversity of the region expresses a great range of such adaptational ideas and, with the loss of such diversity and cultural heritage, our ability to understand the complexity of human interactions with this particular environment is certainly impoverished. But, in seeking to maintain the vitality of the Aboriginal languages of the plateau, this is not an exercise in giving language priority over speakers. Many speakers of Bininj Kunwok dialects on the Arnhem Land Plateau are becoming increasingly aware of the fact that their socioeconomic interests are served by maintaining the vitality of their minority languages in the new land-management economy. The cross-cultural Indigenous-knowledge-recording programs presently in train on the plateau serve as a counter-example to the moralising views of those such as Muehlmann (2007) who argue

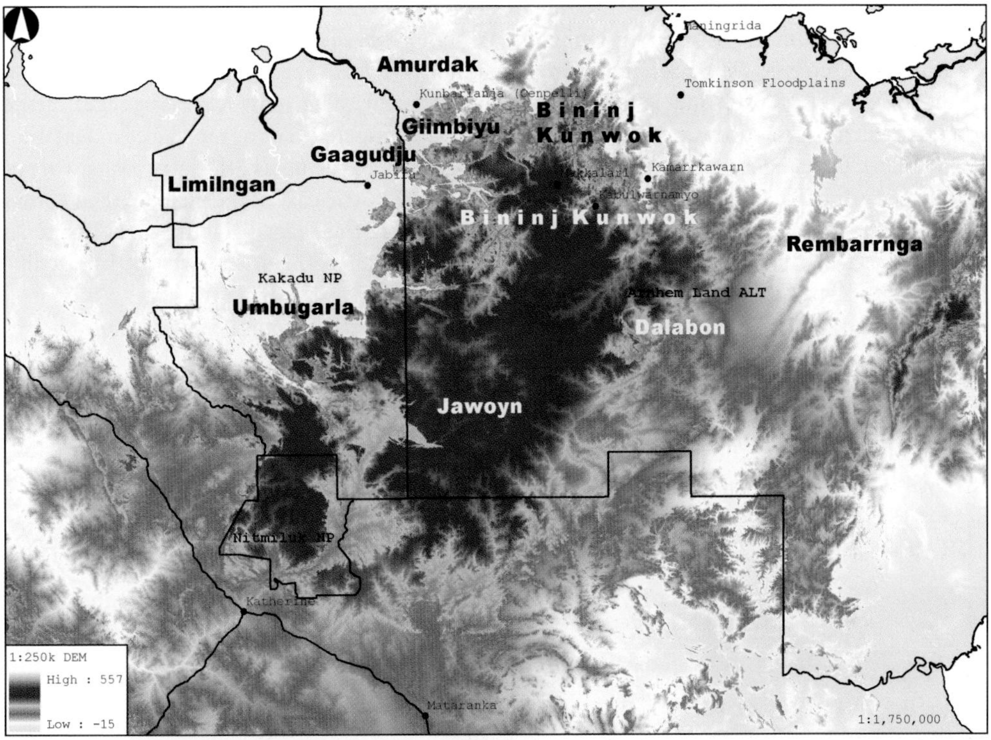

Figure 5.1 Languages of the Arnhem Land Plateau.

that pointing out the links between linguistic and biological diversity in the context of addressing language endangerment is to somehow essentialise both Indigenous people, their languages and nature.

Thirdly, the traditional ecological knowledge presented in this chapter is of further importance beyond notions of intrinsic value in that it can also make a vital contribution to the development of land-management policy. Other contributing authors to this volume outline some of the challenges faced by land managers who, in working with the Aboriginal landowners of the Arnhem Land Plateau, are seeking to address issues such as uncontrollable late-dry-season wildfires, feral animals and plants, and other threats to biodiversity. Traditional ecological knowledge has a contribution to make in the development of new economic opportunities for traditional landowners, some of whom are keen to find ways for younger generations to return to their traditional estates and make a living managing them. As part of this enterprise, and in keeping with the wishes of older generations (and detailed in texts presented here), young people need to learn how their grandparents' generation interacted with landscapes and how management practices were described and referred to.

Languages of the Arnhem Land Plateau

Approximately one-eighth of the continent of Australia – in the north – is home to about 20 language families, which linguists refer to as the northern prefixing languages (Capell 1962) or 'Non-Pama-Nyungan languages' (Evans 2003a). The remaining seven-eighths is occupied by a family that has been named 'Pama-Nyungan'. The names 'Pama' and 'Nyungan' come from the names for 'human' or 'man' in languages in the north-east and south-west of Australia, respectively. One Non-Pama-Nyungan language family, referred to by linguists as 'Gunwinyguan' (Evans 2003a), dominates the Arnhem Land Plateau. The individual languages of the plateau, which are members of this language family, include Rembarrnga to the east, Dalabon to the south, Jawoyn to the south-west and Bininj Kunwok in the north: the latter consisting of a dialect chain of six named varieties stretching from the north-east of the plateau, then across to the north-west area within Kakadu National Park (see the map in Figure 5.1). The texts presented in this chapter are all in Bininj Kunwok (Evans 2003b),[1] particularly the Kundedjnjenghmi, Kunwinjku and Kuninjku dialects.

In the north-west of the plateau around the Alligator Rivers region, there were a number of other languages spoken – Urningangk, Erre and Mengerr(dji) – but these are all now extinct. The neighbouring Gaagadju people referred to this group of languages collectively as 'Giimbiyu' (Birch 2006), which means 'of the rock country'. The descendants of Giimbiyu language speakers now speak Kunwinjku, the Bininj Kunwok dialect now associated with the town of Kunbarlanja (Oenpelli), although most Kunwinjku-speaking clan estates are located to the east of here.

The Aboriginal contributors

Lofty Bardayal Nadjamerrek, Wamud/na-Kodjok subsection, Mok clan.

Bardayal Nadjamerrek's knowledge of the Arnhem Land Plateau, its sites, walking routes, plants, animals and social history is unparalleled in the region today. He has played a central role in encouraging his kinsmen to return and manage their traditional estates and continues to guide the land management and research community of Kabulwarnamyo, on the upper Liverpool River.

Mary Kolkkiwarra, Kodjdjan/ngal-Wamud subsection, Wakmarranj clan.

Mary Kolkkiwarra has also been a key contributor to the Indigenous knowledge recording work currently based at Kabulwarnamyo Outstation. Her knowledge of local plants and women's role in landscape burning has made a valuable contribution to documentation and land-management programs.

Jimmy Kalarriya, Kodjok/na-Wamud subsection, Wurrbbarn clan.

Jimmy Kalarriya is another key contributor to this chapter. He is currently the sole custodian of knowledge about certain sites and walking routes, especially in the northern parts of the plateau. Kalarriya has contributed to a number of knowledge recording projects, including those focused on rock kangaroos, the Oenpelli python and emus.

Jack Namandali Djandjomerr, Kamarrang/na-Bangardi subsection, Bolmo clan.
Jack is a long-term resident of Marlkawo outstation and has contributed much to site survey work, fire knowledge recording and in particular the work on emu ecology.

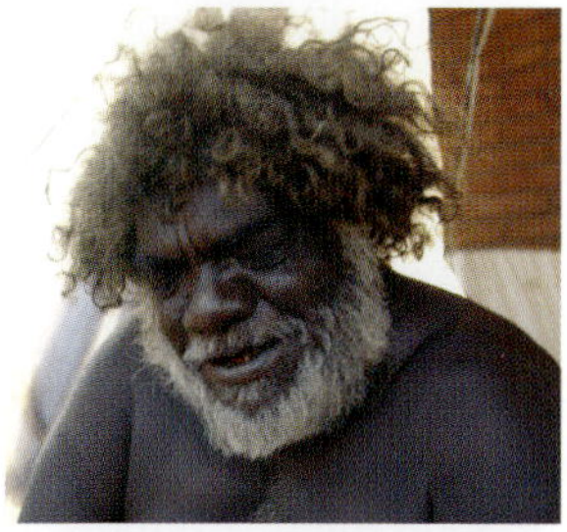

'Big' Bill Birriyabirriya, Bulanj/na-Kangila subsection, Kardbam clan.
Based at Marrkolidjban Outstation in the lower Liverpool River district, Birriyabirriya has comprehensive knowledge of the sites and ecology of this region, which is in the eastern outlier region adjacent to the plateau and includes a number of important neighbouring wetland habitats.

Ruby Bilindja, Ngarridjdjan/ngal-Wakadj subsection, Wakmarranj clan.
Ruby has spent all of her life in the stone country and has extensive knowledge of women's role in landscape burning. Among a variety of other subjects, she also has specialist knowledge about native bees and honey. She was one of the wives of the last great stone tool maker, artist and traditionalist, Wally Mandarrk. Her daughter Jeannie Imangala and son-in-law Romeo Lane also contribute to some of the discussions transcribed in this chapter.

Peter Biless, Ngarridj/na-Wakadj subsection, Djordi clan.
Peter Biless spent most of his earlier life based at Oenpelli, but made frequent trips to popular camping places on the plateau and, in later years, has been based on his country at Kamarrkawarn. He has been instrumental in setting up Warddeken Land Management Ltd that seeks to facilitate new economic opportunities for young Aboriginal people of the plateau through the application of traditional ecological knowledge in contemporary land management.

Mick Kubarkku, (deceased 16 May 2008), Balang/na-Ngarrdij subsection, Kulmarru clan.
Mick Kubarkku lived all of his life in the Mann and Liverpool Rivers districts south-west of Maningrida. He established Yikarrakkal outstation in the early 1980s on the country where he had lived all of his life. An acclaimed artist and traditionalist, Kubarrku grew up in the rock-shelters around the middle Mann River and had extensive environmental knowledge of this region.

Ethnographic background

Chapters 3 and 4 detail the historical context of the lives of the Aboriginal contributors to this chapter. Clearly by the end of the 19th century there had been a major regional breakdown in the healthy functioning and viability of many Aboriginal societies to the west of the Arnhem Land Plateau (Chapter 2). Disease, the availability of destructive recreational drugs and an accompanying exodus of young men had a profound impact on the communities of the plateau. Social changes moved along a trajectory set in train by depopulation rather than by direct foreign invasion. The Arnhem Land Plateau is a vast region of some 34 000 square kilometres with a dramatic topography that limited the number of incursions non-Aboriginal people were able to make into the area. Further, by the late 1920s, the plateau was considered of no economic interest to non-Aboriginal people (Mackay 1929).

Although the destruction of Aboriginal society to the west also affected people on the plateau, it is clear that a small number of groups managed to stay within their traditional ranges of seasonal movement. Certain others travelled back and forth between the plateau and various other settlements and enterprises such as Oenpelli Mission, cattle stations, buffalo camps, mining projects and railway construction camps. In 1939, patrol officer Gordon Sweeney estimated the resident population of the stone country groups (Dedjnjenghmi, Gundjeihmi, Dangbon, Ngalakan and Kunwinjku language groups) to be somewhere around 250 individuals (Sweeney 1939, p. 8–10).[2] Indeed, small groups of Aboriginal people remained on the plateau throughout the 20th century and, in some cases, they continued traditional lifestyles up until the 1960s, having minimal contact with non-Aboriginal people. Evidence for this comes from oral histories, including some of the members of the groups referred to by Sweeney who survive to the present day, but also from the items of contact material culture (glass, steel axes, school slates, ammunition casings, and so on) still found in important rock-shelter camp sites. Additionally, the rock art record – itself a vast historical and artistic treasure – details the recent occupation of many sites across the plateau. Lofty Bardayal Nadjamerrek (b. 1926) has pointed out a number of works of art he created while camping in rock-shelters as a young man, in a variety of locations in the upper Mann and Liverpool Rivers districts, and is able to identify the artists of a number of other paintings throughout the plateau. Other evidence of continued 20th century occupation of the plateau is in the form of the photographs of the surveyor Herbert Basedow. Basedow photographed a number of rock art sites near the Liverpool and Mann Rivers in 1928. Some of these sites have since been located and a comparison of Basedow's photos taken in 1928 with the condition of the galleries today indicates that a number of images have been added since Basedow's visit (Garde and Kohen 2004).

Some of the Aboriginal contributors to this chapter (such as Lofty Bardayal Nadjamerrek) spent much of their youth following traditional walking routes across the plateau, occasionally visiting the Oenpelli Mission, the Maranboy tin mines and the Second World War army camps at Mataranka. Others spent their time moving back and forth to Oenpelli: attending ceremonies further to the east and taking advantage of well-known seasonal resource harvests at key locations across the plateau. To date, ecological knowledge has rarely featured as part of the ethnographies produced by social scientists who have worked in this region (exceptions include Altman 1984; Chaloupka and Giuliani 1984; Lewis 1989) and so the material presented here, despite it appearing quite late in the scheme of things, includes a number of novel additions to our knowledge about the stone country – at least from an ethnographic perspective.[3]

The material presented here was elicited over the past 10 years during discussions and conversations, some of which took place during formal meetings and interviews, and others during casual conversations. The material was digitally recorded, transcribed and translated into

Box 5.1: Land tenure on the Arnhem Land Plateau

The Aboriginal people of the Arnhem Land Plateau own land communally through membership in patrilineal clans known in Bininj Kunwok dialects variously as *kun-mokurrkurr* (Kunwinjku and Gundjeihmi), *kun-nguya* (Kuninjku) or *daworro* (Kune and Mayali). A clan owns one or more named 'clan estates', which are referred to as *kun-ngeybadjan* or *kun-ngeykimuk,* which literally mean 'big names'. Such 'big names' for estates operate as eponyms in the sense that they are usually (but not always) important camping or religious sites within an estate that give rise to the whole estate being referred to by this important individual site within it. There are also many individual named sites within an estate. The term for 'place' is *kun-bolk,* but a site which is a traditional camping place or stopover on a *bininj man-bolh* 'walking route' is known as *kun-red* 'home, hearth, living space'.

A traditional owner of a clan estate is known as *na-redweleng* (if a male), *ngal-redweleng* (if a female) and *birri-redweleng* (plural). Such a person would call an estate their 'father's country' and they collectively exercise the ultimate authority over such an estate. However, people also have certain rights in relation to their mother's country and both a person who calls an estate their mother's country and the country itself are referred to by the noun *karrardwarreken.* An alternative way of referring to one's mother's country is the verb [pronoun.prefix]+*bolkkuykme.* This latter verbal form of reference literally means '+to spray [from the mouth] on the country' – the etymology of which refers to the ritual of a new visitor to a site being introduced to the ancestral spirits of that place by being sprayed with water by a person whose mother is a member of the land owning clan.

e.g. *nga-bolkkuykme:* literally, 'I spray [from my mouth] onto the country' > my mother's land

yi-bolkkuykme: literally, 'you [singular] spray [from your mouth] onto the country > 'your mother's land'

An alternative expression is: *nga-bolk-nunj-yo:* literally 'I-place-saliva-lie' > my mother's land.

Those people who call an estate their 'mother's country' are referred to by the term *djungkay.* The traditional owners or *birri–redweleng* and their *djungkay* 'mother's country kinsmen' must work together to manage an estate. The *djungkay* have particular responsibilities for managing their mother's country and work in consultation with the traditional owners. A person who is a *djungkay* for a particular estate is an important person to be involved in land-management activities such as landscape burning. The traditional owners or *birri-redweleng* [plural] prefer to have their *djungkay* present when undertaking such tasks on their estates or when 'foreigners' are visiting the country.

Another important concept is the division of the cosmos into moieties. Elements of both the natural and supernatural worlds are associated with one of two patrilineal categories known in Bininj Kunwok as *duwa* and *yirridjdja.* All people, clans, places and estates, plants, animals, natural phenomena and supernatural beings are affiliated with either the *duwa* or *yirridjdja* patrimoieties. A *duwa* man must marry a *yirridjdja* woman and vice versa. If a traditional owner or *na-redweleng* is a *yirridjdja* person, their *djungkay* will be a *duwa* person and vice versa. There are also matrilineal moieties known as *–ngarradjku* and *–mardku*.

In some cases, a confederation of neighbouring clans of the same patrimoiety can own an estate in what is referred to as *kabirri-yikadjurren* 'they are in a company relationship'. Should all the members of a clan die without male successors, there are a number of principles at play that determine which clan will succeed to the extinct estate. One of these is that a neighbouring clan of the same moiety can inherit the estate. It is also possible that a group of kinsmen connected through either patrilineal or matrilineal links can join together to succeed to an extinct estate.

English. It is presented as knowledge that is intended to be transmitted orally and has been occasionally edited in order to iron out the untidiness of everyday speech, such as simultaneous utterances and the minimalist and elliptical reference style of Bininj Kunwok speakers (Garde 2003). Traditional ecological knowledge is holistically encyclopaedic in nature and embedded in Indigenous views of the world that do not compartmentalise the natural world (and human relationships with nature for that matter) into divisions in the same way as the knowledge domains of academic disciplines. The presentation of these texts is obviously no attempt to present the entirety of such an encyclopaedia. The texts represent a collection of descriptions, lists, spontaneous explanations and assorted memories as they came to mind during the discussions and interviews that took place over this 10-year period.

Although the material presented here suggests some kind of patterned movement throughout the seasonal cycle, it would be incorrect to suggest that there was no annual variation in the movements of family groups. Ceremonies, invitations, available resources, chance meetings, interpersonal relationship issues, unusual changes in weather and a variety of other factors influenced where people went at the different times of each year. There were patterns of movement, but certainly no immutable itinerary was followed.

SEASONS AND RESOURCES

Research on resource utilisation, seasonality and ecological zones has demonstrated the abundance that Aboriginal people enjoyed around the extensive floodplains and riverine habitats of the Alligator Rivers region to the west of the plateau (Chaloupka 1981; Russell-Smith *et al.* 1997; Brockwell *et al.* 2001). There is no doubt that the population density of the plateau was far lower than that of the adjoining western floodplains because of the comparatively fewer resources available in the rock country. But, of course, there was movement between these two regions. The only archaeological excavation undertaken on the eastern part of the plateau was conducted in 1998 at Ngalirrkewern, a shelter decorated richly with recent phase rock art in a small tributary valley of the Liverpool River (Allen 2002). Although it is only a single site study for the area in question, the findings are possibly consistent with claims of a low population density for the plateau, or at least this particular region of the plateau. It has been suggested that in fact the plateau was only occupied on a seasonal basis (Brockwell *et al.* 2001, p. 375), but this seems an unlikely sweeping generalisation for such a vast region, although this suggestion is qualified with the following rider '... we have to allow for variation within physiographic zones to which the gross label 'plateau' is applied and contemplate a more complex seasonality of human access than the historical reports suggest' (Brockwell *et al.* 2001, p. 375). Other archaeological studies support the view that the plateau had a permanent population throughout the annual seasonal cycle (Guse 1999, p. 58):

> *A broad scale comparison between the archaeological sites of the wetlands and plains to the north and the plateau country to the south reveals a continuum of artefact and raw material richness. There is a general trend towards more diverse archaeological assemblages on the wetland sites along the South Alligator River … There are various reasons for such a trend to exist in these richness scales. A major reason is the importance of the ecological resources of the wetland regions. The pattern does not suggest any abandonment of the plateau for the wetland country on a seasonal basis. The scale highlights the different technological approaches and activities conducted by Aboriginal people in each region.*

Indeed, the variety of ecological habitats and the associated biological diversity have allowed groups of Aboriginal people to base themselves on the high country and to move seasonally within its boundaries in order to profitably exploit available resources. In addition to archaeological arguments, there is also recent ethnographic evidence to support this view. In a conversation with two senior Aboriginal women – Ruby Bilindja and Mary Kolkkiwarra, whose traditional estates are in the north-east of the plateau – they claim that in their youth, their families always stayed in the rock country and they knew very little of the resources of the floodplains:

Murray Garde: *Bale kabbal nawu ngalmangeyi… ngune-yahwurdni ngurri-yawani?*	Which floodplains did you get turtles (*Chelodina rugosa*) from … when you were younger?
Mary Kolkkiwarra: *Kayakki.*	None
Ruby Bilindja: *Kayakki*	None
Mary Kolkkiwarra: *Ngad nani kuwarddewaken arri-ni.*	We were living in the rock country.
Ruby Bilindja: *Kuwarddewaken ngarri-djaldi.*	We were rock country people and we stayed there all the time.

Certainly, some other plateau groups did move down to floodplains in the dry season: to the Alligator Rivers wetlands to the west, and to the Tomkinson River plains in the east.

There are already a number of descriptions of the seasonal cycle for Bininj Kunwok groups in the literature (Altman 1984, 1987, p. 25 for the Kuninjku; Brockwell *et al.* 1995; Russell-Smith *et al.* 1997 for the Gundjeihmi; and Haynes 1985 for the Kune). Both Altman and Russell-Smith *et al.* include comprehensive lists of flora and fauna used respectively by the Kuninjku in the east and the Gundjeihmi in the west and, except for some habitat differences (i.e. the absence of wetlands on the high country) and a few other minor differences, these lists detail basically the same resources used by plateau people. The western studies, however, reveal that groups such as the Gundjeihmi relied solidly on the resources of the Alligator Rivers and Magela Creek floodplains and riverine habitats, but forays into the rocky plateau country were still part of the annual pattern of movement at certain times of the year (Chaloupka 1981). The Kuninjku to the east have territory within tidal river zones and enjoy access to certain estuarine species, in addition to freshwater riverine resources. The material presented here focuses on the resources found in the rock country of the central, central north and north-east parts of the plateau, for people who describe themselves as *warddeken* 'inhabitants of the rock country'. There is evidence in the texts that some central plateau groups (such as the Kundedjnjenghmi) took advantage of visits to floodplains belonging to both their eastern and western neighbours, especially in the early dry season for fishing, or in the later dry season to catch magpie geese.

Bininj Kunwok names for seasons are represented in Table 5.1. These names are the same across the six dialects, except that there are a few differences in the eastern-most dialect, Kune (see Haynes 1985). The names of the seasons in Bininj Kunwok can be reduplicated to indicate 'mid-season' as opposed to a transitional period between seasons:

kudjewk > kudjewhkudjewk 'mid-wet season' also kudjewkbuldjdjarn 'in the middle of the wet season'
(-buldjdjarn = 'centre, middle, depth').

There is also a 'co-occurrence suffix' *–ko* (Evans 2003b, p. 164) which has the sense of 'occurring at the time of', thus:

kudjewk-ko 'in the wet season, at the time of the wet season'
wularri-ko 'in the time of the wet season westerly winds'
wurrkeng-ko 'in the burning season'
an-dudjmiko 'at the time when the green plums (Buchanania obovata) are fruiting'

There is also an alternative construction which takes the form *–keno* (meaning 'time of', Evans 2003b, p. 153). An example is given in the following text by Kune speaker Jack Nawilil:

Jack Nawilil: Namarr ka-bidbun ***kudjewkkeno*** *well kuwardde ka-wendi na-warddewarddeken nungan.*
The male euro climbs up slopes **in the wet season** *and it lives up in the rock country, which is its habitat [at that time].*

Jimmy Kalarriya: wind names

Wind direction is an important seasonal marker. Wind names with a temporal suffix in Bininj Kunwok also serve as seasonal terms as winds blow from particular directions at certain times of the year (as the example *wularri-ko* above illustrates). As can be seen from Table 5.1, there are a large number of named winds and a few of these (*barra, djimurru*) have names which are borrowed from Makassarese (Evans 1992).[4]

Jimmy Kalarriya: *Kun-kurra ka-karrme kun-ngey … Kun-kurra ka-bun bu kune nga-yime ka-rrurndidurndiwe koyek-beh bolkki ka-kurrabun bolkki yekke. Djalkoyekbe ka-wurrabun.*	The winds have names. The [wet season westerly] wind blows and then it reverses and becomes an easterly wind again. We are in the dry season now and that is the wind blowing now. It's blowing from the east.
Ka-djale kune kah-yime ngaleng kudjewk wanjh ka-borledke kunu ngarr nan ka-ngukdowkke korlkkorlmi, ka-ngukdowkke. Wularri ngarrih-yime njamed ka-rralkdadje manimunak, wularri, karrikadbe ka-kurrabun, wularri. Taem	It keeps blowing like that and then in the wet season it changes direction and we see the *korlkkorlmi* winds that make the thunder. The *wularri* wind cuts down the grass for the magpie geese. *Wularri* blows from the west. Then they lay eggs. That *wularri* wind

Table 5.1: Bininj Kunwok names for the seasons and winds.

<table>
<tr><th colspan="3">Dry seasons</th><th colspan="3">Wet seasons</th></tr>
<tr><th>Yekke</th><th>Wurrkeng</th><th>Kurrung, Djurlirrmi</th><th>Kunumeleng</th><th>Kudjewk</th><th>Bangkerreng</th></tr>
<tr><td>early dry season</td><td>cool dry season
wurrk fire > season of fire</td><td>hot dry, build-up season
literally:
Ku- LOCative
dung 'sun'
Djurlirrmi is mid-*kurrung* when humidity and temperatures are both very high.</td><td>first storms of the monsoon</td><td>full wet season
Ku- LOCative
djewk rain</td><td>last rains
knockemdowns
bang strong, powerful.</td></tr>
<tr><th colspan="6">Seasonal wind names</th></tr>
<tr><td colspan="2">**Dalukorro**
strong dry season winds from the east, especially in the mornings.</td><td>**Walirr**
hot south-east winds in *kurrung* season.</td><td></td><td colspan="2">**Kularri** (variant=**Wularri**)
strong westerly winds associated with magpie geese egg laying.</td></tr>
<tr><td rowspan="3"></td><td colspan="2" rowspan="2">**Lidjalidja**
late dry season winds which have moved from easterly to southerly.</td><td colspan="2">**Korlkkorlmi**
westerly winds which bring thunder storms and the rush of cool air just before a thunder storm.
Niyaniya
southerly winds that swing from the south >west and then from the north.</td><td>**Nakul**
'knockemdown storms' which lay down the tall stands of sorghum spear grass.</td></tr>
<tr><td colspan="3">**Barlmarradja** general westerly wet season wind.</td></tr>
<tr><td>**Kun-mayorrk**
gentle south-east winds. This word is also a general term for 'wind' in some dialects.</td><td>**Mahbilil**
evening *kurrung* season winds from the coast. Also known closer to the coast as **Barra** 'northerly/north-west winds' just starting at the end of *kurrung* (Makassarese loan word).</td><td>**Barra**
northerly/ north-west winds continue from *kurrung* to *kunumeleng.*</td><td>**Makkumbu**
westerly winds associated with extended periods of rain. (variant= **Nakkumbu**)</td><td rowspan="2"></td></tr>
<tr><td></td><td></td><td></td><td></td><td>**Djimurru**
a wet season north easterly wind (Makassarese loan word).</td></tr>
</table>

ka-rrabungukdeng. Wularri kunu. Kun-kurra ka-ngeyyo manu wanjh anekke. Wularri ngarri-yime manu kun-kurra an-badjan karrikadbe.	[time]. That's what that wind is called. That big wind from the west we call it *wularri.*
Kudjewk … dalukorro, wanjh ka-berkbakme koyekbe.	These are wet season [westerly winds], … but *dalukorro,* that blows from the east.

...yekke ka-yibukmen wanjh man-djewk ka-ngurdme 'stop' *ka-yime dulukorro. Korlkkorlmi karrikadbeh kudjewk, korlkkorlmi. Makkumbu, karrikadbeh.*	In *yekke* with this wind, the country dries up and then the rain stops, and the *dalukorro* winds start. The *korlkkorlmi* wind blows from the west in the wet season and *makkumbu* wind [also] blowing from the west.
Murray Garde: *Balekeno makkumbu?*	When is the *makkumbu* wind?
Jimmy Kalarriya: *Kudjewk, djal ka-djaldjakdung munguyh en ngalengman. Ka-borledke makkumbu barlmarradja, niyaniya.*	In the wet season, when it keeps raining continuously well that's it then. It changes from *makkumbu* and *barlmarradja* to *niyaniya.*
Niyaniya, karrikadbe kakbibe ka-borledke walembe. Niyaniya ka-borledke walembe karrikadbe, kakbibe. Kakbibe.	*Niyániya,* from the west and then north it turns around to blow from the south. *Niyániya* turns from the south then from the west and around to the north. From the north.[5]
Murray Garde: *Balekeno?*	When?
Jimmy Kalarriya: *Kudjewk. Dalukorro ngarri-yime dalukorro, wanjh ka-bukubukmen man-djewk ka-ngurdme* 'stop' *ka-yime, dalukorro.* 'All stop'. *Ka-ngurdmerren rowk. Wanjh koyekbe ka-berewkberewkme.*	In the wet season. We call it *dalukorro,* a wind that dries up the country and the rain stops. The rain stops and the *dalukorro* wind starts. All rain stops. And then it blows from the east.
Ka-berewkberewkme koyekbe bu an-karnbirr kah-kobun. Ka-berewkberewkme. Dalukorro koyekbe. Wanjh bonj man-djewk ka-yiburnbun.	The wind is blowing from the east when the *Acacia oncinocarpa* shrubs are flowering. It blows [at that time]. The *dalukorro* wind blows from the east. With the arrival of that wind, the rains stop.
...bolkkime ngarr-ni, bolkki ngarr-ni yekke. Koyekbe kah-kurrabun, yekke. Man-djewk wanjh ngurdmeng ngarrh-djalni. Ngarrh-madbun kaluk kun-kudji an-dudjmi ka-kobun wanjh kunukka ka-djakdung nawu darnki wanjh kuridjmidj start *ka-yime bu man-dudjmi ka-kaberrkyo.*	... today we are in the early dry season. The wind is blowing from the east in *yekke* season. We can camp out when the rain has stopped. Then we wait for the green plums (*Buchanania obovata*) to flower and then again it starts to rain and its close to Christmas when the *Buchanania* plums are fruiting all over the place.

Although the higher stone country of the plateau has been described as relatively poor in resources compared with the north-western floodplains, the diversity of habitat types and the range of natural species available has been sufficient to sustain a number of rock-country populations. The descriptions of physiographic zones in the north-western margins of the plateau in Russell-Smith *et al.* (1997, pp. 166–167) also apply in a more general sense for other parts of the plateau except for the absence of any other large wetland habitats higher up in the rock country and the relabelling of 'lowland savanna woodlands' to 'plateau savanna woodlands' and 'lowland jungles' to 'plateau jungles', and so on. A Bininj Kunwok view of landscape and some of the language used to conceptually organise it is discussed in another section below.

The following texts discuss the resources available for each season and some of the sites that people remember visiting at these particular times of the year where such resources are sought. Information about seasonal movement can be interpreted as representative of a range of possible sites that people might visit for a particular time of the year, rather than descriptions of fixed patterns of movement and resource exploitation. We commence with the first rains of the wet season.

Kunumeleng 'First storms of the wet season'

Kunumeleng is a time when the first storms of the coming monsoon start to fall. It is a productive time for many fruit trees and perennial grasses sprout again attracting macropods, which come to eat the soft new shoots. In previous times, *kunumeleng* was a season when people would start to move upstream to rock-shelters on the margin of woodlands in anticipation of the coming wet season.

Bardayal Nadjamerrek: *An-bedje, kunj ka-djalbelkngun kunumeleng ka-djaldalkngun ka-rralkngun An-yokkorol, an-karne makka man-buyika manu njamed an-bedje ka-rralkrohrok manu yoh mayamaya wurlhkarndayh ka-re karndayh ka-ngun djabelno an-mirrhbang.*	In *kunumeleng*, kangaroos eat the new spear grass shoots (*Sorghum spp.*). The large and small varieties of spinifex *Triodia plectrachnoides* are also shooting just like the spear grass, and kangaroos eat grasses such as *Panicum trichoides* and *wurlhkarndayh* (unidentified), the sharp blades of which female Antilopine kangaroos *(Macropus antilopinus)* can eat.
Kunumeleng ku-wardderurrk, arri-re wanjh ku-wardde, dolobbo yika arri-rdakme bu na-kare wanjh arri-ngimeng.	In *kunumeleng* we camp in rock shelters and so we move to the rock country at that time and enclose the shelters we return to annually with stringybark and camp inside.
Murray Garde: *Ku-waddakuken baleh barri-rey bu kunumeleng?*	What are some key *kunumeleng* season camping places?
Bardayal Nadjamerrek: *Bu Kodwalewale maitbi kure Mako Karrang kun-red kure bu Bernberndulh. Kumekke kodbarre arri-yoy dja ku-rurrk. Ka-djalbarndi nomerra kun-wardde kun-wardderurrk karri-yoy, dolobbo arri-rdakmi arri-yoy...yika balabala barri-yoy yika ku-rurrk.*	Maybe in the Kodwalewale estate and places there such as Mako Karrang and Bernberndulh. There we would construct bark shelters and camp inside them. Despite it being rocky, we would camp in the caves and enclose them with stringybark or build platforms inside them which they would sleep on.
Kamarddjabdi arri-yoy kuwardde kunumeleng. Ngaye nga-bimbom rakul, dorddoro, borrobborro. Mawa yi-yime ngudda ngane-yoy an-korrwan arri-nguni.	We also camped at Kamarddjabdi in *kunumeleng*, in the rock country. At that site I painted birds [on the walls of the shelter]; chestnut quilled rock pigeon *(Petrophassa rufipennis)* and partridge pigeon *(Geophaps smithii)*. My father (the one you call grandfather) and I camped there and we used to eat white currants *(Fleuggea virosa)*.

Murray Garde: *Kunumeleng bu ka-djakdung ngaled barri-nguni?*	What else did people eat in *kunumeleng* season?
Bardayal Nadjamerrek: *Kunumeleng an-dudjmi arri-nguni. An-dudjmi, an-dak nawu an-dawk, an-kinjdjek dja kodjbang, kayawal, kayawal ba-rri kolhdeno. Makka worrkimuk ba-wororrmeninj…, yekke wanjh.*	In *kunumeleng* we ate green plums *(Buchanania obovata)*, milky plums *(Persoonia falcata)* and bush cucumber *(Cucumis melo)*, cheeky yams *(Dioscorea bulbifera)*[6], 'water peanuts' *(Aponogeton elongatus)* and long yams *(Dioscorea transversa)*; long yams are just starting to shoot new leaves at this time. When the leaves have grown into maturity it will be *yekke* season.
Mary Kolkkiwarra: *An-dudjmi, an-djarduk, an-kurndalh.*	Green plums *(Buchanania obovata)*, bush apple *(Syzygium suborbiculare)* and black plum *(Vitex glabrata)*.
Bardayal Nadjamerrek: *An-dak ba-yakminj*	The milky plums *(Persoonia falcata)* have finished [by *kunumeleng*].
Mary Kolkkiwarra: *An-dak korroko ba-rradjmeng. An-dak ka-yakmen wanjh* start *ka-yime an-dudjmi.*	Milky plums *(Persoonia falcata)* have finished fruiting. They finish and then the green plums follow *(Buchanania obovata)*.
Murray Garde: *Kunj?*	[What about] kangaroos?
Bardayal Nadjamerrek: *Kunj arri-yami wadda… kuwardde…*	We speared kangaroos around the places where we camped … in the rock country [i.e. rock country macropods]
Mary Kolkkiwarra: *Bu kunumeleng ba-yimi bu ba-djakdungi ba-yimi bu an-dudjmiko* start *ba-mankani wanjh arri-djalni.*	When *kunumeleng* had arrived and the rains had started and the green plums *(Buchanania obovata)* are falling off the trees everywhere, then we would stay in one place.
Bardayal Nadjamerrek: One place!	One place!
Mary Kolkkiwarra: *…yiman ka-yime Kodwalewale kumekke arri-di ba an-dudjmi arri-mang. Arri-djalni wanjh kudjewk ba-rradjmi wanjh arri-wohyimi 'ma karri-derrehmen'. Arri-derrehmi* this side *karri-kukngimeni karrikad yiman ka-yime Mokmek* area *arri-mey wanjh arri-barrarnmangi yika konda arri-rey arri-ni. Ngarri-baley ba-bangmedjewkni wanjh kudjewk ba-yimerranj. kudjewk ka-balyimerran wanjh bangkerreng an-djewk ka-rradjme bu an-bedje ka-kobun.*	For example we would stay in the Kodwalewale estate and collect green plums. We would stay there until the end of the wet season and then say 'OK, let's shift camp'. Then we would move west over on this side [Kabulwarnamyo] to the Mokmek area and into the Liverpool River valley. We'd stay around here until the wet season or maybe stay through the wet season into *bangkerreng* 'last rains' when the spear grass is flowering.

Bardayal Nadjamerrek: *Kure ku-wardde arri-ni ku-ngamed Anbendulkyi, Kodwalewale, koyek arri-yoy Komburr.*	We would camp in the rock country at Anbendulkyi, Kodwalewale is to the east [of Kabulwarnamyo], and there in the east we camped at Komburr and Benbeyibebmeng [in the Marlkawo estate].
Mary Kolkkiwarra: *Ngalengman kilhken Anbendulkyi, bolkbolmoken, yika kudjewk barri-rey arri-wohrey dolobbo arri-rdakmi.*	That is downstream, Abendulkyi is in Bolmo clan country and sometimes they would go there in *kudjewk* (wet season) and we would camp in bark houses or caves enclosed by stringybark.

Further to the west on the floodplains, early *kunumeleng* was also a time to find turtles (*Chelodina rugosa*), which lie buried in periods of aestivation in the mud on billabong margins. Women look for the tiny tell-tale holes in the ground through which the turtles breathe. These are called *mim* and they reveal the location of the turtle beneath the ground. Women also burn grass in *kunumeleng* around the margins of billabongs and water holes to make it easier to find the *mim* as such grass, being close to water, is too green through the dry season and will not burn (Figure 5.2). Once the water has dried further in *kunumeleng*, the grass is burnt making it easier to find the turtles. But back up on the plateau there are few wetlands where turtles are

Figure 5.2 Women burning in *kunumeleng* (first rains) on billabong margins to expose turtle aestivation sites at Kunbarlanja in 2005.

sought by prodding the ground with the long sticks or thin metal rods that women use on the floodplains. In the rock country, turtles are taken out of freshwater streams, but the best season to find them is in *kurrung* and early *kunumeleng* when the water levels are at their lowest.

Mary Kolkkiwarra: *Yekke start kabarri-djalmang bolkki kunumeleng. Kunumeleng kabarri-mang ka-rrudjendi mim.*	They start to find turtles in *yekke* season but now it is *kunumeleng.* In *kunumeleng* they look for them [on the floodplains] by searching for the breathing holes [of buried turtles].

Kudjewk 'Mid-wet season'

Bininj move into the shelter of the rock country during *kudjewk*, especially to sheltered sites that border woodland forest. Obviously people are more sedentary during extended periods of rain, and game is more difficult to find and track with the thickening of vegetation. But, as the discussion below reveals, it was still possible to move around from time to time in the rainy season. As discussed for *kunumeleng* 'early rains', *kudjewk* is a time when there are a variety of edible fruits and tubers available. Some of the most important wet season edible fruits and tubers on the plateau are set out in Appendix 1.

Bardayal Nadjamerrek: *Kudjewkbuldjdjarn ku-wardde arri-yoy kodbarre dolobbo arri-rdakmi na-kare yika arri-rey yika arri-yerrkani kube arri-yun arri-yimi. An-djewk djaldjakdungi nomata. Ba-rradjmeng wanjh arri-wabmeng.*	In the middle of the wet season we camped in rock-shelters or houses made from stringybark and sometimes we would shift camp or sometimes just stay in one place. It didn't matter to us that it rained for long periods of time [and we had to stay in one place]. When the rain stopped falling we could shift camp.
Mary Kolkkiwarra: *Yika ngarri-re wanjh ngarri-dakme dolobbo ngarri-yoy. Bu yiman an-djewk nawu kanh-karrmi yiman ka-yime kan-karrme nawu munguyh an-djewk yiman ka-yime konda ngarri-dakme dolobbo ngamed Kamarrkawarn kure kabarri-djalni kumekke arri-yedbuni* all day *arri-yoy yika.*	Sometimes we go and make bark houses and we camped in those. When the monsoon rains really take hold, falling continuously and limiting our activities, we would go and camp at Kamarrkawarn and just stay in the camp sleeping all day [waiting for the rain to stop].
	Peter Cooke: Any fish still there [at Kamarrkawarn on the Mann River] that time?
Mary Kolkkiwarra: Still *arri-kolungi* samtaem fishing *bikodj darnki bu an-bokimuk wanjh kunukka minj arri-wernhmeninj ba ba-bokolungi little bit wanjh arri-rey arri-weyi kume bu arri-weyi djal mak kun-buyika nawu* fishing line *yakni kunukka.*	We would still walk down to the river to go fishing sometimes but because of the floodwater you can't really fish properly, although we could still do a bit of fishing but another thing is that long ago we didn't have fishing lines.
Bardayal Nadjamerrek: *Wirlmurrng arri-danjbuni.*	We used pronged fishing spears.

Further:

Bardayal Nadjamerrek: *Kumekke [Kunberrkken] arri-yoy. Kunukka arri-bawoni bad konda mak makih arri-yoy ngamed Kamarrkawarn kaluk kure Kamarrkawarn yerre Kunj Ka-djowkke bad ngamed arri-yoy Nakurlkboy.*	[Sometimes] we camped at Kunberrkken near [Makkalarl, Djalbangurrk estate]. We would shift from there and another place where we camped sometimes was near Kamarrkawarn at Kunj Ka-djowkke and at Nakurlkboy [Kodwalewale estate].
An-djanek arri-dongi kun-kare. Ku-djewk bu bangkerreng kure an-kurndalh arri-yoy Yirrirnbuddak	[In the wet season] we would pound *Typhonium* tubers (see Figure 5.3). From the mid to late wet season we might camp in the grove of black plums *(Vitex glabrate)* at Yirrirnbuddak
Bardayal Nadjamerrek: *Kudjewk an-bedje ka-bardkirh ka-djordmen. An-lombeng ka-kobun dja an-darl an-murlmu ka-kobun kudjewk dja an-kolomborr. Djenj med ka-bokimukmen. Djenj bangkerreng wanjh arri-danjbom.*	In the wet season the spear grass (*Sorghum* spp.) is green and growing. *Asteromyrtus symphyocarpa* is flowering in the wet season and so are *Melaleuca viridiflora* and other paperbarks. Certain eucalypts are flowering [*an-kolomborr*, a rock country eucalypt]. It's not yet time for fishing, the water levels must increase first. We spear fish in *bangkerreng* season 'late wet/last rains'.

Figure 5.3 Rock art image at Ngalkombarli [Nakarlwakarlwa estate] of *Typhonium* sp., one of the many yams available in the rock country.

As mentioned above in *kunumeleng*, various grasses are flowering in the wet season including grasses important for kangaroos:

Bardayal Nadjamerrek: *Oh karndakidj, belendjerre. Belendjerre kun-dalk anekke manu ka-dalkngun nawu karndayh yiman kurdubu ka-yime. Ka-balemkimukmen.*	*Themeda* spp. grasses are for Antilopine kangaroos. Male and female Antilopine kangaroos eat *Themeda* grass. They get fat on it.
Mary Kolkkiwarra: *Manekke ka-djaldan ka-djalbelkngun bu bolkki yiman kunumeleng ka-djale yimarnek yekke ka-balyimerran bu kudjewk buldjdjarn wanjh* start *ka-welengkobun* now. *Yiman ka-yime an-bedje ka-rohrok bangkerreng.*	Those grasses are always there, including now in *kunumeleng* and they should be there through *yekke* season but in *kudjewk* they start to flower just like sorghum grasses, which flower through to *bangkerreng.*

The wet season was a good time for hunting flying fox, but as the following text makes clear, they were also hunted at other times of the year. The large jungle of Mererrinj was an important wet season camping place visited for *kuluban* 'flying fox', *kayawal* 'long yams' and the meristems of *marrunj* '*Gronopyllum ramsayi* palm trees'.

Bardayal Nadjamerrek: *Yo kuluban Mererrinj Mererrinj dja Mibirk, yekke kurrung, yekke o kunumeleng, bu kudjewkken, kuluban arri-buni arri-kodmikani arri-yibidbuni arri-kodmikani arri-kodmikani arri-kodmikani, arri-karlangmang arri-bidbidbuniken wanjh kuluban arri-buni kun-dulkyih namalwurru kun-dulk an-kuyekuyeng an-barnba [name] arri-buni arri-buni bow bow kirn ba-yimi barri-berlbakkeyi kirn barri-djurdkodjbakbakeyi. Mererrinj, kunumeleng, kudjewk. Kudjewk kuluban arri-nguniwarre ngalengman dedjno kun-balem! Kukilhken kukabo, ku-ngarre bad arri-yoy karrkad kumekke ku-warddeburrk, kun-kod arri-buni. An-mulmu dja arri-barrhbuni dja an-rurrkyahwurd kumekke ka-rurrkdi. Yika ku-wardderurrk barri-yoy yika kodbarre barri-marnbuni. Kure an-binik arri-yoy.*	Yes, we got flying foxes at Mererrinj and Mibirk in the wet and dry seasons, but we especially used to hunt them in the wet season. We used to climb up the trees, up, up, up, out on to a branch and kill the flying foxes with a long stick called *an-barnba* 'whack whack, screech' they'd break their arms and hit them over the heads and they would break their necks. We went to Mererrinj in *kunumeleng* 'first rains' and *kudjewk* 'full wet season'. We used to eat a lot of flying fox – those plump little rumps! That was down on the creek in the jungle but we camped up higher where there is a traditional camping spot where we made paperbark houses. They used *Melaleuca viridiflora* paperbark to make them and there are also some small rock-shelters there. Sometimes they camped in the rock-shelters and sometimes in paperbark houses. We used to camp in the *Allosyncarpia* forest.

Bangkerreng 'Late wet season'

Mary Kolkkiwarra: … *wanjh arri-balkolungi arri-yime bu bangkerreng* time	… we would go down [downstream, down from the higher rock country] in *bangkerreng* time.

Figure 5.4 Tall stands of Sorghum in *bangkerreng* season (photo: Peter Cooke).

Bangkerreng is the end of the wet season and the final storms and associated winds that knock down the now very tall stands of sorghum grass (Figure 5.4). People are still living in rock-shelters but increasingly moving out into areas of more open savanna woodland or *man-behbe-berrk*. As the waters recede towards the middle and end of *bangkerreng*, fishing becomes a major preoccupation. Dragonflies *kaladjirridj* are seen flying everywhere and are considered a calendar insect, marking the end of the rains.

Mary Kolkkiwarra: *Ka-ngolngurrurdu, bangkerreng burldjdjarn an-djewk ka-yakmen.*	Cirro-cumulus clouds appear in the late wet season of *bangkerreng* and indicate the end of the rain.
An-me arri-yawani ankung arri-nguni, an-wohmi arri-buni karrbilk bangkebangkerreng.	We would look for honey to eat, and we mashed up *Ipomoea graminea* and *Ipomea gracilis* tubers throughout *bangkerreng* season.
An-wohmi manu man-me yiman ka-rohrok biddadda, an-kukyawuyahwurd.	*An-wohmi (Ipomoea graminea)* is edible with a tuber like a potato, but much smaller.
Makka man-lerrelerre ka-kobun manu dja bangkerreng mak bu ngurrurdu-ken kabarri-konguyi djahwodjahwon, ka-mulewan ngurrurdu.	That *Bossiaea bossaeoides* bush flowers in *bangkerreng* and it's the time for emus to come and eat the little flowers. It [the flowering] tells us that emus will start to appear.

An-djamko jarran ka-yo ka-kobun bangkerreng.	*Grevillea dryandri* flowers in *bangkerreng.*

A number of edible tubers are largely only available in *bangkerreng.* One is *Curculigo ensifolia,* called variously *an-burda, an-mulbirrk* or *an-djay.* The tubers are roasted, mashed up and eaten. Another edible tuber available in *bangkerreng* is *Austrodolichos errabundus* called *kurlahbang* or *barddjubarr* in Bininj Kunwok.

An-burda arri-dong arri-bekmarnbun.	We pound the *Curculigo ensifolia* tubers to make a mash.
Mary Kolkkiwarra: *Kurlahbang yimen only djal bangkerreng ngalengarre [bulu bayoy] ba barri-buni kobohbanj. Bangkerreng barri-mangi an-kurlahbang dja anekke an-ngamed … kunan ngurri-yime ngaye barddjubarr nga-ngeybun.*	The old people [the women] used to go and process *an-kurlahbang* in the late wet season. They would get them in the late wet season, the *Austrodolichos errabundus* and whatsit *'kunan [man-kunan]'* which is what you all call it, but I call it *barddjubarr.*
Marlkawo ka-mirnderri an-kurlahbang, karri-karung karri-ngolung karri-kurlahwe.	At Marlkawo there are lots of *Austrodolichos errabundus* plants and we dig them up and roast them and remove the skin.

Not only people come down to the lower more open areas in *bangkerreng,* but so do various macropods:

Jack Nawilil: *Namarr ka-bidbun kudjewkkeno well kuwardde ka-wendi na-warddewarddeken nungan.*	The male euro climbs up slopes in the wet season and it lives up in the rock country, which is its habitat [at that time].
Murray Garde: *Man-me njale ka-ngun?*	What does it eat?
Jack Nawilil: *Man-me man-karrarndalk ka-ngun, djorrkkun dedjmeldurngh ka-ngun and badberrem ka-ngun.*	It eats triodia grass (spinifex), *Ficus platypoda* and *Melodorum rupestre.*
Murray Garde: *Bangkerreng?*	In the late wet season?
Jack Nawilil: *Bangkerreng ka-kolung ka-ngun njamed na… green grass na ka-ngun ka-karurren. Konda badyak* im different *kunj, kalaba. Bad kubad ka-rri nabarlek badborng djukerre barrk.*	In the late wet season, it goes down from the rocks and eats … green grass and digs everywhere. The rock kangaroos are different, as is the antilopine kangaroo. But the nabarlek and the short-eared rock wallabies and black wallaroos live in the rocks [all the time].

In certain places, fish traps of different designs were constructed in *bangkerreng:*

Mary Kolkkiwarra: *Arri-balkolungi kondabeh med Kodwalewale karri-nguni djenj After that, kumekkebeh arri-nani yakmeni Kukadjdjerre arri-rey koyek Dja Kudjam arri-balkolungi karrkad anbu Ankorlod bono kaddum.*	From here [Kabulwarnamyo region] we went to Kodwalewale to eat fish. After that, from there when we saw there were none left, we went to Kukadjdjerre [upstream from Kurrurldul] and down to Kudjam, which is upstream from Ankorlod.
Bardayal Nadjamerrek: *That Kukadjdjerre* im go *raitab ngamed ngamed Kudjam, Kudjam Kardbam djenj arri-dabkeyi djenj, Kawidji ka-bolkduwa.*	From Kukadjdjerre you continue right up to Kudjam in the Kardbam clan country where they blocked [the creek] to get fish. And also at Kawidji, but that is in *duwa* patrimoiety country [Kardbam clan is the opposite *yirridjdja* moiety. There are special rock country fish traps at in these places for catching small fish].
Mary Kolkkiwarra: *Arri-balhmi, ku-warddeyi.*	We made stone fish traps there.
Bardayal Nadjamerrek: *Kunkalewobe.*	Fence fish traps.
Mary Kolkkiwarra: *Kunkalewobe manu ba-balhmi kun-kod kun-dulk ba-djabdjabnami wanjh ba-kirndihnami wanjh ba-balhmi kun-kodyih kun-kod ba-balhmi bonj after that djenj ba-bidbuni ba-yoyo ka-djongi full ba-yimi kume ba-mirnderri wanjh ba-bomunkeweyi ku-buyika kure ka-borey wanjh djenj ba-[?] njamed ba-kurrmi njamed, badjbarra badjbarra nawu yakngarra karri-kadjung kaluk kabarri-marnbun different way bedman old people barri-marnbuni* like *yiman 'net' ka-yime wanjh barri-nganganani barri-kurrmi djenj ba-mankani ba-djalwarrhmi ba-balmi kabarri-burriwe ba-djalyawoyhwarrhmi barri-yawoyhkurrmi badjbarra Yika milil yoh manekke.*	Kunkalewobe are fence fish traps made from paperbark and a fence of stakes which then hold the paperbark used to block the fish. After that the fish having gone upstream come back with the current and are washed through a sluice way into a woven container. A long net called *badjbarra* made from pandanus which the old people made in a different fashion [to how we do weaving today] like a net and the fish would fall into [a basket] and when they saw that it was full they would throw out the fish and set it again and the fish would fall into it again. *Badjbarra,* yes, sometimes it was also made of *Malaisia scandens* vine.
	Peter Cooke: What time of the year do you make those fish traps, off on the side of the river, in shallow water but you make a shade for them?
Bardayal Nadjamerrek: *Arri-kirndihnameng arri-warnamkan. Kun-kalewobe nomo djenj ka-kurduhme* i gan go *ka-ngarrnghmang.*	We put stakes into the ground to make a fence across the creek. This is a trap known as *kun-kalewobe* and fish cannot get around it, they can't continue through, their forward movement is blocked.

Mary Kolkkiwarra: *Nakka* different kind *njamed…*	That's a different kind [of trap], what's it called …?
	Peter Cooke: You make like a little *balabala* [flat roof, shade house] and then cover it up.
Mary Kolkkiwarra: *Yo njamed kabarri-yime bul. Bul kabarri-marnbun bedman kabarri-djalbarrhbun kaberrk kanjdji ka-ngimerren* inside *kanjdji ka-kaberrkdi nakka nabuyika* different *bedman kurih ngayi nga-bengkan barri-yimihwoni wanjh kanjdji ku-djurle ka-welerrme kaluk kabarri-werrhme kabarri-djalwe kabarri-djalwe …wanjh kaban-biddulubun walabi wanjh kabarri-werrhme kabarri-durdmang yekke yiman ka-marnburren bu yiman ka-yime an-bedjekeno ka-bardbakke nakka kuri wanjh ngayi kaban-nang an-kare ngamed ngalengarre kanjdji, yi-bolkngeybu na-Balngarraken, nangamed Djulubbirri kun-red nuye konda* this side, *Kuluban Kayo.*	Yes, they call that *bul* [kind of fish trap]. They make the *bul* and enclose a space and they all go inside [the fish]. This is a different kind [of trap] other people [used to make] but I know about it. The fish used to go inside into the shade in large numbers and then they would scoop them out, throw them out, but the fish would spike them [with their fins/barbs] and so they used hinged-frame fish nets to get them all out. They did this at the start of *yekke* season when the spear grass had fallen down. I remember seeing this a long time ago down on the lowlands there in Balngarra clan country in whatsisname, Jack Nawilil's country, but on this side at Kuluban Kayo.
Bardayal Nadjamerrek: *Ya Korlobidahdah nakka* different *bul duninj ka-mankan.*	Yes and a different trap, a true *bul* fishtrap was made at Korlobidahdah [Cadell River district, eastern plateau].
Mary Kolkkiwarra: *Kun-kod kabarri-barrhbun kabarri-djalkodwe kure kabarri-yimewon an-kimuk kaluk kume djenj ka-mirndengimerren ka-djalngimen ka-barlme kaluk kabarri-welengwerrhme* after that *maitbi kabarri-durdmang walabi ka-bukkendadjme kumekke kanjdji djenj kabarri-bekkan ooooo kabarri-nan rrrrrrr ka-rralak ka-wokdi kanjdji bu* like *ka-njamedme ka-bowokdi djenj nawu yi-burrbun njamedme* like every *yekke.*	They make a paperbark enclosure after collecting a large amount of paperbark and lots of fish go inside [the little house] until its full up and then they scoop all the fish out or maybe they use a hinged-frame net to get them out. Inside they will hear the fish making a noise, making the water rattle rrrr inside [the little house], you know, they did this every *yekke* season.
Bardayal Nadjamerrek: *Dalukorro Yiman ngad karri- nawu, nakka barri-djal karlewobeno barri-kurrmi.*	When the *dalukorro* winds were blowing, but for us here in this area, they used to make fence fish traps.

Yekke 'Early dry season'

Yekke is the start of the dry season. People leave the protection of the rock-shelters and move to lower or downstream areas. Burning starts as soon as the grass will ignite, but many areas are still too green resulting in patches of burnt and unburnt country. This has also been recorded on the eastern Arnhem Land Plateau by Haynes (1991, p. 66), who worked with the Kune in the 1970s:

> *By the time the last rains had fallen, the beginning of the season the Gunei [=Kune] call yegerr [= yekke], burning on the floodplains was well under way and some burning in the woodland had started …*

The reduction of fuel in the early dry season restricts the damage fires can do in the late dry season when men conduct kangaroo fire drives in *kurrung* season. There are few trees with available edible fruit at this time of year, but *yekke* is a time for certain kinds of fishing although fish are only available below the large waterfalls of the major water courses (e.g. Binenge 'Cuthbertson Falls' on the Liverpool River. If you want to go fishing and you are on the upper Liverpool you have to move below Binenge. The lack of large waterfalls on the Mann River, however, means that there are fish in the upper reaches of this river).

Mary Kolkkiwarra: *Yekke wanjh yika arri-bolkbawoni arri-kolungi an-behbeberrk arri-rey wanjh yiman ka-yime koyek djarre. Marlkawo arri-ni, arri-durndengi kondanj [Kabulwarnamyo] arri-ni. Dja karri-waley yakarrikad…*	In *yekke* we would leave the places [where we camped in the wet season, rock-shelters] and go down to the savanna woodlands, walking sometimes a long way to the east. We would camp [in *yekke*] at Marlkawo and then perhaps come back and stay here at Kabulwarnamyo. Or we might travel a long way to the west …
Bardayal Nadjamerrek: *Kunj arri-nguni barrk.*	We would eat black wallaroos [i.e. we would come to this area, Kabulwarnamyo, to hunt black wallaroos in *yekke*].
Mary Kolkkiwarra: *Yekke wanjh ka-bardbakme an-bedje, yekke wanjh an-karnbirr karri-nan ka-kobun an-barndarr karri-welengnan karri-yime yekke.*	In *yekke* the spear grass falls over and in *yekke* when you see *Acacia oncinocarpa* and turkey bush *(Calytrix exstipulata)* flowering, then we say it is *yekke* 'dry season'.
Bardayal Nadjamerrek: [*bangkerreng*] *kayakki ka-bolkkih ka-njihme arri-yime marrek ka-wurlhme yekkeyekke wanjh ka-wurlhme.*	In *bangkerreng* season 'last rains/ knockemdowns' [just before *yekke*] the grass is too green/wet and we say that it will not burn. It will burn throughout *yekke*.
Murray Garde: *Yekke kure ku-wardde barri-bawoni?*	Would they not burn the rock country in *yekke?*
Bardayal Nadjamerrek: *Barri-wurlhkeyi ba-kadjungi ba-rrowen.*	They would burn it [at that time], but it would only creep along and stop [making burnt and unburnt patches].
Murray Garde: *Bininj-wi barri-dokmeni daluk?*	[What about] men and women going ahead [burning]?
Bardayal Nadjamerrek: *Nakka yerre barri-wurlhkeyi barri-rey barri-bulekadjungi 'Buleno karri-ray!' barri-yimi. Daluk o Bininj o dabbarrabbolk o wurdurd kunukka yekke.*	The women came behind [when travelling] burning as they went and following the burnt patches the men had made ahead of them. They used to say 'Follow the burnt patches!' Women and men, old and young [burning], that's what happened in *yekke* 'early dry'.

Mary Kolkkiwarra: *Yika mak daluk arri-wurlhwurlhkeyi, yekke start arri-djalwelengyini Bininj barri-wurlhkeyi ba-djale wurlhkeyi barri-yidangerreni ka-djale kurrung.*	Sometimes we women would burn the country, starting in *yekke* we would keep burning and together with the men we kept burning all the way through to *kurrung* 'hot late dry season'.
Murray Garde: *Kodwalewale yekkekeno baleh ngurri-yoy?*	Would you camp somewhere in Kodwalewale estate in *yekke*?
Bardayal Nadjamerrek: *Oo, Kukarddak Djordi-ken yika mak maki Wurrbbarnbuluken Nabarng.*	Oh at Kukarddak [which is in the Kodwalewale estate] in Djordi clan country or sometimes we would go to the Nabrang estate in Wurrbbarnbulu clan country.
Mary Kolkkiwarra: *Arri-ningihni arri-balkolungi yika kilkken arri-waleyi yika kaddum Kodwalewale yika arri-kolungi yika kure Manmoyi o arri-bale everywhere arri-wam Ngolkwarre arri-bale Kunburray Arri-bale kurih Ngolkwarre Kunburray Kodwalewale, yekke yo.*	We might stay in one area or we might go downstream, sometimes to Kamarrkawarn, sometimes downstream to Manmoyi, but we could go everywhere through the estates of Ngolkwarre, Kunburray and Kodwalewale, in *yekke,* yes.
Murray Garde: *Yekke an-me njale karri-nan?*	What foods do we see in *yekke* season?
Bardayal Nadjamerrek: *An-kung arri-nguni an-wohmi, an-mulkirrk ya burda yi-bengkan kabarri-durrkmang bu kurorrbo ka-dingihdi. An-burda arri-dong arri-bekmarnbun.*	We ate the honey of stingless native bees, *Ipomea* tubers *(Ipomoea graminea, Ipomoea velutina)* as well as *an-mulkirrk*[7] and *Curculigo ensifolia*[8] tubers: you know all those foods which grow in clearings, we pull them out of the ground. We pound the *Curculigo ensifolia* tuber to make a mash.
Murray Garde: *Kunj?*	[What about] kangaroos [in *yekke*]?
Bardayal Nadjamerrek: *Yekke ka-balemdi, yekke nakka bonj.*	In *yekke* they are fat, *yekke* is a good time for them.
… kunj barri-yami barri-nani ka-kinje …Dja an-kung barri-nguni nabiwo, kardderre barri-derdmi yukno barri-dahkendongi barri-barungi.	… they would spear kangaroos and cook them … and they ate various varieties of honey, *nabiwo* and *kardderre* and carried the honey back so they could mix it with other food.
Murray Garde: *Balekeno ngurrim-wam kondanj Wamud?*	What time of the year might you come here [to Kabulwarnamyo] Wamud?
Bardayal Nadjamerrek: *Yekke an-kungken. Kurebeh nga-dolkkang nga-wabmeng [Kabulwarnamyobeh] ngamed Kulnguki, Kodwalewale wanjh arri-wam an-bolh. Kulnguki Djamamburru Ubarr ka-yo. … Djamamburru ka-dolkkan Bonurrkwern kun-*	In *yekke* it's the time for honey. I would move from here [at Kabulwarnamyo] to Kulnguki. Then we might go to Kodwalewale, following the walking route. Then from Kulnguki to Djamamburru where there is an Ubarr ceremonial ground. From Djamamburru you

wadda Kamarrkawarn. That country for *Djordi Kodwalewale.*	go to Bonurrkwern and then home to Kamarrkawarn. That's all in the Kodwalewale estate belonging to the Djordi clan.

There are no major sites for hunting magpie geese on the plateau and so the options are to travel to Gundjeihmi country on the floodplains bordering the plateau to the west or east to billabongs near the lower Liverpool.

Murray Garde: *Ka-bardbakme an-bedje, balekeno kabarri-wurlhke?*	When the spear grass collapses, when do they burn it?
Bardayal Nadjamerrek: *Anekke ka-bardbakme yekkeyekke an-bedje, yekkeken wurlh ba-wolhbarungi, nomo balay ba-rayinj ba-djalkadjukadjungi ba-rroweni ya.*	In the middle of the early dry season *yekke* it has fallen down and in *yekke* the fire would consume it. It would not burn for a long distance, but just follow along in a patch and burn out.
Murray Garde: *Manimunakyak konda Wamud. Yika barri-kolungi bamurruken?*	Are there no magpie geese here Wamud? Would people go down somewhere to get geese?
Bardayal Nadjamerrek: *Kanjdji barri-rayinj kure ngamed Kungorobu koyek Kungorobu dja ngamed Yimayhyirud, bamurru kumekke kulabbarl, yekke.*	They would go down to whatsit, Kungorobu, to the east and to Yimayhyirud to get geese there on those billabongs, in *yekke* 'early dry season'.[9]

Long yams *(Dioscorea transversa)* and 'cheeky yams' *(Dioscorea bulbifera)* are an important perennial food source (except in the wet season when cheeky yams are not available). Certain places in jungle habitats are important yam sites especially in river valleys and gorges. *Yekke* in particular is a time when long yams are easier to find.

Mary Kolkkiwarra: *Kunukka yi-na bu arrih-dja-… kubolkkuken bu arri-yimi* like *kure yiman barri-yikarreni an-me arri-bengkani yiman ka-yime Ngolkwarre, konda kilkken, Kunburray yiman* this side *arrim-kukdurndengi mak kondah mani Mokmek area ngalengarre manekke mane kukadj mirnde ka-rri an-me dja minj na-ngale kayawal ka-bun. Yo yika mahni walem, Morre Birulk Dja marrek na-ngale ka-bun warre bolkkime.*	We have important traditional places where people go to for particular plant foods, such as long yams, which we know such as Ngolkwarre estate and downstream from here [on the Liverpool River] at Kunburray, but also back on this side in the Mokmek area, there are many yams but today no one is collecting them. Yes, also to the south there are important yam sites at Murre and Birulk [in Muruba clan country]. But no one goes there anymore to get the yams.

Another important yam area is Manggabor Creek [i.e. *mankabo* 'creek'] also known as Kumarrirnbang Creek[10] on the north-east of the plateau which has many *Allosyncarpia* forests and other jungles distributed along the gorges and tributaries where the creek has cut through the surrounding rugged sandstone. Jimmy Kalarriya describes sections of this river valley/ gorge as being very important *yekke* long yam sites. The sites he mentions, Ngalbinjdjad and Mibarra, are located along Kumarrirnbang Creek (see the accompanying map):

Jimmy Kalarriya: *Mani karri-wam bolkki, Ngalbinjdjad karrbarda man-kuken. Mani karrbarda an-kuken konda Mibarra. Karrikad manih manekke manih. Djal kawarre, karrbarda. Manih an-barrarn kureh karrih-wam karri-durndi. Mani Mibarra, mani ka-rredjdubbedurndeng, Bumurn manu Bumurn.*	Today we passed through the site Ngalbinjdjad, which is well known for long yams. Long yams were associated with this place Mibarra. There are so many yams here. This gorge is where we travelled back and forth. Here at Mibarra, this gorge comes to a dead end at the site Bumurn.
Mirurr arri-yoy wanjh malahmalayi Ngalurdbirrhmi arri-yoy. Kumekke-beh barrhbuni wanjh Berddolkko arri-wam arri-yoy arri-derrehmi Mibarra. Kumekke arri-yoy karrbarda arri-nguni wanjh arri-derrehmi Kamanemdi. Kumekke-beh arri-dolkkayi arri-wam Yolngbuyken. Malayi arri-djalbolkyurrhkeyi Nadjabba, Kurrbbirnbaleng arri-yoy. Nadjabba bim na-kuken.	We camped at Mirurr and then in the morning we went to Ngalurdbirrhmi and camped there. From there, at first light we walked to Berddolkko and then moved further to Mibarra. We camped there and ate long yams and then continued on to Kamanemdi. From there we set off for Yolngbuyken. The next morning we would walk past Nadjabba and arrive at Kurrbbirnbaleng where we would camp. Nadjabba is a an important rock art site.
Nanebe Berddolkko … ngarri-wakbom. … Karrbarda mani man-wandjad djal ka-warre karrbarda. Karrbarda kukadje.	This area around Berddolkko … where we've been circling [in the helicopter]… long yams *(Dioscorea transversa)* in this river valley are everywhere. There are so many of them here.

Wurrkeng 'Mid-dry season/Fire season'

Wurrkeng is the cool dry season and at this time of year on the higher parts of the plateau, overnight minima can reach as low as 4°C, although the average minimum is 8–15°C. The name for this season is derived from the root *wurrk* 'bushfire' and tells us something of the relationship between seasonality and frequency of fire. The deep significance of landscape burning to plateau Aboriginal people is also reflected in semantic shifts such as the Dalabon language name for 'grass' *wurrhno*, which is also related to the Bininj Kunwok word *–wurrk* 'bushfire'. In this case, the important association between 'fire' and 'grass' has resulted in a semantic shift from 'fire' to 'grass' (the material that burns in a bushfire).[11]

This middle dry season period was the time of year when burning was most frequent, although fires were also lit earlier in *yekke* as the discussion above mentions.

Bardayal Nadjamerrek: *Yekke ka-bolkborledme.*	In *yekke* 'early dry season', the country changes.

Mary Kolkkiwarra: *Wanjh wurrkeng* start *ka-welengbukmen wurrkeng ka-yimerran kurrung darnki. Bininj barri-wurlhwurlhkeyi arri-welengbulekadjungi arri-rey arrban-marnebebmi kuwadda korroko kunj barri-yingkiyami barri-wilkdengi. Ngandi-marnebolkmulewani 'Konda karri-yo' barri-bolkngeybuni. Ma ngarri-yimi wanjh ngarri-welengrey. Barri-yingkihkurrmi kunj barri-kinjeyi.*	Then *wurrkeng* season commences and the country starts to dry up as it moves closer to the hot build up season of *kurrung*. In *wurrkeng*, people were burning everywhere and we would follow the paths of burnt grass [lit by the men ahead of us] until we arrived at the camping place where there would be roast kangaroo waiting for us that they had cooked in ground ovens. At the start of the day the men would say 'we'll camp at such and such a place' and we would say 'OK, see you there' and off we would walk. The men would have kangaroos ready for us, already cooked [by the time the women and children arrived].
Mary Kolkkiwarra: *Yika mak daluk arri-wurlhwurlhkeyi, yekke* start *arri-djalwelengyini Bininj barri-wurlhkeyi ba-djale wurlhkeyi barri-yidangerreni ka-djale kurrung*	Sometimes we women would also burn as we walked starting at *yekke* season the men would be burning all the time continuing all the way through to *kurrung*.
Mary Kolkkiwarra: *An-djoh ka-kobun ka-rrabungundjikan modjarrkki ka-kurrme modjarrkki dabuno* same way *an-djoh. Wurrkeng.*	When the *Acacia difficilis* is flowering, the fresh water crocodiles lay their eggs at the same time, during the flowering of the *Acacia difficilis* in *wurrkeng* season.

Bardayal Nadjamerrek recalls camping at Narralombun in *wurrkeng* to enjoy the abundant honey in this area at this time of year:

Bardayal Nadjamerrek: *Narralombun anmurlmu ka-rri, arri-yoy kuwaddaburrk bu* junction *at Warnkulumbakmeng. Wurrkeng ankung arringuni. Midjohdjodjoh dolobbo arri-rdakmi koyek.*	At Narrolombun there are *Melaleuca leucadendra* trees and we slept at the traditional camping place there at that [river] junction and in *wurrkeng* season we ate honey. To the east, [on southern side of crossing, Manmoyi to Kabulwarnamyo road] is the camp which we made from bark shelters in a grove of *Acacia difficilis.*

Wurrkeng was the time to process the starch in the rhizomes of a jungle fern (possibly *Blechnum indicum*) known as *an-bulkung* in Bininj Kunwok. This food seems to be a specialty of the plateau and its preparation is described by Mary Kolkkiwarra. The interview was recorded in 2001 at Makkalarl in the Djalbangurrk estate.

Murray Garde: *Njale birri-nguni kondanj?*	What did people eat in this place [Makkalarl, Djalbangurrk estate]?

Bardayal Nadjamerrek: *An-me barri-nguni an-bulkung, kunj, ankung, djanay kobbol barri-nguni yok, djanay.*	They ate *an-bulkung* fern rhizome, kangaroos, honey, goanna, bandicoots, and a kind of goanna *(Varanus panoptes)*
Murray Garde: *Yo djanay yiman kalawan.*	*Djanay (Varanus panoptes)* is a kind of goanna.
Bardayal Nadjamerrek: *Ya kalawan bad kuk-buyika.*	Different to *Varanus gouldii.*
Murray Garde: *An-bulkung makka njale?*	What is *an-bulkung?*
Bardayal Nadjamerrek: *An-me.*	(vegetable) food
Murray Garde: *Bale birri-ngalkeng?*	Where did they find it?
Bardayal Nadjamerrek: *Ka-di konda kungarre.*	It's around here in jungles.
Murray Garde: *La birri-durrkmey?*	Did they pull it up?
Bardayal Nadjamerrek: *Barri-karungi, ku-kurlk ka-rri. Barri-rey barri-ngolongi, barri-worrhmi kunak an-kimuk.*	You have to dig it up from the ground. They would roast it on very large fires.
Murray Garde: *Barri-dongi?*	Did they pound it?
Bardayal Nadjamerrek: *Barri-ngolongi med yerre barri-borledkerrinj wanjh barri-dongi wanjh barri-marnbuni kun-djulng mani barri-marnbuni kun-kurlk wanjkih. Ba-marnburreni anbulkulng berlno bad djuy (duyu) makka. Duyu (?) barri-ngoluy. Barri-dongi, kundulk (ngunbulke???), kundulke (?)… yi-bengkan ka-djorlok kabarrimarnbun,* you know that *an-rurrk kunrurrk, that anekke kabarri-dongi, kun-dulk…*	You have to cook it first, turning it over and then pound it, turning it into a pulp in the consistency of soil. This pulp is made from the 'arms' of the rhizome and they pound it on a special piece of wood in which they make a hole and they grind it up in that hole in the wood.
Bardayal Nadjamerrek: *Barri-dongi kabarri-…. barri-djedmi mekem flat ba ba-djorlokkenni that nganme. You seeim that ka-warddedjorlok,* same *barri- barri-dongi.*	They carve the wood until it is flat and then the food sits in a hole (depression) in the wood. Have you ever seen those cupules in the rock-shelters, well they ground it up in those too.
Bardayal Nadjamerrek: *Barri-dongi bad barri-ngolungi korroko barri-dongi ba-yawoyh barri-yawoyhborledkeyi barri-ngolungi barri-yawoyhrungi wanjh ba-bidmeni now kunrung ba-rungi kunak, kay barri-bakkeyi day barri-bakkeyi berlno ya. Barri-marnbuni, barri-dongi anme.* Tharran *ngarri-marnbuni njamed an-kung.*	They cooked it first and then pounded it and pounded it, but they cook it, turning it over again over the coals until the heat from the fire forms it into sticks and they break them off and they pound it then. We used to make it with honey.

Bardayal Nadjamerrek: *Make im ankung* sugar bag, *mekem* sweet one…	Mix it with 'sugar bag [honey] to make it sweet …
Bardayal Nadjamerrek: *barri-ngolungi……. barri-ngolungi, mekem* big fire, *that an-bulkung berlno, ba-rungi wanjh barri-borledkeyi yiken, barri-borledkeyi wanjh 'keyi' barri-bakkeyi now ba-darlehmeninj. Kodjdjan im know marrek ba-rayinj.* I got nother *an-bulkung* there my place.	They roasted it … roast it by making a big fire, those sticks of *an-bulkung* which they roast, turn over again until it would snap 'crack' when you broke it because it had dried out. Kodjdjan [Mary] knows but hasn't been [to collect it for a long time]. I also have *an-bulkung* in my country [Ankung Djang, Kabulwarnamyo].
	Peter Cooke: You can make it anytime of the year or?
Bardayal Nadjamerrek: *Yekke o kurrung* all that, *nomo kudjewk kayakki, tharran ka-boyo* all over.	From *yekke* to *kurrung,* all that, not in the wet season because there was water all over.
Mary Kolkkiwarra: *Ngalkka an-bulkung wurrkeng ngalengarre.*	That *an-bulkung* is [prepared] in *wurrkeng* season.

Kurrung 'Build-up, late dry season'

Kurrung is the hottest time of the year on the plateau, with accompanying high humidity. Storm clouds start to form but don't break until the following *kunumeleng* season. The ground becomes so hot that it is often difficult to walk in the middle of the day. The name of the season derives from the Bininj Kunwok locative prefix *ku-* 'in' and *–dung* 'sun'. When the temperature and humidity has reached a peak, then this time is also known as *Djurlirrmi. Kurrung* is a time when the large macropod fire drives are conducted in specific locations known as *bambarr* 'dead-end fire drives valleys'. These hot fires are designed to injure large kangaroos such as *Macropus antilopinus* so they can be easily speared by hunters working in co-operative teams. The extensive burning that has been conducted throughout the preceding dry season period means that the surrounding fuel has already been reduced, making it less likely for the hot fires of the kangaroo drives to burn out of control and continue outside of the fire drive location. In relation to large kangaroo fire drives, Bardayal Nadjamerrek makes the following comments:

Bardayal Nadjamerrek: *Kurrungburldjdjarn, yekke kudjihno arri-wurlke, yekkeburlhdjarn kudjihno wanjh kurrungburldjdjarn barri-wurlhkeyi an-kimuk an-wurrk kalaba barri-bom kandarlbburru o barrk arri-yami o kanbulerri ku-wardde an-wurrk ba-djirrkkani kunak barri-kengemi, kalaba same way. Kornobolo du ba-birlikehkengemi barri-yami ba-rerrkani ba-berdmaberdmarreni ba-belenghmerreni berdno ba-wolhmerreni*	The middle of *kurrung* [is suitable for kangaroo fire drives], also occasionally in *yekke,* but the biggest drives were in *kurrung,* making fires that would drive out the large male antilopine kangaroos *(Macropus antilopinus),* or black wallaroos *(Macropus bernardus)* and euros *(Macropus robustus)* in the rock country. The fire would push them out and the kangaroos would be afraid of the fire. In the same way, agile wallabies are

ba-darnhnani ba-belenghbelenghmerreni barri-bolkkunibebmi barri-yami.	afraid of the fire and they could spear them when they sat down to hold their tails, licking their wounds, their tails having been burnt in the fire and with their heads close to their tails [distracted] the hunters would sneak up and spear them.

Kurrung is also a good time to look for turtles in creeks as the water levels are now at their lowest point and women would feel around in the sandy creek beds to find them.

Mary Kolkkiwarra: *Marlkawo nakka barri-birrkani kuwukku barri-djuhni ka-djuhkendi barri-djalbirrkani barri-mangi kuwukku barri-birrkani barri-bekkani barri-karrmi kurrungburldjdjarn barri-djuhmi Ngard barri-djalkarrmi makih ngalderrhwo ngalyahwuyahwurd barri-yawoyhdjuhkeyi barri-durndeyi ngale ngalderrhwo ngalkimuk ngalu barri-bebbekeyi o ngalmangeyi barri-mangi barri-bebkeyi kurrung.*	At Marlkawo they would feel around for turtles in the creeks. They get into the water and feel around and grab them in the water when they can feel them. This is in the middle of *kurrung,* [the hot build-up season before the rains] when they go into the creeks. They would get northern snapping turtles *(Elseya dentata)* sometimes getting small ones and then large ones or long neck turtles *(Chelodina rugosa),* getting them [out of the creeks] in *kurrung* season.

While in the creeks looking for turtles in *kurrung,* women and children also gathered fresh-water mussels and the lower water levels in some places allowed men to spear fish:

Bardayal Nadjamerrek: *Kurrung wanjh karnobirr arri-nguni o djenj arri-nguni arri-danjbuni.*	In *kurrung* we ate freshwater mussels or we speared fish.

Low water levels also meant that fish could be killed in rock pools along water courses with plant ichthycides.

	Peter Cooke: What's the really good place for poisoning fish?
Mary Kolkkiwarra: *Arri-weleng kolhbuni kumekke yiman ka-yime Morle Karrang arri-kolhbuni yiman ka-yime Anberd kaddum kure ka-djangdi arri-kolhbuni An-berd Dume Ka-nerrme djenj makka arri-kolhbuni ba-bukmeni kaddum djarre arri-djalkolhbuni Kamarrkawarn area and arri-bodjinihmi yiman ka-yime njale ba-wukdalkani-warre djenj njamed mawurrumburrk mawurrumburrk dja an-balarra arri-rawoni*	We used to poison fish at places such as Morle Karrang and near the penis dreaming at Anberd, upstream and also at Dume Kanerrme we used to poison fish there and much further up in the Kamarrkawarn area we poisoned fish and you could see them float to the surface. We used *Distichostemon hispidulum* and *Owenia vernicosa* sometimes together. We would tie the branches together and drag them through to kill barramundi

arri-dukkani barri-dorrorrokeyi dja yika bedman barri-buni ba-djal[na]rrolkani namarnkol nawu nuk yina dja kuluybirr ngalkka dowerreni dja mardakidjawarre yiman ka-yime all kind *dunbuhmanj* all kind *Bu kurrung Yarangkarr barri-kolhbuni.*

and saratoga and catfish, heaps of them, all kinds of fish, black bream. In *kurrung* season. They also used to poison fish up on the Cadell River and in the easternmost part of the plateau in the Yarangkarr estate [Rembarrnga country].

Another *kurrung* season activity was the collection of white clay (huntite), used for ceremonial, ritual, artistic and medicinal purposes. One of the most famous white clay mines on the plateau is at Madjanngalkku in the Maburrinj estate.

Peter Cooke: When would people collect *delek* [white ochre] from Madjanngalkku?

Mary Kolkkiwarra: *Barri-kolungi yika barri-rey o koyek barri-durndi barri-djalohmangi delek. Kurrung every kurrung. Kurrung time nakka barri-mangi delek. Kulay ba-bukmeninj barri-mangi. Kure yiman Kunbarlanja barri-rey wanjh barri-wohdurndeyi* 'come back time' *and barri-mangi. An-kole bu barri-bunjdadjkeyi. An-kole arri-karnmangi arri-djorrngmi arri-djorrngmi arri-burranbarungi. Yika Kunbarlanja barri-kani bulk barri-worreni. Bandi-woni njalehnjale o middiriyal bandih-woni wirlmurrng djalikiradj bandi-woni.*

They went down or if they were walking to the east they would turn off there [to Madjanngalkku] on their way to get the white clay. That was in *kurrung* season, every *kurrung*. They got the white clay in *kurrung* season. The creek would be dry at that time of year so you could get it. If they went to Kunbarlanja (Oenpelli), on their way back to the east they would collect some when they came back. They would go [west] to get spears [and return via Madjanngalkku]. They would collect the spear shafts, straighten them [with heat] and then paint them with the white clay. Sometimes they would take them [the painted spears and the clay] to Kunbarlanja on the trade route to exchange them. They would give them whatever, cloth or fishing spears or lengths of wire/steel to make spear points.

TALKING ABOUT LANDSCAPES, VEGETATION HABITATS AND ECOLOGICAL ZONES IN BININJ KUNWOK

The topographic and botanical diversity of the Arnhem Land Plateau is reflected in the large number of terms (over 300) in Bininj Kunwok for landscape and vegetation communities. These terms cover the range of scale and specificity from a generic term such as *man-berrk* 'savanna forest' through to a grove of a particular plant *midjohdjodjoh* 'grove of *Acacia difficilis*'. Some terms, however, do not encode scale such that the term *man-dulum* 'hill' can refer to an elevated landform of any size from a bump in the road or a sandcastle through to a major elevated feature covering a square kilometre or larger.

Europeans have brought to the Australian landscape, including the Arnhem Land Plateau, a European-centric taxonomy of the Australian biota. Take for example the definition of the term 'sandstone heath' commonly used by ecologists to refer to a range of vegetation communities on the Arnhem Land Plateau.

A specialist definition:

> *The Arnhem plateau sandstone heath community is a complex of closely interdigitated, intergrading plant formations including* Eucalyptus, Corymbia *or* Callitris-*dominated forests, woodlands, open woodlands, shrublands and hummock grasslands united by the presence of a well-developed sclerophyllous shrub component.*[12]

This definition is based on the English term for the European plant 'heath':

> **1 a:** *a tract of wasteland* **b:** *an extensive area of rather level open uncultivated land usually with poor coarse soil, inferior drainage, and a surface rich in peat or peaty humus* **2 a:** *any of a family (Ericaceae, the heath family) of shrubby dicotyledonous and often evergreen plants that thrive on open barren usually acid and ill-drained soil; especially : an evergreen subshrub of either of two genera (Erica and Calluna) with whorls of needle-like leaves and clusters of small flowers* **b:** *any of various plants that resemble true heaths. (Source: Merriam-Webster online dictionary, www.merriam-webster.com/dictionary/heath)*

An elicited, but composed definition for 'sandstone heath' in Bininj Kunwok is possible, but it would be a descriptive phrase and is therefore not a natural lexical entry in a dictionary:

> *Kun-dulk an-djuhdjumbung kukorlh*
> *'trees and short shrubs associated with bedrock/rocky open zones'*

The point of all this is that although non-Indigenous scientists can identify a particular ecological zone or plant habitat and put a name to it, the same entity is not necessarily similarly identified by Aboriginal people on the plateau as a discrete biome. This is not to say that the plants or the environment of places where 'sandstone heath' is found were of no significance to Aboriginal people. Indeed most of the plants of 'sandstone heath' habitats have Bininj Kunwok names, but this is a reflection of cross-linguistic differences in how landscapes and ecological habitats are ontologised. In the European scientific sense, the classification is motivated by, among other things, intellectual interest in plants, while the lack of a term in Bininj Kunwok is for Aboriginal people on the plateau, possibly motivated by 'utilitarian considerations' (Hunn 1982; Berlin 1992; Burenhult and Levinson 2008) or the fact that the diversity of plants that grow in sandstone heaths is less salient than the landforms on which they grow; that is, *kukorlh* 'open bedrock expanses'.

Therefore, the cross-cultural enterprise of non-Indigenous and Indigenous land managers working together to manage the resources and deal with the threats to the Arnhem Land Plateau will require some mutual cross-cultural understanding of how each group organises its cognitive schemes in relation to the labelling of landscapes and ecological zones. It is also important that the vocabulary and ecological referential systems that have developed over thousands of years in response to human interactions with the plateau's environment continue to be used by younger generations of Bininj. This is no assumption on my part as a sympathetic social scientist, but a clearly stated objective of senior plateau community leaders such as Jimmy Kalarriya and Bardayal Nadjamerrek who, when interviewed for the Australian Broadcasting Corporation television science program *Catalyst* in 2006, made the following comments:

Jimmy Kalarriya: *Yawurrinj ba kabirri-durrkmirri kabarri-wokihme kamak. Ngarrbenbukkan yawurrinj ba bu kabirri-bengkan like ngad maitbi ngarri-danjbik ngarri-dowen.Wanjh yawurrinj kabirri-bolknahnan kun-red. Karri-djarrkdurrkmirri Balanda dorrengh, ba Balanda kun-wok bedberre dja ngad ngarri-bulerri kun-wok ngarri-wokdi.*	This [land management and traditional ecological knowledge recording project] is a good thing and is something we have all wanted. I have really wanted to see this. Our young people working [in this project] is a good thing. We elders need to teach these young people so they will gain this knowledge, because when we three have died [Bardayal Nadjamerrek, Jimmy Kalarriya and Peter Biless] then it's up to young people to look after the country. Then the young people can look after the country. We can work together with non-Aboriginal people so that we use both our Aboriginal language terms and the English words together.
Bardayal Nadjamerrek: *Dja Balanda ngandi-bidyikarrme nawu ka-bengkan kun-wale dja kun-wok ngadberre, birri-wern Balanda minj kabirri-bengkan Kunwinjku. Bu ngayi nga-djare wanjh kunred Bininj ngarri-marnbun, dja Balanda, Balanda law or Bininj law, ngarduk law.*	We need non-Aboriginal people who understand our ways of doing things and our language, as many non-Aboriginal people do not know Kunwinjku. I want to make my place where we Aboriginal people make use of both Aboriginal and non-Aboriginal cultural practice [in managing and referring to the country].

It is no surprise that a hunter-gatherer culture, whose day-to-day survival depends on an understanding of landscape, seasonal variation and biological diversity, should have developed specialised knowledge concerning landscape features, ecological habitats and detailed ways of referring to many aspects of landforms. Cognitive geographers have noted that the linguistic category of landscape terms is universal in all cultures, but the organisational details or cognitive schema for how landscapes are referred to and ontologised varies greatly from culture to culture (Mark and Turk 2003). Seventy years ago, however, the point that landscape terms are a cultural universal needed to be made and anthropologist Donald Thomson working at that time in Arnhem Land made the following remarks (Thomson 1939, p. 211):

> … *'the seasonal factor is recognised by the aborigines themselves, and stressed by the fact that they have classified the types of country, as accurately and as scientifically as any ecologist, giving to each a name, and associating it with specific resources, with its animal and vegetable foods, and its technological products'.*

We turn now to a discussion of the Bininj Kunwok conception of landscape and ecological zones.

BININJ KUNWOK LANDFORM TERMINOLOGY

I wish to sidestep the more abstract question of whether or not landscape in Bininj Kunwok is a self-contained semantic domain and concentrate on a description of terminology, word classes and identify how culture weaves its way into language about landscape. Certainly, the overarching label *kun-bolk* in Bininj Kunwok, can be glossed as 'country, landscape, place' and the interrogative *Bale ka-bolk-yime* 'what is the country like (what is the form of the land)?'[13] could elicit a description containing the names of landscape features. In contrast to the word *kun-bolk* is the term *kun-red*, which means 'home, hearth, traditional camping place', because the site has resources or some cultural significance that drew people to spend time there. All *kun-red* have names and are also described as *kun-redkuken* 'sites of significance in that they are known important historical camping places' (or *waddaburrk* in Kundedjnjenghmi and Dalabon). The adjective *–kuken* can be suffixed to other nouns such as foods and cultural practices and was once translated by a Kunwinjku speaker for me as 'important'.

Basic landscape terms in Bininj Kunwok are nouns. All nouns in Bininj Kunwok have gender agreement (although in some dialects this agreement is breaking down). There are four genders in Bininj Kunwok: masculine, feminine, neuter and vegetable. These classes are often (but not always) marked by the prefixes *na-*, *(ng)al-*, *kun-* and *(m)an-* respectively. Most landscape terms belong in the vegetable class and have the prefix *man-* in some dialects (Kunwinjku, Kuninjku and Kune), while some other dialects drop the initial nasal sound > *an-* (Gundjeihmi and Kundedjnjenghmi).

- *(m)an-dulum* 'convex elevated landform (hill)'
- *(m)an-kabo* 'water course'
- *(m)an-barrarn* 'gorge, valley, lower region with elevated landforms on either side'
- *(m)an-ngarre* 'jungle, thicket'

Not all terms have a noun class prefix e.g. *kabbal* 'plain', *kanjdjikanjdji* 'lowlands', and those with prefixes in some dialects can drop them in others and add a third person possessor suffix *–no* giving *dulumno, kabono* and *barrarnno,* which essentially means the same thing. The locative suffix *ku-* can transform the landform object into a place or location thus *ku-rrulum* (d>rr between vowels) 'in the hill(s), *ku-kabo* 'in the creek', *ku-barrarn* 'in the gorge'.

Elaborations or part features of basic terms are often verbs and seen as processes:

ka-barrarn-djarrinj-mang (intransitive verb)
it gorge.diverge.get
'a gorge which is dissected by another gorge at right angles'
ka-bo-berl-duluburren (reflexive verb)
it water.arm.pierce-reflexive.suffix
'braided stream'
ka-wardde-berre-bame (intransitive verb)
it rock.chest.shine
highly visible scarp or ridge line

Many terms can be described as predicate nouns. This means the root of the term is a noun, but it has a third person singular prefix *ka-* 'it' which is also the kind of prefix found on verbs. Terms in this construction often occur as a combination of subject prefix, noun and adjective:

ka-bo-ngerdken
it water.pregnant [literally, 'it water.pregnant']
'waterhole (on river)'
ka-ngalk-kuyeng
it cliff.long
long and tall escarpment face which runs for some distance.

These terms can then be built up into compound nominal phrases for more specific descriptions, some of which can be conventional terms, while others can deal with translations of English terms such as the rendering of the English word 'watershed':

ka-bolk-rorrbo kunj djurleno
it place.open area kangaroo shade
'open grassland with a few shady trees under which kangaroos like to rest'
ka-bo-larlma-rren ka-bebbe-bo-yarlarrme
it water-divide-itself it each-water-disperse
'water shed'

Landscape terms frequently encode more than just the geomorphology or the form of the land itself but can also include information about vegetation, water and human relationships or interaction with the land. In English, vegetation habitat terms are mostly exclusively concerned with the plant communities they describe, but in Bininj Kunwok this is not so. For example in English, ecologists refer to 'tropical savanna or woodland' or 'tall open forest'. These forests are dominated by eucalypts and other small trees and shrubs with a ground cover dominated by *Sorghum* spp. grasses and various herbaceous species. Generally these environments are called *(m)an-berrk* in Bininj Kunwok but the word also encodes the following semantic features:

- openness and traversibility
- opposition to *kuwarddewardde* 'rock country'
- opposition to riverine environments, such that to retriplicate the stem of the word *(m)an-behbeberrk* denotes a great expanse of woodland with no water. Bininj Kunwok speakers talk about movement across the plateau in terms of walking from one creek, traversing a stretch of *man-berrk* and arriving at the next creek.

These semantic features can be deduced by seeing how the root *-berrk* is used in everyday speech and in compounds as in the following examples:

kanjdji 'inside' versus ku-berrk 'outside'

When you place the locative prefix *ku-* on the stem, the sense becomes 'outside' in opposition to 'inside', but even with the noun class prefix remaining, the word can still be glossed as 'outside':

Ku-berrk *karri-yo wardi karri-ladmen.*
Let's sleep outside otherwise we'll get hot.
Ngarrih-nang ***man-berrk*** *bidbom kinga.*
We saw that the crocodile had climbed up and out in the open (i.e. out of the water).

This sense of openness or accessibility associated with the terms for 'tropical savanna forests' is something that has a human element. The openness of these forests is in opposition to the dominating hindrance of *kuwarddewardde* 'the rock country' or *ka-ngarredubbe* 'impenetrable thickets' and Bininj believe that they have a role to play in keeping these forests 'open and accessible' by undertaking the annual landscape burning as they moved from *kun-redkuken* to *kun-redkuken*, 'important camping location to important camping location'. Areas of open forest are not impersonal tracts that just roll on without delineation. They are divided by patriclan boundaries, and sections of large savanna forest tracts are individually named as the following text illustrating the term *-berrkmarren* 'the meeting up of named tracts of tropical savanna forest' makes clear:

Bardayal Nadjamerrek: ... *ka-balhdurndeng nani kanjdji ka-reh Ngaldaddubbe kondah, Ngaldaddubbe Nakarlwakarlwa* ***kaban-berrkmarren***	... it [the country] goes back down to Ngaldaddubbe here where Ngaldaddubbe and Nakarluwakarluwa meet together in [two areas of named] open forest country.

Encoded in some terms are what might be called 'human affordances', that is, certain terms that index some kind of human interaction with the landscape or facilitate a cultural practice. The term *man-berrk* includes a sense of accessibility. It is a landscape you can walk through with ease (because you keep it that way by burning it regularly) as opposed to the notion of 'hindrance' encoded in *kuwarddewardde* 'rock country' or *ka-ngarredubbe* 'impenetrable thickets'. Other terms which encode human affordances, hindrances or human interaction with the landscape are the following:

bambarr n.

dead end valley, narrow gullies or slopes/screes leading to a cliff, which are the sites chosen for fire drives to trap kangaroos that become injured by fire and can be easily speared.

Makka ngarri-di manekke kunj yiman ba kah-re man-bambarr ka-bidbun, yah ngarri-yame.

We stand there in them (the valleys), and if the kangaroos go up the narrow gully then we spear them.

kurrangmaye n.

the area just outside of the entrance to a rock-shelter, an area just beyond the overhang of a rock-shelter (typically a domestic space associated with a rock-shelter)

-djulngbongun v.t.

prefix = kandi-, ngarrben-/ngarrban-

expression [literally 'drink dust'] signifying relative residence/land ownership on a watercourse; that is, upstream or downstream of the other community

Kandi-djulngbongun.

They drink our dust; that is, their land is downstream from ours.

Ngarrben-djulngbongun.

We drink their dust; that is, our land is downstream from theirs.

ngalkmak prefix = ka-

n. [literally: 'it-cliff.good']

a cliff face that can be negotiated by foot without difficulty

Ka-ngalkwarre karri-kurduhme, ka-ngalkmak karri-kolung.

You have to walk around a sheer cliff face [to get down], but on a broken cliff face/accessible cliff, we can walk down.

antonym = ka-ngalkwarre [literally: 'it-cliff.bad']

Special terms exist for plant communities that are dominated by one species. Again, you can turn the plant name into a location by reduplicating or retriplicating the stem (depending on the number of syllables in the word), removing the noun class prefix *man-* and adding the locative prefix *mi-* so that the name *man-larrh* means 'cypress pine *Callytris intratropica*', but the word *mi-larrhlalarrh* means 'in a grove of cypress pine trees'. This can be done for all plants that grow in stands or groves. Here are some other examples:

mi-rrahdadarl n. [from (m)an-darl]
forest of Melaleuca viridiflora.
mi-djohdjodjoh n. [from (m)an-djoh]
grove of Acacia dificilis trees
mi-bernbehbernbern n. [from (m)an-bernbern]
grove of Eucalyptus alba trees.
mi-binihbinik n. [from (m)an-binik]
grove of an-binik trees, Allosyncarpia ternata
mi-dadjehdadjek n. [from (m)an-dadjek]
grove of fern-leaved grevilleas
synonym = an-dadjek kamirnderri
mi-djamkohdjamko n. [from (m)an-djamko]
grove of sandstone country grevilleas such as Grevillea angulata, Grevillea dryandri, Grevillea formosa
mi-djarrahdjarralda n. [from djarralda]
monsoon vine forest/jungle with large tall trees such as Maranthes corymbosa
mi-djimdjihdjimdjim n. [from (m)an-djimdjim]
grove of water pandanus Pandanus aquaticus.

Certain other terms encode both plant and landform features:

malaka (n).
landscape habitat characterised by flat dry country with numerous eucalypt species.
Andjarrarn malaka-kah.
The an-djarrarn tree [a kind of Eucalypt] is found in flat dry eucalypt country.
malalam (n).
dry land often on gentle slopes or undulating country, typically brown to reddish soils with little grass but covered with a grove of dense but thin tall trees especially Eucalypts.

The Arnhem Land Plateau Indigenous knowledge recording program has to date recorded over 300 terms for landforms and ecological zones. Just to illustrate one landform domain, a selection of terms relating to the rock country and rock formations is given in Appendix 2, and Figures 5.5–5.8.

Figure 5.5 *Ku-warddekimuk an-warddebuldjdjarn*, sandstone outlier massif, extensive plateau country.

Figure 5.6 *Ka-wohbarrarnkodjdubbe*, the dead end 'head' of a gorge.

Figure 5.7 *Ka-warddeyalalyo*, horizontal overhang of rock which provides a shelter below.

Figure 5.8 *Ka-barrarndjarrinjmang (v.i.)*, gorges dissected by other gorges at right angles.

FIRE

This section is a collection of texts that list the various uses of fire in the lives of plateau people. It includes perspectives from both men and women, and again the emphasis is on giving voice to those Aboriginal people of the plateau who are knowledgeable about such matters. Landscape burning was employed by Aboriginal people for a large variety of purposes and for the area with which we are concerned – the tropical north of the Northern Territory – the literature that discusses such variety of purpose is now considerable (e.g. Press 1988; Lewis 1989; Haynes 1991; Russell-Smith 1995; Russell-Smith *et al.* 1997; Yibarbuk 1998; Yibarbuk *et al.* 2001; Whitehead *et al.* 2003). The present compilation, as for the previous sections, consists of transcripts of various audio recordings made over a 10-year period. These include discussions at regional land-management meetings and workshops, formal interviews during field trips and more informal discussions during recreational bush trips or moments of relaxation around the family hearth on outstation residences.

A summary of the landscape burning purposes discussed in the following texts is given in Table 5.2, and a glossary of Bininj Kunwok terms relating to fire is given in the appendices.

Table 5.2: A summary of landscape burning purposes for the Arnhem Land Plateau.

Purpose	Description
Domestic space clearing	Fires are lit to burn an area chosen as a camp site. This also makes it easier to see dangerous animals, such as venomous snakes, centipedes and scorpions.
Fuel reduction and fruit trees	Burning in areas where fruit trees are plentiful ensures that hot late-dry-season fires will not damage fruit trees and makes it easier to collect fallen fruit.
Tuber collecting by women	Burning in areas where various edible tubers are sought makes it easier to find these foods. For some yams the fine tendrils leading to the tuber are followed in the soil and this is easier to do after the surface litter has been burnt.
Green pick encouragement	Areas were burnt in order to encourage game into an area.
Creek-line fire breaks	Creek lines were used as fire breaks and these were purposely created to assist in control of large hot fires.
Kangaroo fire drives	Large coordinated burns were conducted in the late dry season in order to hunt larger macropod species. Other smaller drives were also conducted for smaller macropods.
Walking track clearing	Traditional walking tracks were burnt each year as people moved along them.
Fire and honey	Burning made it easier to find the ground hives of native honey bees.
Fuel reduction	Highly flammable spinifex in the rock country was periodically burnt to keep it under control. Fire in the rock country was also used to flush out rock-dwelling macropods such as the black wallaroo, the short-eared rock wallaby and the nabarlek.
Fire and turtles	Although largely a floodplain-burning technique, fires were lit on the margins of billabongs in the late dry season in order to expose the breathing holes of long neck turtles lying in underground aestivation.
Jungle protection	Closed canopy forests, such as stands of *Allosyncarpia ternata,* were protected from fire by burning fire breaks around them.
Ritual	Although not a form of landscape burning, some aspects of ritual use of fire tell us something about Aboriginal attitudes to hot late-dry-season fires.

kun-rak (var.=kunak) 'fire (generic)'

kun-rak 'domestic camp fire, firewood' (anthropogenic or 'natural')

man-wurrk 'bush fire'

-kinje 'burn something, apply heat or flame to something', ***–rung*** 'to burn (intransitive), ***-wurlhme*** 'to burn (go up in flames)', ***-wurlhke*** 'set fire to something'.

Figure 5.9 Fire nomenclature schema.

The generic term for fire in Bininj Kunwok is *kun-rak*. As for the Dalabon semantic shift mentioned above, the term *kun-rak* (variant *kunak*) also means 'firewood' as used in a domestic fire (Figure 5.9). The word *(m)an-wurrk* refers to 'bushfire' and the fires of landscape burning. There are also specialist terms such as *bilbilk* 'a vigorous fire that burns beneath the matted roots and flotsam of riverbanks, especially in paperbark forests'. There are also sacred terms for fire used in secret men's ceremonies, but these names are not used in the public domain. Furthermore, as for all basic vocabulary, there is a special word for fire that one uses when one addresses affinal kin such as in-laws who are considered 'taboo' (e.g. mother-in-law and this woman's siblings). In such cases, the word *kun-rak* would not be used and the avoidance word for fire *kun-mimal* is substituted[14]. Figure 5.10 shows the Bininj Kunwok terms used for the dynamics of *man-wurrk* 'bushfire'. A more detailed glossary of terms relating to fire in general is given in Appendix 2.

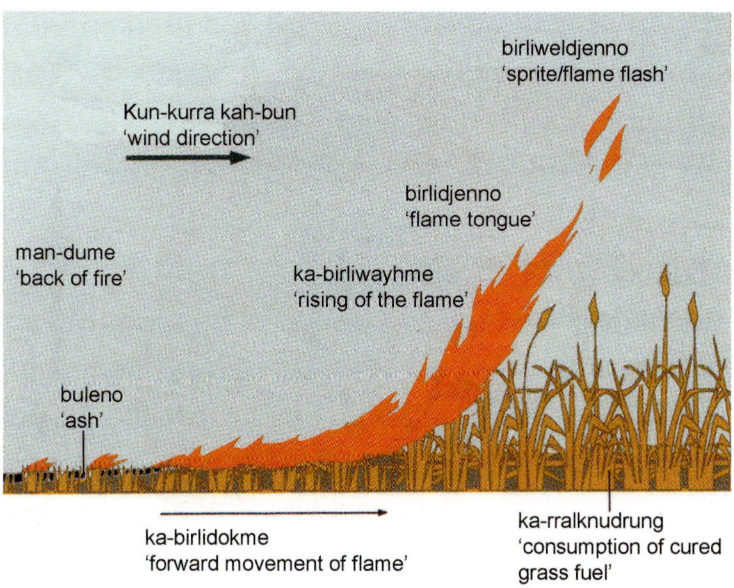

Figure 5.10 Bininj Kunwok terms relating to the dynamics of *man-wurrk* 'landscape fire' (adapted from Cheney and Sullivan 1997).

Burning to clear a living space

On arriving at a camp site, an area around the living space is usually burnt to clean the compound and to reduce the risk from snakes and other dangerous creatures such as scorpions and centipedes.

Jeannie Imangala: *...kubule kareh barri-yerrkani.*	... people used to make camp in burnt areas.
Ruby Bilindja: *Bad kubule barri-werrhwerrhmi barri-yerrkani. Ngarri-werrhwerrhmi kubule wanjh barri-yerrkani*	They would rake away the black ash and camp there. We cleared the black ash and we would sit down there.
Mary Kolkkiwarra: *Yiman ka-yime ngaye na-kohbanj ba-djangkani en nani kam-ngorrkang.*	Like with me, my father would be off hunting and would carry something home [while we were burning the camp ground].
Mary Kolkkiwarra: *Ngan-marnebolknami kumekke ngayo karri-yo that place barri-bolkngeybuni djib. Wanjh ngarri-bolkngeykarrmi kunu arri-djaley an-djad ngarri-yerrkani. Ngarrih-yingkihbuleweyi yika bininj mahki kumekke barri-bolkngeybuni. Wanjh ngarri-djaledjaley ngarri-yerrkani kubule.*	He [father] would name the site they intended walking to and off they'd go. We remembered the name of that place and we would walk off to go and camp there. We would burn off as we walked and when we arrived, we would camp in the area which had been already burnt.
Ruby Bilindja: *Kubule woh.*	In the burnt area (black ash area).

Fire and fruit trees

This text was recorded from Djungkidj Ngindjalarrkku: a young man from Yikarrakkal who has lived all of his life around the middle Mann River region. Here he discusses the relationship between fire and fruit trees, especially those that fruit in the early wet season such as those listed in Appendix 1:

Djungkidj Ngindjalarrkku: *Yika man-dudjmi ngarri-marnebulemarnbun. Man-dudjmi la man-dak.*	Sometimes we make burnt patches for the *Buchanania obovata* trees. *Buchanania obovata* and *Persoonia falcata* trees.
Murray Garde: *Bale ngurri-yime?*	What do you do?
Djungkidj Ngindjalarrkku: *ngarri-wurlhke ngarri-bulemarnbun. Laik ka-marnkan kabbal, ka-kukyo.*	We light fires and make burnt patches. So there's an open and clear place for the fruit to fall onto.
Murray Garde: *Yungki, ngurri-yingkiwurlhke?*	You burn before it fruits?
Djungkidj Ngindjalarrkku: *Ngarri-yingkihwurlhke.*	We burn ahead of time [before flowering].

Murray Garde: *Kukno?*	[When the fruit is] green/unripe?
Djungkidj Ngindjalarrkku: *E'e, kuning ka-yime laik ngarri-wurlhke, wanjh kam-re man-djewk ka-bun, wanjh kune ka-nguybun ka-djordmen.*	No, we burn off and then when some rain starts the trees flower and grow well.
Murray Garde*: Balekeno ngurri-wurlhke?*	When do you burn?
Djungkidj Ngindjalarrkku: *Ngarri-ray ngarri-wurlhke.*	We just go and burn off.
Murray Garde: *Yekkekeno?*	In the early dry?
Djungkidj Ngindjalarrkku: *Yo. Kuneng ngarri-wurlhke wanjh ka-nguybun.*	Yeah. That is when we burn off and then the trees flower.

Women burn to find yams and other tuber foods

Certain edible tubers can be found by identifying the leaves but, at certain times of the year, there will be nothing above the ground to betray the presence of the food that lies beneath the surface. Women use fire to clear the leaf litter allowing them to sift through the soil looking for certain kinds of fine roots that lead them to yams or other tubers.

Mary Kolkkiwarra: *Arri-wurlhke, yiman ka-yime korroko bu arri-rey kobohbanj laik kandi-bukkang bedda* like *ngarri-yahwurdni yiman ka-yime … Arri-rey bu yiman an-yawok dedjrungi karri-wurlhkeyi wanjh. Ngarri-kabbuni manu malahmalayi mani nga-marnbom yi-bengkan?*	We burn the country in the way the old people taught us when we were children such as […]. We also used fire to find the cheeky yam [*an-yawok, 'Dioscorea bulbifera'*]. We cleared away the surface with our hands just like I showed you this morning, remember?
Murray Garde: *Yoh*	Yes
Mary Kolkkiwarra: *Kune arri-yimi wanjh kebmudno arri-ngalkeyi. Mani kebmudno arri-djalkadjuyi ngarri-nani konda ka-rri. Ba-kebmudyirri*	That's what we did until we found the fine roots that grow out of the yam. We then follow them back until we see the tuber. There would be fine tuber roots there.
Jeannie Imangala: *Yawok* and *njale mak karrbarda.*	Cheeky yams *(Dioscorea bulbifera)* and long yams *(Dioscorea transversa).*
Jeannie Imangala: *Ngarri-karungi kebmud ngarri-ngalkeyi.*	We used fire to find the fine tuber roots.
Mary Kolkkiwarra: *Kayawal yerri ani ngarri-yimi ngarri-kebmudyawani ngarri-kebmudngalkeyi. Ngarri-djalkebmudkadjungi ngarri-nani ka-kodj …*	It's the same with long yams *(Dioscorea transversa),* we follow the fine roots that come out of them until we locate the tuber …

Ruby Bilindja: *Kodjno.*	The tuber.
Murray Garde: *bu ngurri-yawani karrbarda nawu daluk...?*	so when women were looking for yams...?
Ruby Bilindja: *Barri-kabbuni.*	They would feel around in the ground with their hands.
Mary Kolkkiwarra: *Buleno barri-wurlhkeyi wanjh.*	They would burn it to turn it to ashes.
Ruby Bilindja: *Kebmudno ngurri-nani.*	You'd see the fine roots.
Mary Kolkkiwarra: *Barri-kabbuni barri-nani laik ngamed kun-yarl yiman wanjki yiman kebmud laik string ngalengarre barri-kabbukabbuni barri-ngalkeng barri-karungi barri-kukkurrmikurrmi barri-mangi barri-yawoyhkabbuni ku-buyika* everywhere *barri-wabwabmi yiman bu barri-nani ba-rungi kaddum ba-yawoyhrungi wanjh barri-kulkweyi kun-bidyih.*	They would run their hands through the top soil to try to find the fine roots, the vine like a fine string that goes down to the yam and they would find it and then dig it out and stack them up, pull one out, and then search for another in another place, everywhere, moving from place to place, looking up where the fire had burnt up into the leaves and then they dig it out with their hands.
Murray Garde: *first time bu ngurri-bebme wanjh ngurri-wurlhke?*	So when you first get there [to look for yams], you burn the place?
Mary Kolkkiwarra: *Yo*	Yes
Mary Kolkkiwarra: *mm, yika mak bu bininj barri-yingkihwurlhkeyi ba-rrordrungi wanjh barri-djale bu mani kaddum ka-rrordbarnhbarndi ku-kurlk konda karri-kabbuni barri-yimi nawu kobohbanj barri-karungi.*	Sometimes too when they had burnt, you can more easily see the seed pods burnt hanging down from above and so then they start digging with their hands, that's how the old people used to dig them out.

Burning for green pick to encourage game

The fact that Aboriginal people burnt landscape to induce new grass growth that attracted game has long been recognised. This text is a conversation with Mick Kubarkku and Lulu Larradjbi.[15]

Murray Garde: *Mak nga-bekkang birri-yimeng birri-wurlhkeng wanjh kun-dalk* like *kun-bule wanjh yimerranj, and ka-rralkbebme and kunj ka-kadjung.*	I've heard that they also burned the country so that kangaroos would be attracted to the new grass shoots which resulted?
Mick Kubarkku: *Yoh, man-nelk. Manbu man-dalknelk, ka-kolhdebebme. Yoh wanjh kolhdeno ka-kadjung ka-ngun wanjh.*	Yes, the soft new green shoots. The new grass shoots. Yes, they follow after this to eat it [the kangaroos].

Lulu Larradjbi: *Ka-kolhdengun.*	It eats the new shoots.
Murray Garde: *Ka-kolhdengun. Kolhdeno njamed kun-dalk?*	It eats the new shoots? The new grass shoots?
Mick Kubarkku: *Yoh ka-rralkbebme, yiman kadni ka-yime ka-rralkbebme.*	Yes, when the grass appears, just like when we plant a vegetable garden, the new grass comes up.
Murray Garde: *Yoh ka-durhdurndulubun.*	It sprouts up everywhere.
Mick Kubarkku: *Yoh wanjh ka-re kunj ka-wake ka-ngun, ka-rralkngun.*	Yes, and the kangaroos crawl across eating the grass.
Murray Garde: *Kunekke mak, bininj birri-wurlhkeng ba ka-kolhdebebmeninj?*	So people also burnt the country to make the new shoots grow up?
Mick Kubarkku: *Yoh, wanjh djurrkmayi kunj kadjuyi bu nguyi.*	Yes, it would attract kangaroos, they follow it to eat it.
Mick Kubarkku: *Yoh nguyi wanjh balemdiwirrinj. Balemno nungkake balemdiwirrinj.*	They eat it and get fat. The fat of the kangaroo, it would get fat.
Murray Garde: *Kunukka-ken mak birri-wurlhkeng?*	So is this also why they made fires?
Mick Kubarkku: *Ng ng kunekke-ken.*	Yes, for this reason.
Mick Kubarkku: *Yoh ka-ngun kunj, nganabbarru warridj. Ka-rralk-ngun. Mane ngarri-marnebulewurlhkemeninj nawu kunj.*	Kangaroos eat it [the new grass] and so do buffaloes. They eat grass. We would make a burnt ash area for the kangaroos.

Burning to make fire breaks along creek lines

The Bininj Kunwok people of the Arnhem Land Plateau are sometimes referred to by other Aboriginal coastal neighbours by the English term 'the creek people'. Small watercourses support a diversity of plants and animals, but they are also important in fire management. Creek lines serve as firebreaks and Bininj take into consideration the location of a creek line that can be used as a fire break when planning burning (Figure 5.11).

Jack Djandjomerr: *An-kabo barri-wurlhkeyi an-kabo yi-bengkan. An-kabo barri-rayi, barri-nayi* too fast *ka-re wanjh barri-wurlhkeyi an-kabo* stopem that natha wan. They didn't want it to get across *kunak nawu,* stoppem *birri-yimi* this side *bu kabohkabo birri-wurlhkeyi wanjh* they stop that *manekke manu kunak rey. Lakadjad dabbarrabbolk birri-yimi nawu kare.*	They used to burn along the creeks, you know. They would go to the creek in the event that they saw a fire was moving too fast and then they would burn a fire break to stop that other fire approaching. They didn't want it to get across the creek, that [dangerous] fire, so on this side they burnt a firebreak to stop that fire from proceeding any further. That's what the old people used to do.

Figure 5.11 A creek-line fire, which creates a firebreak to control approaching larger hot fires (photo: Ira Cooke).

Kangaroo fire drives

Mick Kubarkku: *Makka ngarri-di manekke kunj yiman ba kah-re man-bambarr ka-bidbun yah ngarri-yame.*	We stand there in them (the *bambarr* gullies/ valleys) and if the kangaroos go up the gully/ scree then we spear them.

Large coordinated kangaroo fire drives *man-wurrk kunjken* were conducted usually in the late dry season of *kurrung* but sometimes also earlier in *wurrkeng.* The main targets of this form of hunting were antilopine kangaroos *(Macropus antilopinus)* and euros *(Macropus robustus).* These drives were once an annual event, which required coordination of a large number of people from many groups. An invitation to attend a drive would be made by sending a torch made from bound *Eucalyptus tetrodonta* bark, known as *djalk,* to one's neighbours. Various ritual restrictions had to be observed when staying in the preparation camp at the fire drive location. All spears in the camp had to be laid down flat otherwise antilopine kangaroos would sense their presence and move away. Bardayal Nadjamerrek explains:

Nawu kandakidj an-kole an-kodjdjekurrmi wardi bu baladjabdiwirrinj bu bangmeyurrhkuyuwirrinj, kun-kange ba-kange … ba-bangniwirrinj kunj. Dja	As for the antilopine kangaroo, spears which were going to be used to hunt them had to be placed flat on the ground because if they were stood upright, if they stood up straight

arri-kolekurrmi an-wurrk arri-nami ya arri-kolekurrmi kanjdji borndok an-kole ku-kurlk balayoy bu an-wurrk arri-yikani wardi anbu ba-djabdiwirrinj an-kole ngarri-kurrmeninj ba-kangedarrmeninj yirridjdja nawu bininj nakka, nungka kunj ka-bengkan rowk ba-kangebabanghni 'O ka-kangebabang maitbi Bininj kabarri-re' ba-yimi barri-nami. An-kole arri-kurrmi bu arri-yoy malahmalaywi, nakka yirridjdja yi-bengkan marrngkidj Ubarr nawu o lorrkkon. Yirridjdja na-kudji duwa ngad na-warre kalkberd. Yiman laik kordang marrngkidj bu bad najing nakka yirridjdja yiman Bininj. Yiman ngarri-yime 'Oh djad kunj bi-rredjdjudmeng kabi-dedjdjudme an-bu an-kole arri-djabname. Ka-re balay.	before being used to hunt the kangaroos, the animal would sense the presence of humans and become strong and powerful. If we were placing spears to be used in a kangaroo fire drive, we would put those spears down flat together with the spear thrower on the ground, when we were going off to do a kangaroo fire drive. If we were to put the spears upright in the camp, then the mind of the kangaroo would be alerted because *yirridjdja* moiety kangaroos are like humans.[16] Those kangaroos know everything about our intentions and can think 'Oh I feel energy welling up in my body, there must be humans coming.' When we camped the night before the kangaroo fire drive, [we would be mindful of the fact that] those *yirridjdja* moiety kangaroos are associated with the Ubarr and the Lorrkkon ceremonies and have supernatural power. That's the case for *yirridjdja* moiety kangaroos. But the *duwa* moiety kangaroos, our moiety [speaker is *duwa* moiety], they are not powerful like that. *Yirridjdja* moiety kangaroos are like 'clever men', like human witch doctors. We would say things like this, 'that kangaroo has been [metaphorically] jabbed in the rump by a spear [which was not placed flat on the ground] and it will move away to a distant place now.'

The following text is a discussion with Bill Birriyabirriya and Nancy Kurawalwal at Markolidjban Outstation. They describe how a group of hunters in a fire drive burn grass forcing kangaroos up towards an elevated area or *bambarr* where they cannot escape. Another group of hunters waits here for the kangaroos to arrive. These areas have been previously burnt in preparation, providing refuge from the fire for the second group of hunters and also stopping the large hot blazes from escaping and continuing into neighbouring areas. By this time of year the remaining fuel load is low because people have been burning all through the dry season and the risk of the fire escaping and continuing has been minimised. What Birriyabirriya describes here is also precisely what Altman witnessed at Namilewohwo in 1980 (Chapter 6). The discussion demonstrates coordinated, carefully planned and sequenced activity in fire management.

Murray Garde: *Yoh, but nakka kunj-ken kun-buyika?*	Yeah, but are kangaroo fire drives different? [to other uses of fire]

Bill Birriyabirriya: *Yoh kunj-ken.*	Yes, for kangaroos
Bill Birriyabirriya: *Birri-wurlhkeng birri-borledmeng manu man-dulum and mani birri-borledmeng.*	They lit fires around a hill [like this one over here near Namokardabu] and they encircled it.
Nellie Kurawalwal: *Man-bambarr.*	[Into] valleys/dead end gullies/cliff faces.
Bill Birriyabirriya: *Man-bambarr, and birri-marnbom* [draws a diagram in dirt] … *kunuk' ka-yime, and mane man-dulum, maitbi konda, birri-di kunj namekke kam-wam, kondah and mane bu kureh birri-wurlhkeng kun-rak.*	A valley/gully that rises, and they would make it like this [draws a diagram in dirt] and this is the hill here and they would wait here and the kangaroos would come here and then they would make fire and burn here.
Bill Birriyabirriya: *Kure kam-re kun-rak and namekke kunj olot bad kondah kam-wam.*	The fire would come here and all of the kangaroos would come in here.
Murray Garde: *Minj bale kabirri-yime bu kabirri-bidbun larrk? Kunj*	And so the kangaroos had no where to climb up?
Bill Birriyabirriya: *Ey? Oh kunj ka-bidbun kabirri-bun.*	Oh when they climbed up they would kill them.
Murray Garde: *Aa kabidbun wanjh kumekke …… kabirri-rung?*	Oh they climb up and there … they get burnt?
Bill Birriyabirriya: *Minj ka-rung, la kabirri-djal[?]yame man-kole korroko yameng man-kole, bu korroko birrih-yameng man-kole but bolkki mako.*	They don't get burnt [to death], they just get speared, before they used to use spears, but now we use guns.
Nellie Kurawalwal: *Bolkki mako.*	Guns today.
Murray Garde: *But kabirri-nan kun-kurra, bu kun-kurra ka-bun …*	But what about the wind, when it blows ….
Bill Birriyabirriya: *Kabirri-wurlhke, but konda nawu birrih-di yi-bengkan kure njamed man-bambarr…..* like *kabirri-wurlhke balemane ka-re bu bininj bedberre konda kabirrih-di laik ku-bule kure minj kun-dalk ka-rri. Konda wanjh little bit kah-re maitbi kureh ka-yimen mane kam-re kureh* well block *kabirri-yime manu kun-rak.*	They burn the country and stand there up the valley ….. they burn off and can't go anywhere but just stand there in the ash/burnt area, there's no grass left. And then the others are blocking the other side with the fire.
Murray Garde: *And konda mak bininj kabirrih-ni?*	And there's people standing here also?
Bill Birriyabirriya: *Yo … bininj kabirrih-di because mane manu korroko kabirri-wurhlke like.*	Yes, there's people standing in there because they've already burnt that area.

Murray Garde: *Like boyen birri-wurlhkeng? Yiman ka-yime wolewoleh birri-wurlhkeng wanjh kabirri-durndeng?*	They burnt it beforehand? Like they burnt previously, the day before or whatever and then they go back.
Bill Birriyabirriya: *Yoh bu start laik kunek' kabirri-wurlhke laik konda kabirri-yime kabirri-bulewe, kabirri-wulhke ka-kadjung and kunj kam-re kabid.... kondah ka-rerrkan kabirri-yame, birri-yameng man-kole.*	Yes, at the start they burn this area and make ashes there, and then the kangaroos are forced up there and they sit down there and people spear them.
Bill Birriyabirriya: *bu* start *kabirri-yime laik kabirri-wurlhke man-wurrk yi-bengkan Bulanj?*	At first they burn it, you know a blaze Murray?
Murray Garde: *Yoh*	Yeah.
Bill Birriyabirriya: *Bokenh bininj o bininj kabirri-re bininj kabirri-dawurlhke kun-dalk maitbi raitab kabirri-wurlhke, and birri-buyika mak bininj, konda kabirrih-di.*	A couple of men go and burn right along and another group are standing here.
Murray Garde: *Kure ku-bule?*	Near the ash area?
Bill Birriyabirriya: *Kure ku-rralk.*	In the grass.
Bad kabirri-wurlhke manih kah-kadjung, kureh kah-re, la mane manu kureh kah-re ngalengman, la konda wanjh bedda bininj kabirrih-di.	They burn the country and it moves off itself [the fire], but here the others are standing [waiting].
Bu kabirri-yame and ruy kumekke wanjh yi-bengkan ... kole-dorrengh ruy and birrih-mey.	They spear them and they [the kangaroos] get burnt you know ... burnt, with a spear in them, and they go and grab them.
Yoh man-djewk, like every year, *bu ngarri-wurlhke kunj wanjh karri-wurlhke man-djewk ka-borledme* like, *ku-djewk and like kuning ka-yime karri-wurlhke, munguyh.*	Yes, it's an annual event, we burn off and then another year goes and we burn again, over and over.

Likewise, Mick Kubarkku describes how the kangaroos are often injured by the fire and they lick their burns as they stand in the previously burnt ash area, unaware that hunters are stalking them:

Mick Kubarkku: ... *karri-wurlhke nawu kunj ka-belenghmerren wanjh, ya wanjh karri-danjbun. Man-kole yerrkka, man-kole yerrkka ngarri-yameninj bu belenghmerrimeninj ... kunj,*	... we burn for kangaroos, they stand there licking their wounds, and that's when we spear them. With spears, we used to spear them when they were licking themselves [after being burnt] ... those kangaroos.

Ka-rung wanjh, ka-rung ka-belenghmerren. Minj kan-nan ngarrku, nakka welengh. Ya ka-belenghmerren. Wanjh … ngarri-yame nudno.	They get burned, burned and they lick their burns. They don't see us, they're concerned only with themselves. Yes, they lick their wounds. Then … we spear them while they're wounded with burns.

Conducting kangaroo fire drives at the hottest and driest time of the year is dangerous business and there are numerous stories told of how people were burnt to death in kangaroo fire drive accidents.[17] During the 1980 Namilewohwo fire drive, Jimmy Kalarriya and his classificatory son were nearly burnt. Recollecting this incident during a conversation in 2006, Kalarriya and a group of other elders discuss hot late-dry-season fires and how they need to be controlled by reducing the fuel throughout the early dry season. Patches of grass in certain locations were left unburnt until late in the dry season in preparation for the fire drive, but fire breaks needed to be put in place beforehand, and adjoining areas of dead grass needed to be burnt earlier in the year in order to stop kangaroo fire drives starting fires that might burn out of control for many days.

Jimmy Kalarriya: … *ngad karri-nameng kondah Nimilewohwo bu kurrung-ni.* That time business on *yimihyimi.*	Once we had a kangaroo fire drive at Namilewohwo in the late hot dry season. At that time there was a sacred ceremony [Kunabibi] happening.
Arri-nameng ngayi ngane-kuyinruy na-djakerr Robert ngayi nawu Jabiru kah-ni. Wanjh ngarri-nameng man-wurrk arri-kuyinruy.	We lit the fire and nearly got burnt to death, me and Robert [Namarnyilk], who lives today in Jabiru. When we conducted that [kangaroo] fire drive we nearly got burnt.
But wurdyaw nga-marneyimeng werrk yi-bulewemen kumekke mak matches ngane-kang werrk buleweng.	I said to the boy [Robert], quick light a fire and backburn with these matches we were carrying.
Bulewarlahminj ngane-dolkang ngane-bolkmelmeng …ngane-ruyi.	When the burnt ash area was big enough we jumped into it for safety …we nearly got burnt.
Peter Biless: *Yoh kurrung makka. Makka djal bonj bu birri-wurlhkeyi makka* 'flatout' *djaley*	That's because it was *kurrung* the late hot season, at that time of year the fires will burn 'flat out' with speed and intensity.
Bardayal Nadjamerrek: Flatout *ba-rey bad an-bu yekke yekkeken arri-wurlhkemeninj kuni ba-yimi.*	It burns intensely 'flat out' then, but in the early dry season, that's the time when we used to burn, that's what we used to do.
arri-djalngurdme [kurrung] ba-wurlhmiwurd ba-wurlhmi najing an-bu dalknudyak, djal kun-dalknud makka ka-djale yalhmeninj ka-re ka-dalkraworren yiman.	We would stop burning by *kurrung*, but there might be small fires because all of the dried grass is finished by that time of year, there's no grass left, but if there is a lot of dry grass

	it will burn on ahead with intensity and these fires will then burn fiercely into adjoining unburnt patches of grass [and continue dangerously].
Jeremy Russell-Smith to Jimmy Kalarriya: What, *man-karrarndalk?*	Was it *[Triodia]* spinifex bush?
Jimmy Kalarriya: *Belendjerre*	Kangaroo grass [i.e. *Themeda* spp. grasses]
Jimmy Kalarriya: *Kurrungni kunukka, kurrungni kaluk djalwam manekke.*	It was the late dry hot season and it burnt too far, quickly [where it was not intended that it should burn].
Jimmy Kalarriya: *Mani, nani yi-nan nuye kun-dalk nani.*	[makes hand sign for Antilopine kangaroo] The grass which is eaten by this [the Antilopine kangaroo- in sign language].
Jeremy Russell-Smith: *kanjdjikanjdji?*	OK [was it on the] lowlands?
Jimmy Kalarriya: *Kurrulum*	The hill country [low rolling hills which are a favoured habitat for these kangaroos].

Walking routes and landscape maintenance

When walking from site to site, the men would go on ahead and burn as they went. This might flush out kangaroos, but it also kept the traditional walking routes open and for the women who followed behind, the black burnt patches indicated the path taken by the men. At the end of the day the women and children would arrive at the place mentioned at the beginning of the day and the camp ground would be burnt to clear it.

Bardayal Nadjamerrek: *[bangkerreng] kayakki ka-bolkkih ka-njihme arri-yime marrek ka-wurlhme yekkeyekke wanjh ka-wurlhme.*	In *bangkerreng*, the late wet, we didn't burn much, it was too wet, but in *yekke*, throughout the dry season, it burns.
Bardayal Nadjamerrek: *Barri-wurlhkeyi ba-kadjungi ba-rrowen.*	They burnt in a way such that the fire creeps along and dies out.
Murray Garde: *Bininj-wi barri-dokmeni daluk?*	Are men doing this, going ahead, or women?
Bardayal Nadjamerrek: *Nakka yerre barri-wurlhkeyi barri-rey barri-bulekadjungi 'Buleno karri-ray!' barri-yimi. Daluk o Bininj o dabbarrabbolk o wurdurd kunukka yekke.*	Women would follow us [men] burning behind us and also following the ash areas we had made. 'Let's follow the burnt areas' they would say. Women or men, old people or children. In *yekke*, the dry season.

Mary Kolkkiwarra: *Ngarri-birlikani kuwadda ngarri-yibebmi. Arri-worrhmi, djal anekke kuybuk, kuybuk ngong o njamed mani arri-kani kuydjimbok. Dja yika makih njamed dolobbo nawu yi-bengkan …*	We'd carry the embers (in the banksia cones or pandanus stems) until we got to the camp. We'd light the campfire with the banksia cone or we would carry another plant *koydjimbok*. Or sometimes *Eucalyptus tetrodonta* bark …
Djalk nawu, namekke arri-dukkadukkani arri-birlikani madjdjurn woh. Yika kun-daleh, kun-daleh barri-kani ngadberre kobohbanj korroko bedman nawu barri-marnbuni barri-bordebmi.	We made torches out of rolled up stringybark. Sometimes they used fire drills to light fires, spinning the stick between their hands.
Ruby Bilindja: *Ngarri-wurlhkeyi yika barri-dombuni kun-djalh arri-bakkeyi barri-dombuyi.*	We would light fires, but other times we put them out too, with branches.
Jeannie Imangala: *Kun-malaworr.*	Branches of leaves.
Mary Kolkkiwarra: *Kun-malaworr arri-bakkeyi.*	We'd break off branches.
Ruby Bilindja: *Mm kun-malaworr yiman manih. Barri-dombuni barri-bakkeyi barri-dombuni, barri-dombuni barri-yakwong. Kun-malaworr arri-djalwohwodbuni barri-dombuni rowk. Ka-mak. Bu barri-nani kaddum ba-birlibadwayhmeng. Wanjh barri-dombuni. Barri-djaldombuni barri-yakwoni.*	Mm green leaves, like that. They kept on breaking off branches of leaves until they had extinguished the fires. We would whip with the green branches and put the fire out. It works. They did this if they saw the fire was getting too big. Then they would put it out, until it was completely extinguished.
Jeannie Imangala: *Wanjh kubule kareh barri-yerrkani.*	And then they camp in the ash area.
Ruby Bilindja: *Bad kubule barri-werrhwerrhmi barri-yerrkani. Ngarri-werrhwerrhmi kubule wanjh barri-yerrkani*	They would rake away the black ash and camp there.
Mary Kolkkiwarra: *yiman ka-yime ngaye … na-kohbanj ba-djangkani en nani kam-ngorrkang. Ngan-marnebolknami kumekke nga-yo karri-yo that place barri-bolkngeybuni djib. Wanjh ngarri-bolkngeykarrmi kunu arri-djaley an-djad ngarri-yerrkani. Ngarrih-yingkihbuleweyi yika bininj mahki kumekke barri-bolkngeybuni. Wanjh ngarri-djaledjaley ngarri-yerrkani kubule.*	Like with me, my father would be off hunting and would carry something home. He [father] would name the site they intended walking to and off they'd go. We remembered the name of that place and we would walk off to go and camp there. We would burn off as we walked and when we arrived, we would camp in the area which had been already burnt.
Ruby Bilindja: *Kubule woh.*	In the burnt area (black ash area).

Mary Kolkkiwarra: *Arri-bulekadjungi, barri-wurlhkeyi werrk ngadberre ngandi-marnebulemarnbuni buleno ngarri-rey ngandi-marnebolkngeybuni kure arri-yoy. 'Ma kumekke ngarri-yo ngurri-bebme. Ma! Dja yingkihmarnebulemarnbun!*	We used to follow the burnt patches, the men went on ahead and burnt the country for us and we would follow behind heading towards the place they had named before we all set off. 'We'll camp at place X, you should head for that place! We'll go on ahead and burn the route for you!'
Bardayal Nadjamerrek: *Dja kunj arri-yingkihkinjeyi weleng arri-dow[waddahmik] arri-yami barri-rey arri-yimi 'kuy'*	We would have a kangaroo cooked waiting for the women and children and we would already have set up camp and as they were arriving we would call out to them [so they would know where to go] 'kooooooiiii'.

Patchwork burning

At Kabulwarnamyo Outstation in December 2005, a land-management workshop entitled *Kun-wok Kunbolkken 'The language of land'*[18] was conducted and attended by *Bininj* of all ages and from a number of communities on the plateau. At one point, we discussed the following question, 'How do you describe 'well burnt' savanna in Bininj Kunwok?' Bardayal Nadjamerrek replied with the following comment:

Bardayal Nadjamerrek: *Ka-rung ka-wohdalknjihme ka-kih ka-re hab, ka-re hab ka-re ka-wohnjihnjihme ka-rralkkih.*	It burns resulting in patches of burnt and unburnt grass, it burns into some areas but not others, it burns into other areas [and makes a patchwork pattern] leaving some of the grass unburnt [because] some of it is green.
Murray Garde: *Bu karri-nan ruy rowk, bale karri-yime?*	What is the term for bush that is totally burnt?
Bardayal Nadjamerrek: *Kabarri-yime ba-ruy yerrikahwi ka-yime.*	We use the term *ba-ruy yerrikahwi* 'total area burnt', is what we say.

The Bininj Kunwok term given by Bardayal above *ka-rung ka-wohdalknjihme* is what I have translated as 'patchwork burning' (Figure 5.12). These two words broken into their individual morphemes have the following literal senses:

ka-rung ka-woh-dalk-njihme
it-burns it-partially-grass-is.green/moist

Some moments later, as we were eliciting terms for landscapes by looking at slides on a computer screen, Bardayal excitedly offered the following comments with great animation in response to the photo of *man-berrk* in Figure 5.13 (a site rich in *Buchanania obovata* groves a few kilometres north-west of Marlkawo outstation).

Figure 5.12 *Ka-rung ka-wohdalknjihme*, 'patchwork burning' (photo: Jeremy Russell-Smith).

Figure 5.13 'A good management outcome for farming emus'; *Man-berrk* 'savanna woodland' on the Kabulwarnamyo to Marlkawo road (refer text for details).

Mani dabbarrabbolk barri-djareni, mani ngandi-bukkang, an-dudjmi-ken, ngurrurdu, kunj!	This is [the kind of savanna] the old people wanted, this is what they taught us, [to burn like this] for green plum (*Buchanania obovata*), emus and kangaroos!

The site pictured in Figure 5.13 on the Kabulwarnamyo to Marlkawo road is in some respects a good illustration of Rhys Jones' (1969) term 'firestick farming'. The green plum trees *(Buchanania obovata)* in this region – as in numerous other important green plum sites – are mostly of a noticeably shorter variety (0.5–2 m), which are distinguished by name in Bininj Kunwok from the taller (4–10 m) variety. The tall variety is called *man-dudjmi* or *man-moyi* while the short variety is called *man-wodberr* (see Figure 5.14). The relatively intensive human occupation in the vicinity of this green plum grove and the fire regimes Aboriginal people have instigated here over many generations have most likely resulted in the shorter *man-wodberr* variety. At the same time, the area is known as an important habitat for emus who find the shorter *Buchanania obovata* trees to their liking as the fruit is more easily accessible for them.

The two senior traditional landowners of this region, Jack and George Djandjomerr, have very clear views on the role of fire in attracting emus into the area. However, the hot late-dry-season fires that swept through the area repeatedly up until the 1990s before more traditional regimes were reintroduced, have been blamed for the apparent decline in emu numbers in the region. I asked Jack Djandjomerr about the effects of these repeated hot late-dry-season fires on emu populations. Jack had been busy in 2004–5 trying to reduce fuel loads in the region in

Figure 5.14 Children from Marlkawo collecting *man-wodberr*, the short variety of *Buchanania obovata*.

order to return to a more controllable early dry-season fire regime and avoid the large hot fires of the late dry season. These adjustment fires may have discouraged his emus, but, as he points out, the reintroduction of the proper fire regimes will eventually promote the plant foods sought by emus:

	Murray Garde: Do you think when there are big fires which are not managed properly, it spoils emu habitat?
Jack Djandjomerr: *Woh, manekke nga-wokdi* all the time *bla ngurrurdu nga-wurlhkeng* everywhere from last year im burn lots here *ngurrurdu* no more here now, im gone away but like *ka-marnbun man-me ka-rrurndeng ka-ngungun* then one place *ka-yidjaldurndiwerren kabirri-ngun anme kabirri-worrkmen minj karri-ngalke, bad nuk karri-ngalke karrih-bun.*	Yes, I'm always talking about this in relation to emus, I've been burning everywhere from last year and all these fires have chased away all the emus, but eventually it will make more food for them and they will return to feed and then they will stay around one location to eat the fruit and maybe we will find some of them to hunt or maybe we won't.
	Murray Garde: So emus don't like big fires?
Josie Maralngurra: *Marrek ka-djare ngalkka.*	They don't like it.
Jack Djandjomerr: He [the fires] make him run away from last 2 years he bin burn a lot well im not here *ngurrurdu wam lakidjad* because he bin look *kunak ba-birlinang* well *ba-wam anek ngurrurdu* might be *Wurrbbarn durnderrinj.*	The fire made the emus run away, all the burning, and now they are not here because they saw the fire and so they took off, maybe they all went back to Wurrbbarn [i.e. 'emu'] clan country.
	Murray Garde: So what should people do in respect to fire and emus?
Jack Djandjomerr: no, I like to stop *im,* just one time we just try [to burn] we try *bu anme ka-wernmen anme ka-wernmen wanjh kamak karri-nan wanjh karri-durndeng karri-wurlhke* then *karri-wurlhwurlhke* everywhere like *kamak ka-wernmen anme ka-marneyo* dumaj.	I just want to stop the burning for a while [but] we try to burn so that the fruit trees increase and then when we see that the emus have return then we can continue burning everywhere. When the emu food increases, so will the emus.

I also asked Jack's brother, George Djandjomerr, if the cessation of burning for some time might encourage the return of emus. His reply indicates that the solution is not to stop fire altogether, but to make sure the fires are lit early in the dry season.

George Djandjomerr: *kayakki bu yiman kabirri-wurlhke yi-burrbun nawu boyen kabirri-rengere kabirri-wurlhke* what time *kabirri-wurlhke May bu kabirri-re* April or May helicopter *kabirri-re kabirri-wurlhke*	No [i.e. don't stop burning], you know when they go out burning, what time is that, in May or April in the helicopter those young blokes go out burning, well that is good, it's good to burn early like that. Emus know the

yawurrinj kunukka kamak, early *kunukka kamak Yika mak ka-djalbolkburrbun nawu ngurrurdu ka-djaldurndeng kure man-me ngalengarre ka-dingihdi kumekke ka-rrurndeng ka-ngun yika bu minj kabirri-ngalke kunukka Marlkawo wanjh ku-buyika ka-re kure an-me ka-burrbun kaluk ka-re wardi an-djewk kabirri-re o dird kabirri-re kabirri-yawoyhdurndeng kabirri-marnebebme*	country and they will come back to where they know their food is located and stay there, they'll come back and if there are none around Marlkawo then they will be somewhere else but they know the country and in some years they will come back again

Other kinds of fire drive hunting

Coordinated kangaroo fire drives involving large numbers of people in *bambarr* 'fire drive valleys' were not the only kind of fire drive. Small opportunistic kangaroo fire drives in areas with plentiful fuel were also conducted to flush out macropods exposing them to waiting hunters. Areas with tall grass on the margins of water courses or in rock country were places where fire would be used to flush out macropods.

Bardayal Nadjamerrek: … *karri-wurlhkeyi djamdam (sometimes) barrk, ngarri-nami, ngarri-ngolhngolhkeyi walung, wanjh ngarri-wurlhkeyi yilk yerre, karrimen burldjdjarn. Ya nungan mak kurebe ba-wurlhkeyi bininj nawu nganeh- ngarri-reyi kahwi. La nungan mak bininj kure ba-birlikani, ba-wurlhwurlhkeyi ba-kuyin (disfluent) barrk ngarduk ba-bebmeng bayi(mi), ba-bebmeng ba-yami… ba-wukkurrmi.*	… sometimes we would burn off for the black wallaroos, lighting little fires everywhere (in a ring in the rock country) and then in the middle. Someone else would burn from another side and we went along like this. With people walking along carrying fire, the black wallaroos would come out and sometimes get burnt.
Murray Garde: *Bambarr kuwid nakka…*	You don't hunt them in *bambarr* fire drive gulleys?
Bardayal Nadjamerrek: *Bambarryak, kalaba yirridjdja. Bambarr ka-karrme bad nani kuwarddewaken nakka bambarryak. Ngarri-wurlhkeyi ngarri-nang ka-bebme. Yika* (disfluent) … *'Oh djibowkman ngarduk', ngarri-yimi nga-kadjungi nga-yami.*	No, *bambarr* 'fire drive valleys' are for *yirridjdja* kangaroos. In this landscape there are no fire drive gulleys. We just lit fires and watched them (black wallaroos) come out. I would say 'Oh there's Djibowkman [the name of a mythological black wallaroo] come out for me, I'll go and spear him.'
Murray Garde: *Birri-wurlhkeng bu ka-bebme.*	They lit fires to chase them out?
Bardayal Nadjamerrek: *Wo ka-bebme, anwurrk kabi-bebke, ka-belenghmerren warre, yika ka-belenghmerren djarran anwurrk difren hot kabi-marnbun kabi-kurrnginjbebke. Ka-belbme kuwanjkeno ku-warrwarrku kuworrk ka-belenghmerren. Ngarri-kunibebme ngarri-yame.*	Yes, they'd come out, chased out by the fire and be licking themselves because the fire had made them hot and made them sweat. In the heat their pouches stick to them and so with their heads buried down licking themselves, we would sneak up and spear them.

Honey and fire

On an excursion in 2005 to find *an-kung*, 'honey hives', the discussion turned to the role of fire in helping people locate *nabiwo:* a variety of honey that is found both in the ground (especially termite mounds) and also up in the hollows of trees. Locating the entrance tubes to ground honey hives is much easier to do in areas that have been burnt.

Mary Kolkkiwarra: *Kukurlk nani* its alright *djamtaem kukurlk bonj anyway karridjalyawan* easy one *kangalke kure …*	In the ground it's easier sometimes to find [honey] there …
Mary Kolkkiwarra: *Mani an-kebkali kakbi boyen barri-mey.*	They found some just north of here where that rocky cliff face runs.
Emmanuel: *O baleh?*	Where?
Mary Kolkkiwarra: *Karri-re* next time *ku-bulewali, kure.*	Let's go to where it has been burnt off next time.
Bardayal Nadjamerrek: *Ngurri-wurlhke…*	Burn this area!
Mary Kolkkiwarra: *karri-badjyawan …*	so we can [more easily] see the entrance tubes …
Emmanuel: *La kure nganang, kure nganang kubule nani kure kabirriwurlhke.*	I found a hive over there, there I saw a hive in the ash area, this area where they are doing burning.
Mary Kolkkiwarra: *badjno karrinan kure kayawoyhyerrkan. Ankung, dja kure mak buleyi yika kundalkkuk kanbalhme marrek karringalke.*	We can see the entrance tube where the bees land to go inside. The honey there in the burnt area, because the green grass conceals the hives and stops us from finding them.

Fire, honey, religious rituals and Karrkkanj the brown falcon

People on the Arnhem Land Plateau were well aware of the destructive nature of hot late-dry-season fires. Such fires are particularly destructive of the native bee hives in both the ground and up in the hollows of trees. In the *Ankung Djang* 'honey dreaming' estate of Bardayal Nadjamerrek is a site, Nabiwo Kadjangdi 'Nabiwo honey sacred site', on the banks of the Liverpool River (*nabiwo* is the name of a variety of native stingless honey bee *Trigona mellipes*). At this site there is a small rock of some 30 cm in height that stands up in the ground marking the centre of this totemic site. A honey increase ritual was performed here each year to ensure plentiful honey and also to protect the honey hives from destructive hot late-dry-season fires. The ritual involved clearing the ground around the *nabiwo* stone to make an area where a group of people could sleep. Everyone in the camp joined in circling the camp fire during a part of the ritual called *bordomo.* Wearing paperbark hats, the group would repeatedly circle around the fire until it had died down. Following this ritual the objective was to make sure the fire would burn at a low intensity without large flames throughout the night. To keep the hearth burning gently, pieces of broken black termite mound called *djibdjib* or *djidbinj* were placed on the fire. The subduing of the ritual fire represents the restraint of uncontrolled hot fires that might sweep through and damage the honey resources of a particular area. In the transcript below Bardayal Nadjamerrek, Mary Kolkkiwarra and a number of their grandchildren go to Nabiwo Kadjangdi to visit the site and clear the ground around the sacred rock.

Bardayal Nadjamerrek: [to Kabbindi White, Bardayal's granddaughter] *Mahni manu … yi-djaldjokkok yiray djokkok yiray … beh yikak- … yiwalemwi wurdurd kabarri-di, beh yikodjka, … ngundi-yikarrmen*	This part here … go down and clear this side and then the other side … further towards the south where the children are standing. That's where you make the 'head' of the cleared area. Get the others to help you.
Mary Kolkkiwarra: *Kabbindi yi-birrhmen bikodj yu djungkay djarran,* you *klinimap* there *njamed … ngune-djaldjirrkkan.*	Kabbindi don't you be shy, you are the right person to do this, you are the ritual boss from the correct moiety. Clean it up there, just push all of the debris out of the way.
Mary Kolkkiwarra: You *djungkay* here *makih, ngune-re ngune-yawoyhdjirrkkan kure klinimap ngune-yime clean ngune-bolkbawon. Ma werrk …*	There's more 'correct people' [ritual bosses] here. You two [boys] clean up this area here so that we leave it clean. Come on quick.
Bardayal Nadjamerrek: *Ngune-dulkbakkemen. Beh yi-balderrehmen.*	Break off a branch [as a broom]. Just move further that way.
Mary Kolkkiwarra: *Name kun-wardde nabiwo djang nakka.*	The rock there is the 'dreaming' for *nabiwo* honey.
Bardayal Nadjamerrek: *Beh yi-balray.*	Keep extending it [the cleared area around the rock].
Mary Kolkkiwarra: *Nungan nabiwo ka-djangdi.*	This is the *nabiwo* honey dreaming place.
Mary Kolkkiwarra: *Ngale diyawarra konda koyek ngalengman. Kaluk rerre ngalekke, bad bonj karri-djalbirndulhme.*	That's the *diyawarra* species *(Trigona hockingsi)* honey dreaming just to the east there in that grove of water pandanus. We can go there later, but the splashing we have done in the water will increase that species too.
Mary Kolkkiwarra: *Bedman lorlbban, diyawarra karrimen bobbidj, all kind ka-raworren.*	All the different honey bee species all join together when we do this ritual [*lorlbban, diyawarra, bobbidj*]
Peter Cooke: *Lorlbban, diyawarra, bobbidj?*	[names of honey varieties]
Mary Kolkkiwarra: *All kind ka-djalmarnburren yiman kayime.*	The ritual makes all the honey types increase.
Peter Cooke: *Anyalk?*	Rock honey?

Mary Kolkkiwarra: *Anyalk* all kind *yiman ka-yime ka-mirnderaworren. Nabiwo* and all. *Bad bifo, bu Nabiwo barri-yoy barri-yoy barri-yoy barri-worrhmi kunak kumekke* two sides *barri-worrhmi anyaaaa:hwurd djibdjib* you know that black one black *djibdjib* that one *barri-marnbuni* little bit fire, and they sleep one night before *barridurndengi.*	Including the rock honey too, all of the different types join together in the ritual, including *Nabiwo.* But before, when they used to come and sleep here, they would make campfires on either side of this rock and put pieces of black antbed on the fires either side of the 'dreaming'. You know that black termite mound, well they used that to keep the fires small and they would sleep here for one night and then leave the next morning.
Mary Kolkkiwarra: Don't make this one no … that black *djibdjib.*	Keep the fire small by using the black antbed.
Bardayal Nadjamerrek: *An-njamed, djidbinj*	Called 'djidbinj'.
Mary Kolkkiwarra: *Djidbinj manu yi-bengkan.*	You know that black termite mound 'djidbinj'.
Mary Kolkkiwarra: *Yo wanjh barri-yawworrhmi anekke yiman ba-djalyawworrhmi* inside *kanjdji ba-rungi wanjh barri-budboddi barri-yoy* side side before morning time, *barri-durndengi.*	They made very small fires which kept going a long time, the termite mound chunks would heat up inside and they would sleep on either side of them and in the morning, they would leave.
Mary Kolkkiwarra: *Barri-djalworrhmi minj manu karri-worrhmeninj yiman kunak duninj yi-burrbun manu bolkbaydiwirrinj. Minj ka-bolkbaydi kawarri-wurlhke yiman ka-yime yi-burrbun yinan kah-rung man-kimuk karri-worrhme wanjh ba-badrungi barri-malayidolkkani.*	They used it to make camp fires not with ordinary firewood like this otherwise it would make a bright blaze [and 'burn' the bees of the dreaming place]. They didn't make big bright campfires but they lit small long burning camp fires which kept going until the morning when they got up.
Murray Garde: *Wardi bod kaban-nan o?*	Why, otherwise the dreaming bees would see or something or what?
Mary Kolkkiwarra: *Bod rungi man-kung warridj namekke barri-worrhmi.*	They were ritually 'cooking' the honey and didn't want big fires [in the coming season] to burn the honey bees and ruin the hives.

It is not only humans who use fire for their own benefit. During bushfires in northern Australia, a number of raptors can be seen following the back of the fire to take advantage of the small mammals and insects that are flushed out by the blaze. One raptor in particular, *karrkkanj* the brown falcon *(Falco berigora)*, does more than just wait for the fire to burn into large patches of dry grass. This bird will swoop down, pick up a fire brand and fly off to drop it into another patch of grass. When a fire burns into a creek line and burns out, brown falcons have also been observed collecting fire brands and dropping them on the other unburnt side of the creek in order to continue the fire (Figure 5.15)

Figure 5.15 *Karrkkanj* the brown falcon (*Falco berigora*) carrying and dropping burning embers in order to spread fire to flush out insects and small mammals. Painting by Billy Yalawanga (Kune-Dangbon language group).

This association of brown falcons and fire is celebrated in rituals associated with the hollow log ossuary ceremony known as *lorrkkon*. For a number of nights in succession, sacred songs are sung accompanied by pairs of boomerangs. This singing takes place in public in the midst of the main camp. After a week or so of this evening chanting, there comes a particular night when the men will now leave the public camp and shift to a sacred and restricted location nearby, but out of view of the women and children. As they depart the public camp in the early evening, men of particular patrimoiety subsections (named *Bulanj* and *Kodjok*)[19] line up and in imitation of the brown falcon, they hold a fire brand aloft as they celebrate this special bird in song and ritual.

Burning to protect 'jungles' or closed forests

The restriction on burning in jungle thickets has been well noted in the literature for a number of Aboriginal groups extending from the Alligator Rivers region to the west of the plateau (Russell-Smith *et al.* 1997, p. 177), through to the eastern margins of the plateau around the Cadell River (Haynes 1991, p. 68) and north-east to people from the Blyth River mouth in north-central Arnhem Land (Jones 1975, p. 25, 1980, p. 124). In addition to protecting some of the food resources of jungles (yams, flying foxes, honey and palm tree meristems), jungles on the Arnhem Land Plateau also have cultural significance as *kunredkuken* 'traditional ancestral camping places' where the deep shade was appreciated by their ancestors (Yibarbuk and Cooke 2001, p. 34). Such jungles and forests are sometimes also referred to simply as *kun-djurle* 'shade'. The following text considers the details of how to burn fire breaks around jungles, *Allosyncarpia ternata* forests and cypress pine groves in order to protect them. The terms for firebreak in Bininj Kunwok and mentioned in the text below are based on the idea of 'placing a burnt patch ahead of another fire'. The word *buleno* 'ash, burnt patch' can be incorporated into a compound verb:

-yingki-bule-marnbun *beforehand-ash-make*	>	*Karri-yingkihbulemarnbun.* *'Let's make a fire break!'*

By semantic extention, the word *bule-no* can mean black as in the term *arri-bulerri* 'we black (Aboriginal) people'. Considering the role of fire in Aboriginal culture, this term of self designation takes on an appropriate significance.

There is mention of seasons, time of day to burn, wind intensity and fire behaviour. The text is a transcript of a conversation between Bardayal Nadjamerrek and a younger Aboriginal land-management ranger, Romeo Lane, and was recorded at Kabulwarnamyo in *kunumeleng* in 2005.

Bardayal Nadjamerrek: *... dalukorro ka-birlikimukmen, ... Ka-yakmen afta dalukorro, ka-wurlhme djarran little bit, strong one, an-kimuk kun-kurra, wanjh ka-birliwayhmen djarran kaddum, ...*	... in the *dalukorro* dry season winds, the fires will get bigger ... after the end of *dalukorro* the fires go down and they burn less intensively, but in strong winds the flames lift up high ...
Romeo Lane: *Kun-kurra nawu ka-yurrbirdihme nunganhyak. Wanjh ka-birliwayhke.*	When the wind is 'galloping' along, there should be no fire. The flames will rise up.
Bardayal Nadjamerrek: *Kaddum ka-rrolkan ka-re kaddum ka-birlibarndi.*	They fly up high, the flames rise up.
Romeo Lane: *Karri-nan bu kun-kurra ka-yurrbirdihme en karri-wurlhke ka-birliwayhme karri-nan njamedno ... ka-birlidokme.*	We can see when the wind is blowing strongly and we light a fire, we can that the flames whatsit ... they advance much quicker.
Bardayal Nadjamerrek: *Weldjenno birliweldjenno ka-birliweldjenbebme.*	The flame tongues, there are flame flashes, sprites of flame.
Romeo Lane: *Yo*	Yes

Bardayal Nadjamerrek: *Birlidjenno. Ka-welengre ka-birliwayhme ka-djalkurrabu ka-kolung now, ka-rrowen laik ka-rrowen wanjh ka-djaldowen kandji. Wanjh ka-rrowen an-bolh ka-kadjung ka-re bu wanjh karri-nan. 'Ba-rroweng ba-rroweng karri-yime. Dja kamak karri-yime. Ba barriyimi dabbarrabbolk arri-yimi kandi-marneyimi 'Oh nomo ngurri-wernhwurlhke!' Wanjh barri-yimi 'Ngurri-bawo!' Arri-bawoni. Yawoyhno arrih-ni kaddum yiman konda ba-rrungdi wanjh arri-wurlhkeyi. Arri-nani ba-welengyayahmi djarran ba-rey.*	The tongues of the flame. The flames rise up and burn until they go down and die down to a small creeping fire, down low. When died down like this the fire can crawl along the [walking] track and we can watch it. 'It's died down' we would say. 'That's good', we would say. The old people [previous generation] they would say to us 'Don't make big hot fires! Don't light such fires, leave it.' So we would leave it alone. Again we would sit there [after having watched the behaviour of the fires] and in the middle of the day, we would light a fire. We would watch the fire creep along at low intensity.
Romeo Lane: *Weleng birlirayi.*	Then the flames would move along.
Bardayal Nadjamerrek: *Yoh ba-welengbirlirayi anbu an-dalknud.*	Yes the flames would move through the dried grass.
Romeo Lane: *An-dalknud*	The dried grass.
Bardayal Nadjamerrek: *Ba-ruyi,* all around *ba-ruyi. Arri-djalnahnayi ba-baldungyibmeninj kanjdji ba-birlihkolungi. Laik ba-rroweni barri-yimi 'Ba-rroweng kube kubeh ka-re ba-rroweng.'*	It would burn all around, burning everywhere. We would watch over it until the sun went low in the sky and the flames lowered. The fire was starting to die down and they would say 'It is dying down now, it will not get away, it is going out.'
Romeo Lane: *Wanjh yiman* three four o'clock *wanjh kunukka birlikoluy.*	So three or four o'clock in the in the afternoon the fire would reduce in intensity.
Bardayal Nadjamerrek: Yeah	Yeah
Romeo Lane: *Yeledj birliray.*	And the flames move slowly.
Bardayal Nadjamerrek: *Yalmo ba-rey.*	It would go slowly.
Romeo Lane: *Yiman … en birlirayi en ray wurlhmeng kun-kurra nakka birliwayhmeninj.*	So that … when the flames move off and the winds pick up again then the flames lift up again.
Bardayal Nadjamerrek: *Wanjh derrehmeng.*	Then it moves off again.
Romeo Lane: *Derrehkemeninj beh.*	It moves off.
Bardayal Nadjamerrek: *Yoh. Kumekke ba-djalyimeninj ba-yakmeninj ba-rrowemeninj an-yahwurd an-yahwurd ba-wurlhmeninj ba-kadjuyi wanjh darnki ba-rrowemeninj ka-rrowen.*	Yes. That's what would happen, the fire dies down to a number of small fronts, burning slowly and crawling along until it reduces and dies out.

Bardayal Nadjamerrek: *Kure an-binik dja binik ka-rrulkdi marrek arri-wurlhkemeninj ba-rungi ba-yawoyhrungi ba-rroweni arri-yawoyhwurlhkeyi ba-rey kureh ba-rey bad nomo an-binik kun-djurle makka ba-yakkeni.*	In *an-binik* forests we didn't light fires because repeated fires would kill the forest, so we would burn off [to the side of the forest to make a fire break] and when it died down, we'd light another until it finished the fire break but there was no fire [allowed] in the *an-binik* forest, they are there for shade.
Romeo Lane: *Birri-birlinahnayi* like side and side.	They took care to watch over the fire on all sides [of the jungle].
Bardayal Nadjamerrek: *Arri-birlinahnayi, ba-birlikoluyi arri-nayi, wanjh bu 'Maibi malayi arri-yawoyhbirli … arri-dolkkayi.*	We would watch over the fire until it died down and then [we would say] 'Tomorrow maybe we'll light it again, …when we get up [in the morning].'
Romeo Lane: *Rey wurlhkemeninj*	They would go off and burn again.
Bardayal Nadjamerrek: *Kure arri-bidmarrimeninj kure barri-bawong. Wanjh arri-nayi ka-djalmak.*	We would work together burning and in some places leave it unfired. Then we see this was effective.
Romeo Lane: *Kamak rowk wanjh beh ngurri-birlimunkewemeninj.*	When all was well [with the firebreak] then they sent the fire off to other directions [around the jungle].
Bardayal Nadjamerrek: *Ya, wanjh ba-welengkadjuyi all around.*	Yes, then the fire would burn all around [the jungle].
Romeo Lane: *Bad nawu kurrungburldjdjarn?*	What about in the middle of *kurrung* [hot late dry season]?
Bardayal Nadjamerrek: *Makka larrk! Wardi wurlhmeninj kun-djurle and all yi-bengkan ba-djalruyi nawu kurrungburldjdjarn.*	Not then! Otherwise the shady jungle could be burnt as you know how everything gets burnt in a hot late-dry-season fire.
Romeo Lane: *But early ngurri-wurlhkemeninj ngurri-marnewurlhwurlhkemeninj wanjh yingkihbuleyuwirrinj.*	So you burn early in the year, you burn for the benefit of [the jungle] so that there is a fire break established early [before hot fires can come through].
Bardayal Nadjamerrek: *Ya*	Yes
Bardayal Nadjamerrek: *Buleno ba-yingkibuleyuwirrinj.*	There would be a burnt area, a fire break already in place.
Bardayal Nadjamerrek: *Dja wanjh ba-welenghkadjuyi kurrungken anbu djaid-djaid ba-ruyi.*	So that if a fire should go through in *kurrung* [late dry season], then it would pass to the sides of the jungle.

Romeo Lane: *Wanjh beh bebbebeh birlirayi.*	The fire would burn on either side away from it.
Bardayal Nadjamerrek: *Ya beh ku-buyika ku-buyika ba-welengwurlhmeninj na ba-rayi, dja mak an-binik makka ba-djalngarrediwirrinj ba-ngarredi.*	Yes, the fire would pass somewhere else so that the *anbinik* forest would stand there unaffected.
Romeo Lane: *Yoh*	Yes
Bardayal Nadjamerrek: *Ya… Buleno ka-welengyo.*	Yeah, … there would be a firebreak there.
Romeo Lane: *Ka-yingkihbulerri.*	A fire break beforehand.
Bardayal Nadjamerrek: *Ya*	yes
Romeo Lane: *Wanjh karri-durndeng karri-wurlhke.*	So we can return at some other time and continue burning [without worrying about damaging the jungle].
Bardayal Nadjamerrek: *Minj kurrungburldjdjarn ngurri-wurlhke ka-welengkadjung ka-buleyo yuhyungki ka-yingkihbuleyo kumekke beh ka-djaldowen ka-re.*	Don't burn in the middle of *kurrung* [late dry season] but any fires that do burn then will burn out once they get to the firebreak there.
Romeo Lane: *Wanjh yerrebeh arri-bulemunkewe.*	We should have the fire breaks in place beforehand.
Bardayal Nadjamerrek: *Kure ka-welengre kure ka-re kure ka-re, kurrung nawu.*	So that in *kurrung* season any fires will go here, go there [and die out].
Romeo Lane: *mm*	mm
Bardayal Nadjamerrek: *Ya, that kamak rowk yi-wurlhke. Kune mak ka-yime marrek yi-wurlhke, djamtaem yi-wurlhke yekkeyekke kunukunumeleng… no … dalukorro bu, kunumeleng ka-rralknud.*	Yes that is the proper way to burn. But there are other times of the year from the early dry through to the early wet when you shouldn't burn. Throughout *yekke* there are strong south-easterly winds and at the end of the dry season there can be a build up of large fuel loads [if grass was not burnt].
Romeo Lane: *Yiman mak njamed man-larrh.*	Such as near cypress pines.
Bardayal Nadjamerrek: *Djurleno.*	Shade [The cypress pine groves are also important shade].
Romeo Lane: *Young one, nawu ngarri-nan ka-kaberrkdi mak kumekke ngarri-re ngarri-wurlhke side bedberre…*	We should burn to the side of the young cypress pine trees which are all growing in a grove…
Bardayal Nadjamerrek: *Mm mm*	Mm mm

Romeo Lane: *Yimarnek arri-yingkibulemarnbun kune wanjh kurrungburldjdjarn kamre wardi kunak kamre kubuyikabeh kambirlire yiman Bulmunbeh kamre kabirri-birlimunkewe Bawurrkbarnda wanjh kurebeh kan-djirrkkan.*	So we should burn fire breaks so that in the middle of the late dry season when fires come up from Bulmun which those people have lit or Bawurrkbarnda, as those fires can affect us.
Bardayal Nadjamerrek: ... *an-larrh anbu ka-rrulkrung*	... the cypress pines will be burnt [i.e. are fire sensitive]
Romeo Lane: *Yiman kure an-larrh arri-marnewurlhwurlke arri-yingkihbulebawon.*	So we should also burn fire breaks for the cypress pine groves [to protect them].
Bardayal Nadjamerrek: *Ka-buleyo.*	Fire breaks will be there.
Romeo Lane: *Kure man-binik ka-kaberrkdi.*	And at the *Allosyncarpia ternata* jungles.
Bardayal Nadjamerrek: *Ya*	Yeah
Romeo Lane: *Same way ... wanjh bu kurrungburldjdarn minj worry ngarri-yime.*	Same way ... so that in the hot late dry season, we don't have to worry.
Bardayal Nadjamerrek: *Ng ng, ka-yingkibuleyo, ngurri-djalyime bonj ka-yingkihbuleyo kuberrk ka-yime marrek an-larrh ka-rung wanjh djal woybukkih ka-buleyo ka-welengkadjung kubuyika.*	Mm mm, the fire breaks will be in place before hot fires come through and the cypress pine groves will not get burnt, that's right, and the fires will burn off in another direction.

MAINTAINING ACCESS TO THE HIGHER ROCK COUNTRY

Burning spinifex in the rock country in the late dry season, or in the strong *dalukorro* winds of the early dry, was avoided. In the following text Bardayal Nadjamerrek recommends burning rock country with its highly flammable spinifex grasses only in the late wet season of *bangkerreng*, in the cool of the late afternoon and then not every year. The discussion transcribed below was in the context of walking through spinifex on the margins of higher and denser rock country. In the most remote parts of the plateau, and the most difficult to access, fire was a rare intruder. But certainly some patches of spinifex in rock country needed to be cleared to allow access from time to time.

Murray Garde: *Wamud yi-bengkan nawu an-karrarndalk, yika nawu ka-rralknudkimukmen ...*	Wamud, you know the Triodia spinifex [in the rock country], sometimes it can build up as fuel ...
Bardayal Nadjamerrek: That mean *kaddum ka-rri*, well *nomo yi-djengwurlhke kuni ka-yime* [indicates lower down] *ka-birlikurrmerren ka-mankan ka-rrowen wanjki*, little bit *ka-kolung* now,	That means it's a tall heap, well don't burn it like that, burn it when the flames will be small, down low and it will burn itself out [instead of continuing on], gradually reducing in intensity, crawling along from

ka-balderrehderrehme that mean im finish *maitbi ka-rrungyibme darnki ka-rrowen … bangkerreng yoh.*	one place to another, that means the fire will die out as the sun sets, yes, bangkerreng season [late wet season is the time to burn it].
Murray Garde: *Bu ka-rralknudkimukmen …*	But when the spinifex fuel is big …
Bardayal Nadjamerrek: That mean *ka-birlidjabdi ka-re kondabo [gestures]* 'Oh that *arri-wurlhkeng* but im go down *ka-kolung djarrun ka-nelk, yi-bengkan kun-kurra ka-nelkbarlme* that mean *bangkerreng,* that mean *kamak kabarri-wurlhke.*	That means the flames will stand up high, when they should be down here like this [lowers hands to indicate shorter flames]. [We might say] 'Oh someone has set fire to it, but it will go down, reduce in intensity because [at that time of year] it is only new growth, you know in the last rains of bangkerreng 'knockemdown winds', the spinifex is only young [small] and that means it's good to burn it at that time.
Murray Garde: *Bale ka-yime an-djewk Wamud?*	How many years apart Wamud?
Bardayal Nadjamerrek: *Bokenh arri-djalmarnbun.*	Every couple of years.
Murray Garde: *Bu karri-wurlhke yiddok mayhngong ka-rrowen o kayakki?*	But will burning it kill animals?
Bardayal Nadjamerrek: *Ka-rrowen, bu kurrungburldjdjarn o yekke dalukorro ka-wurlhme* 'all over'. Too much wind *karri-bawo. Ka-mankan kun-kurra ka-yakmen wanjh ngurri-wurlhke ngokkongokkowi wanjh ka-kolung now.*	They will die if you burn [spinifex in the rock country] in the middle of the late dry season [kurrung] or when the south-east winds are blowing in the early dry, then it will burn 'all over'. We should not burn when there is too much wind. When the wind has dropped, when it is finished, then you can burn it late in the afternoon, around dusk, then the flames will be lower.

CONCLUSIONS

Most of the contributors of the texts in this chapter are people who spent a substantial part of their lives walking across the Arnhem Land Plateau before the present day sedentary existence of life on outstations and regional settlements. Their accounts of life on the plateau in terms of how they used fire to manage this vast region tell us something of the nuanced approach to land management. Attention was paid to the conditions of temperature, fuel load, vegetation, habitat and wind intensity in conducting burning. Earlier dry season burning reduced the intensity and extent of fire in the driest and hottest time of the year. The grassy understorey of *man-berrk* 'savanna woodland' was the main focus for burning on the plateau as this was the easiest landscape to traverse. Other areas such as the margins of jungles (as described above) and the *Triodia* spinifex along walking routes on the margins of rock country needed more care.

These anecdotal accounts of landscape burning are largely consistent with what has been recorded from the very first systematic descriptions of Aboriginal landscape burning in Arnhem Land (Jones 1975; Haynes 1985; Russell-Smith 1985b). Burning commenced in the late wet season where possible, and continued throughout the dry season until the rains of the following monsoon, with people seeking to control the intensity and extent of fires at different seasons and in different ecological habitats and landscapes. Burning reached its peak in *wurrkeng* 'the season of fire'. Fires lit after this time were of a specialist nature, such as macropod fire drives or women's turtle collecting fires.

Detailed knowledge of fire regimes for particular places and seasons remains relevant for the development of contemporary land-management policy. The involvement of young Aboriginal people in contemporary land-management work will engender the recreation of experiences that make traditional ecological knowledge relevant in new economic contexts. The main purpose of this chapter has been to give voice to those Aboriginal people whose knowledge about the ecology of the Arnhem Land Plateau is crucial for intergenerational knowledge transfer. No one pretends that there can be a return to the kind of nomadic lifestyles that people on the plateau once conducted. But there is a realisation from younger generations of landowners and managers, that the knowledge of previous generations is a starting point for the development of contemporary management policy. Otto Campion is one such person who values the experience of his elders. During a land-management meeting in 2006, as part of the annual trans-plateau walks that Warddeken Land Management Ltd organises for rock-country communities, Otto made these concluding remarks, which are a fitting conclusion to the texts that have been presented in this chapter.

Otto Campion: *...name na-kare ngarr-durndeng kamak* because *um kun-wok* like *ngarr-karrme* really strong *ngarrku wurdurd* young people *bolkki ka-warrehwarrewon* so really *ngarr-durndeng* back if *ngarr-borledke* new *time-ngarre marrek ngarr-bengkan njale yungki kan-marnbun ngarrku ngarrbin-warkwan yawurrinj, ngarrku kabirri-re kabirri-warkwan* that *kun-wok* and really like *ngad kure ngarri-djare ngarri-durndeng* back	... there are good things in the past we should draw on because like these words we are sharing today [from the knowledgeable elders] they are really our strength. Young people today cannot manage the land well without this knowledge, so really we need to go back to those things if we are entering a new era. We don't know what the future holds for our land and we don't know what the young generation will do without this guidance, so we need to rely on what we know from the past [to manage fire on our country]
Bardayal Nadjamerrek: *Kun-kare*	from the past
Otto Campion: Old ways *ngarr-durndeng ngarr-bokkarren ngarr-ngun mayh,* last one *ngurri-ngun ngudda* that's why *bobo ngarri-yime kun-red namekke ka-yo* last one *maitbi ngarr-bokkarren kunj yi-ngun bobo yi-yime kun-red yi-bobohme*	We need to follow the footsteps of each preceding generation, the old ways when we hunt, you can eat the meat of those animals and then we can say goodbye to the country that sustained us knowing that the knowledge remains, following in the footsteps of others, in the hunting of kangaroos you can say farewell [with confidence] you can say farewell to the country [knowing that the knowledge will be handed on to the next generation].

Bardayal Nadjamerrek: *Bale ka-mak o bale ka-warre?*	Are these words good or not?
Peter Biless: *Yo bu manekke ka-mak makka*	They are of course very good.

ENDNOTES

1 The term *Bininj Kunwok* means 'people's language' and the word *Bininj* has a variety of senses determined by context including: 'Aboriginal', 'human' and 'man'. In this chapter, I use the word *Bininj* to refer to the Aboriginal people of the Arnhem Land Plateau who speak *Bininj Kunwok* dialects.

2 Sweeney (1939) estimated (from interviews with neighbours) that there were 25 Dedjnjengh [Kundedjnjenghmi] and Gundjeihmi people: 50 from the Dangbon and Ngalakan groups, 80 from the upper Kumadderr and Kumarrirnbang groups and he counted 96 people at Marrkolidjban, making a total of at least 250 people living on or nearby the plateau in 1939. There were also an additional 100 Rembarrnga people associated with the camp at Buluhkaduru, but no doubt there were others living in the rock country not far from there.

3 Archaeology in this region has largely focused on the outlier sites of Kakadu National Park and the Alligator Rivers floodplains (e.g. Kamminga and Allen 1973; Schrire 1982; Jones 1985; Hiscock 1991; Brockwell *et al.* 2001). Rock art studies, however, have been more widespread across the region (Chaloupka 1993; Taçon 1993; Garde 2000).

4 The wind name *djimurru* is borrowed from Makassarese *timoro* 'east'. Nicholas Evans (*pers. comm.*) suggests that the wind name *lidjalidja* is also possibly Makassarese in origin: 'Cense 1979, p.402 gives a word *lete'lete'* (which could give *lidjalidja* if filtered through a coastal language) with the meaning 'handelsprauw' and the [Makassarese] example sentence *lete'-lete' kudongkoki antama mae* 'ik ben met een lete'-lete' hier gekomen (bv v d voor de kust liggende eilanden naar Makassar)' suggesting a possible metonymy of season/wind names with the departure of Macassan praus'.

5 Kalarriya is not confused here. The transition from the dry season south-east winds to the northerly and north-westerlies of the monsoon involves this unstable transitional wind *niyaniya* that swings around from one direction to another.

6 These yams are usually finished by *kunumeleng* and not eaten in the wet season, but some will still be available.

7 Scientific name not yet identified.

8 Available mostly in *bangkerreng* but also into early *yekke*.

9 *Yekke* is not the usual season for magpie geese, but geese can sometimes be found at this site on the lower Liverpool River earlier in the year.

10 Aboriginal people do not name complete landscape features such as rivers or water courses. Kumarrirnbang Outstation, (named after the regional Kumarrirnbang estate) is on Manggabor/Kumarrirnbang Creek, and is located at -12° 19' 46", 133° 59' 51". The actual site for Kumarrirnbang outstation is called Kuyahyay.

11 The Bininj Kunwok (and Jawoyn) term *wurrk* 'fire/fire drive' is also widespread in the languages of Arnhem Land beyond the plateau including a number of Yolngu languages from north-east Arnhem Land where the word also means 'fire/kangaroo fire drive' (White 2001, p. 350).

12 This is the definition used in the nomination form to list the 'sandstone heath' of the Arnhem Land Plateau as a threatened ecological community under the *Environment Protection and Biodiversity Conservation Act 1999* (EPBC Act).
13 *Bale ka-bolk-yime?* literally = 'where 3-place-do/say'.
14 *mimal* is also the ordinary register word for 'fire' in the neighbouring Dalabon language (Evans *et al.* 2004, p. 240).
15 This text first appeared in Bowman *et al.* (2001, p. 66).
16 Macropods are associated with either the *yirridjdja* or *duwa* moieties based on species (Garde and Telfer 2006, p. 387). The antilopine kangaroo *(Macropus antilopinus)*, spectacled hare-wallaby *(Lagorchestes conspicillatus)*, and short-eared rock wallaby *(Petrogale brachyotis)* are *yirridjdja* moiety. The euro *(Macropus robustus)*, nabarlek *(Peradorcus concinna)* and agile wallaby *(Macropus agilis)* are *duwa* moiety.
17 See Garde 2003, pp. 479–482 for accounts of such accidents.
18 See the accompanying documentary film of this workshop *Kun-wok Kun-bolkken*, 'The Language of Land' (McKenzie 2006).
19 There are two morphs of the brown falcon: one light, the other dark. The dark morph is considered *Bulanj* subsection and is called *wunwunbu*, the light morph is *kodjok* subsection and is called *kanmilanmila*.

REFERENCES

Altman JC (1984) The dietary utilisation of flora and fauna by contemporary hunter-gatherers at Momega Outstation, north-central Arnhem Land. *Australian Aboriginal Studies* **1**, 35–46.

Altman JC (1987) *Hunter-gatherers Today: An Aboriginal Economy in North Australia.* Australian Institute of Aboriginal Studies, Canberra.

Berlin B (1992) *Ethnobiological Classification: Principles of Categorization of Plants and Animals in Traditional Societies.* Princeton University Press, Princeton.

Birch B (2006) *Erre, Mengerrdji, Urningangk. Three languages from the Alligator Rivers Region of North Western Arnhem Land, Northern Territory, Australia.* Gundjeihmi Aboriginal Corporation, Jabiru, Northern Territory.

Bowman DMJS (1998) The impact of Aboriginal landscape burning on the Australian biota. *New Phytologist* **140**, 385–410.

Bowman DMJS, Garde M and Saulwick A (2001) *Kunj-ken makka manwurrk* – Fire is for kangaroos: interpreting Aboriginal accounts of landscape burning in central Arnhem Land. In *Histories of Old Ages: Essays in Honour of Rhys Jones.* (Eds A Anderson, I Lilley and S O'Connor) pp. 61–78. Pandanus Books, Canberra.

Brockwell C, Levitus R, Russell-Smith J and Forrest P (1995) Aboriginal heritage. In *Kakadu Natural and Cultural Heritage and Management.* (Eds AJ Press, DM Lea, A Webb and AJ Graham) pp. 16–65. Australian Nature Conservation Agency and North Australia Research Unit, Australian National University, Darwin.

Brockwell S, Clarke A and Levitus R (2001) Seasonal movement in the prehistoric human ecology of the Alligator Rivers region, north Australia. In *Histories of Old Ages: Essays in Honour of Rhys Jones.* (Eds A Anderson, I Lilley and S O'Connor) pp. 361–380. Pandanus Books, Canberra.

Burenhult N and Levinson SC (2008) Language and landscape, a cross-linguistic perspective. *Language Sciences* **30**, 135–150.

Capell A (1962) *Some Linguistic Types in Australia.* Oceania Linguistic Monographs, Sydney.

Cense AA (1979) *Makassaars-Nederlands Woodenboek met Nederlands-Makassaars register.* Martinus Nijhoff, Gravenhage, The Netherlands.

Chaloupka G (1981) The traditional movement of a band of Aboriginals in Kakadu. In *Kakadu National Park Education Resources.* (Ed. T Stokes) pp. 162–171. Australian National Parks Wildlife Service, Canberra.

Chaloupka G (1993) *Journey in Time: The World's Longest Continuing Art Tradition. The 50,000-year story of the Australian Aboriginal rock art of Arnhem Land.* Reed Books, Sydney.

Chaloupka G and Giuliani P (1984) '*Gundulk abel gundalg: Mayali flora*'. Unpublished report. Northern Territory Museum of Arts and Sciences, Darwin.

Cheney P and Sullivan A (1997) *Grassfires: Fuel, Weather and Fire Behaviour.* CSIRO Publishing, Melbourne.

Commonwealth of Australia (2005) *Australian Natural Resources Atlas Version 2: Vegetation and Biodiversity.* Accessed at http://www.anra.gov.au/topics/vegetation/index.html in January 2009.

Dyer R, Jacklyn P, Partridge I, Russell-Smith J and Williams RJ (Eds) (2002) *Savanna Burning: Understanding and Using Fire in Northern Australia.* Tropical Savannas Cooperative Research Centre, Darwin.

Ellen R, Parkes P and Bicker A (Eds) (2000) *Indigenous Environmental Knowledge and its Transformations.* Harwood Academic Publishers, Amsterdam.

Evans, N (1992) Macassan loanwords in top end languages. *Australian Journal of Linguistics* **12** (1), 45–91.

Evans N (Ed.) (2003a) *The Non-Pama-Nyungan Languages of Northern Australia: Comparative Studies of the Continent's Most Linguistically Complex Region.* Pacific Linguistics, Canberra.

Evans N (2003b) *Bininj Gun-wok: a pan-dialectal grammar of Mayali, Kunwinjku and Kune* (2 volumes). Pacific Linguistics, Canberra.

Evans N, Merlan F and Tukumba M (2004) *A First Dictionary of Dalabon (Ngalkbon).* Maningrida Arts and Culture, Maningrida, Northern Territory.

Florey M (2001a) Threats to indigenous knowledge: a case study from eastern Indonesia. In *On Biocultural Diversity: Linking Language, Knowledge, and the Environment.* (Ed. L Maffi) pp. 325–342. Smithsonian Institution Press, Washington DC.

Florey M (2001b) Language shift and the demise of indigenous knowledge. In *Sociolinguistic and Psycholinguistic Perspectives on Maintenance and Loss of Minority Languages.* (Eds T Ammerlaan, M Hulsen, H Strating and K Yagmur) pp. 109–125. Waxmann, Münster, Germany.

Garde M (2000) Mick Kubarkku and the rock art of the Mann River district. In *The Oxford Companion to Aboriginal Art and Culture.* (Eds S Kleinert and M Neale) pp. 118–121. Oxford University Press, Melbourne.

Garde M (2003) Social deixis in Bininj Kun-wok conversation. PhD thesis, University of Queensland, Brisbane.

Garde M and Kohen A (2004) Putting Herbert Basedow back in focus: the 1928 expedition to Arnhem Land. *Australian Aboriginal Studies* **1**, 26–36.

Garde M and Telfer W (2006) Indigenous knowledge of rock kangaroo ecology in western Arnhem Land, Australia. *Human Ecology* **34**, 379–406.

Guse D (1999) 'Archaeological survey of the proposed Cameco/Stockdale Work Program 1999, West Arnhem, NT'. Unpublished report. Northern Land Council, Darwin.

Harmon D (1996) Losing species, losing languages: connections between biological and linguistic diversity. *Southwest Journal of Linguistics* **15**, 89–108.

Harmon D (2002) *In Light of Our Differences: How Diversity in Nature and Culture Makes Us Human*. Smithsonian Institution Press, Washington DC.

Haynes CD (1985) The pattern and ecology of *munwag*: traditional Aboriginal fire regimes in north-central Arnhem Land. *Proceedings of the Ecological Society of Australia* **13**, 203–214.

Haynes CD (1991) Use and impact of fire. In *Monsoonal Australia: Landscape, Ecology and Man in the Northern Lowlands*. (Eds CD Haynes, MG Ridpath and MAJ Williams) pp. 61–71. Balkema, The Netherlands.

Hill R and Nowakowski S (2003) *Yalanji-warranga kaban. Yalanji People of the Rainforest Fire Management Book*. Little Ramsay Press, Cairns.

Hiscock P (1991) 'Exploring prehistoric cultural ecology in the Kakadu wetlands: methodological considerations'. Unpublished report to the Australian National Parks and Wildlife service. Archaeological Services, Northern Territory University, Darwin.

Hunn E (1982) The utilitarian factor in folk biological classification. *American Anthropologist* **84**, 830–847.

Jones R (1969) Fire-stick farming. *Australian Natural History* **16**, 224–228.

Jones R (1975) The Neolithic, Palaeolithic and the hunting gardeners: man and land in the antipodes. In *Quaternary Studies*. (Eds RP Suggate and MM Cresswell) pp. 21–34. The Royal Society of New Zealand, Wellington.

Jones R (1980) Hunters in the Australian coastal savanna. In *Ecology in Savanna Environments*. (Ed. D Harris) pp. 107–146. Academic Press, London.

Jones R (Ed.) (1985) *Archaeological Research in the Kakadu National Park*. Australian National Parks and Wildlife Service, Special Publication No. 13. Australian Government Publishing Service, Canberra.

Kamminga J and Allen H (1973) 'Alligator Rivers environmental fact-finding study. Report of the archaeological survey'. Australian Government Publishing Service, Canberra.

Langton M (1998) *Burning Questions: Emerging Environmental Issues for Indigenous Peoples in Northern Australia*. Northern Territory University, Darwin.

Langton M (2000) The fire at the centre of each family: Aboriginal traditional fire regimes and the challenges for reproducing ancient fire management in the protected areas of northern Australia. In *Fire! The Australian Experience: Proceedings of the 1999 Seminar*, pp. 3–32. Australian Academy of Technological Sciences and Engineering Limited, Canberra.

Lewis HT (1989) Ecological and technical knowledge of fire: Aborigines versus park rangers in northern Australia. *American Anthropologist* **91**, 940–961.

Mackay D (1929) An expedition in Arnhem Land in 1928. *The Geographical Journal* **74**, 568–571.

Maffi L (Ed.) (2001a) *Language, Knowledge, and the Environment: The Interdependence of Biological and Cultural Diversity*. Smithsonian Institution Press, Washington DC.

Maffi L (2001b) Linking language and environment: a coevolutionary perspective. In *New Directions in Anthropology and Environment*. (Ed. CL Crumley) pp. 24–48. Altamira Press, Walnut Creek, California.

Maffi L (2005) Linguistic, cultural, and biological diversity. *Annual Review of Anthropology* **34**, 599–617.

Mark DM and Turk AG (2003) Landscape categories in Yindjibarndi: ontology, environment, and language. In *Spatial Information Theory: Foundations of Geographic Information Science*. (Eds W Kuhn, M Worboys and S Timpf) pp. 28–45. Springer-Verlag, Berlin.

McKenzie K (2006) *Kun-wok Kun-bolkken 'The Language of Land'*. DVD film. Centre for Cross-Cultural Research, Australian National University, Canberra.

Muehlmann S (2007) Defending diversity: staking out a common, global interest? In *Discourses of Endangerment: Interest and Ideology in the Defense of Languages.* (Eds A Duchêne and M Heller) pp. 14–34. Continuum, London.

Press AJ (1988) Comparisons of the extent of fire in different land management systems in the Top End of the Northern Territory. *Proceedings of the Ecological Society of Australia* **15**, 167–175.

Russell-Smith J (1985a) A record of change: studies of Holocene vegetation history in the South Alligator River region, Northern Territory. *Proceedings of the Ecological Society of Australia* **13**, 191–202.

Russell-Smith J (1985b) Studies in the jungle: people, fire and monsoon forest. In *Archaeological Research in Kakadu National Park.* (Ed. R Jones) pp. 241–267. Australian National Parks and Wildlife Service, Special Publication No.13. Australian Government Publishing Service, Canberra.

Russell-Smith J (1995) Fire management. In *Kakadu Natural and Cultural Heritage and Management.* (Eds AJ Press, DM Lea, A Webb and AJ Graham) pp. 217–237. Australian Nature Conservation Agency and North Australia Research Unit, Australian National University, Darwin.

Russell-Smith J (2002) Pre-contact Aboriginal, and contemporary fire regimes of the savanna landscapes of northern Australia: patterns, changes and ecological processes. In 'Australian fire regimes: contemporary patterns (April 1998–March 2000) and changes since European settlement'. (Eds J Russell-Smith, R Craig, AM Gill, R Smith and JE Williams). Australia State of the Environment Second Technical Paper Series (Biodiversity). Department of the Environment and Heritage, Canberra, <http://www.ea.gov.au/soe/techpapers/index.html>.

Russell-Smith J, Lucas D, Gapindi M, Gunbunuka B, Kapirigi N, Namingum G, Lucas K, Giuliani P and Chaloupka G (1997) Aboriginal resource utilization and fire management practice in western Arnhem Land, monsoonal northern Australia: notes for prehistory and lessons for the future. *Human Ecology* **25**, 159–195.

Russell-Smith J, Ryan PG, Klessa D, Waight G and Harwood R (1998) Fire regimes, fire-sensitive vegetation, and fire management of the sandstone Arnhem plateau, monsoonal northern Australia. *Journal of Applied Ecology* **35**, 829–846.

Russell-Smith J, Ryan PG and Cheal D (2002) Fire regimes and the conservation of sandstone heath in monsoonal northern Australia: frequency, interval, patchiness. *Biological Conservation* **104**, 91–106.

Russell-Smith J, Yates C, Edwards A, Allan GE, Cook GD, Cooke P, Craig R, Heath B and Smith R (2003) Contemporary fire regimes of northern Australia, 1997–2001: change since Aboriginal occupancy, challenges for sustainable management. *International Journal of Wildland Fire* **12**, 283–297.

Schrire C (1982) *The Alligator Rivers: Prehistory and Archaeology in Western Arnhem Land. Terra Australis 7.* Department of Prehistory, Research School of Pacific Studies, Australian National University, Canberra.

Stepp J, Wyndham F and Zarger R (Eds) (2002) *Ethnobiology and Biocultural Diversity.* University of Georgia Press, Athens.

Sweeney G (1939) 'Report of patrol in Junction Bay, Liverpool River, and Tomkinson River areas, July–August 1939'. Unpublished report from Australian Archives (Northern Territory), Series F1, Item 1949/456.

Taçon P (1993) Regionalism in the recent rock art of western Arnhem Land, Northern Territory. *Archaeology in Oceania* **28**, 112–120.

Thomson DF (1939) The seasonal factor in human culture: illustrated from the life of a contemporary nomadic group. *Proceedings of the Prehistoric Society* **5**, 209–221.

White N (2001) In search of the traditional Australian Aboriginal diet – then and now. In *Histories of Old Ages: Essays in Honour of Rhys Jones.* (Eds A. Anderson, I Lilley and S O'Connor) pp. 343–359. Pandanus Books, Canberra.

Whitehead PJ, Bowman DMJS, Preece N, Fraser F and Cooke P (2003) Customary use of fire by indigenous peoples in northern Australia: its contemporary role in savanna management. *International Journal of Wildland Fire* **12**, 415–425.

Yibarbuk D and Cooke PM (2001) *Bininj mak balanda kunwale manwurrk-ken. Ngoonjook* **20**, 33–37.

Yibarbuk D (1998) Notes on traditional use of fire on the upper Cadell River. In *Burning Questions: Ongoing Environmental Issues for Indigenous Peoples in Northern Australia.* (Ed. M Langton) pp. 1–16. Northern Territory University, Darwin.

Yibarbuk D, Whitehead PJ, Russell-Smith J, Jackson D, Godjuwa C, Cooke P, Choquenot D and Bowman DMJS (2001) Fire ecology and Aboriginal land management in central Arnhem Land, northern Australia: a tradition of ecosystem management. *Journal of Biogeography* **28**, 325–344.

Zent S (2001) Acculturation and ethnobotanical knowledge loss among the Piaroa of Venezuela: demonstration of a quantitative method for the empirical study of TEK change. In *On Biocultural Diversity: Linking Language, Knowledge, and the Environment.* (Ed. L Maffi) pp. 190–211. Smithsonian Institution Press, Washington DC.

APPENDICES

Appendix 5.1: Plant foods of the wet seasons (kunumeleng, kudjewk and bangkerreng)

Species	**Bininj Kunwok Name + short text** Djnj = Kundedjnjenghmi, Gdj =Gundjeihmi, I = Kuninjku	**Which part of wet season?**
Aidia racemosa	Djnj = *anmarredondolem*, Gdj = *djirdu, mulunjmulunj*, I = *kabirlekan* *Anmarredondolem kukabo karri.* The *Aidia racemosa* plant grows near creeks.	*kudjewk, bangkerreng*
Amorphophallus galbra	Djnj, Gdj = *andidjdjanku*, I = *mayhdedj, walangari* (*duwa* moiety), *djedbarlh* (*yirridjdja* moiety, but also the Dalabon name). The tuber is roasted for 1 night. The plant consists of a single stalk of about 1.5 m. with a head of green seeds at the top. Just beneath the base in the ground is the large tuber. Rock wallabies such as the short-eared rock wallaby also eat the tuber. There is considerable variation in shape and size of these plants and, as a result, different sizes of the same species may have different moiety affiliations and names in Bininj Kunwok.	*kudjewk*
Amorphophallus paeoniifolius	Yam with toxins that must be removed before the tuber is edible. W = *kamarn*, Djnj = *morlkalk* *Kamarn arri-ngun an-me manu yiman an-didjdjanku rerri.* The *kamarn* Arum plant is edible and is similar to the *an-didjdjanku* Arum tuber.	All year, but especially *kudjewk, bangkerreng.*
Ampellocissus acetosa	'Wild grape': a low growing and scrambling climber with edible black currant-like fruit *Djn j= an-djakurrkkurrk*, I = *man-djalke*, Gdj = *an-nunjek*, syn. = *anwak* (also the name for a number of plants with similar clustered fruits – *Leea rubra, Melostoma* sp. and *Ampelocissus acetosa).* *Andjakurrkkurrk ngurrurdu ka-ngun, kudjewk.* Emus eat the wild grape in the wet season.	*kudjewk, bangkerreng*
Aneilema siliculosum	Small herb from rock country with edible tubers; the same name is used also for *Murdannia graminea* Djnj = *an-bidkawang* (*man-kawang, man-bidkawang*), Gdj = *an-bidkalkberd* *… ngali ngalengman mak ngarri-karuhkarung. Man-ngalinj en man-kawang, mankawang man-djudjulng. Kunekke ka-dedjmaddi man-kawang.* …we dig other small bush tubers like fi:*Ipomoea*_sp. and fi:*Aneilema_siliculosum* which (the latter) has little small tubers and roots.	*kudjewk, bangkerreng Murdannia graminea* appears in *bangkerreng* and *yekke*
Antidesma parvifolium	Djnj, Gdj = *an-yuku*, I, W = *man-yuku* *An-balinjdja man-me ngarri-nguni, barnemno ngarri-nguni djorlengno, yiman ka-yime yi-bengkan an-djurlukkurn, bad ngalengman ku-warddewaken ngarri-ngeybun an-yuku, an-djorleng* black one. *Anih an-kuken ngarri-ngun, an-yuku.* *Vitex acuminata* has an edible fruit which we eat when ripe and it's like the *Antidesma ghaesembila* fruit, but it grows in the rock country. We call the rock version of *Antidesma ghaesembila* 'an-yuku' *(Antidesma_parvifolium).* When it's ripe it's black. It's a well known traditional food, *Antidesma parvifolium.* The juice is also used in kangaroo hunting sorcery rituals.	*kudjewk, bangkerreng*

Cucumis melo	Bush cucumber: a vine with an edible fruit available in the wet season, also eaten by emus Djnj, Gdj= *an-dawk*, I, W, E= *man-dawk* *Ngale ngal-bengwarr ngalu kurdukadji, kun-mabu warridj ka-ngun. Manekke ngalengarre man-djalke manu ngad Bininj karri-ngun warridj, kan-djenbun and man-dawk warridj, yibengkan man-dawk man-yahwurd, manekke ka-ngun.* She is completely crazy the emu, she even eats charcoal. And *Ampelocissus acetosa* that currant that people also eat, but it burns your tongue and also small bush cucumbers bush cucumber, you know those little things, it eats those.	*kudjewk, bangkerreng*
Cynanchum pedinculatum	Djnj = *an-ngalemerrk*, E = *namelengerrk* Small edible tuber found in sandstone habitats	*kudjewk*
Dioscorea transversa	Long yam: an important staple throughout the year. Djnj = *kayawal*, I, E, W = *karrbarda*	All year
Marsdenia viridifloria	Djnj = *album* *Albun an-me ka-karrme karri-mang bangkerreng, yiman ka-yime Balanda bobbo ngurri-karrme.* *Marsdenia viridifloria* has a fruit that we get in the late wet. It's like that fruit you white people have, pawpaw.	*kudjewk, bangkerreng*
Melodorum rupestre	Djnj, W, I, E, *(m)an-badbirri* *Namekke nadjinem manme ka-ngun man-badbirri, yika ka-kolung ku-wardde ka-kolung man-buyhbuyika ka-ngun.* That black rock kangaroo eats *Melodorum* fruit, sometimes it comes down from the rocks and eats other kinds of food.	*kudjewk*
Persoonia falcata	All dialects = *(m)*an-dak *Man-balmed dja man-dak yi-rrong yi-marnbun yi-ngun karri-rawon.* Pound the *Planchonella arnhemica* and the *Persoonia falcata* fruit together and you can eat it.	*kurrung, kunumeleng, kudjewk*
Pouteria sericea	All dialects = *(m)an-dangnud* Black fruits eaten in wet season	*kunumeleng, kudjewk*
Smilax australis	Djnj = *an-morrodondolem*, Gdj = *an-morrng* *An-morrodondolem kabirri-we ku-wukkukah djenj kangun bamrdek warridj.* They throw *Smilax australis* berries into the water and fish eat them and so do northern snapping turtles *(Elseya dentata)*.	*kudjewk, bangkerreng*
Syzygium armstrongii	Djnj = *an-kardadjirr*, I = *man-kindjilkkindjilk* Edible white fruit *An-kardadjirr kurlahno arri-bun arri-ngolung ka-njerreyhme ka-bangwayhme wanjh arri-djuhke djenj ka-rrowen.* We pound the bark of *Syzygium armstrongii*, warm it over a fire and when it starts sizzling, the poison [ichthycide] comes out and we put it in the water and the fish die.	*kunumeleng, kudjewk, bangkerreng*
Syzygium eucalyptoides ssp. *bleeseri*	Djnj = an-djalbbirdo, larger variety is called *an-djola*, I = *bokorn* *Djalbbirdo, nakka djal kunumeleng nawu* rain time. *Syzygium eucalyptoides* ssp. *bleeseri* is an early wet season fruit, in the rainy season.	*kunumeleng*
Syzygium eucalyptoides ssp. *eucalyptoides*	Djnj = *an-bowngbo, an-bowngbowng*, I = *man-bongbong* *An-bowngbow ku-warddewaken, karri-kukngun ka-rohrok djarduk, man-kukmanjmak.* *Syzygium eucalyptoides* ssp. *eucalyptoides* is from the rock country and is similar to *S. suborbiculare* and tastes delicious.	*kunumeleng, kudjewk*
Syzygium forte	Djnj = *an-boyberre*, I = *kidjdjahdjanj*, E = *madjabuli* *Karrkad nawu karri njamed nakka nawu djalbbirdoh kanjdjikanjdji karri nakka madjabuli nawu kure kuwukku.* That one, the *Syzygium eucalyptoides* ssp. *bleeseri* is up on the higher bushland, but *Syzygium forte* grows down in the creeks, in the water.	*kudjewk*

Syzygium suborbiculare	All dialects (*man-*)*djarduk* *Djarduk nawu yiburrbun nawu djarduk dangbele nakka djal kunumeleng.* You know the red bush apple tree with the white flesh inside, well it fruits in the early wet season.	*kurrung, kunumeleng*
Typhonium spp.	There is considerable variation in shape and size of Arum plants and as a result, different sizes of the same species may have different names in Bininj Kunwok; *madjalkbadj, kubulurr, man-djanek, man-danek.* *Kuwardde kaddum minj bu wernhdjarre darnki, ngarri-bidbom kumekke ngarri-yoy man-danek ngarri-doy.* Up on the rocks not up high but close [to the ground], we climbed up and there we camped and pounded *Typhonium* tubers.	*kudjewk, bangkerreng, yekke*
Vitex acuminata	Djnj = *an-balinjdja, an-dedjbang,* E = *man-dedjbang* *Djabbo ka-ngun an-balindja. Bininj karri-ngun, djabbo nuye duninj anekke an-balindjarr.* The quoll eats the black plum. People eat it (too), but it really belongs to the quoll.	*kunumeleng, kudjewk*
Vitex glabrata	All dialects = (*m*)*an-kurndalh* *Man-kurndalh birri-barndiwirrinj kam-wakemeninj birri-yameninj. Manekke nga-yameng wanjh barnambarl.* They would stand up high in the *Vitex glabrata* tree and when [the emu] came wandering over [to eat the fallen fruit] they would spear it. I speared one from up in a tree, a practice called *barnambarl.*	*kunumeleng, kudjewk*

Appendix 5.2: Bininj Kunwok landscape terms relating to rock country and landforms

Dialect key: Djnj= Kundedjnjenghmi, Gdj= Gundjeihmi, I= Kuninjku, W= Kunwinjku

wardde *prefix =* ***kun-***
n. rock, stone
derivative kun-warddewardde 'escarpment, rocky outcrop'

warddebakabakmeng *prefix =* ***an-, ka-*** *pred. v./ v.i.*
rock rubble, scree rubble, fallen rock debris that has shattered
dialect = Djnj, other dialects = An-warddebakabakmeng an-warddeyahwuyahwurd. Rocky debris broken up into small pieces.

warddebalabala *prefix =* ***an-***
n. 1. flat rock (large scale) 2. any small piece of ledge sandstone, e.g. piece of stone suitable as a grindstone or lala 'grindstone'.

warddebalabala ka-rurrkmarnbun *prefix =* ***ka-***
phrase flat overhang that creates a cave below

warddeberrebame *prefix =* ***ka-***
v.i. scarp or ridge line that runs long and low

warddeboboyo *prefix =* ***ka-***
v.i. sandstone overhang suitable for a shelter
Dabbarrabbolk barri-yonginj kure kawarddeboboyo.
The old people/ancestors used to sleep in sandstone overhanging shelters.
synonym = ka-warddeyalalyo, ka-warddeboddi dialect = Djnj

warddeboddi *prefix =* ***ka-***
v.i., horizontal overhang of rock, which provides a shelter below. synonym = ka-warddeyalalyo, ka-warddeboboyo, dialect = Djnj

warddeburldjdjarn *prefix =* ***ku-***
variant = an-warddeburldjdjarn
n. mass of rock, a sandstone outlier, plateau country
An-warddekimuk an-warddeburldjdjarn.
Rocky plateau massif.

warddedjabdi *prefix* = **ka-**
n. outcrop of rock
Makka mak mimih kumekke ka-reddi. Manih ka-warddedjabdi. Makka wanjh mimih kumekke kah-di.
Mimih spirits also live in that place. There where those boulders are protruding. That's where mimih live.

warddedjudjulngyo *prefix* = **ka-**
v.i. rock rubble from exfoliation or scree deposit, especially finer smaller pieces of sandstone rubble

warddedjuhdjuhkendi *prefix* = **ka-**
rocks lying in a water course
Ka-warddedjuhdjuhkendi ku-barrarn ka-kawadjyirriyo.
'There are rocks in the river gorge with sand lying either side.'

warddedubbe *prefix* = **ka-**
n. impassable rock country, impenetrable rock country

warddedulumkimuk *prefix* = **ka-**
variant =

ka-warddeduluhdulum-kimuk *(reduplicated form)*
n. country with numerous round large boulders (large scale)

warddekarrekarnkudji
prefix = **ka-, an-**
n. rock formation consisting of a narrow column or columns, which may or may not be supporting large boulders on top, slender sandstone pillar or butte.

warddekudjihkudji *prefix* = **an-, man-, ka-**
n. escarpment outlier country with scattered residuals and/or boulders on the plain.
dialect = Djnj
other dialect: I,W = man-warddekudjikudji, kun-warddekudjihkudji.

warddekuyeng *prefix* = **kun-, an-**
n. tall rocky outcrop, butte
Kunwarddekuyeng yiman kayime Nimbuwa.
A tall rock formation such as Nimbuwa.

warddekorlh *prefix* = **ka-, an-**
n. expansive flat rock platform, especially in river valleys and margins
Possible forms = an-warddekorlh, man-warddekorlh, ka-warddekorlh, warddekorlhno.
synonym = ku-korlh, korlhno

warddelobme *n. prefix* = **ka-**
escarpment line.
Ka-warddelobme.
The escarpment line runs along.

warddelungurr *prefix* = **ku-**
n. 1. a large cavity in rock, either in the face of an outlier or under the ground
2. sandstone country with many caves or gorges

warddemirrkbarangbarang *prefix* = **ka-**
predicate noun
expanse of rock of uneven surface with a bright shining appearance (due to variation in reflection on both smooth and uneven surfaces), either as a vertical cliff face or horizontal surface. dialect = Djnj

warddemorrhmorrboy *prefix* = **ka-**
predicate noun
region of flat rocky expanse, often with scattered small boulders and rocky debris on a flat and open area
dialect = Djnj

warddengarlhmi *prefix* = **ba-, ø**
v.i.
rock rubble, scree rubble
Djnj, Gdj = ba-warddengarlhmi

warddengarre *prefix* = **ka-**

warddengarre an-kokbo an-wern *prefix* = **ka-**
phrase a rock country jungle dominated by rock country cycad

wardderurrk *prefix* = **kun-/ ka-, an-**
cave, enclosed space under an overhang
other forms = ka-wardderuhrurrk (reduplicated form) 'many caves'

warddewardde *prefix =* ***kun-/ka-***
n. 1. escarpment,
2. rocky outcrop, very rocky country, land covered in boulders.

warddewardde an-karrarndalkwern *prefix =* ***ka-***
phrase spinifex (Triodia) in rock country

warddewarnamyo *prefix =* ***ka-, ku-***
v.i. parallel ridge lines of sandstone outlier formations or on a smaller scale, small lengths of parallel rock formations on a flat

warddeyalalyo *prefix =* ***ka-***
v.i. horizontal overhang of rock, which provides a shelter below
synonym = ka-warddeboddi, ka-warddeboboyo.
dialect = Djnj

warddeyirriyo *prefix =* ***ka-***
line of rocks or ridgeline (a degree of straightness is implied)

warddeyongoyo *prefix =* ***ka-***
v.i. scattered small boulders

warddeyunkurr *prefix =* ***ka-***
n. variant = ka-djalwarddeyunkurr
a boulder protruding out of the water in a water course as an island.

Appendix 5.3: A glossary of Bininj Kunwok terms relating to fire

bambarr, man-bambarr *n. dead end valley or narrow gully, which are the sites chosen for fire drives to trap kangaroos, which become trapped and injured by fire and can be easily speared.*

bananame *v.i. crackling of a fire*

bilbilk *n. a vigorous fire that burns beneath the matted roots and flotsam of riverbeds, especially in paperbark forests*

birlibang *n. fierce blaze, powerful fire*

birlibarndi *v.i. fire located in a high position (e.g. up in a cliff or in a tree)*

birlibebme *v.i. appearance of flames*

birlibidbun *v.i. the climbing up of a flame or fire, increase in height of a flame, movement of a fire line upstream along a water course or in a drainage basin*

birlidadjme *v.i. the lowering in intensity of a fire/flame*

birlidjenno *n. the tongue of the flame*

birlidokme *v.i. the forward movement of a fire line*

birlidombun *v.t. extinguish a fire*

birlidowen *v.i. extinction of a fire*

birlikadjung *v.t. the movement forward of a fire ('it follows the fuel')*

birlikan *v.t. carry fire or light, take a torch*

birlikimukmen *v.i. the increasing in intensity of a fire/flame*

birlikolung *v.i. the movement of a fire line to a lower elevation or down stream along a water course or drainage basin*

birlilobme *v.i. the quick movement or spreading of a fire through the fuel*

birlimang *v.t. 1.get fire, or a torch. 2. a public section of the lorrkkon ceremony.*

birlimunkewe *v.t. to send fire (with a person taking a brand or a torch)*

birlinahnan *v.t. to supervise a fire, conduct a controlled burn*

birlino, kun-birli *n. birli 'flame' + no '3rd person possessed suffix', flame, fire* *Dalukorro ka-birlikimukmen, bangkerreng wanjh ka-birliyahwurd kun-kurra, ya ka-kolung, ka-rrowen. In the windy part of the dry season, the flames are whipped up, but in the last of the rainy season, the very early dry time, the fires are small because yeah, the wind dies down.*

birlire *v.i. movement of fire through fuel*

birliwarrewon *v.t. interfere undesirably with a flame, to ruin a fire, burn at the wrong time or in a place where one does not have the authority to burn (also figurative 'to be in wrong way marriage')*

birliwayhme *v.i. the rising up of a fire or flame (as in the front of a grass fire)*

birliwe *v.t. light bushfires*

birliweldjenbebme *v.i. the leaping of fire sprites, the flashing of the very tip of a flame*

birliweldjenno *n. (fire) sprite*

birliwerrhme *v.t. the clearing of the bush by a fire, the reduction of vegetation cover through fire*

birliwurlhme *v.i. the burning of a flame, the action of fire taking hold (of the fuel)*

birliyahwurd *n. small fire, small flames*

birliyahwurdmen *v.i. reduction in intensity of a fire*

bule, *prefix* **kun-** *or* **man-,** *also* **buleno** *n. ashes, black area after a fire*

bulewe *v.t. create a burnt area of ground (and by inference, make a fire break)*

bulewurlhke *v.t. create ash by burning grass*

dahbo *prefix =* **man-** *n. red coals of a fire*

dirndelk *n. the new growth of vegetation (after a fire), synonym =* **rakelno**

dume, (m)an- *1. back of a fire (literally 'lower back' but with vegetable noun class prefix) 2. fire drive*

dumekan *v.i. follow behind a moving fire line when using fire to hunt or burn off*

kinje *v.t. burn or cook something*

mardadangkarrk *n. new shoots of grass which appear after fire, new green pick induced by fire*

rak *prefix =* **kun-** *1. generic fire 2. camp-fire 3. fire wood*

rakelno *n. new growth on a branch or stem (after fire), epicormic growth*

-rung *v.i. to burn, be hot, be cooking, give off heat*

-wurlhme *v.i. burn, go up in flames, be hot*

-wurlhke *v.t. set fire to something*

wurrk *prefix =* **man-** *1. bushfire 2. fire drive*

yalhme *v.i. the spreading forth of a bushfire and its attendant noise as fuel is consumed*

-yingkibulemarnbun *v.t. make a fire break literally: 'beforehand-ash-make'.*

-yayahme, ka-birliyayahme *v.i. spreading of fire as it crawls through the bush, movement of a small controlled fire. Also ka-birliyayahme.*

6

Manwurrk (fire drive) at Namilewohwo:

a land-management, hunting and ceremonial event in western Arnhem Land

Jon Altman

SUMMARY

This volume looks at managed fire from ecological, cultural and economic perspectives in northern Australian savannas. I explore this topic from the micro-perspective of one event: a fire drive or *manwurrk* conducted on the edge of the Arnhem Land Plateau at a place with the 'big' or regional name Namilewohwo. This event occurred on 21 September 1980 and, as will be demonstrated, had significant ecological, economic and cultural dimensions that are closely intertwined. This was not the only fire drive that I participated in as a doctoral student undertaking highly participatory fieldwork with the Kuninjku people located in the Mann-Liverpool Rivers region in 1979 and 1980. During the dry seasons, from about June through to October, fire drives were regular events for Kuninjku – often associated with large gatherings at regional ceremonies. But for me this fire drive was exceptional: it was spiritually and emotionally charged and embodied many elements of what I now recognise as highly contested and transforming Kuninjku social norms. More recently, I have learnt that this sentiment is shared by living Kuninjku who were present who somewhat nostalgically regard this as the last great fire drive associated with ceremony on the Arnhem Land escarpment.[1] This ethno-historical account focuses on just one day, which I describe in some detail; I bookend this account initially with some context and then conclude with some contemporary reflections on its significance.

INTRODUCTION

In 1979 and 1980, I lived with a group of Kuninjku people who resided at a cluster of outstations in the Mann-Liverpool Rivers region (see Figure 6.1). At that time very few Kuninjku resided at Maningrida township. I was undertaking research on the structure and functioning of the contemporary Kuninjku economy. In this research my personal residence focused on Mumeka outstation on the eastern bank of the Mann River, but I also accompanied the people

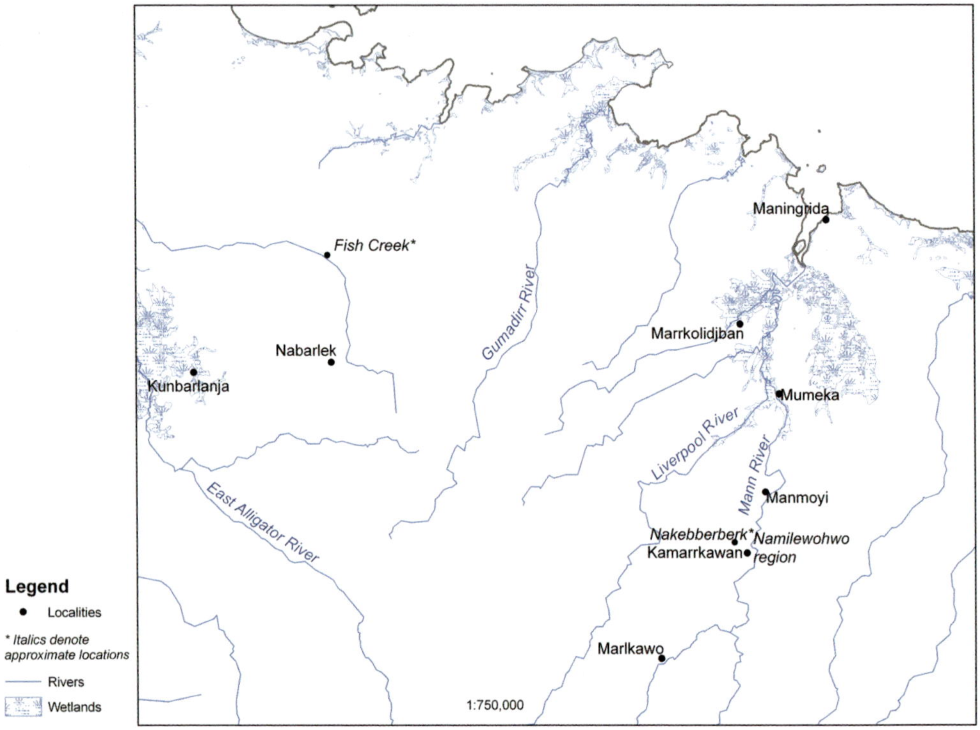

Figure 6.1 Regional map showing outstations.

living at that outstation when they moved elsewhere to seasonal camps or for ceremonies. If I had not, there were times when I would have been the only resident at the outstation locality.

My research methods involved gathering quantitative and qualitative information. Quantitative information was collected on productive activity in three sectors of the local economy: what I termed then subsistence or non-market, arts and crafts and social security (Altman 1987); and what I term now the customary, market and state sectors of the hybrid economy (Altman 2009). Quantitative data were collected in a number of ways:

- taking a daily population census
- recording work effort using time allocation techniques
- recording all hunting, fishing and gathering returns, arts and crafts produced
- recording all earned and unearned (welfare) income received by co-residents.

I observed distribution of resources and also asked people about the social norms, the rules and regulations that governed social relations of production, consumption, distribution and exchange in the cash and non-cash domains.

Much of the story that I recount here was recorded in my field notes. On 25 September 1980, I wrote an eight-page entry in my field diary (No. 6, 25 September–30 October 1980) titled 'Munworrk at Namillehwo-oh' [sic] indicating that even then I regarded this event as significant. I also took photographs, although surprisingly few of this fire drive, primarily I suspect because mixing participation in the hunt and being a photographer was difficult, if not impossible.[2] I also, of course, have my memory of this event: the sights, sounds and smells that are rekindled, especially when looking at photographs of the cooking camp. I am surprised today at some of the observations that were not recorded in my field diary, so writing this short description provides an opportunity to further document this significant event.

As already noted, this was not my first fire drive; like other participants, I had some expectations of what might eventuate in terms of a stampede of macropods (or other species) rushing to escape the fire. Each fire drive that I participated in was different, depending on seasonality and the nature of the physical environment being burnt.

A month earlier, I had participated in a fire drive in a similar 'hills and basins, with rocky outcrops' environment at a place called Ngakebbberrkberrk on the western side of the Liverpool River. A senior Bunungku man from Buluhkaduru outstation, Billy Mirritjawi, had spent time explaining some basics of fire driving to me: issues such as seasonality, wind direction and species targeting. This particular hunt – which involved seven hunters (six armed with guns and old Mirritjawi with four shovel-nose [metal tipped] spears and a spear thrower) – was planned, but far less successful (in terms of hunted game) than the fire drive at Namilewohwo. Only three euros (*Macropus robustus*) were shot and two exploited. This was primarily because we failed to fully encircle the rising ground that we were burning from the bottom up using the south-east Walirr wind that was characteristic of this mid dry season (Kurrung). Mirritjawi explained that to shoot antilopine kangaroos (*Macropus antilopinus*) the encirclement needs to be completed very quickly, while delayed encirclement is more likely to result in exploitation of euros.

MANWURRK AT NAMILEWOHWO

In my fieldwork experience, the fire drive at Namilewohwo was an unusual hunting event because it involved considerable planning and complex cooperation. The event coincided with a large gathering of people at Manmoyi outstation to begin the final month of a major regional ritual cult ceremony called *Kunabibi* (Altman 1987, p. 209–211). There were between 120 and 150 people at Manmoyi, but at that time the outstation's usual population was about 45. This event was called a *Djadjno*, or 'opening' of the ceremony's finale. People mainly came from Mumeka and Marrkolidjban outstations, but also from far away Kunbarlanja where many Manmoyi people had family connections.

There was a strong link between the regional ceremony at Manmoyi, which was 'owned' in this localised context by Peter Mamirriyangu, and organisation of the fire drive. Peter was a Bordoh (patri-clan) Duwa man, who was 'boss' for Manmoyi – located on his clan land – and manager (or *djungkay*) for the rich Yirridjdja hunting grounds that had been intentionally left unburnt for this occasion. Leaving an area unburnt so late in the dry season was itself a masterly feat of land management that had been carefully orchestrated by Manmoyi people who travelled to Namilewohwo on several occasions in previous months to burn patchwork around the elevated ground to be used on this occasion. I was not present for these previous burns, but was informed that they had involved close collaboration between Johnny Didja (Djordi patri-clan) and Manakkali (Djorlam patri-clan) who claimed shared company ownership of Namilewohwo.

The decision to undertake the fire drive on 21 September was influenced by a number of factors and was negotiated by senior men in the context of the *Kunabibi* ceremony. The main motivator seemed to be the need for meat to feed the large gathering: some of the visitors (with whom I was camping) complained that they were 'starving for beef' (or meat). People were about to participate in high-intensity ceremonial activity for much of the night, having been engaged in leisurely, but competitive, card playing (sometimes referred to as gambling) for much of the day. The availability of a number of vehicles and a large group of men made this particular day opportune.

Figure 6.2 Hunters gather at the Marrkolidjban truck, Manmoyi.

There were a number of other reasons for the hunt that only became apparent to me after the event. If successful, the hunt would demonstrate the fecundity of country and the links between giving to the Ancestors in proper ceremony and the country yielding, with Ancestors giving game back. It would also reinforce the authority of elder males – so evident in the ceremonial context – by extending it into profane, every day events. And feeding the large gathering would conversely assist in completing the *Djadjno* ceremony properly because people would be fortified by good sustenance.

Three truck loads of men set out for Namilewohwo. I did not count everyone, but there were 16 armed hunters and probably about 26 people all up, including four initiates from the 'inside business' camp who had to be sneaked out via a back track because they could not be seen by women and children.[3]

The vehicles were a Toyota Landcruiser from Marlkawo, an old Landrover tray back belonging to Tommy Lawikama and the relatively new Marrkolidjban Toyota trayback with roof rack (see Figure 6.2) that had been recently acquired as part of the Nabarlek uranium mine agreement. I was pleased not to even be asked to use my vehicle and to be a passenger in the comfortable front seat of the Marrkolidjban truck.

The drive to the hunting ground along a very rough formed vehicle track was quite spectacular: through rock country with rock art visible from the road at Enamaraway and at Karlba Creek. According to my companions, Sid Kyle-Little – who had been a Native Affairs Branch patrol officer – had marked a tree near here when on a foot patrol in the area in 1949 (see Kyle-Little 1957). We drove through Marebu where people used to camp in rock-shelters.

We arrived at our destination about an hour after leaving Manmoyi, although we travelled only 30 road kilometres.[4] The Marrkolidjban truck was parked at Kunkanin Djuhmeng [place where 'digging stick entered the water'], on the Mann River. The river here has large rocky pools, sometimes 80–100 metres wide, and nearby there is a waterfall (see Figure 6.3), Kamarrkawarn and a *djang* [sacred] hole-in-the-rock called Kanamar-Welbakminj (black cockatoo,

Figure 6.3 Waterfall at Namilewohwo.

wing-broken). This is spectacularly beautiful country. The area to be hunted included two hills of elevation up to about 100 metres above sea level and a valley between them that is easily located using Google Earth today (see Figure 6.4).

We were here to hunt and that is what we did. From what I recall, we did not hang about. We crossed the river at a shallow and narrow crossing downstream from Kankanin Djutmeng and divided into three distinct groups that roughly reflected a combination of our co-residing camping arrangements back at Manmoyi, kinship links, ritual alliances and vehicle groupings. I termed these Groups I, II and III in my field diary and listed the names of the hunters, putting in brackets the weapon that they used for interest and to give a sense of the mode of hunting:

- Group I: Malakkali (.22 rifle), Lofty Bardayal (metal-headed spear and spear thrower/*barrawu* and *borndok*), Peter Binyanbi (shotgun), Peter Nambarradj (metal-headed spear and spear thrower), Morrison Nabordoh (shotgun) and a man of the Narridj sub-section kin term from Oenpelli (shotgun) whose name I did not record. Mamirriyangu was also attached to this group.
- Group II: John Dalngadalnga (.22), James Iyuna (.22) Bobby Wanurr (.22 magnum), Benni Bandawungu (double-barrelled over and under shotgun), Clark Namunjdja (shotgun), Ivan Namirrkki (home-made/bush-fashioned shotgun), Tommy Lawikama (metal headed spear and spear thrower), John Mawurndjul (.44 magnum) and myself (shotgun).
- Group III: Jimmy Kalarriya (shotgun) and four unarmed novices from the 'inside' Kunabibi camp.

I partly list all the weapons because I had never seen so many at one hunt before (in my field diary, on page. 3, I actually had a little table of weapons), but also because in those days people rarely had high-powered rifles. Weapons were usually 12-gauge shotguns, or rifles that fired rim-fired bullets. The spears and spear throwers were items that appeared to me to have

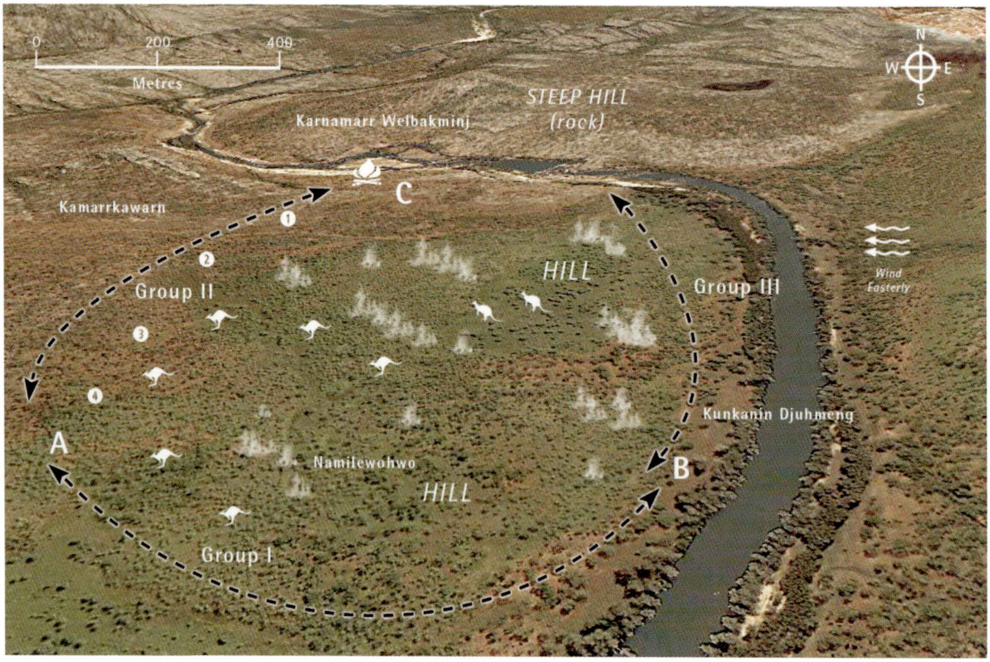

Figure 6.4 Sketch map of hunt from field diary transposed onto Google Earth.

largely symbolic, rather than practical, value and were often associated with ceremonies – with spear throwers in particular often held by authority figures in these contexts.

I try to interpret how the fire drive developed over a period of about 2 hours with the benefit of a sketch map in my field diary that has been transposed onto an image from Google Earth (see Figure 6.4).[5] My perspective was largely as a member of Group II, so I cannot recount exactly what Group I and III did. Point A on the map seemed to have been a pivotal or strategic point: it was from here that at the outset Peter Mamirriyangu stationed the teams, possibly giving Group II (the visitors and ceremonial helpers) the prime position spread along a ridge towards which he would drive the game from the east. Group I was stationed along the southern part of the hill, while Group III was stationed in the grassy valley between the hill on which we were hunting (that on my sketch map was called Namilewohwo) and another smaller area of high ground.

The hunt was skilfully managed by Mamirriyangu: it was unusual to see one man have such authority and everyone moved into position in a relaxed manner at the right time. Hunters did not cover their bodies with *delek* (white pigment) as in previous fire drives I had observed, probably because we were all stationed up wind from the location of the hunted game, something that Mamirriyangu clearly knew.[6]

Mamirriyangu set off to a point nearly diametrically opposite from point A to point B and started making fires using matches, backtracking to point A and joining up with Group I. Our team was spread out along the ridge between point C – our day camp where we had left spare belongings – and point A. Each hunter was accompanied by a non-hunting apprentice – younger men who were not yet using guns – and we waited. Not long after the fire was lit to the east – maybe 20 minutes – kangaroos started bounding in our direction and the hunt began. I suspect that kangaroos also moved in a more southerly direction where the smaller Group I was spread out, while Group III hunted in the area between the two hills where some kangaroos appeared to have taken shelter on some unburnt patches of grassland.[7] After the initial

Figure 6.5 The author struggles to lift up a male kangaroo to show its size.

stampede of game, some hunters who had not been successful went off hunting in a more mobile way.

My time at the hunt was limited to about 30 minutes. After shooting two kangaroos – a male and a female that by good luck had ran towards me – I headed back to the cooking camp. I was the first back at the camp mainly because I thought two kangaroos was enough meat for me to 'contribute'. I was also getting my feet burnt because Jimmy Njiminjuma had borrowed my thongs and I kept stepping on hot embers. I carried the lighter female kangaroo and watched with great admiration as Don Nakadilinj, a man of short stature, struggled with the help of apprentice hunter Murruwardda to get the male kangaroo (that probably weighed as much as Nakadilinj) back to the cooking camp (see Figure 6.5). This journey was approximately one kilometre, thankfully downhill. In terms of work intensity the hunting was easy, but the effort to get whole kangaroo carcasses back to the cooking camp was energy sapping to the point of exhaustion.

I watched the rest of the action from the cool comfort of the cooking camp located on the sandy bank of the Mann River, with some shade provided by paperbark trees. People arrived back at the camp in dribs and drabs over a period of about 1 hour carrying hunted game. Most *karndayh* (female kangaroos) were carried back to the cooking camp whole, but *karndakidj* (larger, male kangaroos) were usually cleaned (gutted) first (or cut up) to reduce the weight by about 20 per cent. Some hunters arrived back empty handed, despite being successful. John Dalngadalnga, for example, shot a large male, but did not carry it to the camp because he said it was too heavy. At the time I could not understand such 'waste', but it needs perhaps to be seen in the context of the overall success of the fire drive and the particular hunter's well-established credentials.

In a table in my field diary (on page 4), I recorded returns for each successful hunter, the species hunted and hunting group affiliation as shown in Table 6.1:

Table 6.1: A record of the hunt from my field diary.

	Male antilopine karndakidj	Female antilopine karndayh	Other*	Total	Group
Kalarriya	1	4		5	III
Bandawungu		2	*Djamunbuk Merlbbe*	4	II
Altman	1	1		2	II
Lawikama	1			1	II
Dalngadalnga	1			1	II
Mawurndjul		1		1	II
Iyuna	1			1	II
Murpunkula		1		1	II
Namunjdja		1		1	II
Malakali	1			1	I
Nambarratj		1		1	I
Wanurr		1		1	I
Narritj		1		1	I
Binyanbi	1			1	I
Total	7	13	2	22	

* *Djamunbuk* is a young male antilopine kangaroo, *merlbbe* is a female agile wallaby (*Macropus agilis*).

I recorded that 19.5 of 22 macropods shot were retrieved and used, or about 90 per cent. Of the unexploited game, one was a male kangaroo shot by Jimmy Kalarriya who had been afflicted by leprosy and so was only dextrous enough to carry a small portion of the animal as well as his gun.

When people brought their kangaroos back they immediately tied them up (front and rear feet as was customary for cooking), singed the fur, cleaned the guts out and cooked and ate the offal. This was a means of marking ownership of hunted game. Even with a large number of kangaroos, there was some debate over distribution of offal – with a lot of claiming going on: some serious, some jocular, some acknowledged and some ignored. The most highly prized part was the abdominal fat, which was cooked lightly in ashes, then soaked in blood and eaten. This is really sweet meat. Kidneys with kidney fat were also popular, as was the heart and liver. Kidneys soaked in blood were placed in undigested stomach contents (to clean and cool for eating) to provide additional flavour.

The kangaroo carcasses were cooked in large underground ovens (see Figure 6.6). As with all macropods, the hind legs were broken at the joint and tied to the front legs with the bark from *mandjoh* (*Acacia difficilis*) trees (see Figure 6.7). With the males, a strip of bark was attached to the testicles, which were placed in stomach blood to soak. A large fire was made and stones from the river shoreline were collected and placed in the fire to heat. When the fire had burnt right down, huge pits were made in the sand – one for the three male kangaroos (cooked in our Team II camp), another for the five female kangaroos. Hot stones were jammed with sticks into the bodies of kangaroos and some hot stones were placed in behind their anuses and then the tails were tied to the front legs. The kangaroos were then placed in the pits and covered with hot sand. There was enough fire to cook everything properly.

Figure 6.6 Preparing the underground oven.

Figure 6.7 Kangaroos ready for cooking.

ANALYSIS OF THE FIRE DRIVE

The fire drive can be analysed from a number of perspectives. Here I focus on four: an economic event; a cultural event; an ecological event; and a demonstration of Kuninjku social norms.

An economic event

This was an extremely successful economic event: in 1 day a group of men went out and secured enough game to feed a population of about 150 people. All told, 22 kangaroos were shot or speared and, of these, 19 and the tail section of another – carried in by Kalarriya – were fully exploited. At the time, I estimated that this totalled about 600 kg of meat that fed everyone for 4–5 days.

Even by the market-replacement prices for fresh meat, this harvest had a value of about $4200 in 1980 dollars: this was a high rate of return from the work effort of 16 hunters and about 10 assistants. There were, of course, costs involved: fuel for the vehicles, ammunition, and so on. But, on the other hand, a very remote community was being supplied with meat and the need to travel to Maningrida for supplies was averted. This aspect of the hunt seemed the most obvious.

A cultural event

The fire drive was a clear demonstration of the ecological knowledge held by elder males and there was a strong sense that this knowledge was being passed on: both to younger hunters and also to apprentices who participated, but did not hunt – such as Arthur Murruwardda and David Murruwardda from Mumeka who were young men in the business camp.

It was Jimmy Kalarriya who most clearly demonstrated an instructional element to the event. He was in charge of the ceremonial initiates who headed off with him and returned last. He was the most successful hunter – shooting one male and four female kangaroos – thus showing the initiates that his ceremonial authority was matched by his hunting prowess. I vividly recall Jimmy and the four youngsters returning in a single file, with Jimmy leading. He was carrying his shotgun and the tail and rump section of a male kangaroo, while each of the initiates was struggling: weighed down carrying a female kangaroo carcass.

This is the first kangaroo hunt I had been on where *barrawu* (spear) and *borndok*)spear thrower) actually increased production possibilities. Tommy Lawigama and Peter Nambarradj each scored with a spear – 13.3 per cent of the exploited meat was in fact 'produced' by spear. It is difficult to speculate how much would have been harvested without guns in pre-colonial times. The only evidence that I am aware of – from Fish Creek in 1948 (McCarthy and McArthur 1960) – suggests that under such close quarter conditions this technology would have been very effective.[8]

An ecological event

The total area that was burnt that day is estimated at 9 km^2, but what struck me as remarkable at the time was that that this area had remained unburnt until late September. My records indicated that there had been no rainfall since April 1980 – a period of about 5 months (Altman 1982, p. 37). When we had driven down south from Manmoyi, there had clearly been landscape clearing on the eastern side of the Mann River, but by cleverly using the river and burning in a patchwork manner back from its eastern bank, land managers had ensured that the hunting grounds had remained unburnt.

I was interested to know to what extent we had exploited the game in this area. Many of the females obviously had joeys that I did not count, but was aware of (see Figure 6.8). Two of my

Figure 6.8 Arthur Murruwardda with a *yaw* (joey).

informants, Jimmy Njiminjuma and Peter Binyanbi, estimated that about half of the antilopine kangaroos present had been shot. Some they suggested would have jumped through the fire: some past the hunters. I saw one escape through our line of hunters between Benny Bandawungu and John Mawurndjul – we had to be careful in what direction one shot at such close quarters with hazy smoke everywhere. Some kangaroos would have hidden on rocky outcrops or unburnt patches.[9]

I went back to Namilewohwo on another hunting expedition a few weeks later: on 23 October 1980 with 16 men, just as the *Kunabibi* was about to finish (Field Notebook No. 9, 4 October–14 November 1980). The motivation was similar: to get meat and blood. On this occasion we did not use fire, but my notes indicate that we did successfully hunt three euros and one young male antilopine kangaroo. This demonstrated to me that our earlier fire drive was very focused (antilopine kangaroos from the valley, not euros from the higher ground). I was also interested to see how the environment had responded positively to the late dry season fire, with green grass shooting despite the absence of rain. In my notebook I also recorded that we saw feral cattle and water buffalo but did not hunt them.

A demonstration of Kuninjku social norms

The fire drive combined many elements of Kuninjku social norms and it was clearly appropriate that these could be demonstrated outside the ceremonial context. The fire drive:

- demonstrated the authority of older men
- created a high degree of camaraderie among the hunters (who were also collaborating in ceremonial performances)
- demonstrated a form of hospitality and provisioning of guests
- broke up a gambling ring that was irritating the ceremonial leaders[10]
- provided opportunity for sharing practices and consumption restrictions (taboos) to be demonstrated.

There was clearly a competitive edge to the hunt and status was accorded to the successful hunters. Because there was widespread hunting success, those who had failed to shoot a kangaroo were clearly disappointed and sought to demonstrate their marksmanship by shooting at fresh water crocodiles swimming in the upper Mann River. On the other hand – as generally observed among Kuninjku at that time – the share that successful hunters received was unrelated to what they themselves had shot, except that hunters received a guaranteed standard share: the head, an arm and half the ribs (see Altman 1987, p. 135). As already noted, some established hunters were not concerned about exerting themselves because, as senior men, they realised they would get a reasonable share in any case.

These sharing rules had some interesting aspects. For example, when kangaroos were taken out of the cooking pits, people could eat anything: it was a free for all and a good time to get a share of others' portions. Also there were limits to sharing that were influenced by some game that was *nadjamun* or restricted. In this particular hunt, the kangaroos shot by Bandawungu and Murrbbunkula were restricted so that people had to keep a close watch on which part came from which kangaroo. The restrictions also allowed young hunters to be extremely generous in accordance with kinship obligations. Bandawungu gave his wallaby to Mamirriya-ngu and a kangaroo to his uncle Nambarradj. Even with an over-abundance of meat, people were fairly competitive about which piece they took and for which kinfolk back at Manmoyi.

CONTEMPORARY REFLECTIONS, WITH THE BENEFIT OF HINDSIGHT

I am not a fire ecologist, and fire just seemed to be such an accepted aspect of everyday life that I probably gave it too little attention. I do remember that we often walked among fire and drove over burning country. Mumeka people liked my truck because it had been sniffer-proofed by the Maningrida Progress Association mechanic David Bond. This meant that the fuel lines did not run anywhere outside the cab and so could not catch fire.[11] With the benefit of hindsight, and with a more recent interest in fire ecology, I offer the following contemporary reflections and generalisations about the particularities of *Manwurrk* at Namilewohwo.

Fire in retrospect

Fire was used continuously by Kuninjku from the time that vegetation was combustible in the late wet (*bangkerreng*) and early dry (*yekke*) seasons. In 1979 and 1980, hunting was done mainly on foot so it was impossible to hunt game unless grass was burnt because visibility was so poor when spear grass was 2 metres high and the crunch of walking through the native grasses would alert game to the hunter's presence.

Some people argue that land-management benefits from fire are epiphenomenonal: that the ecological benefits are unintended spin-offs. Others, from Rhys Jones (1968) onwards, referred to fire-stick farming. Revisiting this particular fire drive has placed me firmly in the latter camp. In 1980, my field diary entry was pretty glib 'The *munworrk* [sic] is an example of good land management and ceremonial management'. With hindsight, and with access to Google Earth, I think this statement was understated. This fire drive was quite remarkable: it was planned, it was neatly executed and we targeted a particular species with outstanding success. Clearly the fire-drive organisers knew exactly what they were doing and it proceeded as they anticipated. I can see now that the area burnt was primarily a valley of native grass that was densely populated by antilopine kangaroos: this was a form of 'kangaroo farming' that I had not previously recognised.

Kangaroo populations in retrospect

I did not get a good sense at the time of how often such major fire drives were undertaken, nor how long populations would have needed to recover from such a large scale drive: I suspect that even under pre-colonial conditions such a hunt would not have occurred at the same place annually. As already noted, it was estimated by informants that we had shot and speared about 50 per cent of the kangaroo population at this site, but just a month later we returned and participated in another successful hunt (for a different species) at the same place.[12]

Peter Cooke (*pers. comm.* 21 May 2008) recalls a conversation about past fire drives that he had with Jimmy Kalarriya in 1992 sitting at the very crossing on the Mann River near where we had our cooking camp in 1980. Kalarriya told Cooke that kangaroos used to be in this area 'like nanny goats' – suggesting that they were very plentiful. 'Nothing now', he suggested, because all the buffalo have eaten all the soft (native) grasses that kangaroo loved'. This is an issue that I will return to, but Cooke is of the view that some of the native grasses may have become locally extinct.

Performance in retrospect

I have participated in many hunting expeditions when the promise of abundant targeted game did not eventuate. But, in this case, the fire-driver organisers were confident of success: before the fire drive the event was talked up, as was knowledge that an area had been saved from fire especially for this event. Expectations were high and reputations were at stake, especially as organising an expedition of this magnitude was extremely rare. It is hypothetical, but I suspect that if the hunt had been unsuccessful, supernatural forces in the form of angry Ancestors would have been blamed – as occurred at an unsuccessful barramundi trapping event that I had witnessed earlier that year.

Part of the performance was the participation of hunters using pre-colonial technology: spears and spear throwers. This was technology that I had regarded as largely symbolic in 1980 because it had been superseded by firearms. Not only did two hunters demonstrate that this technology was still effective, but they clearly impressed younger men with their 'traditional' hunting prowess. This was a part of the theatre associated with the heightened focus on *Kunabibi* ritual: being back in an area that was not permanently occupied, and connecting with Ancestors and an earlier way of living in this now remote part of western Arnhem Land.

Cultural discontinuity

I was a little surprised to hear that Kuninjku themselves now regard this as the last major fire drive on the Arnhem Land escarpment and this has led me to consider what factors might have resulted in the lack of continuity in this practice. The following come to mind.

In 1980, there was no outstation at Kamarrkawarn, so that there was a limited human presence in the immediate vicinity of Namilewohwo. From about 1984, people lived at the reoccupied outstation and in June 1985 an airstrip was completed – I was coincidentally there on 24 June 1985 when the first plane landed. At an ecological level, as noted earlier, there is a distinct possibility that a combination of changed fire regimes and incursion by feral buffalo have significantly impacted on the local occurrence of native grasses on which macropods feed (P. Cooke, *pers. comm.* 21 May 2008). At a technical level, it is unclear if people have the same access to guns that they had in 1980 owing to changes in gun laws following the Port Arthur massacre that made access to firearms so much more difficult for Aboriginal people living in remote Australia (Altman 2003). And at a cultural level, and perhaps more controversially, it seems to me that the organisation of Kuninjku public culture has changed: it is unclear if the motivations, the authority, the sense of theatre associated with grand ceremony, and levels of cooperation all remain strong enough to stage complex major events such as the fire drive at Namilewohwo.

Personal reflections

One of the important factors in participant observation that was rarely discussed in anthropology when I did this fieldwork in 1979 and 1980 was the subjective dimensions of an anthropologist's experience (Marcus and Fischer 1986). My participant observation did have a scholarly purpose, which was to get a sense of work intensity, of how hard one has to work, and also what skill levels are required to be productive. But there are many other dimensions to this experience that might be recalled. So here are a few comments about what was missing in both my ethnography and my field diary.

Heading off to the fire drive in such a large group was very exciting: there was great expectation of hunting success. The hunt itself was over quickly. Once the fire began, the kangaroos raced towards us and, a little like being an English infantryman at Waterloo I suspect, part of the challenge was to hold one's fire until the kangaroos were close. There was luck as well as skill in gaining individual success: depending on where one was positioned in the encircling line, although the usual lauding of the successful did not really reflect this. I cannot overstate how involved I became in the competitive aspects of the experience, although once I had shot two kangaroos at a range of about 30 metres I wanted to get back to the cooking camp quickly.

This was partly because it was hellishly hot in late September, with the temperature increased further by the fire and smoke. It was also because, as already noted, my friend Njiminjuma had borrowed my thongs and I kept on stepping on hot embers. There was no way I could have carried the male kangaroo I shot the 1 kilometre back to the cooking camp: the female was more than enough for me. I remember getting back to the cooking camp, stripping down to my undies and jumping into the Mann River to rehydrate (see Figure 6.5).

As people came back to camp with more and more kangaroos, there was a growing euphoric and jocular high.[13] There were kangaroo carcasses everywhere and ovens being made in the white sand to cook them; highly prized offal was being quickly cooked and shared around and the sweet smell of kangaroo cooking with escarpment herbs was pleasantly overpowering. We had brought only limited supplies of tea and damper, so by the time the first kangaroos were ready to eat, all the hunters were extremely hungry. We engaged in a gluttonous feast, which is so common in such situations of extraordinary surplus.

This euphoria remained with us as we travelled back to Manmoyi. Arriving back in camp, we adopted the Kuninjku cool attitude that accompanies hunting success, while all the time being keenly watchful to protect portions allocated to us and making sure that we delivered them to designated kin in a proper manner. I was caught up in the excitement of the event, but also in the anxiety associated with avoiding excessive claims for my portions. My 'sister' Kaye wanted a kangaroo forearm; a leg designated to Robert Bibora and to be delivered by me disappeared; my 'aunt' stole a kangaroo head that was part of my reward; and I, in turn, 'stole' an apparently unclaimed rump section to substitute for Robert's missing leg.

CONCLUSION

This chapter describes one special day when Kuninjku people engaged in a major fire drive for game. This one day demonstrated a great deal that was working in those days: economic cooperation and productivity, environmental management, a linking of sacred ritual activity with deep cultural significance, and a demonstration of impressive ecological knowledge by ceremonial leaders. People still hunt today, but such large-scale fire drives appear a thing of the past, suggesting that hunting is now undertaken differently. There are many reasons for this transformation: some ecological, some cultural and some economic.

I am pleased to have the opportunity to record this ethnohistorical event that had integrative power for Kuninjku in 1980. In doing so, I am not suggesting that such a major fire drive will not happen again. In the 1960s and 1970s, when Kuninjku people resided at Maningrida and Oenpelli corralled in a government settlement and mission by colonial authorities, no-one would have anticipated an event like *Manwurrk* at Namilewohwo in 1980. Similarly in 2009, although people hunt and live differently, there is no reason to believe that the Indigenous knowledge needed for a similar major fire drive to occur again has been lost. Many of the participants in that fire drive are today's leaders and there is a regional revival in interest in fire ecology that could see similar large-scale land management, hunting and ceremonial events in the future.

ACKNOWLEDGEMENTS

It almost goes without saying that I want to thank and acknowledge all the Kuninjku people with whom I shared the day recalled in this chapter, and many other pleasurable (and some not so pleasurable) days spent together over the past 30 years. It is sad to reflect on how many at that hunt – a number younger than myself – are no longer with us. I would also like to thank: Melinda Hinkson, Jeremy Russell-Smith, Peter Whitehead and Peter Cooke for comments; Murray Garde for comments and assistance with Kuninjku orthography; and John Hughes for expertly drawing Figure 6.4.

ENDNOTES

1 Murray Garde (*pers. comm.*, 24 November 2007, Canberra) recalled that Jimmy Kalarriya who features significantly in this chapter (and at the fire drive) made this comment. In Appendix 7.1 of Garde's PhD thesis *Social Deixis in Bininj Kun-wok Conversation*, University of Queensland, 2002 Kalarriya recounts a hunting story in 1998 that has some elements of the fire drive story told here.

2 The lack of photographs is puzzling because I generally took many. A distinct possibility is that I got caught out with little film on this occasion.

3 Male neophytes who entered a major ritual cult ceremony could not be seen by their uninitiated kin until the completion of the ceremony when they emerged from the secret ceremonial domain back into the secular world.

4 The odometer reading was actually from a subsequent visit in 1985 working on a mapping assignment (Exploration Licence Application 3417) with Peter Cooke for the Northern Land Council. Mapping also indicated that it was only 12 kilometres by river between Manmoyi and Namilewohwo (ELA 3417 Field notebook, June 1985, date of entry 24 June 1985), so clearly our overland route was quite circuitous.

5 Zooming in on the fire drive site with Google Earth is something that I will comment on later. It certainly revived the sentience of the landscape for me: it was almost like being back at Namilewohwo in 1980 again. Note that the vehicle track showing on Google Earth was not there in 1980 because Kamarkawan outstation was only re-established some 4 years after this event. I would like to thank John Hughes for his technical genius in this transposition exercise.

6 The use of pipe clay reduced body odours that can alert hunted game such as kangaroos to human presence and startle them.

7 In conversation with the late Mick Kubarkku, Bowman, Garde and Saulwick (2001, p. 67) record how injured macropods might return to burnt patches and be vulnerable to hunters.
8 Bowman, Garde and Saulwick (2001, pp. 68–69) in translated conversation with the late Jacky Bun.ganiyal relate a discussion about successful hunting with spears at fire drives.
9 There is some description of hunting antilopine kangaroos and euros in 'A hunting story' by Jimmy Kalarriya in Garde (2002, pp. 467–491). Also see Bowman, Garde and Saulwick (2001).
10 There was nothing moralistic about this opposition to gambling: just a concern that it would potentially distract people from proper participation in ceremony if a game with high stakes continued after nightfall when serious ceremonial activity began.
11 At that time, sniffing of petrol was a significant problem in the Maningrida region and these modifications, that included having the fuel inlet inside the vehicle's cab, made it less vulnerable to fuel theft by siphoning. Petrol sniffing has not been a problem in the region for over a decade.
12 I suspect that the estimate of 50 per cent is too high, even though kangaroos have high rates of increase. But, in the absence of an estimate of the regional population, it is impossible to be definitive.
13 This may have been linked in part to a general euphoria associated with ceremonial participation late at night and associated sleep deprivation.

REFERENCES

Altman JC (1982) Hunter-gatherers and the state: the economic anthropology of the Gunwinggu of north Australia. PhD thesis, Department of Archaeology and Anthropology, Australian National University, Canberra.

Altman JC (1987) *Hunter-Gatherers Today: An Aboriginal Economy in North Australia.* Australian Institute of Aboriginal Studies, Canberra.

Altman JC (2003) 'Aboriginal access to firearms for wildlife harvesting: a policy proposal'. *CAEPR Topical Issue 2003/2*, <http://www.anu.edu.au/caepr/Publications/topical/AboriginalAccesstoFirearms.pdf>.

Altman JC (2009) The hybrid economy and anthropological engagements with policy discourse: a brief reflection. *The Australian Journal of Anthropology* (in press).

Bowman DMJS, Garde M and Saulwick A (2001) *Kunj-ken Makka Man-wurrk* fire is for kangaroos: interpreting Aboriginal accounts of landscape burning in central Arnhem Land. In *Histories of Old Ages: Essays in Honour of Rhys Jones.* (Eds A Anderson, I Lilley and S O'Connor) pp. 61–78. Pandanus Books, Australian National University, Canberra.

Garde M (2002) Social Deixis in Bininj Kun-wok Conversation. PhD thesis, School of Social Science, University of Queensland, Brisbane.

Jones R (1968) Firestick farming. *Australian Natural History* **16**, 224–28.

Kyle-Little S (1957) *Whispering Wind: Adventures in Arnhem Land.* Hutchinson, London.

McCarthy FD and McArthur M (1960) The food quest and time factor in Aboriginal economic life. In *Records of the American-Australian Scientific Expedition to Arnhem Land, Volume 2: Anthropology and Nutrition.* (Ed. CP Mountford) pp. 145–194. Melbourne University Press, Melbourne.

Marcus GE and Fischer MMJ (1986) *Anthropology as Cultural Critique: An Experimental Moment in the Human Sciences.* University of Chicago Press, Chicago.

7

Landscape-scale fire research in northern Australia:

delivering multiple benefits in a changing world

Richard J Williams, Damian Barrett, Garry D Cook, A Malcolm Gill, Lindsay Hutley, Adam Liedloff, Bronwyn Myers and John CZ Woinarski

SUMMARY

We describe how fire research in northern Australia, although initially focused on fire behaviour and fire regimes and biodiversity, has diversified to address questions of the role of fire in other sectors of the north Australian economy, such as greenhouse gas emissions, carbon sequestration, the economics of fire management and Indigenous livelihoods. Most fire-management strategies in northern Australia involve the deployment of mixes of burnt and unburnt country across the landscape. Recent research has emphasised that fire frequency and fire severity both matter in the savannas, even though the biome as a whole is characterised by frequent fire and many components of the biodiversity of the savannas show a high degree of resilience to variation in fire regimes. Managing to reduce fire frequency – and hence to provide longer intervals between fires across the landscape – is, and will continue to be, a fundamental objective in landscape fire management. Defining, measuring and delivering heterogeneity in fire regimes has also been a feature of fire research in the past decade in north Australia, and will continue to be important in the coming decades. The management of fire management for optimal carbon benefits is a cross-tenure issue of increasing importance. Approximately 3% of Australia's accountable greenhouse gas emissions result from savanna fires. However, recent ecophysiological research has shown that the northern mesic savannas are a weak sink for carbon, and that sink strength is sensitive to fire regime. However, there is considerable temporal and spatial variation in the sink strength, and further research is needed to quantify such variation. There are considerable economic opportunities to be gained from appropriate fire management in northern Australia, and the technologies developed there have great relevance to the savanna landscapes in our near neighbours, such as eastern Indonesia. Optimal fire regimes for the coming decades will be those that deliver multiple benefits, and will require strong and effective community and governance partnerships to deliver such benefits efficiently and equitably.

INTRODUCTION

Fire is all-pervasive in northern Australia. The evolution of our understanding of the biophysical and social drivers of fire regimes at landscape scales has occurred within a context of a rapidly changing social landscape, with potentially competing objectives, and different views on the role of fire (Williams *et al.* 2002). In 2005, Bond *et al.* published the seminal 'world without fire' paper in which they predicted, based on a simple question, a remarkably different world without fire. Interestingly, the marked structural changes identified by Bond *et al.* (2005) for a variety of global forest biomes were not apparent for northern Australia. Given that fire has been ubiquitous in northern Australia for millennia, the antiquity of the monsoonal climate, perhaps the question is not what sort of landscape a 'no fire world' in northern Australia would look like, but what sort of world does variation in the basic drivers of a fire regime deliver? This question has occupied fire researchers and managers in northern Australia for decades (and no doubt will continue to do so). We wish to further reflect upon this question by considering how fire research has been, is, and might best be, deployed to answer complex questions about how the regional economic futures for northern Australia may be sustained.

Since European settlement of the region, fire research has been a critical component of the research effort devoted to understanding and using the landscapes of northern Australia, and has been expanding rapidly in the past 25 years; it is currently a mature research field. There have been several phases of development, with key ones concerning: the basic weather and fuel drivers of fire; fire and pastoralism; how fire and biodiversity interact; and mapping of fire regimes at landscape scales. Summaries of the major research phases, outputs and outcomes to the turn of 21st century are summarised in Dyer *et al.* (2001); Williams *et al.* (2002); Russell-Smith *et al.* (2003a) and Whitehead *et al.* (2005).

A key outcome of this fire R&D is that the people of northern Australia appreciate the need for managing fire regimes to achieve particular sectoral outcomes. However, it is becoming increasingly apparent that variation in regimes must meet multiple land-management objectives across sectors, which include biodiversity conservation, pastoral production, climate change, the emerging carbon economy, ecosystem services, ecosystem health, human health and Indigenous land management, as part of the process of regional economic development. As such, fire research has come a long way in 30 years: from a focus on the physics, chemistry and biology of combustion to considering the socioeconomic implications of fire, in one of the most flammable, but least populated, landscapes on Earth. The question for fire researchers and managers in the coming decades, then, will be: What fire regimes will optimise resource conservation and economic opportunities in the face of increasingly complex natural resource management?

Most fire-management strategies in northern Australia involve the deployment of mixes of burnt and unburnt country across the landscape. Defining, measuring and delivering such heterogeneity will continue to be important in the coming decades, and there is little doubt that some combination of heterogeneous broadcast of fire will continue to be fundamental to managing the landscape of northern Australia. The issue facing the fire R&D community in the 21st century is therefore identifying the regimes that will bring multiple benefits – wildfire mitigation, biodiversity enhancement, emissions abatement, enhanced carbon sequestration, protection of life and property, and regional community development. The management regimes that result in optimising these complex objectives will necessarily have both spatial and temporal components (including the 'visible' and 'invisible' components of fire regimes; *sensu* Gill *et al.* 2003; Bradstock *et al.* 2005). A critical task for 21st century fire research and management is therefore identifying the scale and grain of broadcast fire that will deliver on these objectives, with equal recognition that 'one size will not fit all'.

Research on fire regimes and biodiversity also has clear implications for other areas of the north Australian economy. It is our contention that defining and delivering heterogeneity in fire regimes is integral to all land-management fields, and we explore this postulate in the remainder of the chapter. The scope of this chapter is not to review all fire research as such. Rather, we illustrate, firstly, how a classical research field – fire ecology – has influenced our understanding of the nature of fire regimes in northern Australia and, secondly, how this in turn has provided a body of understanding that has flexibility, utility and predictive power across sectors and jurisdictions. We briefly review fire research of the last 30 years, illustrate how the main findings are being applied in contemporary fields – with special reference to fire and carbon – and indicate the likely research priorities for landscape fire research in the coming decades. After briefly revisiting work done in the last two decades of the 20th century, we concentrate on work published since about 2002 (i.e. since the publication of the recent review texts listed above).

GRAPPLING WITH COMPLEXITY

Understanding the way in which research on responses of savanna biota to fire can also inform other management questions requires a clear appreciation of the scope of this sort of work. Here we highlight the major findings of two decades of research work and management from around 1980 to 2000. We focus in particular on experimental studies

Fire research for biodiversity conservation

Biodiversity-related research has involved quantification of fire weather, fuels, fire behaviour, the impacts of fire regimes on various plant and animal communities, the mapping of fire regimes, and the biodiversity implications arising from management-imposed fire regimes. This last theme is addressed in Chapter 8 and detailed summaries and historical context are given in: Dyer *et al.* (2001); Williams *et al.* (2002, 2003); Andersen *et al.* (2003, 2005); Russell-Smith *et al.* (2003a); and Whitehead *et al.* (2005). The main features of this research are:

- Fire experiments have a rich tradition; there are numerous well-known experiments at locations such as Annaburroo, Kapalga, Munmarlary and Kidman Springs.
- Internationally, savannas lend themselves to fire experiments (e.g. in South Africa: Kruger National Park, du Toit *et al.* 2003; Hluhluwe National Park, Bond and Archibald 2003).
- The fundamentals of fire – fuel dynamics; fire weather; fire behaviour; and surrogates for intensity and severity – have been well quantified for northern Australia in many technical publications.
- Designed experiments have focused on understanding, firstly, the interactions between fire regime, biodiversity values and conservation management and, secondly, pastoral management.
- Natural experiments have been as important as designed experiments.
- In parallel with this research, there has been increasingly detailed mapping of fire regimes, using remote sensing with ground validation. This has been a critical development in national fire R&D because it has provided quantitative measures of variation in fire regimes that have multiple applications, and are widely accessible to researchers and managers alike (e.g. the North Australia Fire Information website; www.firenorth.org.au/nafi/app/init.jsp).

Major findings from this body of work include:

- Increasingly detailed description of regional savanna fire regime components (*sensu* Gill 1975; Gill *et al.* 1981; Bradstock *et al.* 2002).
- A range of biodiversity responses to variation in fire regime is possible, and no single fire regime can optimise all biodiversity outcomes. For effective conservation of biodiversity in the face of such complexity, conservation goals need to be made explicit.
- Fire frequency and severity both matter in the savannas, even though the biome as a whole is characterised by frequent fire. Therefore, managing to reduce fire frequency – and hence to provide longer intervals between fires across the landscape – is fundamental (Andersen *et al.* 2005).
- The heterogeneity–diversity nexus is complex (Parr and Andersen 2005), but there is an axiomatic need to determine appropriate spatial and temporal variability in fire patterns, at multiple scales including the small (e.g. the home range of small- to medium-sized mammals) through to the large (the home range of mobile fauna; the spread of fire itself under severe fire weather conditions).
- Landscape-scale fire experiments involving formal replicated designs cannot, by their very nature, capture the complexity of fire-landscape-values interactions, nor provide managers with all-encompassing solutions to regional and local problems. Fortunately, the ongoing landscape-scale research generated in response to these foundational studies has been able to both capture some of this ecological complexity, as well as provide ongoing practical frameworks and tools for effective landscape management using fire.
- Notwithstanding the complexity–reality tension that will always bedevil landscape ecological experiments, there is a need for more focused fire experiments. Some additional new, long-term designed experiments have commenced, which vary components of regime (notably frequency and season, and thereby intensity). Examples include the 'Burning for Biodiversity Project' at the Territory Wildlife Park, (http://www.terc.csiro.au/research.asp?Program=FIRE&Project=TWP) and the Boggy Plains Fire project in Kakadu National Park, (http://www.terc.csiro.au/research.asp?Program=FIRE&Project=AFM), and fire and introduced grasses (Rossiter-Rachor 2008).
- Natural experiments, with a focus on documenting variation in landscape features in relation to variation in regime as influenced by land use, will continue to be important, as will other opportunistic, meso-scale experiments. Recent examples include studies undertaken by Woinarski *et al.* (2004) – a study of the impact of long-term fire exclusion near Darwin – and by Fensham *et al.* (2003) – who undertook an experimental manipulation of fire regimes, incorporated extreme drought and examined the differential effects of this combination of treatments on rainforest and savanna tree species in north Queensland.

All of these major findings have profound importance for fire management in all classical and emerging sectors/land tenures of northern Australia. All depend on detecting appropriate temporal and spatial patchiness in the fire regime, which we review briefly below.

HETEROGENEITY IN FIRE REGIMES

The issues of interval and frequency, grain and patchiness are integral to any concept of a fire mosaic. But so are the means of delivery, and the need for evaluation of the effectiveness of the prescribed mosaic against stated aims (Andersen 1999; Keith *et al.* 2002). Fundamental to this is the quantification of regime variation using remote sensing, at local scales (Yates and

Russell-Smith 2003), regional scales (Fisher *et al.* 2003; Felderhof and Gillieson 2006) or continental scales (Russell-Smith *et al.* 2007a).

Gill *et al.* (2009) concluded there was a 'strong role for ... fire, in structuring faunal communities of the tropical savannas of northern Australia' in both experimental (e.g. Woinarski and Ash 2002) and correlative studies (e.g. Kutt and Woinarski 2007). Long-term experimental imposition of different fire regimes produced substantial divergence in fauna (Woinarski 1990; Woinarski *et al.* 1999, 2004; Pardon *et al.* 2003; Corbett *et al.* 2003; Andersen *et al.* 2006), typically with the most notable polarisation in faunal communities being between long-unburnt treatments and those burnt frequently by intense late dry season fires.'

The spatial patchiness of burnt and unburnt features in the landscape is important for conservation management (Russell-Smith *et al.* 2003a; Whitehead *et al.* 2005). Fire-patch size increases with the progression of the dry season, because fires are better able to propagate across the landscape (Yates *et al.* 2008), with a concomitant decrease in the size of unburnt patches (Price *et al.* 2003). However, fire interacts strongly with topography and terrain: numerous unburnt patches occurred within sites in rocky Arnhem Land encompassed by even very intense fires (Price *et al.* 2003). Rockiness therefore provides in-built patchiness 'insurance', with some patches persisting in the landscape to ensure some populations of obligate seeders experience a critical fire-free interval (of about 5 years) such that the populations can persist beyond the primary juvenile period. On flat sites without rocks, the prognosis is very different as extensive, late dry season fires are better able to propagate, with outcomes for interval-sensitive species increasingly dependent on dispersal ability (Price *et al.* 2003).

Intervals between fires also contribute to the heterogeneity of fire in the landscape. The mean 'interval between fires' is the reciprocal of the 'average proportion of the landscape burnt per year' (Gill *et al.* 2003; 2009). Intervals between fires are a critically important component of the fire regime, but are less 'visible' than spatial patches (Gill *et al.* 2003).

Ongoing active management on-country through the use of prescribed burning is therefore needed for most natural resource management (NRM) objectives, because remoteness is of itself no barrier to the evolution of inappropriate fire regimes (Woinarski *et al.* 2007). This raises the critical question of delivering the appropriate, or stated, mix of burnt and unburnt country, and evaluating the effectiveness of such delivery. The science of this is in its infancy (King *et al.* 2006), but northern Australia has international comparative advantage in this field of research given the frequency of fire, especially the regularity with which prescribed burning is deployed, and a relatively simple landscape.

With respect to prescribed burning, there are trade-offs between the proportion of landscape burnt in the early dry season with that in the late dry season, as demonstrated in lowland Kakadu National Park over a 15-year study period (Gill *et al.* 2000). More recently, Price *et al.* (2005) calculated a fire-induced index of heterogeneity at a scale relevant to home ranges of small–medium sized mammals in Kakadu National Park. Fire-induced heterogeneity increased in each successive 5-year period from 1981 to 2000. Importantly, topographic roughness and proximity to both drainage lines and roads increased fire heterogeneity, indicating that existing topography and the road network can be used to deliver biodiversity benefits.

With respect to overall effectiveness, Price *et al.* (2007) analysed the effect of firebreaks created by aerial prescribed burning (APB) in mitigating unplanned fires in western Arnhem Land. They cautioned that such firebreaks, although part of the overall management regime, do not guarantee fire mitigation and there is a continuing need to manage gaps through which late dry season fires can spread.

Further research is needed to identify the optimal mix of linear APB plus patch-scale mosaic to optimise both fire mitigation and habitat heterogeneity outcomes. These are early days, but the types of analyses undertaken by Price *et al.* (2007) in the savannas (and King *et al.*

2006 in temperate systems) will be essential parts of the national and international fire management toolbox.

The above examples are illustrative of the fundamental need for, but difficulty of delivering, broadcast fire for meeting landscape conservation objectives. However, the regimes and their components also have applicability in the fields of climate change research, as will be discussed below.

FIRE RESEARCH AND THE CARBON ECONOMY

Understanding the nature of savanna fire regimes is crucial to understanding the contribution of fire in northern Australia to the nation's greenhouse gas (GHG) budget. This requires an understanding of the contribution of savanna fires to GHG emissions, the capacity of the savannas to take up carbon products from the atmosphere – sequestration – and the interaction between regimes and GHG fluxes. The evolution of carbon markets in northern Australia has the potential to change land use in all sectors, and bring environmental, conservation and commercial benefits (Ockwell and Lovett 2005). Thus, the management of GHGs and biodiversity conservation in Australian savannas via fire management can be considered to be a cross-tenure issue of increasing importance. Below, we outline the predicted changes to the broad climate of northern Australia as a consequence of the enhanced greenhouse effect, and describe the profile of fire-carbon research in northern Australia.

Predicted climate changes for northern Australia

Projected climate changes for northern Australia by 2030 – as a consequence of greenhouse-gas induced global warming – are summarised in Pearce *et al.* (2007). Average temperature is projected to rise by about 1°C and, as a consequence, there will be an increase in the number of days on which temperatures exceed 35°C across the savannas. Projected changes in annual rainfall are relatively small. Other projected changes are that evapotranspiration will increase, but relative humidity is likely to remain unchanged or decrease slightly. Predictions with respect to the main dry season winds that propagate fire across the landscape (the south-east trades) are either for no change or slight increases of 2–5%. The incidence of severe cyclones is likely to increase (Walsh *et al.* 2004) and the proportion of rain that falls during severe storm events is expected to rise by 20 to 30% (Pittock 2003). Lightning may become more frequent (Goldammer and Price 1998).

Atmospheric concentrations of CO_2 are predicted to increase from the current 387 ppm to between 540 ppm to 970 ppm by 2100 depending on emissions trajectories (Solomon *et al.* 2007). This is likely to increase plant productivity, although gains in sequestered carbon may be offset by increased respiration associated with temperature increases. Experimental increases of atmospheric CO_2 in a free-air carbon dioxide experiment (FACE) at a savanna site near Townsville (Stokes *et al.* 2005) demonstrated an increase in the growth of both C4 grasses and C3 woody species. In addition, litter decay rates declined which, combined with an increase in growth, may lead to higher equilibrium fuel loads. The implications of these changes for savanna fire regimes are highly uncertain, because the ways in which climate change will affect: plant, and thus fuel, production; and fire weather, and thence regimes, are inherently uncertain, and increases in both fuel and annual temperatures (especially the number of extremely hot days) may increase the likelihood of fire spreading across the landscape.

Savanna fires and greenhouse gas emissions

The ubiquitous savanna fires of northern Australia produce substantial emissions of GHGs, such as water vapour, carbon dioxide, carbon monoxide (CO), methane (CH_4) and various oxides of nitrogen, especially N_2O. These gases are all released as products of combustion, and enter the atmosphere as constituents of smoke. Nationally, such burning contributes about 3% of Australia's national GHG emissions. Regionally, these emissions may be very significant: for example, they represent, annually, some 35–50% of the Northern Territory's accountable emissions (Russell-Smith *et al.* 2009; Chapter 12).

Emissions of GHGs vary with fire regime. In northern Australia, fires in the late dry season may be extensive and intense, resulting in widespread consumption of fuel – both fine and coarse – and potential death of plants, particularly trees. Tree death results in a long term decline in the above-ground carbon pool and a reduced sequestration potential because evergreen savanna species are capable of carbon uptake all year (Beringer *et al.* 2007). All other things being equal, this scenario of severe late dry season fires results in relatively high GHG emissions. In comparison, fires in the early dry season are relatively low in intensity, consume less fuel and are generally smaller in extent, with lower emissions (Williams *et al.* 2004). The reduction in emissions that results from managing regimes so that the incidence of extensive, late dry season fires is reduced is at the core of the West Arnhem Land Fire Abatement (WALFA) Project (see Chapters 1 and 13).

Savanna fires and carbon sequestration

Ecosystems take up carbon from the atmosphere by photosynthesis, and store it in plants (e.g. tree trunks, branches and roots) and in the soil. Carbon is then re-emitted to the atmosphere via autotrophic and heterotrophic respiration (with small losses via volatile organic carbon emissions from vegetation and organic and inorganic leaching). Carbon is also released to the atmosphere by disturbance, especially fire. The balance between uptake and loss relative to the atmosphere is referred to as the carbon 'sink' or 'sequestration' potential. If more carbon is taken up than lost over time, carbon is said to be sequestered, and the ecosystem is a net carbon sink. Alternatively, if emissions predominate over sequestration, then the ecosystem is a net carbon source to the atmosphere. Given the importance of tropical ecosystems to the world's carbon budget (Stephens *et al.* 2007) and the high incidence of fire in the world's savannas (Justice *et al.* 2003), then knowledge of the carbon budget of tropical savannas in relation to fire is internationally significant. Below, we describe recent research designed to explore these important components of the savanna carbon budget in northern Australia.

Measures of carbon flux – net ecosystem productivity and net biome productivity

There are two sequestration metrics that the international global change community use to assess ecosystem source-sink strength. These are net ecosystem productivity (NEP, sometimes called net ecosystem exchange) and net biome productivity (NBP; Schulze *et al.* 2000). NEP is defined as carbon fixed in the system due to net primary productivity (NPP) minus carbon returned to the atmosphere due to heterotrophic respiration of soil organisms, and autotrophic respiration of fine and coarse roots. NBP is NEP minus carbon returned to the atmosphere due to disturbances such as logging, grazing, fire and cyclones.

Chen *et al.* (2003) addressed the source-sink problem at a *Eucalyptus miniata–E. tetrodonta* site near Darwin. They used an inventory approach by measuring: tree growth and applying allometric equations to calculate biomass increment; shrub increment; and autotrophic and heterotrophic respiration. This produced a point-based estimate of NEP of about -3 t C ha^{-1} y^{-1} (where a negative value indicates net sequestration, i.e. carbon drawn down from the atmosphere, and a positive value indicates emissions to the atmosphere exceed sequestration). Applying a Northern Territory-wide estimate of area burnt to the study area, these authors estimated that NBP was about -1 t C ha^{-1} y^{-1}; that is, a weak sink, even after accounting for fire.

Williams *et al.* (2004) applied the point-source estimate of NEP determined by Chen *et al.* (2003) and (at that time) emerging estimates of NBP from eddy covariance flux towers near Darwin and Katherine to around 32 000 km^2 of *Eucalyptus miniata–E. tetrodonta*-dominated landscapes of western Arnhem Land in the Northern Territory (NT). Those authors scaled their estimates of NEP according to variation in stand basal area and annual rainfall, and derived emissions of C due to burning from remotely-sensed estimates of area burnt, combined with knowledge of fuels, and combustion factors. They concluded that NBP for this tract of country was also about –1 t C ha^{-1} y^{-1}, and that sink strength could be increased by reduced fire frequency.

Beringer *et al.* (2007) published estimates of NEP and NBP from eddy covariance flux towers at Howard Springs, near Darwin. Estimates of NEP were made over a 5-year period, and NBP was estimated by prescribed burning of the savanna within the flux catchment of the tower. Their estimate of NBP over a 2-year period was about –2 t C ha^{-1} y^{-1}.

Barrett *et al.* (2008) applied the VAST (vegetation and soil-carbon transfer) coupled carbon-water model, combined with remotely sensed estimates of area burnt, to the whole continent from 1981–2000 (Barrett 2002). When considering the entire 2M km^2 of savannas, modelled fluxes showed an average NBP value over 6 years from 1995 to 2000 of –0.01 Tg C ha^{-1} yr^{-1} (1.7 Tg C yr^{-1} in total), which ranged from –0.02 to 0.01 Tg C ha^{-1} yr^{-1} (–3.4 to 2.7 Tg C yr^{-1} in total) due to variation in climate on an approximately 5-year interval. (Note that 1 Tg (teragram) = 1 million tonne). This sub-continental scale estimate of sink strength, which included the arid savannas, is two orders of magnitude less than the estimates provided by studies in the mesic savannas of the NT implying that the stronger sink in the more-productive, high rainfall regions is mitigated by episodic fire emissions from semi-arid locales.

Two further studies on permanent monitoring plots, have been recently undertaken in the Top End of the NT. Cook *et al.* (2005) used marked trees and saplings in 0.1 ha plots established as part of the Kapalga Fire Experiment (Andersen *et al.* 2003) to estimate changes in biomass in relation to variation in experimental fire regime. They found NBP of the tree component was dependent on the fire regime. Tree NBP was about –0.2 to –0.5 t C ha^{-1} yr^{-1} for the more benign fire regimes (low-intensity annual early dry season fire for 5 years; unburnt for 5 years). In contrast, tree NBP for the more severe fire regime (5 years of annual moderate–high intensity late dry season fires; infrequent very intense fire) was about +4 t C ha^{-1} yr^{-1}. Murphy *et al.* (Chapter 14) also used a series of permanent plots, established as part of a fire-monitoring system in three NT National Parks (Kakadu, Nitmiluk and Litchfield). These authors found no net sequestration of carbon in the woody components of these plots over a 10-year period, although this assessment was based on coarse measures of tree diameter measured at three sampling times over a decade (0, 5 and 10 years) and a range of fire regimes.

This range of NBP values (~0.0 to –2 t C ha^{-1} yr^{-1}) indicate that the mesic, lowland savannas types of northern Australia are at present a weak sink, but there is spatial variation in the strength of the sink. A key reason for this is the dominant evergreen phenology of these savannas (Williams *et al.* 1997), which feature *Eucalyptus* and *Corymbia* tree species that account for up to 80% of the woody abundance. There is little seasonal variation in canopy fullness (Williams *et al.* 1997) and maximum assimilation rates (Eamus *et al.* 2000) in these species despite the seasonality of rainfall. Beringer *et al.* (2007) observed a consistent net carbon uptake through the dry season – largely attributed to evergreen tree photosynthesis – although this sink reverted to a net source with high-intensity fire (Beringer *et al.* 2007). Elevated atmospheric CO_2 may also enhance this sink capacity (Eamus *et al.* 1995), although there have been limited studies examining responses of Australian savanna eucalypts to elevated CO_2.

The upper range of the reported values above are of the same order of magnitude for reported values of forest NEP and NBP across the tropical and temperate world (–1 to –5 t C ha^{-1} yr^{-1} in the Amazon, Siberia, and Europe; Schulze 2005; Magnani 2007). The global average NBP of the world's savannas is estimated to be about –0.15 t C ha^{-1} yr^{-1} (Grace *et al.* 2006). However, evidence from the studies by Barrett *et al.* (2008) and Murphy *et al.* (Chapter 14) suggest that in some years the savannas will be a mosaic of carbon sources and sinks, determined by prevailing climatic conditions, site quality and the history of both long-term (e.g. cyclones) or short term (e.g. fire) disturbances.

Importantly, the studies of Beringer *et al.* (2007), Cook *et al.* (2005) and Williams *et al.* (2004) indicate that savanna sequestration capacity is sensitive to fire regime. Beringer *et al.* (2007) found that mesic savannas can be a sink, even when burnt by early dry season fires with low to moderate fire intensities. Under increasingly severe late dry season fire regimes, however, the savannas are a net source of carbon for 3 months after a severe fire event (Beringer *et al.* 2007) and, at a sub-continental scale, when low rainfall and heavily grazed systems are included, the savanna region as a whole can be a net source on an annual basis (Barrett *et al.* 2008). This has major implications for international full carbon accounting initiatives, which aim to account for all stores and fluxes of carbon in landscapes.

Under the current Australian Greenhouse Office (now Department of Climate Change) accounting procedures for the national inventory of GHGs, savanna fires are assumed to be CO_2 neutral. That is, it is assumed that carbon lost in fluxes of CO_2 by dry season fires is balanced by CO_2 uptake in the following wet season. The above studies show that this is not necessarily the case, with carbon storage occurring in tree biomass, shrub encroachment and in soils. More research is needed to quantify the sequestration capacity of the entire savanna region (as opposed to mesic, coastal systems that has occurred to date) and its sensitivity to variation in climate, climate change and fire regime.

Estimating long-term sink strength at broad spatial scales in Australia's savannas is subject to considerable uncertainty. However, combining basic inventory data from long-term fire experiments with flux data from various sources has greatly increased our understanding of sink/source relationships in the savannas of northern Australia. This is influencing our understanding of the direction and strength of changes to carbon stocks in relation to land use change, including the management of fire regimes. Further long-term and regional research is needed to reduce the uncertainty concerning carbon sink/source relationships in the Australian savannas.

Fire regimes and invasive grasses

One aspect of global change that has implications for fire regimes – and hence the carbon budget of Australian savannas – is the spread of exotic grass species, particularly Gamba grass (*Andropogon gayanus*). Gamba grass was one of a number of tall grasses introduced from West Africa as an improved-pasture species in the mid–late 20th century (Lonsdale 1994). On pastoral properties, Gamba grass may be reduced by cattle grazing, but on other land tenures such reduction by grazing does not occur, and it is an aggressive colonist of native savanna. It can reach 4–5 m in height, have a standing biomass of up to about 30 t ha^{-1}, which is 5–7 times that of the native grasses, and result in fires an order of magnitude more intense than savanna fires in native grass fuel beds (Rossiter *et al.* 2003). Invasion by this grass initiates a grass-fire cycle in which fires are more intense, thereby decreasing canopy cover of native trees and promoting the standing crop of gamba grass at the expense of native plant species (Douglas and Setterfield 2005; Setterfield *et al.* 2005; Rossiter-Rachor 2008). Its weed potential is such that Western Australia and Queensland Governments banned the sale and planting of the grass in 2008, and all known infestations in Western Australia are to be eradicated. Gamba grass was declared a weed in the Northern Territory in November 2008.

The continued spread of Gamba grass into native savanna will elevate GHG emissions (due to high Gamba grass fuels, increased fire intensity, elevated tree mortality and greater consumption of woody fuels) and reduce carbon sequestration capacity of the native vegetation. There may be a net gain in soil organic carbon under Gamba grass infestations, which has been observed in South American savannas (Fisher *et al.* 1994), but the loss of above- and below-ground woody carbon pools (up to 70 t C ha^{-1}) and associated GHG emissions, is likely to out-weigh this gain in stored soil carbon. More widespread planting of Gamba grass, on the basis that it has the potential to enhance carbon sequestration at landscape scales, is not justified on the basis of current scientific evidence, and would be a poor investment prospect.

THE FUTURE: UNCERTAINTIES, OPPORTUNITIES AND R&D PRIORITIES

Recent research has identified that emerging carbon economy opportunities and better fire management can deliver GHG emissions reduction benefits, potential sequestration enhancement in above- and below-ground C-sources, and other economic and social benefits. However, identifying the strength and diversity of the biophysical stocks and fluxes of carbon, the markets for carbon products, and the rules of engagement for managers and investors, all need critical evaluation.

Uncertainties

Studies on carbon sequestration capacity of the savannas have demonstrated significant spatial variation in sink strength. This is not surprising at the sub-continental scale, given the substantial variation in mean annual rainfall, soils, tree cover and potential productivity across the rainfall gradients of the savanna region (Williams *et al.* 1996). Substantial spatial variation in the pattern of diameter at breast height (DBH) increments within and between different fire regimes on juvenile, sapling and adult savanna trees has been demonstrated by Prior *et al.* (2006). Further work linking detailed flux measurements to plot-based inventories in a range of savanna types, with tree increment measured using precise forestry mensuration methods on permanently marked and geo-located trees – with an assessment of sequestration in soils and assessment of understorey thickening – is required to develop accurate, spatially extensive and rapid assessments of sequestration capacity.

Two key global issues for carbon sequestration that require further research are sink saturation and the permanence of sequestered stocks (in both biomass and soils) of carbon. Sink saturation in living biomass has not been addressed for Australian savannas but, given the evident low standing biomass, rapid sink saturation may occur in these systems. A potential risk to stocks in the vicinity of the coastline concerns tropical cyclones, especially large intense cyclones such as Cyclone Monica, which knocked over more than 50% of trees in 700 km^2 of savanna forest and woodland in Arnhem Land in 2006 (Cook and Goyens 2008). The impact of such infrequent, but extensive, disturbance on both emissions and sequestration potential, and the risks they might pose to GHG abatement projects, also requires further research.

So far, the focus in this chapter has been on carbon fluxes in soil and vegetation. With much of northern Australia burnt each year, fire-derived non-CO_2 fluxes are a significant GHG source in addition to CO_2 (Chapter 13). These GHGs include CH_4, N_2O, NO_x, CO, non-methane hydrocarbons (NMHCs) and volatile organic compounds (VOCs). The Biomass Burning and Lightning Experiment (BIBLE) quantified some of these fluxes (Kondo *et al.* 2003; Russell-

Smith *et al.* 2003b; Shirai *et al.* 2003). In addition to fluxes derived from fire, there may be significant fluxes of CH_4, N_2O and NO_X from soils and termites of the savannas, although currently there are no data describing these fluxes. However, a study of non-CO_2 GHGs derived from savanna soils as a function of land use and fire is underway (Hutley *et al.* unpublished data). Initial data suggest that uncleared savanna is a net sink for CH_4, despite termite mounds producing high levels of methane emissions. N_2O fluxes are low, although these increase when land systems are cleared and converted to fertilised improved pastures for grazing.

Opportunities

In addition to the WALFA region, which is documented in detail in other chapters of this book, there are also other major regions within the savannas of northern Australia where there are excellent opportunities for applying fire abatement to achieve carbon benefits – for example, around the Gulf of Carpentaria, the Kimberley and inland semi-arid savannas. There are also possible opportunities on pastoral land, although the major challenge there is perhaps not fire, but methane emissions due to enteric fermentation from cattle (Cook *et al.* 2009).

There is also offshore potential for the application of the logic and technologies of fire abatement developed in northern Australia. Savannas occupy 12% of the world's land surface (Williams *et al.* 2005), and there are multiple international markets for carbon off-set products (Bayon *et al.* 2007). A particularly enticing prospect concerns fire-prone savannas in eastern Indonesia, where there are extensive grasslands and open eucalypt woodlands that annually carry wildfire (Fisher *et al.* 2006). Closed-canopy gallery and gully forests, which occur in patches surrounded by grassland and savanna, are threatened by widespread, frequent fire. Rural communities depend on these closed forests for a significant part of their livelihoods and, with international carbon off-setting and trading, there is potential for local people to enhance their income, by enhanced fire management of the surrounding savannas in a similar way to that as undertaken for the WALFA program (Djoeroemana *et al.* 2000; Russell-Smith *et al.* 2007b).

Ongoing fire and NRM research issues for the next decade

Biophysical research and modelling

In addition to the uncertainties associated with sink longevity and quanta outlined above, there are a number of additional, ongoing biophysical research needs. One is clearly refining definitions for, and delivery of, appropriate heterogeneity in fire regimes. Heterogeneity will continue to matter to all land tenures and, to this end, we need to better understand the landscape-scale barriers to fire spread, so that the appropriate temporal and spatial distribution of burnt and unburnt country that optimises fire abatement, and flora and fauna requirements, can be delivered in a cost-efficient manner. In addition, and in parallel, the development of better landscape fire and carbon accounting models is required. For example, The National Carbon Accounting System (NCAS) toolkit, developed by the Australian Government's Department of Climate Change, is currently the only nationally recognised method for integrating carbon estimates for a given location under a range of changing management options. The extent to which NCAS can be applied at regional/project scales needs investigation. Research aimed at adequately capturing fundamental savanna ecosystem processes, such as frequent fire and intermittent cyclones within NCAS – and testing this in project-based scenarios – is also required. There is also a growing demand for technical expertise, support and training in the use of NCAS.

One component of this will be to use NCAS in conjunction with other fire-modelling tools, such as FLAMES (fundamentally a savanna fire regimes and vegetation structure dynamics modelling tool; Liedloff and Cook 2007) such that modelling of fire effects using multiple fires with different properties in different seasons – the fire regime of the savannas – can be incorporated into national accounting procedures.

Other pertinent fire information systems include the North Australian Fire Information (NAFI) website that uses satellite (AVHRR and MODIS) information to locate current fires ('hotspots') in near-real-time across the whole of north Australia. Hotspot data are sourced from the Western Australia Department of Land Information and Geoscience Australia, while semi-automated fire scar maps are currently generated by the Tropical Savannas Cooperative Research Centre (TSCRC), Bushfires NT and the Cape York Peninsula Development Association. This system provides an invaluable tool for fire managers, as well as fire ecologists, and is currently managed by the TSCRC. Critical R&D is currently underway to improve the automation of fire scar and severity mapping, and calculation of greenhouse emissions based on fire severity, fire extent and fuel loads. These are significant challenges but, if achieved, the NAFI system could be integrated into systems such as NCAS as a more robust greenhouse-emissions-monitoring tool.

Socioeconomic research

The economics of fire management across sectors in northern Australia is in its infancy (Dyer and Stafford Smith 2003), but international markets for carbon offset products are expanding rapidly. The European Union Emissions Trading Scheme (EU ETS) traded 362 Mt C – worth about US$9 billion – in its first year of operation in 2005–06 (Bayon *et al.* 2007). Knowledge of markets – whether voluntary or regulated – the supply and demand constraints for various carbon products, and the quality of the products (which may be somewhat akin to a credit rating or bond quality) and the governance of trade will require much scrutiny in the years ahead, as major investment consortia entering the carbon market will expect robust verifiable quanta.

Fire management and sustainable development at landscape scales

Savannas cannot be managed purely as carbon sinks as an end in itself: if they were, there would be a risk of them becoming dumping grounds for industrial carbon emissions at the expense of emissions reductions and efficiencies and, acutely, other natural resource management objectives. These trade-offs need to be acknowledged and balanced in land-management decisions, so that perverse outcomes do not result (Cook *et al.* 2009). In addition, sink-saturation issues, and difficulties associated with accurate verification, may limit the attractiveness of offset activities, especially if full carbon accounting is required.

The field is complex, especially given that savanna fire regimes need to be managed not only for biodiversity and greenhouse benefits, but also for: human health (e.g. smoke effects: Bowman *et al.* 2007; Burgess *et al.* 2005; Johnston *et al.* 2002, 2007); primary production, especially cattle grazing: Dyer and Stafford Smith 2003; regulatory issues: Preece 2007; customary Indigenous practice: Whitehead *et al.* 2003; Murphy and Bowman 2007; Indigenous livelihoods, especially via the hybrid economy: Altman 2001; Altman and Cochrane 2005; and, increasingly, life and property. Given this complexity, there is a need for:

1. Landscape-scale fire-management schemes that deliver multiple benefits. That is, combining management of fire for abatement of GHGs, biodiversity benefits (because a number of plant and animal groups in the savannas are likely to benefit from reduced fire extent and frequency: Andersen *et al.* 2003, 2005; Yates *et al.* 2008), enhancement of Indigenous capacity and livelihoods (Luckert *et al.* 2007), and agricultural productivity and regional livelihoods.

2. Effective community and governance partnerships. These need to be enhanced to deliver appropriate tools and outcomes. Partnerships include technical arrangements for delivery and monitoring of, for example, the enhancement of the monitoring and mapping of fires by remote sensing (Pereira 2003) of fire intensity within and between events, and partnerships that deliver on-ground management.
3. Economic analyses. At present, the return on fire R&D investment is high, with a 3.5 multiplier in cost–benefit terms (Centre for International Economics, unpublished data). Fire management involves the whole community, and the multiple benefits that accrue from effective fire management makes fire R&D a worthwhile contributor to sustainable regional economic development in northern Australia. Further economic analyses (e.g. cost–benefit analysis of prescribed burning and development of supply curves for carbon off-set products) will be a fruitful research area.

Fire research in northern Australia has been inter-disciplinary, relevant and productive. The investment prospectus for fire R&D in northern Australia will remain bright for decades to come, and the R&D that will be needed to further our understanding of landscape management will be built upon a very sound foundation that has been accrued by a community of researchers and managers over the past three decades.

ACKNOWLEDGEMENTS

The senior author thanks all his co-authors and the editors for their multiple contributions to the field of fire research in northern Australia. Critical comments on the manuscript were provided by Owen Price.

REFERENCES

Altman JC (2001) 'Sustainable development options on Aboriginal land: the hybrid economy in the twenty-first century'. Discussion Paper 226/201. Centre for Aboriginal Economic Policy Research, Australian National University, Canberra.

Altman JC and Cochrane M (2005) Sustainable development in the indigenous-owned savanna: innovative institutional design for cooperative wildlife management. *Wildlife Research* **32**, 473–480.

Andersen AN (1999) Fire management in northern Australia: beyond command-and-control. *Australian Biologist* **12**, 63–70.

Andersen AN, Cook GD and Williams RJ (Eds) (2003) *Fire in Tropical Savannas: The Kapalga Fire Experiment*. Springer-Verlag, New York.

Andersen AN, Cook GD, Corbett LK, Douglas MM, Eager RW, Russell-Smith J, Setterfield SA, Williams RJ and Woinarski JCZ (2005) Fire frequency and biodiversity conservation in Australian tropical savannas: Implications from the Kapalga fire experiment. *Austral Ecology* **30**, 155–167.

Andersen AN, Hertog T and Woinarski JCZ (2006) Long-term fire exclusion and ant community structure in an Australian tropical savanna: congruence with vegetation succession. *Journal of Biogeography* **33**, 823–832.

Barrett D (2002) Steady state turnover time of carbon in the Australian terrestrial biosphere. *Global Biogeochemical Cycles* **16(4)**, 1108, <doi:10.1029/2002GB001860>.

Barrett D, Renzullo LJ, Guerschman J and Hill MJ (2008) Multi-sensor model-data assimilation for improved modeling of savanna carbon and water budgets. *EOS*

Transactions, American Geophysical Union **88(52)**, Fall Meeting Supplement, Abstract B21B-08.

Bayon R, Hawn A and Hamilton K (2007) *Voluntary Carbon Markets. An International Guide to What They are and How They Work*. Earthscan, London.

Beringer J, Hutley LB, Tapper NJ and Cernusak L (2007) Savanna fires and their impact on net ecosystem productivity in North Australia. *Global Change Biology* **13**, 990–1004.

Bond WJ and Archibald SA (2003) Confronting complexity: fire policy choices in South African savanna parks. *International Journal of Wildland Fire* **12**, 381–389.

Bond WJ, Woodward FI and Midgley GF (2005) The global distribution of ecosystems in a world without fire. *New Phytologist* **165**, 525–538.

Bowman D, Dingle JK, Johnston FH, Parry D and Foley M (2007) Seasonal patterns in biomass smoke pollution and the mid 20th-century transition from Aboriginal to European fire management in northern Australia. *Global Ecology and Biogeography* **16**, 246–256.

Bradstock RA, Williams JE and Gill AM (Eds) (2002) *Flammable Australia: The Fire Regimes and Biodiversity of a Continent*. Cambridge University Press, Cambridge.

Bradstock RA, Bedward M, Gill AM and Cohn JS (2005) Which mosaic? A landscape ecological approach for evaluating interactions between fire regimes, habitat and animals. *Wildlife Research* **32**, 409–423.

Burgess CP, Johnston FH, Bowman D and Whitehead PJ (2005) Healthy country: Healthy people? Exploring the health benefits of indigenous natural resource management. *Australian and New Zealand Journal of Public Health* **29**, 117–122.

Chen X, Hutley LB and Eamus D (2003) Carbon balance of a tropical savanna of northern Australia. *Oecologia* **137**, 405–416.

Cook GD and Goyens C (2008) The impact of wind on trees in Australian tropical savannas: lessons from the impact of Cyclone Monica. *Austral Ecology* **33**, 462–470.

Cook GD, Liedloff AC, Eager RW, Chen X, Williams RJ, O'Grady AP and Hutley LB (2005) The estimation of carbon budgets of frequently burnt tree stands in savannas of northern Australia, using allometric analysis and isotopic discrimination. *Australian Journal of Botany* **53**, 621–630.

Cook GD, Williams RJ, Stokes C, Hutley LB and Ash AJ (2009) Managing sources and sinks of greenhouse gases in Australia's rangelands and tropical savannas. *Rangeland Ecology and Management* (in press).

Corbett LK, Andersen AA and Muller WJ (2003) Terrestrial vertebrates. In *Fire in Tropical Savannas: The Kapalga Fire Experiment*. (Eds AN Andersen, GD Cook and RJ Williams) pp. 126–152. Springer-Verlag, New York.

Djoeroemana S, Semangun H, Saragih B and Sulthoni A (2000) The implications of fire management and reforestation for economic development in East Nusa Tengarra. In *Fire and Sustainable Agricultural and Forestry Development in Eastern Indonesia and Northern Australia*. (Eds J Russell-Smith, G Hill, S Djoeroemana and B Myers) pp. 52–55. Australian Centre for International Agricultural Research, Canberra.

Douglas MM and Setterfield SA (2005) Impacts of exotic tropical grasses: lessons from gamba grass in the Northern Territory. In *Proceedings of the Eighth Queensland Weed Symposium* (Ed W Vogler) pp. 69–73. The Weed Society of Queensland, Brisbane.

du Toit T, Rogers KH and Briggs H (2003) *The Kruger Experience: Ecology and Management of Savanna Heterogeneity*. Island Press, Washington DC.

Dyer R, Jacklyn P, Partridge I, Russell-Smith J and Williams RJ (Eds) (2001) *Savanna Burning: Understanding and Using Fire in Northern Australia*. Tropical Savannas Cooperative Research Centre, Darwin.

Dyer R and Stafford Smith M (2003) Ecological and economic assessment of prescribed burning impacts in semi-arid pastoral lands of northern Australia. *International Journal of Wildland Fire* **12**, 403–413.

Eamus D, Duff GA and Berryman CA (1995) Photosynthetic responses to temperature, light flux-density, CO_2 concentration and vapour pressure deficit in *Eucalyptus tetrodonta* grown under CO_2 enrichment. *Environmental Pollution* **90**, 41–49.

Eamus D, Myers BA, Duff GA and Williams RJ (2000) A cost benefit analysis of leaves of eight Australian savanna tree species of differing leaf life span. *Photosynthetica* **36**, 575–586.

Felderhof L and Gillieson D (2006) Comparison of fire patterns and fire frequency in two tropical savanna bioregions. *Austral Ecology* **31**, 736–746.

Fensham RJ, Fairfax RJ, Butler DW and Bowman D (2003) Effects of fire and drought in a tropical eucalypt savanna colonized by rain forest. *Journal of Biogeography* **30**, 1405–1414.

Fisher MJ, Rao IM, Ayarza MA, Lascano CE, Sanz JI, Thomas RJ and Vera RR (1994) Carbon storage by introduced deep-rooted grasses in the South American savannas. *Nature* **371**, 236–238.

Fisher R, Vigilante T, Yates C and Russell-Smith J (2003) Patterns of landscape fire and predicted vegetation response in the North Kimberley region of Western Australia. *International Journal of Wildland Fire* **12**, 369–379.

Fisher R, Bobanuba WE, Rawambaku A, Hill GJE and Russell-Smith J (2006) Remote sensing of fire regimes in semi-arid Nusa Tenggara Timur, eastern Indonesia: current patterns, future prospects. *International Journal of Wildland Fire* **15**, 307–317.

Gill AM (1975) Fire and the Australian flora: a review. *Australian Forestry* **38**, 4–25.

Gill AM, Groves RH and Noble IR (1981) *Fire and the Australian Biota*. Australian Academy of Science, Canberra.

Gill AM, Ryan PG, Moore PHR and Gibson M (2000) Fire regimes of world heritage Kakadu National Park, Australia. *Austral Ecology* **25**, 616–625.

Gill AM, Allan G and Yates C (2003) Fire-created patchiness in Australian savannas. *International Journal of Wildland Fire* **12**, 323–331.

Gill AM, Williams RJ and Woinarski JCZ (2009) Fires in Australia's tropical savannas: interactions with biodiversity, global warming and exotic biota. In *Fire Ecology and Tropical Ecosystems*. (Ed. M Cochrane) pp. 113–141. Springer, New York.

Goldammer JG and Price C (1998) Potential impacts of climate change on fire regimes in the tropics based on MAGICC and a GISS GCM-derived lightning model. *Climatic Change* **39**, 273–296.

Grace J, San JJ, Meir P, Miranda HS and Montes RA (2006) Productivity and carbon fluxes of tropical savannas. *Journal of Biogeography* **33**, 387–400.

Johnston FH, Kavanagh AM, Bowman DMJS and Scott RK (2002) Exposure to bushfire smoke and asthma: an ecological study. *The Medical Journal of Australia* **176**, 535–538.

Johnston FH, Jacups SP, Vickery AJ and Bowman D (2007) Ecohealth and Aboriginal testimony of the nexus between human health and place. *Ecohealth* **4**, 489–499.

Justice C, Smith R, Gill AM and Csiszar I (2003) A review of current space-based fire monitoring in Australia and the GOFC/GOLD program for international coordination. *International Journal of Wildland Fire* **12**, 247–258.

Keith DA, Williams JW and Woinarski JCZ (2002) Fire management and biodiversity conservation: key approaches and principles. In *Flammable Australia: The Fire Regimes and Biodiversity of a Continent*. (Eds RA Bradstock, JA Williams AM Gill) pp. 402–425. Cambridge University Press, Cambridge, UK.

King KJ, Cary GJ, Bradstock RA, Chapman J, Pyrke A and Marsden-Smedley JB (2006) Simulation of prescribed burning strategies in south-west Tasmania, Australia: effects on unplanned fires, fire regimes, and ecological management values. *International Journal of Wildland Fire* **15**, 527–540.

Kondo Y, Ko M, Koike M, Kawakami S and Ogawa T (2003) Preface to special section on biomass burning and lightning experiment (BIBLE). *Journal of Geophysical Research* **107**, 8397, <doi:10.1029/2002JD002401>.

Kutt AS and Woinarski JCZ (2007) Vegetation and the vertebrate fauna assemblage pattern in response to grazing and fire in a tropical savanna woodland in north-eastern Australia. *Journal of Tropical Ecology* **23**, 95–106.

Liedloff A and Cook GD (2007) Modelling the effects of rainfall variability and fire on tree populations in an Australian tropical savanna using the Flames simulation model. *Ecological Modelling* **201**, 269–282.

Lonsdale WM (1994) Inviting trouble: Introduced pasture species in northern Australia. *Australian Journal of Ecology* **19**, 345–354.

Luckert MK, Campbell B, Gorman JT and Garnett ST (2007). *Investing in Indigenous Natural Resources Management.* CDU Press, Darwin.

Magnani F, Mencuccini M, Borghetti M, Berbigier P, Berninger F, Delzon S, Grelle A, Hari P, Jarvis PG, Kolari P, Kowalski AS, Lankreijer H, Law BE, Lindroth A, Loustau D, Manca G, Moncrieff JB, Rayment M, Tedeschi V, Valentini R and Grace J (2007) The human footprint in the carbon cycle of temperate and boreal forests. *Nature* **447**, 848–850.

Murphy BP and Bowman D (2007) The interdependence of fire, grass, kangaroos and Australian Aborigines: a case study from central Arnhem Land, northern Australia. *Journal of Biogeography* **34**, 237–250.

Ockwell D and Lovett JC (2005) Fire-assisted pastoralism vs. sustainable forestry – the implications of missing markets for carbon in determining optimal land use in the wet-dry tropics of Australia. *Journal of Environmental Management* **75**, 1–9.

Pardon LG, Brook BW and Griffiths AD (2003) Determinants of survival for the northern brown bandicoot under a landscape-scale fire experiment. *Journal of Animal Ecology* **72,** 106–115.

Parr CL and Andersen AN (2005) Patch mosaic burning for biodiversity conservation: a critique of the pyrodiversity paradigm. *Conservation Biology* **20**, 1610–1619.

Pearce K, Holper P, Hopkins M, Bouma W, Whetton P, Hennessy K and Power S (2007) 'Climate change in Australia'. Technical Report 2007. CSIRO/Bureau of Meteorology. Melbourne <http://www.climatechangeinaustralia.gov.au/documents/resources/>.

Pereira JMC (2003) Remote sensing of burned areas in tropical savannas. *International Journal of Wildland Fire* **12,** 259–270.

Pittock B (2003) *Climate Change: An Australian Guide to the Science and Potential Impacts.* Australian Greenhouse Office, Canberra.

Preece N (2007) Traditional and ecological fires and effects of bushfire laws in north Australian savannas. *International Journal of Wildland Fire* **16**, 378–389.

Price O, Russell-Smith J and Edwards A (2003) Fine-scale patchiness of different fire intensities in sandstone heath in northern Australia. *International Journal of Wildland Fire* **12,** 227–236.

Price O, Edwards A, Connors G, Woinarski J, Ryan G, Turner A and Russell-Smith J (2005) Fire heterogeneity in Kakadu National Park, 1980–2000. *Wildlife Research* **32**, 425–433.

Price OF, Edwards AC and Russell-Smith J (2007) Efficacy of permanent firebreaks and aerial prescribed burning in western Arnhem Land, Northern Territory, Australia. *International Journal of Wildland Fire* **16**, 295–305.

Prior LD, Brook BW, Williams RJ, Werner PA, Bradshaw CJA and Bowman DMJS (2006) Environmental and allometric drivers of tree growth rates in a north Australian savanna. *Forest Ecology and Management* **234**, 164–180.

Rossiter NA, Setterfield SA, Douglas MM and Hutley LB (2003) Testing the grass-fire cycle: alien grass invasion in the tropical savannas of northern Australia. *Diversity and Distributions* **9**, 169–176.

Rossiter-Rachor NA (2008) Effects of *Andropogon gayanus* (gamba grass) invasion on ecosystem nitrogen dynamics in an Australian tropical savanna. PhD Thesis, Charles Darwin University, Casuarina Campus, Australia.

Russell-Smith J, Whitehead PJ, Williams RJ and Flannigan M (eds) (2003a) Fire and savanna landscapes in northern Australia: regional lessons and global challenges. *International Journal of Wildland Fire* **12**, 247–440. Special Issue.

Russell-Smith J, Edwards AC and Cook GD (2003b) Reliability of biomass burning estimates from savanna fires: Biomass burning in northern Australia during the 1999 Biomass Burning and Lightning Experiment B field campaign. *Journal of Geophysical Research* **108**(D3), 8405, <doi:10.1029/2001JD000787>.

Russell-Smith J, Yates CP, Whitehead PJ, Smith R, Craig R, Allan GE, Thackway R, Frakes I, Cridland S, Meyer MCP and Gill AM (2007a) Bushfires 'down under': patterns and implications of contemporary Australian landscape burning. *International Journal of Wildland Fire* **16**, 361–377.

Russell-Smith J, Djoeroemana S, Maan J and Pandanga P (2007b) Rural livelihoods and burning practices in savanna landscapes of Nusa Tenggara Timur, eastern Indonesia. *Human Ecology* **35**, 345–359.

Russell-Smith J, Murphy BP, Meyer CP, Cook GD, Maier S, Edwards AC, Schatz J and Brocklehurst P (2009) Improving estimates of savanna burning emissions for greenhouse accounting in northern Australia: limitations, challenges, applications. *International Journal of Wildland Fire* **18**, 1–18.

Schultze E-D, Wirth C and Heimann M (2000) Managing forests after Kyoto. *Science* **289**, 2058–2059.

Schulze E-D (2005) Biological control of the terrestrial carbon sink. *Biogeosciences Discussions* **2**, 1283–1329.

Setterfield SA, Douglas MM, Hutley LB and Welch MA (2005) Effects of canopy cover and ground disturbance on establishment of an invasive grass in an Australian savanna. *Biotropica* **37**, 25–31.

Shirai T, Blake DR, Meinardi S, Rowland FS, Russell-Smith J, Edwards A, Kondo Y, Koike M, Kita K, Machida T, Takegawa N, Nishi N, Kawakami S and Ogawa T (2003) Emission estimates of selected volatile organic compounds from tropical savanna burning in northern Australia. *Journal of Geophysical Research* **108**(D3), 8406, <doi:10.1029/2001JD000841>.

Solomon S, Qin D, Manning M, Alley RB, Berntsen T, Bindoff NL, Chen Z, Chidthaisong A, Gregory JM, Hegerl GC, Heimann M, Hewitson B, Hoskins BJ, Joos F, Jouzel J, Kattsov V, Lohmann U, Matsuno T, Molina M, Nicholls N, Overpeck J, Raga G, Ramaswamy V, Ren J, Rusticucci M, Somerville R, Stocker TF, Whetton P, Wood RA and Wratt D (2007) Technical summary. In *Climate Change 2007: The Physical Science Basis. Contribution of Working Group I to the Fourth Assessment Report of the Intergovernmental Panel on Climate Change.* (Eds S Solomon, D Qin, M Manning, Z Chen, M Marquis, KB Averyt, M Tignor and HL Miller) pp. 19–91. Cambridge University Press, Cambridge, UK.

Stephens BB, Gurney KR, Tans PP, Sweeney C, Peters W, Bruhwiler L, Ciais P, Ramonet M, Bousquet P, Nakazawa T, Aoki S, Machida T, Inoue G, Vinnichenko N, Lloyd J, Jordan A,

Heimann M, Shibistova O, Langenfelds RL, Steele LP, Francey RJ and Denning AS (2007) Weak northern and strong tropical land carbon uptake from vertical profiles of atmospheric CO_2. *Science* **316**, 1732–1735.

Stokes C, Ash A, Tibbett M and Holtum J (2005) OzFACE: the Australian savanna free air CO_2 enrichment facility and its relevance to carbon-cycling issues in a tropical savanna. *Australian Journal of Botany* **53**, 677–687.

Walsh KJE, Nguyen KC and Mcgregor JL (2004) Fine-resolution regional climate model simulations of the impact of climate change on tropical cyclones near Australia. *Climate Dynamics* **22**, 47–56.

Williams RJ, Duff GA, Bowman DMJS and Cook GD (1996) Variation in the composition and structure of tropical savannas as a function of rainfall and soil texture along a large scale climatic gradient in the Northern Territory, Australia. *Journal of Biogeography* **23**, 747–756.

Williams RJ, Myers BA, Muller WJ, Duff GA and Eamus D (1997) Leaf phenology of woody species in a northern Australian tropical savanna. *Ecology* **78**, 2542–2558.

Williams RJ, Griffiths AD and Allan G (2002) Fire regimes and biodiversity in the wet-dry tropical savanna landscapes of northern Australia. In *Flammable Australia: The Fire Regimes and Biodiversity of a Continent.* (Eds RA Bradstock, JA Williams and AM Gill) pp. 281–304. Cambridge University Press, Cambridge, UK.

Williams RJ, Woinarski JCZ and Andersen AN (2003) Fire experiments in northern Australia: lessons for ecology, management and biodiversity conservation. *International Journal of Wildland Fire* **12**, 391–402.

Williams RJ, Hutley LB, Cook GD, Russell-Smith J, Edwards A and Chen X (2004) Assessing the carbon sequestration potential of mesic savannas in the Northern Territory, Australia: Approaches, uncertainties and potential impacts of fire. *Functional Plant Biology* **31**, 415–422.

Williams RJ, Carter J, Duff GA, Woinarski JCZ, Cook, GD and Farrer S (2005) Carbon accounting, land management, science and policy uncertainty in Australian savanna landscapes: Introduction and overview. *Australian Journal of Botany* **53**, 583–588.

Whitehead PJ, Bowman DMJS, Preece N, Fraser F and Cooke P (2003) Customary use of fire by indigenous peoples in northern Australia: its contemporary role in savanna management. *International Journal of Wildland Fire* **12**, 415–25.

Whitehead PJ, Russell-Smith J and Woinarski JCZ (2005) Introduction and overview: fire, landscape heterogeneity and wildlife management in northern Australia. Introduction to Special Issue. *Wildlife Research* **32**, 369–375.

Woinarski JCZ (1990) Effects of fire on the bird communities of tropical woodlands and open forests in northern Australia. *Australian Journal of Ecology* **15**, 1–22.

Woinarski JCZ and Ash AJ (2002) Responses of vertebrates to pastoralism, military land use and landscape position in an Australian tropical savanna. *Austral Ecology* **27**, 311–323.

Woinarski JCZ, Brock C, Fisher A, Milne D and Oliver B (1999) Response of birds and reptiles to fire regimes on pastoral land in the Victoria River District, Northern Territory. *The Rangeland Journal* **21**, 24–38.

Woinarski JCZ, Risler J and Kean L (2004) The response of vegetation and vertebrate fauna to 23 years of fire exclusion in a tropical *Eucalyptus* open forest, Northern Territory, Australia. *Austral Ecology* **29**, 156–176.

Woinarski J, Mackey B, Nix H and Traill B (2007) 'The nature of Northern Australia: natural values, ecological processes and future prospects'. ANU e-press, Canberra.

Yates C and Russell-Smith J (2003) Fire regimes and vegetation sensitivity analysis: an example from Bradshaw Station, monsoonal northern Australia. *International Journal of Wildland Fire* **12**, 349–358.

Yates C, Edwards A and Russell-Smith J (2008) Fires in north Australian savannas: size and frequency matter. *International Journal of Wildland Fire* **17**, 768–781.

8

Fire management and biodiversity of the western Arnhem Land Plateau

John CZ Woinarski, Jeremy Russell-Smith, Alan N Andersen and Kym Brennan

SUMMARY

The sandstone plateau of western Arnhem Land supports an unusually high number of endemic plant and animal species, distinctive vegetation types and many threatened species. Such biodiversity significance is due to the plateau's isolation from other sandstone massifs, its enduring presence within surrounds marked by long-term instability, the diversity of topographic (and resultant microclimatic) features that it offers, and the protection it affords from some otherwise pervasive threats. Some important components of this biodiversity are now in decline, with such trends driven by inappropriate fire regimes and invasive plants and animals. The management priorities required to maintain the biodiversity values of the plateau area are: to reduce the incidence, extent and severity of fire; to control feral animals (particularly cats and buffalo) more effectively; to prevent spread to the plateau of invasive pasture grasses; and to establish or enhance comprehensive monitoring programs that measure trends in biodiversity and management effectiveness.

INTRODUCTION

This chapter describes the biodiversity values of the western Arnhem Land Plateau (the 'stone country') and surrounds, attempts to provide a brief assessment of condition and trends in that biodiversity, describes the responses of some biodiversity components to fire regimes, and gives guidelines for fire management that may provide for biodiversity conservation within a broader management context. This review is constrained by the lack of substantial ecological studies that have been conducted in the stone country proper. In the Top End of the Northern Territory, with some notable exceptions, most of the autecological studies that relate to fire responses of individual plant and animal species, and all of the longer-term ecological experimental studies, have been conducted in the lowlands surrounding the western Arnhem Land Plateau. These lowland studies provide invaluable context and are referred to here, but extrapolations and interpretations derived from such lowland studies should be applied with considerable care to the stone country.

The current fire regimes of the Arnhem Land Plateau are described in detail elsewhere in this volume (e.g. Chapter 9, 10 and 14). The apparent detrimental impacts of those regimes upon biodiversity, as described in this chapter, have recently prompted the establishment of the most detailed and considered ecologically based fire-management strategy yet developed in northern Australia: the *Kakadu National Park Arnhem Land Plateau Fire Management Plan* (Petty *et al.* 2007a). This is an admirable attempt at the resolution of a very significant (and perhaps worsening) conservation management problem. However, its execution and outcomes remain untested, and it focuses specifically only on that 15% of the plateau area encompassed within Kakadu National Park. An analogous management approach that encompasses the rest of the plateau area is described in Chapter 9 (Edwards and Russell-Smith 2009).

BIODIVERSITY VALUES

The broad landscape context and description of the western Arnhem Land Plateau area is provided in Chapters 1 and 4. There has been no consolidated documentation of its biodiversity values, although it has long been recognised as special (Specht 1958; Northern Territory Reserves Board 1972; Christian and Aldrick 1977; Senate Standing Committee on Environment, Recreation and the Arts 1988). Aspects of its conservation significance were described in some of the reports prepared for the land-use planning process that resulted in the establishment and co-habitation of Kakadu National Park and a set of mining ventures (notably Ranger Uranium Environmental Inquiry 1976), subsequently in the description of the conservation values (including World Heritage status) of Kakadu National Park (e.g. Press *et al.* 1995a) and, most recently, in the case for the establishment of Warddeken Indigenous Protected Areas including those parts of the plateau area to the east of the Kakadu boundary.

Intricate knowledge of the biological values and ecological functioning of the plateau area has been held by Indigenous landowners for millennia. This still exists in the keeping of senior custodians (e.g. Russell-Smith *et al.* 1997; Telfer and Garde 2006; Chapter 5), but the generational transfer of this knowledge has been greatly interrupted over the last 50 years as the lands have become increasingly deserted, and the day-to-day need to hold and apply the knowledge has diminished (see Chapter 4). As described elsewhere in this volume, a range of recent initiatives has attempted to reverse this trend by attributing value to this venerable knowledge and seeking to maintain and apply it.

Since European settlement, biologists have been intrigued by the plateau and recognised the distinctiveness of its biota (e.g. Dahl 1897, 1926). Most of the early visits by biologists were brief collecting trips to the plateau margins. The American–Australian Arnhem Land Scientific Expedition of 1948 (Specht 1958) provided the first substantial assessment of the biodiversity of the plateau: particularly in the recognition of the antiquity of its biota, the relatedness of components of its vegetation to nutrient-poor heathlands in temperate Australia, and the distinctiveness of some of its components from the vegetation of the surrounding extensive lowlands. Specht (1958) recognised the melange of floristic elements: co-existing, intermingling or very finely partitioned because of the diversity of environments associated with the marked topographic variability of the plateau, including rock pavements, deeply dissected gorges, wetland systems and deep sandy plains. The elements include:

- relictual rainforest species (mostly in sheltered gorges)
- heathland species (mostly on rock pavements and sandsheets) with a comparably long history in this landscape
- more recent and vagile rainforest species

- components of the eucalypt-dominated savanna systems that form the dominant vegetation type across the surrounding lowlands.

Plant diversity is high: partly because of this fine-scale mix of environments and vegetation elements, but also because the highly dissected plateau allows for extensive localised speciation of groups with limited dispersal ability, such as the *Stylidium* trigger plants (Carlquist 1979).

Many of the relictual species are of outstanding evolutionary and conservation significance. The plateau has remained an unbroken (but often isolated) feature of the landscape for more than 100 million years, while the surrounding lowlands have intermittently been inundated (Needman 1988; Duretto and Ladiges 1997). The range of landscape elements (and hence microclimates) across the plateau has offered shelter to species faced with climate changes to which they would otherwise be ill-suited. The persistence of some of these species has been remarkable. In some cases, the persistence is now extraordinarily precarious. Figure 8.1 illustrates some examples of the distributions of some species for which the plateau provides refuge. From at least the Jurassic (136–195 million years ago), much of the Gondwana continent was dominated by conifers including the genus *Podocarpus* (Adam 1994); fossil pollen samples from sediment cores near the source of the South Alligator River attest to the presence of *Nothofagus* and *Podocarpus* in this region around 50 million years ago (Truswell 1982; Wilson *et al.* 1996). *Podocarpus* persists in Australia now mostly in high-altitude rainforests of the Great Dividing Range. In the Northern Territory, it exists only in a small number of highly incised gully systems draining into the East Alligator River, with the species present, *P. grayae*, separated by more than 1500 km from its nearest other populations, in northern Queensland (Figure 8.1, 8.2a). Another example of a relictual species is the heathland plant *Drummondita calida*, which is known in the Northern Territory only from the western escarpment of the Arnhem Land Plateau. This population is remarkably disjunct from its only other occurrence in the Einasleigh Uplands area of north Queensland, which is itself separated from all other species in the genus, which are all restricted to heathlands in south-western Australia (Figure 8.1). A contrasting example is the spectacular tree *Hildegardia australiensis* (Figure 8.2b; Figure 8.4e), which is restricted to a few sites of the Arnhem Land Plateau; the other 10 members of this genus occur in Africa, south-eastern Asia and the Caribbean. A feature of many of the most localised plant species of the Arnhem Land Plateau is their current restriction to some of the most rugged habitats, such as cliffs, waterfalls, gorges and defiles. Such sites offer the greatest possible protection from fire and/or buffering from otherwise largely unfavourable climatic conditions. Examples include the threatened plants *Boronia viridiflora* and *Sauropus filicinus*, which are both now restricted to crevices in a small set of sandstone cliff faces (Figure 8.2c).

The international significance of the Arnhem Land Plateau as a centre for plant biodiversity was explicitly recognised through the inclusion of a 'Kakadu–Alligator Rivers Region' in a listing of the world's centres for plant diversity (Ingwersen 1995), in part on the basis that this area included about 40 endemic plant species. This region was also recognised in a more systematic analysis of Australian sites of plant endemism by Crisp *et al.* (2001). More recently, and with access to a far more comprehensive information base, Woinarski *et al.* (2006) analysed the distributions of all plant species endemic to the Northern Territory, and established that the Arnhem Land Plateau was by far the most significant Northern Territory site for endemic plants (Figure 8.3a), with 172 plant species restricted to the plateau and a further 25 species mostly restricted (>90% of their range) to it. Although some sites in the Arnhem Land Plateau area are botanically well explored, many other areas remain largely unvisited and the number of plant species (including endemic plant species) known from the plateau continues to increase substantially (Brennan 1996; Harwood and Dessein 2005).

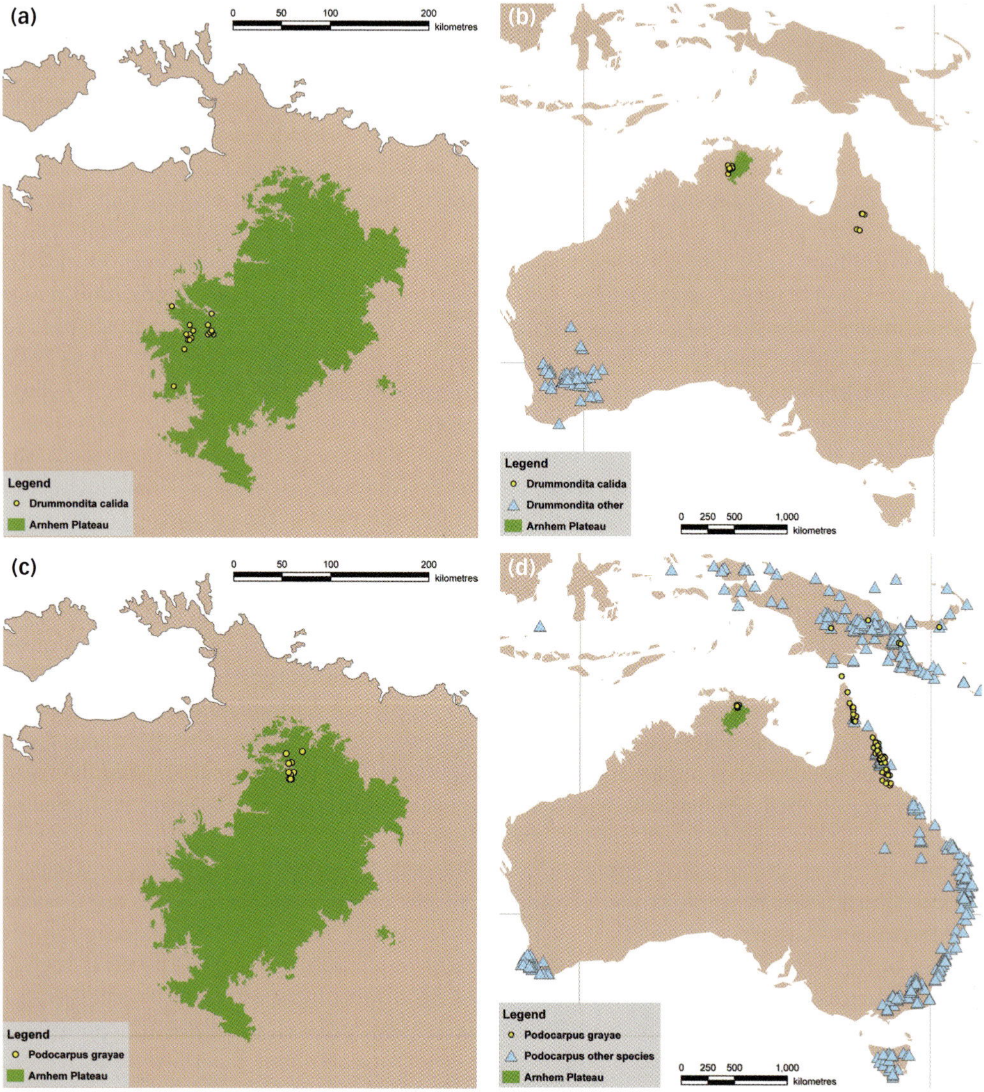

Figure 8.1 Maps of the current locations of (a) *Drummondita calida* in the western Arnhem Land Plateau, (b) all *Drummondita* species, (c) *Podocarpus grayae* in the western Arnhem Land Plateau, (d) all *Podocarpus* species (but noting that records beyond Australia are not comprehensive). (Source: Australian Virtual Herbarium).

As with plants, the Arnhem Land Plateau is of international significance for endemic, relictual and threatened fauna. To date, there has been no comprehensive assessment of the richness and endemicity of the invertebrate fauna of the western Arnhem Land Plateau. Our knowledge of the terrestrial invertebrate fauna is especially poor, and is based primarily on opportunistic collections from that part of the Arnhem Land Plateau and escarpment in Kakadu National Park (Press *et al.* 1995b). Two of the best known groups are ants and grasshoppers. The ant fauna of Australia's monsoonal tropics is exceptionally rich (Andersen 2000), and there are likely to be several hundred species in the western Arnhem Land Plateau. The ant fauna of the sandstone country associated with the western Arnhem Land escarpment is highly distinctive,

Figure 8.2 Examples of some plants and animals endemic to – or with relictual distribution including the – Arnhem Land Plateau: (a) *Podocarpus grayae* (photo: Jeremy Russell-Smith), (b) *Hildegardia australiensis* (photo: A. Fisher), (c) *Boronia viridiflora* (photo: D. Liddle), (d) Leichhardt's grasshopper on one its main host plants, *Pityrodia jamesii* (photo: P. Barrow). This specimen carries a plastic disk used as marker during a recent study of dispersal.

with perhaps 30% of species not occurring in the lowland savanna to the west. Many of these are undoubtedly endemic to the immediate region, and this endemicity occurs at higher taxonomic levels, including an undescribed species group of *Meranoplus* that is most unlike other members of the genus (Andersen 2006). Such endemicity is clearly true more generally for the western Arnhem Land Plateau, with, for example, *Aphaenogaster reicheli* known only from Podocarpus Canyon (Shattuck 2008), and a species of *Rhytidoponera* recently collected from the Mann River region representing a new and highly unusual species group (AN Andersen, unpublished). The western Arnhem Land region is also notable for the occurrence of species

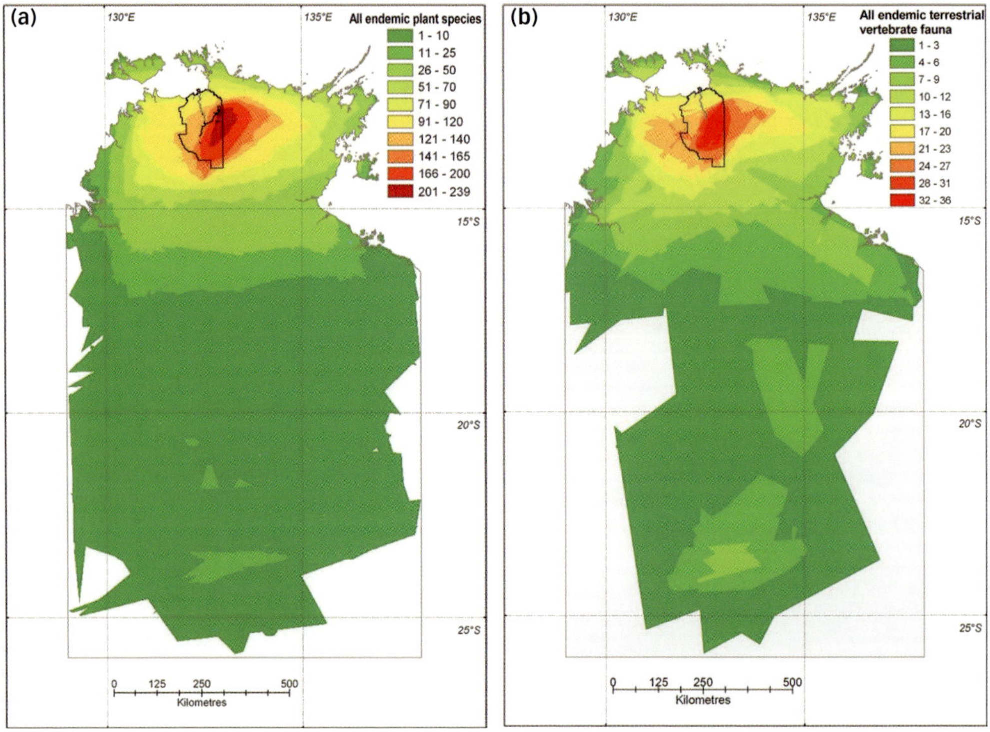

Figure 8.3 Contour maps showing the richness of Northern Territory endemic (a) plant species and (b) terrestrial vertebrate species, with the boundary of Kakadu National Park superimposed. (Sources: Woinarski *et al.* (2006) for plants; Woinarski and Hempel (unpublished) for vertebrate species)

with highly disjunct distributions. For example, the only record of the infamous bull-ant genus *Myrmecia* in the Top End is an undescribed species from the Marrawal plateau in the southwest of the region, which also occurs on the Mitchell plateau in Western Australia's northern Kimberley (AN Andersen, unpublished.). The coastal Queensland species *Dolichoderus scrobiculatus* has also been collected from the Marrawal plateau – about 2000 km from its previously known range; this is the only record of the genus in the Northern Territory. The only record of the ant genus *Prionopelta* in the Northern Territory is from a rainforest patch associated with the Arnhem Land Plateau (Reichel and Andersen 1996).

The Kakadu region supports a highly diverse and significant grasshopper fauna, with at least 11 species endemic to sandstone country of the Arnhem escarpment and plateau (Andersen *et al.* 2000). One notable plateau species is Leichhardt's grasshopper *Petasida ephippigera* (Figure 8.2d), which is restricted to the plateau of western Arnhem Land and a few other sandstone isolates in the Top End and Victoria River region (Lowe 1995). It is notable for its dependence on a narrow set of host food plants: primarily two aromatic shrub species of *Pityrodia* (Verbenaceae), and for its bright and spectacular orange and blue aposematic coloration. It has special spiritual significance for the local Jawoyn people, who relate it to the 'lightning man' (Namarrgon) responsible for the wet season storms.

The biogeographic significance of the west Arnhem Land Plateau for ants and grasshoppers is also shown in other insect groups for which there is sufficient information. For example, several species of dragonfly are endemic to streams associated with the plateau (Press *et al.* 1995b). Some of these, such as *Hemigomphus magela*, are most closely allied to species from temperate southern Australia. All other species of *Hemigomphus* occur in coastal and montane

Table 8.1: List of vertebrate species endemic to the Arnhem Land Plateau.

Fish	
Hypseleotris barrawayi	
Craterocephalus marianae	Mariana's hardyhead
Pingalla midgleyi	Midgley's grunter
Frogs	
Uperoleia arenicola	Jabiru toadlet
Litoria personata	masked rock-frog
Reptiles	
Chelodina burrungandjii	sandstone long-necked turtle
Gehyra pamela	Arnhem Land spotted dtella
Oedura gemmata	jewelled velvet gecko
Pseudothecadactylus lindneri	giant cave gecko
Ctenotus arnhemensis	Arnhemland ctenotus
Ctenotus coggeri	Cogger's ctenotus
Ctenotus gagudju	Kakadu ctenotus
Ctenotus kurnbudj	Alligator Rivers ctenotus
Egernia obiri	Arnhemland egernia
Menetia concinna	neat menetia
Morelia oenpelliensis	Oenpelli python
Birds	
Petrophassa rufipennis	chestnut-quilled rock-pigeon
Ptilinopus cinctus	banded fruit-dove
Amytornis woodwardi	white-throated grass-wren
Meliphaga albilineata	white-lined honeyeater
Mammals	
Macropus bernardus	black wallaroo
Zyzomys maini	Arnhem Land rock-rat

streams of eastern Australia (Watson *et al.* 1991), with closely allied genera occurring in south-western Australia and South America.

The freshwater invertebrate fauna of the western Arnhem Land Plateau also includes a substantial and significant endemic component, although quantification of this is currently hampered by very limited collection and taxonomic scrutiny (such limitations are now being addressed by a range of collaborative studies involving the Environmental Research Institute of the Supervising Scientist). The fauna includes an endemic family of shrimps – the Kakaducarididae, comprising two endemic genera (Bruce 1993, Bruce and Short 1993, Page *et al.* 2008) – as well as an endemic genus of phreatoicidean isopod (*Eophreatoicus)* that has exceptional species-level diversity (Dr G Wilson, Australian Museum *pers. comm.*). Most of these macro-crustacean species have very restricted distributions: often limited to single streams, seeps or springs.

Twenty-two vertebrate species (three fish, two frog, eleven reptile, four bird and two mammal) are endemic to the Arnhem Land Plateau (Table 8.1) – by far the highest concentration in the Northern Territory (Figure 8.2b), and very high by Australian standards. The list includes one of Australia's largest snakes (the Oenpelli python *Morelia oenpelliensis*) and a large macropod (the black wallaroo *Macropus bernardus*). Other vertebrate species thought to be restricted to the plateau (notably a small carnivorous marsupial, *Planigale* sp.) remain undescribed.

Some of the plateau endemics form disjunct populations of more wide-ranging superspecies, whose level of taxonomic divergence provides some indication of the duration of their current isolation and the extent to which they can or cannot disperse through less-suitable lowland habitats. For example, the white-lined honeyeater *Meliphaga albilineata*, chestnut-quilled rock-pigeon *Petrophassa rufipennis*, white-throated grass-wren *Amytornis woodwardi*, Arnhem rock-rat *Zyzomys maini* and giant cave gecko *Pseudothecadactylus lindneri* all have closely related species (allospecies) occurring in comparable sandstone environments in the Kimberley, Western Australia (*M. fordiana, P. albipennis, A. housei, Z. woodwardi* and *P. cavaticus*, respectively). There are also comparable 'sister' species for the grass-wren (Carpentarian grass-wren *A. dorotheae*) and rock-rat (Carpentarian rock-rat *Z. palatalis*) in the smaller isolated sandstone ranges of the Gulf of Carpentaria hinterland. Other species occur broadly across almost all of the disjunct ranges of northern Australia (e.g. common rock-rat *Z. argurus*), or patchily across some (e.g. nabarlek *Petrogale concinna*, sandstone antechinus *Pseudantechinus bilarni*: Fisher *et al.* 2000). More idiosyncratic is the banded fruit-dove *Ptlinopus cinctus*, which is restricted in Australia to rainforests of the Arnhem Land Plateau, but variably considered specifically distinct from a closely related form occurring from Bali to the Lesser Sunda Islands (including Timor) in Indonesia (Christidis and Boles 2007).

The plateau supports not only the Northern Territory's largest concentration of endemic species but also one of the largest concentration of threatened species (Table 8.2; Department of Natural Resources Environment and The Arts 2005). This latter factor reflects two somewhat opposing factors:

1. that some species have declined substantially across their broader (lowland) range, but are now finding their last refuge in the stone country: in part because there may be some moderation in the plateau region of some otherwise pervasive threats (such as weeds and feral animals)
2. that some species largely restricted to the plateau are now declining in the area (e.g. white-throated grass-wren).

In fact, these two variants are probably just truncated versions of a process that has been operating over very long timeframes: that species ebb and flow from the plateau to the lowlands, retreating to the plateau during inhospitable periods and expanding out in favourable times. Where the periods are particularly inhospitable, there will also be retreat to the most rugged (insulated) parts of the plateau and, at times of particular stress, even these may prove inadequate.

The plateau supports not only threatened and relictual species but also distinctive assemblages of species (Figure 8.4). The most notable of these are sandstone rainforests and sandstone heathlands. There is a range of rainforest assemblages in and around the Arnhem Land Plateau that are floristically dissimilar to the rainforests of the lowlands (Russell-Smith 1991). The most extensive and notable are closed forests dominated by the ancient proto-eucalypt *Allosyncarpia ternata*. This majestic tree is restricted to the Arnhem Land Plateau, and its closest relatives comprise similarly old relictual genera isolated in mountain tops in Queensland (*Stockwellia*), eastern Malesia (*Eucalyptopsis*) and New Caledonia (*Arillastrum*) (Ladiges *et al.* 2003; Crisp *et al.* 2004). On the Arnhem Land Plateau, *Allosyncarpia ternata* can form monospecific forests (especially on slopes and sandsheets) or mix with a diverse set of rainforest trees (especially in gullies, watercourses and springs) (Bowman 1991; Russell-Smith 1991), with a total extent of 1138 km^2 (Russell-Smith *et al.* 1993).

Table 8.2: List of threatened species occurring in the Arnhem Land Plateau. Trend assessment (for the Arnhem Land Plateau area) is based on interpretation from Woinarski (2004a) and Woinarski et al. (2007). Codes for conservation status: CR=critically endangered; EN=endangered; VU=vulnerable; DD=data deficient; – = not listed. Northern Territory (NT) status is as defined in regulations of the *Territory Parks and Wildlife Conservation Act* 2000; Australian (Aust.) status is under the *Environment Protection and Biodiversity Conservation Act* 1999.

Species		Status (NT)	Status (Aust.)	Trend
Plants				
Acacia sp. Graveside Gorge		CR	CR	unknown
Boronia quadrilata		VU	VU	unknown
Boronia viridiflora		VU	VU	unknown
Cephalomanes obscurum		EN	–	unknown
Eleocharis retroflexa		DD	VU	unknown
Freycinetia excelsa		VU	–	unknown
Hibiscus brennanii		VU	VU	unknown
Lithomyrtus linariifolia		VU	–	unknown
Sauropus filicinus		DD	VU	unknown
Toechima sp. East Alligator		EN	EN	unknown
Utricularia singeriana		VU	–	unknown
Reptiles				
Varanus mertensi	Mertens' water monitor	VU	–	declining
Varanus panoptes	yellow-spotted monitor	VU	–	declining
Egernia obiri	Arnhemland egernia	EN	EN	declining
Morelia oenpelliensis	Oenpelli python	VU	–	unknown
Birds				
Dromaius novaehollandiae	emu	VU	–	declining
Erythrotriorchis radiatus	red goshawk	VU	VU	stable
Ardeotis australis	Australian bustard	VU	–	declining
Geophaps smithii smithii	partridge pigeon	VU	VU	unknown
Tyto novaehollandiae kimberli	masked owl (north Australian mainland)	VU	VU	unknown
Amytornis woodwardi	white-throated grasswren	VU	–	declining
Falcunculus frontatus whitei	crested shrike-tit (northern)	VU	VU	unknown
Erythrura gouldiae	gouldian finch	EN	EN	stable
Mammals				
Dasyurus hallucatus	northern quoll	CR	EN	declining
Phascogale pirata	northern brush-tailed phascogale	VU	–	declining
Isoodon auratus	golden bandicoot	EN	VU	declining, probably extinct
Hipposideros diadema inornatus	Arnhem leaf-nosed bat	VU	–	unknown
Conilurus penicillatus	brush-tailed rabbit-rat	VU	–	declining
Mesembriomys macrurus	golden-backed tree-rat	CR	VU	declining
Notomys aquilo	northern hopping-mouse	VU	VU	unknown
Zyzomys maini	Arnhem Land rock-rat	VU	VU	declining

Figure 8.4 Characteristic landscapes and ecological communities of the stone country: (a) fissures and gullies offering shelter for fire-sensitive plants and rainforests, (b) escarpment, (c) edge of sandstone heathland with woodland dominated by *Callitris intratropica*, (d) rainforest, (e) rainforest patch on rocky hilltop, here dominated by *Hildegardia australiensis* (photos by Jeremy Russell-Smith with the exception of (b), by Cameron Yates).

Heathland communities are scattered across Australia, especially on nutrient-poor sandy substrates. They are typically dominated by woody shrubs belonging to the families Myrtaceae, Fabaceae and Ericaceae. Many of the heathland species reproduce only by seed, and characteristically require a disturbance-free interval of several to many years before they are sufficiently mature to produce seeds. Across the Arnhem Land Plateau, open heathlands and shrublands occur on rock platforms, skeletal sandy soils and deeper sand sheets – occupying a total area of between 18 920 and 24 510 km^2, depending upon the inclusion, or not, of sandstone isolates outside the periphery of the core plateau area (Blake 2004). These formations may be floristically diverse, and many of the constituent plant and animal species are endemic. More species-poor versions of these communities extend from the Arnhem Land Plateau to smaller and less-dissected sandstone isolates elsewhere in northern Australia, or to sandsheets and coastal dunes of the lowlands; these attenuated heathlands typically include lower representation of obligate-seeding species.

The Arnhem Land Plateau supports an unusually diverse set (13 species) of hummock grasses *Triodia* spp.: in part mixed among the sandstone heathlands, as the dominant component of understorey in some woodlands, or as one of the few plants growing in boulder fields and rock pavements. As with hummock grasses elsewhere (Morton and James 1988), these may be ecologically pivotal: providing key habitat, shelter and food resources for many fauna species. For the western Arnhem Land Plateau, the white-throated grass-wren is probably one such example. The plateau is also the centre of diversity (11 species) of resurrection grasses *Micraira* spp. – all with unusual moss-like appearance, spirally arranged leaves, and mostly occurring on rock pavements with no or skeletal soil. These typically highly localised species are particularly adapted to the extreme seasonal challenge of such sites, notably by showing very rapid response of dried-off foliage to rehydration ('revival': Lazarides 1979).

Some plateau habitats are far less distinctive. As with the lowlands, eucalypt open forests and woodlands occur over extensive areas (about 75% of the Western Arnhem Land Fire Abatement project area: Edwards and Russell-Smith 2009) of the plateau. While there is some degree of floristic variation between upland and lowland eucalypt woodlands (e.g. Taylor and Dunlop 1985; Wilson *et al.* 1990; Dunlop and Webb 1991), they also show many structural and floristic similarities. Dominant plateau woodland trees include *Eucalyptus tetrodonta, E. miniata* (and its endemic close relative *E. gigantangion*), *E. tintinnans, Corymbia dichromophloia, C. arnhemensis, C. bleeseri, C. kombolgiensis, C. ferruginea, C. setosa, C. latifolia, Erythrophleum chlorostachys* and *Xanthostemon paradoxus*, with overstorey and understorey species composition and forest/woodland structure varying subtly or markedly in response to variation in soil depth and texture, geology and landscape position (Burgman and Thompson 1982; Bowman *et al.* 1990, 1993; Wilson *et al.* 1990, 1996). These woodlands support many endemic understorey species, particularly in the genera *Acacia, Boronia, Calytrix, Gardenia, Goodenia, Grevillea, Hibiscus, Jacksonia, Lithomyrtus* and *Spermacoce*, with many of these understorey plants also forming dominant components in adjacent or intermingling heathlands (Wilson *et al.* 1996; Woinarski *et al.* 2006). As scattered copses, isolated individuals or occasionally dominant in patches, the northern cypress-pine *Callitris intratropica* is a notable feature of these woodlands, and provides a signal of the complex evolutionary mix in the flora of this region. As with many of the heathland plant species, *Callitris* is an obligate seeder, with, in at least some habitats, new plants requiring a fire-free period of at least 10 years to mature sufficiently to set seed (Russell-Smith 2006).

CONDITION AND TRENDS IN BIODIVERSITY

As evident in the assessment of trends for threatened species given in Table 8.2, there is little or no information on trends in the status of many of the plateau species of most conservation significance. Even for those species for which some trend is provided, the monitoring or other data from which this assessment is based is notably meagre. There is no substantial or sustained targeted monitoring program for any threatened species in the plateau area, although some limited monitoring programs have been established recently, with baseline sampling (Kerrigan 2003, 2004; Woinarski 2004a; Cowie 2005; see also Chapter 10).

Limited trend information on some components of biodiversity is available from a range of sources, including Indigenous knowledge, targeted monitoring, incidental by-product from other monitoring, and some ecological studies. In some cases, trends on the plateau reflect more general trends across the broader region (e.g. golden bandicoot *Isoodon auratus*, northern quoll *Dasyurus hallucatus*: Woinarski *et al.* 2001); in other cases, trends on the plateau are more muted or delayed than sharper trends in the lowlands (e.g. *Callitris intratropica*); but in no case is the plateau unequivocally defying trends in the lowlands.

Species have been lost from the plateau. Spectacular depictions of thylacines *Thylacinus cyanocephalus* and Tasmanian devils *Sarcophilus harrisi* in rock art (and, in the latter case, as subfossils) attest to the regional extinction of these species sometime in the last 20 000 years, and perhaps as recently as 3000 years BP (Calaby and White 1967; Brandl 1972; Calaby and Lewis 1977; Lewis 1977). Bones from surficial archaeological sites demonstrate the presence of northern hopping-mouse *Notomys aquilo*, probably as recently as post-European contact (Jones 1985). This species appears to be now absent from the plateau area, with the sole exception of a 1973 record from the upper Caddell River (Woinarski *et al.* 1999). The golden bandicoot was probably present across the plateau (there are a few isolated records from the nearby lowlands, including as recently as 1967, near Goodparla: Woinarski 2004a), but appears now to have disappeared from the entire Top End mainland. The golden-backed tree-rat *Mesembriomys macrurus* is recorded in the Northern Territory from only three specimens: the most recent being in 1969 from Deaf Adder Gorge (Woinarski 2004a). It has not been trapped since, notwithstanding substantial survey effort including in many areas of the plateau. However, a recent study using stuffed mammals as props to record traditional knowledge (M. Ziembicki *pers. comm.*) has indicated that this species may persist in at least one site on the plateau.

Several species have declined precipitously over the last decade following the colonisation of the plateau by cane toads *Rhinella marina* (mostly from 2000–01). The most extreme of these declines has been for the northern quoll (Watson and Woinarski 2003), which has suffered consequential local extinction over most of its range. Bucking the general trend, a population of quolls has persisted (so far) following toad colonisation near the East Alligator ranger station at Kakadu National Park (J. Woinarski unpublished). This persistence mirrors some examples of local populations of quolls persisting with toads in parts of Queensland, typically in the most rugged areas, which are the highest habitat quality for northern quolls (Woinarski *et al.* 2008). Other species known to have suffered at least short-term declines in the plateau (and throughout the broader Top End region) because of toads include Mertens' water monitor *Varanus mertensi*, yellow-spotted goanna *V. panoptes*, freshwater crocodile *Crocodylus johnstonei* and king brown snake *Pseudechis australis* (e.g. Watson and Woinarski 2003; Doody *et al.* 2006; Smith and Phillips 2006; Griffiths and McKay 2007). Declines are also suspected for some small dasyurid marsupials (e.g. northern brush-tailed phascogale *Phascogale pirata*), some agamid lizards and some other snake species (Watson and Woinarski 2003). It is possible that the significant endemic invertebrate fauna of the plateau wetlands may also have been detrimentally affected by toads (Dr C. Humphries *pers. comm.*).

Across much of the Top End, there has been substantial decline of the mammal fauna over the last century, with declines continuing over the last decades (e.g. Woinarski *et al.* 2001). Anecdotal information from Indigenous landowners and other experienced observers suggest that there has also been a substantial decline of medium-sized mammals in the plateau, which has been reported for some other conspicuous and significant animal species notably including emu *Dromaius novaehollandiae* (Yibarbuk and Cooke 2001) and nabarlek *Petrogale concinna* (D. Pearson *pers. comm.*). Results from fauna sampling at 63 fire monitoring plots at Kakadu (including 26 plots in the Arnhem Land Plateau and sandstone outliers) indicate a substantial decline between the baseline sampling (in 2001–02) and most recent sampling (2007–08), with decline occurring both in lowlands and uplands, although perhaps more muted in uplands (Figure 8.5). Such downward trends are also evident from more-intensive sampling at particular sites (e.g. Nawurlandja: Watson and Woinarski 2003).

In contrast, the numbers of some feral animals are increasing in the plateau area, although there are few population estimates available. Gorman *et al.* (2007) reported substantial recent increases in the abundance of feral water buffalo *Bubalus bubalis* from central Arnhem Land, which have been building up again following a significant reduction in their numbers during the Brucellosis and Tuberculosis Eradication Campaign between 1985 and 1989. The cane toad invaded the entire plateau area over a 3–5 year period after 2000. The first records of the exotic black rat *Rattus rattus* have been reported (from several areas) in the stone country over the last 5 years. There is some anecdotal evidence for increases in feral cats *Felis catus* and pigs *Sus scrofa*; and feral cattle *Bos taurus/indicus* and horses *Equus caballus* remain at least locally abundant, despite some localised hunting and more systematic aerial shooting programs (Franklin *et al.* 2008).

There is far more information available for trends in (at least some) plants, most particularly for some key charismatic plant species considered indicators of fire regimes – *Allosyncarpia ternata* and *Callitris intratropica* – with adequate information also from specific studies for some other fire-sensitive plant species (e.g. *Petraeomyrtus punicea*: Russell-Smith 2006). More generally, there is a very considerable body of data also available for nearly 1000 plant species from the large set of established fire-monitoring plots at Kakadu and Nitmiluk National Parks (Edwards *et al.* 2003; see also Chapter 10), although only about 10% of these have been recorded in a sufficient number of plots to allow for robust monitoring. Trends in the status and condition of some key vegetation communities are also determinable from a range of studies including fire-plot monitoring (Edwards *et al.* 2003; Chapter 10), remotely-sensed and field-based measurements of boundaries and extent (Bowman 1994; Banfai and Bowman 2006; Bowman and Dingle 2006; Prior *et al.* 2007), correlative distributional and fire-response studies (Bowman and Panton 1993; Russell-Smith *et al.* 1998, 2002; Yibarbuk *et al.* 2001; Bowman *et al.* 2001a, 2004; Bowman and Prior 2004), small-scale experiments (Russell-Smith *et al.* 2002), and demographic modeling (Price and Bowman 1994; Prior *et al.* 2007).

The pattern for broad-scale decline of *Callitris* in the plateau area is clear (Haynes 1985, 1991; Bowman and Panton 1993; Bowman *et al.* 2001a; Bowman and Prior 2004; Russell-Smith 2006; Prior *et al.* 2007), although the prevailing decline may be reversed or stabilised at sites having benign fire regimes (Yibarbuk *et al.* 2001; Edwards *et al.* 2003; Bowman and Prior 2004; Edwards and Russell-Smith 2009). Dead *Callitris* trees remain conspicuous in the landscape for decades, and hence allow for an unusually good assessment of trends from aerial survey or ground-based monitoring (Bowman and Panton 1993; Bowman and Prior 2004; Edwards and Russell-Smith 2009), although the diffuse distribution of *Callitris* across the plateau makes it difficult to map extent and boundary changes effectively, with the exception of some larger and sharply edged copses (e.g. Round Jungle: Bowman and Dingle 2006; Prior *et al.* 2007; Edwards and Russell-Smith 2009). Demographic modelling studies have demonstrated highly significant (current and) expected declines for *Callitris*, under most prevailing fire regimes (e.g. Price and Bowman 1994; Prior *et al.* 2007).

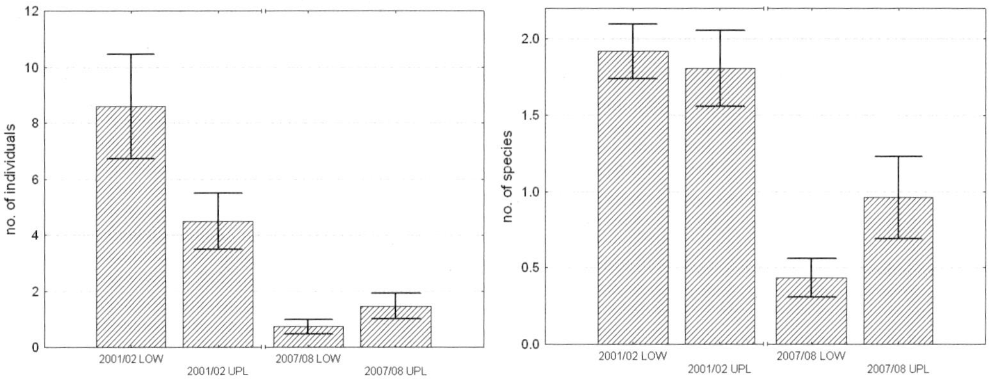

Figure 8.5 Trends in native mammal populations (a) number of individuals (b) number of species in Kakadu National Park, derived from repeat sampling at 63 fixed monitoring plots. Columns represent means, whiskers represent standard errors; 'LOW' indicates lowland sites; 'UPL' indicates upland sites.

In contrast, forests dominated by *Allosyncarpia* are more tractably mapped and monitored, and changes over time in the position of their edges – and of the demographic fingerprint of these edges – provide robust and explicit measures of their trend and condition. However, long-term monitoring of the boundaries and (hence) extent of *Allosyncarpia* forests has suggested some inconsistent trends (Bowman and Dingle 2006), probably resolvable with reference to site-specific fire histories. Across much of northern Australia, there are general trends for expansion of rainforests (and increased 'thickening' in some savanna woodlands), probably due to some increases in rainfall over the last few decades and to the effects of increased atmospheric CO_2 (Sharp and Whittaker 2003; Sharp and Bowman 2004; Vigilante and Bowman 2004a; Banfai and Bowman 2005, 2006; Bowman and Dingle 2006; Brook and Bowman 2006), albeit modulated (or in some cases reversed) by local factors (notably unfavourable fire regimes).

Beyond estimation of trends in the boundaries of *Allosyncarpia* patches, there have also been assessments of the condition of those patches, and of other plateau rainforests. Russell-Smith and Bowman (1992) reported that: 67% of the sandstone rainforests were 'severely disturbed' by fire; 47% 'severely disturbed' by feral buffalo and/or cattle; and 27% 'severely disturbed' by feral pigs. These measures of health are unlikely to have shown substantial improvement in the 16 years hence.

The overall condition and trend of sandstone heathlands is poor: the general trend has been for diminution of area, and local-scale loss of plant species richness, particularly of the functionally significant obligate seeder species (Russell-Smith *et al.* 1998, 2002; Edwards *et al.* 2003; Edwards and Russell-Smith 2009). This pattern has been variable, and is closely associated with the prevailing fire regime and site location. Overall decline in sandstone heathlands has been of such concern that they have recently been nominated for listing as an endangered ecological community under the Australian *Environment Protection and Biodiversity Conservation Act.* If approved, it would be the first listed community in the Northern Territory, and only the second of the 38 listed Australian ecological communities whose qualification for listing has not been linked to threats from intensive development: the only other exception being for a 5 km^2 community in south-western Australia threatened primarily by the cinnamon fungus *Phytophthora cinnamomi*. It would also be one of very few listed threatened ecological communities with such a large proportion included in protected areas (23% of its extent is included in Kakadu National Park, 3.5% in Nitmiluk National Park, and almost all of the remainder is within the proposed Warddeken Indigenous Protected Area: A. Edwards *pers. comm.*).

There is some information focused on the spread of undesirable plants into, around and in the plateau, including some consideration of the ongoing increase in extent and dominance of exotic weeds (including mission grass *Pennisetum* spp.: P. Cooke *pers. comm.*). Weed incidence in the plateau area is generally low relative to the surrounding lowlands and pastoral lands (Franklin *et al.* 2008); however, there is increasing incidence or risk of incursions, mostly along road corridors and from current or abandoned sites of more intensive development (such as mine sites), with feral animals (and macropods) responsible for some dispersal. The current incidence of some weeds is so localised that eradication and prevention of further incursions may be possible, given appropriate resourcing for targeted management (P. Cooke *pers. comm.*).

RESPONSES TO FIRE REGIMES

From the discussion above about trends in biodiversity on the plateau, it is evident that many environmental values are in decline. Much of this decline is attributable to inappropriate fire regimes. To some extent, there is a contrast between plants and animals in this pattern. Decline has been far more marked for many animal species than for plant species. Fire is the principal agent for decline for plants and vegetation communities, but there may be a broader range of threatening factors that have complex and compounding impacts for animals. Furthermore, the response to fire regimes may be more intricate and contextual for animals than for plants.

In terms of conservation management, the most straightforward responses of species to fire regimes are those of obligate seeder plant species. For many such species, adult plants are readily killed by a single fire event, and the survival of the species at a burnt location is totally dependent upon regeneration from seed shed before (or in some cases, after) the fire event. This life history strategy will work only when there is no more fire at that site until after the new plants have reached maturity and produced seeds. (However, the system can be tweaked a little by variable seed dormancy and survival periods, variable susceptibility of adult plants to fires of varying intensity and timing, and by variability in the rate and mode of dispersal of propagules between burnt and unburnt patches: Vigilante and Bowman (2004a).) For many heathland plants – and some plants from other vegetation types – such a required fire-free interval is generally at least 3 years, and may more optimally be 5–8 years. For some of these species, life history also demands another fire requirement – because of shading or other resource exploitation by adult plants, and because plants have a limited lifespan, recruitment may require fire to remove the mature (or senescent) generation; that is, there may be a maximum fire-free period. The models developed for *Callitris intratropica* (Price and Bowman 1994; Russell-Smith 2006; Prior *et al.* 2007) provide the most informed such fire requirements currently available for obligate seeder species in the plateau, and suggest a preferred regime with shortest interval between successive fires of at least 10 years (Edwards and Russell-Smith 2009).

Plant species that can reproduce vegetatively after fire (such as most eucalypts and *Allosyncarpia*) typically have greater flexibility and resilience in persisting through a range of fire regimes. However, these species can still be severely disadvantaged by fires that are so intense that the mature plant is killed, or by fires that are so frequent that regeneration is fatally compromised. For rainforest patches, losses can be extreme (from intense fire); or gradual accretion may occur over decades from retreat at the margins or through openings in the canopy that allow for invasion by fire-promoting grasses (Russell-Smith and Bowman 1992; Prior *et al.* 2007). More broadly, in the savanna woodlands, frequent 'hot' fire may decrease the woody basal area and increase grass fuels, leading to irrevocable change in overall vegetation structure – and fire regimes – and render fire management increasingly challenging (Bowman *et al.* 2001a,

2004; Russell-Smith *et al.* 2003; Bowman and Prior 2004; Liedloff and Cook 2007). For the Arnhem Land Plateau, such a fire–grass cycle currently involves almost entirely native grasses (particularly spinifex *Triodia* spp. and sorghums *Sarga* spp.: Franklin *et al.* 2008); but any further incursions by the exotic pasture grasses of the surrounding lowlands may ratchet up the pace and extent of this detrimental change (Rossiter *et al.* 2003; Rossiter-Rachor *et al.* 2008).

To the extent that floristic composition and vegetation structure contribute to habitat suitability for most fauna species, the responses of vegetation to fire regimes also directly affect fauna. But the responses of fauna to fire may be far more complex. In general, the greater mobility of fauna than flora species allows them to work around the highly deterministic site-specific responses to fire of plants. Animals may be able to move from (fire-dictated) unsuitable to suitable patches, although this capability varies depending upon the scale of fire and the dispersal ability of the animals. Leichhardt's grasshopper, whose individuals may have a dispersal range of less than hundreds of metres, may be far more readily overcome by a 1 km^2 patch of recently burnt habitat than would a masked owl *Tyto novaehollandiae*, which may see such a patch as simply another layer of heterogeneity readily encompassed within a far larger territory.

Most information on faunal responses to, and requirements for, fire are derived from studies in the lowlands adjacent to the western Arnhem Land Plateau. But, with due caveats, the conclusions from such studies are broadly applicable to the plateau. In the next few paragraphs, we briefly describe some of the fire responses demonstrated in these studies. Most of these studies describe species that are in decline, with a generally inappropriate fire regime a factor contributing to that decline.

In lowland open forests, the partridge pigeon *Geophaps smithii* prefers habitat that has both patches of dense grassy understorey in which to provide shelter for the nest (little more than a scrape in the ground), but also open (burnt) areas in which to forage most efficiently (for fallen seeds) (Fraser *et al.* 2003). The optimal fire regime for this threatened species thus appears to be cool patchy fires at a scale smaller than its territory (1–10 ha), with fires early in the dry season (in order to provide extensive access to the resource of fallen seeds), although fires at this time come with the risk of destruction of nests.

The frilled lizard *Chlamydosaurus kingii* forages most effectively in open (more recently burnt) areas, but suffers greater mortality from fires in the late dry season (Griffiths and Christian 1996). As with the partridge pigeon, its optimal fire regime is probably spatially variable patchy fine-scale and early dry season fire, with most detriment arising from frequent late dry season fires (Brook and Griffiths 2004).

The brush-tailed rabbit-rat *Conilurus penicillatus* dens mostly in fallen hollow logs and hollows in large old trees (Firth *et al.* 2006a). Such sites are likely to be susceptible to intense fires (Williams *et al.* 2003), particularly those late in the dry season. The preferred diet of rabbit-rats comprises grass seeds, particularly those of perennial grass species (Firth *et al.* 2005). These are more likely to dominate in areas that are not exposed to annual fires; and indeed, broad-scale correlative distributional studies have demonstrated a negative association of rabbit-rats with areas that have been most recently burnt (Firth *et al.* 2006b). Thus a preferred fire regime for this species is likely to be one of infrequent fire, with any such fire preferably being relatively fine scale and of low intensity.

The related, but larger, black-footed tree-rat *Mesembriomys gouldii* also dens in tree hollows (and the tangled fronds of pandanus), but its diet comprises a higher proportion of fruits, and it is likely to find it difficult to forage on the ground when the ground layer is encumbered with dense grasses. Its habitat quality is likely to be greatest when the woodlands support a dense tall shrub layer (for these provide the greatest abundance and diversity of fruits) and sparse grass layer (Rankmore 2006). Such a structure occurs only where fire is infrequent (e.g. intervals of 5 or more years) (Taylor and Dunlop 1985; Russell-Smith *et al.* 2003; Vigilante and Bowman 2004b; Woinarski *et al.* 2004).

In an intensive radio-tracking study of the northern quoll in the Kakadu lowlands (before the arrival of cane toads), Oakwood (2000) found that the main cause of mortality was predation, and that this rate of predation increased proportionally to the extent of the home range that was burnt. Consistent with this assessment, more broad-ranging distributional studies have demonstrated that the occurrence of northern quolls was most likely in longer-unburnt sites (Woinarski *et al.* 2008). The management of fire for this species in the Arnhem Land Plateau area may be a moot point anyway, given its recent broad-scale demise following cane toad invasion.

The Gouldian finch *Erythrura gouldiae* occurs particularly on stony hillsides, including around the margins of the western Arnhem Land Plateau. Like the partridge pigeon, it forages most efficiently in open (burnt) areas, but has a special requirement for a sequential set of seeds of differing grass species to eke it through a resource bottleneck in the early wet season (Dostine *et al.* 2001). This bottleneck arises because the first rains trigger the loss (through germination) of the fallen grass seed resource on which this species depends (Garnett and Crowley 1994). The gouldian finch then works on an ecological knife-edge: relying on the local spatial and temporal juxtaposition of both seeds from grass species that are slower to germinate and fast-growing species that most quickly produce the new season's seeds. Perennial grasses are critical to this strategy, and these are most likely to persist in the landscape when fires are relatively infrequent (Woinarski *et al.* 2005). Finally, this species nests in tree hollows, particularly on hill slopes. The availability of such hollows and hollow-bearing trees is likely to decrease with frequent fire.

There have been no substantial management-related studies on an iconic endemic bird of the Arnhem Land Plateau: the white-throated grass-wren. However, it is thought to be closely associated with large and extensive (i.e. longer unburnt) hummocks of spinifex *Triodia* spp. and, given the current regime of frequent fire (Edwards and Russell-Smith 2009), likely to be in considerable decline (Woinarski 1992; Woinarski *et al.* 2007: but cf. Noske 1992). A closely related species, the Carpentarian grass-wren, appears to have undergone a very major decline over the last few decades because of too frequent and extensive fire in similar habitat in the Gulf of Carpentaria hinterland (Perry 2005; Woinarski *et al.* 2007).

There is a diverse macropod fauna in the Arnhem Land Plateau and surrounds. Indigenous people have long used fire to manipulate habitat suitability for (and hunt) these 'game' species, with some burning designed specifically to stimulate grass production ('green pick') to increase the abundance (or concentrate the distribution) of macropods (Bowman *et al.* 2001b; Telfer and Garde 2006; Murphy and Bowman 2007). Although frequent patchy fires probably do enhance habitat suitability for some of the larger macropod species (including the endemic black wallaroo) in open forests and sandsheets, it is not clear that this is beneficial for smaller macropods (e.g. short-eared rock-wallaby *Petrogale brachyotis*) and in rocky areas (Murphy and Bowman 2007).

Experimental manipulation of fire regimes has provided valuable insight into the responses of plant and animal species, particularly those of the savanna lowlands. The most substantial of these studies was that at Kapalga in the Kakadu lowlands, where four different experimental fire regimes (ranging from fire exclusion to annual, high-intensity fires in the late dry season) were imposed on replicated sub-catchments over a 5-year period (Andersen *et al.* 2003, 2005). The responses of most insect, reptile and bird species were muted (perhaps because of the short timeframe), but there were pronounced responses by small mammals, with particular reductions in abundance in areas burnt annually in the late dry season. For vertebrate fauna, the most detailed analysis of the Kapalga study was for northern brown bandicoot *Isoodon macrourus*

(Pardon *et al.* 2003). This demonstrated that bandicoots declined under all uniformly applied fire regimes – most particularly under that of annual late dry season fire – with least decline in unburnt and regular early dry season fires. Pardon *et al.* (2003) concluded that fine-scale patchy fires were most beneficial for bandicoots, and frequent intense fires were most detrimental. A conclusion from the Kapalga fire experiment was that the overall proportion of land burnt each year should be reduced. Such reduction could be achieved through spatially explicit long-term protection from fire of some areas and/or more generally through landscape-wide reduction in the annual extent of fire (Andersen *et al.* 2005).

For the longer-lasting (26 years), but unreplicated and smaller-scale, fire 'experiment' (unburnt versus annually burnt) at Solar Village, Woinarski *et al.* (2004) reported substantial change in the vegetation structure and floristic composition of savanna woodlands (towards rainforest elements), with consequential changes in assemblages of vertebrates (Woinarski *et al.* 2004) and ants (Andersen *et al.* 2006). This study demonstrated that a set of species including *Glaphyromorphus* skinks, bar-shouldered dove *Geopelia humeralis*, white-gaped honeyeater *Lichenostomus unicolor*, white-throated honeyeater *Melithreptus albogularis*, dusky honeyeater *Myzomela obscura*, northern fantail *Rhipidura rufiventris*, yellow oriole *Oriolus flavocinctus*, common brushtail possum *Trichosurus vulpecula* and black-footed tree-rat is clearly disadvantaged by frequent fire, and some of these species are in decline (and/or are occurring at abundances far below their potential) because frequent fire in lowland woodlands has left little long unburnt habitat (Woinarski 2004b). Conversely, other species were more common in the frequently burnt savanna woodlands at Solar Village, and most of these species are flourishing broadly across the landscape.

The only experimental fire study in the plateau is that by Begg *et al.* (1981) at Little Nourlangie Rock (Nawurlandja): a plateau outlier in Kakadu. This study examined the abundance and distribution of four mammal species (sandstone antechinus, Arnhem rock-rat, common rock-rat and northern quoll) before and for a year after a single (intense) fire. For the first three species, abundance decreased markedly following fire. Based on these results, Begg *et al.* (1981) recommended that fire regimes should comprise management burns in the early to mid dry season, at intervals of at least 3–5 years.

There have been some studies that have compared biodiversity at sites subject to ongoing traditional fire management with sites without such management. The forerunner of such studies was that of Haynes (1985, 1991), who noted that there was more successful regeneration of *Callitris* in clan estates near Maningrida that were traditionally managed than in areas not subject to such management. This theme – with particular reference to *Callitris* – has been reported in several subsequent studies (e.g. Yibarbuk *et al.* 2001; Bowman *et al.* 2001a, 2004; Bowman and Prior 2004), and reinforces and corroborates demographic-based modelling. From such studies, mixed-age *Callitris* stands, with successful recruitment, are recognised as the product of, and a marker of, landscape health and benign fire management.

But healthy *Callitris* stands are not a principal objective of traditional management. Rather, they happen to be a collateral consequence of such management that aims to promote the availability of useful resources and to serve a range of other utilitarian ends. The extent to which such management provides the optimal regime for biodiversity more generally remains unresolved (as indeed is the composition of such a biodiversity nirvana). The most assertive claim for such a conjunction has been made by Yibarbuk *et al.* (2001), who reported that one traditionally managed clan estate had unusually high biodiversity values. The generality of such a claim is arguable, given its basis on but one site and the lack of explicit quantitative comparisons with comparable areas subjected to contrasting fire regimes. But although maybe not yet proven, there is obvious sense in such an assertion: millennia of Indigenous management presumably produced some sort of equilibrium for contemporary patterns of biodiversity, and withdrawal of that management will destabilise that equilibrium: inevitably with some species finding the new regime less suitable.

Indigenous knowledge of fire is intricate (see Chapter 5), and its application provides considerable opportunity for improving the management – and conservation outlook – for the plateau region. However, although such application of traditional fire management is necessary, it may no longer be sufficient. Across most of the plateau area, there has been a 50–70 year hiatus in traditional management (Bowman *et al.* 2001a; Chapter 4). This may have shifted the ecological state to a condition (characterised particularly by increase in flammable grass fuels: Bowman *et al.* 2001a) that no longer responds in the same way to customary management. Such a shift may have been catalysed or further compounded by the spread of invasive weeds and changes in vegetation structure, floristics and fuel loads caused by feral animals (Werner 2005; Petty *et al.* 2007b). The difficulty of, or advantage in, (re-)imposing traditional management on such an altered state may be further blurred in at least some areas by changes in the rationale and practice of that management: away from traditional to 'corrective' fire management (Lewis 1994; Bowman *et al.* 2004). The broad-scale application of traditional fire management is also compromised by the reduced population of Aboriginal people across the plateau area, the reduced need to manage the land to provide sustenance, and the erosion of traditional knowledge – although recent initiatives aim to counter all of these constraints (Chapter 11).

CONCLUSION

This chapter demonstrates the extraordinarily high value of the Arnhem Land Plateau for biodiversity conservation, and notes that this value is being diminished. There is a need to identify these values more comprehensively (particularly for poorly known invertebrate groups, and more generally for biota in less-well inventoried areas of the plateau), and to establish more systematic monitoring that will provide more robust data on population trends and more insight into the factors that drive those trends.

Notwithstanding the incompleteness of our knowledge, it is clear that current fire regimes are a major driver of detrimental change in the plateau's biodiversity. To some extent, this is typical of environments in the Northern Territory or northern Australia generally, but there is also exceptionality about fire and biodiversity in the plateau. We examined this issue by developing a matrix of threatened species by threatening factors for all listed species in the Northern Territory (from Woinarski *et al.* 2007), and then tallying, for every subregion, the number of species affected by each threat. The matrix revealed that fire was by far the most frequently listed threatening factor for threatened species (particularly plants) in the plateau area (see Chapter 10), and that fire affected unusually many threatened species in the plateau area relative to most other regions in the Northern Territory (Figure 8.6). That is, fire is the most significant problem for threatened species in the plateau, and that this problem is worse in the plateau area than almost anywhere else in the Northern Territory.

Most species suffering decline in the plateau are disadvantaged by the high frequency and large extent of the current fire regime. A management goal of a substantial reduction in fire frequency, especially the frequency of large (e.g. >10 km^2) fires should be defined and implemented. Fire frequency data (from 1990) presented in Chapter 9 provide such an opportunity. Even with the limited knowledge base, this would be a 'no regrets' policy. This goal may be put into operation either through setting aside some substantial fire refuge areas, or reducing the overall fire frequency across the entire landscape. While noting the need for lower fire frequency and impact, other chapters consider the institutional and practical mechanisms for achieving this. In part, this can be done through resourcing Indigenous landholders to manage fire at landscape scales, informed by traditional practice.

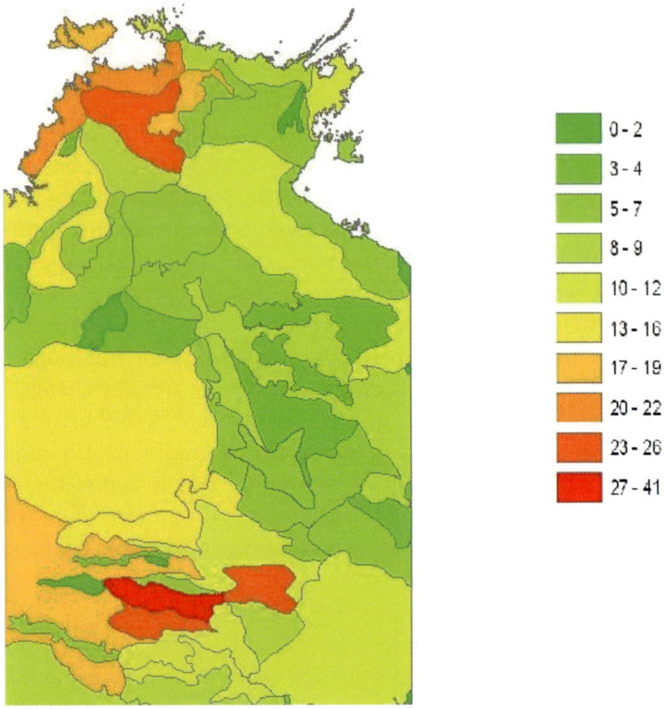

Figure 8.6 The number of threatened species for which fire is a recognised threat, for Northern Territory IBRA subregions (tallied from information in Woinarski *et al.* 2007).

There may be a happy conjunction in conflating the virtue of traditional fire management (with its associated implication for Indigenous employment and management direction) with optimal management for explicit biodiversity priorities. But uncritical acceptance or assertion of such symmetry will ultimately lead to disappointment for all interests, because:

- Significant components of faunal biodiversity are declining for reasons other than inappropriate fire regimes (alone)
- It is not yet clear that re-imposition of traditional modes of fire management will be sufficient to re-establish the ecological equilibrium that existed prior to European settlement
- At least some fauna species probably have habitat requirements that are poorly met by either the currently prevailing or more traditional fire regimes.

For these reasons, we recommend that management for biodiversity conservation in the plateau area complements enhanced fine-scale fire management (with targets and thresholds of reduced fire frequency) with management of other threatening factors (such as control of feral buffalo and cats, and elimination of any outbreaks of invasive pasture grasses), and that management resourcing is explicitly linked to biodiversity outcomes and is of a magnitude commensurate with the area's international significance for biodiversity.

ACKNOWLEDGEMENTS

We thank the very many landowners and others who have facilitated our work in, and educated us about, this beautiful area for much of our lives. We thank Matthew Fegan and Damian

Milne for preparation of some figures. We thank Chris Humphrey for information about freshwater invertebrates. Information presented here on fauna monitoring in Kakadu National Park derives from a study supported by Parks Australia.

REFERENCES

Adam P (1994) *Australian Rainforests.* Oxford University Press, Oxford.

Andersen AN (2000) *The Ants of Northern Australia: A Guide to the Monsoonal Fauna.* CSIRO Publishing, Melbourne.

Andersen AN (2006) A systematic overview of Australian species of the myrmicine ant genus Meranoplus F. Smith, 1893 (Hymenoptera: Formicidae). *Myrmecologische Nachrichten* **8**, 157–170.

Andersen AN, Lowe LM and Rentz DCF (2000) The grasshopper (Orthoptera: Acridoidea, Eumastacoidea and Tettigonioidea) fauna of Kakadu National Park in the Australian seasonal tropics: biogeography, habitat associations and functional groups. *Australian Journal of Zoology* **48**, 431–442.

Andersen AN, Cook GD and Williams RJ (Eds) (2003) *Fire in Tropical Savannas: The Kapalga Experiment.* Springer-Verlag, New York.

Andersen AN, Cook GD, Corbett LK, Douglas MM, Eager RW, Russell-Smith J, Setterfield SA, Williams RJ and Woinarski JCZ (2005) Fire frequency and biodiversity conservation in Australian tropical savannas: implications from the Kapalga fire experiment. *Austral Ecology* **30**, 155–167.

Andersen AN, Hertog T and Woinarski JCZ (2006) Long-term fire exclusion and ant community structure in an Australian tropical savanna: congruence with vegetation succession. *Journal of Biogeography* **33**, 823–832.

Banfai DS and Bowman DMJS (2005) Dynamics of a savanna-forest mosaic in the Australian monsoon tropics inferred from stand structures and historical aerial photography. *Australian Journal of Botany* **53**, 185–194.

Banfai DS and Bowman DMJS (2006) Forty years of lowland monsoon rainforest expansion in Kakadu National Park, northern Australia. *Biological Conservation* **131**, 553–565.

Begg RJ, Martin KC and Price NF (1981) The small mammals of Little Nourlangie Rock, N.T. V. The effects of fire. *Australian Wildlife Research* **8**, 515–527.

Blake G (2004) An object-orientated approach to mapping the distribution of sandstone heath vegetation on the Arnhem plateau. B.Sc. (Hons.) thesis. Charles Darwin University, Darwin.

Bowman DMJS (1991) Environmental determinants of *Allosyncarpia ternata* forests that are endemic to Western Arnhem Land, Northern Australia. *Australian Journal of Botany* **39**, 575–589.

Bowman DMJS (1994) Preliminary observations on the mortality of *Allosyncarpia ternata* stems on the Arnhem Land Plateau, northern Australia. *Australian Forestry* **57**, 62–64.

Bowman DMJS and Dingle JK (2006) Late 20th century landscape-wide expansion of *Allosyncarpia ternata* (Myrtaceae) forests in Kakadu National Park, northern Australia. *Australian Journal of Botany* **54**, 707–715.

Bowman DMJS and Panton WJ (1993) Decline of *Callitris intratropica* R.T. Baker and H.G. Smith in the Northern Territory: implications for pre- and post-European colonization ire regimes. *Journal of Biogeography* **20**, 373–381.

Bowman DMJS and Prior LD (2004) Impact of Aboriginal landscape burning on woody vegetation in *Eucalyptus tetrodonta* savanna in Arnhem Land, northern Australia. *Journal of Biogeography* **31**, 807–817.

Bowman DMJS, Wilson BA and Fensham BS (1990) Sandstone vegetation pattern in the Jim Jim Falls region, Northern Territory, Australia. *Australian Journal of Ecology* **15**, 163–174.

Bowman DMJS, Woinarski JCZ and Menkhorst KA (1993) Environmental correlates of tree species diversity in Stage III of Kakadu National Park, northern Australia. *Australian Journal of Botany* **41**, 649–660.

Bowman DMJS, Price O, Whitehead PJ and Walsh A (2001a) The 'wilderness effect' and the decline of *Callitris intratropica* on the Arnhem Land Plateau, northern Australia. *Australian Journal of Botany* **49**, 665–672.

Bowman DMJS, Garde M and Saulwick A (2001b) *Kunj-ken Makka Man-wurrk*. Fire is for kangaroos: interpreting Aboriginal accounts of landscape burning in central Arnhem Land. In *Histories of Old Ages: Essays in Honour of Rhys Jones*. (Eds A Anderson, I Lilley and S O'Connor) pp. 61–78. Australian National University, Canberra.

Bowman DMJS, Walsh A and Prior LD (2004) Landscape analysis of Aboriginal fire management in central Arnhem Land, north Australia. *Journal of Biogeography* **31** 207–223.

Brandl EJ (1972) Thylacine designs in Arnhem Land rock paintings. *Archaeology and Physical Anthropology in Oceania* **7**, 24–30.

Brennan K (1996) 'An annotated checklist of the vascular plants of the Alligator Rivers Region, Northern Territory, Australia'. Supervising Scientist Report no. 109. Supervising Scientist, Darwin.

Brook BW and Bowman DMJS (2006) Postcards from the past: charting the landscape-scale conversion of tropical Australian savanna to closed forest during the 20th century. *Landscape Ecology* **21**, 1253–1266.

Brook BW and Griffiths AD (2004) Fire management for the Frillneck Lizard *Chlamydosaurus kingii* in northern Australia. In *Species Conservation and Management: Case Studies Using RAMAS® GIS*. (Eds HR Akçakaya, MA Burgman, O Kindvall, P Sjögren-Gulve, J Hatfield and MA McCarthy) pp. 312–325. Oxford University Press, Oxford.

Bruce AJ (1993) *Kakaducaris glabra* gen. nov., sp. nov., a new freshwater shrimp from the Kakadu National Park, Northern Territory, Australia, Crustacea: Decapoda: Paleamonidae with the designation of a new subfamily Kakaducaridinae. *Hydrobiologia* **268**, 27–44.

Bruce AJ and Short JW (1993) *Leptopalaemon gagadjui* gen. nov., a new freshwater palaemonid shrimp from Arnhem land, and a re-evaluation of *Palaemontes holthuisi* Strenth, with a designation of a new genus, *Calathaemon*. *Hydrobiologia* **257**, 73–94.

Burgman MA and Thompson BS (1982) Cluster analysis, ordination and dominance-structural classification applied to diverse tropical vegetation at Jabiluka, Northern Territory. *Australian Journal of Ecology* **7**, 375–387.

Calaby J and Lewis D (1977) The Tasmanian devil in Arnhem Land rock art. *Mankind* **11**, 150–151.

Calaby J and White C (1967) The Tasmanian devil (*Sarcophilus harrissii*) in northern Australia in recent times. *Australian Journal of Science* **29**, 473–475.

Carlquist S (1979) *Stylidium* in Arnhem Land: new species, modes of speciation on the sandstone plateau, and comments on floral mimicry. *Alsio* **9**, 411–461.

Christian CS and Aldrick JM (1977) 'Alligator Rivers Study: a review report of the Alligator Rivers Region Environmental Fact-finding study'. AGPS, Canberra.

Christidis L and Boles WE (2007) *Systematics and Taxonomy of Australian Birds*. CSIRO Publishing, Melbourne.

Cowie I (2005) 'Kakadu threatened flora report. Volume 3. Results of a threatened flora survey'. Department of Infrastructure Planning and Environment, Darwin.

Crisp MD, Laffan S, Linder HP and Monro A (2001) Endemism in the Australian flora. *Journal of Biogeography* **28**, 183–198.

Crisp MD, Cook LG and Steane DA (2004) Radiation of the Australian flora: what can comparisons of molecular phylogenies across multiple taxa tell us about the evolution of present-day communities? *Philosophical Transactions of the Royal Society of London. Series B. Biological Sciences* **359**, 1551–1571.

Dahl K (1897) Biological notes on north-Australian mammalia. *Zoologist*, Series 4, **1**, 189–216.

Dahl K (1926) *In Savage Australia: An Account of a Hunting and Collecting Expedition to Arnhem Land and Dampier Land*. Philip Allan and Co., London.

Department of Natural Resources Environment and The Arts (2005) 'Northern Territory Parks and Conservation Masterplan'. Department of Natural Resources Environment and The Arts, Darwin.

Doody JS, Green B, Sims R, Rhind D, West P and Steer D (2006) Indirect impacts of invasive cane toads (*Bufo marinus*) on nest predation in pig-nosed turtles (*Carettochelys insculpta*). *Wildlife Research* **33**, 349–354.

Dostine PL, Johnson GC, Franklin DC and Hempel C (2001) Seasonal use of savanna landscapes by the Gouldian finch, *Erythrura gouldiae*, and co-existing finches in the Yinberrie Hills area, Northern Territory. *Wildlife Research* **28**, 445–458.

Dunlop CR and Webb LJ (1991) Flora and vegetation. In *Monsoonal Australia. Landscape, Ecology and Man in the Northern Lowlands*. (Eds CD Haynes, MG Ridpath and MAJ Williams) pp. 41–60. Balkema, Rotterdam.

Duretto MF and Ladiges PY (1997) Morphological variation within the *Boronia grandisepala* group (Rutaceae) and the description of nine taxa endemic to the Northern Territory, Australia. *Australian Systematic Botany* **10**, 249–302.

Edwards A and Russell-Smith J (2009) Ecological thresholds and the status of fire-sensitive vegetation in western Arnhem Land, northern Australia: implications for management. *International Journal of Wildland Fire* **18**, 127–146.

Edwards A, Kennett R, Price O, Russell-Smith J, Spiers G and Woinarski J (2003) Monitoring the impacts of fire regimes on biodiversity in northern Australia: an example from Kakadu National Park. *International Journal of Wildland Fire* **12**, 427–440.

Firth RSC, Jefferys E, Woinarski JCZ and Noske RA (2005) The diet of the brush-tailed rabbit rat (*Conilurus penicillatus*) from the monsoonal tropics of the Northern Territory, Australia. *Wildlife Research* **32**, 517–523.

Firth RSC, Woinarski JCZ and Noske RA (2006a) Home range and den characteristics of the brush-tailed rabbit-rat (*Conilurus penicillatus*) in the monsoonal tropics of the Northern Territory, Australia. *Wildlife Research* **33**, 397–407.

Firth RSC, Woinarski JCZ, Brennan KG and Hempel C (2006b) Environmental relationships of the brush-tailed rabbit-rat, *Conilurus penicillatus*, and other small mammals on the Tiwi Islands, northern Australia. *Journal of Biogeography* **33**, 1820–1837.

Fisher A, Woinarski JCZ, Churchill S, Trainor C, Griffiths AD, Palmer C and Cooper N (2000) Distribution of the rock-dwelling dasyurids *Pseudantechinus bilarni* and *Pseudantechinus ningbing* in the Northern Territory. *Northern Territory Naturalist* **16**, 1–13.

Franklin DC, Petty AM, Williamson GJ, Brook BW and Bowman DMJS (2008) Monitoring contrasting land management in the savanna landscapes of northern Australia. *Environmental Management* **41**, 501–515.

Fraser F, Lawson V, Morrison S, Christopherson P, McGreggor S and Rawlinson M (2003) Fire management experiment for the declining Partridge Pigeon, Kakadu National Park. *Ecological Management and Restoration* **4**, 93–101.

Garnett ST and Crowley GM (1994) Wet season feeding by four species of granivorous birds in the Northern Territory. *Australian Bird Watcher* **15**, 306–309.

Gorman JT, Saalfeld K and Griffiths AD (2007) Population estimates of Asian Water Buffalo and wild cattle in the Arafura Swamp, central Arnhem Land. *Northern Territory Naturalist* **19**, 1–8.

Griffiths AD and Christian KA (1996) The effects of fire on the frillneck lizard (*Chlamydosaurus kingii*) in northern Australia. *Australian Journal of Ecology* **21**, 386–398.

Griffiths AD and McKay JL (2007) Cane toads reduce the abundance and site occupancy of Merten's water monitor (*Varanus mertensi*). *Wildlife Research* **34**, 609–615.

Harwood R and Dessein S (2005) Australian *Spermacoce* (Rubiaceae: Spermacoceae). I. Northern Territory. *Australian Systematic Botany* **18**, 297–365.

Haynes CD (1985) The pattern and ecology of *munwag*: traditional Aboriginal fire regimes in north-central Arnhem Land. *Proceedings of the Ecological Society of Australia* **13**, 203–214.

Haynes CD (1991) Use and impact of fire. In *Monsoonal Australia: Landscapes, Ecology and Man in the Northern Lowlands.* (Eds CD Haynes, MG Ridpath and MAJ Williams) pp. 61–71. Balkema, Rotterdam.

Ingwersen F (1995) Kakadu-Alligator Rivers region, Northern Territory, Australia. In *Centres of Plant Diversity: A Guide and Strategy for Their Conservation. Volume 2. Asia, Australasia and the Pacific.* (Eds SD Davis, VH Heywood and AC Hamilton) pp. 471–475. IUCN Publications Unit, Cambridge UK.

Jones R (Ed.) (1985) 'Archaeological research in Kakadu National Park'. Special publication no. 13. Australian National Parks and Wildlife Service, Canberra.

Kerrigan R (2003) 'Kakadu threatened flora report. Results of a threatened flora survey, 2003'. Department of Infrastructure Planning and Environment, Darwin.

Kerrigan R (2004) 'Kakadu threatened flora report. Volume 2. Results of a threatened flora survey, 2004'. Department of Infrastructure Planning and Environment, Darwin.

Ladiges PY, Udovicic F and Nelson G (2003) Australian biogeographic connections and the phylogeny of large genera in the plant family Myrtaceae. *Journal of Biogeography* **30**, 989–999.

Lazarides M (1979) *Micraira* F. Muell. (Poaceae, Micrairoideae). *Brunonia* **2**, 67–84.

Lewis D (1977) More striped designs in Arnhem Land rock paintings. *Archaeology and Physical Anthropology in Oceania* **12**, 98–111.

Lewis HT (1994) Management fires vs. corrective fires in Northern Australia: an analogue for environmental change. *Chemosphere* **29**, 949–963.

Liedloff AC and Cook GD (2007) Modelling the effects of rainfall variability and fire on tree populations in an Australian tropical savanna with the FLAMES simulation model. *Ecological Modelling* **201**, 269–282.

Lowe L (1995) Preliminary investigations of the biology and management of Leichhardt's Grasshopper, *Petasida ephippigera* White. *Journal of Orthoptera Research* **4**, 219–221.

Morton SR and James CD (1988) The diversity and abundance of lizards in arid Australia: a new hypothesis. *American Naturalist* **132**, 237–256.

Murphy BP and Bowman DMJS (2007) The interdependence of fire, grass, kangaroos and Australian Aborigines: a case study from central Arnhem Land, northern Australia. *Journal of Biogeography* **34**, 237–250.

Needman RS (1988) *Geology of the Alligator Rivers Uranium Field, Northern Territory.* Australian Government Publishing Service, Canberra.

Northern Territory Reserves Board (1972) 'Notes on a proposal for a National Park in the Alligator River Area of the Far North of the Northern Territory'. Alice Springs Printing and Publishing Co., Alice Springs.

Noske R (1992) The status and ecology of the white-throated grass-wren *Amytornis woodwardi*. *Emu* **92**, 39–51.

Oakwood M (2000) Reproduction and demography of the northern quoll, *Dasyurus hallucatus*, in the lowland savanna of northern Australia. *Australian Journal of Zoology* **48**, 519–539.

Page TJ, Short JW, Humphrey CL, Hillyer MJ and Hughes JM (2008) Molecular systematics of the Kakaducarididae (Crustacea: Decapoda: Caridae). *Molecular Phylogenetics and Evolution* **46**, 1003–1014.

Pardon LG, Brook BW, Griffiths AD and Braithwaite RW (2003) Determinants of survival for the northern brown bandicoot under a landscape-scale fire experiment. *Journal of Animal Ecology* **72**, 106–115.

Perry J (2005) Habitat and status of the Carpentarian grasswren (*Amytornis dorotheae*) in the Northern Territory. B.App.Sci. thesis. Charles Darwin University, Darwin.

Petty A, Alderson J, Muller R, Scheibe O, Wilson K and Winderlich S (2007a) 'Kakadu National Park. Arnhemland plateau draft fire management plan'. Kakadu National Park and the Tropical Savannas Cooperative Research Centre, Darwin.

Petty AM, Werner PA, Lehmann CER, Riley JA, Banfai DS and Elliott LP (2007b) Savanna responses to feral buffalo in Kakadu National Park, Australia. *Ecological Monographs* **77**, 441–463.

Press AJ, Lea DM, Webb A and Graham AJ (Eds) (1995a) 'Kakadu: natural and cultural heritage management'. Australian Nature Conservation Agency and North Australia Research Unit, Darwin.

Press T, Brock J and Andersen A (1995b) Fauna. In *Kakadu: Natural and Cultural Heritage and Management.* (Eds T Press, D Lea, A Webb and A Graham) pp. 167–216. Australian Nature Conservation Agency and North Australia Research Unit, Darwin.

Price O and Bowman DMJS (1994) Fire-stick forestry: a matrix model in support of skilful fire management of *Callitris intratropica* R.T. Baker by north Australian Aborigines. *Journal of Biogeography* **21**, 573–580.

Prior LD, Bowman DMJS and Brook BW (2007) Growth and survival of two north Australian relictual tree species, *Allosyncarpia ternata* (Myrtaceae) and *Callitris intratropica* (Cupressaceae). *Ecological Research* **22**, 228–236.

Rankmore BR (2006) Impacts of habitat fragmentation on the vertebrate fauna of the tropical savannas of northern Australia; with special reference to medium-sized mammals. PhD thesis. Charles Darwin University, Darwin.

Ranger Uranium Environmental Inquiry (1976) 'First report'. AGPS, Canberra.

Reichel H and Andersen AN (1996) The rainforest ant fauna of Australia's Northern Territory. *Australian Journal of Zoology* **44**, 81–95.

Rossiter NA, Setterfield SA, Douglas MM and Hutley LB (2003) Testing the grass-fire cycle: exotic grass invasion in the tropical savannas of northern Australia. *Diversity and Distributions* **9**, 169–176.

Rossiter-Rachor NA, Setterfield SA, Douglas MM, Hutley LB and Cook GD (2008) *Andropogon gayanus* (Gamba grass) invasion increases fire-mediated nitrogen losses in the tropical savannas of northern Australia. *Ecosystems* **11**, 77–88.

Russell-Smith J (1991) Classification, species richness and environmental relations of monsoon rain forest in northern Australia. *Journal of Vegetation Science* **2**, 259–278.

Russell-Smith J (2006) Recruitment dynamics of the long-lived obligate seeders *Callitris intratropica* (Cupressaceae) and *Petraeomyrtus punicea* (Myrtaceae). *Australian Journal of Botany* **54**, 479–485.

Russell-Smith J and Bowman DMJS (1992) Conservation of monsoon rainforest isolates in the Northern Territory, Australia. *Biological Conservation* **59**, 51–63.

Russell-Smith J, Lucas DE, Brock J and Bowman DMJS (1993) *Allosyncarpia*-dominated rain forest in monsoonal northern Australia. *Journal of Vegetation Science* **4**, 67–82.

Russell-Smith J, Lucas D, Gapindi M, Gunbunuka B, Kapirigi N, Namingum G, Lucas K, Giuliani P and Chaloupka G (1997) Aboriginal resource utilization and fire management practice in western Arnhem Land, monsoonal northern Australia: notes for prehistory, lessons for the future. *Human Ecology* **25**, 159–195.

Russell-Smith J, Ryan PG, Klessa D, Waight G and Harwood R (1998) Fire regimes, fire-sensitive vegetation and fire management of the sandstone Arnhem plateau, monsoonal northern Australia. *Journal of Applied Ecology* **35**, 829–846.

Russell-Smith J, Ryan PG and Cheal DC (2002) Fire regimes and the conservation of sandstone heath in monsoonal northern Australia: frequency, interval, patchiness. *Biological Conservation* **104**, 91–106.

Russell-Smith J, Whitehead PJ, Cook GD and Hoare JL (2003) Response of *Eucalyptus*-dominated savanna to frequent fires: lessons from Munmarlary, 1973–1996. *Ecological Monographs* **73**, 349–375.

Senate Standing Committee on Environment, Recreation and the Arts (1988) 'The potential of the Kakadu National Park region'. AGPS, Canberra.

Sharp BR and Bowman DMJS (2004) Patterns of long-term woody vegetation change in a sandstone-plateau savanna woodland, Northern Territory, Australia. *Journal of Tropical Ecology* **20**, 259–270.

Sharp BR and Whittaker RJ (2003) The irreversible cattle-driven transformation of a seasonally flooded Australian savanna. *Journal of Biogeography* **30**, 783–802.

Shattuck SO (2008) Australian ants of the genus Aphaenogaster (Hymenoptera: Formicidae). *Zootaxa* **1677**, 25–45.

Smith J and Phillips BL (2006) Toxic tucker: the potential impact of cane toads on Australian reptiles. *Pacific Conservation Biology* **12**, 40–49.

Specht RL (1958) The geographical relationships of the flora of Arnhem Land. In *Records of the American-Australian Scientific Expedition to Arnhem Land. 3. Botany and Plant Ecology.* (Eds RL Specht and CP Mountford) pp. 415–478. Melbourne University Press, Melbourne.

Taylor JA and Dunlop CR (1985) Plant communities of the wet-dry tropics of Australia: the Alligator Rivers region, Northern Territory. *Proceedings of the Ecological Society of Australia* **13**, 83–128.

Telfer WR and Garde MJ (2006) Indigenous knowledge of rock kangaroo ecology in western Arnhem Land, Australia. *Human Ecology* **34**, 379–406.

Truswell EM (1982) 'Palynology of a Tertiary core sample from South Alligator area, E.L. 1287'. Bureau of Mineral Resources, Canberra.

Vigilante T and Bowman DMJS (2004a) Effects of fire history on the structure and floristic composition of woody vegetation around Kalumburu, North Kimberley, Australia: a landscape-scale natural experiment. *Australian Journal of Botany* **52**, 381–404.

Vigilante T and Bowman DMJS (2004b) Effects of individual fire events on the flower production of fruit-bearing tree species, with reference to Aboriginal people's management and use, at Kalumburu, North Kimberley, Australia. *Australian Journal of Botany* **52**, 405–415.

Watson JAL, Theischinger G and Abbey HM (1991) *The Australian Dragonflies: A Guide to the Identification and Habitats of Australian Odonata*. CSIRO Publishing, Melbourne.

Watson M and Woinarski J (2003) 'Vertebrate monitoring and resampling in Kakadu National Park, 2002'. Parks and Wildlife Commission of the Northern Territory, Darwin.

Werner PA (2005) Impact of feral water buffalo and fire on growth and survival of mature savanna trees: an experimental field study in Kakadu National Park, northern Australia. *Austral Ecology* **30**, 625–647.

Williams RJ, Muller WJ, Wahren C-H, Setterfield SA and Cusack J (2003) Vegetation. In *Fire in Tropical Savannas. The Kapalga Experiment.* (Eds AN Andersen, GD Cook and RJ Williams) pp. 79–106. Springer-Verlag, New York.

Wilson BA, Brocklehurst PS, Clark MJ and Dickinson KJM (1990) 'Vegetation survey of the Northern Territory, Australia'. Technical report no. 49. Conservation Commission of the Northern Territory, Darwin.

Wilson BA, Russell-Smith J and Williams R (1996) Terrestrial vegetation. In *Landscape and Vegetation Ecology of the Kakadu Region.* (Eds CM Finlayson and I von Oertzen) pp. 57–79. Kluwer, Dordrecht, The Netherlands.

Woinarski JCZ (1992) The conservation status of the White-throated Grass-wren *Amytornis woodwardi*, an example of problems in status designation. *Northern Territory Naturalist* **13**, 1–5.

Woinarski J (2004a) 'Threatened plants and animals in Kakadu National Park: a review and recommendations for management'. NT Department of Infrastructure Planning and Environment, Darwin.

Woinarski JCZ (2004b) The forest fauna of the Northern Territory: knowledge, conservation and management. In *Conservation of Australia's Forest Fauna.* 2nd edn (Ed. D Lunney) pp. 36–55. Royal Zoological Society of New South Wales, Sydney.

Woinarski JCZ, Gambold N, Wurst D, Flannery TF, Smith AP, Chatto R and Fisher A (1999) Distribution and habitat of the Northern Hopping Mouse *Notomys aquilo. Wildlife Research* **26**, 495–511.

Woinarski JCZ, Milne DJ and Wanganeen G (2001) Changes in mammal populations in relatively intact landscapes of Kakadu National Park, Northern Territory, Australia. *Austral Ecology* **26**, 360–370.

Woinarski JCZ, Risler J and Kean L (2004) The response of vegetation and vertebrate fauna to 23 years of fire exclusion in a tropical *Eucalyptus* open forest, Northern Territory, Australia. *Austral Ecology* **29**, 156–176.

Woinarski JCZ, Williams RJ, Price O and Rankmore B (2005) Landscapes without boundaries: wildlife and their environments in northern Australia. *Wildlife Research* **32**, 377–388.

Woinarski JCZ, Hempel C, Cowie I, Brennan K, Kerrigan R, Leach G and Russell-Smith J (2006) Distributional patterns of plant species endemic to the Northern Territory, Australia. *Australian Journal of Botany* **54**, 627–640.

Woinarski J, Pavey C, Kerrigan R, Cowie I and Ward S (2007) 'Lost from our landscape: threatened species of the Northern Territory'. NT Government Printer, Darwin.

Woinarski JCZ, Oakwood M, Winter J, Burnett S, Milne D, Foster P, Myles H and Holmes B (2008) 'Surviving the toads: patterns of persistence of the northern quoll *Dasyurus hallucatus* in Queensland'. Department of Natural Resources, Environment and the Arts, Darwin.

Yibarbuk D and Cooke P (2001) Fire, fruit and emus. In *Savanna Burning: Understanding and Using Fire in Northern Australia.* (Eds R Dyer, P Jacklyn, I Partridge, J Russell-Smith and D Williams) p. 40. Tropical Savannas Cooperative Research Centre, Darwin.

Yibarbuk D, Whitehead PJ, Russell-Smith J, Jackson D, Godjuwa C, Fisher A, Cooke P, Choquenot D and Bowman DMJS (2001) Fire ecology and Aboriginal land management in central Arnhem land, northern Australia: a tradition of ecosystem management. *Journal of Biogeography* **28**, 325–343.

9

Ecological thresholds and the status of fire-sensitive vegetation in western Arnhem Land, northern Australia:

implications for management

Andrew C Edwards and Jeremy Russell-Smith (adapted from the International Journal of Wildland Fire, *reproduced with permission)*

SUMMARY

This chapter examines the application of the ecological thresholds concept to fire-management issues concerning fire-sensitive vegetation types associated with the remote, biodiversity-rich, sandstone Arnhem plateau, in western Arnhem Land, monsoonal northern Australia. In the absence of detailed assessments of fire regime impacts on component biota such as those for the adjoining Nitmiluk and World Heritage Kakadu National Parks, the chapter builds on validated 16-year fire history and vegetation structural mapping products derived principally from Landsat-scale imagery, to apply critical ecological thresholds criteria as defined by fire regime parameters for assessing the status of fire-sensitive habitat and species elements. Assembled data indicate that the 24 000 km^2 study region today experiences fire regimes characterised generally by high annual frequencies (mean = 36.6%) of large (>10 km^2) fires that occur mostly in the late dry season under severe fire-weather conditions. Collectively, such conditions substantially exceed defined ecological thresholds for significant proportions of fire-sensitive indicator rainforest and heath vegetation types, and the long-lived obligate seeder conifer tree species, *Callitris intratropica*. Thresholds criteria are recognised as an effective tool for informing ecological fire management in a variety of geographic settings.

INTRODUCTION

The design and implementation of ecologically sustainable, adaptive fire management is a key issue facing reserve and landscape managers in fire-prone landscapes throughout the world. This is particularly true of tropical and subtropical savanna systems which, in many regions, are fire-affected on an annual-biennial basis, and globally constitute the largest annually recurring source of pyrogenic emissions (Andreae 1997). It is widely recognised that fire regimes play a significant role in regulating the vegetation structure and functioning of savanna systems, including influences on the dynamic balance between woody plants and flammable grasses (Scholes and Archer 1997; Higgins *et al.* 2000).

In Australia, an annual mean of around 19% of the 1.9 M km^2 tropical savannas region (i.e. 25% of the continental landmass) has been fire-affected over the period 1997–2005, making this by far the most fire-prone extensive landscape type in that continent (AGO 2007; Russell-Smith *et al.* 2007). Given that fire-prone savanna woodlands of northern Australia are characterised by very low human population densities (<0.01 persons km^{-2} outside of towns and settlements) and associated infrastructure development, a reliably long dry season and mostly subdued terrain, individual fires (although often developed from multiple ignition sources) may burn over tens of thousands of square kilometres (Luke and McArthur 1978; Yates *et al.* 2008).

Fire activity in northern Australian savannas is unevenly distributed: (a) spatially – occurring mostly in higher rainfall, biodiversity-rich, remote and rugged terrain in northern coastal/sub-coastal areas, often under Indigenous (Aboriginal) land tenure arrangements where landowners typically have limited resources and management capacity, and (b) temporally – occurring today mostly as uncontrolled wildfires in the latter half of the 7+ month dry season under relatively severe fire weather conditions (Russell-Smith *et al.* 2003a; Whitehead *et al.* 2003). In most regions, fire regime patterns today differ markedly from those under Aboriginal land stewardship at the time of European colonisation (e.g. Russell-Smith *et al.* 2003a). Contemporary fire patterns have significant regional implications for biodiversity and soil conservation, greenhouse emissions, pastoral production and broader social issues (e.g. Townsend and Douglas 2000; Dyer *et al.* 2001; Williams *et al.* 2002; Russell-Smith *et al.* 2003a, 2007; Whitehead *et al.* 2003; Ellis *et al.* 2004; AGO 2007; Preece 2007).

Assessment of the impacts of northern Australian fire regimes on biodiversity values at landscape scales has been made with reference to the status of:

1. vulnerable habitats (e.g. rainforests – McKenzie and Belbin 1992; Russell-Smith and Bowman 1992; heaths – Russell-Smith *et al.* 1998, 2002)
2. key indicator plant species such as the long-lived, fire-sensitive (obligate seeder) conifer, *Callitris intratropica* (Bowman and Panton 1993; Price and Bowman 1994; Graham 2002; Yates and Russell-Smith 2003)
3. putative impacts on vertebrate fauna (e.g. Franklin 1999; Woinarski *et al.* 2001, 2005).

Such ecological assessments have been augmented with spatial analyses derived from recent Landsat or Advanced Very High Resolution Radiometer (AVHRR) imagery of fire regime parameters: particularly seasonality, extent, frequency, interval, patchiness and heterogeneity (e.g. Russell-Smith *et al.* 1997, 2002, 2003a; Edwards *et al.* 2001; Fisher *et al.* 2003; Yates and Russell-Smith 2003; Price *et al.* 2005; Yates *et al.* 2008). Collectively, these studies support the contention that relatively fire-sensitive elements (species, habitats) occurring within the broader savanna matrix are increasingly vulnerable to contemporary north Australian fire regimes.

Assessment of the status of, and development of, appropriate management responses concerning fire-sensitive elements in many fire-prone regional northern Australia landscapes is hampered given remoteness (i.e. lack of tracks and roads), and the absence of reliable and high-resolution vegetation mapping and fire-mapping spatial datasets. Where such datasets exist, however, landscape-scale assessments based on assessment of defined critical ecological threshold criteria (e.g. minimum and maximum fire-free intervals to ensure effective replenishment of seed banks in perennial obligate seeder taxa; minimum patch size requirements for sedentary animals and poorly dispersed plant taxa) can be highly informative (Keith *et al.* 2002). Comparable 'thresholds of concern' have been developed to inform management responses in South African savanna conservation reserves, albeit aimed more at delivering variable fire regimes (especially fire-return intervals and patch sizes) to promote faunal diversity (van Wilgen *et al.* 1998, 2003, 2004; Bond and Archibald 2003).

In this chapter we apply critical fire regimes thresholds criteria to the rapid appraisal of the status of fire-sensitive habitats and species in the western Arnhem Land region of northern Australia. Although this extensive region is recognised internationally for its biodiversity values, is contiguous with two major conservation reserves (Nitmiluk and World Heritage Kakadu National Parks) and is known to be very fire-prone, limited ecological assessment of the status of component biota is available with respect to contemporary fire regimes. Surveys of the status of the long-lived obligate seeder conifer, *Callitris intratropica*, report substantial population decline in at least portions of that region (Bowman and Panton 1993; Bowman *et al.* 2001a). We take advantage of recent compilation of detailed 16-year Landsat-based fire mapping and associated vegetation structural mapping (~1:100,000) to assess the likely condition of fire-sensitive habitat (sandstone rainforest, heath) and species (*Callitris*) elements with respect to critical thresholds as defined by a number of fire regime parameters. The chapter provides:

1. a fire regime baseline against which current and future fire management activities in this recognised high biodiversity project area can be assessed
2. a clear example of the utility of applying critical threshold criteria for informing landscape-scale ecological assessment and management issues (see Groffman *et al.* 2006).

REGIONAL CONTEXT

The study focuses on the 23 893 km^2 Western Arnhem Land Fire Abatement (WALFA) project area (Figure 9.1). This area comprises remote, rugged and largely unpopulated terrain which, as for Arnhem Land generally, is under Aboriginal freehold tenure. Year-round human settlement of the WALFA area amounts to around 200 persons, who are concentrated mostly in a few small communities ('out-stations'), particularly along the northern fringe. Like Kakadu National Park immediately to the west (Figure 9. 1), the WALFA area is recognised nationally (Crisp *et al.* 2001; Woinarski *et al.* 2006) and internationally (Ingwersen 1995) for its biodiversity values.

In the north-west, the WALFA area is dominated by the Arnhem plateau – mostly at less than 400 m elevation – comprising resistant, flat-bedded Middle Proterozoic quartzose sandstones, criss-crossed by tensional joints that have been deeply weathered and eroded to form a maze of narrow valleys and gorges (Needham 1988). Northern and western rims of the plateau are bounded by sheer and spectacular escarpments. Soils, where present, are typically skeletal and infertile sands (Aldrick 1976). In restricted situations, deep sand plains are developed on plateau summits. In the south-east, the physiography tends to be more subdued: characterised by gently undulating plains and rises derived from Cretaceous sediments (Lynch and Wilson 1998).

Following Wilson *et al.* (1990), vegetation cover of sandstone formations in the north-west is mostly a low open-woodland: a complex of shrubby heath communities, comprising many fire-sensitive, obligate seeder shrub species (Russell-Smith *et al.* 1998) with scattered emergent trees (e.g. *Eucalyptus, Gardenia* and *Terminalia*), interspersed with a substantial cover of highly

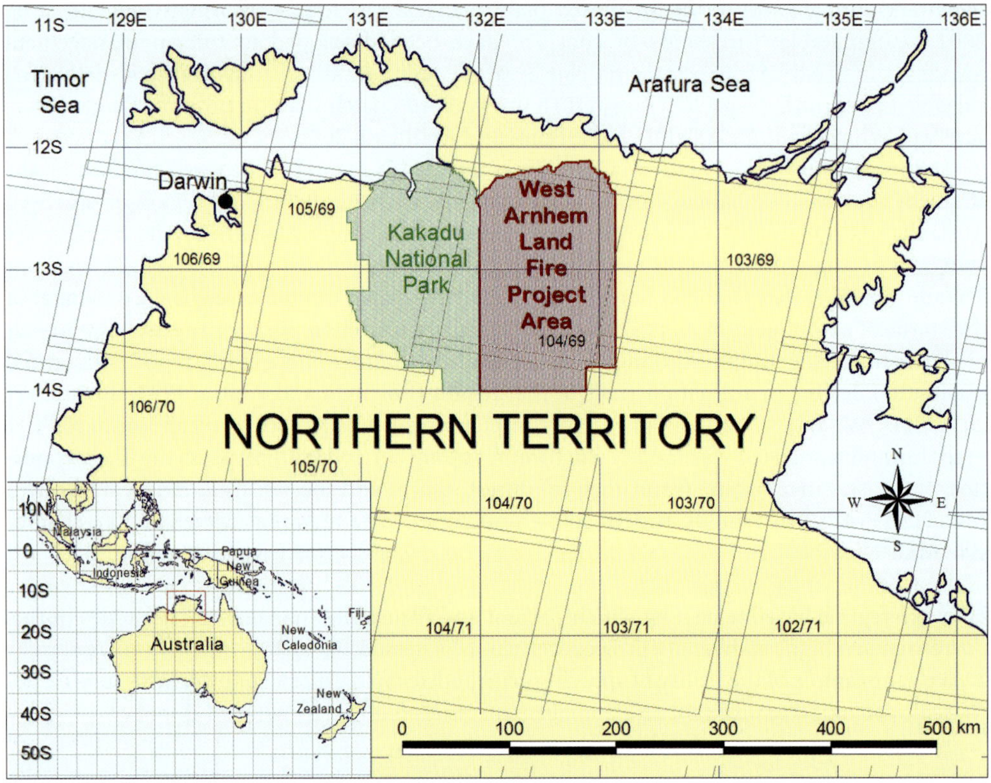

Figure 9.1 Location of the West Arnhem Land Project Area (WALFA) in relation to Kakadu National Park, and the 'Top End' of the Northern Territory, Australia.

flammable hummock spinifex (*Triodia*) grasses. On deeper sands, the vegetation takes an open-forest form dominated by *Eucalyptus* over a range of shrubs and slender perennial and annual grasses. The fire-sensitive conifer, *Callitris intratropica*, occurs widely, either singly or in copses, on these sandy soils (Bowman and Panton 1993). On shallower soils, particularly in the south-east, the vegetation is of a characteristic savanna woodland form dominated by *Eucalyptus* typically over flammable annual grasses, especially *Sarga timorense*. Seasonally inundated floodplains support mostly *Melaleuca* and *Eucalyptus* over a range of vigorous perennial grasses. Rainforest – dominated by the myrtaceous sclerophyll *Allosyncarpia ternata* – occurs in a variety of situations, but typically in rugged, broken terrain affording some level of fire protection (Russell-Smith *et al.* 1993).

As for northern Australia generally, the regional climate is characterised by marked rainfall seasonality, with over 90% falling in the summer months from November to April. Mean annual rainfall in the region is likely to be between 1000 and 1500 mm, based on records from regional weather stations outside of the WALFA area. Although the amount of rainfall is highly variable from year to year, the rainy or wet season is a highly reliable event (Taylor and Tulloch 1985). Daily maximum temperatures average above 30°C over the year. Winds in the dry season period are dominated by dry south-easterlies, turning to increasingly from the north-west in association with the arrival of the Asian monsoon.

These climatic conditions are conducive for rapid development of grassy fuels following a fire event, such that fires may recur at any one site on an annual–biennial basis (Williams *et al.*

2002). Accumulation of fine grass and litter fuels is sufficient to support intense fires (>5000 kW m^{-1}) in all but exposed rocky situations under late dry season climatic conditions within 1–3 years of having been burnt previously (Russell-Smith *et al.* 1998). Fire severity generally increases with the progression of the dry season given the increasingly severe fire weather (stronger winds, higher temperatures and lower humidity), and lower fuel moisture conditions (Gill *et al.* 1996, Williams *et al.* 1998, 2003; Russell-Smith *et al.* 2003b; Russell-Smith and Edwards 2006).

Fire management

Contemporary fire patterns in the WALFA area date from the early 1900s and are associated with rapid regional Aboriginal depopulation, following movement away into newly developing missions and settlements, and the impacts of disease (Keen 1980). Accumulating ethnographic evidence indicates that customary Aboriginal landscape burning in WALFA and surrounding areas was undertaken extensively and systematically throughout the dry season for a variety of livelihood (e.g. hunting and gathering) and secular (e.g. 'cleaning' of country) purposes (Haynes 1985, 1991; Yibarbuk *et al.* 2001; Chapter 5).

Frequent burning, especially in areas regularly frequented by mobile family groups such as along creekline corridors, doubtless significantly reduced fuel loads over extensive areas and instigated a seasonal network of burnt firebreaks – thereby reducing the risk of extensive, intense wildfires burning, especially in late dry season conditions (Chapter 5). For example, as noted by Bowman and Panton (1993) and Bowman *et al.* (2001a), an artefact of this management pattern was that, at the time of European colonisation, stands of the long-lived obligate seeder, *Callitris intratropica*, were relatively widespread on the Arnhem plateau. Today, however, in the face of contemporary fire regimes dominated by extensive and frequent wildfires (see Results), many populations of *C. intratropica* in the WALFA area are showing signs of significant decline (Bowman and Panton 1993; Bowman *et al.* 2001a).

In recognition of the landscape-scale impacts of contemporary fire regimes on regional biodiversity values, including down-wind impacts on adjoining Kakadu and Nitmiluk National Parks (Figure 9.1), a concerted program to develop more ecologically and economically sustainable fire management has been undertaken from the mid-1990s, using public (Australian and Northern Territory Governments) and community funding and resources. This program has a number of components:

- to build capacity, including training, in local Indigenous communities and organisations towards the delivery of landscape-scale natural resource management
- to implement prescribed landscape-scale fire management through the development of collaborative partnerships between key regional stakeholders (Aboriginal landowners; adjoining National Parks; Bushfires NT – the regional rural fire management agency)
- the development of an economically sustainable basis for regional fire management through the development of a commercial greenhouse gas emissions abatement offsets program (as described in Chapter 12.

METHODS

The assessment of ecological thresholds presented here is based on:

1. vegetation structure mapping, particularly of fire-sensitive sandstone heath and rainforest communities

2. associated aerial survey of the distribution of fire-sensitive *Callitris intratropica*
3. fire-history mapping derived from Landsat imagery for the period 1990–2005.

As detailed below, we then assess the current status of respective fire-sensitive communities in the WALFA area with reference to fire thresholds (frequency, interval, patch size and seasonality/intensity) as defined from available literature.

Vegetation structure mapping and distribution of Callitris intratropica

In the absence of regional vegetation mapping at scales larger than 1:1 000 000, mapping of major vegetation structural types (following Specht 1981) in the WALFA area at 1:100 000 was derived from Landsat TM imagery using an object-based classification approach. Additionally, an aerial survey of the extant distribution of the long-lived obligate seeder, *Calltris intratropica*, was undertaken. The applied methodology for these components was as follows:

Sandstone heath/low open-woodland

Vegetation was mapped initially for a mask of the whole of the Arnhem plateau region, as described by Blake (2005). Classification of Arnhem plateau heath vegetation was undertaken using eCognition's standard nearest neighbour (NN) classifier algorithm following methodological details as given in Definiens Imaging (2004). Classification was conducted using training sites for 13 vegetation structural classes. Importantly, these land-cover classes included four heath types (sandstone heath, heath/rainforest, heath/woodland and burnt heath). NN classification at an appropriately fine segmentation/landscape scale was undertaken using at least two image feature descriptors (selected from – mean, standard deviation, band ratio, mean difference to neighbours and texture) for each of the six Landsat TM bands. A total of 21 image features was used to classify the image (Blake 2005).

Closed forest

Preliminary closed forest mapping was available from interpretation of 1:25 000 aerial-photographs taken in 1991 (J Russell-Smith and D Lucas, unpublished). This digitised coverage was used as a guide in an object-oriented classification approach based on Landsat TM imagery as for mapping of sandstone heath. Closed forest on the Arnhem plateau occurs as two distinct classes and is readily discerned on Landsat imagery: (a) linear riparian communities dominated typically by *Melaleuca* intermixed with evergreen rain forest taxa; (b) generally small (>1 ha) patches of evergreen forest dominated by *Allosyncarpia ternata* occupying dissected sandstone terrain, typically in fire refugia. These closed forest classes are readily separable from less-dense vegetation, and were mapped as part of the process of mapping non-heath vegetation as described below.

Open forest, Woodland and Open woodland communities

After masking out heath and low open woodland vegetation structural types, object-oriented classification was undertaken on remaining Arnhem plateau vegetation with respect to approximately 10 training sites per class derived from: (a) rainforest patches (see above); (b) 147 vegetation structure plots sampled across the region between 2003–2005; and (c) an aerial survey of Arnhem plateau vegetation structural types undertaken in 2005. NN classification of non-heath vegetation was conducted at a segmentation scale of about1 ha, using image feature descriptors (mean and standard deviation) for each of three Landsat TM bands (3, 4 and 7), the normalised difference of bands 3 and 4 (NDVI), a 100 × 100 m pixel digital elevation model (DEM) and a slope surface derived from the DEM in ArcInfo (ESRI 1982–1997) using 1:250 000 drainage as a local lowest point constraint.

The final vegetation map used here comprised five vegetation structural classes. Patch size distributions were calculated for respective vegetation map units, where individual patches were defined as areas of contiguous pixels, including those on diagonals, constrained within the study region. Reliability assessment of vegetation mapping products was undertaken with respect to: (a) an aerial survey (196 km) of Arnhem plateau vegetation structural types, undertaken in 2005; and (b) a foliage projective cover (FPC) surface derived from Landsat TM imagery (Meakin *et al.* 2001) covering the Top End of the Northern Territory.

Callitris intratropica

An aerial survey of the WALFA region was undertaken in early 2007, flying at approximately 500 ft (~150 m) and 80 knots (~150 km h^{-1}). Mature trees are readily surveyed from the air, and dead stems are also highly visible and may remain standing for some decades after death as they are generally resistant to termite attack (Bowman and Panton 1993). Counts of live and standing-dead stems were made each 30 seconds along a 250 m swath, with waypoints logged using real-time mapping software connected to a global positioning system (GPS) device. Random-start stratified east–west traverses were around 100 km long and 20 km apart. A linear total of approximately 1300 km was surveyed. Semi-log population rank estimates were made separately for live and dead stems as follows: 0 = no stems; 1 = 1 stem; 2 = 2–5 stems; 3 = 6–10; 4 = 11–20; 5 = 21–50; 6 = 51–100; 7 = 101–200; 8 = 201–500; 9 = 501–1000; 10 > 1000. For analysis, the mid-point values (counts) of respective classes were used, with rank 10 = 1000 individuals.

Within a Geographic Information System, polygon segments were created and attributed with counts of living and dead stems by buffering line segments representing the shortest distance between successive GPS points. Polygons were converted to a raster file with 50 m pixels, where each segment was given a unique identifying number and then intersected with vegetation mapping. The mean count of living or dead stems per unit area in each of the five major habitats was derived as the product of: (a) the areal proportion of each habitat in each unique segment and (b) the areal proportion of the count of living or dead stems in that segment.

Fire history mapping

Mapping of burnt (fire affected) areas was undertaken over the WALFA area for the period 1990–2005, derived from at least three satellite images each dry season (Table 9.1). For convenience, burnt-area mapping for the seasonal period from April to May to the end of July is defined here as occurring in the early dry season (EDS), and from August to November–December as the late dry season (LDS). 'Mid' and 'very late' dry season periods referred to in Table 9.1 are for scenes acquired in the middle (June–July) and as late in the year as possible, respectively, before the onset of cloudy conditions. Fires in the EDS period are observed typically to be of low severity/intensity, whereas fires in the LDS are observed to be typically much more severe/intense (Russell-Smith and Edwards 2006).

For the period 1990–1995, mapping of burnt area was derived mostly from Landsat MSS false colour images assembled for the adjacent Kakadu National Park as part of its fire-mapping program (Russell-Smith *et al.* 1997). These images were printed on photo-quality media, which were then scanned to create a digital image. Given that this Landsat MSS coverage was neither temporally nor spatially complete for the WALFA region, supplementary digital Landsat images were acquired from the Australian Centre for Remote Sensing (ACRES) as degraded thumbnails (or 'quicklooks') at about 250 × 250 m spatial resolution (https://acs.auslig.gov.au/intro.html).

From 1996–2005, Landsat ETM+ or TM scenes were used as the basis for burnt-area mapping. However, given incomplete coverage due to significant cloudiness, especially in the latter part of the dry season associated with the imminent onset of the wet season, it was found

Table 9.1: Landsat 'quicklook', MSS, TM and ETM+, MODIS (250m) and AVHRR satellite image acquisition dates procured to create the west Arnhem Land Fire Project area fire history, 1990–2005. Different coloured fonts represent data type as follows: Landsat ETM+; Landsat TM; Landsat MSS; *Landsat 'quicklook'*. The shaded cells are predominantly Landsat MSS but also Landsat quicklooks for the same date were used to augment the spatial coverage for the standard Landsat (104/69) scene (see Figure 9.1).

	Season			
Year	Early	Mid	Late	Very late
2005	20050602	20050720	20051024	MODIS
2004	20040604	20040802	20041005	*20041208*/MODIS
2003	20030731	–	20031003	*20031104*/MODIS
2002	20020602	20020805	20020922	*20021227*/MODIS
2001	20010530	20010702	20010801	*20011122*/AVHRR
2000	20000628	20000815	20001103	
1999	19990517	19990821	19990930	*19991016/19991117*
1998	19980701	19980818	19980926	*19981122*
1997	19970425	19970730	19971002	*19971103*
1996	19960609	19960727	19960913	*19961015/19961202*
1995	19950623	19950725	19950911	*19951013*/AVHRR
1994	19940620	19940807	19941010	*19941111*
1993		19930802	19931023	*19931108*
1992	19920527	19920716	19920916	*19921004*/AVHRR
1991	19910612	19910703	19911018	*19911103*
1990	*19900609*	*19900727*	*19901031*	

necessary to augment Landsat with more frequent, but lower-resolution AVHRR and, from 2001, Moderate Resolution Imaging Spectroradiometer (MODIS), imagery. A summary of the imagery sources used in this study is given in Table 9.1.

The fire-mapping procedure was as follows. A base image from 1999 (Landsat ETM+) was rectified in ER Mapper image-processing software (ER Mapper 2006), using 100 ground control points (GCPs) selected from available 1:50 000 topographic mapping (source: Geoscience Australia) with the WGS84 datum. In every instance, other images used in this assessment were co-registered to the base image using invariant bright targets (usually exposed rock), given the absence of permanent infrastructure and changes of drainage lines with time. The image was enhanced using histogram stretches of bands 3, 4, 5 and 7 and the thermal band 6 (rescaled). In any single year, difference images were used for mapping of burnt areas in subsequently acquired scenes. Unburnt areas were removed by first digitising vectors roughly around the fire-affected areas.

Enhanced spatially digitised images were classified with ER Mapper (2006) using an unsupervised classification procedure (ISOCLASS), generating at least 50 classes. Each generated class was individually painted over the enhanced image and attributed as 'burnt' or 'unburnt'.

'Burnt' classes were combined into a single image, filtered with a 3 × 3 median filter, converted to an ArcInfo GRID raster file, and further filtered to achieve homogeneous contiguous burnt and unburnt areas of a minimum size of 1 ha. The raster files throughout a given annual dry season were overlayed so that each cell was only affected once, and always attributed to the earliest fire (i.e. LDS fires burnt EDS also, are low severity). Finally, burnt areas were attributed as occurring in EDS or LDS periods.

Patch size distributions of fire-affected areas for respective seasons were assessed in similar manner as for major vegetation structural types. Aerial surveys were conducted for the purpose of collecting ground-validation points to conduct an error analysis of the mapping of the annual fire-affected areas, following procedures outlined by Russell-Smith *et al.* (1997), and Edwards *et al.* (2001). Validation data are presented here for 1998–2004. In summary, validation was undertaken using random-start aerial transects in a helicopter flown at approximately 100 m above the ground and at approximately 150 km h^{-1} (80 knots). Waypoints were collected for an area directly beneath the aircraft every 15 seconds, and attributed as 'burnt', 'unburnt' or 'patchy' (at least 20% of area unburnt). For assessment, points were buffered 50 m, and intersected with burnt area mapping. In the case of 'patchy' waypoints, if any part of the buffered point intersected a mapped burnt area the result was considered positive. Standard omission/ commission matrices, and calculation of the Kappa coefficient of agreement (Congalton 1991) were then derived.

Fire-regime thresholds

Fire-regime thresholds were defined for respective fire-vulnerable *Allosyncarpia* closed forest, sandstone heath and *Callitris intratropica* vegetation types based on published information sources (Table 9.2), and then assessed against respective fire-mapping surfaces, for the period 1990–2005. For conservation of fire-sensitive biodiversity purposes, it is assumed that threshold values exceeding 10% of the extent of respective fire-sensitive vegetation types over the 16 years of records are unsustainable, or at least of concern (Table 9.2). Collectively, these assessments required the creation of the following datasets using standard GIS approaches:

- *Allosyncarpia closed forest* – extent and frequency of LDS fires on closed forest boundaries/ perimeters, where mapped boundaries were buffered either side of the boundary by 50 m, to create an inner and outer buffer area.
- *sandstone heath* – (1) shortest-interval (SI) between recurring fires, in years, for each 100 × 100 m pixel, where unburnt and once-only-burnt pixels were not used in SI calculations; (2) SI between large fires, defined by patch sizes of contiguously burnt pixels greater than 100 ha extent.
- *Callitris intratropica* – (1) extent and frequency of LDS fires, intersected with population density estimates for mature (and dead) *C. intratropica* in each of the four extensive open-forest and open-woodland (including heath) vegetation classes; (2) SI between large LDS fires as for sandstone heath, for each extensive open vegetation type.

RESULTS

Reliability of mapping surfaces

Overall reliability of vegetation structural mapping was assessed as 89% (Kappa coefficient = 0.8), as derived from independent vegetation structure mapping data obtained through 196 km aerial survey in complex terrain in the north-west of the study region (Table 9.3a). Relatively

Table 9.2: Fire regime vulnerability components, and associated threshold parameters for Arnhem plateau fire-vulnerable vegetation types.

Vegetation type	Fire regime vulnerability	Threshold parameters assessed	Critical threshold
Allosyncarpia ternata closed forest	Comprising mostly resprouter species (Russell-Smith *et al.* 1998), this vegetation type is nonetheless vulnerable, especially on exposed margins, to LDS fires given that trunks of the dominant canopy tree (*Allosyncarpia ternata*) are often scarred from previous fires, and/or hollowed by termites (Russell-Smith *et al.* 1993; Bowman 1994; Prior *et al.* 2006).	(1) frequency and proportion of *Allosyncarpia* closed forest perimeter affected by LDS fires	≥10% of *Allosyncarpia* boundaries affected by one or more LDS fires
Sandstone heath	Comprising many obligate seeder shrub taxa, mostly with juvenile periods ≤3 years, but others ≤5 years, and a few (including some *Acacia* spp., and the serotinous *Petraeomyrtus punicea*) c.10 years (Russell-Smith *et al.* 1998, 2002; Russell-Smith 2006).	(1) shortest-interval (SI) between recurring fires, for ≤3, ≤5, ≤10 years	≥10% of sandstone heath patches affected by one or more recurring fires ≤3 years
	Given limited dispersal capacity in many component obligate seeder taxa (e.g. Russell-Smith 2006), recruitment in these species (especially those possessing non-dormant seed banks) also vulnerable to frequency of large fires.	(2) SI between large fires, in contiguously burnt patch sizes ≥100 km^2	≥10% of sandstone heath patches affected by one or more large (≥100 km^2) recurring fires ≤3 years
Callitris intratropica	Long-lived obligate seeder tree occurring in stands within woodland matrix (Stocker 1966; Bowman and Panton 1993), with juvenile periods likely to substantially exceed 10 years in many situations (Russell-Smith 2006). Although juveniles and adults afforded some protection from non-grassy, low fuels in dense, larger *Callitris* stands (Stocker 1966), adults vulnerable to intense LDS fires.	(1) frequency of LDS fires	≥10% of woodland habitat occupied by *Callitris* affected by at least two LDS fires
	Given limited dispersal capacity and non-serotinous, non-dormant seed banks (Stocker 1966), *Callitris* is vulnerable to frequency of large LDS fires	(2) SI between large LDS fires, in contiguously burnt patch sizes ≥100 km^2	≥10% of woodland habitat occupied by *Callitris* affected by at least two large (≥100 km^2) LDS fires

high commission errors were observed for the closed forest class, which were attributable mostly to the typically small patch size of this habitat type (see below) in conjunction with apparent consistent geo-locational error. Intersection between vegetation structural mapping and an available foliage projective cover (FPC) surface (Meakin *et al.* 2001) indicates:

1. substantial canopy gaps in closed forest (mean FPC = 57.6%) associated with typically highly rocky terrain
2. considerable overlap in FPC values for open forest, woodland, and sandstone woodland mapping units
3. generally sparse vegetation cover associated with the heath mapping unit (Table 9.2b).

Mean overall accuracy of fire mapping was 87% (Kappa co-efficient = 0.72) based on assessment against available independent ground validation data, for 1999–2004 (Table 9.4). This level of reliability is consistent with similar fire-mapping studies employed using Landsat imagery elsewhere in northern Australia (e.g. Russell-Smith *et al.* 1997; Edwards *et al.* 2001; Fisher *et al.* 2003; Yates and Russell-Smith 2003). Fire mapping prior to 1998 was undertaken

Table 9.3: Reliability of vegetation structure mapping as assessed through (a) aerial survey, heath, closed forest, woodland/open forest of heath, (b) all five vegetation structure types in relation to a Foliage Projective Cover (FPC) surface derived from Landsat imagery (from Meakin et al. 2001); S.E.M. = Standard Error of the Mean.

(a) Aerial survey assessment

Habitat	Producer's accuracy (omission)		User's accuracy (commission)		Overall accuracy	Kappa coefficient
	In class (%)	Not in class (%)	In class (%)	Not in class (%)	89%	0.80
Heath	84	16	96	4		
Closed forest	87	13	74	26		
Woodland/open forest	94	6	85	15		

(b) Foliage projective cover assessment

Habitat class	Mean FPC	S.E.M.
Closed forest	57.6	2.2
Open forest	36.9	2.0
Woodland	30.7	2.0
Sandstone woodland	35.8	2.7
Sandstone heath	24.6	0.8

with Landsat MSS. Although no validation data were available for assessing earlier mapping accuracy, equivalent methods were employed. Fire mapping for earlier years was undertaken upon completion of mapping 1998–2005, and we note proportions of early and late dry season fires were consistent over both mapping periods (Figure 9.2).

Vegetation and fire regimes patterning

The distribution of mapped major vegetation structural types is given in Figure 9.3. Notably, sandstone woodland and heath vegetation types occur principally in the north-western sector, with closed forest occurring as scattered patches associated particularly with rugged terrain in northern and north-western sectors. Conversely, open forest and woodland types occur generally throughout the more gently undulating south-east, but also fringing the north-west perimeter. Despite individual maximum patch sizes much larger than 1000 ha for all respective vegetation structural types, mean and median patch sizes are substantially smaller (Table 9.5). Characteristic closed forest (rainforest) patch sizes are shown to be especially small (Table 9.5).

Table 9.4: Accuracy assessment derived from aerial ground 'truthing' surveys conducted over the WALFA project area for the purposes of validating fire history mapping.

Date	No. of points	Overall accuracy (Kappa coefficient)
1999	1362	84% (0.68)
2000	1496	84% (0.67)
2001	1275	87% (0.70)
2002	860	89% (0.77)
2003	250	90% (0.80)
2004	444	85% (0.67)
Mean		**87% (0.72)**

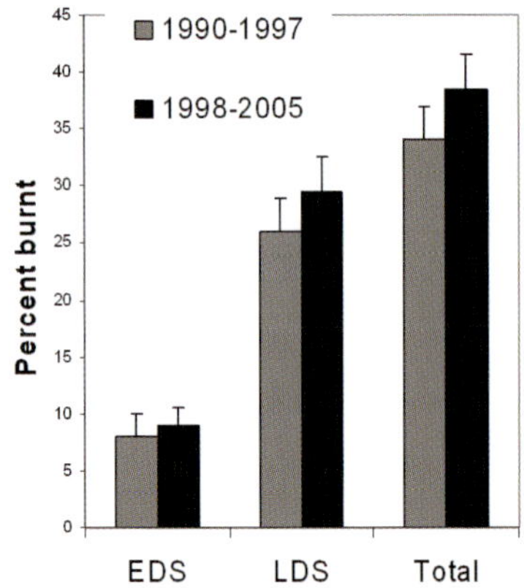

Figure 9.2 Comparison of fire mapping between two mapping periods 1990–1997 (Landsat MSS) and 1998–2005 (Landsat TM/ETM+); error bars = S.E.M.

Figure 9.3 Vegetation structure mapping for the West Arnhem Land region.

Table 9.5: Patch size attributes of major vegetation structural types in the West Arnhem study region.

Vegetation structure type	Total area (km^2)	Total no. of patches	Average patch size (ha)	Median patch size (ha)	Maximum patch size (ha)
Closed forest	571	8325	7	2	3305
Open forest	6809	8461	80	3	159 424
Woodland	6175	9957	62	3	315 410
Sandstone woodland	5024	11 442	44	2	93 274
Heath	5314	8754	61	3	69 055
Total	**23 894**				

Callitris intratopica was observed to occur generally throughout the WALFA region in a highly clumped distribution pattern (Figure 9.4). Although mean sampled density of live stems was 99.8 stems km^{-2} (Table 9.6a), no live *Callitris* stems were observed in 56% of sampled segments, and sampled densities of 100 stems km^{-2} or more occurred in only 8% of segments. By vegetation type, *Callitris* was observed to occur at greatest densities in non-sandstone open forest and woodland, and least in sandstone vegetation types, especially in association with

Table 9.6: (a) Mean density (± S.E.M.) of live and dead Cypress Pine (*Callitris intratropica*) stems per respective vegetation types. (b) Proportion of dead: dead + live stem counts for respective population rank categories of live stems. Total area surveyed respectively for live and dead stems = 325 km^2 from 1300 km of aerial transect.

(a) Mean density

Vegetation type	Sampling intensity	Mean density (stems/km^2)	
	(%)	Live stems	Dead stems
Closed forest	1.2	69 ± 37	3 ± 1
Open forest	1.2	137 ± 43	20 ± 4
Woodland	1.5	128 ± 43	19 ± 4
Sandstone woodland	1.3	85 ± 18	17 ± 4
Sandstone heath	1.6	81 ± 18	15 ± 4
Total area	1.4	99.8 ± 15	14.8 ± 1.6

(b) Ratio dead : dead + live stems

Rank count of live stems (n = number of survey segments)	Proportion dead:live stems
0 (n = 211)	100
1 (n = 101)	89.2
5 (n = 139)	63.9
15 (n = 100)	50.8
35 (n = 83)	30.6
75 (n = 65)	24.4
150 (n = 9)	31.2
350 (n = 13)	12.5
750 (n = 37)	1.6
1000 (n = 13)	1.6

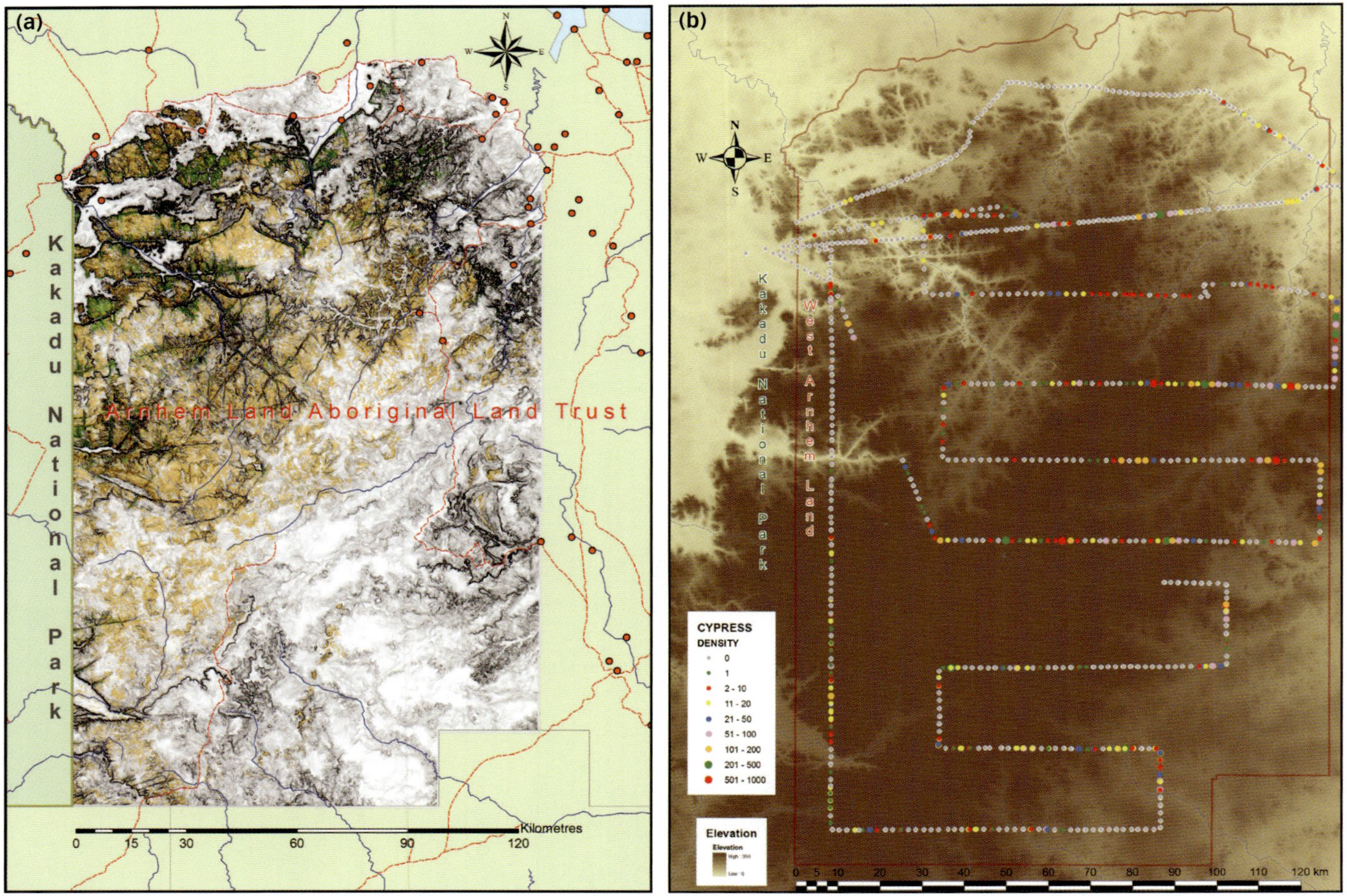

Figure 9.4 Distribution of: (a) outstations and settlements (red dots), tracks and roads (red lines), major drainage lines (blue lines), variability in topographic relief (gray shading), heaths (orange shading), rainforests (green shading); (b) live Cypress Pine (*Callitris intratropica*) stems based on aerial survey of 650 km^2 of the WALFA region. Note that larger region of western Arnhem Land shown in (a) to depict wider spread of outstations and settlements.

closed forest (Table 9.6a). Mean densities of live *Callitris* stems were consistently greater than dead stems in all vegetation types and, proportionately, the ratio of dead:live stems was least in closed forest (1:23) and greatest in open canopy sandstone (1:5) vegetation types (Table 9.6a). However, the ratio of dead:live stems was found to be very substantially greater in aerial survey segments supporting small population counts of live *Callitris*, and conversely was very low in those few segments supporting dense stands of live *Callitris* (Table 9.6b).

Over the 16 years of records, a mean 36.6% p.a. of the total WALFA study region area was burnt, comprising 9.8% in the EDS and 26.7% in the LDS. These data translate to fires recurring every 2.7 years for the WALFA region as a whole, and every 3.7 years for LDS fires. High fire-prone areas occur generally in the south-east, and also fringing the north-west (Figure 9.5a,b,c). Based on fire frequency distribution data for different vegetation types (Figure 9.6), mean fire return periods range from 5 years for closed forest, 3.9 years for sandstone heath, 3.3 years for sandstone woodland, to 2.3 years and 2.2 years, respectively, for open forest and woodland.

Assessment of minimum fire interval data indicates that, overall, 55.1% of the entire WALFA area experienced at least one repeat fire in a year following, with 82.6% experiencing at least one repeat fire within 2 years (Figure 9.7); note that a minimum fire interval of 0 in Figure 9.7 describes unburnt pixels, and those which burnt only once, over the 16 year study period. By vegetation type, the majority of woodland (77.4%) and open forest (67.2%) types experienced repeat fires within one year, whereas the majority of sandstone woodland (74.6%) and sandstone heath (62.2%) experienced repeat fires within 2 years (Figure 9.7). The majority of closed forest (64.3%) vegetation experienced a repeat fire within 3 years, with a further 22.7% remaining either unburnt or burnt only once (i.e. no interval) over the study period. The associated spatial patterns are given in Figure 9.5d.

Over the study period, on average more than 10 000 individual fire patches were recorded each year for each vegetation type, except for closed forest (mean 4362 patches). More patches were observed to occur in the LDS in each of five logarithmic-scale patch-size classes in respective vegetation types (Figure 9.8). In contrast to closed forest, which was burnt mostly (51.5%) by smaller fires <0.1 km^2, sandstone heath (63.9%) was burnt mostly by fires 1 km^2 or more, and open forest (62.9%), woodland (58%) and sandstone woodland (50.8%) types were burnt mostly by fires 10 km^2 or more (Figure 9.8). Mean fire patch sizes ranged from 2.6 ha for closed forest to 25 ha for woodland.

Thresholds assessment

Based on assembled fire regime data, and with reference to thresholds criteria for respective fire-sensitive vegetation components given in Table 9.2, our assessment indicates:

- *Allosyncarpia closed forest* – 41.1% of the 472 km^2 inner 50 m perimeter buffer was affected three or more times by LDS fires, while 47.2% of the 726 km^2 outer 50 m buffer was similarly fire-affected (Table 9.7). Such LDS fire frequencies substantially exceed critical threshold values (Table 9.2). By contrast, only 3% of inner and 4% of the outer perimeter buffers, respectively, were affected three or more times by EDS fires. Overall, 17% and 13% of inner and outer buffered areas, respectively, were not burnt by LDS fires over 16 years.
- *Sandstone heath* – 83.2% was affected by fires with minimum return intervals of 1–3 years; 6% was affected by fires with minimum intervals of more than 5 years; and 6% was either unaffected by fire or was burnt once over the 16 years of fire records (Figure 9.6). For large fires (≥100 km^2), 66% of heath habitats were burnt at least once within 1–3 years of having been burnt previously, including 34% having been burnt at similar short intervals on two

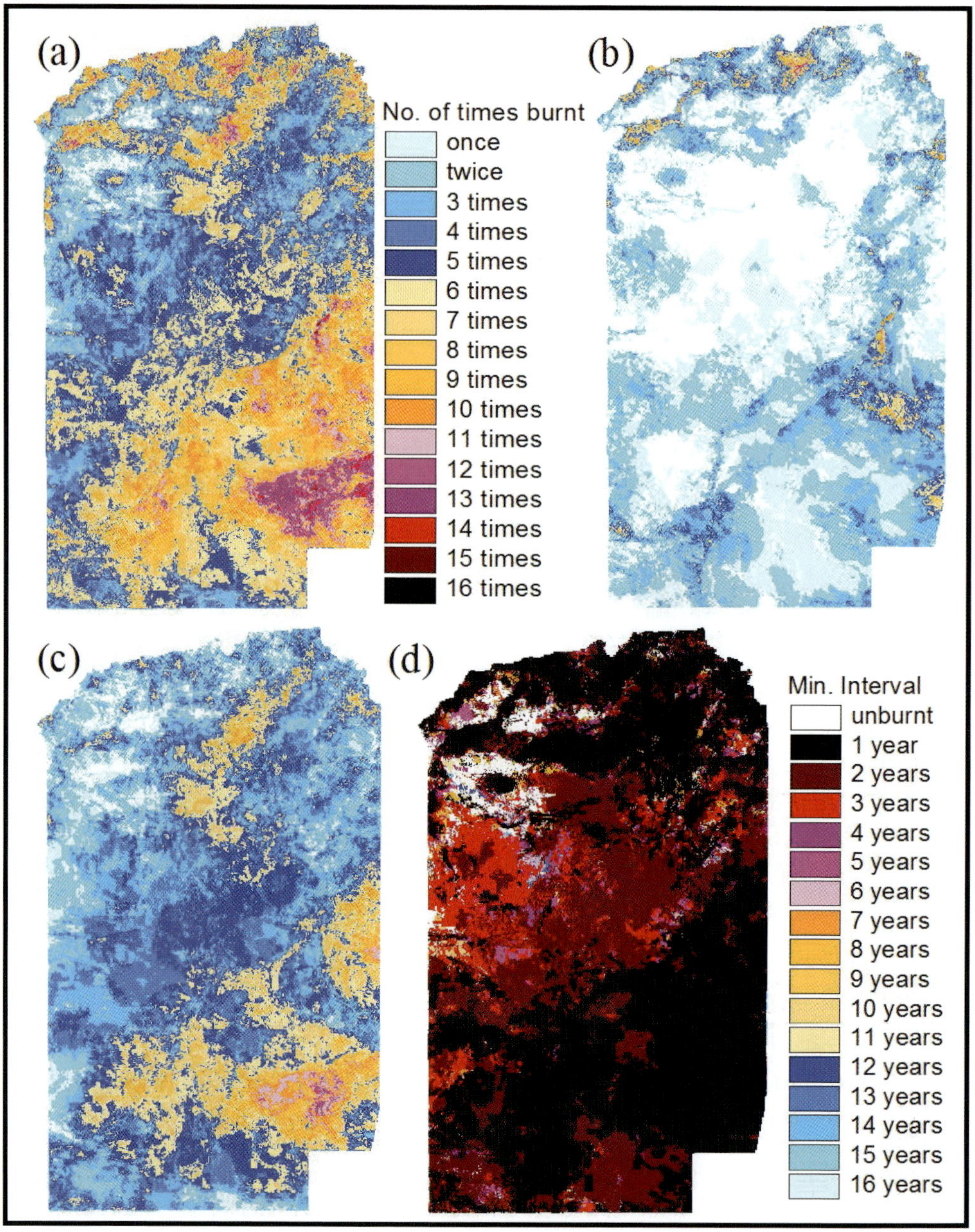

Figure 9.5 Fire frequency and minimum interval between fires, 1990–2005: (a) fire frequency, (b) frequency of early dry season (EDS) fires, (c) frequency of late dry season (LDS) fires, (d) shortest interval between fires. Note that a minimum fire interval of 'unburnt' describes unburnt pixels and those that burnt only once, over the 16-year study period.

or more occasions (Figure 9.9). Such fire frequencies substantially exceed critical threshold values (Table 9.2).

- *Callitris intratropica* – given that this species occurs generally throughout the WALFA region, but especially in open forest and woodland vegetation (Figure 9.4), stands were

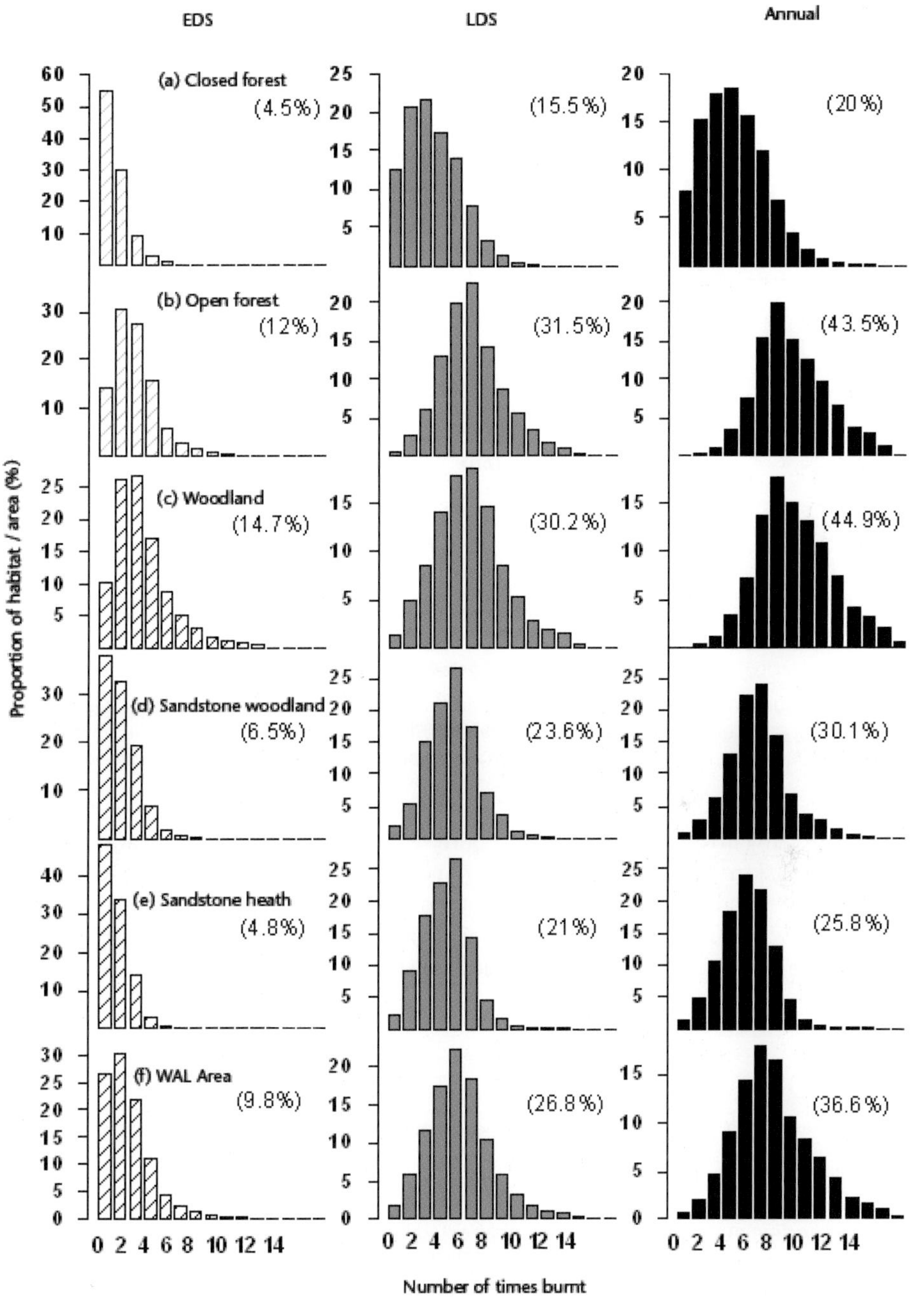

Figure 9.6 Seasonal (EDS, LDS, annual) frequency distribution of burning, 1990–2005, for (a) closed forest, (b) open forest, (c) woodland, (d) sandstone woodland, (e) heath, (f) entire West Arnhem Land study area. Numbers in parentheses are annual average of percentage burned.

shown to be exposed to relatively intense LDS fires on average every 3.7 years in the WALFA region overall, and 3.2 and 3.3 years in open forest and woodland, respectively (Figure 9.6). Given the coincidence between the relatively high proportion of the WALFA

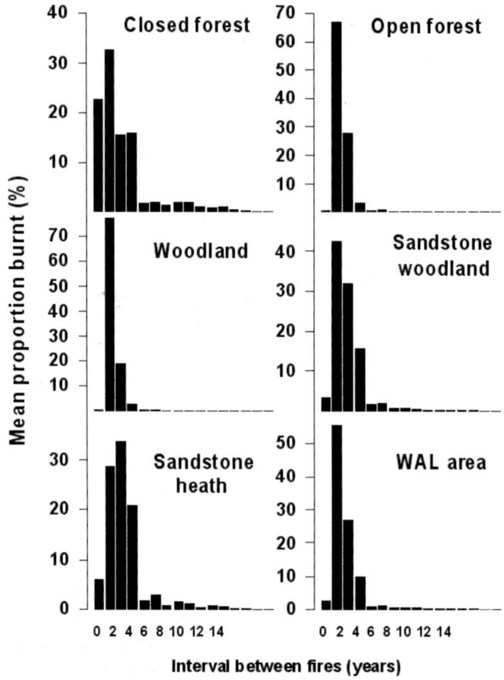

Figure 9.7 Proportions of respective habitats, and the entire West Arnhem Land study area, burnt by fires reoccurring at different minimum intervals (years), 1990–2005. Note that a minimum fire interval of 0 describes unburnt pixels, and those which burnt only once, over the 16 year study period.

region annually affected by LDS fires (Figure 9.6), and the very significant contributions of large (≥100 km^2) fire patches, especially in open forest, woodland and sandstone woodland (Figure 9.8), it follows that *Callitris* stands in at least these habitat types are affected by frequently recurring large LDS fires, as depicted in Figure 9.9 for recurrence of large fires generally. Such LDS fire frequencies substantially exceed critical threshold values (Table 9.2).

DISCUSSION

Assembled data indicate that this remote biodiversity-rich region of the Australian tropical savannas:

Table 9.7: Proportion of closed forest perimeter affected by late dry season (LDS) fires, 1990–2005, where total inner perimeter = 472 km^2, total outer perimeter = 726 km^2.

Burning frequency	Inner 50m perimeter buffer fire affected (%)	Outer 50m perimeter buffer fire affected (%)
Unburnt	17.0	13.4
Burnt 1–2 times	41.9	39.4
Burnt 3–5 times	35.8	40.9
Burnt 6–12 times	5.3	6.3
Burnt 13–16 times	0	0

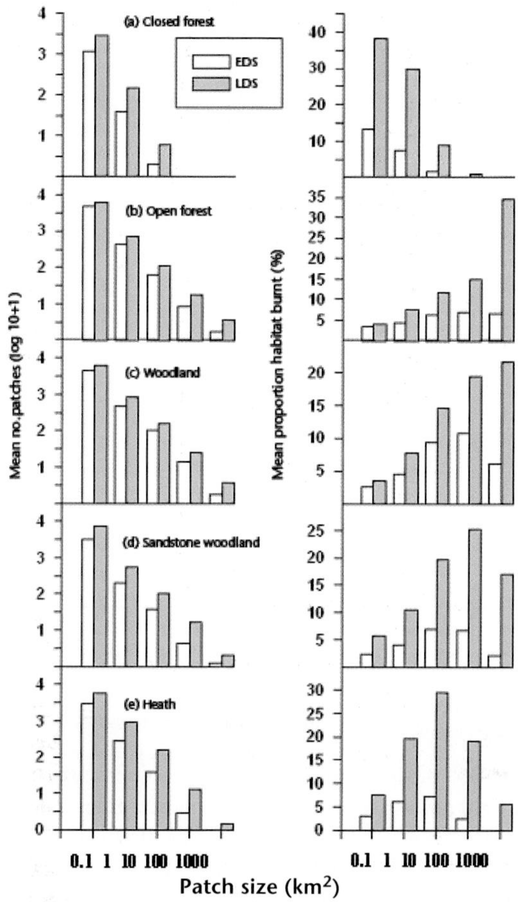

Figure 9.8 Early dry season (EDS) and late dry season (LDS) mean fire patch-size distributions for respective habitat types, and mean areal proportions of respective habitat types burnt by fires of different patch sizes, 1990–2005. Fire patch-size classes on the x-axis are as follows: 0.1 describes fires less than 0.1 km^2, 1 describes fires between 0.1 and 1 km^2 ... 1000 describes fires greater than 1000 km^2. Note that both left and right panels refer to the same vegetation structure type.

1. includes an array of fire-sensitive (both severity- and interval-dependent) vegetation types and constituent taxa, including long-lived obligate seeders, which are scattered widely, if unevenly, within the savanna matrix
2. today experiences fire regimes characterised generally by high annual frequencies of extensive, large fires that occur typically in the late dry season under severe fire-weather conditions
3. collectively substantially exceed defined ecological tolerances or thresholds over significant proportions of fire-sensitive indicator vegetation types and species.

For an 8000 km^2 region including extensive areas of the north-eastern Arnhem plateau, Bowman *et al.* (2004) likewise found that, on the basis of irregular sampling of Landsat imagery over the period 1988–1997, fire activity occurred very predominantly in the LDS. The extent by which critical fire regimes are exceeded in respective vegetation types over just 16 years of records sets imposing fire management and conservation challenges for future decades.

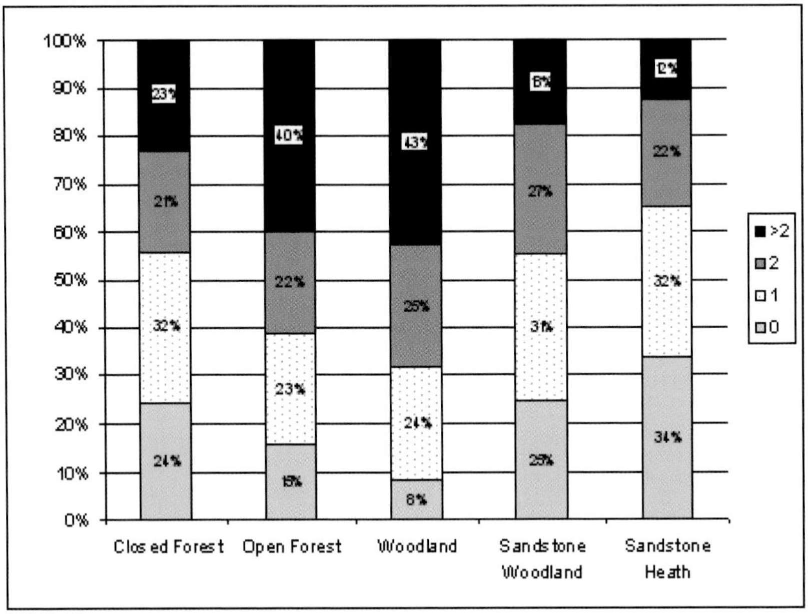

Figure 9.9 Areal proportion of short intervals (SI – intervals of 1, 2 or 3 years only) between large (≥100 km²) fires from 1990–2005, for major vegetation structure types, where: '0' = no short intervals, '1' = one SI only, '2' = two SIs, '> 2' = three or more SIs.

These observations reflect similar concerns expressed for sandstone vegetation types in the contiguous Kakadu and Nitmiluk National Parks. For example, ecological assessments of fire regimes undertaken mostly in or adjacent to both parks illustrate significant:

1. fire severity impacts on *Allosyncarpia* closed forest boundaries (e.g. Russell-Smith 1991, Russell-Smith *et al.* 1993; Bowman 1994; Bowman and Dingle 2006; Prior *et al.* 2006)
2. frequent fire impacts on sandstone heaths (Russell-Smith *et al.* 1998, 2002; Russell-Smith 2006)
3. fire severity and frequency impacts on *Callitris intratropica* (Bowman and Panton 1993; Price and Bowman 1994; Prior *et al.* 2006; Russell-Smith 2006).

Published fire history information derived principally from Landsat imagery for Kakadu indicates that, for the period 1980–2000, an annual average of 21% of sandstone heath and 27% of sandstone woodland habitat was burnt (Edwards *et al.* 2003), mostly in the LDS period (Russell-Smith *et al.* 1997). These quanta are very similar to those reported here for the WALFA region (Figure 9.6).

Recent aerial photo assessments of multi-decadal-scale change in regional rainforest assemblages report net rainforest patch expansion in lowland and hillslope situations in Litchfield National Park (Bowman *et al.* 2001b), and lowlands and sandstone uplands in the contiguous Kakadu National Park (Banfai and Bowman 2006; Bowman and Dingle 2006). However, it is possible that results reported in the latter sandstone uplands study reflect substantial canopy infilling (associated with decadal trends in increasing rainfall) and issues of scale and sampling methodology, rather than expansion per se (J. Freeman *pers. comm.* April 2007). Regardless, Bowman and Dingle (2006) found that 'forest expansion was negatively correlated with fire activity', and that exposed forest margins were those most at risk.

For a narrow east–west transect survey undertaken in the north of the present study area, Bowman *et al.* (2001a) suggested that low proportions of dead: live *Callitris* stems were attributable to proximity to Aboriginal outstations and associated fine-scale fire management. However, perusal of their Figure 9.3 indicates that, as observed in this study, low proportions of dead stems are also associated typically with larger live *Callitris* populations. Given the marginal distribution of Aboriginal outstations on the northern fringe, and to the south-east of the WALFA area (Figure 9.4a), the large live *Callitris* populations sampled in this study are typically very remote from Aboriginal settlement (Figure 9.4b). Collectively, these observations illustrate that contemporary fire regimes are exerting significant pressures on the savanna matrix in the WALFA region generally. Populations of *Callitris* are observed to be in collapse over much of northern and north-western Australia as a result of frequent, intense late dry season fires (e.g. Bowman and Panton 1993; Graham 2002; Yates and Russell-Smith 2003).

Whereas vegetation/habitat patch sizes typically are very small – with median patch sizes of 2–3 ha (Table 9.5) – at least 66% of the extent of all habitat types has been burnt by repeat large fires (>100 km^2) within 1–3 years of previously having been burnt (Figure 9.9). The influence of small numbers of fires contributing a large proportion of fire extent is widely reported in other northern Australian savannas (Yates and Russell-Smith 2002; Yates *et al.* 2008), and internationally (Kasischke and French 1995; Keeley *et al.* 1999; Stocks *et al.* 2003).

The extent and frequency of such large fires has significant implications generally for regional conservation of obligate seeder species with limited dispersal capacities, and fauna with restricted home ranges (Woinarski *et al.* 2005; Yates *et al.* 2008). Approximately 50% of Arnhem plateau sandstone heath shrub taxa are reported as being obligate seeders, with as many as 10% exhibiting primary juvenile periods (i.e. the time taken from germination to first seed production) of 5 years and more (Russell-Smith *et al.* 1998). Although many of these taxa possess dormant soil-borne seed banks (e.g. *Acacia*, legumes generally), and thus may persist in the face of at least some repeat short-interval fires, others such as the long-lived serotinous shrub, *Petraeomyrtus punicea*, do not exhibit this capacity (Russell-Smith 2006). For the long-lived obligate seeder taxa *Callitris intratropica* and *Petraeomyrtus punicea*, typical propagule dispersal distances range from tens of metres (Stocker 1966; Russell-Smith 2006), to within metres (Russell-Smith 2006) of parent canopies, respectively.

Likewise, observations also implicate impacts of regional fire regimes on endemic, locally restricted sandstone heath and sandstone woodland fauna, including the spectacular Leichhardt's Grasshopper, *Petasida ephippigera* (Lowe 1995) and the White-throated Grass-wren, *Amytornis woodwardi* (Noske 1992; Woinarski 1992). The latter species relies on patches of long unburnt Spinifex (*Triodia* spp.) hummock grasses in sandstone habitats. In a complementary assessment to that presented here, but at a broader savanna scale, Yates *et al.* (2008) provide a comprehensive review of the impacts of fire size and frequency on a variety of birds and small mammals with specialist habitat and/or small (<1 ha – <100 ha) home-range requirements.

It is salutary to note that fire-patch-size distributions reported here underestimate significantly the 'true' extent of large fire patches, given GIS analytic procedures that constrained estimation of patch sizes to (1) the study region, (2) imposed seasons and (3) the sizes of habitat patches. The importance of the third constraint is illustrated by parallel assessment (Yates *et al.* 2008) of a temporal subset (1997–2005) of the same fire extent data as used here, where fire-patch-size distributions were calculated by fire extent only (i.e. irrespective of habitat type). Whereas Yates *et al.* (2008) describe mean fire patch sizes as being 1.2 km^2 and 4 km^2 for the EDS and LDS, respectively, mean fire patch sizes for individual habitat types described here range from 0.02–0.2 km^2 in the EDS, to 0.07–0.55 km^2 in the LDS.

Methodological caveats to this assessment include, firstly, recognition of the limitations of remotely sensed imagery: especially for mapping of fine-scale fire patchiness, which, in turn, has significant implications for longer-term population dynamics (Price *et al.* 2003). On the basis of an assessment of fire patchiness in contiguous 5 × 5 m quadrats on the Arnhem plateau, Price *et al.* (2003) found that, in relatively rocky (~40–50%) terrain, a mean 25% of transect areas remained unburnt in the EDS, whereas 15% remained unburnt in the LDS. While acknowledging that our particular reliance on Landsat imagery means that sub-pixel, metre-scale fire patchiness is not effectively addressed, we note that even very patchy fires recurring at short intervals have been documented to incur significant deleterious impacts on local populations of regional obligate seeder shrubs (Russell-Smith *et al.* 2002; Russell-Smith 2006).

Secondly, fire seasonality is used here as a surrogate for fire severity, following other regional studies in northern Australian savanna systems (Williams *et al.* 1998, 2003; Russell-Smith *et al.* 2003b). However, we acknowledge that fire severity is likewise seasonally heterogeneous. For example, based on 10 years of observations of fire plot data ($n = 178$) from Kakadu and Nitmiluk National Parks, Russell-Smith and Edwards (2006) recorded that 21% of EDS fires were of moderate–high severity (scorch height typically ⩾2 m), and 24% of LDS fires were of low severity (scorch height <2 m).

In the absence of direct empirical observations of fire regime impacts on vegetation components over the vast WALFA study region, the application of detailed, verified fire history and vegetation mapping data derived from relatively high-resolution satellite imagery, coupled with defined ecological thresholds based on published observations, clearly provides for a robust, powerful and practicable means for assessing and ongoing monitoring of the status of regional fire-sensitive vegetation types. Thresholds criteria are recognised as an effective tool for informing ecological fire management in a variety of geographic settings (Keith *et al.* 2002; Bradstock and Kenny 2003; van Wilgen *et al.* 1998, 2003, 2004). As noted by van Wilgen *et al.* (2004), however, the ecological thresholds criteria adopted here do not necessarily apply to, nor have currency in, other savanna systems – especially those where resprouting is the predominant fire- response trait of woody species.

The broader issue raised by these assembled data is how to effectively respond to this assessment. Unlike the conservation reserves situated immediately to the west, the Indigenous custodians and landowners of the WALFA region have very limited financial and infrastructural resources and today are only occasional visitors at best – being resident in surrounding settlements and communities. There is next to no road or track infrastructure. And yet the region is acknowledged for its international biodiversity and cultural values. In a very real sense, the reinstatement of effective fire management in this remote area is required not just for sustaining the values of the WALFA region itself, but also for those nationally and internationally acclaimed conservation reserves that are situated down-wind of those large fires that erupt regularly outside their borders. As explored more fully in Chapter 12 the development of economically sustainable bases for regional fire management is critical to long-term conservation of the WALFA region's internationally acclaimed biodiversity values.

ACKNOWLEDGEMENTS

Cypress survey data were assembled by Felicity Watt, Cameron Yates, Brett Murphy, Brian Lynch (all Bushfires NT), and Ian Munro (Bawinanga Aboriginal Corporation). This work, which was undertaken for a over a decade, has been supported by the Natural Heritage Trust and associated community, the Northern Territory Government, and Commonwealth Government resources, Bushfires NT and the Tropical Savannas Management Cooperative Research Centre based Charles Darwin University, Darwin. It is a pleasure to acknowledge the contributions of many Indigenous and non-Indigenous colleagues involved in the WALFA project, but especially the inspirational leadership and commitment of Wamud Bardayal Nadjamerrek and Gamerang Peter Cooke.

REFERENCES

AGO (2007) 'National greenhouse gas inventory 2006'. Australian Greenhouse Office, Canberra.

Aldrick JM (1976) Soils of the Alligator Rivers area. In *Lands of the Alligator Rivers Area, Northern Territory.* (Ed. R Story) pp. 71–111. Land Research Series No. 38. CSIRO, Melbourne.

Andreae MO (1997) Emissions of trace gases and aerosols from southern African fires. In *Fires in Southern African Savannas; Ecological and Atmospheric Perspectives.* (Eds BW van Wilgen, MO Andreae, JG Goldammer and JA Lindsay) pp. 161–184. Witwatersrand University Press, Johannesburg.

Banfai DS and Bowman DMJS (2006) Forty years of lowland monsoon rainforest expansion in Kakadu National Park, northern Australia. *Biological Conservation* **131**, 553–565.

Blake G (2005) Object oriented mapping of sandstone heath vegetation on the Arnhem Plateau. Unpublished BSc (Hons.) thesis. Charles Darwin University, Darwin.

Bond WJ and Archibald S (2003) Confronting complexity: fire policy choices in South African savanna parks. *International Journal of Wildland Fire* **12**, 381–389.

Bowman DMJS (1994) Preliminary observations on the mortality of *Allosyncarpia ternata* stems on the Arnhem Land Plateau, northern Australia. *Australian Forestry* **57**, 62–64.

Bowman DMJS and Dingle JK (2006) Late 20th century landscape-wide expansion of *Allosyncarpia ternata* (Myrtaceae) forest in Kakadu National Park, northern Australia. *Australian Journal of Botany* **54**, 707–715.

Bowman DMJS and Panton WJ (1993) Decline of *Callitris intratropica* in the Northern Territory: implications for pre- and post-colonisation fire regimes. *Journal of Biogeography* **20**, 373–381.

Bowman DMJS, Price O, Whitehead PJ and Walsh A (2001a) The 'wilderness effect' and the decline of *Callitris intratropica* on the Arnhem Land Plateau, northern Australia. *Australian Journal of Botany* **49**, 665–672.

Bowman DMJS, Walsh A and Milne DJ (2001b) Forest expansion and grassland contraction within a *Eucalyptus* savanna matrix between 1941 and 1994 at Litchfield National Park in the Australian monsoon tropics. *Global Ecology and Biogeography* **10**, 535–548.

Bowman DMJS, Walsh A and Prior LD (2004) Landscape analysis of Aboriginal fire management in central Arnhem Land, northern Australia. *Journal of Biogeography* **31**, 207–223.

Bradstock RA and Kenny BJ (2003) An application of plant functional types to fire management in a conservation reserve in southeastern Australia. *Journal of Vegetation Science* **14**, 345–354.

Congalton RG (1991) A review of assessing the accuracy of classifications of remotely sensed data. *Remote Sensing of Environment* **37**, 35–46.

Crisp MD, Laffan S, Linder HP and Monro A (2001) Endemism in the Australian flora. *Journal of Biogeography* **28**, 183–198.

Definiens Imaging (2004) 'eCognition User Guide 4'. Definiens Imaging, Munich, Germany.

Dyer R, Jacklyn P, Partridge I, Russell-Smith J and Williams RJ (Eds) (2001) *Savanna Burning: Understanding and Using Fire in Northern Australia.* Tropical Savannas Cooperative Research Centre, Darwin.

Edwards A, Kennett R, Price O, Russell-Smith J, Spiers G and Woinarski J (2003) Monitoring the impacts of fire regimes on vegetation in northern Australia: an example from Kakadu National Park. *International Journal of Wildland Fire* **12**, 427–440.

Edwards A, Hauser P, McCartney J, Armstrong M, Thackway R, Allan G, Hempel C and Russell-Smith J (2001) A tale of two parks: contemporary fire regimes of Litchfield and Nitmiluk National Parks, monsoonal northern Australia. *International Journal of Wildland Fire* **10**, 79–89.

Ellis S, Kanowski P and Whelan R (2004) 'National inquiry on bushfire mitigation and management'. Commonwealth of Australia, Canberra. See <www.coag.gov.au>.

ER Mapper (2006) 'ER Mapper Professional Version 7.1'. Earth Resource Mapping Pty Ltd., Perth, Australia.

ESRI (1982–1997) 'ARC/INFO Version 7.2.1'. Environmental Systems Research Institute Inc., Redlands, USA.

Fisher R, Vigilante T, Yates CP and Russell-Smith J (2003) Patterns of landscape fire and predicted vegetation response in the North Kimberley region of Western Australia. *International Journal of Wildland Fire* **12**, 369–379.

Franklin DC (1999) Evidence of disarray amongst granivorous bird assemblages in the savannas of northern Australia, a region of sparse human settlement. *Biological Conservation* **90**, 53–68.

Gill AM, Moore PHR and Williams RJ (1996) Fire weather in the wet-dry tropics of the World Heritage Kakadu National Park, Australia. *Australian Journal of Ecology* **21**, 302–308.

Graham G (2002) 'An assessment of the distribution and status of Arnhem Cypress Pine *Callitris intratropica* (R. T. Baker and H. G. Sm.) in the Kimberley Region, Western Australia'. Report to the National Reserves System Program, Environment Australia, Canberra. Western Australia Dept. Conservation and Land Management, Perth.

Groffman PM, Baron JS, Blett T, Gold AJ, Goodman I, Gunderson LH, Levinson BM, Palmer MA, Paerl HW, Peterson GD, Poff NL, Rejeski DW, Reynolds JF, Turner MG, Weathers KC and Wiens J (2006) Ecological thresholds: The key to successful environmental management or an important concept with no practical application? *Ecosystems* **9**, 1–13.

Haynes CD (1985) The pattern and ecology of *munwag*: traditional Aboriginal fire regimes in north-central Arnhem Land. *Proceedings of the Ecological Society of Australia* **13**, 203–214.

Haynes CD (1991) Use and impact of fire. In *Monsoonal Australia: Landscape, Ecology and Man in the Northern Lowlands.* (Eds CD Haynes, MG Ridpath and MAJ Williams) pp. 61–71. Balkema, Amsterdam.

Higgins SI, Bond WJ and Trollope WSW (2000) Fire, resprouting and variability: a recipe for grass-tree coexistence in savanna. *Journal of Ecology* **88**, 213–229.

Ingwersen F (1995) Kakadu-Alligator Rivers Region, Northern Territory, Australia – CPD Site Au4. In *Centres of Plant Diversity: A Guide and Strategy for their Conservation. Vol. 2. Asia, Australasia and the Pacific.* (Eds SD Davis, VH Heywood and AC Hamilton) pp. 471–475. WWF and IUCN, Cambridge, UK.

Kasischke KS and French NHF (1995) Locating and estimating the aerial extent of wildfires in Alaskan boreal forests using multiple-season AVHRR NDVI composite data. *Remote Sensing of Environment* **51**, 263–275.

Keeley JK, Fotheringham CJ and Morais M (1999) Re-examining fire suppression impacts on brushland fire regimes. *Science* **284**, 1829–1832.

Keen I (1980) The Alligator Rivers Aborigines – retrospect and prospect. In *Northern Australia: Options and Implications.* (Ed. R Jones) pp. 171–185. Australian National University, Canberra.

Keith DA, Williams JE and Woinarski JCZ (2002) Fire management and biodiversity conservation: key approaches and principles. In *Flammable Australia.* (Eds RA Bradstock, JE Williams and AM Gill) pp. 401–425. Cambridge University Press, Cambridge, UK.

Lowe L (1995) Preliminary investigations of the biology and management of Leichhardt's grasshopper, *Petasida ephippigera* White. *Journal of Orthopera Research* **4**, 219–221.

Luke RH, McArthur AG (1978) *Bushfires in Australia*. Australian Government Publishing Service, Canberra.

Lynch BT and Wilson PL (1998) 'Land systems of Arnhem Land'. Technical Report No. R97/1. Northern Territory Department Lands Planning and Environment, Darwin.

McKenzie NL and Belbin L (1991) Kimberley rainforest communities: reserve recommendations and management considerations. In *Kimberley Rainforests Australia*. (Eds NL McKenzie, R B Johnston and PG Kendrick.) pp. 453–468. Surrey Beatty & Sons, Sydney.

Meakin C, Owen G, Brocklehurst P and Lewis D (2001) 'Mapping the forest cover of the Northern Territory, Document 1: field methodology and FPC indexing'. Report RCAB35. Northern Territory Department of Lands, Planning and Environment, Darwin.

Needham RS (1988) 'Geology of the Alligator Rivers Uranium Field, Northern Territory'. Bulletin 224. Bureau of Mineral Resources, Canberra.

Noske R (1992) The status and ecology of the White-throated Grasswren *Amytornis woodwardi*. *Emu* **92**, 39–51.

Preece N (2007) Traditional and ecological fires and effects of bushfire laws in north Australian savannas. *International Journal of Wildland Fire* **16**, 378–389.

Price O and Bowman DMJS (1994) Fire-stick forestry: a matrix model in support of skilful fire management of *Callitris intratropica* R.T. Baker by north Australian Aborigines. *Journal of Biogeography* **21**, 573–580.

Price O, Edwards A, Connors G. Woinarski J, Ryan G, Turner A and Russell-Smith J (2005) Fire heterogeneity in Kakadu National Park, 1980–2000. *Wildlife Research* **32**, 425–433.

Price O, Russell-Smith J and Edwards A (2003) Fine-scale patchiness of different fire intensities in sandstone heath vegetation in northern Australia. *International Journal of Wildland Fire* **12**, 227–236.

Prior LD, Bowman DMJS and Brooke BW (2006) Growth and survival of two north Australian relictual tree species, *Allosyncarpia ternata* (Myrtaceae) and *Callitris intratropica* (Cupressaceae). *Ecological Research* **22**, 228–236, <doi:10.1007/s11284-006-0011-2>.

Russell-Smith J (1991) Classification, species richness, and environmental relations of monsoon rain forest in northern Australia. *Journal of Vegetation Science* **2**, 259–278.

Russell-Smith J (2006) Recruitment dynamics of the long-lived obligate seeders *Callitris intratropica* (Cupressaceae) and *Petraeomyrtus punicea* (Myrtaceae). *Australian Journal of Botany* **54**, 479–485.

Russell-Smith J and Bowman DMJS (1992) Conservation of monsoon rainforest isolates in the Northern Territory, Australia. *Biological Conservation* **59**, 51–63.

Russell-Smith J and Edwards AC (2006) Seasonality and fire severity in savanna landscapes of monsoonal northern Australia. *International Journal of Wildland Fire* **15**, 541–550.

Russell-Smith J, Lucas DE, Brock J and Bowman DMJS (1993) *Allosyncarpia*-dominated rain forest in monsoonal northern Australia. *Journal of Vegetation Science* **4**, 67–82.

Russell-Smith J, Ryan PG and Cheal D (2002) Fire regimes and the conservation of sandstone heath in monsoonal northern Australia: frequency, interval, patchiness. *Biological Conservation* **104**, 91–106.

Russell-Smith J, Ryan PG and DuRieu R (1997) A Landsat MSS derived fire history of Kakadu National Park, monsoonal northern Australia, 1980–94: seasonal extent, frequency and patchiness. *Journal of Applied Ecology* **34**, 748–766.

Russell-Smith J, Ryan PG. Klessa D, Waight G and Harwood R (1998) Fire regimes, fire-sensitive vegetation and fire management of the sandstone Arnhem plateau, monsoonal northern Australia. *Journal of Applied Ecology* **35**, 829–846.

Russell-Smith J, Yates C, Edwards A, Allan GE, Cook GD, Cooke P, Craig R, Heath B and Smith R (2003a) Contemporary fire regimes of northern Australia, 1997–2001: change since Aboriginal occupancy, challenges for sustainable management. *International Journal of Wildland Fire* **12**, 283–297.

Russell-Smith J, Whitehead PJ, Cook GD and Hoare JL (2003b) Response of *Eucalyptus*-dominated savanna to frequent fires: lessons from Munmarlary, 1973–1996. *Ecological Monographs* **73**, 349–375.

Russell-Smith J, Yates CP, Whitehead P, Smith R, Craig R, Allan G, Thackway R, Frakes I, Cridland, Meyer CP and Gill AM (2007) Bushfires 'down under': patterns and implications of Australian landscape burning. *International Journal of Wildland Fire* **16**, 361–377.

Scholes RJ and Archer SR (1997) Tree-grass interactions in savannas. *Annual Review of Ecology and Systematics* **28**, 517–544.

Specht RL (1981) Project foliage cover and standing biomass. In *Vegetation Classification in Australia*. (Eds AN Gillison and DJ Anderson) pp. 10–21. CSIRO, Canberra.

Stocker GC (1966) 'Aspects of the seeding habits of *Callitris intratropica*'. Forestry and Timber Bureau Leaflet No. 97. Commonwealth of Australia Department of National Development, Canberra.

Stocks BJ, Mason JA, Todd JB, Bosch EM, Wotton BM, Amiro BD, Flannigan MD, Hirsch KG, Logan KA, Martell DL and Skinner WR (2003) Large forest fires in Canada, 1989–1997. *Journal of Geophysical Research* **107**, 8149, <doi:10.1029/2001JD000484>.

Taylor J and Tulloch D (1985) Rainfall in the wet-dry tropics: extreme events at Darwin and similarities between years during the period 1870–1973. *Australian Journal of Ecology* **10**, 281–295.

Townsend SA and Douglas MM (2000) The effect of three fire regimes on stream water quality, water yield and export co-efficients in a tropical savanna (northern Australia). *Journal of Hydrology* **229**, 118–137.

van Wilgen BW, Biggs HC and Potgieter ALF (1998) Fire management and research in the Kruger National Park, with suggestions on the detection of thresholds of potential concern. *Koedoe* **41**, 69–87.

van Wilgen BW, Govender N, Biggs HC, Ntsala D and Funda XN (2004) Response of savanna fire regimes to changing fire-management policies in a large African National Park. *Conservation Biology* **18**, 1533–1540.

van Wilgen BW, Trollope WSW, Biggs HC, Potgieter ALF and Brockett BH (2003) Fire as a driver of ecosystem variability. In *The Kruger Experience: Ecology and Management of Savanna Heterogeneity*. (Eds JT du Toit, KH Rogers and HC Biggs) pp. 149–170. Island Press, Washington DC.

Whitehead PJ, Bowman DMJS, Preece N, Fraser F and Cooke P (2003) Customary use of fire by indigenous peoples in northern Australia: its contemporary role in savanna management. *International Journal of Wildland Fire* **12**, 415–425.

Williams RJ, Gill AM and Moore PHR (1998) Seasonal changes in fire behaviour in a tropical savanna in northern Australia. *International Journal of Wildland Fire* **8**, 227–239.

Williams RJ, Gill AM and Moore PHR (2003) Fire behaviour. In *Fire in Tropical Savannas: The Kapalga Experiment*. (Eds AN Andersen, GD Cook and RJ Williams) pp. 33–46. Springer-Verlag, New York.

Williams RJ, Griffiths AD and Allan GE (2002) Fire regimes and biodiversity in the wet-dry tropical landscapes of northern Australia. In *Flammable Australia: The Fire Regimes and Biodiversity of a Continent*. (Eds RA Bradstock, JE Williams and AM Gill) pp. 281–304. Cambridge University Press, Cambridge, UK.

Wilson BA, Brocklehurst PS, Clark MJ and Dickinson KJM (1990) 'Vegetation survey of the Northern Territory'. Technical Report No. 49. Conservation Commission of the Northern Territory, Darwin.

Woinarski JCZ (1992) The conservation status of the White-throated Grasswren *Amytornis woodwardi*, an example of problems in status designation. *Northern Territory Naturalist* **13**, 1–5.

Woinarski JCZ, Milne D and Wanganeen G (2001) Changes in mammal populations in relatively intact landscapes of Kakadu National Park, Northern Territory, Australia. *Austral Ecology* **26**, 360–370.

Woinarski JCZ, Hempel C, Cowie I, Brennan K, Kerrigan R, Leach G and Russell-Smith J (2006) Distributional patterning of restricted range plant species in the Northern Territory, Australia. *Australian Journal of Botany* **54**, 627–640.

Woinarski JCZ, Williams RJ, Price O and Rankmore B (2005) Landscapes without boundaries: wildlife and their environments in northern Australia. *Wildlife Research* **32**, 377–388.

Yates C and Russell-Smith J (2002) An assessment of the accuracy of DOLA's Northern Australia NOAA-AVHRR Fire Affected Area (FAA) map products. In 'Australian fire regimes: contemporary patterns (April 1998–March 2000) and changes since European settlement'. (Eds Russell-Smith J, Craig R, Gill AM, Smith R and Williams J). Australia State of the Environment Second Technical Paper Series (Biodiversity). Department of the Environment and Heritage, Canberra. <http://www.ea.gov.au/soe/techpapers/index.html>

Yates CP and Russell-Smith J (2003) Fire regimes and vegetation sensitivity analysis: an example from Bradshaw Station, monsoonal northern Australia. *International Journal of Wildland Fire* **12**, 349–358.

Yates CP, Edwards AC and Russell-Smith J (2008) Big fires and their ecological impacts in Australian savannas: size and frequency matters. *International Journal of Wildland Fire* **17**, 768–781.

Yibarbuk D, Whitehead PJ, Russell-Smith J, Jackson D, Godjuwa C, Cooke P, Choquenot D and Bowman DMJS (2001) Fire ecology and Aboriginal land management in central Arnhem Land, northern Australia: a tradition of ecosystem management. *Journal of Biogeography* **28**, 325–344.

10

Fire and biodiversity monitoring for conservation managers:

a 10-year assessment of the 'Three Parks' (Kakadu, Litchfield and Nitmiluk) program

Jeremy Russell-Smith, Andrew C Edwards, John CZ Woinarski, John McCartney, Sarah Kerin, Steve Winderlich, Brett P Murphy and Felicity A Watt

SUMMARY

The chapter describes key results and observations from the first 10 years (1995–2004) of a fire regime and biodiversity monitoring program established in Kakadu, Litchfield and Nitmiluk National Parks. Although the major purposes for establishing the 'Three Parks' program were to inform park managers of the consequences of prescribed management actions – and associated development of a long-term reference for the benefit of future managers – the program has yielded significant insights of benefit to fire management and related ecological issues in the Top End region generally. The program has two complementary components: fire-extent mapping derived from Landsat imagery; and annual monitoring of fire occurrence and severity, and 5- yearly quantitative sampling of vegetation and vertebrate fauna at 220 permanent plots. Plots were selected to represent all major flammable habitat types. We describe fire histories of the three parks over the 10-year assessment period derived from remote sensing studies and, based on monitoring plot assessments, selected analyses of vegetation (woodland, the fire-sensitive conifer *Callitris intratropica*, rainforest) and vertebrate fauna responses to fire regime characteristics. We show that: (a) mean annual extent of burning has declined in Kakadu, and increased in Litchfield and Nitmiluk; (b) there has been concomitant increase in burning in the early dry season period on all parks; (c) changes in savanna vegetation structure generally have been only weakly related to fire regime characteristics; although (d) severe fires have had deleterious impacts on some savanna structural attributes, *Callitris* populations, and sandstone rainforests; and (e) high fire frequency is associated with significantly reduced numbers of small native mammal species. Although assessment of the impacts of fire frequency on fire-sensitive sandstone heaths is yet to be undertaken, high fire frequencies observed at many plots – and on the three parks generally – are cause for concern. We consider the utility of the program in meeting the information requirements of ranger staff, the power of the program to detect change and the practical effectiveness of the program in helping inform adaptive management responses to identified fire-management challenges.

INTRODUCTION

Fire is a recurring element of the mesic (>1000 mm rainfall p.a.) savannas of northern Australia where, on average, over one-third and, in some places, more than half the landscape is burnt annually. At the savanna-wide scale, such observations have been informed from the early 1990s through mapping of burnt (or, more properly, fire-affected) areas from relatively coarse resolution (~1.1 km^2 pixels) Advanced Very High Resolution Radiometer (AVHRR) imagery on the United States' National Oceanic and Atmospheric Administration (NOAA) series of satellites. At more local scales, fine resolution LANDSAT (~0.1–0.5 ha pixels) imagery has been used to map fires and characterise fire regimes in a large number of regional studies since the 1980s.

As described elsewhere in this volume, such satellite-based observations have complemented an extensive research effort (commencing from the 1960s) focusing on describing and understanding savanna fire behaviour, and the ecological responses of regional vegetation and faunal components to different fire regimes (for recent reviews see Williams *et al.* 2002, 2003a; Woinarski *et al.* 2005a). Fire ecology research has targeted savanna systems and component species especially, but substantial other work has examined the fire ecology of floodplains, and relatively fire-sensitive rainforest and sandstone heath habitats.

A parallel development since the mid-1990s has been joint establishment of a long-term fire monitoring program on three of the Top End's major regional National Parks: the 19 092 km^2 World Heritage Kakadu National Park, jointly managed by its Aboriginal traditional owners and Parks Australia – an agency of the Australian Government; the contiguous 2924 km^2 Nitmiluk National Park, jointly managed by its Aboriginal traditional owners and the Northern Territory Government's Parks and Wildlife Service; and the 1464 km^2 Litchfield National Park, also managed by the Parks and Wildlife Service (Figure 10.1). Impetus for development of this joint program has had a number of drivers, including requirements for:

- maintaining an up-to-date fire history coverage, based on interpretation of satellite imagery for each park, accompanied by systematic on-ground verification
- recognition of the ecological significance of fire as agent of biotic change, particularly in relation to the effects of more-intense fires late in the dry season
- providing a means by which changes in status and condition of habitats and species (including vulnerable flora and vertebrates) can be assessed with respect to different fire regime parameters (e.g. seasonality, frequency, interval and intensity)
- requirements for quantifying management success with respect to stated objectives in management plans
- providing a means by which park managers and traditional owners can themselves assess the long-term effects of their management actions, as part of an adaptive management process.

Importantly, the program was borne out of recognition that long-term experimental fire research plot studies established in lowland savanna at Munmarlary (1972–1996) and Kapalga (1990–1994) in present-day Kakadu, did not adequately meet the multi-factorial needs of regional managers given: (a) the highly restricted set of habitat and fire treatments involved in those experiments; as well as (b) disenchantment with a research process that did not fully engage regional managers in either its design or implementation (Andersen and McKaige 1997; Russell-Smith *et al.* 2003). This negative experience with a research-imposed agenda led to a key philosophical standpoint being taken by park staff concerning the purpose of establishing

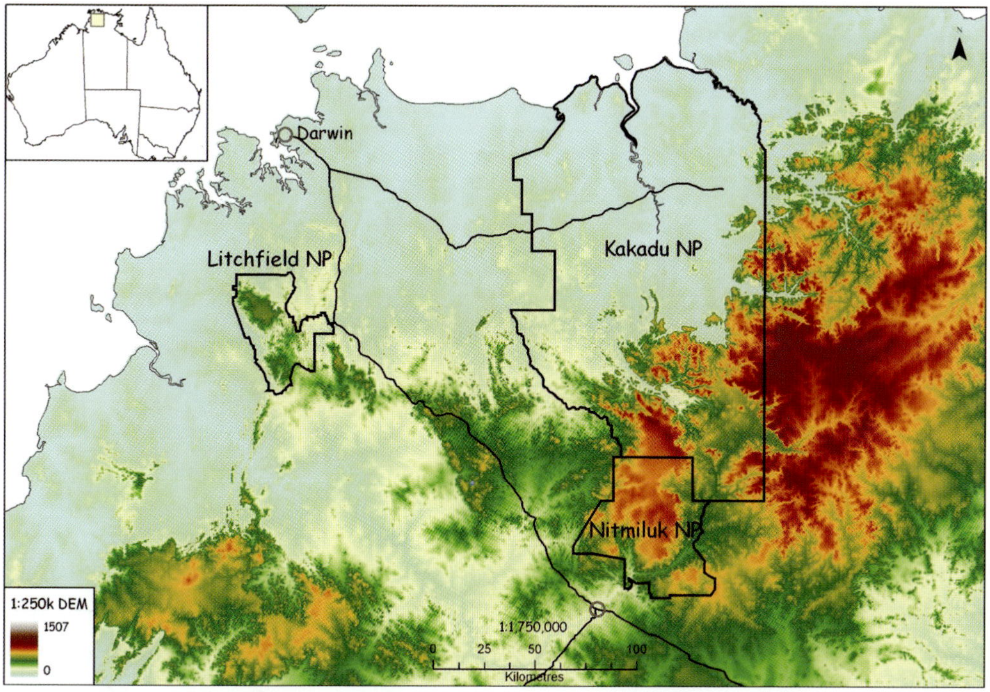

Figure 10.1 Location of Kakadu, Litchfield and Nitmiluk National Parks.

this fire-monitoring program – the program must primarily serve the long-term information interests of park managers and traditional owners (Russell-Smith and Ryan 1996). Nevertheless, the program continues to provide fundamentally important fire history, vegetation and faunal ecological perspectives (Edwards *et al.* 2001, 2003; Woinarski *et al.* 2004; Price *et al.* 2005) that are relevant to the three parks, as well as northern Australian savanna systems generally. Various assessments presented elsewhere in this volume, for example, have been informed by datasets assembled through the Three Parks program.

This chapter first outlines methodological aspects concerning monitoring program components, then considers representative results obtained from assessment of various datasets (satellite-derived fire histories, flora and fauna responses) assembled over the first 10 years (1995–2004) of the program, and key assessment of the value of the applied program to the land managers it was designed primarily to assist. Purposefully, our presentation of results entails use of simple descriptive statistics; more sophisticated modelling treatments (embedding various assumptions) are, at the time of writing, under review in various journals.

Regional context

The three parks are located in the north-western 'Top End' of the Northern Territory (Figure 10.1). The entire region shares a broadly similar monsoonal climate, with rainfall ranging from around 1500 mm in the vicinity of Litchfield and the northern sub-coastal portions of Kakadu, to around 1000 mm in Nitmiluk in the south-east. Although annually highly variable in quantity, the wet season (predominantly Nov–Mar) reliably delivers substantial rainfall to the entire study region.

The parks also share various physiographic features in common: notably extensive tracts of infertile, typically rugged sandstone uplands and associated plateau surfaces, surrounded by

Table 10.1: Area of major landscape units and proportion occupied by vegetation types in Kakadu, Litchfield and Nitmiluk National Parks. Note floodplain data not repeated (given only under 'Landscape Unit' heading).

Landscape unit	Kakadu	Litchfield	Nitmiluk
Total area (km²)	19 092	1464	2924
Landcape unit (%)			
Lowlands	67	37	29
Sandstone uplands	22	63	71
Floodplains	11	–	–
Vegetation type (%)			
Lowland woodland/open forest	65	36	27
Lowland rainforest	0.6	0.5	0.2
Sandstone woodland/open forest	14	63	65
Sandstone heath	6.5	<0.1	7
Sandstone rainforest	2	0.4	<0.1
Lancewood (*Acacia shirleyi*)	–	–	0.8

typically laterised undulating lowlands and plains. Kakadu is famous also for its extensive estuarine and freshwater wetland floodplain systems. By park, Kakadu encompasses mostly lowland formations (67%), then sandstone uplands (22%), and floodplains (11%). Conversely, Litchfield and Nitmiluk comprise mostly sandstone uplands (Table 10.1). Apart from floodplain systems, vegetation types are broadly similar across the three parks and are dominated by extensive eucalypt-dominated open forest and woodland savannas, with lesser components of sandstone heaths, monsoon rainforest and, in Nitmiluk, small patches of closed-canopy lancewood (*Acacia shirleyi*) thicket (Table 10.1; Figure 10.2).

Prevailing climatic conditions are conducive to rapid development of grassy fuels following fire, such that fires may recur at any one site on an annual–biennial basis (Walker 1981; Williams *et al.* 2002). Even in rugged sandstone formations, accumulation of fine grass and litter fuels is sufficient to support intense fires (>5000 kW m^{-1}) in all but exposed rock situations under late dry season climatic conditions within 1–3 years of having been burnt previously (Russell-Smith *et al.* 1998). Fires in the region are almost invariably anthropogenic in origin, and ground-borne (Stocker and Mott 1981).

FIRE MONITORING PROGRAM

The Three Parks fire-monitoring program comprises two complementary components: satellite-based mapping of fire events and on-ground assessment of changes in biota at a large series of permanent plots. The value of, and requirements and support for, the undertaking of this program on each of the three parks is set out in respective planning documents (Litchfield National Park 1998; Nitmiluk National Park 2002; Kakadu National Park 2007).

Satellite monitoring

Fire histories have been assembled for all three parks derived from interpretation of sub-hectare resolution LANDSAT MSS (in early years for Kakadu) and, principally, LANDSAT TM and ETM imagery. For Kakadu, the available fire history dates from 1980. For Litchfield and Nitmiluk, fire histories are available from 1990 and 1989, respectively. For all parks, annual

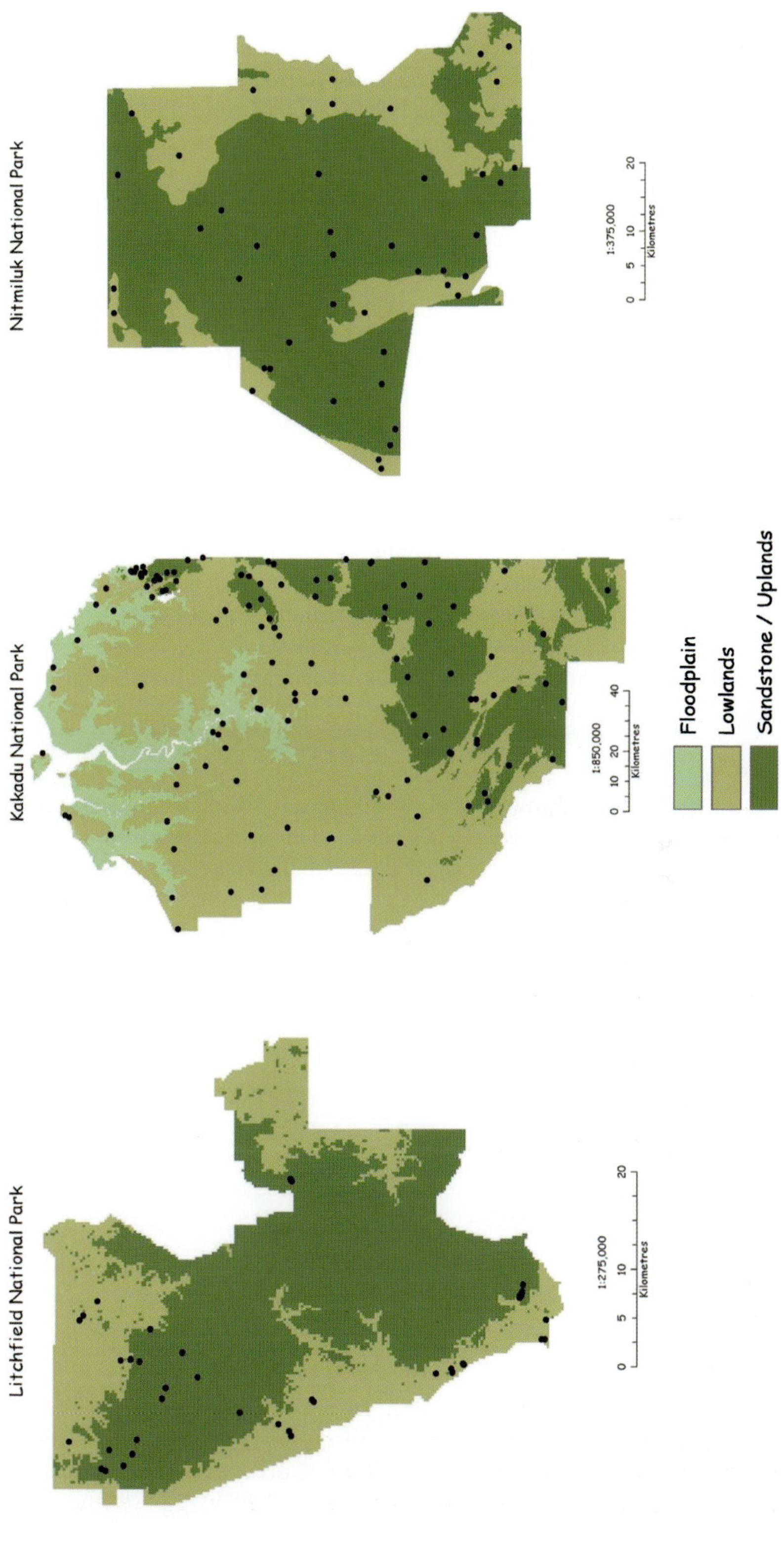

Figure 10.2 Locations of permanent monitoring plots with respect to major landform units, Kakadu, Litchfield and Nitmiluk National Parks.

fire mapping has been derived from at least three sampled scenes as follows: a first image obtained ideally early in the dry season (around late May/early June); a second obtained around the end of the main burning early season period (late July/early August); and a third image obtained as late in the year as possible before the onset of extensive cloudy conditions associated with the developing wet season. More frequent sampling of imagery is required early in the dry season period given regrowth of perennial grasses under still relatively favourable soil moisture conditions (Russell-Smith *et al.* 1997; Bowman *et al.* 2003). To account for cloudiness issues in the late dry season period, fire mapping from LANDSAT imagery is regularly augmented with coarser-resolution, daily imagery available from AVHRR and, in recent years, MODIS sensors. Fire-mapping data have been validated annually since the mid-1990s for all three parks, based on stratified aerial transect assessments. Methodological details concerning the satellite-based fire-mapping program, including validation assessments, are given in Russell-Smith *et al.* (1997) and Edwards *et al.* (2001, 2003).

An allied component of the imagery-derived monitoring program has been development of the North Australia Fire Information (NAFI) website (http://www.firenorth.org.au). This site provides daily mapping of fire hotspots detected from AVHRR and MODIS sensors, as well as regular (e.g. weekly in the peak fire season) manual updating of fire extent mapping derived principally from MODIS. To date, NAFI has been developed under the umbrella of the Tropical Savannas Management Cooperative Research Centre based in Darwin, with fire information, weather and mapping inputs from a number of contributing organisations including Landgate (Western Australia Government), Geoscience Australia, Bureau of Meteorology, Cape York Peninsula Development Association, and Bushfires NT. No fee is currently charged for this service.

Permanent plots

A total 220 permanent monitoring plots (40 m × 20 m) was established on the three parks in 1994–1995 to monitor biotic change. Of these, 133 plots are located in Kakadu, with 41 and 46 in Litchfield and Nitmiluk, respectively (Table 10.2). These sample a variety of landform and

Table 10.2: Numbers of plots in Kakadu, Litchfield and Nitmiluk National Parks.

	No. plots			
Landscape unit	Kakadu	Litchfield	Nitmiluk	Total
Total	133	41	46	220
Landscape unit				
Lowlands	72	17	14	103
Sandstone uplands	49	20	30	99
Floodplains	12	4	2	18
Vegetation type				
Lowland woodland/open forest	55	16	13	84
Lowland rainforest	17	1	1	19
Sandstone woodland/open forest	13	19	21	53
Sandstone heath	16	1	3	20
Sandstone rainforest	20	0	3	23
Lancewood (*Acacia shirleyi*) thicket	0	0	3	3

Notes: (i) floodplain plots, typically focusing on *Melaleuca* spp, not repeated under Vegetation type heading; (ii) many plots are situated across ecotones, and in data below; (iii) numbers of rainforest and lancewood plots include all plots ecotonal with other vegetation types; (iv) numbers of heath plots include all plots ecotonal with sandstone woodland but not rainforest.

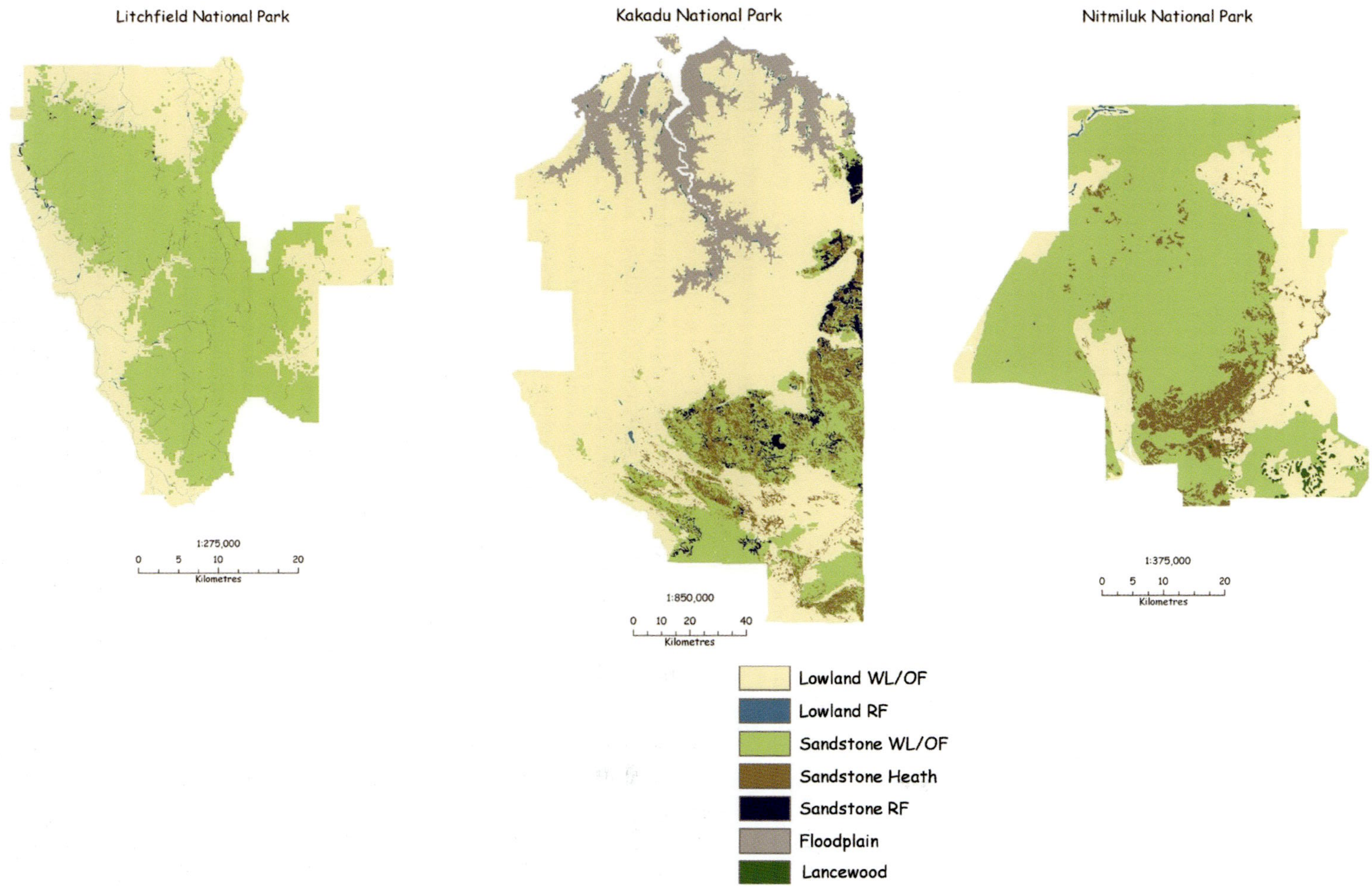

Figure 10.3 Major vegetation/habitats, Kakadu, Litchfield and Nitmiluk National Parks.

vegetation type/habitat conditions (Table 10.2; Figure 10.3). A substantial proportion of plots was positioned deliberately at sites likely to most efficiently reveal environmental dynamics, especially at ecotones and in patches of fire-sensitive vegetation (e.g. stands of *Callitris* or sandstone heaths). As well, many plots are located at, or in the near vicinity of, intensively managed sites such as camp-grounds and other tourist destinations.

The major purposes for establishing the program were to inform park managers of the consequences of prescribed management actions (including leaving areas unburnt for many years), and associated development of a long-term reference for the benefit of future managers. As such, the program was not designed with statistical elegance in mind (e.g. as in a replicated experimental design with randomised plot selection and regularly applied fire treatments). Rather, park managers requested a robust monitoring program to inform them of the consequences of different management actions across the spectrum of major habitat types occurring in respective parks (Russell-Smith and Ryan 1996; Edwards *et al.* 2003). Re-sampling of plot vegetation attributes across all three parks is planned to be undertaken every 5 years, with the second re-sampling having been completed in 2004–2005. Monitoring plot establishment and re-sampling exercises have been conducted at the end of the wet season period to facilitate flora identification (i.e. before herbaceous plants wither, or the vegetation is burnt). While following generally the same sampling regimen, fauna sampling has not been undertaken so consistently at each of the three parks for reasons described below.

Vegetation sampling

Detailed information on vegetation and other environmental factors is recorded at each plot. For trees (woody stems ≥5 cm dbh), all individuals in the plot are counted and tagged; for taller shrubs (>0.5 m), all individuals are counted within a fixed 40 m × 10 m sub-plot; counts of small shrubs (≤0.5 cm) are recorded in two 40 × 1 m fixed transects; and for herbaceous ground-layer species, cover is recorded in each of 40 fixed 1m × 1m quadrats. All species occurring within each of three height strata (trees, shrubs and ground-layer) are counted, such that any species may be recorded in more than one class. Importantly for practical assessment purposes (see below), for each plot target species and/or habitat conditions were defined at establishment. Full details of the vegetation sampling design are given in Russell-Smith and Ryan (1996) and Edwards *et al.* (2003).

Vegetation sampling has been undertaken jointly by park ranger staff with expert technical assistance, especially from Northern Territory Government agencies: the Parks and Wildlife Service (formerly Parks and Wildlife Commission) and Bushfires NT (formerly Bushfires Council). For Litchfield and Nitmiluk, monitoring plot assessments have been undertaken as part of major ranger-training exercises, where rangers have been assembled from throughout the Northern Territory, including from Indigenous ranger groups. These exercises typically have been conducted over 2-week periods, and have involved:

1. plot sampling and associated data entry (in a Microsoft Access database), analysis (including assessment of the status of target species and/or habitat condition), interpretation
2. editing of remotely-sensed fire-mapping products and associated GIS analyses (primarily using Arcview)
3. report writing
4. presentation of findings to training camp participants and invited senior park management and scientific personnel.

Although a broadly similar process is undertaken in Kakadu, the larger number of plots involved means that sampling is undertaken over a larger number of weeks in 2 years. In addition, given staffing issues (see below), there is less formal involvement of Kakadu staff in the entry, assessment and presentation of assembled datasets. Detailed reports summarising establishment and re-sampling programs are available for each park; for example, for the 2000–2001 re-sampling period, see O'Leary *et al.* (2000); PWCNT (2000); Turner *et al.* (2002).

Fauna sampling

The monitoring program was designed originally to sample vegetation. Its applicability for the monitoring also of fauna was recognised subsequently. As a consequence, the monitoring of fauna has generally lagged behind the vegetation monitoring program. Whereas all 133 monitoring plots in Kakadu have been sampled for vegetation three times over the period 1995–2004, sampling for fauna has been far less substantial and regular: 15 of the Kakadu plots have been monitored for fauna three times, 45 plots twice, 68 plots only once, and 6 plots have not been sampled at all. Sampling intensity has been better for Litchfield, where 40 of the 41 plots have now been sampled three times, more or less at the same time as the vegetation sampling. The 46 plots in Nitmiluk have been sampled only once (in 2005). This more fragmentary monitoring performance means that it is not yet feasible to undertake detailed analyses of fauna trends and responses in a manner comparable to that summarised here for plants.

Fauna sampling is restricted to terrestrial vertebrates. Invertebrates have not been included, partly because of the taxonomic impediment of a generally very poorly known invertebrate fauna. The sampling of terrestrial vertebrates follows a protocol now widely used for fauna survey in northern Australia (e.g. Woinarski *et al.* 2004), allowing comparative linkage of trends and inventory to a far broader base than simply this set of monitoring plots.

At (or immediately adjacent to) the fixed monitoring plots, this fauna sampling protocol comprises searches and trapping over a 72 hr period, including:

1. eight 'instantaneous' counts (predominantly in the early morning) of birds present in a 1 ha plot
2. two nocturnal spotlighting searches, each for 10 min, of a 50 m × 50 m plot within the 1 ha plot, for frogs, reptiles, birds and mammals
3. three 10-min day-time searches for reptiles, frogs and mammals in the 50 m×50 m plot
4. an array of 20 Elliott traps (metal box traps 30 cm × 7 cm × 7 cm) and cage traps (65 cm × 15 cm × 15 cm) evenly spaced along the perimeter of the 50 m × 50 m plot (mostly to sample small mammals), baited with a mixture of peanut butter, honey and oats, and checked early every morning
5. two pitfall traps (20 L plastic buckets dug into the ground), with 8 m of 30 cm high flywire drift-line netting.

For each species, an abundance value is calculated as the sum of all individuals captured or reported in searches.

This sampling protocol is notably more demanding than the vegetation sampling. Sampling requires the monitoring plot to be visited repeatedly over a 3-day period, and the requirement to spotlight search at night obviously requires nocturnal access (typically involving camping near the site). Further, the equipment required for sampling is far more substantial than that for vegetation monitoring. Hence, the fauna sampling tends to be more expensive (especially so for plots requiring helicopter access) and to take much more time. Partly because of this logistical constraint, fauna sampling, at least in Kakadu, has been spaced over more months of the year than vegetation sampling, albeit with the requirement that subsequent sampling of any plot should occur at approximately the same time of year as any previous sampling of the plot (+/– one month).

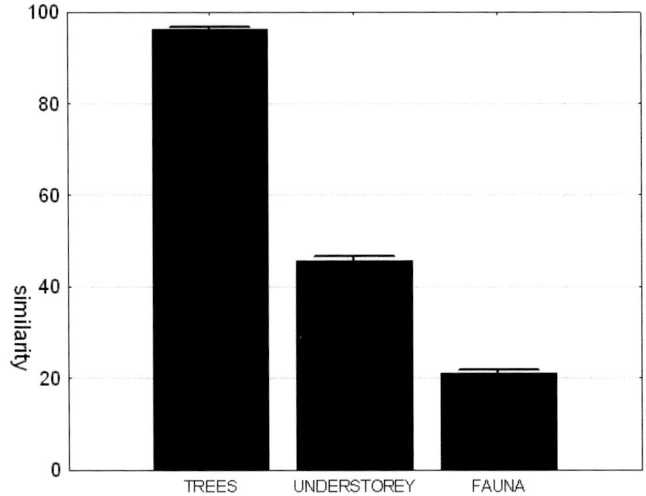

Figure 10.4 Similarity between sampling episodes of the same plot in tree species presence, understorey plant species presence and vertebrate fauna species presence, where 'similarity = (the number of species reported in a plot at both time 1 and time 2)/ (the cumulative tally of different species recorded in that plot across times 1 and 2)'.

Fauna sampling has additional complications:

- some fauna species (e.g. burrowing skinks) present in a plot may not be detected readily
- some fauna species may disperse in and out of the plot during sampling (and happen to be over- or under-recorded by chance during the sampling events)
- there may be some differences between observers in observation skills and identifications
- there may be seasonal and year-to-year variation between years in regional occurrence and 'trapability' or detectability of some species, unrelated to fire or other threatening factors.

For this set of reasons, there may be substantially more unwanted 'noise' between monitoring episodes for fauna than for vegetation, and hence far less statistical power and interpretational acuity. As an illustration of this, we compared a simple measure of inter-sample similarity in species composition [(the number of species reported in a plot at both time 1 and time 2)/ (the cumulative tally of different species recorded in that plot across times 1 and 2)], across monitoring plots for the set of tree species, understorey plant species and vertebrate species (Figure 10.4). Unsurprisingly, the occurrence of tree species was relatively invariant; there was substantially more variation in the inventory of understorey plant species at the same plot across samples 5 years apart; but there was remarkably little similarity in the species complement of vertebrates at the same plot in 5 year samples. Although part of this variability in fauna assemblages may be attributable to a meaningful trend, another, probably more substantial, part may be largely measurement error or other muddying chaos.

In assessing the impact of this variability on the ability to detect trend, Woinarski *et al.* (2004) determined that for adequate power (in this case, the ability to be 90% certain of detecting a 20% change in our index of relative abundance, and with a 10% chance of accepting a Type I error), the Litchfield fauna monitoring program should be expanded from its current 47 plots to more than 1000. This is obviously an impractical ask. But the integration of monitoring across the three parks (here, to a tally of 220 plots) allows a substantial advance towards this ideal.

Fire occurrence and severity

The occurrence of fire at each plot is assessed at least bi-annually by on-ground visits by ranger staff, and this record is used – as well as the undertaking of aerial surveys – to assess the reliability of the fire history developed from the satellite imagery. For Kakadu and Nitmiluk, the photo record assembled for these visits for the period 1995–2004 has also been used to develop a practical field-based index of fire severity (low severity = leaf scorch height <2 m, and/or fires patchy with <20% of ground cover remaining unburnt; moderate severity = leaf scorch height >2 m, but upper canopy unscorched; high severity = upper canopy scorched), and as a basis for assessing seasonal fire severity characteristics in different vegetation types (Russell-Smith and Edwards 2006).

THE FIRST 10-YEAR ASSESSMENT, 1995–2004

After the first 10 years of the program, it is salutary to consider aspects of the data assembled to date and, perhaps even more importantly, to assess whether the program has been meeting its primary objectives, especially the information needs of management staff.

Fire history

Fire histories derived from LANDSAT imagery for each of the three parks are broken down by season in Table 10.3 and Figure 10.5, where the early dry season (EDS) period covers fires generally before August, and the late dry season (LDS) period describes fires after that date. Although these seasonal distinctions are somewhat arbitrary, various regional studies illustrate that fires in the EDS period *tend* to be of lower intensity (Williams *et al.* 1998, 2003b; Russell-Smith *et al.* 2003) and more patchy (Price *et al.* 2003; Williams *et al.* 2003b). Using data assembled for Kakadu and Nitmiluk National Parks as part of the monitoring plot program described in this chapter, it was observed that for 719 fires recorded from 178 plots over the period 1995–2004,

Table 10.3: Proportion (%) of Kakadu, Litchfield and Nitmiluk National Parks burnt by season 1995–2004, derived from LANDSAT satellite imagery. EDS = early dry season (pre-August); LDS = late dry season (August onwards). Note floodplain vegetation type data given only under Landscape unit heading.

Landscape unit	Proportion burnt (%)								
	Kakadu			Litchfield			Nitmiluk		
	EDS	LDS	Annual	EDS	LDS	Annual	EDS	LDS	Annual
Total area	28.6	13.1	41.7	49.4	16.1	65.5	29.3	21.6	50.9
Landscape unit									
Lowlands	35.8	14.0	49.8	44	14.6	58.7	29.9	23.7	53.6
Sandstone uplands	12.9	13.2	26.1	54.2	15.9	70.2	29	20.8	49.8
Floodplains	16.6	7.4	24.1	–	–	–	–	–	–
Vegetation type									
Lowland woodland/open forest	35.8	14.2	50.0	44.2	14.7	58.9	30	23.9	53.9
Lowland rainforest	21.6	8.2	29.8	33.6	11.6	45.2	16.1	13	29.1
Sandstone woodland/open forest	15.3	14.0	29.3	52.6	17.01	69.6	29.5	21.5	51
Sandstone heath	13.3	12.5	25.7	–	–	–	25	14.6	39.6
Sandstone rainforest	5.2	7.3	12.5	46.4	14.4	60.8	26.5	12	38.4
Lancewood (*Acacia shirleyi*)	–	–	–	–	–	–	32.3	24.8	57.1

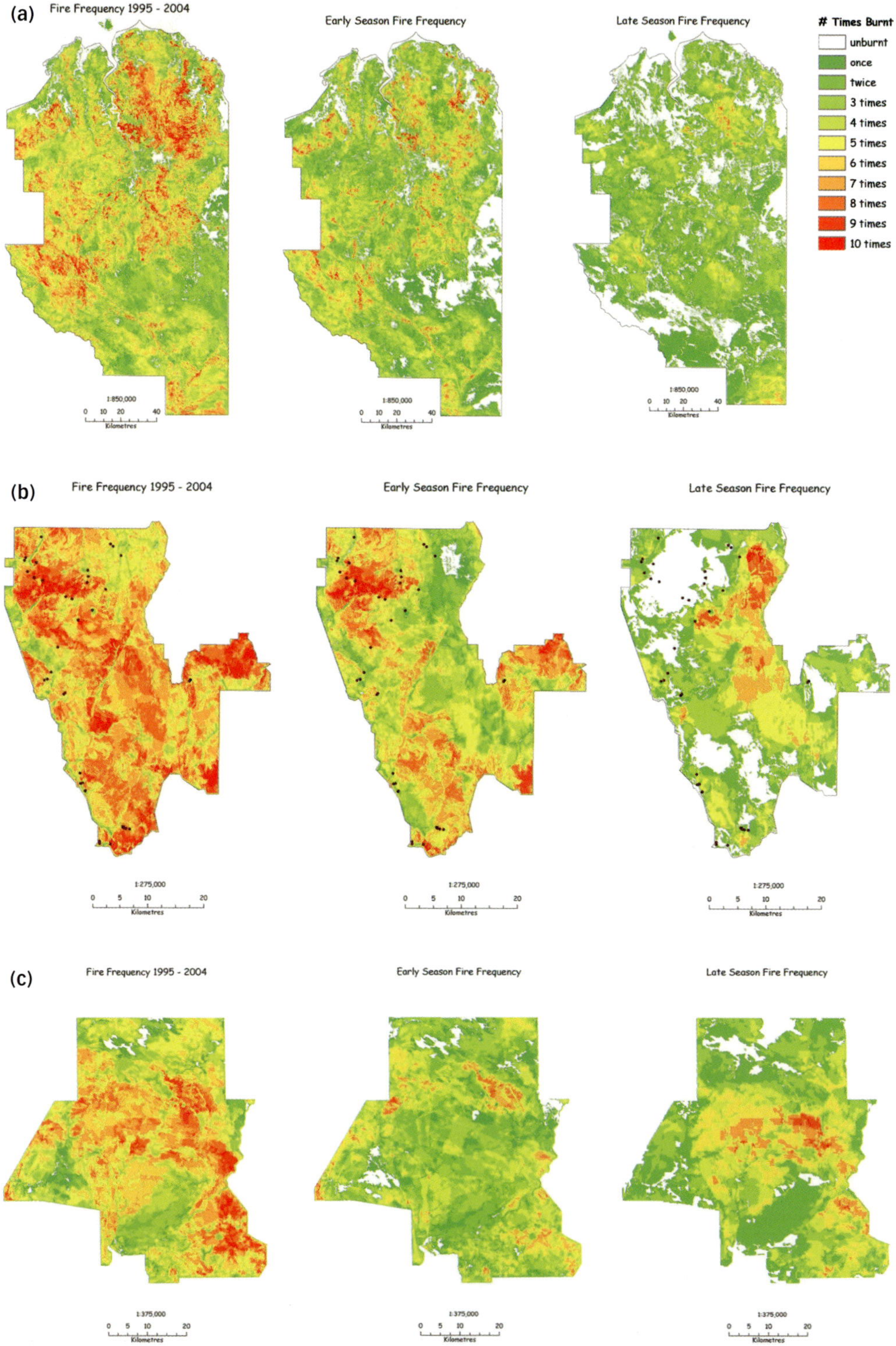

Figure 10.5 Fire frequency 1995–2004. (a) Kakadu, (b) Litchfield, (c) Nitmiluk National Parks.

80% of EDS fires were of very low severity (fire-line intensities $\leqslant$1000 kW m^{-1}, with leaf scorch heights typically <2 m), whereas fires later in the dry season were typically of substantially greater severity. Similar trends were evident for vegetation occupying all landform types (Russell-Smith and Edwards 2006). Also, as described by Garde *et al.* (Chapter 5), the EDS period effectively covers the main landscape burning period of as undertaken traditionally in western Arnhem Land generally.

By park, an annual average of 65.5% of Litchfield, 50.9% of Nitmiluk and 41.7% of Kakadu was burnt between 1995 and 2004 (Table 10.3). In all parks, fires burnt most frequently in the EDS period, by a factor of 3.1 in Litchfield, 2.2 in Kakadu and 1.4 in Nitmiluk. Approximately 50% of lowland landform units were burnt in all parks: particularly fire-prone open forest and woodland savannas. By contrast, the annual extent of burning of sandstone uplands varied markedly between parks: for example, on average, 70% of sandstone savanna was burnt in Litchfield, 51% in Nitmiluk and 29% in Kakadu (Table 10.3). Sandstone heath communities in Kakadu and Nitmiluk were burnt at slightly lesser frequencies than for open canopied sandstone vegetation communities generally; note that sandstone heath is very restricted in Litchfield (Table 10.1).

The annual frequency of burning observed in rainforest and lancewood thicket (the latter in Nitmiluk only) communities likewise varied markedly between parks: being least in sandstone rainforest in Kakadu (12.5%) and substantially greater (ranging between 30 and 61%) in all other situations (Table 10.3). It is noted, however, that, with the singular exception of relatively extensive tracts of endemic sandstone rainforest in Kakadu (Table 10.1), rainforest patches (and lancewood thickets) in these other situations are typically very small (often $\leqslant$5 ha) and often linear (e.g. along creek-lines or ridge-lines). Although they are thus particularly fire-prone along their external margins, such large perimeter: area ratios probably exaggerates estimation of fire incidence, given pixel sizes of around 30 m and associated mapping registration issues.

As a means for assessing progress in the delivery of fire-management programs in these parks, we are able to compare the above fire history data for 1995–2004 with published fire history data available for earlier periods for all three parks. Data for Kakadu are available for the period 1980–1994 (Russell-Smith *et al.* 1997). Overall, the mean annual extent of burning in Kakadu has declined 4.1%, particularly in the LDS period (20.8% previously to 13.1% here) and across all three major landform units, especially lowlands. There have been concomitant slight increases in the mean extent of EDS burning. Fire history data for Litchfield and Nitmiluk are available from 1990 to 1997, and 1989 to 1997, respectively (Edwards *et al.* 2001). Mean fire extent in the 1995–2004 assessment period is 9.6% greater in Litchfield, and 10.2% greater in Nitmiluk. These increases have occurred particularly in the EDS in both parks, with a concomitant slight decline in mean LDS fire extent in Nitmiluk, and a slight increase in Litchfield. Other notable shifts in fire regime on these parks include a large increase in the extent of burning (both EDS and LDS) in lowland situations in Nitmiluk (31.6% previously to 53.9% now), and marked increase in (EDS) burning both in lowland and sandstone upland situations in Litchfield.

Vegetation responses

As described in Table 10.1, the 220 permanent monitoring plots sample a variety of key vegetation/habitat types in each of the three parks. Given: (a) the detail with which vegetation attributes have been audited on each of three occasions over the 10-year history of the program; (b) the accompanying detailed recording of fire occurrence, seasonality and severity; and (c) the very considerable efforts that have been given to curating these data (e.g. maintaining taxonomic consistency); the assembled data provide a powerful means for assessing the responses of respective vegetation types and constituent species to different fire-regime parameters (e.g. fire frequency, severity, time-since-last-fire and fire interval).

Although it is not our purpose here to present a comprehensive assessment of vegetation-fire responses in data assembled to date, we provide below assessments of broad-scale fire effects on two vegetation types (woodland/open-forest and rainforest) and one iconic fire-sensitive species, *Callitris intratropica*, in two landscape settings (lowlands and sandstone uplands). These assessments combine data from all three parks, and illustrate the practical value of this program to inform park managers of the effects of their fire-management program(s).

Savanna woodland

A total of 137 monitoring plots are established in woodland/open forest vegetation, dominated principally by eucalypts, with 84 plots situated in lowland and 53 plots in sandstone upland situations. Although approximately half the plots occur in Kakadu, there are disproportionately more sandstone woodland plots both in Litchfield and Nitmiluk (Table 10.1). None of these plots samples an ecotonal situation. For the assessment presented here, data from 14 plots (mostly lowland woodland) have been excluded given incomplete fire severity data for 11 plots, apparent major vegetation changes not related to fire regime in two plots, and incomplete sampling for a further plot. A more detailed statistical assessment is being undertaken on this latter 123 plot dataset at the time of writing.

Major vegetation trends for the three parks over the 10-year period are summarised in Table 10.4. For lowland woodland plots, overall there has been mean net increase in ground cover and tree species richness, shrub and tree stem density, and stem basal area. Conversely, on sandstone woodland plots, there has mostly been decline in all these same parameters except for ground-cover species richness. These trends are generally consistent across all three parks, although declines in stem basal area of lowland woodland plots were observed only for Kakadu.

Overall, woodland plots have been burnt on average at an annual frequency of 0.55 (almost every second year), mostly (0.42) in the early dry season (pre-August). Burning of lowland woodland plots has occurred at a slightly higher frequency (0.56) than for sandstone woodland plots (0.52). However, sandstone woodland plots have burnt marginally more frequently at moderate and high fire severities (0.18), than lowland woodland plots (0.17). By park, sandstone plots have been burnt most frequently in Litchfield (0.67), then Nitmiluk (0.46) and Kakadu (0.39) (Figure 10.6). High-severity fires, however, occur most frequently in Nitmiluk, then Litchfield and Kakadu (Figure 10.6).

The overriding importance of fire severity vis-à-vis other fire frequency parameters on woodland vegetation response is illustrated in Table 10.5, where:

1. fire frequency *per se* is not significantly ($\alpha = 0.05$) correlated with any vegetation parameter
2. frequency of moderate and severe fires is significantly negatively correlated only with tall shrub density
3. frequency of severe fires is significantly negatively correlated with nine parameters, including tall shrub density, tree stem density and basal area.

The importance of fire severity on woody vegetation structural diversity suggested in these data supports observations made generally from north Australian regional savanna fire experiments conducted at Kapalga and Munmarlary (e.g. Williams *et al.* 1999, 2003a; Russell-Smith *et al.* 2003; Prior *et al.* 2006). However, it is interesting to note that fire frequency *per se* is not implicated as being a significant driver of woody structural change in these data,

Table 10.4: Change in lowland and sandstone woodland vegetation attributes over ten years at long-term monitoring plots in Kakadu, Litchfield, Nitmiluk NPs, 1995–2004.

Attribute	Woodland vegetation type							
	Lowland				Sandstone			
	Kakadu (n = 44)	Litchfield (n = 16)	Nitmiluk (n = 13)	Total (n = 73)	Kakadu (n = 11)	Litchfield (n = 18)	Nitmiluk (n = 21)	Total (n = 50)
Change in ground cover species richness								
Forbs	5.43 (16.02)	0.69 (17.50)	1.85 (13.85)	3.75 (15.96)	4.36 (15.27)	–1.12 (20.17)	4.24 (10.05)	2.32 (14.84)
Graminoids	2.23 (5.30)	0.38 (12.13)	1.77 (8.54)	1.74 (7.37)	3.18 (6.00)	–1.18 (9.50)	3.43 (8.71)	1.76 (8.40)
Change in shrub layer density (ha^{-1})								
Shrubs overall	709 (9633)	8111 (14 122)	654 (2738)	2321 (9389)	–832 (10 180)	–8447 (16 958)	–2783 (8058)	–4393 (11 729)
Shrubs <50 cm	444 (6699)	8273 (10 914)	673 (1327)	2201 (6666)	121 (7254)	–7861 (13 792)	–1756 (5321)	–3541 (8796)
Shrubs 50–200 cm	98 (2765)	–114 (2725)	–50 (1275)	25 (2491)	–830 (2513)	–650 (2879)	–1071 (2646)	–867 (2701)
Shrubs >200 cm – <5 cm DBH	166 (170)	–48 (483)	31 (137)	95 (233)	–123 (413)	64 (288)	44 (90)	14 (232)
Change in tree layer								
Species richness overall	0.49 (6.82)	–0.31 (7.88)	0.15 (4.77)	0.25 (6.68)	–0.18 (7.45)	–0.83 (9.39)	–0.95 (5.33)	–0.74 (7.26)
Fruiting trees	0.26 (2.36)	0.13 (2.31)	0.31 (0.69)	0.24 (2.05)	–0.09 (3.00)	–0.17 (3.72)	–0.38 (1.29)	–0.24 (2.54)
Deciduous trees	0.05 (2.75)	0.13 (3.13)	0.15 (2.92)	0.08 (2.86)	0.18 (2.64)	0.17 (2.72)	–0.33 (1.48)	–0.04 (2.18)
Evergreen trees spp. richness	0.44 (4.07)	–0.44 (4.75)	0.00 (1.85)	0.17 (3.82)	–0.36 (4.82)	–1.00 (6.67)	–0.62 (3.86)	–0.70 (5.08)
Stem density overall (ha^{-1})	42.1 (342.2)	–19.5 (461.7)	21.2 (248.1)	24.6 (248.1)	–27.7 (311.7)	–80.6 (438.9)	–55.4 (265.5)	–58.3 (338.1)
Stems 5–<10 cm DBH	48.5 (157.3)	0.0 (201.6)	21.2 (122.1)	32.8 (160.7)	–29.2 (111.7)	–48.6 (228.5)	–28.6 (113.1)	–35.9 (154.3)
Stems 10–30 cm DBH	–7.2 (156.7)	–21.1 (241.4)	–1.0 (116.3)	–9.1 (168.1)	1.5 (175.0)	–37.5 (191.0)	–28.0 (129.8)	–24.9 (161.8)
Stems >30 cm DBH	0.7 (28.2)	1.6 (18.8)	1.0 (9.6)	0.9 (22.8)	0.0 (25.0)	5.6 (19.4)	1.2 (22.6)	2.5 (22.0)
Basal area (m^2 ha^{-1})	–0.18 (8.27)	0.18 (8.79)	0.45 (3.95)	0.01 (7.61)	–0.99 (8.93)	–0.42 (8.24)	–0.73 (6.69)	–0.68 (7.74)

n = no. of monitoring plots. Numbers in parentheses are mean values of attributes at establishment (t1).

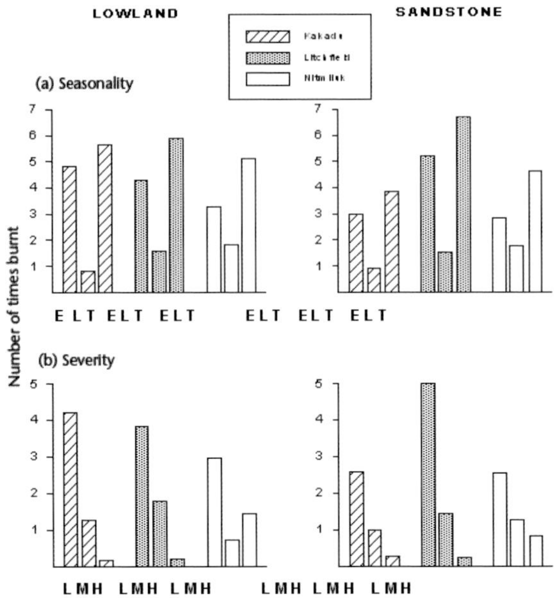

Figure 10.6 Mean fire frequency for lowland and sandstone woodland plots in Kakadu, Litchfield and Nitmiluk National Parks for (a) seasonality of burning (E = Early dry season; L = Late dry season; T = total) and (b) fire severity (L = Low; M = Moderate; H = High), over the period 1995–2004.

perhaps given that most of these fires happened to occur early in the dry season period and were typically of low intensity (Figure 10.6), and thereby more patchy (Price *et al.* 2003; Williams *et al.* 2003b).

In contrast, using various data from different fire treatments (of 4–7 years duration) at Kapalga, Prior *et al.* (2006) found that frequent fires early in the dry season resulted in reduced height growth of juvenile trees, whereas late-season fires resulted in increased juvenile height increment. Those authors suggested that differential fire seasonality effects probably occur because juveniles are physiologically active early in the dry season, but are effectively dormant in the late dry season. Such effects may have been exacerbated by the treatments themselves because elsewhere it has been noted that both early and late dry season fire intensities at Kapalga were significantly greater than those observed both at Munmarlary and in regional savannas more generally (Russell-Smith and Edwards 2006). At Munmarlary, little change was evident in population structures of dominant eucalypts after 22 years of imposed low–moderate severity annual early and late dry season treatments: a response interpreted to reflect that, as suggested in other studies, recruitment of eucalypts into the canopy appears to require significantly reduced root competition through death of dominant eucalypts (Russell-Smith *et al.* 2003).

Furthermore, using other observations from the same woodland sites as described here, recent complementary statistical analyses of the effects of fire regime (e.g. fire frequency, seasonality and severity) parameters on various vegetation structure, growth increment, stem recruitment and mortality responses, illustrate a diversity of often subtle relationships. For example, although fire seasonality and severity are clearly strongly correlated, fire severity was shown to have stronger influences on herbaceous species richness (Russell-Smith *et al.* unpublished data), stem growth increment (Chapter 14) and mortality (Prior *et al.* 2009), whereas tree-stem recruitment was shown to be affected particularly by frequency of early dry season fires (Prior *et al.* 2009). Regardless, fire frequency *per se* has significant implications for main-

Table 10.5: Spearman rank correlations between woodland vegetation response variables change with fire frequency variables, 1995–2004.

Response variable	*n*	Fire frequency		Frequency moderate and severe fires		Frequency severe fires	
		r_s	*P*	r_s	*P*	r_s	*P*
Forb species richness change (t3–t1):	123	0.02	NS	–0.05	NS	–0.12	NS
annuals	123	0.07	NS	–0.20	NS	**–0.28**	**0.002**
perennials	123	0.07	NS	–0.07	NS	–0.06	NS
Graminoid species richness (t3–t1):	123	–0.09	NS	–0.19	NS	–0.09	NS
annuals	123	–0.04	NS	–0.19	NS	**–0.20**	**0.03**
perennials	123	–0.07	NS	–0.01	NS	–0.09	NS
Tree species richness (t3–t1):	122	–0.14	NS	–0.10	NS	**–0.29**	**0.001**
fruiting tree species richness	122	–0.04	NS	–0.05	NS	–0.15	NS
deciduous tree species richness	122	–0.08	NS	<–0.01	NS	**–0.18**	**0.05**
evergreen species richness	122	–0.10	NS	–0.10	NS	**–0.25**	**0.005**
Shrub density change (t3/t1):	123	0.13	NS	0.03	NS	0.10	NS
small shrubs (<50 cm)	118	0.10	NS	–0.01	NS	0.09	NS
medium shrubs (50–200 cm)	122	0.12	NS	–0.04	NS	–0.03	NS
tall shrubs (>200 cm – <5 cm DBH)	104	–0.13	NS	**–0.41**	**<0.0001**	**–0.24**	**0.02**
Tree stem density change (t3/t1):	122	<0.01	NS	–0.09	NS	**–0.21**	**0.02**
small stems (5–10 cm DBH)	120	<–0.01	NS	–0.08	NS	**–0.20**	**0.03**
medium stems (>10 – <30 cm DBH)	122	–0.04	NS	–0.02	NS	–0.05	NS
large stems (≥30 cm DBH)	83	–0.03	NS	0.02	NS	–0.19	NS
Basal area (m^2 ha^{-1}) change (t3/t1)	122	<0.01	NS	–0.02	NS	**–0.19**	**0.03**

n = number of plot observations; NS = not significant (α = 0.05); P ≤ 0.01 given in bold. Differences in sample size (n) due to absence of records of some attributes in certain plots

tenance of populations of fire-sensitive obligate seeder species, especially in regional heath communities (Russell-Smith *et al.* 1998), and, as discussed also below, relatively fire sensitive components of savanna fauna (Woinarski *et al.* 2001, 2005a; Andersen *et al.* 2005.

Rainforest

A total of 42 rainforest plots was monitored across the three parks, with the great majority (37) occurring in Kakadu. With two exceptions, all plots sample boundary ecotones with other vegetation types. In this assessment, we present data only from the 40 ecotonal plots. Over the 10 years of records, based on assessment along fixed transects of changes in location of canopy cover conditions, rainforest boundaries in lowland situations were observed to expand slightly (mean 1.2 m), and those in sandstone uplands to contract very slightly (mean –0.2 m), as reflected by changes in measurements of dense canopy cover in quadrats along fixed transects. Similarly, stem basal area was observed to increase by 8% overall, and by 34% for rainforest species in lowland plots, whereas stem basal area declined slightly (–3.5%) for all species, including rainforest species, in sandstone plots (Table 10.6). Boundary expansion and marked increase of rainforest stem basal area at lowland plots occurred despite high frequencies of mostly low severity fires, whereas decline in basal area at sandstone plots was significantly associated with more severe fires (Table 10.7).

Table 10.6: Rainforest plot vegetation change and fire regime characteristics (± S.E.M), 1995–2004.

Variable	Lowland rainforest plots (*n* = 18)	Sandstone rainforest plots (*n* = 22)
Mean change in boundary, *t*3–*t*1 (m)	1.24 ± 0.56 (*n* = 15)	−0.17 ± 0.44 (*n* = 18)
Mean change in Canopy Cover Index, *t*3/*t*1 (%)	138.4 ± 12.6 (*n* = 16)	235.1 ± 85.2 (*n* = 22)
Mean basal area, *t*1 ($m^2.ha^{-1}$)		
– rainforest species	9.7 ± 4.0 (*n* = 17)	10.5 ± 1.5 (*n* = 21)
– non-rainforest species	7.4 ± 2.2 (*n* = 17)	3.0 ± 0.7 (*n* = 21)
– all species	17.1 ± 3.9 (*n* = 17)	13.5 ± 1.6 (*n* = 21)
Mean change in basal area, *t*3/*t*1 (%)		
– rainforest species	134.4 ± 6.5 (*n* = 17)	97.2 ± 5.6 (*n* = 21)
– non-rainforest species	98.4 ± 5.6 (*n* = 17)	10.3.0 ± 7.8 (*n* = 21)
– all species	108.3 ± 5.2 (*n* = 17)	96.5 ± 5.7 (*n* = 22)
Mean fire frequency	0.58 ± 0.08 (*n* = 18)	0.2 ± 0.04 (*n* = 22)
Mean frequency moderate and severe fires	0.12 ± 0.03 (*n* = 17)	0.13 ± 0.02 (*n* = 19)

The above observations follow generally the trends observed at woodland plots over the 10-year monitoring period. In common, (a) woody vegetation structural components on our monitoring plots have tended to increase in lowland situations and decline in sandstone uplands (Tables 10.4 and 10.6), and (b) the main apparent driver of these changes is the frequency of severe fires (Tables 10.5 and 10.7). In both datasets, fire frequency per se apparently has had relatively little affect on woody change. The importance of severe fires on regional rainforest vegetation, especially on patch margins, is widely reported (e.g. Russell-Smith and Bowman 1992; Bowman 1994; Prior *et al.* 2007).

Recent aerial photo assessments of multi-decadal-scale change in regional rainforest assemblages report net rainforest patch expansion in lowland and hillslope situations in Litchfield (Bowman *et al.* 2001), and lowlands and sandstone uplands in Kakadu (Banfai and Bowman 2006; Bowman and Dingle 2006). However, it is also possible that results reported in the latter sandstone uplands study reflect substantial canopy infilling (associated with decadal trends in increasing rainfall) and issues of scale and sampling methodology, rather than expansion per sc (J. Freeman *pers. comm.*, April 2007). Regardless, Bowman and Dingle (2006) found that 'forest expansion was negatively correlated with fire activity', and that exposed forest margins were those most at risk.

Table 10.7: Spearman rank correlations between vegetation change with fire regime variables, 1995–2004.

	Fire regime variables											
	Fire frequency						Frequency moderate and severe fires					
	Lowland			Sandstone			Lowland			Sandstone		
Response variable	*n*	r_s	*P*	*n*	r_s	*P*	*n*	r_s	*P*	*n*	r_s	*P*
Change in boundary (t3–t1)	15	0.39	NS	17	−0.29	NS	15	0.16	NS	14	−0.39	NS
Change in basal area (t3/t1)	17	0.16	NS	22	−0.23	NS	17	0.13	NS	19	**−0.50**	**0.02**

n = no. of plot observations; NS = not significant (α = 0.05).

Callitris intratropica

A total of 30 monitoring plots across the three parks sample *Callitris* populations, with 24 of these occurring in sandstone uplands, and 22 plots containing *Callitris* in Kakadu. Overall, while mean stem basal area across plots has increased both on lowland (0.84 $m^2 ha^{-1}$) and sandstone (0.21 $m^2 ha^{-1}$) plots over the 10-year period, both mean tree-stem density and density of juveniles has declined markedly – lowland plots: –13.3 tree stems ha^{-1}, –320 juveniles ha^{-1}; sandstone plots: –11 tree stems ha^{-1}, –647 juveniles ha^{-1}. Evidently, contemporary fire regimes are having differential impacts on smaller size classes. The population and fire frequency data for 17 woodland (i.e. non-ecotonal) plots presented in Figure 10.7 illustrate the above observations;

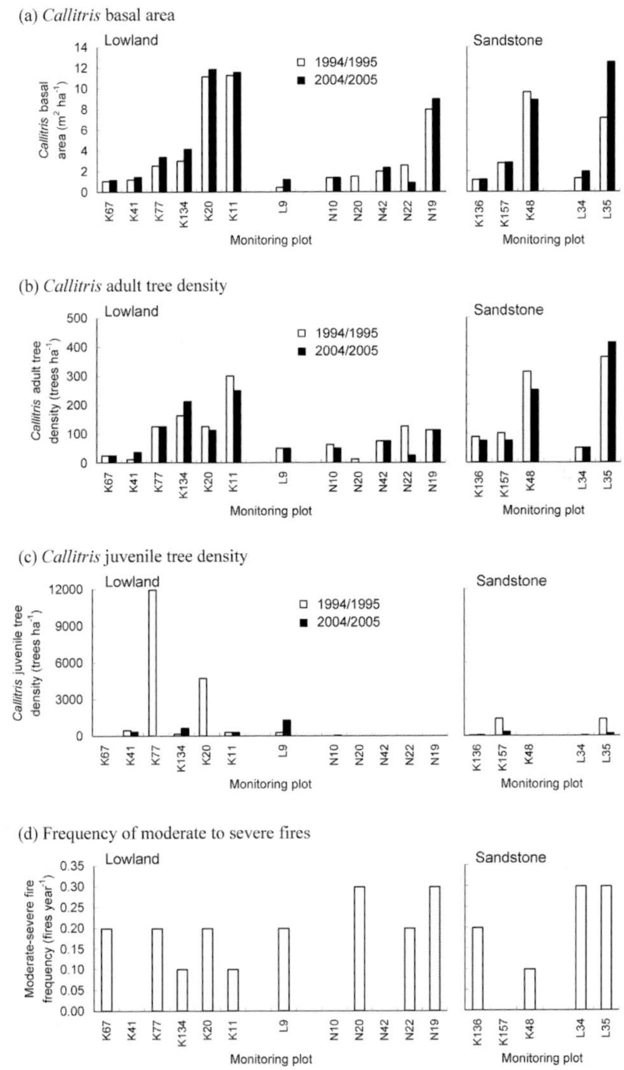

Figure 10.7 Change in status of *Callitris intratropica* at respective lowland and sandstone upland woodland monitoring plots in Kakadu, Litchfield and Nitmiluk National Parks, 1995–2004 for (a) tree basal area, (b) density of adult stems (≥5 cm D.B.H.), (c) juvenile <5 cm D.B.H.) stem density, and (d) frequency of moderate and high-severity fires.

particularly concerning is the collapse and/or conspicuous absence of juveniles at many of these plots (Figure 10.7c). Such observations mirror those made in many other regional studies concerning the impact of contemporary fire regimes on stands and population structures of this previously widespread, fire-sensitive obligate seeder species (e.g. Bowman and Panton 1993, Bowman *et al.* 2001; Prior *et al.* 2007; Russell-Smith 2006; Chapter 9).

FAUNA

As described above, it is premature to provide a detailed analysis of the results of fauna monitoring across the three parks. To date, there has been only one formal publication of the fauna monitoring program, describing the comparison of benchmark (1995–96) and the next subsequent monitoring (2001–02) of the Litchfield plots (Woinarski *et al.* 2004). That analysis demonstrated some substantial shifts in species composition between these two sampling periods, and significant changes in the abundance of 18 of the 92 fauna species reported sufficiently often to analyse. However, change in the faunal composition of individual plots between the baseline and subsequent monitoring periods showed no significant association with the fire history of those plots over this period. Woinarski *et al.* (2004) suggested a number of factors that may have led to this lack of connection:

1. that the fire history preceding the benchmark sampling was not considered in the analysis
2. that the period between baseline and subsequent monitoring (6 years) may have been insufficient to cause appreciable vegetation change and hence appreciable change in habitat quality
3. that fauna species may be influenced at least as much by the pattern of fire in areas surrounding the plot than by the fire patterning within the plot itself
4. that the imprecision in fauna sampling associated with variable detectability and other factors may have muted the response signal to fire
5. that the faunal assemblages are so dynamic that it is very difficult to recognise a significant fire effect.

Woinarski *et al.* (2005b) reported on the baseline fauna sampling of Nitmiluk, and used these inventory data to relate faunal abundance with the preceding fire history (number of fires in the 10 years prior to sampling) of the plots. Across the 46 monitoring plots sampled, the richness of small native mammal species was strongly and inversely related to the number of fires experienced (estimate = –0.71, Wald statistic = 8.7, $p<0.01$, deviance explained = 35%): that is, the more frequently a plot had been burnt, the fewer the number of small native mammal species present. The Nitmiluk baseline fauna sampling was notable in reporting extremely low numbers of small native mammals (mean of 0.17 species per plot). Given the detected relationship between the richness of small native mammals and the preceding incidence of fire, the high fire frequency in the Nitmiluk plots (mean of 4.1 fires in the 10 years preceding sampling) may have contributed to this impoverishment. For the Nitmiluk baseline sampling, there was no significant relationship for richness of native frogs, reptiles, birds, larger native mammals or feral mammals with fire frequency.

A small set (15 plots) of the fire-monitoring plots in Kakadu have now been sampled for fauna three times. The data from these plots serve to corroborate, and indeed provide objective evidence of, a reported regional-wide decline in the small native mammal fauna (Figure 10.8). The extent to which fire has contributed (and/or continues to contribute) to this decline is not yet clearly defined, but inter-plot variation in fire regimes should provide a powerful tool for partitioning this impact and response from other putative threatening factors.

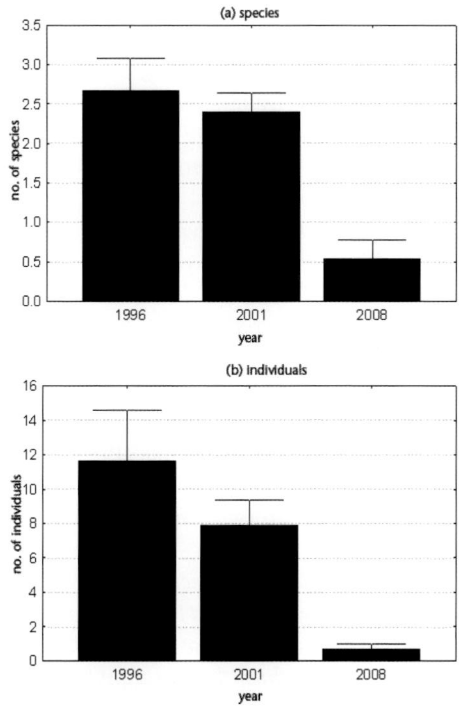

Figure 10.8 Trends over a 12-year monitoring period for small native mammal fauna at 15 fire monitoring plots in Kakadu National Park: (a) number of mammal species per plot, (b) number of individuals of mammal species per plot.

Staff engagement

As outlined in the introduction, a key design feature of the program was to address the fire-information needs of management staff. Although the extent to which this has been achieved is admittedly difficult to quantify, over the 10 years of the program a series of reports has been prepared by park staff as part of 5-year assessments undertaken on the participating parks. In Litchfield and Nitmiluk, such assessments have been undertaken as part of formal 'training camp' programs undertaken over 10-day periods, involving ranger staff from around the Northern Territory. The report of the 10-year assessment for Litchfield, for example, notes that as well as Northern Territory Parks and Wildlife rangers, the camp formally involved Aboriginal ranger groups from Anindilyakwa, Dhimurru, Yirralka Layunhapuy, Wagaman and Larrakia (PWCNT 2005a). For Kakadu, the large number of plots has meant that, to date, plot assessments have needed to be undertaken over two wet season periods and, given the dispersed ranger infrastructure in that park, separately within the five ranger districts rather than as a single integrated program.

For all park assessments, a requirement has been that participating staff undertake a formal quantitative analysis of the program for the 5-year period in question, involving assessment both of remotely sensed fire history and plot-based data. In the case of Litchfield and Nitmiluk, such assessments have involved reporting findings to senior staff at a formal presentation at the conclusion of respective training camps. For Kakadu, findings have been reported by ranger representatives to the park's Board of Management. Each assessment report is available digitally.

Collectively, the reports affirm the utility of those assessment programs:

> *Overall, the [Kakadu] workshop participants agreed that the project to date has been an outstanding success. The project has opened an avenue for field staff with experience in fire and land management to analyse actual data from the country they manage, instead of just having feelings about how the country is responding to prescribed fire management. The original goal of the project as a training exercise is still very much a reality, with all participants gaining valuable training from the report analysis sessions. (Turner* et al. *2002).*
>
> *We were exposed to a wide range of activities, such as collecting, entering and analysing floristic data – this was the first experience for many of us in this type of monitoring work. Learning how to use GIS to digitise fire scars on maps from satellite imagery was also of great benefit. Overall, rangers will be able to apply their newly acquired skills in their own parks and contribute to improved park management practices ... Through meeting with other rangers, a greater appreciation of how other parks operate was achieved. The relationships developed between the participants during the entire camp will be lasting. (PWCNT 2005b)*

Moreover, staff have taken the opportunity to comment on related training issues:

> *This type of data collection and analysis project is long overdue in Kakadu National Park, and, with recent impacts such as weeds like Gamba Grass and Mission Grass that are likely to impact on fire regimes, these baseline data will become more and more important. The use of Geographical Information Systems (GIS) and database programs are well utilised in other situations outside the Park, and the value of these management tools cannot be understated. A recommendation of this workshop is that specific training be given to staff in aspects of GIS and database use, design, and maintenance... (Turner* et al. *2002)*

Although participant feedback from these assessment and training exercises reinforces the practical value of the monitoring program to respective park management authorities, it remains an issue of significant concern to the authors that – along with cultural and resource management monitoring programs in general and despite the significance of fire management in particular to regional ecological processes (see below) – fire-monitoring programs on these parks continue to require repeated justification in seemingly ever-constrained conservation management budgets. To date, the program has been driven largely by the ongoing commitment of concerned individuals as opposed to structured institutionalised support. Edwards *et al.* (2003: Table 2) provide an estimate that fire monitoring in Kakadu constitutes about 1% of the annual park budget when annualised over 5-year periods. We argue that, in the absence of rigorous monitoring data and defined quantitative objectives, it is not possible to assess the effectiveness or otherwise of fire-management programs and to respond to management challenges in an adaptive process.

VALUE OF THE MONITORING PROGRAM – STRENGTHS AND WEAKNESSES

In the introduction, we indicated a number of requirements that the monitoring program was designed to address. In preceding sections we have established that the program effectively provides for ongoing development of: (a) satellite imagery-derived fire history databases of the parks and (b) photographic and numerical databases describing vegetation and vertebrate fauna responses to ambient fire regimes, as quantitative means for (c) assessing changes in status and condition of habitats and at least common species, and (d) potentially, assessing management success with respect to stated objectives. As well, we have seen that the program generally addresses the daily (e.g. NAFI website) and ecological (e.g. monitoring plot database) fire information needs of park management staff, as well as incorporating formal, albeit occasional, training associated with plot re-assessment exercises. Below we discuss a number of components and issues requiring further consideration.

Monitoring framework

The developing fire-mapping and ecological-plot databases already provide powerful tools for monitoring key aspects of management-imposed fire regimes and their ecological effects on the three parks. This applies especially to assessment of the responses of major regional vegetation/habitat types and widespread common plant species. In this chapter, we have illustrated the utility of the program for assessing the fire responses of regional savannas and rainforest communities, and one fire-sensitive long-lived obligate seeder, *Callitris intratropica*. More targeted statistical analyses are currently being undertaken for the assembled woodland dataset and, over the next year or two, a detailed assessment of floristic change on heath plots is due to be undertaken. While comprising only a small proportion of the vegetation of the three parks (Table 10.1), regional heaths include many obligate seeder shrub species and have been shown previously to be particularly susceptible to frequent fires (Russell-Smith *et al.* 2002). Heaths are disproportionately represented with 31 plots: 11 of these describing ecotonal situations principally with sandstone rainforest. These assessments of site-scale processes usefully complement studies addressing multi-decadal landscape-scale change in vegetation cover, where the fire regime is but one of the major drivers along with climatic and grazing/browsing effects (Bowman *et al.* 2001; Banfai and Bowman 2006; Bowman and Dingle 2006; Lehmann *et al.* 2009; Petty *et al.* 2007b).

Despite these benefits, the monitoring program does not cater particularly well for fauna, nor address rare or threatened taxa. Probably more so than for vegetation, the fauna of the tropical savannas exhibits complex responses to fire, and is known also to be substantially affected by a range of other threatening factors. Much of the mammalian fauna, and some species in other fauna groups, is suffering regional-level decline, through various and unresolved combinations of inappropriate fire regimes, feral predators, vegetation change wrought by pastoralism or feral stock, poisoning by cane toads, and/or disease (Franklin 1999; Woinarski 2000, 2004a; Woinarski *et al.* 2001, 2007; Franklin *et al.* 2005).

Nonetheless, inappropriate fire regime is the factor that has been identified as a threat to by far the largest number of listed threatened species in the Kakadu area, so there is obvious merit in using fire as the foundation for an integrated monitoring program that considers vegetation and vertebrate fauna (Chapter 8). With appropriate design and analytical tools, the impacts of other threats may be possible to disentangle or compartmentalise: allowing such a monitoring program to provide collateral benefit in the form of insight into the effects of factors other than fire.

A primary objective of management in conservation reserves is to retain biodiversity, with special attention directed at the conservation of the most threatened species. In the three parks considered here, threatened species are particularly susceptible to inappropriate fire regimes (Table 10.8), and hence trends in the status of threatened species should be a primary focus in evaluating the efficacy of fire management. However, most threatened species in these parks are – almost axiomatically – highly localised and/or uncommon, and hence unlikely to be adequately sampled in monitoring plots that have been sited based on other criteria. Although the Three Parks fire monitoring program targeted some monitoring sites to include examples

Table 10.8: Tallies of threats recognised to affect listed threatened species occurring in Kakadu (and near Kakadu, for some plants). Threats are listed in decreasing order of frequency. Marine and aquatic fauna excluded. Note that threats are recognised for the entire range of the threatened species rather than Kakadu alone (e.g. 'clearing' is not generally a threat in Kakadu but may be an issue for elsewhere in the species' range). Data derived from Woinarski *et al.* (2007).

	Plant species		Animal species	Total
Threat	Kakadu (n = 9)	Nearby (n = 9)	(n = 20)	
Fire	5	5	15	25
Weeds	2	1	8	11
Feral herbivores	3	2	4	9
Feral predators	0	0	8	8
Clearing	3	0	5	8
Pastoralism	0	0	7	7
Changes in hydrology	4	1	1	6
Climate change	1	0	3	4
Cane toads	0	0	3	3
Harvesting	0	0	3	3

of high profile fire-sensitive plants (notably *Callitris intratropica*), occurrence of listed threatened plants or animals was generally not considered in the siting of monitoring plots.

As a consequence, the Three Parks monitoring program provides very little information on threatened species. For example, Edwards *et al.* (2003) reported that 'of the 11 threatened plant species recorded from Kakadu, only one (*Hildegardia australiensis*) was recorded in the [133] monitoring plots, and this on only one plot'. It is a perennial issue with monitoring programs that although they may prove useful for the purpose for which they were designed (in this case, the assessment of landscape-wide responses of vegetation to fire regimes), they are not necessarily the right instrument for other purposes (and indeed it may be unreasonable to expect them to be so). However, for common plant species, a recent assessment of the power of the monitoring program to detect 10% change in population size after three visits (10 years) indicated that, as a rule of thumb, at least 10 plots is required to provide a useful longer-term monitoring framework for persistent long-lived taxa, about 20 plots for shorter-lived perennials, and about 50 plots for ephemeral species (Price *et al.* unpublished data). Recognising the insufficiency of the Three Parks monitoring program to answer questions about the fate of threatened species, some specific complementary monitoring projects for selected threatened species have been attempted at Kakadu (e.g. Fraser *et al.* 2003; Kerrigan 2004; Woinarski 2004b), although these are not yet firmly entrenched in the park's management program.

Adaptive management, ecological thresholds and management responses

In contemporary management settings, we expect that fire management should be planned, implemented and monitored within adaptive-management frameworks (Walters 1986; Andersen 1999; Whitehead 1999). Increasingly, we recognise also the utility of setting management targets or, better still, thresholds to provide flexible, but objective, criteria against which to measure management success, and/or prompt management responses, including remedial action where required. For example, 'thresholds of potential concern' have been formally established to inform fire management in Kruger National Park (van Wilgen *et al.* 1998, 2003). Those authors proposed defining thresholds relating to various fire-regime characteristics (e.g. fire-return interval, fire size-class distribution and extent of unplanned fire) as 'surrogate measures' for status of biodiversity generally. Consistent exceeding of the unplanned fire threshold resulted in a substantial change in fire-management policy and practice in 2002 (van Wilgen *et al.* 2004).

Despite evident problems with contemporary fire regimes on each of the three parks, particularly as these affect fire-sensitive flora and fauna (e.g. results presented here for *Callitris* and small mammals), the management response has mostly been muted. Although to some genuine extent that lack of response has been dictated by limited available resourcing, it serves to underline the inherent mismatch between expectations that park staff emphasise management of infrastructure and tourists, while also protecting the very values that the reserves were established to maintain in perpetuity. Nonetheless, it is salutary to reflect on recent fire-management experience in Kakadu where – in 2001, 2004 and 2006 – very large fires burnt through biodiversity-rich areas of the sandstone uplands. In response, Kakadu has developed a fire-management plan that incorporates thresholds criteria specifically for the Arnhem plateau (Petty *et al.* 2007a). Assessments included within that planning document, and independently (Brennan 2008), establish that 'ecological management thresholds' relating to average fire frequency in fire-sensitive habitats have been very substantially exceeded both over the period 1990–2006, and especially so in the period 2001–2006. For example, 42% of fire-sensitive sandstone heath communities were burnt at least twice within 6 years (Brennan 2008) whereas the minimum fire-return period threshold is given as 5 years in the plan. Contemporary fire regime impacts on heaths of the Arnhem plateau, including those in Kakadu and Nitmiluk, are considered so dire as to warrant them being nominated as an 'endangered community' under Australian environmental legislation (WWF 2005).

Importantly, Kakadu has responded constructively by resourcing a potentially effective plateau fire-management program that complements the endeavours of Western Arnhem Land Fire Abatement (WALFA) partners over the remainder of the plateau. This example is perhaps the first instance since the establishment of the Three Parks fire monitoring program where the program itself has played a significant role in informing and eliciting an adaptive management response. In Chapter 8, John Woinarski and others outline the need for further changes in fire-management practice based on substantial, albeit incomplete, evidence from vertebrate-monitoring data. Rather than see fire/biodiversity monitoring programs as an expensive hindrance, we set a challenge to those responsible for park management budgets – and the national ecological bureaucracy specifically – to better understand the essential value of long-term monitoring programs in assisting staff, traditional owners and respective park management programs to learn from, adapt and deliver the very outcomes espoused in worthy plans of management and lofty vision statements.

ACKNOWLEDGEMENTS

This chapter is dedicated to our ranger colleagues, especially those not listed as authors who were instrumental in establishing the program – Phil Hauser (Nitmiluk) and Greg Spiers (Kakadu), and Paul Ryan who developed the original database and remote sensing capability. Assessments have been undertaken by large numbers of staff from all three parks, and more broadly from respective participating organisations (Parks Australia, Parks and Wildlife Service of the Northern Territory). Many thanks go to contributing botanists including Kym Brennan, Ian Cowie, Judy Egan, Bob Harwood, Raelee Kerrigan, Diane Lucas, Carrie Michell and Jenny Risler. Fauna sampling of fire plots has been a major logistical challenge, involving many people. Among many others, we'd like to thank particularly Alaric Fisher, Kym Brennan,

Damian Milne, Martin Armstrong, Michelle Watson, Jenni Low Choy, Gay Crowley, Tony Griffiths, Mani Berghout, Mark Ziembicki, Riikka Hokkanen, Lindley McKay, Tida Nou, Rod Kennett, and Scott Morrison.

REFERENCES

Andersen AN (1999) Fire management in northern Australia: beyond command-and-control. *Australian Biologist* **12**, 63–70.

Andersen AN, Cook GD, Corbett LK, Douglas MM, Eager RW, Russell-Smith J, Setterfield SA, Williams RJ and Woinarski JCZ (2005) Fire frequency and biodiversity conservation in Australian tropical savannas: implications from the Kapalga fire experiment. *Austral Ecology* **30**, 155–167.

Andersen AN and MacKaige BJ (1997) Burning issues: communicating fire research in Australia. In *Ecology for Everyone: Communicating Ecology to Scientists, the Public and the Politicians*. (Eds RT Wills and RJ Hobbs) pp. 88–96. Surrey Beatty and Sons, Sydney.

Banfai DS and Bowman DMJS (2006) Forty years of lowland monsoon rainforest expansion in Kakadu National Park, northern Australia. *Biological Conservation* **131**, 553–565.

Bowman DMJS and Dingle JK (2006) Late 20th century landscape-wide expansion of *Allosyncarpia ternata* (Myrtaceae) forest in Kakadu National Park, northern Australia. *Australian Journal of Botany* **54**, 707–715.

Bowman DMJS and Panton WJ (1993) Decline of *Callitris intratropica* in the Northern Territory: implications for pre- and post-colonisation fire regimes. *Journal of Biogeography* **20**, 373–381.

Bowman DMJS, Price O, Whitehead PJ and Walsh A (2001) The 'wilderness effect' and the decline *of Callitris intratropica* on the Arnhem Land Plateau, northern Australia. *Australian Journal of Botany* **49**, 665–672.

Bowman DMJS, Walsh A and Milne DJ (2001) Forest expansion and grassland contraction within a *Eucalyptus* savanna matrix between 1941 and 1994 at Litchfield National Park in the Australian monsoon tropics. *Global Ecology and Biogeography* **10**, 535–548.

Bowman DMJS, Zhang Y, Walsh A and Williams RJ (2003) Experimental comparison of four remote sensing techniques to map tropical savanna fire-scars using LANDSAT-TM imagery. *International Journal of Wildland Fire* **12**, 341–348.

Brennan K (2008) 'An aerial assessment of the incidence of fire in rocky (and associated) environments in Kakadu National Park during the 2006 dry season'. Unpublished report to Kakadu National Park. Northern Territory Parks and Wildlife Service, Berrimah.

Edwards A, Hauser P, Anderson M, McCartney J, Armstrong M, Thackway R, Allan GE and Russell-Smith J (2001) A tale of two parks: contemporary fire regimes of Litchfield and Nitmiluk National Parks, monsoonal northern Australia. *International Journal of Wildland Fire* **10**, 79–89.

Edwards A, Kennett R, Price O, Russell-Smith J, Spiers G and Woinarski J (2003) Monitoring the impacts of fire regimes on vegetation in northern Australia an example from Kakadu National Park. *International Journal of Wildland Fire* **12**, 427–440.

Franklin DC (1999) Evidence of disarray amongst granivorous bird assemblages in the savannas of northern Australia, a region of sparse human settlement. *Biological Conservation* **90**, 53–68.

Franklin DC, Whitehead PJ, Pardon G, Matthews J, McMahon P and McIntyre D (2005) Geographic patterns and correlates of the decline of granivorous birds in northern Australia. *Wildlife Research* **32**, 399–408.

Fraser F, Lawson V, Morrison S, Christopherson P, McGregor S and Rawlinson M (2003) Fire management experiment for the declining Partridge Pigeon, Kakadu National Park. *Ecological Management & Restoration* **4**, 94–102.

Kakadu National Park (2007) *Kakadu National Park Plan of Management*. Director of National Parks, Australian Government, Canberra.

Kerrigan R (2004) 'Kakadu threatened flora report. Volume 2: Results of a threatened flora survey 2004'. Northern Territory Department of Infrastructure Planning and Environment, Darwin.

Lehmann C, Prior LD and Bowman DMJS (2009) Multi-decadal dynamics of tree cover in an Australian tropical savanna. *Austral Ecology*, in press.

Litchfield National Park (1998) *Litchfield National Park Fire Management Plan*. Parks and Wildlife Commission of the Northern Territory, Darwin.

Nitmiluk National Park (2002) 'Nitmiluk (Katherine Gorge) plan of management'. Parks and Wildlife Commission of the Northern Territory, Darwin.

O'Leary P, Sharpe C, McMahon K, Russell-Smith J and Edwards A (2001) 'Litchfield National Park fire monitoring plot survey and analysis'. Technical Report No. 67. Parks and Wildlife Commission of the Northern Territory, and Bushfires Council of the Northern Territory, Darwin.

Petty AM, Alderson J, Muller R, Scheibe O, Wilson K and Winderlich S (2007a) *Kakadu National Park, Arnhem Land Plateau Fire Management Plan*. Kakadu National Park, Jabiru, Northern Territory.

Petty AM, Werner PA, Lehmann CER, Riley JE, Banfai DS and Elliott LP (2007b) Savanna responses to feral buffalo in Kakadu National Park, Australia. *Ecological Monographs* **77**, 441–463.

Price O, Russell-Smith J and Edwards A (2003) Fine-scale patchiness of different fire intensities in sandstone heath vegetation in northern Australia. *International Journal of Wildland Fire* **12**, 227–236.

Price O, Edwards A, Connors G, Woinarski J, Ryan G, Turner A and Russell-Smith J (2005) Fire heterogeneity in Kakadu National Park: 1980–2000. *Wildlife Research* **32**, 425–433.

Prior LD, Bowman DMJS and Brooke BW (2007) Growth and survival of two north Australian relictual tree species, *Allosyncarpia ternata* (Myrtaceae) and *Callitris intratropica* (Cupressaceae). *Ecological Research* **22**, 228–236.

Prior LD, Brook BW, Williams RJ, Werner PA, Bradshaw CJA and Bowman DMJS (2006) Environmental and allometric drivers of tree growth rates in a north Australian savanna. *Forest Ecology and Management* **234**, 164–180.

Prior LD, Murphy BP and Russell-Smith J (2009) Environmental and demographic correlates of tree recruitment and mortality in north Australian savannas. *Forest Ecology and Management* **257**, 66–74

PWCNT (Parks and Wildlife Commission of the Northern Territory) (2000) 'Nitmiluk fire study ranger training camp, 20–31 March 2000'. Parks and Wildlife Commission of the Northern Territory and Bushfires Council of the Northern Territory, Katherine.

PWCNT (2005a) 'Litchfield National Park fire monitoring plot survey and analysis'. Parks and Wildlife Commission of the Northern Territory, and Bushfires Council of the Northern Territory, Darwin.

PWCNT (2005b) 'Nitmiluk National Park fire monitoring plot survey and analysis: Ranger training camp 4–15 April 2005'. Parks and Wildlife Commission of the Northern Territory WCNT and Bushfires Council of the Northern Territory, Katherine.

Russell-Smith J (2006) Recruitment dynamics of the long-lived obligate seeders *Callitris intratropica* (Cupressaceae) and *Petraeomyrtus punicea* (Myrtaceae). *Australian Journal of Botany* **54**, 479–485.

Russell-Smith J and Bowman DMJS (1992) Conservation of monsoon rainforest isolates in the Northern Territory, Australia. *Biological Conservation* **59**, 51–63.

Russell-Smith J and Edwards A (2006) Seasonality and fire severity in savanna landscapes of monsoonal northern Australia. *International Journal of Wildland Fire* **15**, 541–550.

Russell-Smith J and Ryan PG (1996) Establishing a photo-point monitoring system in Kakadu National Park: rationale and protocol. In *Proceedings of the Fire Ecology Workshop, Darwin, August 1994*. pp. 80–91. Parks and Wildlife Commission of the Northern Territory, Darwin.

Russell-Smith J, Ryan PG and DuRieu R (1997) A LANDSAT MSS-derived fire history of Kakadu National Park, monsoonal northern Australia, 1980–1994: seasonal extent, frequency and patchiness. *Journal of Applied Ecology* **34**, 748–766.

Russell-Smith J, Ryan PG, Klessa D, Waight G and Harwood R (1998) Fire regimes, fire-sensitive vegetation, and fire management of the sandstone Arnhem Plateau, monsoonal northern Australia. *Journal of Applied Ecology* **35**, 829–846.

Russell-Smith J, Ryan PG and Cheal D (2002) Fire regimes and the conservation of sandstone heath in monsoonal northern Australia: frequency, interval, patchiness. *Biological Conservation* **104**, 91–106.

Russell-Smith J, Whitehead PJ, Cook GD and Hoare JL (2003) Response of *Eucalyptus*-dominated savanna to frequent fires: lessons from Munmarlary, 1973–1996. *Ecological Monographs* **73**, 349–375.

Stocker GC and Mott JJ (1981) Fire in the tropical forests and woodlands of northern Australia. In *Fire and the Australian Biota*. (Eds AM Gill, RM Groves and IR Noble) pp. 425–439. Australian Academy of Science, Canberra.

Turner A, Fordham B, Hamann S, Morrison S, Muller R, Pickworth A, Edwards A and Russell-Smith J (2002) 'Kakadu National Park fire monitoring plot survey and analysis'. Technical Report No. 72. Kakadu National Park, and Bushfires Council of the Northern Territory, Darwin.

van Wilgen BW, Biggs HC and Potgieter ALF (1998) Fire management and research in the Kruger National Park, with suggestions on the detection of thresholds of potential concern. *Koedoe* **41**, 69–87.

van Wilgen BW, Govender N, Biggs HC, Ntsala D and Funda XN (2004) Response of savanna fire regimes to changing fire-management policies in a large African National Park. *Conservation Biology* **18**, 1533–1540.

van Wilgen BW, Trollope WSW, Biggs HC, Potgieter ALF and Brockett BH (2003) Fire as a driver of ecosystem variability. In *The Kruger Experience: Ecology and Management of Savanna Heterogeneity*. (Eds JT du Toit, KH Rogers and HC Biggs) pp. 149–170. Island Press, Washington DC.

Walker J (1981) Fuel dynamics in Australian vegetation. In *Fire and the Australian Biota*. (Eds AM Gill, RM Groves and IR Noble) pp. 101–127. Australian Academy of Science, Canberra.

Walters C (1986) *Adaptive Management of Renewable Resources*. Macmillan, New York.

Whitehead PJ (1999) Promoting conservation in landscapes subject to change: lessons from the Mary River. *Australian Biologist* **12**, 50–62.

Williams RJ, Cook GD, Gill AM and Moore PHR (1999) Fire regime, fire intensity and tree survival in a tropical savanna in northern Australia. *Australian Journal of Ecology* **24**, 50–59.

Williams RJ, Gill AM and Moore PHR (1998) Seasonal changes in fire behaviour in a tropical savanna in northern Australia. *International Journal of Wildland Fire* **8**, 227–239.

Williams RJ, Griffiths AD and Allan GE (2002) Fire regimes and biodiversity in the wet-dry tropical landscapes of northern Australia. In *Flammable Australia: The Fire Regimes and Biodiversity of a Continent*. (Eds RA Bradstock, JE Williams and AM Gill) pp. 281–304. Cambridge University Press, Cambridge, UK.

Williams RJ, Woinarski JCZ and Andersen AN (2003a) Fire experiments in northern Australia: contributions to ecological understanding and biodiversity conservation in tropical savannas. *International Journal of Wildland Fire* **12**, 391–402.

Williams RJ, Gill AM and Moore PHR (2003b) Fire behaviour. In *Fire in Tropical Savannas: The Kapalga Experiment*. (Eds AN Andersen, GD Cook and RJ Williams) pp. 33–46. Springer-Verlag, New York.

Woinarski JCZ (2000) The conservation status of rodents in the monsoonal tropics of the Northern Territory. *Wildlife Research* **27**, 421–435.

Woinarski JCZ (2004a) The forest fauna of the Northern Territory: knowledge, conservation and management. In *Conservation of Australia's Forest Fauna* (second edition). (Ed. D Lunney) pp. 36–55. Royal Zoological Society of New South Wales, Sydney.

Woinarski JCZ (2004b) 'Threatened plants and animals in Kakadu National Park: a review and recommendations for management'. Unpublished report to Parks Australia North. Parks and Wildlife Commission of the Northern Territory, Darwin.

Woinarski JCZ, Milne D and Wanganeen G (2001) Changes in mammal populations in relatively intact landscapes of Kakadu National Park, Northern Territory, Australia. *Austral Ecology* **26**, 360–370.

Woinarski JCZ, Armstrong M, Price O, McCartney J, Griffiths AD and Fisher A (2004) The terrestrial vertebrate fauna of Litchfield National Park, Northern Territory: monitoring over a 6-year period and response to fire history. *Wildlife Research* **31**, 587–596.

Woinarski JCZ, Williams RJ, Price O and Rankmore B (2005a) Landscapes without boundaries: wildlife and their environments in northern Australia. *Wildlife Research* **32**, 377–388.

Woinarski J, Milne D, Palmer C, Fisher A, Ward S, Risler J, Brennan K and Berghout M (2005b) 'Fauna monitoring at Nitmiluk fire plots: baseline sampling, 2005'. Department of Natural Resources, Environment and the Arts, Darwin.

Woinarski J, Pavey C, Kerrigan R, Cowie I and Ward S (2007) *Lost From Our Landscape: Threatened Species of the Northern Territory*. Northern Territory Government Printer, Darwin.

WWF (World Wildlife Fund) (2005) 'Nomination for listing Arnhem Plateau sandstone heath as an Endangered Ecological Community under Australia's *Environment Protection and Biodiversity Conservation Act 1999*'. Unpublished document lodged with the Commonwealth of Australia Dept Environment and Heritage, Canberra. WWF Australia, Darwin.

11

The West Arnhem Land Fire Abatement (WALFA) project:

the institutional environment and its implications

Peter J Whitehead, Paul Purdon, Peter M Cooke, Jeremy Russell-Smith and Stephen Sutton

SUMMARY

The WALFA project was developed on a foundation of long-term ecological studies on the Arnhem Land Plateau, many of them relating to the impacts of fire on vegetation dynamics. The insights provided by this work permitted a very lateral connection of the region's fire-management problems with development of a major natural gas processing facility on Darwin Harbour, some 300 km west. To offset the gas plant's environmental impacts, developers contracted to support for 17 years the region's Indigenous people to re-establish and maintain control over fire regimes. The goals were to protect important areas of monsoon forest and reduce emissions of greenhouse gases.

The project also drew on many other earlier initiatives that had built some of the institutions needed for fire management operating over a large area. A new Arnhem Fire Control Region was created in 2004. Operational capability had been built through creation and consolidation of regional Ranger groups under the Northern Land Council's Caring for Country Program. Regional resource management organisations (e.g. Bawinanga Aboriginal Corporation) offered many forms of operational support, especially transport. The Tropical Savanna Management Cooperative Research Centre built – with the assistance of government – technology for real-time detection of fire 'hotspots' and prompt mapping of fire scars that is essential for both planning and day-to-day management responses. The 'glue' to join these disparate components into an effective system was provided by committed individuals with a mix of scientific and organisational skills and a long-term commitment to the region and its people. The continued availability of these individuals was, in turn, dependent on funding from a wide range of sources, including government and philanthropic organisations.

In this chapter we describe the array of activities generated by the project, and the social and environmental benefits. Against this background, we examine strengths and weaknesses of the present arrangements and, in particular, identify gaps and the new institutions needed to reduce dependence on interested individuals and other goodwill. We build our treatment on the proposition that entering markets in the provision of ecosystem services – in particular trading in carbon credits – will be an essential component of more robust arrangements capable of consolidating realisation of benefits.

BACKGROUND

Preceding chapters in this volume have shown how management of fire in Australia's tropical north has critical implications for environments and human relationships with savanna landscapes. Choices made, and management performance in acting on choices, will influence, among many other things, vegetation patterns, plant and animal diversity, the status of threatened species, emissions of greenhouse gases and other (including particulate) pollutants and, ultimately, the availability and quality of ecosystem services on which all users of the savannas depend.

Many of these insights are far from new. Despite evidence that present fire-management policy and practice is often misguided or poorly executed (Russell-Smith *et al.* 2003; Whitehead *et al.* 2005), concerns about biodiversity, status of native vegetation and even risks to life and property or threats to pastoral production have, until recently, mostly failed to drive effective responses. The desirability of engaging the regions' Indigenous people in dealing with threats and realising opportunities (Whitehead 1999; 2000) was also slow to gain traction. But demonstration of the sheer scale of greenhouse gas emissions from savanna fires – as part of Australia's National Greenhouse Gas Inventory and its reporting obligations under the Kyoto Protocol (Chapter 1) – has contributed to a remarkable change in perception of this issue.

That shift has been dictated by two important aspects of emerging regulatory frameworks for greenhouse gas management. On the one hand, there is potential for an emissions trading scheme to impose financial liabilities for greenhouse gas (GHG) emissions on all users of fire (of whom there are many in the northern savannas), or collectively on the nation as it strives to meet challenging targets in GHG mitigation. In addition to widespread customary use by Indigenous people in 'firestick farming', non-Indigenous pastoralists and other agricultural enterprises regularly use fire to foster production ends. On the other hand, established fire users who can show that they have reduced their emissions below business-as-usual levels may be positioned to generate formally recognised 'offsets' as tradable credits for sale in voluntary or compliance markets for carbon (Whitehead *et al.* 2008).

Costs to the Australian economy and the potential for substantial new financial obligations for individuals and options for enterprise have focused minds in a way that more abstract or non-monetary losses and gains failed to achieve. But this new view of the place of fire in the north Australian economy has yet to be translated into effective policy and practice, or even clear statements of expectations. At all geographical and institutional levels – from international treaties, through national and regional governments to local organisations – policy settings and approaches are in flux. In this chapter, we seek to mark a potential path through that developing policy terrain, drawing on experience from the WALFA Project and, in turn, suggesting how that experience can be used to achieve multiple benefits for the savannas region.

In particular, we address the questions:

- What positions should Australia take to international negotiations in regard to treatment of savanna fire under the successor agreement to the Kyoto Protocol?
- How should Australia treat savanna fire under its own emissions trading scheme?
- What actions should state and territory governments take to complement international and national settings and extract optimum regional benefit from management of fire-related emissions?
- How might the region's Indigenous people organise to optimise socioeconomic benefits from engagement in improved regional fire management?

Although our focus in addressing these questions and providing tentative answers – or at least pointers to their resolution – is primarily on Indigenous interests, we also touch on implications of emissions trading for non-Indigenous fire managers in industry and government.

INTERNATIONAL AGREEMENTS

All international agreements are, by their nature, more or less unsatisfactory, lowest-common-denominator compromises that deliver incremental benefits rather than optimal solutions to the problems they address. The Kyoto Protocol is no exception to this general rule. The primary intent is to reduce addition to the atmosphere of huge volumes of GHGs from fossil fuels sequestered over millions of years. The means is to set binding caps on emissions from developed and developing countries, to which they are expected to respond by reducing energy use – especially energy provided by combustion of fossil fuels – or by finding alternative energy sources that produce lower net emissions.

This conceptually simple equation is complicated by the fact that, as highlighted in earlier chapters, many GHGs are integral to natural and agricultural cycles that influence the amounts of gases reaching the atmosphere and being removed from it. In addition to differences among nation states in the way they use fossil fuels, there are also great variations in their use of other resources connected to natural and semi-natural processes, including treatment of forests (Gullison *et al.* 2007). Some of the world's largest natural stores of carbon are held by developing nations who, by definition, have yet to enjoy the socioeconomic benefits of industrial development and intensive land use.

Initial targets for emissions reductions by developed nations under the Kyoto Protocol were negotiated mostly on reductions in use of fossil fuels and took little account of the potential role of other carbon sources and sinks. Consequently, recognition of the role of natural processes and land use influencing GHG emissions was at best partial and clumsy (Schlamadinger *et al.* 2007) or, less charitably, unfair and damaging. Most significant was a failure to deal effectively with options for avoiding further deforestation (Grace 2004). Methods for dealing with capture of carbon by reforestation (of areas previously cleared of woody vegetation) or afforestation (of areas that did not in recent history support forests) are complex and, perversely, capped to prevent large claims in contributions to net reductions in emissions (Hohne *et al.* 2007).

Critics of the notion of putting greater emphasis on the role of the biosphere in setting goals and managing emissions point to the risk of some nations receiving 'unwarranted' credits for carbon captured through changed patterns of land use unrelated to reductions in energy use. Accessing such credits might allow them to prolong high rates of fossil fuel consumption. They also cite measurement difficulties that confound verification of national, or even project-based emissions outcomes from 'natural' or semi-natural systems (see Canadell *et al.* 2007 for a discussion of related issues). Australia's national emission accounts are a striking realisation of the fear that fossil fuel use will continue as usual despite commitments to binding targets. A negotiated 8% **increase** in total emissions, combined with the last minute addition of the 'Australia' clause on emissions from clearing native vegetation in Article 3.7 of the Protocol, made the nation's task relatively easy. Australia was able to maintain quite rapid growth in industrial and domestic emissions and yet approach its weak targets by claiming credits from rates of land clearing lower than a high 1990 base rate (see Department of Climate Change 2008a).

Although this sort of departure from the spirit of the United Nations Framework Convention on Climate Change (UNFCCC) may foster continued scepticism about the role of the land use, land use change and forestry sectors (LULUCF) in changing behaviour and levels of dependence on fossil fuels, there were some signals of a freeing of attitudes at the 2007 Bali

Climate Change Conference. One of few significant decisions of the Conference set goals for reducing deforestation in developing nations (Decision 2/CP13 available at http://unfccc.int/resource/docs/2007/cop13/eng/06a01.pdf#page=8).

Savanna fire is treated under IPCC guidelines as an agricultural emission, on the grounds that most fires are lit deliberately (Russell-Smith *et al.* 2007) to meet a variety of Indigenous (Whitehead *et al.* 2003) or non-Indigenous (Dyer and Stafford-Smith 2003) production goals. The preceding digression into the LULUCF sector is, however, important because it illustrates the sort of opposition to 'quirky' sources of credit that management of savanna fire may confront, if Australia seeks to treat it as a source of emissions that can be credibly managed to mitigate national GHG emissions. To illustrate this: in discussions of the difficulties confronting credits for reforestation, fire is often cited as a threat to 'permanence' of carbon sequestration and hence a source of uncertainty rather than of well-validated credits.

Agriculture as a whole is regarded as creating special difficulties for attribution of emission liabilities, given the many small enterprises involved and the difficulties of measuring many emission sources and sinks. The Australian Government's White Paper (Department of Climate Change 2008b) for an emissions trading scheme, to be known as the Carbon Pollution Reduction Scheme (CPRS), proposes a work program to determine by 2013 whether or not to include agriculture in the scheme from 2015. Later in this chapter, we consider further the implications of this decision assuming that savanna fire continues to be treated as an agricultural emission.

NATIONAL POLICY, ESPECIALLY IN REGARD TO EMISSIONS TRADING

Australia has examined policy settings for GHG management through two high-level processes, brought together during 2008. The states and territories established the Garnaut Climate Change Review, in which the national Labor Government also chose to participate soon after its election in late 2007. The Australian Government – in addition to being informed by the Garnaut Review – undertook parallel policy analysis through its agencies, including economic modelling by the Australian Treasury, and concurrent public consultation processes.

In September 2008, the Garnaut Review (2008a) issued its final report. The Australian Treasury (Commonwealth of Australia 2008) released modelling of the economic costs of emissions mitigation in October 2008. The White Paper, drawing on all of this analysis, appeared in December 2008 (Department of Climate Change 2008b). Taken together, these documents present the most authoritative statements of likely national direction. Basic features of all analyses are reinforcement of the desirability of acting decisively and acting early on unambiguous reduction targets, and reliance on an emissions trading scheme covering as many sectors as possible.

On savanna fire, Garnaut, in a suite of estimates of potential reductions in agricultural emissions, estimated that a 50% reduction in emissions from savanna fire was plausible and the steps necessary to realise such a goal were summarised in a case study (Northern Territory Government 2008a). The White Paper (Department of Climate Change 2008b) makes a number of important statements about savanna fire. First, in a discussion of emission sources that will probably remain outside the CPRS, it uses savanna fire as an example of sources that are too difficult to attribute to individuals or corporate entities for assigning liability. Second, it notes that offsets (as saleable credits) can only come from emission sources outside the CPRS. Third, and perhaps most important from the perspective of this discussion, it repeats commitments to support Indigenous people financially and otherwise to facilitate their participation in carbon markets, including producing offsets from savanna fire. The Australian Government released a draft of proposed law to implement the CPRS in March 2009.

Garnaut and the White Paper also promote the potential for Australia to take an important role in promoting effective global responses to climate change, by acting decisively as a developed nation with some international credibility. But a convincing position will depend on more than the rigour of the proposed CPRS. The White Paper explicitly recognises the importance of adopting national policy consistent with the attitudes Australia puts to others in negotiating post-Kyoto agreements. In addition to commitments to better management of savanna fire, it also notes the need for good domestic policy and performance on deforestation and acknowledges the need for domestic incentives to avoid deforestation. Connections between the deforestation issue and savanna burning are explored later in this paper.

NORTHERN TERRITORY POLICY

The Northern Territory Government, in its strategy for greenhouse action (Northern Territory Government 2006), committed to work with the Australian Government and states to examine the feasibility of recognising savanna fire emissions as a source of GHG offsets. It has also supported related research. In its submission to the national government's Green Paper on the CPRS (Department of Climate Change 2008b), the Northern Territory supported investigation of offsets from reductions in savanna burning, while recognising that that further work was required on scientific and methodological issues (Northern Territory Government 2008b).

Because savanna fire is one of few emission sources in the Northern Territory that appears amenable to prompt reduction at reasonable cost (Whitehead *et al.* 2008), serious support for efforts to reduce unnecessary emissions from savanna fire will almost certainly feature in the Territory Government's response to the national climate change agenda. A number of important steps have already been taken in this direction through support for a globally unique 'experiment' in regional-scale intervention in fire regimes.

THE WALFA PROJECT

The great significance of the WALFA experiment is that it demonstrates the feasibility of substantially reducing fire-generated emissions in Australia's tropical savannas. But it did not start with that intent.

Initiated in 1997, WALFA was developed on a technical foundation of ecological studies on the western side of the Arnhem Land Plateau, initially focused in Kakadu National Park, and assessing impacts of fire on vegetation dynamics. Many of those studies were prompted by concerns about conservation of diverse assemblages of often endemic plants and animals as summarised in earlier chapters. At the same time, senior Indigenous landowners, regional Indigenous organisations, the Northern Land Council (the peak body for Indigenous landowners in the northern half of the Northern Territory), and Bushfires NT (a rural fire management authority), were motivated to address concerns identified by landowners regarding chronic fire management issues. They sought to re-establish effective management of fire at a regional scale, to be achieved by strong coordination of complementary local actions. Importantly, Aboriginal partners saw an opportunity to bring Aboriginal managers and management back to 'orphaned country' (Chapter 4).

The insights provided by this work permitted a very lateral connection to development of a major natural gas processing facility on Darwin Harbour, some 300 km west of the WALFA site. The plant was developed by Darwin Liquefied Natural Gas PL (DLNG): a subsidiary of the transnational energy company ConocoPhillips. To offset loss of monsoon forest removed to accommodate plant and related infrastructure – and also to offset to some extent the GHG emissions from the LNG plant – the developers agreed to support Indigenous people to restore fire management over 2.8 million ha of western Arnhem Land, which became the WALFA project area. Negotiations were concluded with DLNG in 2005.

Contracts included a target for an average annual reduction of 100 000 tonnes CO_2-equivalent relative to a 10 year (1995 to 2004) baseline of the region's fire-derived GHG emissions. Under an agreement extending over 17 years the developer pays AU$1 million per annum (indexed to 2006) for greenhouse-effective fire-management services. The Australian and Northern Territory governments recognised the potential of the project to deliver GHG abatement, and formally accepted that DLNG's commitment met obligations under environmental law. The company does not, however, purchase tradeable carbon credits from the project. Developing systems to produce tradeable credits will be an important step in the project's evolution. An opportunity to take this step may be available through a fifth year review of the contract between DLNG and the Northern Territory Government scheduled for 2010.

WALFA operations

The project drew capacity and strength from earlier initiatives that had built some of the institutions needed for collaborative fire management over large areas. A new Arnhem Land Bushfires Council Region – comprising solely Aboriginal-owned land and advised by an entirely Indigenous advisory committee – was created in 2004 (see Box 11.1). Operational capability was built incrementally by Indigenous groups now collaborating in the project through the Northern Land Council's Caring for Country program (Northern Land Council 2006). Regional resource management organisations (e.g. Bawinanga Aboriginal Corporation) offered many forms of operational support, especially transport. The Tropical Savanna Management Cooperative Research Centre (created in 1992 under the Australian Government's Cooperative Research Centres Program), with the particular assistance of the Western Australian Government, developed and refined technology for real-time detection of fire 'hotspots' and near real-time mapping of fire scars that is essential for both planning and day-to-day fire management responses. The 'glue' to join these disparate components into a workable system was provided by committed individuals with a mix of scientific and organisational skills and a long-term commitment to the region and its people. Their continued availability to work on the project was, in turn, dependent on funding from a wide range of sources, including government and philanthropic organisations.

As shown in preceding chapters (especially Chapter 5), exercise of customary skills in fire management is associated with a rich vocabulary and knowledge base describing fire behaviour, the circumstances under which different fire behaviours are observed, and environmental outcomes (Yibarbuk *et al.* 2001; Chapter 5). This sort of knowledge and practice is critical for reliable delivery of fire regimes calculated to protect forests and reduce emissions, while also protecting other natural and cultural values. But because the wider landscape is no longer under intensive traditional management, additional layers of management are needed.

Under the WALFA project design, aerial burning (incendiary pellets dropped from helicopters or fixed wing aircraft) is used to establish boundaries around and within the project area to prevent intrusion of fires from neighbouring lands, and to contain other fires lit within the project area. Strategic aerial burns are complemented by finer-scale burning from motor vehicles or by walkers to create effective barriers to subsequent fire. Within these barriers, objectives involving tighter control over timing and extent of individual fires can be addressed. Some ancient, but more recently disused, walking tracks through the rugged sandstone landscape have been re-established to facilitate operations (Chapter 4), and more tracks will be established in future. This 'two toolkits' approach is a key feature of the project. Modern technology is used early in the annual burning program to establish a favourable

Box 11.1 Creation of the Arnhem Fire Control Region

In 1965, the Legislative Council of the Northern Territory established the *Bush Fires Control Ordinance,* which provided an organisational system for control of bushfires. Two key elements of the system were the Bushfires Council and Regional Committees. The Bushfires Council was envisaged as a non-operational body, chiefly required to administer the Ordinance and advise the Administrator (equivalent to a state governor) on bushfire matters.

The Regional Committees were intended to be the operational fire management bodies. This has flowed through to current Northern Territory law which charges Regional committees, among other things, to:

- prevent and control bushfires in its fire control region
- exercise and discharge such powers, duties and functions as are conferred or imposed on it by or under this or any other Act.

By 1967, seven regional committees – the Northern, Southern, Alice Springs East, Alice Springs West, Victoria River, Elliott/Wauchope and Barkly – were established and had conducted their inaugural meetings.

Although it was clearly the intent to place decision making and operational capacity in the hands of landholders 'on the ground', two public sector Fire Control Officers – one for the north of the Territory and one for the south – were appointed in May 1966 to facilitate coordination of resources across the enormous areas of Territory. By May the following year, statutory and operational measures were put in place to control bushfires. These included the declaration of the first Fire Danger Period (2 weeks in May south of the 21st parallel) and the first use of Prohibited Fire Days.

During subsequent years, the boundaries of these regions varied depending upon demographic, economic and technological changes, such that by the year 2000 nine Bushfire Control Regions had been established with Regional Committees.

For all of the 35 years of operation of the Regional Committees up until 2001, resourcing, and setting priorities for, fire prevention and suppression was guided by a broad three-part categorisation of the Territory. In 1963, it was determined by the Legislative Council that the three categories of land for the purpose of fire control should be:

- extensive areas where little or no development has taken place and bushfires in these areas are not of high economic importance
- intensive areas where development is taking place and bushfires are of economic importance
- areas of fire prohibition where bushfires are likely to cause heavy damage to investments and improvements (Northern Territory Legislative Council Hansard 14 November 1963).

Although Arnhem Land was incorporated within a region (Arafura) little or no effort was made to contain wildfire other than in Kakadu and as undertaken by traditional owners. And Indigenous fire control was in a state of marked decline from the 1950s onward (see Chapters 3 and 4). With the exception of a few nodal and notable mineral deposits, little of Arnhem Land could be found to conform to a characterisation of 'high economic importance' to the Commonwealth or to the Northern Territory. That is not to say the lands and waters of the region were not of economic importance to its inhabitants (Chapters 5 and 6).

This situation changed after 2001 when traditional owners from Arnhem Land attended a meeting of the Arafura Region Bushfires Committee. At that time, the Arafura Region stretched from the mouth of the Fitzmaurice River on the Joseph Bonaparte Gulf around the northern Top End (not including Darwin and its rural area) to the mouth of the Roper on the Gulf of Carpentaria: over 171 000 km^2 and nearly 13% of the Territory.

The deputation to the Arafura Regional Committee sought to establish fire control in the Arnhem Land: something that was not possible at the time with a single Fire Control Officer, working within the parameters of the 1963 prioritisation of fire land management. The Arafura Regional Committee agreed to take the matter forward and, in March 2002, the Bushfires Council resolved to recommend to the relevant minister of a now self-governing Northern Territory that a new fire control region be established to include Arnhem Land and Kakadu National Park. This recommendation gained momentum due to the particularly bad 2002 fire season, which devastated 38.4 million hectares of the Territory and highlighted the degree to which the current fire management arrangements were stretched. The creation of a new fire control region would relieve some of this tension.

A period of consultation between the Northern Territory Government and traditional owners and other stakeholders ensued to determine the most appropriate boundaries for the region and how it should be staffed. This latter issue was approached with some sensitivity recognising that employment of an Aboriginal Fire Control Officer with a traditional moiety affiliation may lead to concerns regarding improper representation of the alternative moiety, and there may in fact need to be two Officers, representing *Dhuwa* and *Yirritja* (see Chapter 5).

These discussions continued until 2004 when the Territory Government approved the establishment of the new Fire Control Region. The Arnhem Fire Control Region was declared by the Minister under the *Northern Territory Bushfires Act* on 8 May 2004 and was gazetted on 26 May 2004.

Michael Carter was appointed as the Regional Fire Control Officer (RFCO) in July 2004 and moved quickly to promote fire management capacity in the region. A Regional Committee, consisting of Otto Campion, George Djanggawangga, Djawa Yunipingu, Jessie Alderson, Bobbie Nunggumajbar, Curtis Andrews and Michael Carter held its inaugural meeting on 5 October 2004 in Katherine and appointed Otto Campion as Chair. The region now hosts the West Arnhem Land Fire Abatement project (WALFA): the world's first GHG abatement fire-management project, which has seen a reversal in fire regime over an area of 28 000 km^2. The Committee continues to promote better fire management in the region and is actively seeking to develop fire-management projects that generate income, as well as protecting the traditional economic resources that most inhabitants of the region still rely on.

environment within which to apply knowledge and labour-intensive finer-scale management through the remainder of the fire season. This combination of technologies greatly reduces the risk inherent in any use of fire in such flammable landscapes.

Integration of the perspectives and contributions of five different Indigenous organisations (and many more clans and families) and the views of government regulators and ecological researchers is achieved through regular consultations. Participating groups meet to develop annual plans organised into several sub-regions. Budgets are distributed according to the work program. Senior Indigenous knowledge custodians – including a highly respected traditional owner, Lofty Bardayal Nadjamerrek AO – play critical roles in planning and subsequent Indigenous authorisation and implementation of the annual work program. Different groups take responsibility for designated areas that include their traditional lands, but coordinate their work through a headquarters base at Kabulwarnamyo on the north-central margin of the plateau (Figure 5.1). Seasonal employment for up to 30 Indigenous people has been created.

Despite not being initially designed to deliver tradeable GHG credits, many technical and operational issues relevant to trade have been explored and are now reasonably well understood. In particular, the WALFA experience indicates that it is feasible to deliver well-validated credits. Indeed, knowledge and practice in measuring and reporting emissions from remote sensing and mapping of fires, and estimation of emissions by application of established National Greenhouse Gas Inventory methodology is regarded as world's best practice (Chapter 13), with emissions modelling being based typically on parameters measured directly in the project area.

Over the 3 years of operation to 2007, the project has exceeded expectations by delivering reductions in GHG emissions totalling 450 000 tonnes CO_2-equivalent (50% above target) from a 1995–2004 baseline. This performance was recorded while methods and capacity were being built and also depended on logistic, technical and financial support from a number of sources. Performance under more settled circumstances may differ substantially.

Influences on WALFA development

In another paper (Whitehead *et al.* 2008), we briefly examined influences on WALFA's present and future capacity to engage in fire management for sustained reductions in GHG emissions. We summarised external economic, technical, and environmental factors, how Indigenous land managers are affected by, and respond to, those influences and how policy-makers might take account of the particulars of Indigenous lives and norms to promote positive environmental and social outcomes. However, that treatment was necessarily broad and warrants further examination.

Here we attempt a deeper treatment of a subset of inter-related issues that are likely to have decisive influences on performance and hence capacity to engage productively with carbon markets. The key issues are:

- the nature and scale of partnerships and their relationship to performance
- Indigenous leadership
- modes of Indigenous engagement
- institutions for sharing obligations and benefits.

Existing partnerships

We have argued that the WALFA project is successful because over 3 years it has delivered change in fire regimes and non-CO_2 GHG emissions bettering contracted targets – and done this while methods and capacity were still being developed. That achievement, and the ability to demonstrate it, has depended on a complex mix of activities supported and coordinated by a large array of individuals and organisations. Cataloguing all linkages is too cumbersome a task to be attempted here. However, a summary of the roles of the various partners is sufficient to illustrate. These are:

- research providers from Commonwealth, state and territory governments (at least six different agencies) and four universities
- direct and indirect funding support from more than 20 individual sources, including governments, research cooperatives, national and international philanthropic or conservation organisations, Indigenous organisations and commercial interests
- operational support from at least eight organisations encompassing government and Indigenous institutions and
- coordination and leadership from individuals sometimes working outside partner organisations.

Importantly, despite the large array of participants, there is presently little redundancy. Each group makes a financially, functionally or geographically discrete contribution. The project is therefore sensitive to loss or scaling back of the contribution of any one of them.

In addition to inherent complexity, a striking feature is particular dependence on the goodwill or grant-dependent capacity of many groups (especially research providers) rather than contractual or other commercial arrangements. Formal and long-term contractual partnerships exist only between the Northern Territory Government (through the Department of Natural Resources Environment, the Arts and Sports and Bushfires NT) and DLNG and between the Northern Territory Government and the Northern Land Council. In essence, government has extended its role as environmental regulator to act as intermediary between the private company DLNG and the Northern Land Council, representing Indigenous landowners.

Indigenous partnerships among five participating Ranger groups and the Indigenous associations or corporations that support them, are informal collaborations brokered through the Northern Land Council and the regional Indigenous organisations themselves. Research partnerships have been coordinated through the Tropical Savannas Management CRC under an umbrella savanna fire-management project with many diverse elements. Connections of researchers to Indigenous groups have been facilitated by the North Australian Indigenous Land and Sea Management Alliance (NAILSMA).

Australian Government involvement has to date included generic support to Indigenous Ranger groups through Community Development Employment Projects (CDEP) and, since 2007, the Working on Country program that provides for full time jobs outside the CDEP or welfare systems, as well as research support from the (now) Department of Climate Change. Northern Territory Government support has been channelled through the Arnhem Fire Control Region and less directly by various forms of support for Indigenous Ranger groups. A number of committed individuals have acted as intermediaries between these groups to assist in the alignment of expectations and support.

With such diversity comes a rich mix of expectations. The elements of this mix presently appear compatible – at least in principle. Government seeks active fire management for a range of environmental reasons and to protect life and property, and also to promote development in remote regions and reduced dependence on welfare (Northern Territory Government 2007). Indigenous people wish to re-assert control over their lands, and reduce reliance on government welfare programs through engagement in enterprise that allows people to benefit from land ownership and presence on country (Northern Land Council 2006). They wish to see customary knowledge and skills respected and applied (Yu 2007). Research providers seek opportunities to refine understanding of vegetation and carbon dynamics in savannas and the biophysical and socioeconomic implications of Indigenous involvement in delivery of environmental services (this volume; Luckert *et al.* 2007). Contributing NGOs seek conservation of cultural and biological diversity (e.g. The Christensen Fund 2005).

But specific expectations from this project – aside from the contractual abatement target – have rarely been formally put, so that supportive individuals within various partner organisations may experience difficulty in making and defending decisions about levels and nature of involvement. Their goodwill has provided 'glue' to hold the project together, and realise its successes to date. But such arrangements are inherently fragile and cannot provide an adequate basis for secure long-term operations. Moreover, the present arrangements provide no mechanisms or access to skills for interacting effectively with markets in the future. It is important that roles and expectations be better defined and agreed before more overtly commercial activities are contemplated. Here we focus on those areas of cooperation in which clarification of roles is most required to consolidate the WALFA project and its potential successors.

Changing roles: Indigenous groups, governments and private interests

Indigenous individuals and groups presently play dominant roles in planning and implementing annual fire-management programs. They decide who will be involved in on-ground work on the Indigenous lands that make up the whole of the project area, and how available budgets will be distributed among participating groups. Respected custodians of Indigenous knowledge oversee both plans and implementation. Non-Indigenous perspectives, including coordination and regulatory issues, are considered by Indigenous participants during planning and operations. Connections with formal regulatory processes are facilitated through the Arnhem Fire Control Region, established under Territory law.

Present tangible benefits from the project are principally additional employment for members of participating Ranger groups, who are paid at about the same rates as government land managers doing equivalent work. Indigenous knowledge custodians and advisers are paid a higher rate for time spent in planning and supervision. Intangible benefits relate to satisfaction that management of country is being improved, and that improvements are secured largely through application of Indigenous knowledge and methods. Engagement of young people and the opportunities offered for transfer of knowledge and skills are also regarded as important (Chapter 4).

Although the relatively informal structures built around regular meetings have worked smoothly enough to meet targets and maintain participation, some tensions and unresolved issues exist. The Northern Land Council, representing traditional landowners, has a particular obligation to ensure that landowners share incomes derived from management of and use of resources from their lands. But the prevailing ethos among many participants is that income should be distributed entirely on active contributions to meeting targets in fire mitigation. Indeed, it may be argued that landowners are receiving services that may exceed in value any 'royalties' (rent) that might otherwise be payable for any associated commercial activity on their lands.

Although presently unresolved, this issue has not threatened effective operations. However, should the project lead to more overtly commercial arrangements offering a number of products – including tradable credits in GHG emissions – then it will be essential that 'rules' for decisions about participation and benefit-sharing be entirely clear. Given that commercial viability of WALFA, and any similar projects, is likely to require long-term cooperation across large areas, and hence different clan and language groups (see also the discussion below of the scale issue in regard to options for engaging with markets), rules must be accepted as fair and be enforceable by a wide range of participants. Internal mechanisms for negotiation and resolution of disputes about participation and benefit-sharing will be required. Basing internal rules on customary assignments of land ownership, roles in land management and rights of access to resources (see Chapters 5 and 6) would appear a reasonable first step, because customary practice is in part about achieving coordination of some aspects of fire use across clan boundaries (Yibarbuk *et al.* 2001). The same Indigenous leaders who presently legitimise application of Indigenous knowledge to 21st century goals in management of the global atmosphere can also provide direction on participation and benefit-sharing issues.

Moves to more-commercial operations and delivery of private (including corporate) benefits will require sustained operation at carbon prices within a range that is presently difficult to predict. In 2008, WALFA estimates of costs in delivering non-CO_2 emissions abatement fall in the range $7–30 per tonne CO_2-equivalent, depending on the range of establishment costs taken into account and assumptions made about ongoing dependence on existing human and other capital originally built around – and still needed for – servicing other needs. These estimates overlap the carbon price ($23 per tonne in 2005 dollars) that the Garnaut Review (2008b) identifies as likely to be generated by modest interim targets (10% reduction of 2000 emissions by 2020). The fire-abatement work and associated expenditures also generate additional (so far unquantified) biodiversity and human social benefits for which demand in private markets or from governments is uncertain.

Assuming that new structures and institutions are built around, or depend at least in part on, existing organisations, it will be essential to show profitability or, at the very least, that operations do not degrade the capacity of those organisations to deliver services in other areas. Doing this raises two sets of closely linked issues on which Indigenous individuals and organisations must provide direction.

First, robust accounting systems will be needed to record sources of finances, contributions of effort, and use of shared equipment or other infrastructure. Second, rules will be required to identify activities that can legitimately be treated as contributing to creation of fire-abatement credits. Discrimination of contributing from non-contributing activity is particularly challenging because fire use is so tightly connected to Indigenous livelihoods and spiritual matters. For example, should participation in the increase ceremonies for native bees (and the honey they produce) and associated knowledge transfers involving fire use (Chapter 5) be treated as a contribution to the commercial fire abatement operation? Similarly, would the sequence of activities done to protect sites from early burning for later use in hunting (Chapter 6) be treated as contributing effort? On the one hand, such activity contributes to the general capacity to prevent unwanted burns but, on the other, later deliberate burning for hunting in the protected site may reduce incomes that could be realised from abatement if the decision was taken to forgo the hunt.

Decisions about 'eligibility' of activity are chiefly the province of Indigenous participants, but those 'cultural' decisions will be best made with access to comprehensive information about financial implications. Forums, or other processes for working through these questions, should be established as soon as possible, so that institutional arrangements are agreed before fully commercial operations are attempted.

Resolution will influence the corporate structures through which Indigenous service deliverers ultimately seek to work. Altman (2007) and Altman and Cochrane (2005) argue that hybrid organisations delivering a mix of private and public benefits are plausible – and indeed indispensible – in sparsely populated remote regions where economies of scale demand shared use of available human and other capital. Whatever decisions are taken about organisational structures, both 'internal' Indigenous and external non-Indigenous interests are likely to demand high standards of accounting and accountability. It will be important to draw on the experience of regional organisations with a history of successfully operating commercial enterprise while also delivering basic community services. The Bawinanga Aboriginal Corporation at Maningrida and the Jawoyn Association in Katherine provide important examples of such organisations.

Obviously, governments play key roles in fire management. They specify in law the obligations of fire users to act to protect the public interest. Under the Australian Federal system, fire use in non-urban landscapes has been treated as a land-management and agricultural issue. So, regulation of use falls predominantly to state or territory governments. But the states and territories have agreed that GHG emissions should be managed nationally. Although the two roles of prescribing required behaviour (states and territories) and influencing behaviour indirectly by imposing costs through permits and emissions trading (national government) are superficially distinct, they intersect in critical ways.

Fire users wishing to adopt emission reduction strategies to generate recognised offsets will need to show that their actions go beyond meeting basic obligations to protect human life and health and property, or maintain land condition and biodiversity: that they have gone beyond compliance. State and territory bushfires legislation is acknowledged to be deficient in its treatment of environmental and Indigenous issues in the savannas (Hughes 1995; Preece 2007), and these deficiencies should be corrected. But, in setting the non-GHG compliance bar, state and territory governments determine the scope for fire users to generate GHG credits at costs that permit trading. This relationship exists irrespective of whether savanna fire sits entirely within a compliance framework under which all fire attracts financial liabilities, or nominally sits outside but allows trades of legitimate credits from better fire management into compliance or voluntary markets. It will be essential that state and territory legislators are clear about the intent of their non-GHG compliance regimes and do not inadvertently compromise identification and national and international recognition of 'beyond compliance' work done specifically to reduce GHG emissions. This issue is further explored below in examining options for Indigenous engagement with markets.

Important additional government roles are ensuring that all those who may be interested and positioned to contribute to mitigation of emissions have reasonable opportunities to do so and, in particular, plausible pathways for accessing carbon-trading markets. Clearly, Indigenous people, as major landowners in the tropics and holders of rights to use for hunting and foraging much of the remaining pastoral and conservation land (Chapter 1; Figure 1.7), are positioned to influence north Australia's fire-derived emissions very substantially. And, as WALFA has demonstrated, they have the will and the capacity to make quantifiable emission reductions. But optimising Indigenous peoples' contributions to GHG mitigation and the generation of income to reduce entrenched socioeconomic disadvantage will require active support.

The Northern Territory and relevant state governments have played, and should continue to play, technical roles in applying remote sensing and land-monitoring capabilities to fire detection and mapping (Chapter 13). These capabilities are essential for measuring and reporting emissions and validating credits. But there must be a shift from present dependence on support from relatively short-term research consortia to robust and properly resourced, ongoing public sector activity. Because monitoring and mapping facilities are needed to meet national needs in reporting emissions, there may be a role for the Australian Government to support enduring arrangements.

This issue connects closely to development of methods and standards for estimating emissions and for demonstrating achievement of reductions that have the potential to be offered as tradeable credits. There is scope for improved parameters for seasonal variation in burning efficiency and resultant levels of non-CO_2 emissions, as well as the potential for automated mapping of fire. The Australian and state or territory governments should collaborate in the inter-related tasks of defining standards that would allow trade into the national emissions trading scheme, as well as meeting the most stringent standards applied in international voluntary and compliance markets, and undertaking related research to increase confidence in quality of estimates.

Equally important will be support to consolidate gains in the relevant capacity of regional people. Operational capacity in natural resource management recognised by the wider Australian community has been built over more than a decade of Caring for Country work (Northern Land Council 2006). A shift in emphasis is now required to support development of Indigenous businesses engaging in the voluntary and compliance markets. Financial support is available through the Australian Government's 2007 commitment to assist Indigenous communities to enter emerging carbon markets, but territory and state agencies engaged in promoting economic and regional development might also contribute.

Although support from government is critical, expertise in such specialised markets as global carbon trading does not reside in government. Private enterprise and skills must be enlisted. However, private interests wishing to form business relationships with Aboriginal land managers face many obstacles. Important challenges include:

- identification of Indigenous partners with
  - genuine interest and commitment
  - appropriate authority
  - existing institutions presently supporting or readily modifiable to support business operations
  - track records in delivering contracted or otherwise agreed outcomes
- integration across land areas large enough to support reliable delivery by 'smoothing' local variation in performance and spreading risk
- substantial variation in experience and capacity of participating individuals and groups
- cultural differences in views of viable and sustainable arrangements.

There are many non-exclusive options for dealing with these very substantial challenges, a number of which are presently being examined in one way or another through WALFA and other initiatives (e.g. Hill *et al.* 2006). It is useful for this discussion to consider three somewhat different approaches.

First, individual entrepreneurs working in carbon and other environmental services markets may choose to finance development of appropriate institutional arrangements in selected Indigenous groups, subject to agreement that they will have priority access to opportunities developed by those groups. This option has the virtue of fostering a diversity of approaches and responses. But, in addition to the obvious financial risks of failure of such open-ended arrangements, a laissez faire approach may also result in fragmentation of effort and so compromise the regional-scale cooperation needed to produce reliable abatement. Each separate arrangement may also require considerable time to develop shared understanding and trust. In bilateral arrangements, there is a risk that neither party will have sufficient understanding of the operational demands, regulatory frameworks and legal implications of integrated fire management. Given a long history of business failure in remote and Indigenous Australia, missteps in individual efforts, for whatever reason, may cause Indigenous participants to be leery of future regional collaborations.

Second, Indigenous groups or bodies working independently of expert business partners may develop entirely new institutions or modify existing institutions to provide and market environmental services. In this case, business expertise would come from existing Indigenous businesses or through consultancy or similar arrangements. As noted above, some elements of this independent approach will be required anyway to derive culturally appropriate operating methods matched to different biophysical conditions, but this option will provide a greater measure of Indigenous control over all features of design. The obvious risk is that expectations may be unrealistic and, in particular, the fit to the market and regulatory demands may be compromised by weak connections to those with exposure to complex and evolving carbon markets.

A third approach is for Indigenous groups, government and private interests to cooperate to develop and test 'model' business arrangements to meet the demands of existing and emerging markets. Collaboration across these sectors may help to improve the match of regulatory regimes, Indigenous institutions and market expectations and hence provide superior models for general application. The disadvantage is that such collaborations may be clumsy and, if rigidly formalised, may be process-oriented and slow.

We prefer a version of the third option under which the lead is taken by one or more Indigenous bodies, who determine arrangements for involvement of private interests in a number of regional initiatives, and where government facilitates building of regional institutions, structuring interactions and provides advice on regulatory issues. If done well, such arrangements will ensure that all participants have access to all relevant technical information and expertise (built over a decade, mostly at public expense), while retaining flexibility to explore diverse approaches matched to the regional context. Government support for regional tests of different approaches to resource management has worked well in other contexts, including management of migratory fauna requiring coordination across jurisdictional, community and natural boundaries. We consider the shape that such collaborations might take at the end of the chapter.

Indigenous leadership

As we have noted at various points, Indigenous leadership is critical in a number of arenas:

- **intellectual** – in developing and articulating Indigenous approaches to application of customary knowledge to fire management for non-customary purposes
- **institutional** – in building structures and processes for planning, reaching agreement and implementing plans in ways that work for individuals, families, clans and language groups
- **political** – for framing and presenting policy options to decision-makers in government and in peak Indigenous bodies
- **operational** – in organising and coordinating inputs to on-ground work.

This volume presents some of the outcomes of existing Indigenous leadership in all of these areas, especially on the intellectual and theoretical underpinning of culturally informed fire management and in operational delivery of services. NAILSMA, as the land and resource management facility of north Australia's major Indigenous Land Councils, is increasingly engaged in putting policy on fire management and related issues to national, state and territory governments and is increasingly treated as an authoritative voice by government and other interested parties. NAILSMA also plays a role in identifying and supporting present intellectual leaders in Indigenous resource management and in promoting mentoring of future leaders.

Less developed to date is Indigenous leadership on the institutional arrangements needed for robust engagement with markets. At present there are no arrangements to bring together the combined experience and knowledge of Indigenous corporations with business experience to develop templates or models for application by others lacking such experience. NAILSMA might play a particularly important role in this arena.

Identifying the most productive potential approaches is challenged by present (early 2009) ambiguity about international and national policy, and hence the array of opportunities and challenges that Indigenous interests should position themselves to meet. In an effort to provide some sense of direction, we turn now to an analysis of options for treatment of savanna fire within emissions trading markets, and identify the most plausible approaches from the full array of potential options.

OPTIONS FOR ENGAGING WITH MARKETS

WALFA is the first example of commercial support for large-scale delivery of ecosystem services in northern Australia. The project continues to draw on government and philanthropic support for contributing Ranger groups, but a long and comparatively large financial commitment by major industry is a critical feature of WALFA's success to date. This private investment has helped consolidate earlier initiatives and generated confidence for others to invest both financially and non-financially. Non-government support, with its strong contractual underpinning, is perhaps seen to be more robust than the sometimes over-defined and too-often re-defined government programs that wax and wane in response to political or competing budgetary pressures (Altman and Whitehead 2003).

The emergence of markets in GHG emissions has added to interest among Indigenous groups in strengthening links among land ownership, the management skills and obligations associated with land ownership, and payments based on recognition of the economic value of Indigenous management activity. Despite signals in the White Paper (Department of Climate Change 2008b) that savanna fire will not attract liabilities under the proposed CPRS, but may be treated as a source of credits for trade with covered sectors, it is important to consider the implications of different decisions. How should Indigenous fire managers connect their environmental management products with the proposed Australian CPRS, and in turn how that scheme might be designed to facilitate Indigenous peoples' contributions to national targets in emission reductions through commercially rewarding management? We consider here five classes of arrangements that, in one form or another, would appear to cover the entire field.

Option 1: direct coverage of fire-related emissions

It is generally agreed that covering all significant sources of emissions within the CPRS will optimise efficiency and so help keep down total economic costs of meeting a given reductions target. However, agriculture presents many difficulties. Estimating emissions and attributing liability to individuals or selecting other appropriate points for imposing liability (producer, processer, distributor, wholesaler or retailer) presents particular problems. For these reasons, agriculture will not be a covered sector on initiation of the Australian emission trading arrangements in 2010, but may be considered for coverage from 2015 (Department of Climate Change 2008b). Nonetheless, because savanna fire is presently treated internationally as an agricultural emission, it is necessary to consider the implications for fire management when agriculture is covered or, conceivably, savanna fire is considered for direct coverage separately from agriculture.

As we have shown, it is practicable to measure fire-related emissions with reasonable accuracy and precision at large spatial scales (Chapter 13), including the regional scale at which WALFA shows good coordination of ground-based operations and reliable delivery on targets is achievable. But, at finer scales, attribution of fires to individuals faces the same classes of difficulties faced by agriculture more generally, plus a suite of additional difficulties. It is not only difficult to describe individual fires with the accuracy needed to estimate emissions, but attribution faces the particular difficulty of identifying the ignition source and the person lighting individual fires. And individual fires with a single ignition source often straddle property boundaries. Which individual or group should accept responsibility for emissions from multi-property fires originating outside those properties and ignited for purposes unrelated to production on them?

Special problems are created for fire users who generate no income from their fire use but nonetheless use fire for legitimate purposes such as conservation management, protection of life and property, or to meet other culturally dictated obligations. Is it reasonable or feasible to require Indigenous landowners and land managers to meet the carbon costs of activities that are entirely non-commercial, but rather an indispensible part of Indigenous living? How will fire use associated with the exercise of native title rights such as hunting be handled, especially if it occurs outside Indigenous-owned lands? How will government-mandated and conducted prescribed burning liabilities be measured and met?

It is difficult to conceive of a plausible monitoring, accounting and enforcement framework for attributing and obtaining payment for the many thousands of individual fires that affect the savannas annually. Effective fire management along WALFA lines will actually increase the number of individual, often smaller, fires. The potential benefits in GHG mitigation of improved management of savanna fire are potentially large because the areas affected are so large; but potential gross annual returns from WALFA (assuming a carbon price of \$23 per tonne CO_2-equivalent equates to about \$1.00 per hectare). The costs of administering a system based on individual attribution could exceed such returns.

Perhaps most importantly, focus on individual fires and assignments of personal liability for them run counter to the integration and collaboration needed to achieve reliable GHG mitigation. Because preventative burns are the most effective tool for asserting control, some individual landowners may actually need to increase their use of fire to promote regional reductions. This is necessary because the optimal configuration of preventative burns is concentration in carefully selected sites rather than uniform or random application across the entire landscape. The protection provided by spatially focused and optimally timed fires enables other landowners to assert tighter control at finer spatial scales – to achieve shared reduction in total emissions at a regional scale.

In addition to the basic conflict between individual attribution and the operational demand to concentrate rather than reduce fire in some parts of the landscape, individual attribution is also challenged by the scale of operations needed for commercial viability. Annual variance in fire regimes is high at spatial scales smaller than the average size of planned or unplanned fires: in a given year smaller areas may be burned entirely or not at all. Extreme year-to-year variation in performance, and hence income, is likely to threaten the viability of small (say) family-based businesses, even if their average performance over several years meets targets and is otherwise commercially sustainable. Costs to both purchasers and suppliers in many bilateral interactions may also be prohibitively high. Demonstrating consistent and relevant levels of performance to markets to protect the reputation and price of the fire abatement product is plausible at large spatial scales, but much more difficult at smaller spatial scales. WALFA illustrates how smoothing out variation in performance, income and costs might be achieved by actively pooling activities over large areas in a similar fashion to carbon pooling arrangements for sequestration projects (e.g. AGO 2005).

We consider that the barriers to full attribution and individual liability for emission that would be required for full and direct coverage by the proposed CPRS are insurmountable.

Option 2: direct coverage but with special conditions for Indigenous fire users

If fire were to be directly covered by the CPRS, the particular circumstances of Indigenous people in northern Australia – as owners of large areas of land often unsuitable for orthodox production – could be recognised in part by total exemption from liabilities, or by issue of free permits to a ceiling at or below present average emissions.

A total exemption would offer no incentive for Indigenous landholders, land managers and residents to contribute to mitigation. Given the large proportion of savanna lands under Indigenous ownership, this would clearly be an undesirable situation. Retaining potential for liability but issuing some free permits to individual Indigenous landowners – perhaps together with the option to trade permits not used – would confront all of the attribution problems identified above, which we have concluded rule out this option unless permit issue and surrender is rolled up into large spatial units. But, achieving large scale accounting of many individual liabilities would then shift costs of administration to Indigenous groups, without offering compensating benefits.

The act of accounting in this way would also conflict with maintenance of customary practice and important norms of Indigenous society. Consider the example of fire use for hunting and associated ceremony described in Chapter 6. Would the owner(s) of the land be required to surrender permits, or those land managers who lit the fires? How could the distribution of benefits (in both food availability and discharge of social obligations) from the activity be aligned with liabilities for fire use and surrender of permits? Would those sharing in hunting returns be liable for a portion of the costs?

Forms of liability being differentiated on ethnicity or culture would also work against the potential for Indigenous groups to partner with non-Indigenous neighbours to achieve regional-scale emission reductions. Organisations such as the Northern Territory Cattlemen's Association are interested in exploring options to participate in regional fire management activities (I. McLean *pers. comm.*).

We conclude that Indigenous exemptions or access to free permits do not overcome the many important barriers to direct coverage, but may add new layers of complexity and intractability.

Option 3: savanna fire outside the CPRS, but with options to trade emission credits with covered sectors

For the reasons already given, it may be desirable to avoid the large transaction costs of having savanna fire directly covered by Australia's CPRS. But, to provide incentives for groups to combine their efforts to mitigate fire-related emissions, it may be useful to allow them to trade credits as offsets for industry sectors covered by the national CPRS. Garnaut (2008a) has suggested that no particular limits should be placed on options for covered sectors to access well-validated domestic offsets to supplement purchase of permits issued under an emissions trading scheme. The White Paper on the CPRS (Department of Climate Change 2008b) repeats commitments to facilitate Indigenous engagement with carbon markets, including by management of savanna fire.

The WALFA experience illustrates the potential to generate offsets at prices comparable with the unit cost considered likely early in the life of Australia's CPRS (Garnaut Review 2008b). The quantities of credits potentially available will, however, be small compared with total demand. For example, a suite of five WALFA style projects spread across the savannas and performing at similar levels might generate credits averaging about 1.0 million tonnes per year. Even with ambitious emission reduction targets, total permits issued to covered sectors early in the proposed CPRS will be several hundred million tonnes per year.

Although the income gained from savanna fire credits could be highly significant in the weak regional economies of northern Australia, trading offsets into such a large national pool is unlikely to distort the compliance market. Perhaps the greatest risk, given a prevailing view of fire as a source of uncertainty, may be to the reputation of the national CPRS and consequent constraints on capacity to link it to international trading arrangements. This reputational risk can be managed by ensuring that fire-related credits meet the highest standards of verification and that institutional arrangements supporting their creation and marketing are comprehensive and demonstrably sound.

Whatever decision is taken by Australia, the approach adopted should be clearly and comprehensively documented and details made available to other parties who may have interests in accessing credits generated in Australia or allowing trade between their national schemes and Australia's.

Option 4: savanna fire outside the CPRS and credits tradable only with voluntary markets

If, for ideological or other reasons, a 'pure' compliance emissions trading scheme is desired, it may be decided that no offsets from uncovered sectors should be available to sectors covered by the CPRS. In this situation, north Australian fire managers may still have incentives to reduce emissions if they are able to trade into voluntary markets, which currently accept credits subject to a range of accreditation mechanisms of varying rigour (see Ribón and Scott (2007) and Downie (2007). Price is influenced by the quality of accreditation, so it will be important that savanna fire managers entering such markets seek and achieve the highest standards. There appear to be no technical obstacles to savanna fire credits doing so (see Chapter 13).

Indeed, because generation of savanna fire credits will often also entrain biodiversity conservation and other benefits for landscape function – plus social benefits in regional and Indigenous communities – they may command a premium in voluntary markets, which may be accessed by industry in part as demonstrations of good corporate citizenship. Multiple benefits delivered by Indigenous people may be an important attraction in such marketplaces.

Unfortunately, in framing standards for the voluntary carbon market, the Australian Government has questioned the validity of domestic offsets originating in a sector not covered by the CPRS, where that sector's emissions are nonetheless counted towards the Kyoto target (Department of Climate Change 2009). Savanna burning falls into this category.

The argument is based on considerations of additionality: the requirement that abatement is real, measurable and permanent. It is argued that offsets from activities like savanna fire management, no matter how effective in reducing real emissions at a project or regional scale, may not be truly additional – at a national scale and hence to Kyoto obligations – because they allow Australia still to meet those obligations if emissions from counted sources are increased by an amount equivalent to the reduction in burning emissions. Interpretation of the additionality concept in this way has the potential to place a number of classes of domestic offsets outside Australian standards and remove an array of options to deliver important benefits. In addition to linked social and biodiversity conservation outcomes, these benefits include fostering emissions abatement in sectors not covered by the CPRS and using such reductions to set more ambitious total national abatement targets and trajectories than would otherwise be plausible.

There appear to be many ways to resolve the additionality challenge, including surrender of emission units (assigned amount units or AAUs) that define the total emissions permitted under the Kyoto Protocol or successor agreements. But so far there has been no commitment to take appropriate actions. We trust that the minutiae of emissions accounting will not be allowed to deny opportunity for Indigenous fire managers to engage productively with voluntary markets.

Option 5: savanna fire outside the CPRS with no trading into any market, but direct support of fire-management activity to meet emission -reduction targets

The Australian and Northern Territory Governments both support Indigenous land managers in various ways to deliver services that maintain the condition of landscapes and natural resources. The Australian Government's Working on Country program and the Territory's legislated initiative for joint management of national parks are important examples. Other north Australian states are also engaged in work of this sort. None of these arrangements seeks reductions in fire use. However, there is no reason in principle why similar programs of support should not be developed to deliver specific national and sub-national targets in GHG mitigation across the savannas (Luckert and Whitehead 2007).

However, governments and many Indigenous commentators have urged greater involvement of Indigenous people with the mainstream economy as a critical step in redressing Indigenous socioeconomic disadvantage (Pearson 1999). Other sectors, including the pastoral industry, are also interested in exploring commercial opportunities associated with the carbon economy (I. MacLean *pers. comm.*). Developing enterprises to enter carbon markets could be an important pathway to wider economic activity or – in remote regions where the market economy is largely absent – the only option. Furthermore, Indigenous individuals and organisations are suspicious of the durability of commitments from government and concerned at the high costs of complying with over-prescribed reporting and other program requirements (Altman and Whitehead 2003). Such concerns would be only partially allayed if government chose to purchase GHG mitigation services through tender or similar arrangements outside a legislated emissions-trading scheme.

Direct support under grants or similar programs for a specified period could, and probably should, be part of a longer term strategy to build capacity in Indigenous organisations: ultimately leading to additional opportunities in either or both compliance and voluntary carbon markets and commercial delivery of other environmental services. But neither direct program-based support nor government acting as sole purchaser of GHG mitigation services in savannas would be seen by many present and potential participants as the best long-term arrangement.

REALISING THE FAVOURED OPTION

We consider that the most plausible arrangement that generates the greatest net benefits is for savanna fire to attract no liabilities under the CPRS but nonetheless be treated as a source of offsets available to both covered and uncovered sectors (Option 3). This option provides incentives to reduce emissions by well-coordinated, regional-scale fire-management arrangements, but avoids the disabling burden of individual attribution. The White Paper (Department of Climate Change 2008b) strongly and consistently suggests that the Australian Government is prepared to consider this position and hence seek compatible treatment under future international agreements replacing the Kyoto Protocol. Acting on the assumption that this position will be endorsed nationally and internationally, we turn to the additional actions that should be taken now to consolidate WALFA and to apply its lessons in other places. In approximate priority sequence, we have identified seven actions, which are discussed below.

1. Derive a business plan for WALFA's transition to markets

WALFA's Indigenous participants have indicated their long-term interest in engaging more directly with carbon markets (Whitehead *et al.* 2008). However, present contractual arrangements with DLNG PL extend to 2022. A review of the contract scheduled for 2010 may provide an opportunity to explore options for direct access to carbon markets, but will obviously be subject to negotiation with DLNG. The company's involvement has been critical in getting the project to the position where such options have become plausible.

To inform negotiations, Indigenous participants must drive development of a robust business plan based on conservative estimates of incomes. A reasonable base for projecting incomes from markets may be the carbon price (approx $23 per tonne in 2005 values) estimated by the Garnaut Review from models of relative modest mitigation targets and trajectories (Garnaut 2008b). The plan will need to:

- take account of present commitments and how they will continue to be met or modified
- set rules for participation and benefit sharing
- identify required technical and operational support
- deal with issues in risk management (*ex post* credits, need to ensure continuity of supply through accruing reserves and the like)
- identify opportunities for new or linked products.

The plan will include key details of the design of products and show how they will be packaged for markets as meeting all relevant standards. It should also include an integrated program for ongoing workforce development, including links to supporting government and non-government programs.

Arrangements should be made to access high-quality business and financial advice through consultancy or, if compromise to quality and timeliness can be avoided, through *pro bono* contributions from other well-credentialed sources.

2. Design governance arrangements capable of delivering on the business plan

Governance arrangements and business planning are inseparable, but we treat them individually here to emphasise the importance of Indigenous partners setting clear and plausible business goals before focusing on the complexities of governance structures or processes and the large array of options available to them.

A key issue in governance is the spatial scale necessary for managing risk and realising economies of scale: the nature of structures and processes necessary to coordinate activities of multiple groups will vary with spatial scale. Although the large scale of WALFA has been defined by the geography of the Arnhem Land Plateau, participants need to reaffirm their commitment to work collectively and determine the conditions under which they will do so. Despite the relative cultural homogeneity of the key WALFA interests, formalising arrangements for coordinating activity and sharing benefits is a fundamental requirement for ensuring the long-term stability of the collaboration.

It will be important to engage Indigenous commercial and quasi-commercial organisations in shaping governance arrangements that draw heavily on their experience. This could perhaps be best achieved through a forum to summarise the experience and insights of such organisations, facilitated by business advisers. In addition to contractual or other formalisation of cooperatives, governance issues to be considered will include comprehensive systems for recording relevant inputs, rules for determining benefits based on levels of inputs or other legitimate interest, and associated financial systems matched to business goals.

3. Develop standards for validation of credits

If they are to engage effectively with commercial carbon markets, Indigenous providers will need to build and maintain credibility. Technical issues to be addressed include acceptance of methods for estimating non-CO_2 emissions and length and timing of baselines. Existing methods are broadly compatible with Australia's Greenhouse Friendly standards, which are likely to form the basis for national standards for the CPRS and standards for the voluntary market (Department of Climate Change 2009). Setting standards is principally the responsibility of the Australian Government, but coordinated input from the Northern Territory and relevant state governments, research providers and Indigenous organisations such as NAILSMA will be useful. It will be particularly important that methods that reduce costs without compromising accuracy or precision of estimates are fully explored, including automated fire mapping. Additional, well-focused research will also be necessary to refine emissions factors and their seasonal variation (see Russell-Smith *et al.* 2008).

Work to explore independent 'top down' surrogates for non-CO_2 emissions to validate present and refined empirical models should continue (C.P. Meyer *pers. comm.*). All present participants have agreed that meeting the highest standards of measurement and validation is a key goal for the project.

The currency for measurement of emissions abatement is obvious (tonnes CO_2-equivalent per year), but the parameters for measuring other biophysical and social benefits from fire management are far less clear. Biodiversity is particularly problematic. It is desirable that projected biodiversity and other benefits be considered now, and methods for measuring their attainment or otherwise be developed. In part, these may include the status of endemic and fire-sensitive animals (Chapter 8) and plants (Chapter 9).

4. Design systems for independent monitoring and evaluation

Markets are likely to require systems for monitoring delivery of GHG abatement and other products that are demonstrably independent of the suppliers of credits. Remote sensing appears to be the only method available to measure variables such as fire extent and timing at the scales and frequencies required. At present, systems for delivering data and analysis of this type at the quality and frequency required reside principally in government or in research facilities.

For building credibility of savanna fire products, it is important that arrangements for independent validation are agreed and implemented soon. Given the present location of expertise and facilities, this is likely to require agreement among the national government, the relevant states and the Northern Territory.

5. Adding sequestration

It is well understood that fire can influence the tree–grass balance in savannas and so affect quantities of carbon sequestered in woody vegetation (Chapters 7 and 14). However, the specific relationships between fire regime and woody biomass in different vegetation types are presently relatively poorly known. Nonetheless, it would be useful to give consideration to demands on management structures if additional claims for carbon sequestration in natural forests and woodlands were shown to be plausible and significant in scale.

The challenges here are fundamentally different from the annual abatement in non-CO_2 gas emissions provided by present WALFA arrangements. International protocols for dealing with sequestration are well-developed and restrictive, and participation could require quite different operational and business models. For example, the size of the potential fire-managed sink, its 'durability' (permanence) and the costs and risks of maintaining it indefinitely (or at least for very long periods) require careful evaluation before associated credits could be validated and offered commercially.

6. Identify connected issues for government

Bushfire laws in all north Australian jurisdictions are deficient in their treatment of savanna fire. Although it is desirable that deficiencies are corrected – particularly to legitimise active use of fire to manage fire risk – it is also important to ensure that changes do not compromise the effectiveness of potential financial incentives to reduce emissions from savanna fire. For example, new laws should not confound operational responses to protection of life and property, biodiversity or other environmental conservation, and abatement of GHGs. The processes and incentives needed to achieve in each area may be different.

Governments at all levels play important roles in management of risk. Two classes of risk are particularly relevant. The first involves year-to-year variation in delivery of abatement credits that may be experienced even when operating better than competently. These could arise through unusually adverse weather events or a low probability, but plausible, sequence of reasonably well-understood risks such as the failure of fire-breaks or incidents of arson. The second type of risk flows from either or both incompetent performance or the collapse of a business for other reasons that could compromise the reputation of fire-related abatement products. In either case, it is desirable that potential roles for Indigenous peak bodies and the Northern Territory or Australian Government in restoring operations or redeeming failures are identified and agreed. Commercial alternatives may involve various forms of insurance, but, given the novelty of the proposed operations, coverage may be problematic until greater experience is gained.

7. Write templates for governance and other business arrangements for other potential participants

As we have noted, the size of the markets into which savanna fire credits may be sold is very large. Because potential demand is so high compared with the likely supply of domestic offset credits, suppliers in northern Australia will not be in competition with each other in regard to price, which will be set by larger markets. Indeed, there may be benefits in close collaboration to ensure that all participants maintain very high standards and so boost the reputation of their similar products. This issue is taken up again in Chapter 15.

CONCLUSION

Global and national responses to the threat of severe anthropogenic climate change remain uncertain. Ambiguity is generated by unwillingness to accept the immediate social costs of reducing high and increasing levels of fossil energy use for potential long-term benefits that could be lost to others who fail to change their behaviours. Under such circumstances, evidence of national commitment to take all reasonable steps to reduce emissions, and to promote equitable sharing of costs and benefits, is essential to foster conditions for serious negotiation. Real demonstrations of such commitment have an important role in building national and international confidence that no nation or sector of society will bear a disproportionate burden (Garnaut 2008b).

It is in that context that the management of emissions from savanna fire assume some significance. Globally, improved management of savanna fire can make a proportionally small contribution to the substantial cuts in emissions that appear necessary to reduce risk. But, in the sparsely populated Northern Territory, where per capita emissions are among the highest in the world, savanna fire is the single largest sectoral contributor to the emissions profile and the only emissions source that local people have the opportunity to cut substantially. The WALFA experience has shown the feasibility of asserting control at a relatively modest cost and in ways that bring environmental and socioeconomic benefits to otherwise disadvantaged regions. If Australia is to seriously influence the actions of others, it is inconceivable that its national response will not include strong commitment to reduce emissions from all sources that are amenable to relatively low-cost control, including savanna burning.

But the challenge of managing fire in infrastructure-poor regions, where highly flammable landscapes are uninterrupted across huge areas, depends on much more than an accumulation or more-or-less independent actions from individuals. It requires consistent high levels of collaboration and coordination over large areas and sharing of costs and benefits. It is in this arena that the major challenges remain. There are no precedents for harnessing activities that are central to cultural and social identity to also capture commercial benefits, or to do so across such large regions owned and managed by groups with limited or no business experience.

We have concluded that, like the technical and operational challenges posed by WALFA, arrangements for commercial governance will also be best addressed by testing working arrangements with the active support of Indigenous organisations and government. Government involvement should continue as long as is necessary to build demonstrably robust models that have the potential to be extended to new areas or new classes of enterprise that demand high levels of collaboration to overcome infrastructure or other deficits. In many remote regions, these deficits are likely to continue to present important barriers to enterprise, unless systematically addressed by sharing insights, effort and resources.

REFERENCES

AGO (2005) 'Planning forest sink projects. A guide to carbon pooling and investment structures'. Australian Greenhouse Office, Department of Environment and Heritage, Canberra. <http://www.climatechange.gov.au/land/publications/pubs/carbonpooling.pdf in July 2008>.

Altman JC (2007) Alleviating poverty in remote Indigenous Australia: the role of the hybrid economy. *Development Bulletin* **72**, 47–51.

Altman JC and Cochrane M (2005) Sustainable development in the indigenous-owned savanna: innovative institutional design for cooperative wildlife management. *Wildlife Research* **32(5)**, 473–480.

Altman J and Whitehead PJ (2003) 'Caring for country and sustainable Indigenous development: opportunities, constraints and innovation'. CAEPR Working Paper: No. 20/2003. Centre for Aboriginal Economic Policy Research, Australian National University, Darwin.

Canadell J, Kirschbaum M, Kurz W, Sanz M-J., Schlamadinger B and Yamagata Y (2007) Factoring out natural and indirect human effects on terrestrial carbon sinks and sources. *Environmental Science and Policy* **10**, 370–384.

Commonwealth of Australia (2008). 'Australia's low pollution future: the economics of climate change mitigation'. The Treasury, Canberra.

Department of Climate Change (2008a) 'National greenhouse gas inventory 2006: Australia's national greenhouse gas accounts'. Department of Climate Change, Canberra.

Department of Climate Change (2008b) 'Carbon pollution reduction scheme: Australia's low pollution future'. White paper. Two volumes. Department of Climate Change, Canberra.

Department of Climate Change (2009) 'National carbon offset standard: discussion paper'. Department of Climate Change, Canberra.

Downie C (2007) 'Carbon offsets; saviour or cop-out'. Research Paper No. 48, August 2007. Australia Institute, Canberra.

Dyer R and Stafford Smith M (2003) Ecological and economic assessment of prescribed burning impacts in semi-arid pastoral lands of northern Australia. *International Journal of Wildland Fire* **12**, 403–413.

Garnaut R (2008a) 'The Garnaut climate change review: final report'. Cambridge University Press, Melbourne.

Garnaut Climate Change Review (2008b) 'Targets and trajectories: supplementary draft report'. Garnaut Climate Change Review, Canberra. <http://www.garnautreport.org.au>.

Grace J (2004) Understanding and managing the global carbon cycle. *Journal of Ecology* **92**, 189–202.

Gullison R, Frumhoff P, Canadell J, Field C, Nepstad D, Hayhoe K, Avissar R, Curran L, Friedlingstein P, Jones C and Nobre C (2007) Tropical forests and climate policy. *Science* **316**, 985–986.

Hill R, Golson K, Lowe PA, Mann MK, Hayes S and Blackwood JE (Eds) (2006) *Kimberley Appropriate Economies Roundtable Forum Proceedings.* Convened 11–13 October 2005, Fitzroy Crossing, WA, by the Kimberley Land Council, Environs Kimberley and Australian Conservation Foundation. Australian Conservation Foundation, Cairns.

Hohne N, Wartmann S, Herold A and Freidbauer A (2007) The rules for land use, land use change and forestry under the Kyoto Protocol – lessons learned for the future climate negotiations. *Environmental Science and Policy* **10,** 353–369.

Hughes C (1995) One land, two laws: Aboriginal fire management. *Environmental and Planning Law* **12**, 37–49.

Luckert MK, Campbell B, Gorman JT and Garnett ST (Eds) (2007) *Investing in Indigenous Natural Resources Management.* Charles Darwin University, Darwin.

Luckert MK and Whitehead PJ (2007) A general case for natural resource management: market failures and government policy. In *Investing in Indigenous Natural Resources Management* (Eds MK Luckert, B Campbell, JT Gorman and ST Garnett). pp. 11–18. Charles Darwin University, Darwin.

Northern Land Council (2006) *Celebrating Ten Years of Caring for Country.* Northern Land Council, Darwin.

Northern Territory Government (2006) 'The Northern Territory strategy for greenhouse action'. Department of Natural Resources, Environment, the Arts and Sports, Northern Territory Government, Darwin.

Northern Territory Government (2007) 'Closing the gap of Indigenous disadvantage: generational plan of action'. Department of the Chief Minister, Northern Territory Government, Darwin. <http://action.nt.gov.au/plan_of_action/docs/closing_the_gap.pdf>.

Northern Territory Government (2008a) 'Abating greenhouse gas emissions through strategic management of savanna fires – Northern Territory'. Available at <http://www.garnautreview.org.au/CA25734E0016A131/pages/all-reports--resources>.

Northern Territory Government (2008b) 'Northern Territory Government submission to the Carbon Pollution Reduction Scheme Green Paper'. <http://www.nt.gov.au/dcm/legislation/climatechange/docs/NTG_Response_to_Carbon_Pollution_reduction.pdf>.

Pearson N (1999) *Our Right to Take Responsibility.* Noel Pearson and Associates, Cairns.

Preece N (2007) Traditional and ecological fires and effects of bushfire laws in north Australian savannas. *International Journal of Wildland Fire* **16**, 378–389.

Ribón L and Scott H (2007) 'Carbon offset providers in Australia 2007'. Global Sustainability at RMIT University, Melbourne.

Russell-Smith J, Whitehead P, Williams R and Flannigan M (2003) Fire and savanna landscapes in northern Australia: regional lessons and global challenges – preface. *International Journal of Wildland Fire* **12,** v–ix.

Russell-Smith J, Yates CP, Whitehead PJ, Smith R, Craig R, Allan GE, Thackway R, Frakes I, Cridland S, Meyer CP and Gill M (2007) Bushfires 'down under': patterns and implications of contemporary Australian landscape burning. *International Journal of Wildland Fire* **16**, 361–377.

Schlamadinger B, Bird N, Johns T, Brown S, Canadell J, Ciccarese L, Dutschke M, Fiedler J, Fischling A, Fearnside P, Forner C, Freibauer A, Frumhoff P, Hoehne N, Kirschbaum M, Labat A, Marland G, Michaelowa A, Montanarella L, Moutinho P, Murdiyarso D, Pena N, Pingoud K, Rakonczay Z, Rametsteiner E, Rock J, Sanz M, Schneider U, Shvidenko A, Skutsch M, Smith P, Somogyi Z, Trines E, Ward M and Yamagata Y (2007) A synopsis of land use, land-use change and forestry (LULUCF) under the Kyoto Protocol and Marrakech Accords. *Environmental Science and Policy* **10**, 271–282.

The Christensen Fund (2005) 'The Christensen Fund vision statement'. <http://www.christensenfund.org/frame_about.html>.

Whitehead PJ (1999) Is it time to fill the north's empty landscapes? *Savanna Links*, September-October 1999, 6–7.

Whitehead PJ (2000) The clever country: where cows manage wildlife. In *Business as Usual? Local Conflicts and Global Challenges*. (Ed. R Dixon) pp. 155–168. Australian National University, Darwin.

Whitehead PJ, Bowman D, Preece N, Fraser F and Cook P (2003) Customary use of fire by indigenous peoples in north Australia: its contemporary role in savanna management. *International Journal of Wildland Fire* **12**, 415–425.

Whitehead P, Russell-Smith J and Woinarski J (2005) Introduction and overview: fire, landscape heterogeneity and wildlife management in northern Australia. Introduction to Special Issue. *Wildlife Research* **32**, 369–375.

Whitehead P, Cooke P, Fenton J, Purdon P, Russell-Smith J and Sutton S (2008) Indigenous people in northern Australia: an emerging contribution to the global management of climate change through prescribed savanna burning. *Public Administration and Development* **28**, 374-385.

Yibarbuk DM, Whitehead PJ, Russell-Smith J, Jackson D, Godjuwa C, Fisher A, Cooke P, Choquenot D and Bowman DJMS (2001) Fire ecology and Aboriginal land management in central Arnhem Land, northern Australia: a tradition of ecosystem management. *Journal of Biogeography* **28**, 325–344.

Yu P (2007) *Growing the Alliance. National Indigenous Land and Sea Management Conference*, Cardwell Queensland. 10 October 2007. <http://www.nailsma.org.au/nailsma/forum/downloads/Cardwell-speech-07.pdf>

12

Fire, fuels and greenhouse gases

Garry D Cook and CP (Mick) Meyer

SUMMARY

Savanna fuel loads comprise mostly fine fuels consisting of cured grasses and leaf and twig litter from trees and shrubs, and coarse woody debris. Seasonal accessions of leaf litter result in this component becoming increasingly significant with the progression of the dry season. Fuel loads reach a steady state when the annual amount of decomposition equals that of accumulation; rates of decomposition of fine fuel in the Australian tropical savannas are substantially higher than those of drier and more temperate regions. Savanna burning emissions occur through two processes: flaming combustion, where fine fuels are readily combusted; and smouldering combustion, where heavier woody fuels are more slowly combusted. These two combustion processes result in different proportions of gaseous emissions, with flaming combustion producing more oxidised gaseous species (e.g. CO_2), and smouldering combustion relatively more reduced species (e.g. CO, CH_4). CO_2, CH_4 and N_2O that are relatively long lived in the atmosphere and are the main 'greenhouse gases' considered in evaluating the effect of biomass burning on global warming. However, to date, CO_2 typically has not been accounted for in savanna burning emissions inventories on the assumption that CO_2 produced by savanna burning during the dry season is re-absorbed by plant growth the following wet season. Reducing fire frequency reduces emissions of greenhouse gases from savannas because more of the litter is decomposed biologically following pathways that, compared with savanna fires, produce fewer emissions per unit of litter consumed. For example, savanna fires emit substantially more emissions of the greenhouse gases methane (CH_4) and nitrous oxide (N_2O) than occurs through biological pathways.

FIRES AND GREENHOUSE GASES IN NORTHERN AUSTRALIA

In common with savanna lands worldwide, fires are frequent across Australia's wet–dry tropics (Bond 2005; Russell-Smith *et al.* 2007), with more than four fires per decade occurring across much of area (Russell-Smith *et al.* 2003a). These fires produce 2 to 4% of the accountable emissions across Australia and about 40% of the Northern Territory's emissions (National Greenhouse Gas Inventory Committee 2007b, c). Biomass burning in Australia contributes about 6–8% of global carbon emissions from biomass fires (Meyer *et al.* 2008).

Over the past 50 to 100 years, depopulation of extensive areas of the Arnhem Land in the Northern Territory has led to land being unmanaged, with consequent increases in the severity of regional fire regimes. In the savanna lands of the Northern Territory, traditional knowledge of fire management is still extant and fire is used as part of land management across many tenures (Dyer *et al.* 2001; Yibarbuk *et al.* 2001). There has been extensive research into fire ecology and emissions (e.g. Hurst *et al.* 1994a, 1994b; Andersen *et al.* 2003, 2005; Russell-Smith *et al.* 2003b; Whitehead *et al.* 2005; Luhar *et al.* 2008; Meyer *et al.* 2008; Chapters 7 and 13). Thus, land-management actions to reduce the severity of fire regimes at a landscape scale on Aboriginal lands in Arnhem Land (Gorman *et al.* 2007) is taking place against a background of:

1. considerable potential for improvements to fire management
2. considerable expertise in traditional and non-traditional management of savanna fires
3. a well-developed understanding of the impacts of fire on vegetation, fauna, carbon budgets and greenhouse gas emissions.

Nevertheless, several important knowledge gaps in the estimation of greenhouse emissions budgets remain: fuel dynamics across the range of vegetation types; the contribution of woody debris to emissions; variations in emission factors in time and space; the role of emissions of CO_2, ozone precursors and aerosols in atmospheric forcing; and the consequences of changed management for greenhouse gas budgets. This chapter aims to provide a background to these issues.

ACCOUNTABLE EMISSIONS: OVERVIEW AND RATIONALE

Savanna fires produce a great diversity of emissions, which are rapidly transferred to the troposphere and dispersed (Andreae and Merlet 2001). Of these gases, carbon dioxide (CO_2), water vapour (H_2O), nitrous oxide (N_2O) and methane (CH_4) directly contribute to global warming. Emissions of oxides of nitrogen (NO and NO_2, collectively termed NO_x) and volatile organic compounds contribute to the production of tropospheric ozone (O_3), which is also greenhouse active. The fates of these gases in the atmosphere vary considerably. Water vapour, for example, has a residence time of days, but CO_2, CH_4 and N_2O are long lived and are the main gases considered in evaluating the effect of biomass burning on global warming (Forster *et al.* 2007).

Under the 1996 Intergovernmental Panel on Climate Change (IPCC) guidelines for greenhouse gas inventories, it was assumed that CO_2 produced by savanna burning during the dry season is re-absorbed by plant growth the following wet season, and therefore savanna burning does not alter carbon stocks. Thus savanna burning was assumed to produce no net CO_2 flux (IPCC 1996) and only N_2O and CH_4 emissions were accounted for under savanna burning in Australia's inventory. Although savanna fires emit much greater amounts of CO_2 than N_2O and CH_4, the global warming potentials of N_2O and CH_4 are respectively 290 and 21 times that of CO_2.

Notwithstanding the default assumption in the 1996 inventory guidelines that savanna burning causes no net change in carbon stocks, an allowance was made for the case where fires caused substantial degradation of the savanna and a loss of carbon (IPCC 1996). In Australia, this provision could have been applied to situations where savanna fires were causing the loss of fire-sensitive vegetation within the savanna matrix (e.g. Bowman 1992), but such effects were minor compared with the extensive burning of highly fire tolerant savanna vegetation, and the Australian inventory of savanna burning made no account of carbon flux (National Greenhouse Gas Inventory Committee 1994). Nevertheless, it has now been well established that fire regimes can have substantial impacts on carbon fluxes in Australian savannas, despite the high fire tolerance of these ecosystems (Williams *et al.* 1999, 2005; Russell-Smith *et al.* 2003a; Cook *et al.* 2005; Beringer *et al.* 2007; Liedloff and Cook 2007). Consequently, the 2006 revision to the IPCC guidelines, which provides for accounting of both increases and decreases in carbon stocks, has significant implications for estimation and formal accounting of the magnitude of emissions abatement that might result from improved fire management.

THE COMBUSTION PROCESS

When a fire front passes through a savanna landscape, much of the fine fuel load – defined as particles having a diameter less than 6 mm – is burnt rapidly by flaming combustion. Grasses are often the most obvious component of the fine fuel, and therefore savanna fires are frequently referred to as grass fires (Cheney and Sullivan 1997). However, in the *Eucalyptus*-dominated savannas of northern Australia, tree litter, made up of leaves and twigs, often comprises more than half the fine fuel load (Cook 1994, 2003).

Shrubs are another important component of fine fuels on the Arnhem Land Plateau (Figure 12.1), but have rarely been considered in other studies of savanna burning in Australia. The shrubs represent many families of plants and include re-sprouted lignotubers of savanna trees. Across four habitats in Arnhem Land, the contribution of grass and shrubs each ranged from about one-tenth to one-third of the total fine fuels (Russell-Smith *et al.* 2009).

Figure 12.1 Combustion of a *Grevillea formosa* shrub.

Box 12.1: The difference between flaming and smouldering combustion

Andreae and Merlet (2001) describe the range of different combustion types that occur during biomass fires, and this overview is taken from their work. Flaming and smouldering combustion dominate in savanna fires. As a fire front approaches, heat from the fire causes a complex mixture of tar and gas products to be released from the fuel. Following dilution of this mix with air and its subsequent ignition, flaming combustion produces a range of simple highly oxidised molecules including CO_2, H_2O, NO, NO_2, and SO_2. Depending on the nature of the fire, intermediate products of flame chemistry – such as CO, N_2O, CH_4, H_2, C_2H_4, C_2H_2, polycyclic aromatic hydrocarbons and soot particles – are also released during this stage. Smouldering combustion begins when most of the volatile compounds have been released from the surface of the fuel, and flaming combustion has ceased. This phase of combustion is characterised by a much greater proportion of carbon being released as CO and CH_4, and is accompanied by a great diversity of other partially oxidised compounds similar to the products of the initial heating. Flaming dominates during the passage of a fire front and smouldering dominates the later phases.

The dynamics of grassy fuels in northern Australian savannas differ considerably from that of tree litter. Whereas grass grows during the wet season and cures in the early dry season – and is thereafter available to burn – the peak litter fall from trees does not occur until mid to late dry season (Cook 2003). Thus, fuel loads in the early dry season are less than those later in the year.

In addition to the fine fuel, coarse woody debris (CWD) comprises a substantial fuel load (Russell-Smith *et al.* 2009). This debris results from branch fall and the death of trees. The passage of a fire front causes the ignition of the CWD, but combustion occurs over many days. Unlike fine fuels, most fires burn less than one-third of the CWD, but the proportion of the CWD burnt increases with fire severity (Russell-Smith *et al.* 2009). Combustion of CWD mostly occurs through smouldering rather than flaming (see Box 12.1).

EMISSIONS MEASUREMENTS IN NORTHERN AUSTRALIA

In studies of biomass burning emissions, the amounts of trace gases can be expressed as a ratio against either carbon dioxide or carbon monoxide. The difference between the ratios in the smoke plume and those in the background air are reported as **excess mixing ratios**. Measurements of the excess mixing ratio of $\Delta CO/\Delta CO_2$ at Kapalga in Kakadu National Park showed that smouldering combustion releases about three times more CO relative to CO_2 than flaming combustion (Table 12.1). On-ground measurements of mixed combustion indicated a lower proportion of flaming combustion than was measured in the smoke plume using an aircraft to collect samples.

Emission factors are calculated from the excess mixing ratios and take account of the fuel characteristics. Here, emission factors are defined as the proportion of the fuel carbon or nitrogen emitted as each species of gas. The emission factors from northern Australian savanna fires indicate that they produce more CO and less CO_2 than fires in savannas in Brazil and Southern Africa (Table 12.2) – indicating a greater proportion of smouldering combustion in

Table 12.1: Differences between combustion types and sampling methodology on the excess mixing ratio of $\Delta CO/\Delta CO_2$. The percentage of flaming combustion in the on-ground mixed and aircraft samples was calculated from the on-ground measurements of flaming and smouldering combustion.

	$\Delta CO/\Delta CO_2$	Percentage flaming
Flaming	0.055	
Smouldering	0.168	
Mixed – collected on ground	0.115	47
Plume collected by aircraft	0.09	69

Australian savannas. In contrast with the overseas savannas, Southern Australian forest fires have even more smouldering combustion than Australian savanna fires. These differences probably reflect variation in the relative proportions of woody fuel being consumed, with increasing amounts of woody fuel contributing to greater proportions of smouldering combustion.

FUEL DYNAMICS AND ITS IMPLICATIONS FOR EMISSIONS

The dynamics of fine fuels in the region's savannas have been described by Cook (2003) and incorporated into a model of savanna dynamics (Liedloff and Cook 2007). Briefly, with increasing time since fire, fine litter in the form of grass, as well as leaves and twigs of trees and shrubs, is added to the litter layer where it also decomposes. Decomposition of the fine fuel occurs mainly through the action of ultra-violet light, microbial and fungal decay and termite activity. Fuel loads reach a steady state when the annual amount of decomposition equals that of accumulation (Olson 1963). The rates of decomposition of fine fuel in the Australian tropical savannas are substantially higher than those of drier and more temperate regions (Walker 1981). With a decomposition rate of 0.8 – as measured in the lowland savannas of Kakadu National Park – 90 % of the steady state loads of fine fuel would accumulate in about 3 years (Cook 2003).

CWD is produced by trees. For a system in steady state, the net annual increase in wood and bark in live standing trees should be balanced by losses to the dead pool of CWD: either standing or lying. In a typical *Eucalyptus miniata/E. tetrodonta* savanna, the annual production of above-ground wood is about 0.87 t C ha^{-1} yr^{-1}, of which nearly one quarter, or about 0.21 t C ha^{-1} yr^{-1}, is consumed by termites within live standing trees (Cook *et al.* 2005). Thus, the net annual increase in the stock of wood in live trees should be about 0.66 t C ha^{-1} yr^{-1}. For bark, the corresponding data are 0.51, 0.38 and 0.13 t C ha^{-1} yr^{-1}, although some of the bark loss is also due to combustion. Thus, the annual net production of wood and bark, and therefore the annualised mean loss from fire from the CWD pool, should be about 0.79 t C ha^{-1} yr^{-1}.

Table 12.2: A comparison of emission factors from fires in savannas in Australia, Brazil, and southern Africa and forest fire in southern Australia.

	Northern Australian savanna[1]	Southern Australian forest[2]	Brazilian cerrado[3]	Southern African savanna[4]
CO_2	0.87	0.85	0.94	0.93
CO	0.078	0.091	0.05	0.065
CH_4	0.0035	0.0054	0.002	0.00375

1: Hurst *et al.* 1994a; 2: Hurst *et al.* 1996; 3: Ward *et al.* 1992; 4: Ward *et al.* 1996

The mean stock of dead CWD at steady state (F_c) is about 4.5 t C ha^{-1} (Russell-Smith *et al.* 2009). From this value, and the mean annual rate of addition of coarse woody debris (L = 0.79 t C ha^{-1} yr^{-1}), the rate of decomposition (k) from all sources for CWD can then be calculated following the approach of Olson (1963) and Walker (1981) as:

$$k = \frac{L}{F_c}$$

The rate of decomposition of CWD from all sources including fire is thus about 0.18. A time period of 3/k, or 17 years, would be required to reach 95% of the steady state load of CWD, if it was all removed at one time. However, unlike fine fuels, which can be largely consumed by one fire, CWD is never all consumed by a single fire. Thus, a neat relationship between time since fire and the load of coarse woody debris should not be expected. An independent estimate of the consumption of CWD by fires can be derived from the measurements of emissions ratios of CO/CO_2 and CH_4/CO_2 from fires at Kapalga (Hurst et al. 1994a). This work ignored in its calculations any contribution of CWD to the release of trace gases. However, if it is assumed that all of the fine fuel was burnt with flaming combustion, and all of the CWD was burnt with smouldering combustion, then the estimated 69% contribution of flaming combustion (Table 1) means that about 1.1 t C ha^{-1} of CWD had been consumed by the time the plume was sampled by aircraft. This estimate is roughly consistent with the estimate of 0.79 t C ha^{-1} derived from consideration of tree growth rates and stocks of CWD.

HOW DOES FIRE ABATEMENT REDUCE GREENHOUSE GAS EMISSIONS?

Reducing fire frequency reduces emissions of greenhouse gases from savannas because more of the litter is decomposed biologically following pathways that, compared with savanna fires, produce fewer emissions per unit of litter consumed, or that reduce the rate of emissions production. In the savannas of northern Australia, termites are major consumers of litter. Similar to cattle and other bovines, termites have anaerobic, fermentative microbial communities in their alimentary tract and so can be significant emitters of CH_4 (Khalil *et al.* 1990; Brauman *et al.* 1992). However, unlike cattle, most termite activity occurs underground, and soils are usually substantial sinks of CH_4 because of the activity of methanotrophic bacteria. Thus, studies in southern Australia, Africa and South-east Asia have shown that CH_4 emissions from high levels of termite activity are usually countered by the soil sink (Eggleton *et al.* 1999; MacDonald *et al.* 1999; Galbally *et al.* 2008).

Thus, although biological decomposers and consumers of litter also can emit CH_4, the proximity of the soil, which is a strong CH_4 sink, means that net emissions from biological decomposition are close to zero. In contrast, CH_4 release by fires is rapidly carried hundreds and thousands of metres into the atmosphere, where it is dispersed and can contribute to global warming.

What happens to the nitrogen contained in the fuel if it is not released by combustion? The unburned fuel decomposes, releasing organic nitrogen (N) compounds into the soil. Soil organic N is mineralised by soil microorganisms to a reduced inorganic form (NH_4^+), which is either absorbed by plants, leached out of the root zone or metabolised by nitrifying bacteria to nitrate (NO_3^-). Nitrification produces N_2O which diffuses to the atmosphere. Nitrate is also absorbed by roots and, under anaerobic conditions – mostly during rain events – is further denitrified to molecular nitrogen (N_2) releasing additional N_2O in the process. Australian rangeland soils are characteristically nitrogen deficient and plant roots absorb most of the NH_4^+, leaving little available for nitrification and denitrification. Also, these soils have N:C ratios greater than 25 (Bowman and Cook 2001; Cook 2001), which inhibits mineralisation and nitrification activity (Parnas 1975; Attiwill and Leeper 1987). Given the relatively low concentrations of inorganic N, and the low rates of nitrification activity, it is not unexpected that Galbally *et al.* (2008) found N_2O emission rates from rangeland soils to be low; globally, rangeland N_2O emission rates vary from 0 to 0.7 kg N ha^{-1} yr^{-1}, with a mean of 0.25 kg N ha^{-1} yr^{-1}. Livesey (*pers. comm.*) reports comparable emission rates in savanna woodland soils in Northern Australia.

Organic nitrogen exists in a range of forms: some reactive and some not. For convenience, the soil is frequently approximated to two organic N pools: a small and reactive pool of labile N and a larger recalcitrant pool. The change in N_2O flux resulting from the addition of organic N from decomposed litter depends on the way in which it is partitioned between these pools. The size of these pools is rarely quantified by direct measurement; however, we can estimate upper and lower limits to the N_2O flux change. If we assume that all the soil organic N is labile, then the change in N_2O flux will be proportional to the relative increase in the total soil pool caused by the addition of organic N from unburned fuel. This is the lower limit. From Cook (2001), the organic N concentration in surface soil is 0.09% or 4.1 t N ha^{-1} in a root zone depth of 0.3 m and the mass of fuel N burned in the WALFA domain in 2004 was 0.02 t N ha^{-1} (Russell-Smith *et al.* 2009). If this mass of N was returned to the soil organic N pool, then an additional 1.6 g N ha^{-1} would be emitted to the atmosphere as N_2O by soil microbial metabolism. This is <1% of the 0.3 kg N ha^{-1} N_2O emitted by burning the same amount of fuel.

The upper limit is estimated by assuming that all the fuel nitrogen passes through a very small, but very active, labile pool, which is isolated from the large recalcitrant pool. This is equivalent to applying the same mass of fuel nitrogen as an ammonium fertiliser. There have been no studies of N_2O emissions from fertiliser additions to non-agricultural soils in savanna woodland. However, there have been two comprehensive studies of annual N_2O emissions from urea-fertilised dryland wheat production in Australia; at Rutherglen, northern Victoria and at Cunderdin, Western Australia (Galbally *et al.* 2005; Barton *et al.* 2007). At Rutherglen, 0.05% to 0.1% of fertiliser N was emitted as N_2O, while at Cunderdin it was less than 0.02%. Both wheat systems are extremely nitrogen limited, similar to the tropical savanna woodlands. In contrast, when savanna fuel is burned, 0.8% of fuel nitrogen is emitted as N_2O (Hurst *et al.* 1994b).

In summary, if fuel is not burned, but decomposes, then it would release between 1% and 12% of the N_2O emitted by the biomass combustion pathway.

The high decomposition rate of fine fuels in savannas means that increasing the average fire-free interval causes substantial increases in the proportion of litter consumed biologically, thus causing substantial decreases in the proportion exposed to combustion (Figure 12.2). Reductions in fire frequency will thereby substantially reduce the annualised amount of fuel consumed by fire.

Figure 12.3 shows the effect of fire frequency and seasonality on mean annual emissions of CH_4 and N_2O expressed as CO_2 equivalents. This was calculated using the fine fuel model of Cook (2003) and measurements of the effect of fire season on fire severity, and of fire severity on CWD consumption (Russell-Smith and Edwards 2006; Russell-Smith *et al.* 2009). It was assumed that all CWD was burnt with smouldering combustion and all fine fuel was burnt with flaming combustion, and that 5% of CWD was consumed by termites each year. The emissions were calculated using currently available knowledge of emission factors. Although the variation in CH_4 emission factors with degree of smouldering combustion is reasonably well defined, it has not been determined for N_2O emission factors. The current view is that N_2O emission is less in smouldering combustion than in flaming combustion. However, in the

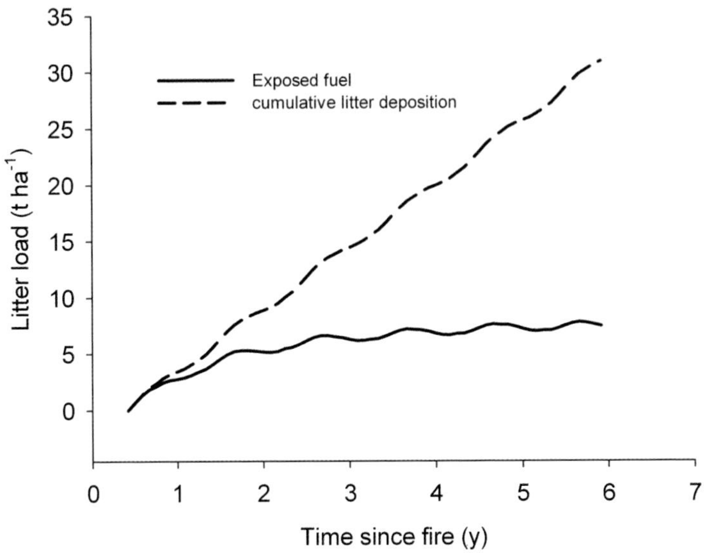

Figure 12.2 The effect of time since fire on the difference between the amount of litter exposed to combustion and the cumulative litter deposition. This difference represents the litter decomposed biologically. Based on the work of Cook (2003).

absence of reliable observations, in this calculation the N_2O emission factor is assumed to be the same in both combustion phases. The results show that emissions increase strongly with increasing fire frequency, while seasonality has a much weaker effect. N_2O emissions represent about 35% of the combined accountable emissions.

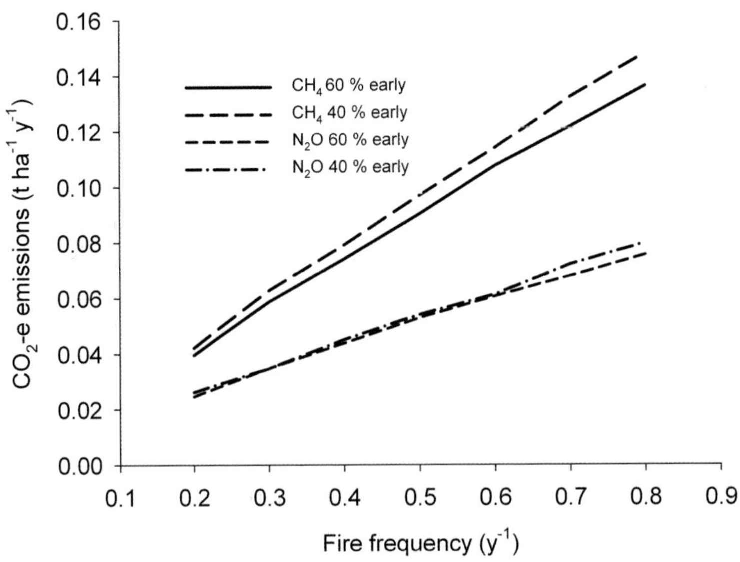

Figure 12.3 The effect of fire frequency and seasonality on pyrogenic emissions of CH_4 and N_2O expressed as CO_2 equivalents.

With decreasing mean annual rainfall, tree density tends to decrease in north Australian savannas (Williams *et al.* 1996; Cook *et al.* 2002) and thus the amounts of fine fuel derived from trees should also decline. Unfortunately, little is known about the effect of declining rainfall within the savanna biome on decomposition rates. Given the importance of termites to decomposition – and the ubiquitous nature of termites in the region – and the lower variability in the duration of the rainy season than in the amount of rain received (Cook and Heerdegen 2001), decomposition could be expected to remain high for much of the biome. With this assumption, fine fuel loads derived from trees are calculated to decline about 10-fold with a decline in mean annual rainfall from 1600 mm in Darwin savannas to about 400 mm in inland savannas.

IMPROVING EMISSIONS ESTIMATES

Fuels

Fuel loads, and the proportion of that fuel combusted, represent one of the major uncertainties in the estimation of emissions from savanna burning. From its inception, the Australian National Greenhouse Gas Inventory (NGGI) relied upon expert opinion to estimate fuel loads at the resolution of Australian states. Consequently, estimates varied considerably. In the 1994 NGGI Workbook for Agriculture, for example, the mean fuel loads estimated for savannas fires in Western Australia, Northern Territory and Queensland were respectively 10, 5.0 and 10.3 t ha^{-1} (National Greenhouse Gas Inventory Committee 1994), while in the 1996 NGGI the respective values were 8.3, 4.1 and 3.0 t ha^{-1}. Across these three jurisdictions, savanna burning occurs in similar landscapes so three-fold variation across jurisdictions is improbable. Such uncertainty derives from the widely varying definitions of fuel in the scientific literature resulting from differing reasons for conducting measurements. For example, most studies have only focused on the fine fuels – and many only on the grassy component. A synthesis of studies of fine fuel loads in savannas dominated by native grasses (Table 12.3) shows the variation across the north Australian jurisdictions to be far less than indicated by the 1994 and 1996 values agreed by the expert panel. The most recent estimates for fuel loads used in the NGGI are 3 t ha^{-1} for fine fuels in savanna grasslands across the three jurisdictions, and 12.7 t ha^{-1} for total fuels in savanna woodlands in Northern Territory and Western Australia (National Greenhouse Gas Inventory Committee 2007a).

Emission factors

The emission factors used currently for all Australian savanna burning are those determined from measurements of savanna fires at Kapalga and nearby (Hurst *et al.* 1994a, 1994b, 1996). As shown in Table 12.2, emission factors vary significantly across vegetation types, mainly because of the different proportions of flaming and smouldering combustion driven by

Table 12.3: The biomass of fine fuels in savanna lands of Northern Territory, Queensland and Western Australia, as reported in published literature and synthesised by Cook and Eager (2005).

Jurisdiction	Mean (sd)	Maximum	N
NT	5.4 (2.0)	9.0	46
Qld	5.0 (3.1)	12.7	33
WA	6.3 (2.8)	12.5	11
All	5.8 (3.2)	12.7	84

variations in fuel characteristics and fire behaviour. Because the relative proportions of grassy and woody fuels vary across the dry season and between the diverse fire-prone habitat types in northern Australia, it could be expected that the emission factors also vary considerably. Better estimates of the variation in emissions factors between late and early dry season fires – and between heathland, woodland and grassland habitats – would considerably improve emissions estimates.

The role of trees

In response to guidelines from the IPCC, it had been assumed under the National Greenhouse Gas Inventory accounting procedures that savanna burning did not alter carbon stocks at landscape scales (National Greenhouse Gas Inventory Committee 1994). Thus, to date, the inventory has only considered the direct emissions of N_2O and CH_4. Nevertheless, the emissions of CO_2 from Australian savanna fires are substantial, and approximately of the same order of magnitude as all Australian fossil fuel emissions (Hurst *et al.* 1994a). At the scale of one hectare, direct emissions of CO_2 are more than 20 times the CO_2-equivalent emissions of CH_4 and N_2O.

The 2006 IPCC guidelines (IPCC 2006) for both forests and grasslands allow for estimation of changes in biomass due to management practices. In Australian savannas, which could be defined under the guidelines as either forests or grasslands, changes in fire and grazing animal management can have marked effects on carbon stocks. Northern Australian savannas have been shown to be carbon sinks in common with many of the world's ecosystems under the rising levels of atmospheric CO_2: most likely because of increasing woodiness (Williams *et al.* 2004, 2005, Chapter 7; Beringer *et al.* 2007; Chapter 14). This is true even under frequent fires, but fire management can serve to increase their sink strength. Nevertheless, most ecosystems of the world are carbon sinks most of the time simply because major disturbance often occurs on short time spans, but recovery can take decades (Gifford and Howden 2001). In an environment where individual fires can kill up to 50% of the basal area of trees – and cyclones can destroy vast areas (Williams *et al.* 1999; Cook and Goyens 2008) – caution is required in interpreting the sink strength of savannas. Thus, small changes in the net flux of CO_2 will have great impact on the overall greenhouse gas budget.

WHAT ARE THE CONSTRAINTS TO FIRE ABATEMENT EMISSIONS-REDUCTIONS PROJECTS?

Fire frequency must be greater than desirable as defined by multiple stakeholders in the project

This condition is most likely to be met in extensive lands in which the spread of fire is unmanaged and largely unmitigated by breaks in fuel continuity, either through natural barriers or anthropogenic barriers such as cultivation, urban development or heavy grazing of livestock. In Arnhem Land, habitats such as sandstone heathlands, monsoon forests and *Callitris* woodlands are threatened by frequent intense fires. Thus, reducing fire frequency has multiple benefits. However, in landscapes where fire-dependent ecosystems could be jeopardised by reducing fire frequency, emissions reductions could still be achieved, but at a cost to biodiversity conservation. In north-east Queensland, for example, reducing fire frequency is having a deleterious effect on biodiversity by allowing the invasion of rainforest into wet sclerophyll habitats on the western fringe of the wet tropics (Trott 1995).

Reductions in fire frequency are both conceptually and practically feasible

Reductions in fire frequency require that both ignition sources and fire spread can be controlled. The prevalence of anthropogenic fires in the north of the continent (Russell-Smith *et al.* 2007), and the ability to impose burnt fire breaks that connect with natural barriers to fire spread, makes this region highly suitable for emissions-abatement projects.

Reductions in fire frequency lead to reduced emissions that are verifiable and compliant with the regulatory environment

Currently, estimates of emissions from savanna burning in Australia combine a bottom-up approach to fuel load estimation with a top-down approach through use of satellite imagery to estimate burnt area. In the future, direct assessments of emissions via satellite should be possible and calibrated against bottom-up estimates to provide robust, routine and verifiable measurements of greenhouse gas emissions from fires.

The financial benefit of emissions reductions is greater than the cost of mitigating fire frequency

Fire abatement projects to reduce emissions are unlikely to be viable where the costs of reducing fire frequency are high, and the reductions in emissions are low. In far northern Australia, where fire frequencies are often biennial, the benefits of fire-abatement schemes would be more quickly realised and verified than in central Australia where fires occur at decadal or greater frequencies.

Fire abatement strategies are perceived as low risk

Unlike tree-planting schemes, which aim to sequester carbon and suffer the risk of loss of that carbon through natural or anthropogenic disturbance, fire-abatement schemes aim to manage natural landscapes to minimise emissions. Major disturbances, such as Cyclone Monica, which uprooted or snapped thousands of square kilometres of trees (Cook and Goyens 2008), do not constitute a risk to an emission-abatement project, but rather an opportunity to reduce the rate of transfer of the felled carbon stock to the atmosphere below that which would occur in the absence of management intervention.

CONCLUSION

Better management of savanna fires, leading to reduced fire frequency, provides the opportunity for remote communities in northern Australia to contribute to the reduction in Australian greenhouse gas emissions in a way that also provides benefits for biodiversity conservation and livelihoods. Fire abatement can be undertaken in this region, supported by a sound cultural, practical and scientific background. Much progress has been made towards defining emissions estimates at the project scale, and further research will reduce uncertainties in estimates. Fire-abatement projects for minimising landscape scales emissions of greenhouse gases are qualitatively different from sequestration projects involving tree planting.

REFERENCES

Andersen AN, Cook GD, Corbett LK, Douglas MM, Eager RW, Russell-Smith J, Setterfield SA, Williams RJ and Woinarski JCZ (2005) Fire frequency and biodiversity conservation in Australian tropical savannas, implications from the Kapalga fire experiment. *Austral Ecology* **30**, 155–167.

Andersen AN, Cook GD and Williams RJ (Eds) (2003) *Fire in Tropical Savannas: The Kapalga Experiment.* Springer-Verlag, New York.

Andreae MO and Merlet P (2001) Emission of trace gases and aerosols from biomass burning. *Global Biogeochemical Cycles* **15**, 955–966.

Attiwill PM and Leeper GW (1987) *Forest Soils and Nutrient Cycles.* Melbourne University Press, Melbourne.

Barton L, Kiese R, Gatter D, Butterbach-Bahl K, Buck R, Hinz C and Murphy D (2007) Nitrous oxide emissions from a cropped soil in a semi-arid climate. *Global Change Biology* **14**, 177–192.

Beringer J, Hutley LB, Tapper NJ and Cernusak LA (2007) Savanna fires and their impact on net ecosystem productivity in North Australia. *Global Change Biology* **13**, 990–1004.

Bond WJ (2005) Large parts of the world are brown or black: A different view on the 'Green World' hypothesis. *Journal of Vegetation Science* **16**, 261–266.

Bowman DMJS and Cook GD (2001) Can stable carbon isotopes (^{13}C) in soil carbon be used to describe the dynamics of Eucalyptus savanna–rainforest boundaries in the Australian monsoon tropics? *Austral Ecology* **27**, 94–102.

Bowman DMJS (1992) A burnt out case? Reply to Lonsdale and Braithwaite (1991) *Austral Ecology* **17**, 103–106.

Brauman A, Kane MD, Labat M and Breznak JA (1992) Genesis of acetate and methane by gut bacteria of nutritionally diverse termites. *Science* **257**, 1384–1387.

Cheney P and Sullivan A (1997) *Grassfires: Fuel, Weather and Fire Behaviour.* CSIRO Publishing, Melbourne.

Cook GD (1994) The fate of nutrients during fires in a tropical savanna. *Australian Journal of Ecology* **19**, 359–365.

Cook GD (2001) Effects of frequent fires and grazing on stable nitrogen isotope ratios of vegetation in northern Australia. *Austral Ecology* **26**, 630–636.

Cook GD (2003) Fuel dynamics, nutrients and atmospheric chemistry. In *Fire in Tropical Savannas: The Kapalga Experiment.* (Eds AN Andersen, GD Cook and RJ Williams) pp. 47–58. Springer, New York.

Cook GD and Eager RW (2005) 'Savanna burning emissions across north Australia: a review of data'. Unpublished report to Australian Greenhouse Office, Canberra. CSIRO Sustainable Ecosystems, Darwin.

Cook GD and Goyens CMAC (2008) The impact of wind on trees in Australian tropical savannas: lessons from the impact of Cyclone Monica. *Austral Ecology* **33**, 462–470.

Cook GD and Heerdegen RG (2001) Spatial variation in the duration of the rainy season in monsoonal Australia. *International Journal of Climatology* **21**, 1831–1840.

Cook GD, Liedloff AC, Eager RW, Chen X, Williams RJ, O'Grady AP and Hutley LB (2005) The estimation of carbon budgets of frequently burnt tree stands in savannas of northern Australia, using allometric and isotopic discrimination. *Australian Journal of Botany* **53**, 621–630.

Cook GD, Williams RJ, Hutley LB, O'Grady AP and Liedloff AC (2002) Variation in vegetative water use in the savannas of the North Australian Tropical Transect. *Journal of Vegetation Science* **13**, 413–418.

Dyer R, Jacklyn P, Partridge I, Russell-Smith J and Williams RJ (Eds) (2001) *Savanna Burning: Understanding and Using Fire in Northern Australia.* Tropical Savannas Cooperative Research Centre, Darwin.

Eggleton P, Homathevi R, Jones DT, MacDonald JA, Jeeva D, Bignell DE, Davies RG and Maryati M (1999) Termite assemblages, forest disturbance and greenhouse gas fluxes in Sabah,

East Malaysia. *Philosophical Transactions of the Royal Society of London. Series B, Biological Sciences* **354**, 1791–1802.

Forster P. Ramaswamy V, Artaxo P, Berntsen T, Betts R, Fahey DW, Haywood J, Lean J, Lowe DC, Myhre G, Nganga J, Prinn R, Raga G, Schulz M and van Dorland R (2007) Changes in atmospheric constituents and in radiative forcing. In *Climate Change 2007: The Physical Science Basis.* (Eds S Solomon, D Qin, M Manning, Z Chen, M Marquis, KB Averyt, M Tignor and HL Miller) pp. 129–234. Contribution of Working Group I to the Fourth Assessment Report of the Intergovernmental Panel on Climate Change. Cambridge University Press, Cambridge, UK.

Galbally I, Meyer CP, Bentley S, Weeks I, Leuning R, Kelly K, Phillips F, Barker-Reid F, Gates W, Baigent W, Eckard W and Grace W (2005) A study of environmental and management drivers of non-CO_2 greenhouse gas emissions in Australian agro-ecosystems. *Environmental Sciences* **2**, 133–142.

Galbally IE, Kirstine WV, Meyer CP and Wang YP (2008) Soil–atmosphere trace gas exchange in semiarid and arid zones. *Journal of Environmental Quality* **37**, 599–607.

Gifford R and Howdon M (2001) Vegetation thickening in an ecological perspective: significance to national greenhouse gas inventories. *Environmental Science and Policy* **4**, 59–72.

Gorman JT, Williams RJ, Russell-Smith J, Hutley L and Cook GD (2007) A case for indigenous fire management. In *Investing in Indigenous Natural Resources Management.* (Eds MK Luckert, B Campbell, JT Gorman and ST Garnett) pp. 22–30. Charles Darwin University Press, Darwin.

Hurst DF, Griffith DWT and Cook GD (1994a) Trace gas emissions and biomass burning in tropical Australian savannas. *Journal of Geophysical Research* **99**, 16 441–16 456.

Hurst DF, Griffith DWT, Carras JN, Williams DJ and Fraser PJ (1994b) Measurements of trace gas emitted by Australian savanna fires during the 1990 dry season. *Journal of Atmospheric Chem*istry **18**, 33–56.

Hurst DF, Griffith DWT and Cook GD (1996). Trace gas emissions from biomass burning in Australia. In *Biomass Burning and Global Change.* (Ed. JS Levine) pp. 787–792. MIT Press, Cambridge, Massachusetts.

IPCC (1996) *Revised 1996 Intergovernmental Panel on Climate Change (IPCC) Guidelines for National Greenhouse Gas Inventories.* 3 vols. IPCC/OECD/IEA, Paris, France.

IPCC (2006) *2006 IPCC Guidelines for National Greenhouse Gas Inventories.* (Eds S Eggelston, L Buendia, K Miwa, T Ngara and K Tanabe). Institute for Global Environmental Strategies (IGES) for the Intergovernmental Panel on Climate Change, Kanagawa, Japan.

Khalil MAK, Rasmussen RA, French JRJ and Holt JA (1990) The influence of termites on atmospheric trace gases: CH_4, CO_2, $CHCl_3$, N_2O, CO, H_2 and light hydrocarbons. *Journal of Geophysical Research* **95(D4)**, 3619–3634.

Liedloff AC and Cook GD (2007) Modelling the effects of rainfall variability and fire on tree populations in an Australian tropical savanna using the Flames simulation model. *Ecological Modelling* **201**, 269–282.

Luhar AK, Mitchell RM, Meyer CP, Qin Y, Campbell S, Gras JL and Parry D (2008) Biomass burning emissions over northern Australia constrained by aerosol measurements: II–Model validation, and impacts on air quality and radiative forcing. *Atmospheric Environment* **42**, 1647–1664.

MacDonald JA, Jeeva D, Eggleton P, Davies R, Bignell DE, Fowler D, Lawton J and Maryati M (1999) The effect of termite biomass and anthropogenic disturbance on the CH_4 budgets of tropical forests in Cameroon and Borneo. *Global Change Biology* **5**, 869–879.

Meyer CP, Luhar AK and Mitchell RM (2008) Biomass burning emissions over northern Australia constrained by aerosol measurements: I–Modelling the distribution of hourly emissions. *Atmospheric Environment* **42**, 1629–1646.

National Greenhouse Gas Inventory Committee (1994) *Agriculture: Workbook for Non-Carbon Dioxide Gases from the Biosphere*. Department of the Environment, Sport and Territories, Canberra.

National Greenhouse Gas Inventory Committee (2007a) *Australian Methodology for the Estimation of Greenhouse Gas Emissions and Sinks 2006: Agriculture*. Australian Greenhouse Office, Canberra.

National Greenhouse Gas Inventory Committee (2007b) *National Greenhouse Gas Inventory 2005*. Australian Greenhouse Office, Canberra.

National Greenhouse Gas Inventory Committee (2007c) *State and Territory Greenhouse Gas Inventories 2005*. Australian Greenhouse Office, Canberra.

Olson JS (1963) Energy storage and the balance of producers and decomposers in ecological systems. *Ecology* **44**, 322–331.

Parnas H (1975) Model for decomposition of organic material by microorganisms. *Soil Biology and Biochemistry* **7**, 161–169.

Russell-Smith J and Edwards A (2006) Seasonality and fire severity in savanna landscapes of monsoonal northern Australia. *International Journal of Wildland Fire* **15**, 541–550.

Russell-Smith J, Yates CP, Edwards A, Allan GE, Cook GD, Cooke P, Craig R, Heath B and Smith R (2003a) Contemporary fire regimes of northern Australia: change since Aboriginal occupancy, challenges for sustainable management. *International Journal of Wildland Fire* **12**, 283–297.

Russell-Smith J, Whitehead PJ, Cook GD and Hoare JL (2003b). Response of *Eucalyptus*-dominated savanna to frequent fires: lessons from Munmarlary, 1973–1996. *Ecological Monographs* **73**, 349–375.

Russell-Smith J, Yates CP, Whitehead PJ, Smith R, Craig R, Allan G, Thackway R, Frakes I, Cridland, Meyer CP and Gill AM (2007) Bushfires 'down under': patterns and implications of Australian landscape burning. *International Journal of Wildland Fire* **16**, 361–377.

Russell-Smith J, Murphy BP, Meyer CP, Cook GD, Maier S, Edwards AC, Schatz J and Brocklehurst P (2009) Improving estimates of savanna burning emissions for greenhouse accounting in northern Australia. *International Journal of Wildland Fire* **18**, 1–18.

Trott P (1995) A rare habitat feels the squeeze. *Ecos* **84**, 27–31.

Walker J (1981) Fuel dynamics in Australian vegetation. In *Fire and the Australian Biota*. (Eds AM Gill, RH Groves and IR Noble) pp. 101–127. Australian Academy of Science, Canberra.

Ward DE, Susott RA, Kauffman JB, Rasmussen RA and Setzer AW (1992) Smoke and fire characteristics for cerrado and deforestation burns in Brazil: BASE B experiment. *Journal of Geophysical Research* **97**, 14 601–14 619.

Ward DE, Hao WM, Susott RA, Babbitt RE, Shea RW, Kauffman JB and Justice CO (1996) Effect of fuel composition on combustion efficiency and emission factors for African savanna ecosystems. *Journal of Geophysical Research* **101**, 23 569–23 576.

Whitehead PJ, Russell-Smith J and Woinarski JCZ (2005) Fire, landscape heterogeneity and wildlife management in Australia's tropical savannas: introduction and overview. *Wildlife Research* **32**, 369–375.

Williams RJ, Duff GA, Bowman DMJS and Cook GD (1996) Variation in the composition and structure of tropical savannas as a function of rainfall and soil texture along a large-scale climatic gradient in the Northern territory, Australia. *Journal of Biogeography* **23**, 747–756.

Williams RJ, Cook GD, Gill AM and Moore PHR (1999) Fire regime, fire intensity and tree survival in a tropical savanna in northern Australia. *Australian Journal of Ecology* **24**, 50–59.

Williams RJ, Hutley LB, Cook GD, Russell-Smith J, Edwards A and Chen XY (2004) Assessing the carbon sequestration potential of mesic savannas in the Northern Territory, Australia: approaches, uncertainties and potential impacts of fire. *Functional Plant Biology* **31**, 415–422.

Williams RJ, Carter J, Duff GA, Woinarski JCZ, Cook GD and Farrer SL (2005) Carbon accounting, land management, science and policy uncertainty in Australian savanna landscapes: introduction and overview. *Australian Journal of Botany* **53**, 583–588.

Yibarbuk D, Whitehead PJ, Russell-Smith J, Jackson D, Godjuwa C, Cooke P, Choquenot D and Bowman DMJS (2001) Fire ecology and Aboriginal land management in central Arnhem Land, northern Australia: a tradition of ecosystem management. *Journal of Biogeography* **28**, 325–344.

13

Improving estimates of savanna burning emissions for greenhouse accounting in northern Australia:

limitations, challenges and applications

Jeremy Russell-Smith, Brett P Murphy, Mick CP Meyer, Garry D Cook, Stefan W Maier, Andrew C Edwards, Jon Schatz and Peter Brocklehurst (adapted from the International Journal of Wildland Fire, *reproduced with permission)*

SUMMARY

Although biomass burning of savannas is recognised as a major global source of greenhouse gas emissions, quantification remains problematic, with resulting regional emissions estimates often differing markedly. Here we undertake a critical assessment of Australia's National Greenhouse Gas Inventory (NGGI) savanna burning emissions methodology. We describe the methodology developed for, and results and associated uncertainties derived from, a landscape-scale emissions abatement project in fire-prone western Arnhem Land, northern Australia. The methodology incorporates (a) detailed fire history and vegetation structure and fuels type mapping derived from satellite imagery, (b) field-based assessments of fuel load accumulation, burning efficiencies (patchiness, combustion efficiency, ash retention) and N:C composition, and (c) application of standard, regionally derived emission factors. Importantly, this refined methodology differs from the NGGI by incorporation of fire seasonality and severity components, and substantial improvements in baseline data. We consider how the application of a fire-management program aimed at shifting the seasonality of burning (from one currently dominated by extensive late dry season wildfires to one where strategic fire management is undertaken earlier in the year) can provide significant project-based emissions abatement. The approach has wider application to fire-prone savanna systems dominated by anthropogenic sources of ignition.

INTRODUCTION

It is widely recognised that biomass burning of tropical savanna biomes is a globally significant driver of CO_2 cycling and a source of the greenhouse gases CO_2, CH_4, N_2O, other chemically reactive atmospheric trace gases and aerosol particles (Scholes and Andreae 2000; Kondo *et al.* 2003). However, accurate quantification of such emissions is problematic, being reliant on reliable estimation of various parameters including the spatial and temporal distribution of burning, appropriate fuel-load estimates and gaseous emission factors for different fuel fractions and fire types. The application of different assumptions and approaches – particularly estimation of the extent of biomass burning and fuel-load estimates – has resulted in various global or regional estimates of savanna biomass burning and derived emissions (Seiler and Crutzen 1980; Crutzen and Andreae 1990; Hao and Liu 1994; Scholes *et al.* 1996a, b; Shirai *et al.* 2003; Streets *et al.* 2003; Liousse *et al.* 2004). Although there are considerable uncertainties in greenhouse gas (GHG) emissions estimates, the extent and contribution of biomass burning from Australian savannas is typically ranked second or third after Africa (e.g. Hao and Liu 1994; Carmona-Moreno *et al.* 2005).

Under the provisions of the Kyoto Protocol, participating Tier 1 (developed economy) countries are required, where pertinent, to account for emissions of GHGs (specifically CH_4 and N_2O) from 'prescribed burning of savannas' (UNFCCC 1998: Article 3, Annex A). Although Australia only recently ratified the Kyoto Protocol, Australia's National Greenhouse Gas Inventory (NGGI) reports on savanna burning emissions; typically, accountable GHG emissions annually contribute between 1 and 3% of Australia's NGGI (Meyer 2004; AGO 2007a). The Kyoto Protocol also establishes a framework for developing market-based instruments to address anthropogenic sources and sinks of GHG emissions (UNFCCC 1998: Article 6). Accountable GHG emissions from Australian savanna burning are predominantly associated with anthropogenic sources (Russell-Smith *et al.* 2007).

Contemporary fire regimes across the sparsely settled 1.9M km^2 northern Australian tropical savannas region have significant implications for biodiversity and soil conservation, GHG emissions, pastoral production and broader social issues (e.g. Dyer *et al.* 2001; Williams *et al.* 2002; Russell-Smith *et al.* 2003, 2007; Whitehead *et al.* 2003). Key issues include developing economically viable solutions for implementing ecologically sustainable landscape-scale fire management, particularly for vast, remote, biodiversity-rich regions where few employment opportunities exist, especially for indigenous (Aboriginal) communities (Whitehead *et al.* 2003). In 2005, the first major program aimed at substantially reducing GHG emissions from savanna burning in northern Australia commenced over 24 000 km^2 of the Western Arnhem Land Fire Abatement (WALFA) project area. Although that project operates under the authority of Australian and Northern Territory Government approvals, it is essentially a 'voluntary' offset arrangement (*sensu* Bayon *et al.* 2006), given current absence of formal, regulated market mechanisms in Australia.

In this chapter, we address ongoing development and regional refinement of the pyrogenic emissions assessment methodology established under guidelines developed by the Intergovernmental Panel on Climate Change – IPCC (1997). Those guidelines underpin current methodologies adopted both in Australia's NGGI accounting of savanna burning, as well as informing WALFA project emissions abatement assessments. We begin with description of the current NGGI methodology (AGO 2007b). We then describe details of remotely sensed- and field-based methodologies developed for, and key results arising from, ongoing biomass-burning emissions assessments for the WALFA region. In particular, we address the implications of fire seasonality on biomass burning and resultant emissions, and provide an assessment of uncertainties associated with respective parameters. Our broad purpose is to identify limitations and challenges in current Australian approaches, especially with reference to extending the current WALFA methodology to other prospective GHG-abatement projects in fire-prone northern Australia, and potentially to fire-prone savanna systems generally.

SAVANNA BURNING AND AUSTRALIA'S NGGI METHODOLOGY

Australia's NGGI accounts emissions for Sector 4E, Prescribed Burning of Savannas, using a country-specific methodology (AGO 2007b). Inventory accounting of GHG emissions uses an essentially bottom-up approach in which the emissions from each source, or subsector, is estimated as the product of an activity and an emission factor. The emission from major sectors and regions is the aggregate of all the sources and subsectors. Parties to the UNFCCC compile their national GHG inventories using reporting categories defined in the Revised 1996 Guidelines for Greenhouse Gas Inventories (IPCC 1997) following the procedures outlined in the IPCC Good Practice Guidelines (IPCC 2000, 2003). The methodologies used for each subsector may be the IPCC default methodologies or verified country-specific methodologies.

In this accepted IPCC methodology, for each region *j*, the emission of a trace species *i* (E_{ij}) is determined from savanna fire activity; that is, the mass of fuel pyrolised (FP_j) and the emission factor for each trace species (EF_{ij}).

$$E_{ij} = FP_j \times EF_{ij} \tag{1}$$

The mass of fuel pyrolised is the product of the area exposed to fire (A_j), the fuel load (FL) and the burning efficiency factor (BEF). BEF is defined as the mass of fuel that is exposed to fire that is pyrolised. It is calculated from the mass of fuel (M_{fuel}) prior to combustion and the mass of ash and unburned fuel residue remaining after combustion (M_{ash}),

$$BEF = 1 - \frac{M_{ash}}{M_{fuel}}$$

As a first approximation, fuels in the NGGI are divided into two size classes: fine fuels comprising leaf litter, grass and twigs less than 6 mm diameter; and heavy fuels. In relatively intense fires, fine fuels typically burn completely (i.e. with very high efficiencies), while heavy fuels tend to burn with lower efficiency. If *k* is the fuel class then:

$$FP_j = A_j \sum_k (FL_{jk} BEF_{jk})$$

The area exposed to fire is the area of the fire scar A_j^1 corrected for the patchiness of the fire (P_j), that is:

$$A_j = A_j^1 P_j$$

The emission factors (EFs) can be defined either relative to fuel mass pyrolised, or relative to the fuel elemental content. The NGGI methodology uses the latter. For carbon species, CH_4 and CO, and volatile organic compounds (VOC), EFs are expressed relative to fuel carbon, and the nitrogen species N_2O and NO_x are expressed relative to fuel nitrogen. Fuel carbon mass is determined from fuel mass by the fuel carbon content (CC_{jk}), while fuel nitrogen is derived from the fuel mass by the product of CC_{jk}, and the fuel nitrogen to carbon ratio NC_{jk}.

Combining these equations, for CH_4, CO and VOC, the emission E_{ij} is then:

$$E_{ij} = EF_{ij} A_j^1 P_j \sum_k (FL_{jk} BEF_{jk} CC_{jk}) M_i \tag{2}$$

And for N_2O and NO_x is:

$$E_{ij} = EF_{ij} A_j^1 P_j \sum_k (FL_{jk} BEF_{jk} CC_{jk} NC_{jk}) M_i \tag{3}$$

where M_i is the ratio of the molecular mass to the elemental mass for each species. By convention, the mass of NO_x is expressed as NO_2.

Under IPCC guidelines, the NGGI methodology is Tier 2, that is, disaggregated spatially, with country-specific emission factors and parameters. It contrasts with Tier 3 methodologies, which are spatially highly disaggregated with emissions estimated by using complex models. For example, for savanna fires in a Tier 3 methodology, fuels and GHG emissions would be calculated from a carbon cycle model that incorporates disturbance from fire and is driven by monthly varying inputs that include climate parameters, radiation, vegetation indices, fire areas and hotspots. Both approaches have their strengths. Tier 3 methodologies, when based on an appropriate model, can account for complex nonlinear processes and interactions, and can provide estimates of emissions at fine spatial and temporal resolution. Calibration and verification of the models is achieved by comparison with detailed measurements of processes and outputs at a small number of key locations in the landscape. However, the cost of Tier 3 sophistication can be a loss of transparency, and difficulty and expense in sourcing accurate spatial input data.

Tier 2 methodologies are largely empirical. Their reliability and strength depend on the accuracy of the classification into strata that capture most of the system variance between strata, and minimise the variance within strata. The process of determining the parameter values across all strata, if complete, fully characterises the behaviour and accuracy of the methodology. However, empirical models are valid only within the domain for which they are calibrated and therefore it is important that the classification produces strata that are stable to external drivers such as changing climate, land management or other sources of disturbance. If this is achieved, then the parameter values for each stratum (i.e. FL_j, BEF_j, etc) should also remain stable, with overall system behaviour occurring through changes in activity (fire area) or the population or size of the individual strata (vegetation class, seasonality, etc).

Tier 2 methodologies with the properties described above should be suitable for project-based GHG accounting for carbon offsets or carbon credits. In the WALFA project, the GHG emissions offset derives from increasing the residence time of carbon and nitrogen in the landscape by reducing the average fuel consumption per year, through altering the seasonality and extent of savanna fires by early season prescribed burning. To account for this, the regional stratification of the NGGI (j) is expanded to the vegetation classes used in the fire management planning; j = *eucalypt closed forest, eucalypt open forest, eucalypt woodland, sandstone woodland, sandstone heath, closed forest and riparian*. Fuel-load classes are expanded to k = *fine fuel, coarse, heavy, shrub*. Two additional strata are required to account for seasonality in burning efficiency; season (l = *early, late*) and fire severity (m = *low, moderate and high*). A third stratum describing fire regime (n = *years since last burned*) extends the accuracy of fuel loads through accounting for fuel accumulation with time.

Including these extensions in equations 2 and 3 produces, for CH_4, CO and VOC:

$$E_i = EF_i \sum_{jln} ((A_{jln}^1 P_i) \sum_{km} (FL_{jkn} BEF_{jkl} S_{lm} CC_{jk})) M_i \tag{4}$$

and for N_2O and NO_x:

$$E_i = EF_i \sum_{jln} ((A^1_{jln} P_i) \sum_{km} (FL_{jkn} BEF_{jkl} S_{lm} CC_{jk} NC_{jk})) M_i \quad (5)$$

and where S_{lm} is the fraction of fires of severity class m in season l

WALFA GREENHOUSE GAS EMISSIONS ASSESSMENT METHODOLOGY

This section describes evaluation of algorithm parameter values, based on extensive field- and GIS-based studies undertaken for the WALFA project over the past 5 years. Background information detailing the regional context of contemporary fire patterns in the fire-prone, biodiversity-rich 23 893 km^2 WALFA region, and putative impacts on fire-sensitive vegetation components, are described in Edwards and Russell-Smith (2009) and Chapter 9. That paper also provides details of applied vegetation structure and fire extent mapping methodologies, including validation of respective derived surfaces. Here, we focus on the broader methodology applied to, and substantive results arising from, ongoing assessment of GHG emissions derived from savanna burning in that same region.

The methodology and results presented here build substantially on preliminary work (Russell-Smith *et al.* 2004) that has informed development of Australia's current NGGI methodology for accounting of GHG emissions from savanna burning, especially with respect to regional parameter values for fuel loads and burning efficiencies (Meyer 2004; AGO 2007b). The current WALFA regional assessment methodology includes major refinements on the NGGI, namely: (1) finer-resolution fire extent mapping (with Landsat as opposed to AVHRR imagery), especially accounting for seasonal differences in fire characteristics and properties; (2) substantial further field assessment, enabling more robust parameterisation of habitat- and seasonal-specific relationships for different FL and BEF components; and (3) further detailed assessments of C and N concentrations of different fuel components. In particular, the consideration of seasonality issues in the WALFA methodology is an important addition in that it provides a further opportunity for accounting for GHG emissions abatement via strategic fire management in the early dry season (EDS; see below, and in Discussion), other than by simply reducing the annual mean extent of burning.

For simplicity of presentation, we describe the methods and results for individual major parameters as considered previously in Equations 1–5. In subsequent sections we: (1) summarise and consolidate these individual observations; (2) compare recent results with those as given in Meyer (2004), Russell-Smith *et al.* (2004); and as incorporated into the current NGGI methodology (AGO 2007b), (3) identify key gaps and required further refinements; and (4) present a detailed statistical assessment of uncertainties associated with derived parameters.

BURNT AREA

The land area exposed to fire is the combination of the fire scar area; that is, the area defined by the outer boundary of the scar and the degree of patchiness within the fire scar.

Scar area

Validated fire history mapping for the WALFA area is available from 1990 to the present: mostly from Landsat imagery sampled three times through the 7+ month dry (burning) season, as outlined in Chapter 9. For convenience, but based also on Indigenous fire management

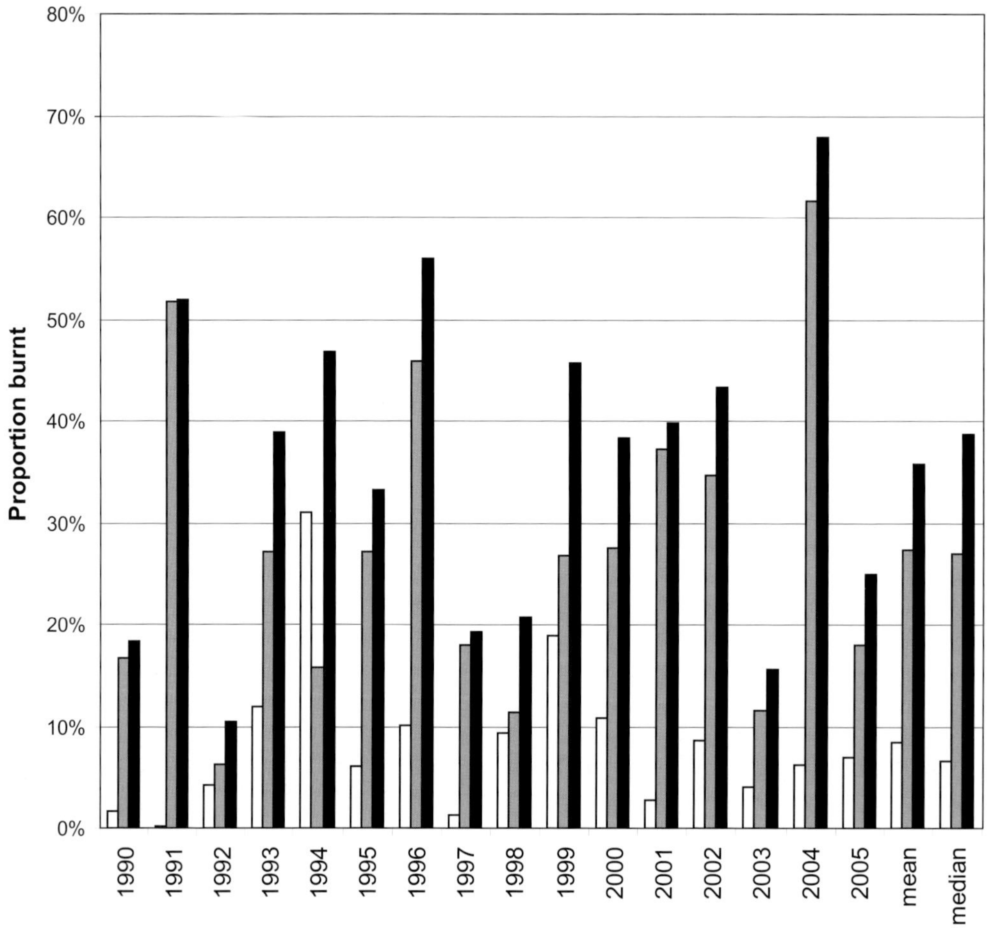

Figure 13.1 Fire extent in the 23 893 km^2 WALFA region, by season, 1990–2004 (from Edwards and Russell-Smith 2009), where fire extent for early dry season = clear bars, late dry season = grey bars, annual = black bars. Mean and median values also shown.

seasonality (Haynes 1985; Russell-Smith *et al.* 1997a), burnt area mapping for the period from April to the end of July is defined as occurring in the EDS, and from August to December as the late dry season (LDS). Based on fire severity observations made from 178 plots over 10 years in adjacent Kakadu and Nitmiluk National Parks, fires in the EDS period are observed typically to be of low severity/intensity, whereas fires in the LDS are observed to be typically much more severe/intense (Russell-Smith and Edwards 2006). In the absence of fire severity mapping, currently the subject of a PhD program being undertaken by AC Edwards, we apply these fire severity observations proportionately by season to assess post-fire fuel consumption (see below).

Over the period 1990–2005, a mean of 35.7% p.a. of the WALFA region was burnt (or, more properly, fire affected), comprising 8.3% in EDS and 27.3% in LDS periods, respectively (Figure 13.1).

Patchiness

Not all the area within a fire scar is burnt; however, optical satellite sensors typically lack sufficient resolution to detect this fine scale, that is, within-pixel patchiness. Patchiness is

Table 13.1: Fine-scale patchiness of fires on the Arnhem Land Plateau in early and late dry seasons. Patchiness is defined here as the average proportion of Landsat pixels identified as burnt that are actually burnt based on ground-truthing assessments.

Season	Source	Replicates	Patchiness (% burnt)	Weighted mean
Early	Price *et al.* (2003)	1300	74.9	70.9
	This study	669	63.3	
Late	Price *et al.* (2003)	556	84.6	88.9
	This study	280	97.6	

measured using ground surveys and, in regional savannas, is related both to seasonality and severity (Williams *et al.* 1998; Price *et al.* 2003; Russell-Smith and Edwards 2006). For present purposes, we consider separate estimates of fine-scale patchiness for the EDS and LDS, based on data assembled for the Arnhem Plateau as follows: (1) the transect study of Price *et al.* (2003); and (2) more recent post-fire assessments made both on assessment plots, and along transects adapting the methodology of Price *et al.* (2003). Price *et al.* (2003) demonstrated that EDS fires are significantly more patchy than LDS fires, and the more recent post-fire assessments support this (Table 13.1).

VEGETATION/FUEL TYPE

Validated vegetation/fuels mapping was undertaken for the WALFA region based on five vegetation structure classes (closed forest – 571 km^2, open forest – 6809 km^2, woodland – 6175 km^2, sandstone woodland – 5024 km^2, sandstone heath – 5314 km^2), applying the methodology as outlined in Edwards and Russell-Smith (2009: Chapter 9). This mapping surface is used as the basis for stratifying (1) derived fire history parameters (e.g. fire frequency and time-since-burnt) for application in biomass burning calculations, and (2) informing sampling efforts for assessment of fuel component accumulation with time-since-burnt. For present purposes, the relatively small closed forest class is combined with open forest.

FUEL-LOAD ACCUMULATION

Fuel-load accumulation of four fuel components (fine fuels, coarse fuels, heavy fuels and shrubs) has, to date, been assessed at 72 sites – in most cases each with three replicate plots (n = 219) – spread widely throughout the WALFA study area. Sites were stratified by vegetation/fuel type and time-since-burnt. At each plot a permanent 100 m transect of variable lateral dimension, depending on tree (≥5 cm diameter at breast height DBH) density, was established. Originally, it was intended to treat each site as an individual observation. However, it became readily apparent that, given the large plot size and scattered distribution of replicate plots, 'replicate' plots exhibited both substantial between- and within-plot heterogeneity with respect to recent fire history attributes. As such, for fuel accumulation purposes we have treated plots as discrete observations. In other instances, however (e.g. carbon and nitrogen sampling of fuel components), results are reported by site – comprising the bulking of samples from plot 'replicates'.

Fuels sampling at each plot was undertaken in the middle of the year (dry season) as follows:

- *Fine fuels* – sampling undertaken consistently at five equidistant 1 × 1 m quadrats. Fine fuel (<0.6 cm diameter) samples were collected separately for grass and litter components, weighed in the field using a digital balance, and subsequently corrected for oven dry weight based on one sub-sample for grass and litter, respectively. For presentation purposes, litter and grass components are aggregated and described here as 'fine fuels'.

- *Coarse fuels* – defined as sticks and other mid-sized, mostly woody debris (≥0.6 cm – <5 cm diameter), sampling undertaken consistently at ten equidistant 1 × 1 quadrats, corrected for dry weight.
- *Heavy fuels* – sampling undertaken consistently in a 5 × 100 m swath, recording the length, diameter and hollowness of all fuel sections >5 cm diameter. Assuming each piece was cylindrical in shape, we estimated total volume of heavy litter and total mass assuming a specific gravity of 0.995 tonnes m^{-3} (approximating that of eucalypt wood; Eamus *et al.* 2000).
- *Shrubs* – live shrubs were counted by species, and dead shrubs cumulatively, consistently in five 1 × 20 m sections, in each of four height classes: <0.5 m; 0.5–1.0 m; 1–2 m; and >2 m–<5 cm DBH. To estimate mean live shrub mass for each plot, we harvested and weighed the above-ground parts of at least three individuals of 107 live shrub samples (representing 46 common species) per respective height class. Species were allocated to one or more vegetation types in which they commonly occur and, for each vegetation type, mean shrub mass per height class was calculated. For 37 samples from 21 common species, we also measured the ratio of leaf mass to stem mass, and a mean value calculated for each height class. These values were used to estimate the mean mass of dead shrubs.

Simple linear regression was used to examine relationships between the accumulation of fuel-load components (i.e. grass, litter, coarse fuels, heavy fuels and shrubs) with time-since-burnt, stratified by vegetation type. A natural log transformation was applied both to fuel-load response (given strong positive skews for most observations), as well as to time-since-burnt (given apparent non-linearity). In those cases where significant relationships ($p<0.05$) were not observed between fuel-load components with time-since-burnt, we assumed that fuel load was best described by a simple mean, calculated using natural log transformed values. Based on a previous assessment of fine fuels (Russell-Smith *et al.* 1998), fuel loads for closed forest were assumed to be the same as open forest.

Fine fuel loads were positively related to time-since-burnt in all vegetation types, whereas coarse and heavy fuels, and shrub fuel loads were only related to time-since-burnt in sandstone woodland (Figure 13.2). Given general paucity of fine fuel samples greater than 4 years, we have also included in Figure 13.2 an additional 72 fine fuel samples from adjacent Kakadu National Park, which were sampled in a comparable manner to fine fuels data reported here (Russell-Smith *et al.* 1998). These additional samples are particularly pertinent to sandstone heath vegetation, but provide some increased temporal sampling coverage also to open forest and sandstone woodland vegetation types.

A significant seasonal feature not addressed in sampled fine fuel accumulation data concerns additional inputs of leaf and twig litter in the LDS. For example, 5 years of annual observations from an experimental fire treatment concerning lowland woodland in Kakadu National Park indicated that mean fine fuel loads were 1.7 t ha^{-1}, or 53%, greater in LDS treatments owing to substantial litter fall from trees during the dry season (Cook 2003; Williams *et al.* 2003). This has a significant bearing on seasonally available fuel loads and, as considered later, provides a major opportunity for substantially reducing GHG emissions through strategic EDS savanna burning. To account for seasonal variation in fine fuel loads, we have corrected LDS fine fuel loads by 1.7 t ha^{-1}, relative to the fuel loads predicted on the basis of our collected data (shown in Figure 13.2).

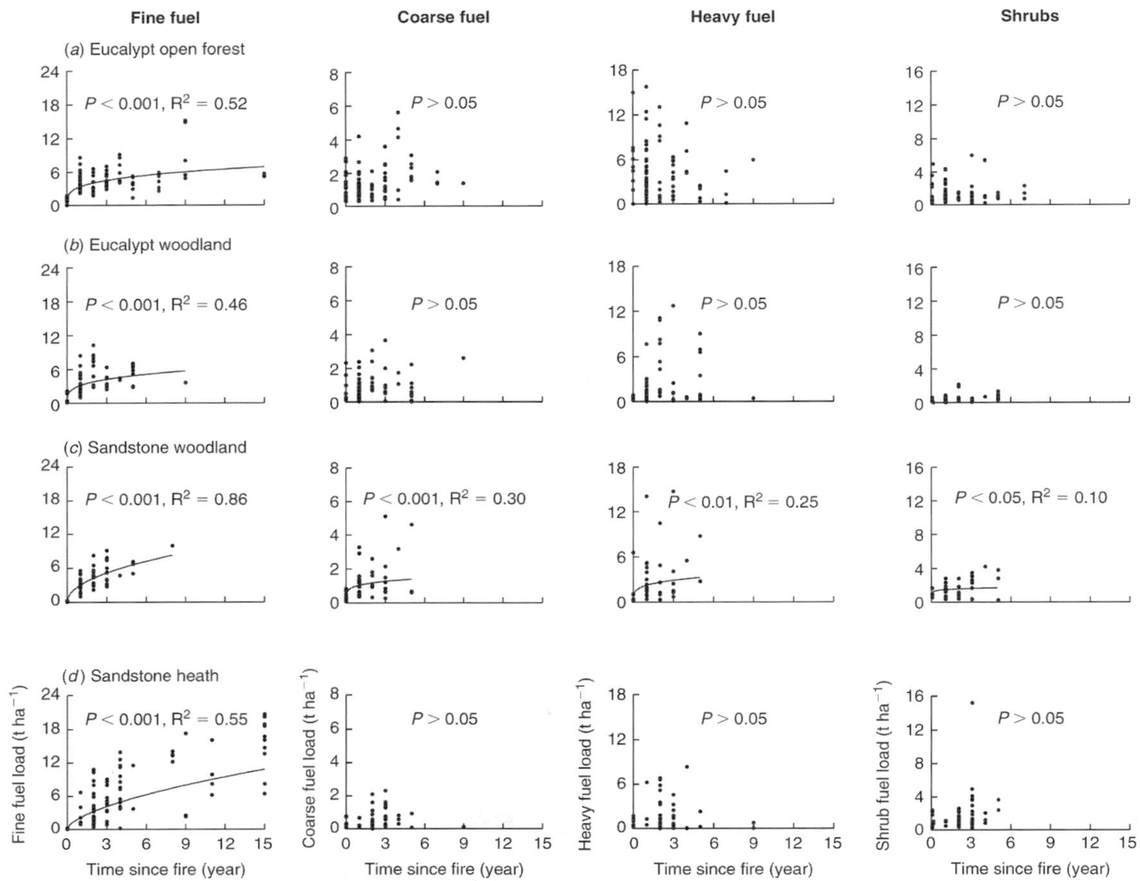

Figure 13.2 Relationship between accumulation of respective fuel components and time-since-burnt, in respective vegetation types: (a) eucalypt open forest; (b) eucalypt woodland; (c) sandstone woodland; and (d) sandstone heath. The regression lines, p-values and R^2 values refer to the results of linear regressions of fuel load (natural log transformed) versus time-since-burnt (natural log transformed).

BURNING EFFICIENCY FACTORS (BEFS)

We measured the proportion of each fuel component consumed (i.e. either volatilised or converted to ash) – referred to as fuel consumption – following exposure to fires of varying severity where fuel loads had been measured the previous day. Although some very fine ash may have been lost in this process, it was (a) typically impractical to measure ash residue immediately following fire treatments given more pressing fire management priorities and (b) anyway, it is likely that some fine ash may have redistributed from surrounding burnt areas. Regardless, these errors are likely to be minimal compared with the sampling process (dustpan and brush, vacuum cleaner). The severity of 42 imposed fire treatments was assessed on the basis of leaf-scorch height, following the methodology of Russell-Smith and Edwards (2006) where: low severity fires = leaf scorch height <2 m, and/or fires patchy with <20% of ground cover remaining unburnt; moderate severity = leaf scorch height >2 m, but upper canopy unscorched; high severity = upper canopy scorched.

Burning efficiency is defined as the mass of fuel exposed to fire that is pyrolised. For fine fuels, which can be sampled using small quadrats, BEF is usually determined directly from the mass of fuel (M_{fuel}) prior to combustion and the mass of ash and unburnt fuel residue remaining after combustion (M_{ash}); that is, BEF = $1 - M_{ash}/M_{fuel}$. However, generally this is impractical for coarse and heavy fuel because large area quadrats would be required to ensure homogeneous samples. In this study an alternative approach was applied.

In the case of fine fuels and shrubs, we compared pre-fire and post-fire measurements of fuel load to calculate fuel consumption. Post-fire fuel loads were assessed using an identical methodology as that used for pre-fire fuel loads. In the case of coarse and heavy fuels, mean estimates of consumption – based on the proportion of individual sticks and logs consumed by the fire (as evidenced from ash trails) – were made along 100 × 1 m transects. This approach does not allow for residual ash and therefore, unless subsequently corrected, fuel-load consumption tends to overestimate BEF.

We compared fuel consumption by fires of varying severity using analysis of variance. Given that fuel consumption was strictly bounded by zero and one, we applied a logit transformation prior to analysis (Crawley 2002). For each fuel component, fire severity had a significant effect on fuel consumption: being consistently greater in fires of high versus low severity (Figure 13.3). Using the respective EDS and LDS frequencies of low, moderate and high severity fires as reported in Russell-Smith and Edwards (2006), we derived estimates of fuel consumption for each fuel component for each season (Table 13.2).

Table 13.2: Consumption rates for each fuel component for low, moderate and high severity fires. The frequency of low, moderate and high severity fires in the early and late dry seasons (from Russell-Smith and Edwards 2006) is used to derive estimated fuel consumption rates for each season. Refer text for details.

	Frequency (%)		Fuel consumption (%)			
Fire severity	Early dry season	Late dry season	Fine fuel	Coarse fuels	Heavy fuels	Shrubs
Low	72	19	69	5	12	26
Moderate	20	47	85	32	19	24
High	8	34	97	58	58	68
Early dry season			74	15	17	29
Late dry season			86	36	31	39

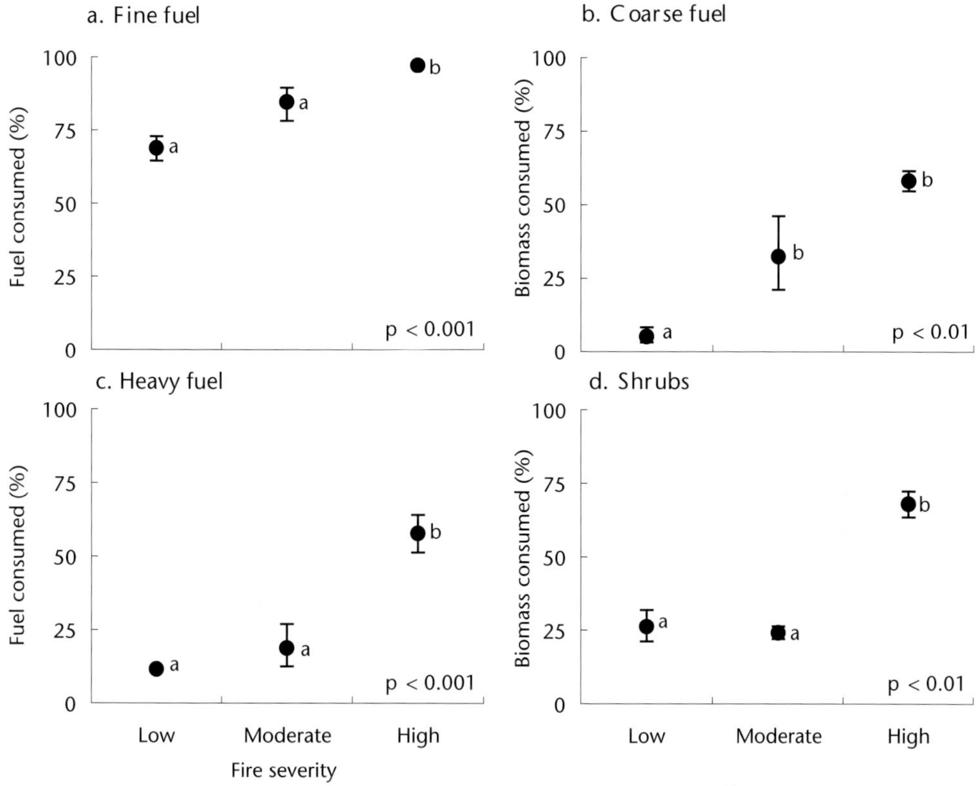

Figure 13.3 The mean proportion of each fuel component consumed (fuel consumption) in relation to fire severity. Each p-value refers to the results of an analysis of variance. The means marked by the same letter are not significantly different ($p<0.05$, Tukey HSD). The error bars represent ±1 standard error.

Of the fuel that was consumed in each fire, we assumed that some was pyrolysed (volatilised) and the remainder was converted to ash. Some of the ash remained *in situ*, and some was entrained within the smoke plume. To estimate the proportion of fuel that was pyrolysed, rather than just consumed, we required an estimate of the proportion of consumed fuel that was converted to ash.

At seven sites where we estimated the proportion of fuel consumed, we also measured the post-fire abundance of ash. Dry weight estimates of ash were undertaken in five 20 × 20 cm quadrats per plot, sampled within 1 day of fire treatments. Using the mass balance approach of Fearnside *et al.* (1993), we estimated the proportion of consumed fuel (including all fuel components) that was converted to ash and remained *in situ*. To use this approach, it was necessary to estimate the carbon content of ash (35.1 ± 4.9% [S.E.M.], $n = 7$) and of each fuel component (see carbon content below). The mean figure for *in situ* ash conversion was 11.6 ± 2.3% (S.E.M.) of consumed fuel. Given that we could not determine which fuel components contributed to the ash, we assumed that the *in situ* ash conversion rate was the same for all fuel components.

Fuel consumption (Table 13.2) was converted to BEF by reducing each by the average fraction of fuel remaining after combustion as ash (11.6%). In the absence of more comprehensive data, we applied this rate equally to fires of low, moderate and high severity.

Table 13.3: Carbon content and nitrogen to carbon (N:C) ratios measured for (a) vegetation components. These values were used to estimate values for the (b) fuel components. Standard errors are indicated for the means for each vegetation component.

Component	Carbon content (%)	N:C ratio	*n*
(a) Vegetation (measured values)			
Grass	43.9 ± 0.3	0.0087 ± 0.0002	68
Non-grassy leaves	50.6 ± 0.3	0.0122 ± 0.0003	66
Twigs	50.1 ± 0.2	0.0081 ± 0.0004	38
(b) Fuel (estimated values)			
Grass	43.9	0.0087	
Fine fuels[1]	49.4	0.0101	
Coarse and heavy fuels	50.1	0.0081	
Shrubs[2]	50.2	0.0093	

[1] Calculated assuming an average litter composition of 46% non-grassy leaves, 39% twigs and 15% grass.
[2] Calculated assuming an average litter composition of 69% twigs and 31% non-grassy leaves.

CARBON AND NITROGEN CONTENT OF FUEL COMPONENTS

We measured carbon content and N:C ratio of standing grasses from 68 sites and non-grassy leaf litter from 66 sites. Measurements were also made for twigs collected from 38 sites. Using these measured values, we derived estimates of carbon content and N:C ratio for each fuel component (Table 13.3). To do this, we made the following assumptions: (1) fine litter had a composition equivalent to 46% non-grassy leaves, 39% twigs and 15% grass (based on mean litter composition at 50 sites); (2) coarse and heavy fuels were equivalent to twigs; and (3) shrubs had a composition equivalent to 69% twigs and 31% non-grassy leaves (based on mean leaf:stem ratio for 37 individual shrub samples).

DIFFERENCES BETWEEN 2004 AND CURRENT PARAMETERS

Compared with the original 2004 methodology developed for the WALFA project region (Meyer 2004; Russell-Smith *et al.* 2004), revised parameters that have a considerable affect on estimated CO_2-equivalent emissions substantially include various fuel-load estimates and, to a much lesser extent, pyrolysis efficiencies (Table 13.4). Fine fuel load estimates have decreased slightly in the EDS, but increased markedly for the LDS given additional leaf litter inputs. Other fuel-load components have also generally declined in both seasons, with the exception of a marked decrease in heavy fuels; for a typical year (average of 2000–2004), the 2004 methodology estimate of heavy fuel load is five times the current estimate. As noted in Table 13.4, this is due to the use of a different calculation method, especially the taking into account of the hollowness of most woody debris due to termite activity. We note, nevertheless, that woody debris is likely to be highly spatially variable due to site productivity (i.e. related to stand basal area), past severity of fire regimes and especially the influence of cyclonic and localised very strong wind events. In combination, these factors afford significant challenges for modelling.

Current estimates of pyrolysis efficiency have both increased and decreased, depending on the fuel component (Table 13.4). Importantly, however, LDS pyrolysis efficiencies for fine fuels – the source of the majority of CO_2-equivalent emissions – have decreased substantially (from an assumed 100% in 2004 estimates, to 86% as measured; Table 13.1). The differences between the 2004 methodology and current parameters have a considerable effect on emissions estimates. For

Table 13.4: Summary of parameters used in emission estimates, based on data available from Russell-Smith et al. (2004), and current methodology as presented here. Average fuel loads (2000–2004) and estimated emissions for the study area are also shown.

Parameter/variable	2004	Current	Comments/refinement required
(a) FLA			
Average fuel load (t ha^{-1})			• Further fuels sampling at longer-unburnt sites (>5 yrs), especially for open forest, woodland, sandstone woodland • Further fuels sampling at sites both in early and late dry seasons, especially to refine leaf litter inputs • Given lack of relationship for both coarse and heavy fuel components with time-since-burnt, explore relationships with vegetation density parameters (e.g. basal area, foliage projective cover) • Large difference between heavy fuel estimates in 2004 and current values due to different calculation method
Early dry season			
– fine fuels	4.60	4.16	
– coarse fuels	1.12	0.69	
– heavy fuels	7.72	1.35	
– shrubs	0.83	0.62	
Late dry season			
– fine fuels	4.40	5.95	
– coarse fuels	0.99	0.71	
– heavy fuels	8.29	1.45	
– shrubs	0.75	0.84	
Fire patchiness (proportion of area burnt)			
Early dry season	71%	71%	• Further sampling of fire patchiness, particularly in LDS • Modelling of patchiness with rockiness and other terrain features, and fire severity
Late dry season	84%	89%	
(b) BEF			
Pyrolysis efficiency (proportion of fuel pyrolysed)			• Further post-fire combustion sampling effort with respect to fire severity • Ultimately, application of fire severity as derived from remote sensing to entire study region, rather than use of seasonality as surrogate
Early dry season			
– fine fuels	70%	74%	
– coarse litter	3%	15%	
– heavy litter	5%	17%	
– shrubs	45%	29%	
Late dry season			
– fine fuels	100%	86%	
– coarse fuels	25%	36%	
– heavy fuels	25%	31%	
– shrubs	75%	39%	
Residual ash (proportion of consumed biomass)			• Further assess post-fire ash deposition and ash entrainment, especially with respect to fire severity
entrained ash	6%	–	
ash remaining *in situ*	–	11.6%	
(c) Carbon and nitrogen content			
Fuel carbon content	46% (for all fuels)		• Adequate sampling undertaken
– fine fuels		49%	
– coarse and heavy fuels		50%	
– shrubs		50%	
Fuel N:C ratio	0.0106 (for all fuels)		• Adequate sampling undertaken
– fine fuels		0.0101	
– coarse and heavy fuels		0.0081	
– shrubs		0.0093	
(d) Estimated emissions (t)			
CH_4	11 126	9772	• Assess relationships between emission factors, fire severity and fuel moisture (seasonality) • Refine emissions assessment in light of undertaking further work on above parameters
N_2O	303	216	
CO_2-equivalent	327 354	272 040	

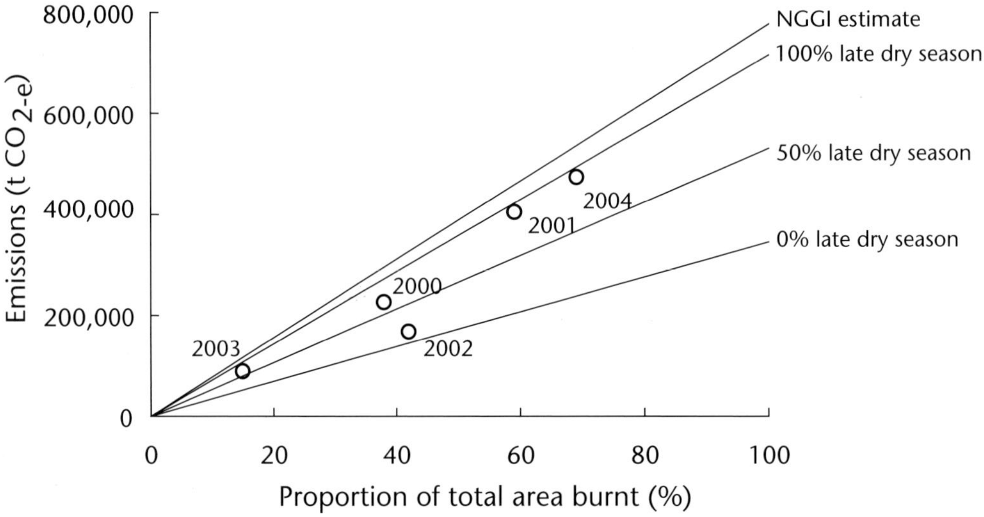

Figure 13.4 Estimated CO_2-equivalent emissions from the WALFA study area in each year in the period 2000–2004. The lines indicate the relationship between emissions and the proportion of the total area burnt, and the proportion of burning taking place during the late dry season. Emissions estimated using the current NGGI methodology (AGO 2007b) are also shown.

a typical year (average of 2000–2004), estimates of CO_2-equivalent emissions using the current parameters are 83% of those using the earlier 2004 parameters (Table 13.4; Figure 13.4).

UNCERTAINTY IN EMISSIONS

Good practice guidance for GHG inventory preparation (IPCC 1997, 2006) requires that emission estimates include an assessment of their uncertainties. Our determination of parameter values permits uncertainty analysis following the approach used in the Australian NGGI. This is a Monte Carlo analysis in which activity and input parameter values are replaced by probability density functions (PDFs) and the probability distributions of the outputs following the Monte Carlo simulation. The 95% uncertainty ranges are defined by the 2.5 and 97.5 percentiles, which avoids any assumptions about the symmetry of the distributions.

The uncertainty model was based on algorithms described in Equations 4 and 5. The PDFs are listed in Table 13.5. Mean fire severity, burning efficiency, carbon content, N:C ratio, and CH_4 and N_2O emission factors were fitted by normal distributions. Seasonal fire extent areas were fitted by lognormal distributions. Fuel loads were calculated using the regressions shown in Figure 13.2, with the parameter values replaced by the appropriate normal distributions. Patchiness was not adequately described by either normal or lognormal distributions. LDS patchiness varied from 0.9 to 1.0, whereas the frequency distribution of EDS patchiness was distributed between 0 and 1: increasing from a minimum at 0.3 to a maximum at 1 (Figure 13.5). In the uncertainty analysis, both distributions were approximated by triangular PDFs. For this study, all parameters were assumed to be independent; however, it is likely that correlations do exist between both parameters and strata. In future studies these should be investigated, quantified and included in the analysis.

The total CO_2-equivalent emissions and their uncertainties are presented in Table 13.6. Most of the emission (~65%) comes from LDS combustion of fine fuels, predominantly in eucalypt and sandstone woodland, with a smaller contribution (11%) from EDS fine fuels. The

Table 13.5: Probability density functions (PDFs) used in the Uncertainty Analysis.

Parameter		Unit	PDF	min	max	Most likely	CV[1] or SE[2]	
Area		ha	Lognormal	–	–	annual activity	0.1	CV
Patchiness								
	early		Triangular	0	1	1		
	late		Triangular	0.8	1	1		
Fuel load								
ln(FL)= a0 + a1*t								
	Fine-EOF	a0	normal			1.22	0.06	SE
	Fine-EW	a0	normal			1.15	0.07	SE
	Fine-SH	a0	normal			0.77	0.10	SE
	Fine-SW	a0	normal			1.11	0.06	SE
	Shrub-SW	a0	normal			0.36	0.11	SE
	Fine-EOF	a1	normal			0.26	0.03	SE
	Fine-EW	a1	normal			0.27	0.04	SE
	Fine-SH	a1	normal			0.60	0.06	SE
	Fine-SW	a1	normal			0.48	0.03	SE
	Shrub-SW	a1	normal			0.09	0.04	SE
Mass	Coarse-EOF	t ha^{-1}	lognormal			1.43	0.07	CV
	Coarse-EW	t/ha	lognormal			0.90	0.11	CV
	Coarse-SH	t/ha	lognormal			0.58	0.13	CV
	Coarse-SW	t/ha	lognormal			1.23	0.14	CV
	Heavy-EOF	t/ha	lognormal			4.81	0.11	CV
	Heavy-EW	t/ha	lognormal			2.18	0.12	CV
	Heavy-SH	t/ha	lognormal			1.68	0.14	CV
	Heavy-SW	t/ha	lognormal			3.42	0.15	CV
	Shrub-EOF	t/ha	lognormal			1.46	0.76	CV
	Shrub-EW	t/ha	lognormal			0.49	2.28	CV
	Shrub-SH	t/ha	lognormal			1.77	0.64	CV
BEF								
Low severity	fine		Logistic			0.69	0.06	CV
	coarse		Logistic			0.06	0.49	CV
	heavy		Logistic			0.12	0.07	CV
	shrub		Logistic			0.27	0.20	CV
Moderate severity	fine		Logistic			0.84	0.07	CV
	coarse		Logistic			0.34	0.36	CV
	heavy		Logistic			0.20	0.38	CV
	shrub		Logistic			0.24	0.09	CV
High severity	fine		Logistic			0.97	0.003	CV
	coarse		Logistic			0.58	0.06	CV
	heavy		Logistic			0.58	0.11	CV
	shrub		Logistic			0.68	0.06	CV
Fire severity distribution								
Early season	low		Truncated normal	0	1	0.722	0.20	CV

Table 13.5 (Continued)

Parameter		Unit	PDF	min	max	Most likely	CV[1] or SE[2]	
	moderate		Truncated normal	0	1	0.198	0.20	CV
	high		Truncated normal	0	1	0.079	0.20	CV
Late season	low		Truncated normal	0	1	0.187	0.20	CV
	moderate		Truncated normal	0	1	0.473	0.20	CV
	high		Truncated normal	0	1	0.341	0.20	CV
Carbon content								
	Coarse		Normal	0	1	0.49	0.02	CV
	Coarse		Normal	0	1	0.50	0.02	CV
	heavy		Normal	0	1	0.50	0.02	CV
	Shrub		Normal	0	1	0.50	0.02	CV
N:C ratio								
	Coarse		Normal	0		0.0096	0.1	CV
	Coarse		Normal	0		0.0081	0.1	CV
	heavy		Normal	0		0.0081	0.1	CV
	Shrub		Normal	0		0.0093	0.1	CV
EF	CH_4		Normal	0		0.0035	0.02	CV
	N_2O		Normal	0		0.0076	0.02	CV

[1] Coefficient of variation
[2] Standard error of the mean

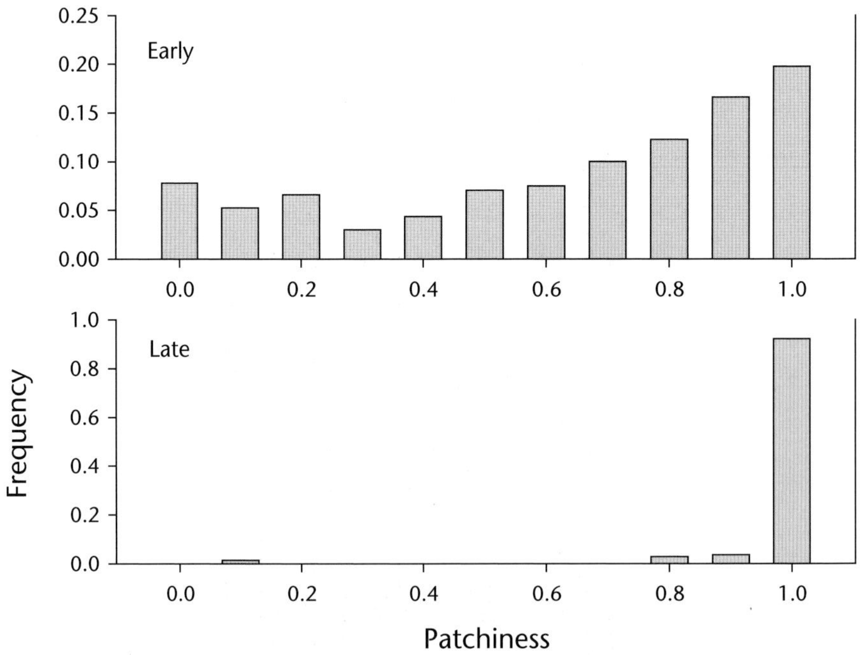

Figure 13.5 Frequency distribution of fire patchiness during early and late dry seasons.

Table 13.6: The 95% confidence ranges for total CO_2-equivalent emissions for 2000–2004 for respective fuel components. Note that the range in the total uncertainty estimate is not symmetrical around the mean emissions estimate of 272 Gg CO_2-equivalent (as presented in Table 13.4), given various transformations for determining probability density functions as described in Table 13.5. CV = coefficient of variation.

	Emission (Gg CO_2-equivalent)									
Vegetation class	Early season				Late season				Total	
	Fine	Coarse	Heavy	Shrub	Fine	Coarse	Heavy	Shrub		CV
Eucalypt open forest	0.6–3.9	0.0–0.6	0.0–0.1	0.0–1.1	10–19	0–4	0–15	0–7	**16–38**	0.25
Eucalypt woodland	5.8–40.8	0.1–5.1	0.0–0.0	0.3–8.6	65–138	2–22	3–60	1–16	**108–228**	0.20
Sandstone heath	0.9–6.5	0.0–0.5	0.0–0.0	0.0–0.9	27–57	0–6	0–19	0–26	**38–88**	0.23
Sandstone woodland	0.7–5.1	0.0–0.4	0.0–0.0	0.1–0.6	39–72	2–6	4–11	4–8	**56–96**	0.13
Uncertainty	**8.0–56.2**	**0.2–5.8**	**0.0–0.1**	**0.5–9.8**	**142–284**	**6–31**	**11–85**	**8–45**	**225–424**	
CV	0.4	0.8	1.2	0.7	0.2	0.5	0.7	0.5	0.16	

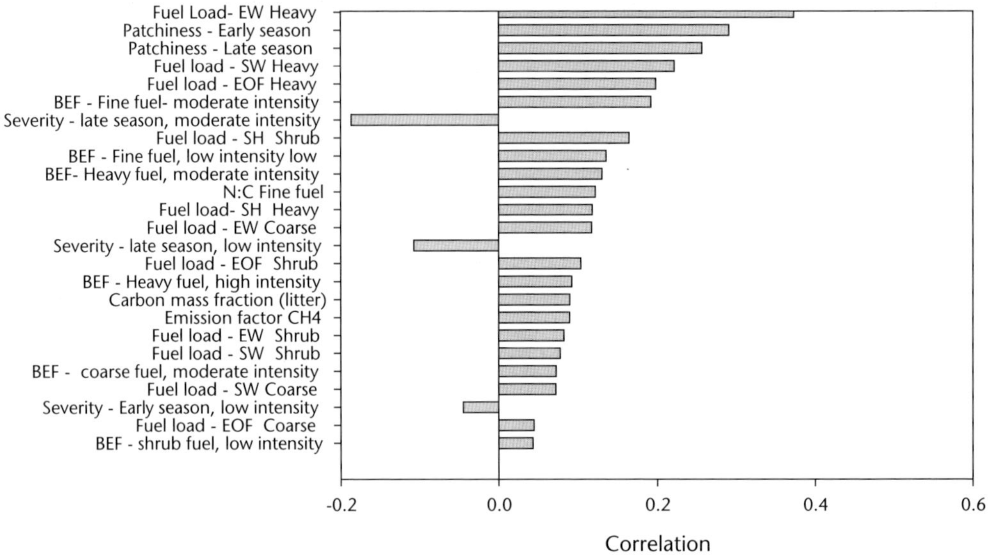

Figure 13.6 The sensitivity of total greenhouse gas emissions (CO_2-equivalent) to input parameters.

upper and lower 95% uncertainty limits for emissions from fine fuels are a factor of approximately 0.3 of the mean emission. However, this increases to a factor of ±2.5 in the coarse and heavy fuels in the early fire season. Across vegetation classes, uncertainty factors range from ±0.5 to ±0.7. Overall, these are substantially lower than the factor of ±2 for the savanna woodlands and arid grasslands reported in the Australian NGGI for 2006 (AGO 2007a).

The sensitivity of total CO_2-equivalent emissions to individual parameters is determined from a multiple regression between inputs (parameters and activities) and the total emissions. The analysis, presented in Figure 13.6, describes the relative change in total emissions caused by variation in the input parameters. This analysis shows that emissions are highly dependent on the heavy fuel loads, the early and late-season fire patchiness, and the burning efficiencies – particularly of the fine and heavy fuels. Notably, uncertainty in fire area has little leverage on the total emissions, because it is assumed that fire-scar area estimates between strata are independent, and consequently their variability tends to average out.

The total emission also appears relatively insensitive to uncertainty in emission factor (in this case for CH_4). However, this may be a consequence of a simplification in the inventory model, due to a lack of data, which assigns an annual mean emission factor to all strata. Emission factors probably vary with season, fuel class and fire severity and therefore, when stratified emission data become available, it is likely that greater sensitivity to emission factor will become apparent.

DISCUSSION

Emissions estimates and uncertainties

The current study indicates that in a typical year (average of 2000–2004) GHG emissions attributable to savanna burning from the WALFA project area were 272 Gg (Table 13.4) CO_2-equivalent, with a 95% confidence range of around ±100 Gg CO_2-equivalent (Table 13.6). The confidence limit is based solely on the uncertainties of the parameter estimates, and it is assumed the inventory structure (i.e. the emissions model) is sound. This is a characteristic of all bottom-up inventories that are not constrained by independent measurements of emissions or emission surrogates such as concentrations of trace species (e.g. CO or particulate matter). The reliability of the emissions estimates relies on the accuracy of the inventory model and the determination of the model parameters and the input (i.e. activity) data. Reliability is emerging as a significant issue among the growing number of bottom-up global and regional spatial inventories of biomass combustion emissions. For example, for the Australian region, the emissions estimates of Horowitz *et al.* (2003), Shirai *et al.* (2003), Hoelzemann *et al.* (2004), Ito and Penner (2004) and van der Werf *et al.* (2006) differ by up to a factor of six, particularly due to differences in estimation of fire area and fuel load (Kasischke and Penner 2004). Each of these estimates is considered by their authors to be reliable, so the question is where do the differences arise? From our uncertainty assessment, the WALFA domain emissions should be accurate at the 95% confidence level to within a factor of 30–35% of the mean. Uncertainties associated with WALFA parameter estimates – and their broader applicability to other (especially Australian) savanna regions – are considered below.

Fire extent mapping

We consider this generally highly reliable for the WALFA region, given the application of multi-temporal annual mapping based on fine-scale Landsat imagery (from 1990–present) and associated independent validation (Edwards and Russell-Smith 2009). However, apart from a number of regional fire histories derived from Landsat imagery – primarily for savanna conservation properties (e.g. Kakadu National Park) – consistent (semi-automated) fire extent mapping at the Australian savanna-wide scale is available only from coarse-resolution NOAA-AVHRR imagery since 1997 (Craig *et al.* 2002), and more recently from MODIS. An assessment of the accuracy of fire extent mapping in northern Australian savannas based on comparison with mapping from Landsat and associated field validation found that, apart from one Landsat scene, fire mapping from AVHRR consistently under-estimated the 'true' extent of burning by as much as 10–20% (Yates and Russell-Smith 2002). Australia's NGGI currently uses the AVHRR-derived fire-mapping product as the basis for calculations.

For regional GHG emission abatement project purposes, however, reliable estimation of seasonal fire extent needs to be based preferably on multi-temporal annual fire mapping using fine-resolution imagery, or at least calibration of fire mapping from coarser-scale imagery against fine-scale imagery spatio-temporal subsets. For the latter, it is a requirement that the error of the coarse scale mapping is predictable. This is difficult to achieve with the manual or semi-automated mapping currently used operationally. It requires the use of fully automated mapping, which is independent of a human operator. Luckily these fully automated mapping systems are becoming available (e.g. Roy *et al.* 2005). Some of these algorithms also provide confidence estimates that would be important inputs for accuracy assessment. Beyond this, there is the potential to develop algorithms to estimate the fraction of a pixel affected by fire. So far, development of algorithms for mapping fire extent has focused on binary ('hard') classification of each pixel using usually only the direction of change of reflectance. By using quantitative values of change of reflectance, there is the possibility to distinguish sub-classes (at least classes like 'partly burnt' and 'completely burnt').

Vegetation structure/fuel type mapping

We also consider the fuel type cover to be reliable for the WALFA region based on robust digital canopy cover classification (1 ha scale) of Landsat imagery and other spatial data sources, and associated independent validation (Edwards and Russell-Smith 2009). By contrast, consistent vegetation mapping for Australian savannas is available at 1:2 000 000 (Fox *et al.* 2001), whereas the NGGI uses an 'agro-ecological' regionalisation comprising five zones (Meyer 2004; AGO 2007b). For regional GHG emission abatement project purposes, however, reliable vegetation/fuels mapping needs to be based on validated fine-scale vegetation structural mapping.

Fuel loads

Fine fuel load accumulation for different vegetation structural types in the WALFA study area was found to occur at broadly similar rates and in similar quantities to those reported in other comparable regional studies (Russell-Smith *et al.* 1998; Cook 2003). Given the frequency of fire in the WALFA area, mean fine fuels (4.16 t.ha^{-1}; Table 13.4) were well below equilibrium (long unburnt) levels observed in above studies. Although not addressed specifically by field sampling in the present study, it is widely observed that accumulation of fine fuels in savannas is substantially greater in the LDS, associated with increased litter (leaf and twig) drop as the dry season progresses (e.g. Hopkins 1966; Williams *et al.* 1998; Cook 2003; Hennenberg *et al.* 2006). Cook (2003) has shown that the amount of litter drop is proportional to stand basal area in regional savannas. Kauffman *et al.* (1994) also note the diminishing contribution of grassy fuels to fine fuel loads along a gradient of increasing woody biomass in Brazil. We contend that estimation of fine fuel loads in different regional vegetation structure types, and under different environmental settings (e.g. rainfall and soil conditions), can be described satisfactorily by simple relationships. As noted previously, fine fuels provide the major contribution to emissions from savanna burning in the WALFA region, and presumably for Australian savannas generally.

Conversely, statistically significant relationships describing accumulation of other fuel components (coarse, heavy and shrub fuels) were not apparent in our assembled data, with the exception of sandstone woodland, and then at much reduced explanatory power relative to fine fuels (Figure 13.2). The overall contribution of coarse and heavy fuels combined (~2 t.ha^{-1}; Table 13.4) is higher than that reported generally in African and South American studies (Kauffman *et al.* 1994; Shea *et al.* 1996; Barbosa and Fearnside 2005), but substantially less than the 4.4 t.ha^{-1} reported by Rose (2006) from an assessment of heavy fuel accumulation in Australian eucalypt savanna open forest. This latter study also found no clear relationship between heavy fuel accumulation with time-since-burnt. Rather, other observations would suggest that both dead standing stems and heavy ground fuels are more likely to accumulate in response to severe fires/fire regimes (Williams *et al.* 1998; Cook *et al.* 2005), as well as (probably highly significant) effects of localised and sub-regional strong windy events (e.g. pre-monsoon storms and cyclones). GHG emissions from burning of tropical forests are, conversely, far more dependent on combustion of heavy fuels (Pereira *et al.* 1999).

The variable lack of response exhibited by shrub fuel accumulation with time is also likely to reflect very substantial inter-site variability in historical processes (Russell-Smith *et al.* 1998). In the absence of simple predictive relationships describing accumulation in these fuel components, we have assumed that, at large spatial scales, arithmetic means are the most useful descriptors.

Burning efficiency factors (BEFs)

In Australian savanna studies, fire patchiness is recognised to reflect aspects of seasonality, fire severity and rockiness (Price *et al.* 2003; Williams *et al.* 2003) and, with further data collection (see Table 13.4), is potentially readily modelled. More limited and uncertain data are available, however, for post-fire assessments of fuel consumption and aspects associated with ash entrainment and retention.

In the present study, consumption of different fuel components was found to be dependent on fire severity. Based on the observed distribution of mild, moderate and severe fires throughout the year, we estimated mean consumption of fine fuels to be 74% in the EDS and 86% in the LDS, and 29% and 39% for heavy fuels in the respective seasons (Table 13.2). Studies elsewhere report consumption of savanna grassland/woodland fine fuels ranging generally between around 90 and 100% (Kauffman *et al.* 1994; Shea *et al.* 1996; Ito and Penner 2004; Barbosa and Fearnside 2005), although Hoffa *et al.* (1999) note that, in Zambian grassland, pyrolysis increases substantially as the dry season progresses (from about 50% in early June to 90% in August). Greater variability is evident in consumption of heavy fuels; for example, with reported values ranging from 9 to 31% in studies presented by Kauffman *et al.* (1994), Shea *et al.* (1996), Ito and Penner (2004), Barbosa and Fearnside (2005), Rose (2006). Clearly, for modelling purposes, the effects of fire severity and seasonal variation in fuel moisture need to be more consistently described.

We did not directly measure the production of entrained ash during a fire; that is, ash transported away from the site of a fire within a smoke plume. However, the results of Cook (1994) suggest that during lowland woodland fires in Kakadu National Park, entrained ash represents between about 3 and 9% of consumed biomass. However, ash that remains *in situ* following a fire will comprise at least some entrained ash that has resettled from surrounding areas. Although we do not account for entrained ash specifically in this study, it is consistent with the methodology that has been used to calculate pyrolysis efficiency elsewhere; namely, entrained ash is generally not accounted for (e.g. Kauffman *et al.* 1994; McNaughton *et al.* 1998). A further important factor not accounted for is the effect of fire severity on the production of ash, given that severe fires are likely to result in more efficient volatilisation of consumed biomass (Stronach and McNaughton 1989). For our present purposes, we have assumed that residual ash encompasses entrained ash redeposited from nearby sources. With these qualifications in mind, we note that the *in situ* ash conversion rate calculated here (11.6%; Table 13.4) is generally consistent with studies from Kakadu National Park (7%: Cook 1994) and other tropical savanna and forest ecosystems, which typically range from around 3% (e.g. Kauffman *et al.* 2002; Barbosa and Fearnside 2005) to 11% (e.g. Kauffman *et al.* 1994). Clearly, issues concerning ash production and entrainment require further research.

Emission factors (EFs)

Although most of above parameters have been evaluated for all or most strata, the emissions factors are taken from the Australian NGGI methodology. These are annual averages measured in Kakadu National Park by Hurst *et al.* (1994a, b) using grab samples of air collected in flasks either at ground level or from aircraft. Although they sampled both early and late dry season fires, the variability between samples collected within-fires was greater than the differences between fires, and they were unable to draw any conclusions about seasonal variation in EFs. This is a standard problem when sampling heterogeneous plumes. A second series of airborne and ground-based measurements made in 1999 during the late fire season by Shirai *et al.* (2003) closely support the results of Hurst *et al.* (1994a, b). The Australian CO EFs (~0.05 to 0.08) are slightly higher compared with 0.05 for other savanna regions (Andreae and Merlet 2001), and indicate a small bias towards mixed and smouldering combustion – probably associated with the higher fraction of tree-leaf litter in the fine fuels than occurs in the grass-dominated savannas on other continents. The Australian savanna CH_4 emission factors, however, are similar to those reported from southern Africa, but higher than South America (Andreae and Merlet 2001; Shirai *et al.* 2003).

The CO emission factor is closely related to modified combustion efficiency (MCE = $[CO_2]/([CO]+[CO_2])$), which is strongly correlated with the EF for trace species (Hao *et al.* 1996; Ward *et al.* 1996). MCE varies with fuel class: grasses tend to have high MCE (~0.95), whereas forest fuels have lower MCE (~0.9: Ward *et al.* 1996; Andreae and Merlet 2001) and therefore the fuel class mixture is likely to be a determinant of CH_4 emission, as is fuel moisture. The relationship between MCE and EF was exploited by Korontzi *et al.* (2003, 2004) to explore the seasonality of emissions from fires in southern Africa. These authors found early-season fires in southern Africa tend to have lower MCE. Hence, methane emission is likely to be disproportionately higher in the early fire season than indicated by fire area; in fact, it was suggested that the bulk of CH_4 emissions might come from early-season fires (Korontzi 2005). In contrast, NO_x emissions are positively correlated with MCE and therefore are likely to be disproportionately weighted toward LDS fires. The relationship between MCE and N_2O emissions is less clear, although there is some indication that it is weighted to high MCE. This might also apply in Australia; however, given that the more severe fires of the LDS are more likely to burn shrubs, coarse and heavy fuels, which will burn with a greater proportion of smouldering combustion (Andreae and Merlet 2001), the emission factors of these reduced species are likely to be greater. Hurst *et al.* (1994a) found that smouldering combustion in north Australian savannas produced substantially more CO and CH_4 than flaming combustion.

It should be noted that, to date, all analyses of seasonality rely on derived relationships between a greenness index, which describes the moisture content of the fuel, and MCE; there are currently no comprehensive measurements of the seasonality of emissions composition. However, it is likely that interaction between the mix of grass, forest litter, coarse and heavy fuels, and fuel moisture on MCE is expected to produce EF seasonality that is regionally specific. In Australian savanna, fuels tend to be fully cured before burning commences in the EDS and therefore the strong seasonality reported by Korontzi (2005) probably does not occur to the same extent in Australia. However, quantifying the seasonality of emission factors is obviously a priority for improving emissions accounting.

Bottom-up and top-down approaches

Although verification of a bottom-up inventory model structure and parameters lends confidence to the accuracy of model predictions, independent validation is required to prove them. Validation of emissions at regional or larger scales is a major challenge – seldom undertaken – and, to date, has given mixed results. A study of emissions in northern Australia used an emission model coupled with a dispersion model to calculate regional hourly concentrations of PM2.5 produced by biomass burning throughout 2004, and found good agreement between observed and predicted concentrations in Darwin, and observed and predicted aerosol optical depth (AOD) at two locations (Meyer *et al.* 2008; Luhar *et al.* 2008). A similar approach involving coupling the Global Fire Emissions Database emissions model (Giglio *et al.* 2006) to a Lagrangian dispersion model was used by Saarikoski *et al.* (2007) to evaluate the contribution of several fire events in northern Europe to local air quality, with partial success. However, studies using CO concentrations from MOPITT (measurement of pollution in the troposphere sensor on NASAs TERRA satellite) to estimate biomass burning emissions in southern Africa using atmospheric inversion techniques (Arellano *et al.* 2004; Petron *et al.* 2004) concluded that bottom-up inventories substantially underestimate total emissions. These studies also found substantial differences in the seasonality of emissions from the inventory predictions.

What then are the options for improving the accuracy of emissions accounting? Looking from the bottom up, there is an increasing requirement to define the dependence of GHG emissions on fuel class, fuel properties and seasonality. This can be achieved to a degree through field measurement, but there are alternatives. Increasingly, data assimilation techniques are being used to constrain weather-model predictions with observations both at the land surface and from satellites. Similar approaches are used to constrain parameters in land-surface models (e.g. Wang and Barrett 2003; Barrett *et al.* 2005). Data that could be used to constrain emissions models include: satellite measurements of fire severity (Smith and Wooster 2005; Smith *et al.* 2005); vegetation curing index; fuel moisture estimates (e.g. Hao and Qu 2007; Verbesselt *et al.* 2007); and both surface and satellite observations of aerosol and CO concentration.

Fire severity can be used directly to predict aerosol and gaseous carbon and nitrogen emission rates, while curing and greenness indices – as well as fuel moisture estimates – can be used to estimate MCE, and hence EF, for a wide range of trace species (Hao *et al.* 1996, Korontzi *et al.* 2004). There is evidence that fuel type (cured dead grasses versus green alive shrubs or trees) can be derived from satellite remote sensing (Maier 2008). Finally, measured concentration fields from satellite or ground-based approaches can be used to constrain both emission estimates and model parameters.

Satellite-based thermal measurements of active fires (fire radiative power) provide another independent estimate of fuel combustion rates (Wooster *et al.* 2005). These can be used for calibration or validation of the other approaches, although the method still needs to be tested for fires in Australia.

The WALFA offset model

The underlying premise of WALFA as a GHG emissions offset project is that substantial annual emissions abatement can be achieved, and quantified, through the implementation of strategic prescribed fire management and associated monitoring and accounting. As described previously, the vast WALFA project area is a remote, rugged, biodiversity-rich, today mostly unpopulated landscape where, between 1990 and 2005, an average of 36% p.a. of the region was burnt – predominantly by LDS wildfires. It is the contention of project partners that reinstatement of strategic fire management applying both traditional knowledge and using contemporary practices (e.g. aerial prescribed burning) and tools (e.g. GIS for project planning, implementation and monitoring purposes) can (a) substantially reduce GHG emissions associated with wildfires, (b) help to address chronic contemporary fire regime impacts on biodiversity values, and (c) provide culturally appropriate employment opportunities for regional Indigenous communities.

Achieving significant GHG emissions abatement requires, firstly, the undertaking of a prescribed burning program that is strategic in its implementation to reduce the current annual extent of wildfire. To that end, project partners have commenced critical assessment of using landscape features (watercourses, terrain and tracks) and aerial burning to deliver more-effective management (Price *et al.* 2007). Edwards *et al.* (2003) document the reduction in area burnt using a more strategic management approach in the adjoining 20 000 km^2 Kakadu National Park, from a mean of 45% for 1980–1995 compared with 40% for the period 1996–2000.

Associated substantial emissions reductions can be achieved also through implementing a fire-management program that focuses on delivering strategic burning in the EDS. As detailed in the present study, burning in the EDS provides increased patchiness and reduced fuel consumption, associated with typically less-severe fires. Overall, our data allow us to estimate that emissions from early season fires typically emit 48% of late-season fires per unit area (compare lines for 0% and 100% LDS fires in Figure 13.4). An assessment of the fire history of Kakadu National Park from 1980 to 1994 (covering the 15-year period after its creation) found a pronounced shift from a fire regime dominated by LDS fires until the mid-1980s, to one dominated by EDS fires subsequently (Russell-Smith *et al.* 1997b). In the more recent fire history assessment undertaken by Edwards *et al.* (2003, see above) for Kakadu, the 5% reduction achieved in annual burning during the second 5 year period, compared with the first 15 years, was attributable predominantly to reduction in late-season fires. The Kakadu example illustrates that, with adequate resources, it is feasible to implement a more benign, GHG-friendly fire regime.

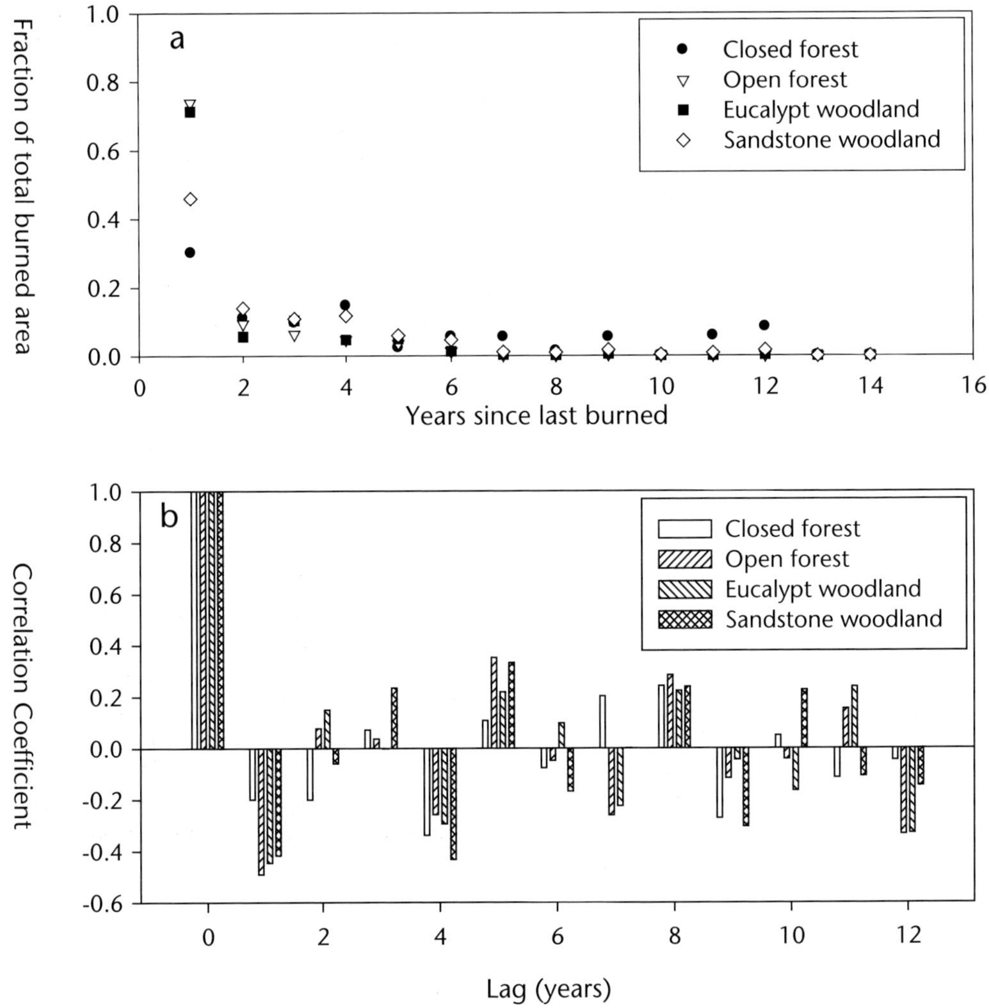

Figure 13.7 Analysis of fire regimes in the WALFA domain: (a) fire return interval for WALFA vegetation classes; (b) the autocorrelation functions of annual fire area in each of the vegetation classes.

The final issue that we canvass here concerns determining the length of the pre-project (business-as-usual) emission baseline against which WALFA (and other similar savanna fire-abatement projects) need to measure their abatement outcomes. This is particularly pertinent given that current IPCC guidelines (IPCC 1997, 2000) do not specify a temporal baseline horizon. If the annual baseline emission has substantial interannual variability, then ideally the sources of the variability – typically climatic drivers – should be determined and their impacts quantified, so that their effects can be separated from the management impacts. Although at regional scales, the impacts of climate may be apparent (e.g. Meyer 2004), at fine spatial scales, climate impacts are often swamped by other sources of variability and therefore cannot easily be quantified (Russell-Smith *et al.* 2004). However, climate trends typically occur on a timescale of decades, and therefore shorter-term interannual variation can often be treated as random noise around a mean. The issue to resolve is the length of an appropriate time window that averages out within-fire-cycle variation, without removing slow trends due to climate shifts.

The fire regimes in the vegetation strata of the WALFA domain can be characterised by their fire return frequency. Figure 13.7a shows the distribution of fire return frequency in the area burned in 2005. In all cases, the median return period is 13 years or less and in all classes more than 85% of the area was reburned less than 10 year previously. In eucalypt woodland, sandstone woodland and open forest, the median return period is closer to 2 years. The short return period is reflected in the autocorrelation functions of the fire history (Figure 13.7b), which show negative correlations at lags of 1, 4, 7, and so on, years, which are presumably consistent with the time required to rebuild the fuel loads in the absence of prescribed strategic fire management. An averaging period of at least 10 years (i.e. covering a number of fuel accumulation–fire cycles) is required to define an emissions baseline that is not compromised by natural variability. The established WALFA emissions baseline covers the pre-project decadal period 1995–2004.

CONCLUSION

The regional GHG emissions inventory methodology presented here is shown to be generally robust: delivering emissions estimates congruent with those of Australia's NGGI. The regional methodology is enhanced by the explicit incorporation of terms for fire seasonality and severity. Addressing reduction in estimate uncertainties requires further work, especially on coarse and heavy fuels, and seasonality and severity components of burning efficiency (patchiness, ash retention and entrainment) and emission factors. More generally, emission estimates derived from this project-based, bottom-up approach need to be qualified and constrained with respect to independent top-down assessments; for example, taking advantage of measurements of CO and aerosol particulates derived from satellite-based sensors coupled with dispersion modelling. With regionally specific adaptations of emissions inventory parameters, the developed methodology has significant potential for emissions accounting of savanna burning offset and abatement projects in other regions. Such opportunities would appear to exist especially in situations where there is a need to implement more strategic burning to reduce anthropogenically dominated, LDS emissions.

ACKNOWLEDGEMENTS

The authors acknowledge the contribution of WALFA partners to support for and the undertaking of this work. We acknowledge also the close reading of this complex manuscript, and many constructive comments made, by reviewers and the Associate Editor. Funding was provided by the Tropical Savannas Management Cooperative Research Centre, Natural Heritage Trust, and the Australian Greenhouse Office.

REFERENCES

AGO (2007a) *National Greenhouse Gas Inventory 2005.* Australian Greenhouse Office, Canberra.

AGO (2007b) *Australian Methodology for the Estimation of Greenhouse Emissions and Sinks 2005 – Agriculture.* Australian Greenhouse Office Canberra.

Andreae MO and Merlet P (2001) Emission of trace gases and aerosols from biomass burning. *Global Biogeochemical Cycles* **15**, 955–966.

Arellano AF Jr, Kasibhatla PS, Giglio L, van der Werf GR and Randerson JT (2004) Top-down estimates of global CO sources using MOPITT measurements. *Geophysical Research Letters* **31**, L01104, <doi:10.1029/2003GL018609>.

Barbosa RI and Fearnside PM (2005) Above-ground biomass and the fate of carbon after burning in the savannas of Roraima, Brazilian Amazonia. *Forest Ecology and Management* **216**, 295–316.

Barrett DJ, Hill MJ, Hutley LB, Beringer J, Xu JH, Cook GD, Carter JO and Williams RJ (2005) Prospects for improving savanna biophysical models by using multiple-constraints model-data assimilation methods. *Australian Journal of Botany* **53**, 689–714.

Bayon R, Hawn A and Hamilton K (2006) *Voluntary Carbon Markets: An International Business Guide, What They Are and How They Work*. Earthscan, London.

Carmona-Moreno C, Belward A, Malingreau J-P, Hartley A, Garcia-Alegre M, Antonovskiy M, Buchshtaber V and Pivovarov V (2005) Characterizing interannual variations in global fire calendar using data from Earth observing satellites. *Global Change Biology* **11**, 1537–1555.

Cook GD (1994) The fate of nutrients during fires in a tropical savanna. *Australian Journal of Ecology* **19**, 359–365.

Cook GD (2003) Fuel dynamics, nutrients and atmospheric chemistry. In *Fire in Tropical Savannas: The Kapalga Experiment*. (Eds AN Andersen, GD Cook and RJ Williams) pp. 47–58. Springer, New York.

Cook GD, Liedloff AC, Eager RW, Chen X, Williams RJ, O'Grady AP and Hutley LB (2005) The estimation of carbon budgets of frequently burnt tree stands in savannas of northern Australia using allometric analysis and isotopic discrimination. *Australian Journal of Botany* **53**, 621–630.

Craig R, Heath B, Raisbeck-Brown N, Steber M, Marsden J and Smith R (2002) The distribution, extent and seasonality of large fires in Australia, April 1998–March 2000, as mapped from NOAA–AVHRR imagery. In 'Australian fire regimes: contemporary patterns (April 1998–March 2000) and changes since European settlement'. (Eds J Russell-Smith, R Craig, AM Gill, R Smith and JE Williams). Australia State of the Environment Second Technical Paper Series (Biodiversity). Department of the Environment and Heritage, Canberra. <http://www.ea.gov.au/soe/techpapers/index.html>

Crawley MJ (2002) *Statistical Computing: An Introduction to Data Analysis Using S-Plus*. John Wiley and Sons, Chichester, UK.

Crutzen PJ and Andreae MO (1990) Biomass burning in the tropics: impact on atmospheric chemistry and biogeochemical cycles. *Science* **250**, 1669–1678.

Dyer R, Jacklyn P, Partridge I, Russell-Smith J and Williams RJ (Eds) (2001) *Savanna Burning: Understanding and Using Fire in Northern Australia*. Tropical Savannas Cooperative Research Centre, Darwin.

Eamus D, McGuiness K and Burrows W (2000) 'Review of allometric relationships for estimating woody biomass for Queensland, the Northern Territory and Western Australia'. Report to the Australian Greenhouse Office, Canberra. Charles Darwin University, Canberra.

Edwards AC and Russell-Smith J (2009) Ecological thresholds and the status of fire-sensitive vegetation in western Arnhem Land, northern Australia: limitations, challenges, applications. *International Journal of Wildland Fire* **18**, 127–146.

Edwards A, Kennett R, Price O, Russell-Smith J, Spiers G and Woinarski J (2003) Monitoring the impacts of fire regimes on vegetation in northern Australia an example from Kakadu National Park. *International Journal of Wildland Fire* **12**, 427–440.

Fearnside PM, Leal N and Fernandes FM (1993) Rainforest burning and the global carbon budget: biomass, combustion efficiency, and charcoal formation in the Brazilian Amazon. *Journal of Geophysical Research* **98**, 16733–16743.

Fox ID, Neldner VJ, Wilson GW and Bannink PJ (2001) *The Vegetation of the Tropical Australian Savannas*. Queensland Environment Protection Agency, Brisbane.

Giglio L, van der Werf GR, Randerson JT, Collatz GJ and Kasibhatla P (2006) Global estimation of burned area using MODIS active fire observations. *Atmospheric Chemistry and Physics* **6**, 957–974.

Hao WM and Liu MH (1994) Spatial and temporal distribution of tropical biomass burning. *Global Biogeochemical Cycles* **8**, 495–503.

Hao WM, Ward DE, Olbu G and Baker S (1996) Emissions of CO_2, CO and hydrocarbons from fires in diverse African savanna ecosystems. *Journal of Geophysical Research* **101**, 23 577–23 584.

Hao X and Qu JJ (2007) Retrieval of real-time live fuel moisture content using MODIS measurements. *Remote Sensing of Environment* **108**, 130–137.

Haynes CD (1985) The pattern and ecology of *munwag*: traditional Aboriginal fire regimes in north-central Arnhem Land. *Proceedings of the Ecological Society of Australia* **13**, 203–214.

Hennenberg KJ, Fischer F, Kouadio K, Goetze D, Orthmann B, Linsenmair KE, Jeltsch F and Porembski S (2006) Phytomass and fire occurrence along forest-savanna transects in the Comoe National Park, Ivory Coast. *Journal of Tropical Ecology* **22**, 303–311.

Hoelzemann JJ, Schultz MG, Brasseur GP, Granier C and Simon N (2004) Global Wildland Fire Emission Model (GWEM): evaluating the use of global area burnt satellite data. *Journal of Geophysical Research* **109**, D14S04, <doi:10.1029/2003JD003666>.

Hoffa EA, Ward DE, Hao WM, Susott RA and Wakimoto RH (1999) Seasonality of carbon emissions from biomass burning in Zambian savanna. *Journal of Geophysical Research* **104,** D11, 13 841–13 853.

Hopkins B (1966) Vegetation of the Olokemeji Forest Reserve, Nigeria IV. The litter and soil with special reference to their seasonal changes. *Journal of Ecology* **54**, 687–703.

Horowitz LW, Walters S, Mauzerall DL, Emmons LK, Rasch PJ, Granier C, Tie X, Lamarque JF, Schultz MG, Tyndall GS, Orlando JJ and Brasseur GP (2003) A global simulation of tropospheric ozone and related tracers: description and evaluation of MOZART, version 2. *Journal of Geophysical Research* **108**(D24), 4784, <doi:10.1029/2002JD002853>.

Hurst DF, Griffith DWT and Cook GD (1994a) Trace gas emissions and biomass burning in tropical Australian savannas. *Journal of Geophysical Research* **99**, 16 441–16 456.

Hurst DF, Griffith DWT, Carras JN, Williams DJ and Fraser PJ (1994b) Measurements of trace gas emitted by Australian savanna fires during the 1990 dry season. *Journal of Atmospheric Chem*istry **18**, 33–56.

IPCC (1997) *Revised 1996 Intergovernmental Panel on Climate Change (IPCC) Guidelines for National Greenhouse Gas Inventories.* 3 vols. IPCC/OECD/IEA, Paris, France.

IPCC (2000) *Good Practice Guidance and Uncertainty Management in National Greenhouse Gas Inventories.* (Eds J Penman, D Kruger, I Galbally, T Hiraishi, B Nyenzi, S Emmanul, L Buendia, R Hoppaus, T Martinsen, J Meijer, K Miwa and K Tanabe). IPCC National Greenhouse Gas Inventories Programme. Published for the IPCC by the Institute for Global Environmental Strategies, Kanagawa, Japan.

IPCC (2003) *Good Practice Guidance for Land Use, Land-Use Change and Forestry.* (Eds J Penman, M Gytarsky, T Hiraishi, T Krug, D Kruger, R Pipatti, L Buendia, K Miwa , T Ngara, K Tanabe and F Wagner). Institute for Global Environmental Strategies (IGES) for the Intergovernmental Panel on Climate Change, Kanagawa, Japan.

IPCC (2006) *2006 IPCC Guidelines for National Greenhouse Gas Inventories.* (Eds S Eggelston, L Buendia, K Miwa, T Ngara and K Tanabe). Institute for Global Environmental Strategies (IGES) for the Intergovernmental Panel on Climate Change, Kanagawa, Japan.

Ito A and Penner JE (2004) Global estimates of biomass burning emissions based on satellite imagery for the year 2000. *Journal of Geophysical Research* **109**, D14S05, <doi:10.1029/2003JD004423>.

Kasischke ES and Penner JE (2004) Improving global estimates of atmospheric emissions from biomass burning. *Journal of Geophysical Research* **109,** D14S01, <doi:10.1029/2004JD004972>.

Kauffman JB, Cummings DL and Ward DE (1994) Relationships of fire, biomass and nutrient dynamics along a vegetation gradient in the Brazilian cerrado. *Journal of Ecology* **82,** 519–531.

Kauffman JB, Steele MD, Cummings DL and Jaramillo VJ (2002) Biomass dynamics associated with deforestation, fire, and conversion to cattle pasture in a Mexican tropical dry forest. *Forest Ecology and Management* **176**, 1–12.

Kondo Y, Takegawa N, Miyazaki Y, Ko M, Koike M, Kita K, Kawakami S, Shirai T, Ogawa T, Blake DR, Liley B and Russell-Smith J (2003) Effects of biomass burning and lightning on atmospheric chemistry over Australia and South-east Asia. *International Journal of Wildland Fire* **12**, 271–281.

Korontzi S (2005) Seasonal patterns in biomass burning emissions from southern African vegetation fires for the year 2000. *Global Change Biology* **11**, 1680–1700,

Korontzi S, Roy DP, Justice CO and Ward DE (2004) Modeling and sensitivity analysis of fire emissions in southern Africa during SAFARI 2000. *Remote Sensing of Environment* **92,** 376–396.

Korontzi S, Justice CO and Scholes RL (2003) Influence of timing and spatial extent of savanna fires in southern Africa on atmospheric emissions. *Journal of Arid Environments* **54**, 395–404, <doi:10.1006/jare.2002.1098>.

Liousse C, Andreae MO, Artaxo P, Barbosa P, Cachier H, Gregoire J-M, Hobbs P, Lavoue D, Mouillot F, Penner J, Scholes M and Schultz MG (2004) Deriving global quantitative estimates for spatial and temporal distributions of biomass burning emissions. In *Emissions of Atmospheric Trace Compounds.* (Eds C Granier, P Artaxo and C Reeves) pp. 77–120. Kluwer Academic, Dordrecht, the Netherlands.

Luhar AK, Mitchell RM, Meyer CP (Mick), Campbell S, Gras JL and Parry D (2009) Biomass burning emissions over northern Australia constrained by aerosol measurements: II – Model validation and impacts on air quality and climate. *Atmospheric Environment* (in press).

McNaughton SJ, Stronach NRH and Georgiadis NJ (1998) Combustion in natural fires and global emissions budgets. *Ecological Applications* **8**, 464–468.

Maier SW (2009) Fire induced changes in surface reflectance on the Australian continent as measured with MODIS. *International Journal of Remote Sensing* (in preparation).

Meyer CP (Mick) (2004) 'Establishing a consistent time-series of greenhouse gas emission estimates from savanna burning in Australia'. Report to the Australian Greenhouse Office, Canberra. CSIRO Publishing, Melbourne.

Meyer CP (Mick), Luhar AK and Mitchell RM (2009) Biomass burning emissions over northern Australia constrained by aerosol measurements: I – Modelling the distribution of hourly emissions. *Atmospheric Environment* (in press).

Pereira JMC, Pereira BS, Barbosa PM, Stroppiana D, Vasconcelos MJP and Grégoire J-M (1999) Satellite monitoring of the overall fire activity in the EXPRESSO study area during the dry season experiment: active fires, burnt areas, and atmospheric emissions. *Journal of Geophysical Research – Atmospheres*, **104** (D23), 30 701–30 712.

Petron G, Granier C, Khattatov C, Yudin V, Lamarque JF, Emmons L, Gille J and Edwards DP (2004) Monthly CO surface sources inventory based on the 2000–2001 MOPITT satellite data. *Geophysical Research Letters*, **31**, L21107, <doi:10.1029/2004GL020560>.

Price O, Edwards A and Russell-Smith J (2007) The efficacy of permanent firebreaks and aerial prescribed burning in western Arnhem Land, Northern Territory, Australia. *International Journal of Wildland Fire* **16**, 295–307.

Price O, Russell-Smith J. and Edwards A (2003) Fine-scale patchiness of different fire intensities in sandstone heath vegetation in northern Australia. *International Journal of Wildland Fire* **12**, 227–236.

Rose D (2006) The dynamics of coarse woody debris in the mesic tropical savannas of the Darwin region: abundance, decomposition and consumption by fire. BSc (Hons) thesis. Charles Darwin University, Darwin.

Roy DP, Jin Y, Lewis PE and Justice CO (2005) Prototyping a global algorithm for systematic fire-affected area mapping using MODIS time series data. *Remote Sensing of Environment* **97**, 137–162.

Russell-Smith J and Edwards AC (2006) Seasonality and fire severity in savanna landscapes of monsoonal northern Australia. *International Journal of Wildland Fire* **15**, 541–550.

Russell-Smith J, Lucas D, Gapindi M, Gunbunuka B, Kapirigi N, Namingum G, Lucas K, Giuliani P and Chaloupka G (1997a) Aboriginal resource utilization and fire management practice in western Arnhem Land, monsoonal northern Australia: notes for prehistory and lessons for the future. *Human Ecology* **25**, 159–195.

Russell-Smith J, Ryan PG and DuRieu R (1997b) A LANDSAT MSS-derived fire history of Kakadu National Park, monsoonal northern Australia, 1980–1994: seasonal extent, frequency and patchiness. *Journal of Applied Ecology* **34**, 748–766.

Russell-Smith J, Ryan PG. Klessa D, Waight G and Harwood R (1998) Fire regimes, fire-sensitive vegetation and fire management of the sandstone Arnhem Plateau, monsoonal northern Australia. *Journal of Applied Ecology* **35**, 829–846.

Russell-Smith J, Yates C, Edwards A, Allan GE, Cook GD, Cooke P, Craig R, Heath B and Smith R (2003) Contemporary fire regimes of northern Australia, 1997–2001: change since Aboriginal occupancy, challenges for sustainable management. *International Journal of Wildland Fire* **12**, 283–297.

Russell-Smith J, Edwards AC, Cook GD, Brocklehurst P and Schatz J (2004) 'Improving greenhouse emissions estimates associated with savanna burning in northern Australia'. Report to Australian Greenhouse Office, Canberra. Tropical Savannas Management Cooperative Research Centre, Darwin.

Russell-Smith J, Yates CP, Whitehead P, Smith R, Craig R, Allan G, Thackway R, Frakes I, Cridland, Meyer CP and Gill AM (2007) Bushfires 'down under': patterns and implications of Australian landscape burning. *International Journal of Wildland Fire* **16**, 361–377.

Saarikoski S, Sillanpaa M, Sofiev M, Timonen H, Saarnio K, Teinila K, Karppinen A., Kukkonen J and Hillamo R (2007),Chemical composition of aerosols during a major biomass burning episode over northern Europe in spring 2006: Experimental and modelling assessments. *Atmospheric Environment* **41**, 3577–3589.

Scholes M and Andreae MO (2000) Biogenic and pyrogenic emissions from Africa and their impact on the global atmosphere. *Ambio* **29**, 23–29.

Scholes RJ, Kendall J and Justice CO (1996a) The quantity of biomass burned in southern Africa. *Journal of Geophysical Research* **101** (D19), 23 667–23 676.

Scholes RJ, Ward DE and Justice CO (1996b) Emissions of trace gases and aerosol particles due to vegetation burning in southern hemisphere Africa. *Journal of Geophysical Research* **101** (D19), 23 677–23 682.

Seiler W and Crutzen PJ (1980) Estimates of gross and net fluxes of carbon between the biosphere and atmosphere from biomass burning. *Climatic Change* **2**, 207–247.

Shea RW, Shea BW and Kauffman JB (1996) Fuel biomass and combustion factors associated with fires in savanna ecosystems in South Africa and Zambia. *Journal of Geophysical Research* **101** (D19), 23 551–23 568.

Shirai T, Blake DR, Meinardi S, Rowland FS, Russell-Smith J, Edwards A, Kondo Y, Koike M, Kita K, Machida T, Takegawa N, Nishi N, Kawakami S and Ogawa T (2003) Emission estimates of selected volatile organic compounds from tropical savanna burning in northern Australia. *Journal of Geophysical Research – Atmospheres*, **108**(D3), 8406, <doi:10.1029/2001JD000841>.

Smith AMS and Wooster MJ (2005) Remote classification of head and backfire types from MODIS fire radiative power and smoke plume observations. *International Journal of Wildland Fire* **14**, 249–254.

Smith AMS, Wooster MJ, Drake NA, Dipotso FM, Falkowski MJ and Hudak AT (2005) Testing the potential of multi-spectral remote sensing for retrospectively estimating fire severity in African Savannahs. *Remote Sensing of Environment* **97**, 92–115.

Streets DC, Yarber RF, Woo J-H and Carmichael GR (2003) Biomass burning in Asia: annual and seasonal estimates and atmospheric emissions. *Global Biogeochemical Cycles* **17**(4), 1099, <doi:10.1029/2003GB002040>.

Stronach NRH and McNaughton SJ (1989) Grassland fire dynamics in the Serengeti ecosystem, and a potential method of retrospectively estimating fire energy. *Journal of Applied Ecology* **26**, 1025–1033.

UNFCCC (1998) 'Kyoto Protocol to the United Nations Framework Convention on Climate Change'. United Nations. <http://unfccc.int/resource/docs/convkp/kpeng.pdf>.

van der Werf GR, Randerson JT, Giglio L, Collatz GJ, Kasibhatla PS and Arellano AF Jr. (2006) Interannual variability in global biomass burning emissions from 1997 to 2004. *Atmospheric Chemistry and Physics* **6**, 3423–3441.

Verbesselt J, Somers B, Lhermitte S, Jonckheere I, van Aardt J and Coppin P (2007) Monitoring herbaceous fuel moisture content with SPOT VEGETATION time-series for fire risk prediction in savanna ecosystems. *Remote Sensing of Environment* **108**, 357–368.

Wang YP and Barrett DJ (2003) Estimating regional and terrestrial carbon fluxes for the Australian continent using a multiple constraint approach: I. using remotely sensed data and ecological observations of net primary production. *Tellus* **55B**, 270–289.

Ward DE, Hao WM, Susott RA, Babbitt RE, Shea RW, Kauffman JB and Justice CO (1996) Effect of fuel composition on combustion efficiency and emission factors for African savanna ecosystems. *Journal of Geophysical Research* **101**(D19), 23 569–23 576.

Whitehead PJ, Bowman DMJS, Preece N, Fraser F and Cooke P (2003) Customary use of fire by indigenous peoples in northern Australia: its contemporary role in savanna management. *International Journal of Wildland Fire* **12**, 415–425.

Williams RJ, Gill AM and Moore PHR (2003) Fire behaviour. In *Fire in Tropical Savannas: The Kapalga Experiment*. (Eds AN Andersen, GD Cook and RJ Williams) pp. 33–46. Springer-Verlag, New York.

Williams RJ, Gill AM and Moore PHR (1998) Seasonal changes in fire behaviour in a tropical savanna in northern Australia. *International Journal of Wildland Fire* **8**, 227–239.

Williams RJ, Griffiths AD and Allan GE (2002) Fire regimes and biodiversity in the wet-dry tropical landscapes of northern Australia. In *Flammable Australia: The Fire Regimes and Biodiversity of a Continent*. (Eds RA Bradstock, JE Williams and AM Gill) pp. 281–304. Cambridge University Press, Cambridge, UK.

Wooster MJ, Roberts G, Perry GLW and Kaufman YJ (2005) Retrieval of biomass combustion rates and totals from fire radiative power observations: FRP derivation and calibration relationships between biomass consumption and fire radiative energy release. *Journal Geophysical Research* **110**, D24311, <doi:10.1029/2005JD006318>.

Yates C and Russell-Smith J (2002) An assessment of the accuracy of DOLA's Northern Australia NOAA-AVHRR Fire Affected Area (FAA) map products. In *Australian fire regimes: contemporary patterns (April 1998–March 2000) and changes since European settlement.* (Eds J Russell-Smith, R Craig, AM Gill, R Smith and J Williams). Australia State of the Environment Second Technical Paper Series (Biodiversity). Department of the Environment and Heritage, Canberra, <http://www.ea.gov.au/soe/techpapers/index.html>.

14

Fire management and woody biomass carbon stocks in mesic savannas

Brett P Murphy, Jeremy Russell-Smith, Felicity A Watt and Garry D Cook

SUMMARY

Northern Australia's savannas are highly productive, yet fire-prone, ecosystems, and it has been suggested that reductions in fire frequency could substantially increase the amount of carbon they store, especially in woody biomass. Using data from 136 vegetation and fire-monitoring plots throughout the savannas of three Top End national parks (Kakadu, Nitmiluk and Litchfield), we examined changes in live woody biomass carbon stocks over a 10-year period in relation to fire frequency. Contrary to widely held perceptions that Top End mesic savannas are thickening, there was no net change in woody biomass carbon stocks between 1994 and 2004. However, we found that high frequencies of fires, especially severe fires, resulted in reductions in woody biomass carbon stocks, suggesting that appropriate fire management could lead to increased carbon storage. A model we developed to explain changes in tree carbon stocks generally agreed with the changes reported in the published literature from Top End savannas. For this reason, we were confident our model could provide crude, but still useful, predictions of the likely effects of different management scenarios. Using western Arnhem Land as a case study, we evaluated the likely effects of improved fire management on woody biomass carbon stocks. Based on models derived from the three parks dataset, we suggest that a substantial, yet plausible, reduction in fire frequency – especially late dry season fires – may increase woody biomass carbon stocks by around 6.1 t C ha^{-1} over a century, which equates to a carbon offset of 22 t CO_2-equivalent ha^{-1}. This crude estimate is larger than that generated by non-CO_2 greenhouse gas abatement over a century. Thus, improvements in fire management aimed at non-CO_2 greenhouse gas abatement are likely to offer additional benefits in terms of carbon sequestration.

INTRODUCTION

The dynamics of tropical savannas have received enormous attention in recent years, with much of the focus on factors that determine woody biomass (Scholes and Archer 1997;

Sankaran *et al.* 2004; Sankaran *et al.* 2005, 2008; Bucini and Hanan 2007). Much of the interest in savanna dynamics has been driven by the prominent role this biome plays in the global carbon cycle, contributing around 30% of global terrestrial net primary production (NPP) – roughly equivalent to the NPP of tropical forests (Grace *et al.* 2006). However, because of the very high frequency of fires in savannas – with around 20% of these areas burnt annually (Dwyer *et al.* 2000) – much of the carbon captured is quickly released back to the atmosphere, such that savannas sequester and store much less carbon than other highly productive biomes, such as tropical forests (Grace *et al.* 2006). Recent work generally suggests that northern Australia's mesic savannas are a weak carbon sink, with several estimates of the rate of sequestration in the order of 1–2 t C ha yr^{-1}, despite very frequent, and often severe, fires (Williams *et al.* 2004; Beringer *et al.* 2007). The nature of the carbon sink remains poorly defined, with little evidence to support, or refute, the notion that the sink is due to increasing woody biomass, possibly as a result of increasing CO_2 concentrations: a process suggested by Grace *et al.* (2006). Given that fires result in large losses of carbon to the atmosphere, it has been suggested that reducing savanna fire frequencies – especially the frequency of severe fires – could lead to a substantial increase in rates of carbon sequestration (Williams *et al.* 2004; Grace *et al.* 2006; Beringer *et al.* 2007). It is likely that an increase in carbon sequestration following reductions in fire frequency would be largely the result of an increase in woody biomass, although the effects of fire on soil carbon stocks – by far the largest carbon pool in savannas – remain poorly understood (Bird *et al.* 2000; Chen *et al.* 2005; Ansley *et al.* 2006).

There is very strong evidence in the published literature to support the notion that reducing the frequency or severity of fires, or both, leads to an increase in woody biomass in savannas. Numerous studies, from various parts of the world, have demonstrated dramatic increases in woody biomass following complete fire exclusion (e.g. South America: San Jose *et al.* 1998; North America: Tilman *et al.* 2000; Africa: Bond *et al.* 2005). However, studies such as these have examined an extreme treatment that is effectively impossible on a large scale in fire-prone savannas; that is, complete fire exclusion. More valuable insights into the negative relationship between fire frequency and the abundance of woody biomass have been provided by a number of long-term, large-scale fire experiments. Perhaps the best example is the 40-year study undertaken in the semi-arid savannas of Kruger National Park, South Africa, where: complete fire exclusion resulted in very large positive changes in woody biomass; triennial and biennial fires resulted in smaller positive changes; and annual fires resulted in negative changes (Higgins *et al.* 2007). Two similar fire-experiments have been undertaken in northern Australia: Munmarlary (Russell-Smith *et al.* 2003) and Kapalga (Williams *et al.* 1999), both in present-day Kakadu National Park. These two studies generally, although not always consistently, demonstrated that fire exclusion leads to an increase in woody biomass and frequent severe fires lead to a decrease. Frequent mild fires appear to have only a relatively minor negative effect on woody biomass.

Another approach to examining the effect of fire regime on woody biomass is the use of 'natural experiments', whereby vegetation changes are examined in relation to the ambient fire regime, rather than imposed fire treatments. A recent example of such an approach is the work of Lehmann *et al.* (2008, 2009). Examining changes in tree cover in the late 20th century in present-day Kakadu National Park, these authors found a negative relationship between change in tree cover and fire frequency, especially the frequency of late dry season fires. One of the advantages of such an approach is that it can produce results that are more readily useful to land managers seeking to predict the effects of lesser changes in fire regimes – changes that may plausibly be within their reach. In contrast, traditional fire experiments (e.g. Kruger, Kapalga and Munmarlary) focus, by necessity, on a limited number of fire treatments that are often extreme and unrealistic (e.g. no fires or annual severe fires), and the results are therefore of relatively little immediate use to land managers (Russell-Smith *et al.* 2003).

Recently, a large vegetation and fire-monitoring dataset, spanning 10 years, has been compiled for three large Top End national parks: Kakadu; Nitmiluk; and Litchfield (Chapter 10). Though based on a monitoring program designed to detect changes in biodiversity rather than woody biomass, this dataset provides a good opportunity to carry out another 'natural experiment' and examine the magnitude of changes in woody biomass carbon stocks over a 10-year period, in relation to the characteristics of the ambient fire regime. In this chapter, we discuss a recent analysis of the Three Parks dataset and use it to address two questions:

1. Are the Top End's mesic savannas thickening?
2. To what extent can fire management be used to increase woody biomass carbon stocks?

We then use western Arnhem Land as a case study, and provide a preliminary assessment of the effects that recent improvements in fire management may have on woody biomass carbon stocks.

THE THREE PARKS VEGETATION AND FIRE DATASET

The Three Parks vegetation and fire-monitoring program is described in detail in Chapter 10. Throughout Kakadu, Litchfield and Nitmiluk National Parks, a total of 220 monitoring plots were established in 1994–1995 in a range of vegetation types, with detailed vegetation assessments being carried out in 1994–1995, 1999–2000 and 2004–2005. As well as the vegetation assessments, a detailed fire history has been assembled for each of the plots. In this chapter, we only examine the plots located in savanna woodland vegetation: in either lowland or sandstone plateau landform types (136 plots in total). Although these plots were not located on a strictly random basis, but often positioned in areas of interest to park rangers, the large number of plots and their widespread distribution (see Fig. 10.2), suggest that they are likely to be quite representative of Top End savanna woodlands.

Estimation of live woody biomass carbon stocks

Live woody biomass is made up of two components: trees and shrubs. During each detailed vegetation assessment, a complete inventory of trees was made at each monitoring plot, and this was used to estimate tree carbon stock. The diameter at breast height (DBH) of each live adult tree (trees with DBH $\geq$5 cm) within 40 $\times$ 20 m plots was recorded, such that live tree basal area could be calculated. Using the allometric equations provided by Cook *et al.* (2005), live tree basal area (m^2 ha^{-1}) was converted to tree carbon stock, including both above-ground and below-ground tree biomass (t C ha^{-1}).

Inventories of shrubs were also made during the vegetation assessments, and these were used to estimate shrub carbon stock. The inventories involved counting all shrubs within three height classes: small (<50 cm); medium (50–200 cm); and large (>200 cm). Small shrubs were counted in two 40 $\times$ 1 m plots, whereas medium and large shrubs were counted in a single 40 $\times$ 10 m plot. For each size class, we estimated the above-ground biomass of a typical shrub, using data collected in western Arnhem Land (J. Russell-Smith unpublished). To estimate total shrub biomass, including below-ground biomass, we assumed a root:shoot ratio of 0.49, based on a number of published studies in savannas or tropical forests (Menaut and Cesar 1979; Becker *et al.* 1999; Valladares *et al.* 2000; Grainger and Becker 2001; Werner and Murphy 2001). The biomass of a typical shrub was converted to carbon stock of a typical shrub, assuming that plant biomass has a carbon content of 49% (Gifford 2000). Finally, the estimated carbon stock per shrub was multiplied by shrub density to produce an areal estimate of shrub carbon stocks (t C ha^{-1}).

Fire history

The Three Parks dataset also contains detailed fire histories for each monitoring plot. Throughout the study period, plots were usually visited, and photographed, once or twice each year, such that we have a record of whether each plot was burnt in any given year and the severity of the fire. Fires were categorised as mild, moderate or severe, using the severity index developed by Russell-Smith and Edwards (2006) – based largely on the scorch height of leaves. Mild fires were those with scorch heights of less than 2 m, suggesting Byram fire-line intensities of much less than 1000 kW m^{-1}. Moderate fires were those where the canopy was scorched to mid-height, suggesting intensities of 1000–2000 kW m^{-1}. Severe fires were those where the canopy was scorched to its full height, suggesting intensities of more than 2000 kW m^{-1}. A detailed description of the compilation of the fire histories is provided by Russell-Smith and Edwards (2006).

CHANGES IN WOODY BIOMASS CARBON STOCKS 1994–2004: ARE TOP END SAVANNAS THICKENING?

Between 1994 and 2004, there was virtually no net change in woody biomass carbon stocks in savannas throughout the three parks. Carbon stocks in tree and shrub biomass were 25.1 and 0.64 t C ha^{-1}, respectively, in both 1994 and 2004 (Figure 14.1) – representing a net change of 0.0 t C ha^{-1} ±1.1 (95% CI). There was a slight increase between 1994 and 1999, but this was offset almost exactly by a subsequent decrease between 1999 and 2004. A similar pattern was detected in both lowland and sandstone plots, with very little evidence of a net change in either case (0.8 t C ha^{-1} ±1.4 and -1.2 t C ha^{-1} ±1.6, respectively).

Not only was there no net change in carbon stocks, but very few individual sites experienced large rates of change (Figure 14.2). For example, only three sites (<3%) experienced either positive or negative changes as large as 1.7 t C ha^{-1} yr^{-1} in magnitude; that is, the rate of woody thickening suggested by Beringer *et al.* (2007) for their Howard Springs study site.

Clearly the absence of woody thickening during the study period was not due to below average rainfall, because the decade 1994–2004 was the wettest on record. During this period, annual rainfall averaged 1396 mm across the three parks, compared with a long-term average of 1227 mm (Australia Bureau of Meteorology). If anything, we would have expected elevated rates of thickening during this period.

These findings are in stark contrast to several previous reports of high carbon sequestration rates in northern Australian savannas. Beringer *et al.* (2007) estimated a sequestration rate of 2.0 t C ha^{-1} yr^{-1} at Howard Springs and suggested that 1.2 t C ha^{-1} yr^{-1} (60%) and 0.5 t C ha^{-1} yr^{-1} (15%) of this was due to increases in tree and shrub biomass, respectively (Table 14.1). Clearly, there is the potential for the Howard Springs study site to be thickening at a relatively high rate, especially given that it was damaged by Cyclone Tracy in 1974 and is probably still recovering (O'Grady *et al.* 2000). If this is the case, then caution should be used when extrapolating observations from this site across the wider savanna landscape.

Other than the work at Howard Springs and other sites near Darwin (Chen *et al.* 2003; Beringer *et al.* 2007) there seems to be little empirical evidence in the published literature of substantial woody thickening in Top End savannas that is greater than about 0.5 t C ha^{-1} yr^{-1}. Several studies have examined changes in tree basal area, from which changes in tree carbon stocks can be inferred. However, the vegetation examined was subject to experimental fire treatments (e.g. Kapalga: Williams *et al.* 1999; Munmarlary: Russell-Smith *et al.* 2003), rather

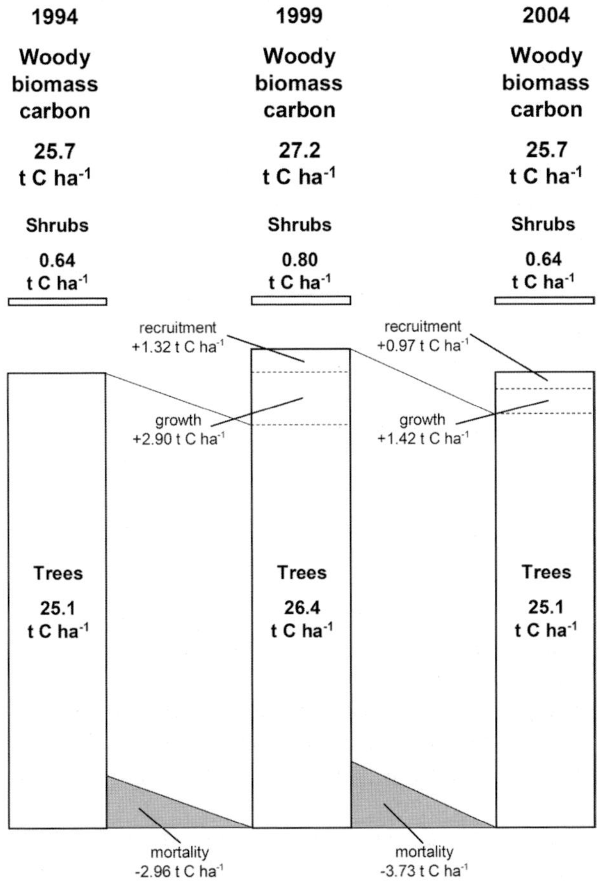

Figure 14.1 Schematic representation of changes in woody biomass carbon stocks over a 10-year period (1994–2004). The relative contributions of growth, recruitment and mortality to change in carbon stocks are indicated.

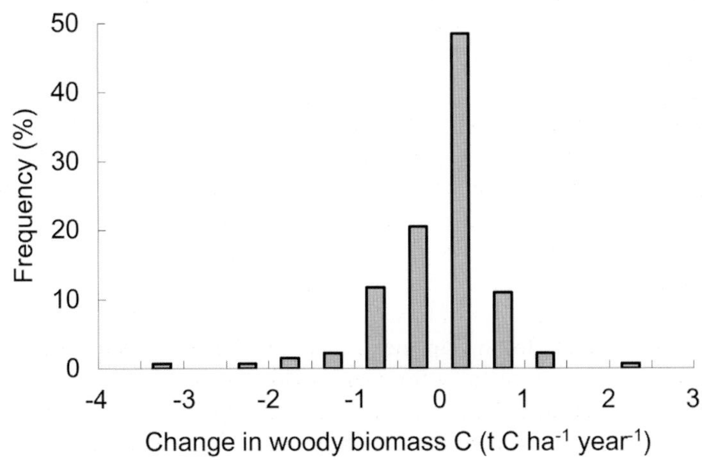

Figure 14.2 Frequency distribution of annualised changes in woody biomass carbon stocks in 136 savanna woodland sites over a 10-year period (1994–2004).

Table 14.1: Summary of previous studies that have examined (a) changes in biomass carbon stocks and (b) ecosystem productivity in northern Australian savannas. Where the authors provided changes in tree basal area, we converted these to tree carbon stocks using the allometric equations of Cook *et al.* (2005). To allow easier comparison with changes in carbon stocks, positive values of ecosystem productivity indicate carbon sinks. The studies are listed in chronological order.

Reference	Location	Disturbance regime	Change parameter	Value ($t\ C\ ha^{-1}\ yr^{-1}$)	Method	Study period
(a) Changes in biomass carbon stocks						
Williams *et al.* (1999)	Eucalypt open forest, Kapalga, Kakadu National Park, NT	No fire	Change in tree biomass carbon stocks	0.2	Successive inventories	5 years
		Annual early fires		0.0		
		Annual late fires		–1.9		
Burrows *et al.* (2002)	Eucalypt woodland, Queensland (ranging from latitudes of Brisbane to Cairns)	Intensification of cattle grazing, presumably accompanied by decrease in fire frequency	Change in tree biomass carbon stocks, including dead biomass	0.5	Successive inventories	14 years
				0.9		2 years
Russell-Smith *et al.* (2003)	Eucalypt woodland, Munmarlary, Kakadu National Park, NT	No fire	Change in tree biomass carbon stocks	1.4	Successive inventories	23 years
		Biennial early fires		0.1		
		Annual early fires		1.6		
		Annual late fires		1.0		
	Eucalypt open forest, Munmarlary, Kakadu National Park, NT	No fire		1.4		21 years
		Biennial early fires		0.7		
		Annual early fires		0.4		
		Annual late fires		-0.6		
Chen *et al.* (2003)	Eucalypt open forest, four sites within 65 km of Darwin, NT	Vegetation at some sites recovering from a severe tropical cyclone in 1974 Ambient fire regime	Change in biomass carbon stocks (including both woody and herbaceous components)	3.1	Successive inventories	≤2 years

Reference	Location	Disturbance regime	Change parameter	Value (t C ha^{-1} yr^{-1})	Method	Study period
Cook *et al.* (2005)	Eucalypt open forest, Kapalga, Kakadu National Park, NT	No fire	Change in tree biomass carbon stocks	0.5 to 1.0	Simulation study. Single basal area measurement and an assumed DBH increment for each of six species. The effect of fire estimated using a modelled relationship between stem survival, DBH and fire intensity.	Single basal area measurement only. Mean DBH increment measured over 13 years. Stem survival measured over 5 years.
		Annual early fires (intensity 2.1 MW m^{-1})		–0.1 to 0.1		
		Annual late fires (intensity 7.7 MW m^{-1})		–3.3 to –2.5		
		Annual very high severity fires (intensity 15 MW m^{-1})		–9.8 to –5.2		
Lehmann *et al.* (2009)	*Eucalyptus tetrodonta* woodland and open forest, Kakadu National Park, NT	Ambient fire regime	Change in tree biomass carbon stocks	0.1	Assessments of tree cover change from aerial photography. Calibrated to basal area change.	40 years
(b) Ecosystem productivity						
Eamus *et al.* (2001)	Eucalypt open forest, Howard Springs, NT	Vegetation recovering from a severe tropical cyclone in 1974 No fire	Net ecosystem productivity[1]	2.8	Eddy covariance flux towers	2 years
Chen *et al.* (2003)	Eucalypt open forest, four sites within 65 km of Darwin, NT	Vegetation at some sites recovering from a severe tropical cyclone in 1974 Ambient fire regime	Net ecosystem productivity[1]	3.8	Carbon balance	≤2 years
Williams *et al.* (1999)	Eucalypt open forest and woodland, Top End, NT	Fire regime typical of western Arnhem Land, 25–57% burnt annually	Net biome productivity[2]	1.0 to 1.5	NEP estimated from literature review. NBP then estimated using assumed C losses due to fire.	
Beringer *et al.* (2007)	Eucalypt open forest, Howard Springs, NT	Vegetation recovering from a severe tropical cyclone in 1974 Annual fires moderate–severe (intensity 1.0–3.5 MW m^{-1})	Net ecosystem productivity[1]	3.6	NEP measured using eddy covariance flux towers. NBP then estimated using assumed C losses due to fire.	5 years
			Net biome productivity[2]	2.0		

[1] Net ecosystem productivity is the carbon fixed by ecosystems due to net primary productivity, minus the carbon losses due to heterotrophic respiration.
[2] Net biome productivity is net ecosystem productivity minus the carbon losses due to disturbance (e.g. fire, grazing).

than the ambient fire regime, making it very difficult to extrapolate from these results to the wider savanna landscape. Even in treatments subject to complete fire exclusion, tree carbon stocks only increased by between 0.2 and 1.4 t C ha^{-1} yr^{-1} at Kapalga (5 years) and Munmarlary (21–23 years), respectively (Table 14.1). Cook *et al.* (2005) used tree growth and mortality data from Kapalga to estimate that, under a regime of biennial moderate severity fires (typical intensity 2.1 MW m^{-1}), which is reasonably typical of the wider savanna landscape, tree carbon stocks would increase by only 0.2–0.5 t C ha^{-1} yr^{-1}, depending on soil depth (Table 14.1). Very recently, Lehmann *et al.* (2009) have used historical aerial photography to estimate that tree cover in Kakadu increased by just 5% between 1964 and 2004. Using a relationship between tree cover and basal area ($R^2 = 0.64$) provided by those authors, we estimate that this change in tree cover represents a change in tree carbon stocks of around 0.1 t C ha^{-1} yr^{-1} (Table 14.1).

In conclusion, the published literature contains very little empirical evidence that northern Australian savannas are thickening substantially. The Three Parks vegetation dataset represents the Top End's only longer-term (10-year) inventory of woody biomass carbon stocks in areas subject to an ambient fire regime, and this dataset strongly suggests that thickening is not occurring, despite coinciding with an exceptionally wet decade. The oft-quoted fact that northern Australian savannas are a substantial carbon sink appears to be based on evidence from a few sites near Darwin, where vegetation is thought to be recovering from tropical cyclone damage sustained in 1974. The apparent discrepancy in thickening observed at a single site (Howard Springs) and at a large number of sites across the Top End (the Three Parks dataset) highlights the question of how to accurately monitor woody biomass carbon stocks if carbon storage in savannas ever becomes a marketable commodity. It is clear that a large array of monitoring sites will be required to capture adequately the spatial variation in carbon stocks. It is important to note that the issue of whether or not Top End savannas are a carbon sink is not especially relevant to efforts to manage savannas in a way that can generate carbon offsets. The relevant issue is whether management can substantially shift carbon fluxes in the desired direction; the starting point – whether source or sink – is essentially moot. We will now examine the relationship between fire frequency and carbon fluxes: in this case, changes in woody biomass carbon stocks.

EFFECTS OF FIRE ON WOODY BIOMASS CARBON STOCKS

Examination of the raw changes in tree and shrub carbon stocks suggests an effect of frequent, severe fires (Figure 14.3). Generally, plots that were subject to one or more severe fires, or repeated moderate fires, experienced declines in tree and shrub carbon stocks.

These initial observations were confirmed using a multivariate modelling approach. Using six predictor variables – including the frequency of mild, moderate and severe fires – we constructed a set of 12 candidate models to explain the proportional change in either tree or shrub carbon stocks, per plot, in each 5-year monitoring period (i.e. 1994–1999, 1999–2004; Table 14.2). Because repeated measures were made for each plot, models were constructed as linear mixed effects models with plot specified as a random effect. Model selection was based on AIC_c – a robust form of Akaike's Information Criterion – and model predictions were made by averaging the entire candidate set, weighted according to w_i (Burnham and Anderson 2002). All models were constructed using the computer program R (version 2.6.2; R Development Core Team 2008).

The model-selection procedure provided strong support for an effect of fire frequency on changes in both tree and shrub carbon stocks. Models incorporating fire frequency were most likely to be the best in the candidate set (w_i of 0.94 and 0.98 for tree and shrub carbon stocks,

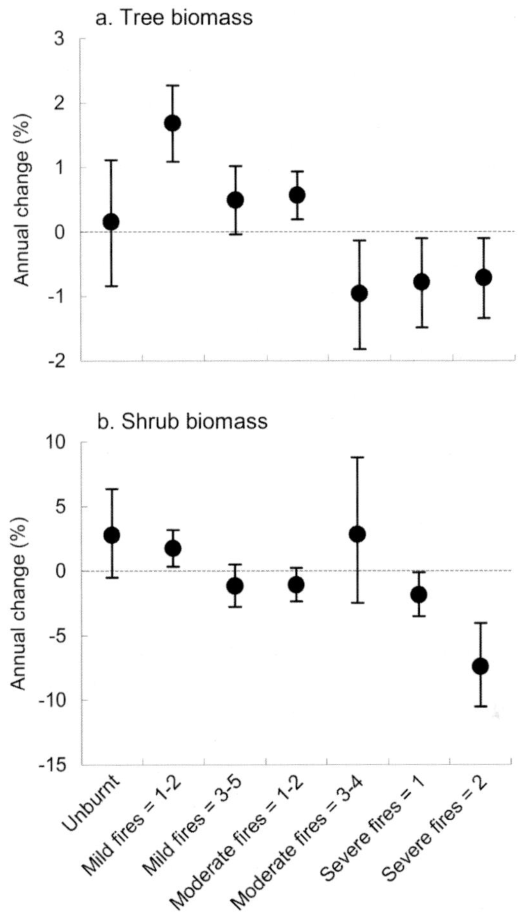

Figure 14.3 Mean annual change in carbon stocks in (a) tree and (b) shrub biomass, in relation to fire regimes of varying fire frequency and severity. From left to right along the x-axis, fires are increasingly frequent and severe. Standard errors are shown.

respectively; Table 14.2). The relationship between fire frequency and changes in both tree and shrub carbon stocks was negative (Figure 14.4). In the case of severe fire frequency, the relationship was highly significant for both trees ($p<0.005$) and shrubs ($p<0.01$). The effect of mild and moderate fire frequency on changes in carbon stocks was less clear, and smaller in magnitude (Fig. 4); only the relationship between mild fire frequency and change in shrub carbon stocks was statistically significant ($p<0.005$). The best models explained 11% and 23% of the variation in change in tree and shrub carbon stocks, respectively.

Our results for tree carbon stocks, which dominate woody biomass carbon stocks overall, are consistent with the savanna literature from northern Australia that suggests that frequent fires, especially severe fires, lead to reductions in tree biomass, and an absence of fires leads to an increase in tree biomass (Williams *et al.* 1999; Russell-Smith *et al.* 2003). The Kapalga fire experiment involved three fire treatments (no fires, early annual fires and late annual fires), applied over 5 years. The no fire and annual early fire treatments resulted in very minor changes in tree basal area (and hence tree carbon stocks), whereas the late annual fire treatment resulted in a large decrease in basal area (Williams *et al.* 1999). The changes in tree

Table 14.2: Candidate models to explain variation in changes in (a) tree and (b) shrub biomass carbon stocks between 1994 and 2004, and results of the model selection procedure. Models are shown ranked in ascending order of the model selection criterion AICc. Δi is the difference between the model's AICc value and the lowest AICc value in the candidate set; wi is the probability of the model being the best in the candidate set. Models with a high level of empirical support (Δi < 2) are shown in bold.

Model	AIC_c	Δ	w_i	R^2
(a) Proportional change in tree biomass carbon stocks				
initial biomass[1] + landform[2] + mild fires + moderate fires + severe fires[3]	**–24.6**	**0.0**	**0.46**	**11**
initial biomass + mild fires + moderate fires + severe fires	**–23.0**	**1.6**	**0.21**	**10**
initial biomass + MAR[4] + landform + mild fires + moderate fires + severe fires	**–22.9**	**1.7**	**0.19**	**11**
initial biomass + MAR + mild fires + moderate fires + severe fires	–21.1	3.5	0.08	10
initial biomass + landform	–19.0	5.7	0.03	7
initial biomass	–17.7	6.9	0.01	6
initial biomass + MAR + landform	–16.9	7.7	0.01	7
initial biomass + MAR	–15.7	9.0	0.01	6
landform	–6.0	18.6	0.00	1
mild fires + moderate fires + severe fires	–5.5	19.1	0.00	3
null model	–4.9	19.7	0.00	0
MAR	–3.6	21.0	0.00	0
(b) Proportional change in shrub biomass carbon stocks				
initial biomass + MAR + landform + mild fires + moderate fires + severe fires	**–426.0**	**0.0**	**0.75**	**23**
initial biomass + MAR + mild fires + moderate fires + severe fires	–423.2	2.8	0.19	22
initial biomass + landform + mild fires + moderate fires + severe fires	–420.2	5.8	0.04	21
initial biomass + MAR + landform	–417.2	8.7	0.01	19
initial biomass + MAR	–416.2	9.7	0.01	18
initial biomass + mild fires + moderate fires + severe fires	–415.9	10.0	0.00	19
initial biomass + landform	–411.8	14.2	0.00	17
initial biomass	–409.1	16.9	0.00	15
landform	–370.6	55.4	0.00	2
null model	–367.5	58.5	0.00	0
mild fires + moderate fires + severe fires	–366.0	60.0	0.00	2
MAR	–365.7	60.3	0.00	0

[1] 'initial biomass' is the natural logarithm of initial tree or shrub biomass.
[2] 'landform' is either lowland or sandstone.
[3] 'mild fires', 'moderate fires' and 'severe fires' represent fire frequency.
[4] 'MAR' is mean annual rainfall, derived from the computer programme, ANUClim (version 5.1, Australian National University, Canberra).

carbon stocks observed at Kapalga agree remarkably well with the predictions of our model based on the Three Parks dataset (Figure 14.5).

The Munmarlary fire experiment produced broadly similar results to those from Kapalga. At Munmarlary there were four treatments (no fires, early annual fires, biennial fires and late annual fires) applied over 21 years in the case of open forest areas, and 23 years in the case of woodland areas (Russell-Smith *et al.* 2003). Fire intensities were generally much lower than those at Kapalga.

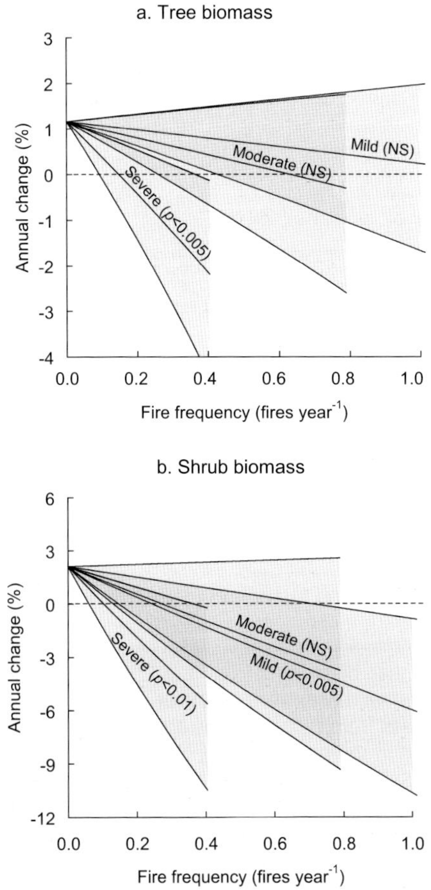

Figure 14.4 The modelled relationship between annual change in carbon stocks in (a) tree and (b) shrub biomass and the frequency of mild, moderate and severe fires. The predictions are based on averaging of the entire candidate sets shown in Table 14.2, weighted according to w_i. The shaded areas represent the 95% confidence interval associated with fire frequency.

In open forest areas, plots subject to less-severe fire treatments experienced increases in tree basal area, with the greatest increase in the unburnt plots, followed by the biennial fire and early annual fire plots (2.3%, 1.9% and 0.9% annually, respectively). In contrast, late annual fire plots experienced declines in tree basal area of 1.6% annually. These observations are consistent with the view that frequent severe fires result in declines in tree biomass.

In the woodland areas at Munmarlary, there were no significant differences between fire treatments in terms of their effect on tree basal area, despite the fires in the woodland areas typically being of greater intensity than those in the open forest areas. Strikingly, tree basal area increased markedly in nearly all treatments, including late annual fires. In the unburnt, early annual and late annual fire plots, basal area increased by 4.0%, 4.9% and 3.7% annually, respectively, whereas the biennial fire plot only increased by 0.3% annually. Rates of increase as high as 4–5% annually, doubling to tripling basal area over the duration of the study, cannot be explained in terms of the fire treatments alone. It is possible that historical site factors are responsible, such as severe disturbance prior to the establishment of the fire experiment. Possible sources of major disturbance include buffalo, which were effectively eradicated from the area during the study period (Russell-Smith *et al.* 2003), storm damage, or a very severe

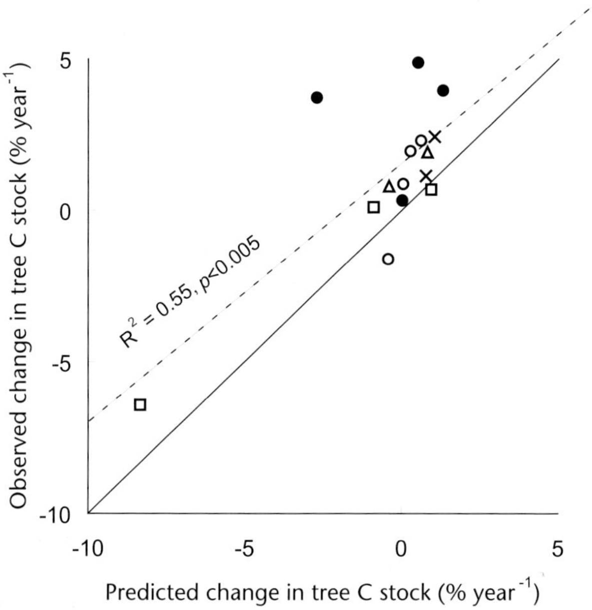

Figure 14.5 Results of a number of studies from Top Ends savannas, examining the effect of fire regimes on tree abundance (either basal area or cover, which is converted to biomass here). The observed changes in tree biomass from those studies are plotted against the changes predicted by the regression equations developed in the present study. The data are from Kapalga (Williams *et al.* 1999: squares), Munmarlary (Russell-Smith *et al.* 2003: empty circles = open forest, filled circles = woodland), greater Kakadu (Lehmann *et al.* 2009: triangles) and Solar Village near Darwin (Woinarski *et al.* 2004: crosses). The dashed line represents a line of best fit for predicted versus observed changes in tree carbon stocks, and the solid line represents an ideal, one-to-one relationship.

prior fire regime. It is also possible that the effects of elevated rainfall and atmospheric CO_2 concentration during the study period, both of which are likely to increase tree biomass, may have had a particularly pronounced effect in woodland areas, possibly due to greater water limitation in these areas.

Two other datasets provide insights into the effect of fire on tree biomass in the Top End. Near Darwin, Woinarski *et al.* (2004) compared tree basal area in an area subject to fire exclusion for 23 years with an adjacent area subject to the ambient fire regime, characterised by frequent annual fires. Relative to the frequently burnt area, tree carbon stocks in the unburnt area thickened by between 1.1% and 2.4% per year. Lehmann *et al.* (2009) found that changes in tree cover in Kakadu over a 20-year period, were related to fire frequency. If we convert tree cover change to basal area change using the equation provided by those authors, their modelling results suggest that unburnt areas had tree carbon stocks increasing at a rate of about 1.9% annually, whereas areas burnt annually had tree carbon stocks increasing at a rate of only 0.8% annually.

In summary, the changes in tree carbon stocks suggested by our model – based on the Three Parks dataset – are remarkably consistent with three published studies of tree biomass in Top End savannas, including the Kapalga fire experiment (Williams *et al.* 1999; Woinarski *et al.* 2004; Lehmann *et al.* 2009) (Figure 14.5). However, our predictions seem to be much less consistent with the results from Munmarlary (Russell-Smith *et al.* 2003), primarily due to the anomalously high rates of thickening in the woodland plots. We suggest that this thickening is probably related to the recent history of that particular site. Despite the limited agreement with the Munmarlary woodland results, there was still good overall agreement between our model's predictions and the changes reported in the published literature from Top End savannas ($R^2 = 0.55$, $p<0.005$; Figure 14.5). Given this agreement – and the strong theoretical basis for a relationship between severe fire frequency and changes in tree biomass – we conclude that our model of change in tree carbon stocks is robust, and a suitable tool for suggesting the likely response of tree carbon stocks to fire management in Top End savannas. In contrast, we have not been able to validate our model of change in shrub carbon stocks, due to a lack of published results, and we therefore have less confidence in its predictions. However, given the very minor contribution of shrubs to total woody biomass carbon stocks (Figure 14.1), we consider shrub biomass to play a relatively unimportant role in the woody biomass carbon budget, and consider the greater uncertainty associated with our shrub carbon stock model to be acceptable.

A CASE STUDY OF WESTERN ARNHEM LAND

Until very recently, despite the work of a number of groups and individuals, fire management in western Arnhem Land has been characterised by a lack of the coordinated effort needed to deal with the threat of frequent large, late dry season fires. It is clear that the fire regime that prevailed had adverse impacts on biodiversity values, especially relatively fire-sensitive obligate seeder plants and relatively immobile vertebrate taxa (Chapters 10 and 8, respectively), and negative impacts on carbon stocks also seem likely (Williams *et al.* 2004; Cook *et al.* 2005; Beringer *et al.* 2007). Recently, the area has also supported the West Arnhem Land Fire Abatement (WALFA) project that aims to use fire management to reduce emissions of non-CO_2 greenhouse gases such as methane and nitrous oxide (Chapters 12 and 13). It seems likely that the improvements in fire management associated with the WALFA project will have additional benefits in terms of increasing carbon stored in woody biomass, yet there has been little opportunity to estimate the likely magnitude of these benefits. For these reasons, western Arnhem Land presents a good opportunity to consider the likely effects of different fire-management scenarios on woody biomass carbon stocks, using the models we have developed from the Three Parks dataset.

Management scenarios

Here we compare the likely effects of two management scenarios on woody biomass carbon stocks in western Arnhem Land. The first scenario is one where limited, and mostly localised, fire management is carried out, and fire regimes are dominated by late dry season (LDS) fires. This is effectively the scenario that has occurred in recent decades. For example, during the period 1995–2004, when little fire management occurred, around 44% of the savanna vegetation of western Arnhem Land was burnt each year, mostly (31%) during the late dry season (Chapter 9).

The second scenario is one we consider to be a realistic target of fire management in western Arnhem Land, given the current availability of resources including industry support for non-CO_2 greenhouse gas abatement. Specifically, we consider a modest reduction in overall fire frequency (from 44% to 33%) and, more importantly, a larger reduction in LDS fire frequency (from 31% to 17%) to be a substantial improvement in fire management that – based on practical experience from 2005 to 2007 – can plausibly be achieved in the near future.

Likely effects on carbon stocks

Under a scenario of limited fire management, our model suggests very little change in woody biomass carbon stocks in the long-term (100 years), with an increase of just 0.6 t C ha^{-1} expected (Figure 14.6). Under the improved fire management scenario, however, our model suggests a

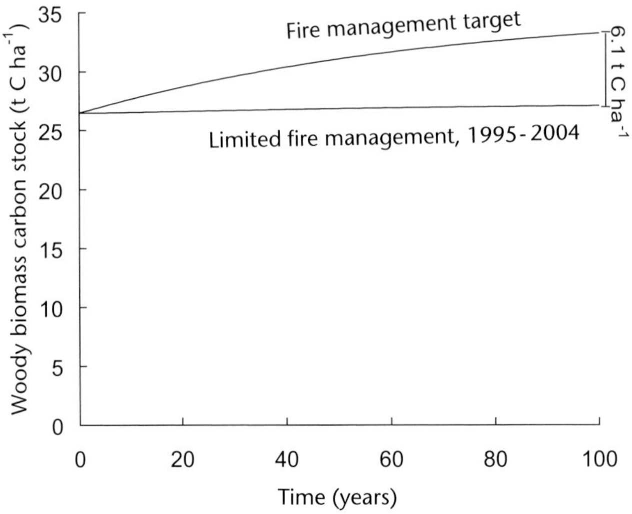

Figure 14.6 The predicted change in woody biomass carbon stocks in savanna woodlands in the West Arnhem Land Fire Abatement project area, in response to two management scenarios: no fire management, typical of the period 1995–2004; and a realistic target of fire management (see text for details). Initial tree and shrub carbon stocks (25.8 t C ha^{-1} and 0.72 t C ha^{-1}, respectively) were the mean values for the Kakadu monitoring plots in 2004.

substantial increase in carbon stocks, around 6.8 t C ha^{-1}. The difference between these two outcomes (6.1 t C ha^{-1}: discrepancy due to rounding) represents the additional carbon that would be sequestered into woody biomass because of the improvement in fire management.

It is important to put this figure of 6.1 t C ha^{-1} into an appropriate context. Expressed as a carbon dioxide equivalent (CO_2-e) – the standard unit of carbon trading – this figure would be about 22 t CO_2-e ha^{-1}, or 0.22 t CO_2-e ha^{-1} yr^{-1}, for the next 100 years. Although this may seem relatively small, it is worth considering that the improved fire management is likely to result in a non-CO_2 greenhouse gas abatement – that is, the type of abatement that underpins the WALFA project – of just 0.04 t CO_2-e ha^{-1} $year^{-1}$, calculated using the method outlined in Chapter 13. Thus, it seems likely that in the first century of a greenhouse gas abatement project such as WALFA, carbon sequestration by woody biomass could be several times greater than the non-CO_2 greenhouse gas abatement that is currently allowed under current greenhouse gas accounting rules.

We should point out that, although the models we have used to predict changes in woody biomass carbon stocks in response to fire management scenarios seem generally robust, they have an inherent flaw. That is, they are imprecise phenomenological models based on observations taken during a period of higher than average rainfall. In addition to low precision and resultant high levels of uncertainty about quantitative estimates, there is a risk of bias. They are not mechanistic or theory based models and hence unable to take into account the major environmental changes expected in the coming decades, such as increased atmospheric CO_2 concentration and temperature. It is difficult to predict how such changes will affect the relationships between fire management and woody biomass. However, it has been suggested that increased CO_2 concentrations are likely to increase the difference in woody biomass between areas with severe fire regimes (i.e. unmanaged) and relatively benign regimes (i.e. well managed) (Bond *et al.* 2003). Similarly, we could expect a temperature increase to result in more extreme fire weather, and hence a greater frequency of severe fires, such that enhanced fire management would become even more important for maintaining woody biomass. In summary, it seems that our predictions of the carbon sequestration benefits of fire management are likely to be conservative in the face of increased CO_2 concentrations and increased temperatures.

Although we have not specifically considered soil carbon stocks – the largest carbon pool in savannas – it is likely that improvements in fire management would also affect these. However, the nature of the relationship between soil carbon stocks and fire frequency and severity remains unclear, with soil carbon stocks having been demonstrated to both increase (Bird *et al.* 2000) and decrease (Chen *et al.* 2005; Ansley *et al.* 2006) in response to reductions in fire frequency. In order to validate the notion that improvements in fire management can result in increased carbon storage by savannas, understanding the dynamics of soil carbon stocks must be a research priority.

CONCLUSION

The Three Parks vegetation and fire monitoring dataset provides an excellent opportunity to examine changes in woody biomass carbon stocks across a large expanse of savanna vegetation over a decade. From our analysis of this dataset, it is clear that: (1) Top End savannas, in general, are not thickening; although (2) reductions in the frequency of severe fires can lead to a substantial increase in woody biomass carbon stocks over time.

Our conclusion that Top End savannas are not thickening contradicts a widely held view that has emerged in northern Australian savanna ecology in recent years. Northern Australia's mesic savannas are generally considered to be sequestering 1–2 t C ha^{-1} yr^{-1}, with most of this carbon presumably going into woody biomass. We suggest that this misconception has arisen primarily because published measurements of NEP have been confined to several sites near Darwin. It is likely that these sites are still recovering from tropical cyclone damage sustained in 1974 and may therefore have atypically high rates of NEP. Determination of NEP for a wider range of mesic savanna sites throughout northern Australia will lead to a better understanding of the true variation of NEP values, and hence rates of carbon sequestration. Such research is currently being undertaken by a team led by Lindsay Hutley at Charles Darwin University.

We consider discussions about whether northern Australia's savannas are presently a carbon source or sink irrelevant to attempts to manage fire regimes in a way that generates carbon offsets. The more important issue is whether fire management can shift carbon fluxes in a desired direction, regardless of the starting point. The Three Parks dataset suggests this is possible, with changes in woody biomass carbon stocks related to the frequency of fires, especially severe fires. This finding is certainly not ground-breaking, with essentially identical conclusions drawn from the Kapalga and Munmarlary fire experiments. The advantage of the Three Parks dataset, however, is that it allows us to examine the response of woody structure to a diverse, ambient fire regime, rather than a limited number of unrealistic fire treatments. For this reason, inferences about the effects of fire management on woody biomass carbon stocks are more readily drawn using the Three Parks analysis.

Although our statistical models are imprecise, application to our case study of western Arnhem Land suggests that active fire management may offer substantial benefits in terms of increasing woody biomass carbon stocks. Over a century, the carbon offset associated with this increase may be larger than that generated by abatement of non-CO_2 greenhouse gases. Although current greenhouse gas accounting arrangements don't incorporate sequestration by savannas, such arrangements may change in the future – especially if the relationship between fire management and carbon sequestration can be more confidently measured. Given the indication that carbon offsets could be generated by carbon sequestration in savannas flowing from plausible improvements in fire management, further research in this area should be a priority.

ACKNOWLEDGEMENTS

Thanks go to the staff of the three parks, who undertook the vegetation and fire assessments, as well as the numerous botanists who assisted. Lynda Prior provided helpful comments on an earlier version of the manuscript.

REFERENCES

Ansley RJ, Boutton TW and Skjemstad JO (2006) Soil organic carbon and black carbon storage and dynamics under different fire regimes in temperate mixed-grass savanna. *Global Biogeochemical Cycles* **20**, GB3006. <doi:10.1029/2005GB002670>.

Becker P, Sharbini N and Yahya R (1999) Root architecture and root : shoot allocation of shrubs and saplings in two lowland tropical forests: Implications for life-form composition. *Biotropica* **31**, 93–101.

Beringer J, Hutley LB, Tapper NJ and Cernusak LA (2007) Savanna fires and their impact on net ecosystem productivity in North Australia. *Global Change Biology* **13**, 990–1004.

Bird MI, Veenendaal EM, Moyo C, Lloyd J and Frost P (2000) Effect of fire and soil texture on soil carbon in a sub-humid savanna (Matopos, Zimbabwe). *Geoderma* **94**, 71–90.

Bond WJ, Midgley GF and Woodward FI (2003) The importance of low atmospheric CO_2 and fire in promoting the spread of grasslands and savannas. *Global Change Biology* **9**, 973–982.

Bond WJ, Woodward FI and Midgley GF (2005) The global distribution of ecosystems in a world without fire. *New Phytologist* **165**, 525–538.

Bucini G and Hanan NP (2007) A continental-scale analysis of tree cover in African savannas. *Global Ecology and Biogeography* **16**, 593–605.

Burnham KP and Anderson DR (2002) *Model Selection and Multimodel Inference: A Practical Information-theoretic Approach*. Springer-Verlag, New York.

Burrows WH, Henry BK, Back PV, Hoffmann MB, Tait LJ, Anderson ER, Menke N, Danaher T, Carter JO and McKeon GM (2002) Growth and carbon stock change in eucalypt woodlands in northeast Australia: ecological and greenhouse sink implications. *Global Change Biology* **8**, 769–784.

Cernusak LA, Hutley LB, Beringer J and Tapper NJ (2006) Stem and leaf gas exchange and their responses to fire in a north Australian tropical savanna. *Plant Cell and Environment* **29**, 632–646.

Chen XY, Hutley LB and Eamus D (2003) Carbon balance of a tropical savanna of northern Australia. *Oecologia* **137**, 405–416.

Chen XY, Hutley LB and Eamus D (2005) Soil organic carbon content at a range of north Australian tropical savannas with contrasting site histories. *Plant and Soil* **268**, 161–171.

Cook GD, Liedloff AC, Eager RW, Chen X, Williams RJ, O'Grady AP and Hutley LB (2005) The estimation of carbon budgets of frequently burnt tree stands in savannas of northern Australia, using allometric analysis and isotopic discrimination. *Australian Journal of Botany* **53**, 621–630.

Dwyer E, Pinnock S, Grégoire J-M and Pereira JMC (2000) Global spatial and temporal distribution of vegetation fire as determined from satellite observations. *International Journal of Remote Sensing* **21**, 1289–1302.

Eamus D, Hutley LB and O'Grady AP (2001) Daily and seasonal patterns of carbon and water fluxes above a north Australian savanna. *Tree Physiology* **21**, 977–988.

Gifford RM (2000) *Carbon Contents of Above-ground Tissues of Forest and Woodland Trees.* Australian Greenhouse Office, Canberra.

Grace J, San Jose J, Meir P, Miranda HS and Montes RA (2006) Productivity and carbon fluxes of tropical savannas. *Journal of Biogeography* **33**, 387–400.

Grainger J and Becker P (2001) Root architecture and root:shoot allocation of shrubs and saplings in a Bruneian heath forest. *Biotropica* **33**, 363–368.

Higgins SI, Bond WJ, February EC, Bronn A, Euston-Brown DIW, Enslin B, Govender N, Rademan L, O'Regan S, Potgieter ALF, Scheiter S, Sowry R, Trollope L, Trollope WSW (2007) Effects of four decades of fire manipulation on woody vegetation structure in savanna. *Ecology* **88**, 1119–1125.

Lehmann CER, Prior LD and Bowman DMJS (2009) Multi-decadal dynamics of tree cover in an Australian tropical savanna. *Austral Ecology,* in press.

Lehmann CER, Prior LD, Williams RJ and Bowman DMJS (2008) Spatio-temporal trends in tree cover of a tropical mesic savanna are driven by landscape disturbance. *Journal of Applied Ecology* **45**, 1304–1311.

Menaut JC and Cesar J (1979) Structure and primary productivity of Lamto savannas, Ivory Coast. *Ecology* **60**, 1197–1210.

O'Grady AP, Chen X, Eamus D and Hutley LB (2000) Composition, leaf area index and standing biomass of eucalypt open forests near Darwin in the Northern Territory, Australia. *Australian Journal of Botany* **48**, 629–638.

R Development Core Team (2008) *R: A Language and Environment for Statistical Computing.* R Foundation for Statistical Computing, Vienna.

Russell-Smith J and Edwards AC (2006) Seasonality and fire severity in savanna landscapes of monsoonal northern Australia. *International Journal of Wildland Fire* **15**, 541–550.

Russell-Smith J, Whitehead PJ, Cook GD and Hoare JL (2003) Response of Eucalyptus-dominated savanna to frequent fires: lessons from Munmarlary, 1973–1996. *Ecological Monographs* **73**, 349–375.

San Jose JJ, Montes RA and Farinas MR (1998) Carbon stocks and fluxes in a temporal scaling from a savanna to a semi-deciduous forest. *Forest Ecology and Management* **105**, 251–262.

Sankaran M, Ratnam J and Hanan NP (2004) Tree-grass coexistence in savannas revisited – insights from an examination of assumptions and mechanisms invoked in existing models. *Ecology Letters* **7**, 480–490.

Sankaran M, Hanan NP, Scholes RJ, Ratnam J, Augustine DJ, Cade BS, Gignoux J, Higgins SI, Le Roux X, Ludwig F, Ardo J, Banyikwa F, Bronn A, Bucini G, Caylor KK, Coughenour MB, Diouf A, Ekaya W, Feral CJ, February EC, Frost PGH, Hiernaux P, Hrabar H, Metzger KL, Prins HHT, Ringrose S, Sea W, Tews J, Worden J, Zambatis N (2005) Determinants of woody cover in African savannas. *Nature* **438**, 846–849.

Sankaran M, Ratnam J and Hanan N (2008) Woody cover in African savannas: the role of resources, fire and herbivory. *Global Ecology and Biogeography* **17**, 236–245.

Scholes RJ and Archer SR (1997) Tree-grass interactions in savannas. *Annual Review of Ecology and Systematics* **28**, 517–544.

Tilman D, Reich P, Phillips H, Menton M, Patel A, Vos E, Peterson D and Knops J (2000) Fire suppression and ecosystem carbon storage. *Ecology* **81**, 2680–2685.

Valladares F, Wright SJ, Lasso E, Kitajima K and Pearcy RW (2000) Plastic phenotypic response to light of 16 congeneric shrubs from a Panamanian rainforest. *Ecology* **81**, 1925–1936.

Werner PA and Murphy PG (2001) Size-specific biomass allocation and water content of above- and below-ground components of three Eucalyptus species in a northern Australian savanna. *Australian Journal of Botany* **49**, 155–167.

Williams RJ, Cook GD, Gill AM and Moore PH (1999) Fire regime, fire intensity and tree survival in a tropical savanna in northern Australia. *Australian Journal of Ecology* **24**, 50–59.

Williams RJ, Hutley LB, Cook GD, Russell-Smith J, Edwards A and Chen XY (2004) Assessing the carbon sequestration potential of mesic savannas in the Northern Territory, Australia: approaches, uncertainties and potential impacts of fire. *Functional Plant Biology* **31**, 415–422.

Woinarski JCZ, Risler J and Kean L (2004) Response of vegetation and vertebrate fauna to 23 years of fire exclusion in a tropical *Eucalyptus* open forest, Northern Territory, Australia. *Austral Ecology* **29**, 156–176.

15

Fire management futures:

new options for environmental and socioeconomic benefit

Peter J Whitehead, Jeremy Russell-Smith and Peter M Cooke

SUMMARY

In the preceding chapters, we traced the factors that led to present poor standards of fire management in northern Australia and explored issues affecting capacity to meet our collective obligations to care for country. Our many contributors covered: the social history that disrupted long established modes of fire use; the persistence of commitment and survival of knowledge among Indigenous people despite this history; developments in the science and technology of greenhouse gas emissions abatement; options for trading carbon credits; opportunities for enhanced conservation of biodiversity; pathways for realising social benefits and promoting cultural renewal from enhanced fire-management activity; and institutional arrangements to support all of these outcomes. Here, we consider interactions among these issues and with rapidly evolving national and international policy and legal settings. We show how these interactions could be shaped to optimise outcomes for northern Australia. Our analysis is based on the expectation that Australia will seize the opportunity to apply national and international responses to the climate change challenge in ways that generate multiple environmental and social benefits for northern Australia. This optimistic view requires creative approaches from government to improved management of the biosphere for greenhouse gas (GHG) abatement and biosequestration. The global carbon cycle connects centralised industrial development to the remotest parts of the northern Australia. To date, those connections appear to have brought more costs than benefits. The new carbon markets should be deployed to bring some of the benefits of development to those who are motivated to generate new products in novel ways, while also contributing to national well-being. The greatest challenges to realising this vision lie not in the formal science of GHG mitigation, but in governance arrangements for new business systems that must be fit for application at regional scales, and associated arrangements for Indigenous people to gain equitable access to financial returns from those systems.

SOCIAL HISTORY AND CONTEMPORARY CONDITIONS

Over the last 150-odd years, the decisions of Indigenous people to move to centres of settler activity offering access to goods and services, population decline through disease, violence and other social dysfunction, plus active government-supported programs to remove people from their lands, all contributed to loss of fire managers from large areas, including western Arnhem Land (Chapters 2, 3 and 4). The creation of what Cooke (Chapter 4) has called a 'modern wilderness' weakened capacity to manage land and resources. Subsequent 'return to country' through the outstations movement brought some benefits (e.g. Bowman *et al.* 2001), but did not itself fully restore this capacity, because the return was incomplete and inadequately resourced. Moreover, the use of space and patterns of movement had changed to include regular interaction with the mid-sized regional settlements (Taylor *et al.* 2006) and reduced use of traditional routes for accessing seasonally abundant resources.

It is also clear that the creation of and residence in, or regular access to, these larger settlements did not bring major social benefits or connections to the mainstream economy. The Northern Australian society and economy are characterised by relatively strong economic growth (Access Economics 2008), with good prospects for future minerals and energy production, but dysfunctional socioeconomic development. In effect, there is a dual socio-economy, with a strong export sector and a few advantaged major urban centres existing alongside, and contrasting with, considerable social disadvantage in the remote, mostly Indigenous, domain (Gerritsen 2008).

The contemporary socioeconomic situation of the Indigenous domain can be illustrated by reference to the 2006 Australian Census (ABS 2008a) and in particular data from the Statistical Local Areas (SLAs) approximating the WALFA boundaries (Figure 15.1). WALFA straddles large parts of the West Arnhem and Nyirranggulung Mardrulk Ngadberre SLAs. The project draws most of its participants from the combined area of these SLAs and the associated Gunbalanya township (Kunbarllanjnja SLA).

The total population (2006 Census, based on place of usual residence) of these areas (62 432 km^2) combined is 5016, with 92.7% identifying themselves as Indigenous. The average population density for the project region is very low at 0.08 persons.km^{-2}. The largest settlement is Maningrida, which is within the West Arnhem SLA but outside the north-east margin of the WALFA project area, with a population of 2063 – 92.4% being Indigenous. Relatively high levels of Indigenous representation in the resident population are typical of remote Australia, but for this region the representation is much higher than the average for other parts of the nation also classified by the Australian Bureau of Statistics as remote (15%) or very remote (48%) (ABS 2008b). It is important to note that the Census suffers from substantial undercounts, especially of Indigenous people in remote areas (Taylor *et al.* 2006), so the Indigenous representation is also likely to be under-estimated.

Indigenous population structure

A large proportion of the region's Indigenous population is under 15 years of age (35.1%), and a small number over 55 (7.0%). For non-Indigenous Australia, the corresponding average figures are 14.2% and 22.0% respectively. These differences reflect relatively high fertility among Indigenous females and premature mortality in both sexes. The total life-span of the Indigenous population is about 17 years shorter than for non-Indigenous Australians. Health is often poor across the life-span (Burgess *et al.* 2005).

Although high mortality presently suppresses overall rates of growth in the Indigenous population, in the foreseeable future the relative rate of population growth will continue to exceed non-Indigenous rates of population growth in remote areas. More importantly, from the perspective of this study, the Indigenous population of working age appears likely to

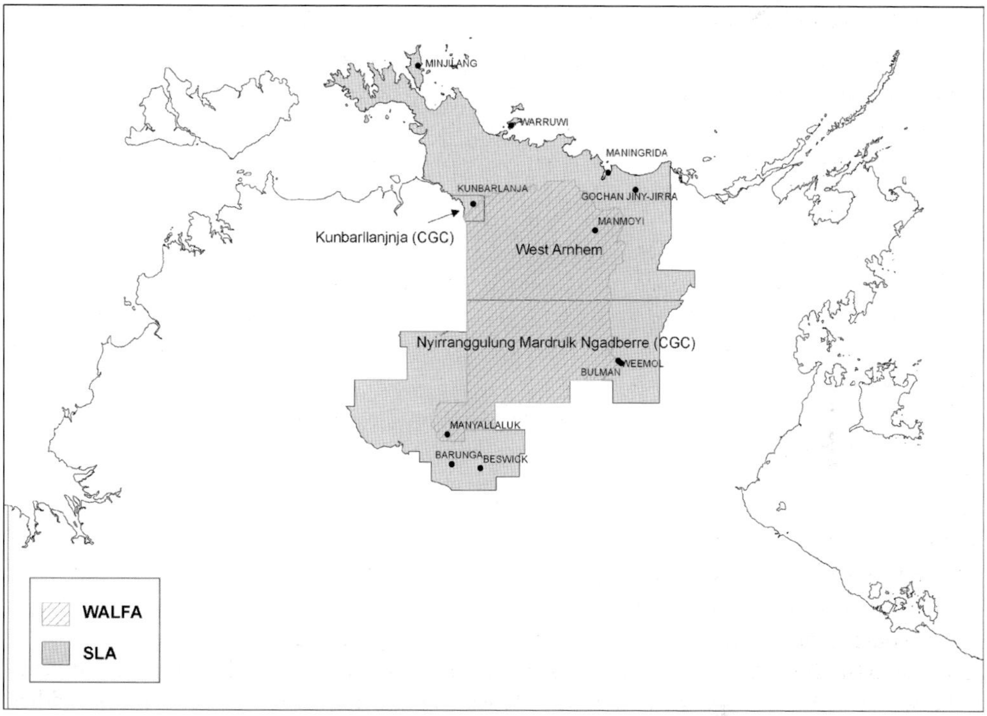

Figure 15.1 Boundaries of statistical local areas (SLAs) that encompass the WALFA project area and places of residence of most of the Indigenous people engaging with it.

continue to increase in remote areas at a much higher rate than plausible rates of increase in employment opportunities of a sort accessible to Indigenous residents – adding to the large proportion of the community already unable to access jobs (e.g. Taylor 2006).

Education

Systemic failure of education systems in remote and very remote regions has particularly disadvantaged Indigenous people. In the WALFA 'catchment', the proportion of Indigenous people who report no formal schooling at all (7.8%) substantially exceeds the number completing secondary schooling (6.8%): an additional 34.4% have not progressed beyond year 8 (first year of high school). None of the region's non-Indigenous population reported having no schooling, and 55.2% had completed year 12. Clearly, much of the Indigenous population will experience great difficulty in accessing the limited available mainstream employment.

Housing

Housing used by Indigenous people in the region is often overcrowded. Of households occupying dwellings with three or fewer bedrooms, 45.2% accommodated more than seven and 21.5% more than 10 persons. The average number of resident people per bedroom in Indigenous households in SLAs within the WALFA area ranged from 2.1 to 2.9.

Employment and incomes

The dominant employment categories reported in the 2006 Census for Indigenous residents were labourer (44.6%) and community/personal service worker (18.6%). Non-indigenous employment was concentrated in management (16.5%) and professional roles (38.4%). The industry categories in which Indigenous employment was most concentrated were public administration and safety (49.2%) and health care and social assistance (23.9%). For non-Indigenous people, 40.2% were in public administration and 18.5% in education and training.

These figures reflect the limited presence of private employers in the region. There are no major mines, and the livestock industry is confined to harvesting feral buffalo in part of the region. Harvesting wildlife (e.g. royalties for crocodiles eggs going to crocodile ranches located outside the region) generates some incomes. Tourism is limited principally to 'safari' camps for sports fishing and hunting managed from the major settlements. Significant incomes are derived from the sale of Indigenous art and crafts (e.g. MAC 2007). The Indigenous customary economy continues to be important in some parts of the region, especially outside the larger centres (Altman 2004), but details are not captured in the Census.

The modal income category for Indigenous people in the region was $150–249 per week and 57.1% of persons aged over 15 years were in this category. The median Indigenous wage varied slightly among SLAs, ranging from $185 to $207 per week. A large proportion of Indigenous people also reported no income (11.1%). The Australian average weekly earnings were about $1180 at the time of the 2006 Census (ABS 2008b). The great majority (92.6%) of those in the region earning more than $1000 per week were non-Indigenous.

The modal Indigenous income coincides with rates payable under the Community Development Employment Scheme (CDEP): an employment program that underpins basic service delivery in communities and, less frequently, activities such as land (and fire) management. Many Indigenous Rangers (see Northern Land Council (2006 for a description of the Caring for Country Ranger groups) working on the WALFA project had all, or a substantial portion of, their income paid under this scheme.

This snapshot of conditions for the Indigenous population illustrate both the need for new sources of income and meaningful employment and – because of the failure of educational systems and low standards of health – the difficulties Indigenous residents face in accessing any opportunities that may become available in mainstream activities.

Role of outstations

Altman (2006) shows for a portion of the WALFA area that distribution of indices of socioeconomic condition for residents of the larger regional settlements differ somewhat from residents of outstations. Employment and health status appear better at outstations, and education and home ownership better at townships. Higher formal employment probably reflects greater involvement in the CDEP scheme for people on outstations. The median adult cash income at outstations and townships is the same. It would appear that aggregation into two larger settlements, occupied by up to 60% of the total population, itself does little to lift economic or social outcomes. But aggregation has obvious advantages for government in lowering the cost of delivering basic services.

Despite the equivocal evidence base, outstations have been described as cultural museums that condemn residents to permanent, institutionalised disadvantage, and smaller aggregations as non-viable (Vanstone 2005). Costs and deficiencies in service delivery, especially in education and health, have motivated attempts to rationalise services to outstations (Altman 2006). The Northern Territory Government has sought comment on an outstations discussion paper (Northern Territory Government 2008), which makes clear that no new outstations will be funded, and explores options for matching present service delivery to outstation population. Discussion will occur against a background of calls for a return to a policy of remote depopulation, with some decision makers and community leaders arguing that outstations should be abandoned in favour of relocation to larger regional townships.

We suggest that the WALFA experience presents an alternative view of the value of small communities, such as the Kabulwarnamyo outstation, in northern Australia: as an important investment in infrastructure to promote delivery of land-management services for which there is growing demand both nationally and internationally. Focus on aggregating people in larger settlements may compromise capacity to deliver ecosystem services that demand familiarity with, and regular presence on, country. Recent Northern Territory policy on homelands argues 'that the future ... lies in ... innovation and ... emerging economic opportunities' (see www.workingfuture.nt.gov.au). It remains to be seen how support will be matched to innovation and economic opportunity.

INDIGENOUS KNOWLEDGE AND SKILLS

The persistence of Indigenous knowledge and skills regarding fire in the western Arnhem Land area appears to owe much to the small number of people who spent a large part of their lives outside the major settlements and in the plateau (e.g. Yibarbuk *et al.* 2001). These people were much less constrained by settler and other interests in their approaches to management of land than Indigenous owners or residents of other lands (Chapter 2), and probably more strongly dependent on resources managed with, or accessed using, fire (Chapters 5 and 6). Whatever the explanation, it is clear that fire managers working in the region have access to a rich vein of sophisticated Indigenous knowledge and related skills.

Those skills extend to capacities to predict the behaviour of fire under different conditions and landscape types, and the effects those fires will have on vegetation and resources used by wildlife over the short and long term (Chapter 5). Obviously, many of these skills are essential for effective design and implementation of fire-management plans, irrespective of the goals. But the comprehensiveness of knowledge is particularly important for ensuring that techniques adopted to reduce GHG emissions do not damage other values. The record provided by collaborators with Murray Garde (Chapter 5) provide an invaluable framework for facilitating dialogue between traditional owners and other Indigenous and non-Indigenous interests for setting shared objectives and working jointly to achieve them.

This group of Indigenous informants also remains available to transfer knowledge actively to other Indigenous participants who will possess varying levels of knowledge about fire use in the stone country. Younger people in particular have much to learn. Oral history and ethnographic studies of the sort presented here may also offer an adjunct to direct contact with senior informants, especially if integrated with formal educational programs. There are opportunities to exploit the many facets of the fire management discipline as a vehicle for learning basic skills in numeracy and literacy, as well as applied knowledge in chemistry, physics and technology (remote sensing and computing). Otherwise abstract concepts can be grounded and rendered relevant though linkage to Indigenous knowledge in fire behaviour, climate and ecology and connected to social obligations to look after land and resources.

It is difficult to know how many other regions in northern Australia have sustained such rich sources of Indigenous fire knowledge. However, experience in other disciplines (ethno-botany, ethno-zoology) suggests that provided effort is made to engage with people recognised for their expertise within their communities, it will be possible to bring valuable locally relevant knowledge to bear on design and execution of equivalent programs in other regions.

FORMAL SCIENCE AND REGIONAL FIRE MANAGEMENT

The formal science supporting the WALFA project is obviously of more recent origin than the Indigenous, although it draws on a relatively long tradition of high-quality fire research in

Australia (Chapter 7). Major accomplishments include: cataloguing the fire sensitivity of most plant species occurring in the savannas; ongoing enhancement of capacity to detect and map fire using remote sensing technology; descriptive and sometimes quantitative models of the impact of fire of different seasonality and intensity on vegetation dynamics (composition and structure); long-term studies on relationships between fire regimes (frequency, timing and intensity) and vegetation dynamics in many landscape types, based on permanent plots; understanding of GHG emissions from savanna fire and their variation with season; and integration of relevant information in comprehensive methodologies to estimate GHG emissions at regional scales.

Gaps, or areas needing refinement, relate particularly to improved methods for (automated) mapping of fire incidence and area, improved understanding for seasonal variation in emissions factors, and hence model parameters, and better information on above and below-ground bio-sequestration and its variation with fire regime.

Although existing methods for estimating emissions from savanna fire in the plateau represent best practice (Russell-Smith *et al.* 2009), it is desirable that accuracy and precision of estimates be improved so that practice can be refined to optimise environmental benefits and increase confidence in fire-related offsets that may be offered for sale in commercial markets (Whitehead *et al.* 2008). Over the longer term, with experience in a wider range of environments, it may be possible to identify patterns in quantitative relationships between emissions behaviour and environmental conditions and vegetation types. Such capability would facilitate identification of additional sites where substantial reductions in emissions are plausible, and help to design efficient interventions to achieve mitigation.

We have also identified better understanding of the potential for enhanced bio-sequestration in savannas through fire management as an important issue (Chapters 7 and 14). In part, this question will be addressed by continuation of studies in permanent fire plots, but it is desirable that additional studies be done to explore the mechanics of increased carbon storage in both woody biomass and soils and in a wider range of landscapes and land-management situations than are encompassed by the existing fire plots.

BIODIVERSITY CONSERVATION THROUGH FIRE MANAGEMENT

It will also be necessary to link such work to measures of biodiversity outcomes. In the Arnhem Land Plateau, it appears likely that the fire regimes that reduce annual emissions (and perhaps enhance bio-sequestration) will also generate substantial biodiversity benefits because they reduce frequency of intense fires below ecological thresholds (Chapter 9). Early in the process of redressing very poor contemporary fire management, gains in both biodiversity and GHG emissions reduction are readily achieved. But adding increments may become progressively more challenging and encounter diminishing returns on additional effort.

Indeed, in mature managed systems it is possible that biodiversity decrements will be encountered at some point; if, for example, additional reductions in emissions result in excessive increase in density of woody stems that compromise some features of habitat important for fauna, or see the displacement of disturbance-dependent plants. The presently available information suggests that in the stone country such change will be slow because the rate of change in stem density and plant biomass is low (Chapter 14). However, many purchasers of fire-related credits in compliance and (especially) voluntary markets are likely to seek robust demonstration that generating benefits from reduced emissions do not create costs in other areas. Moreover, separate markets are likely to develop for biodiversity credits (e.g. Mansfield 2008), which will demand high standards of verification, and hence development of robust metrics. Optimal balance between GHG benefits and biodiversity may vary among ecosystem types. For example, in the absence of fire or other disturbance, some wetlands can become dominated by a single vegetation type in just a few years (Whitehead and McGuffog 1997).

OPTIONS FOR TRADING CARBON CREDITS

We have concluded that the most favourable treatment of credits from abatement of fire-related GHG emissions will be to trade them as offsets in both compliance and voluntary markets. As noted above, achieving this goal will require the highest standards of independent validation of credits claimed, and robust institutions for managing other forms of risk. But perhaps the greatest risk is that government policy will be determined on other grounds and the potential gains never realised.

The Australian Government White Paper provides some insights into what those other grounds might be. There is a strong pre-occupation with accounting and the challenges it creates, particularly in regard to the dynamics of the biosphere and in industries such as agriculture that involve strong interactions with the biosphere. Agriculture has been deferred from direct inclusion in the Carbon Pollution Reduction Scheme (CPRS) – with a decision to be made in 2013 – primarily because of difficulties in measurement and attribution of liability. Deforestation and deforestation avoided have not been considered for inclusion pending development of robust international mechanisms for ensuring mitigation benefits are actually achieved. In regard to savanna fire, concerns have been raised at the complexity of property rights on Indigenous land and the associated difficulty of identifying single commercial entities to take on scheme obligations for those emissions.

In our view, both the accounting and property rights concerns are over-stated. The WALFA experience points to ways of managing collaborative research to solve complex measurement problems and to coordinate management actions over large areas. Many Indigenous institutions, including those participating in WALFA, have long experience of resolving competing claims regarding rights. There are numerous examples of mechanisms for distribution of financial benefits in the face of competing claims. In the Indigenous domain, they may sometimes be complex, and will often be contested, but it seems extraordinarily pessimistic to suggest that these sorts of difficulties should be treated as disabling, when there are opportunities to make significant contributions to mitigation and to realise important benefits for Australia's most disadvantaged people.

The national climate change response, and the CPRS as the primary emissions mitigation vehicle, involves a number of major risks that must be carefully managed. Difficulties of distributing benefits from collaborative management of fire surely pale into insignificance compared with these national challenges. We have considered (Chapter 11) institutions that will be needed to allow WALFA participants to move from arrangements with a single 'purchaser' to fruitful interactions with complex markets.

The space being given to these and other issues in government papers shows that it will be important for Indigenous interests to develop and put forward positive solutions to such concerns if they are to foster good, balanced decisions. The Australian Government has undertaken to support development of Indigenous capacity to make contributions to national mitigation efforts through allocation of $10 million to a 3-year program. But it is to be hoped that the Government will not pre-empt this work by taking decisions on accounting or other grounds that knowingly or inadvertently close options in fire management, or any related areas potentially accessible to Indigenous people, especially in bio-sequestration and deforestation avoided. The reality of such risks is illustrated by suggestions – in documents exploring development of standards for voluntary offsets – that credits from abatement of emissions from savanna burning may be ineligible for consideration (Chapter 11). The stakes for Indigenous people are too high for opportunities to be squandered for want of imagination and some additional effort.

We urge that, internationally, Australia should position itself – in negotiations at and leading to Copenhagen in 2009 – to press for recognition of abatement opportunities based on improved management of predominantly natural, but actively managed, systems. Benefits of better fire management in Australian and global savannas should not be lost in definitional dead-ends concerning illusions of permanence or hair splitting about the naturalness of systems from which credits are derived. Australia should particularly emphasise the synergies that these and related opportunities bring to national responses to obligations under a large array of other treaties. Positive treatment of well-validated fire-abatement measures will, for example, also help deliver under the Convention on Biological Diversity, and commitments in other international agreements relating to the role and status of Indigenous people in resource management.

PATHWAYS FOR SOCIAL DEVELOPMENT

Implying that getting policy for fire management right is a key issue for Indigenous people may strike some readers as over-stated. But its importance is not just its value as an enterprise in its own right, but as an exemplar of an array of possibilities, and a test of the willingness of decision makers to make serious efforts to connect policy and rhetoric on Indigenous social and economic development to other policy arenas. Without a commitment to lateral connections, then the present focus on Indigenous affairs through initiatives such as the Northern Territory intervention (see Yu *et al.* 2008) risks repeating the failures of the past. Present attention on issues of law and order, physical health and protection of children is essential, but long-term solutions to social disadvantage will require action on many fronts, including development of regional enterprise and more robust self-governing communities.

The global carbon cycle and its effects on climate connect the remotest parts of Arnhem Land with centres of industrial activity – but the connections are presently entirely negative. Regional communities feel the costs, but cannot capture the benefits. Development of global (or at least national) carbon markets offers a unique and perhaps unrepeatable opportunity to connect remote and Indigenous communities to the benefits of industrial society: and to do so in ways that other forms of development that occasionally penetrate to the regions have been unable to achieve (e.g. Kakadu Region Social Impact Study 1997; Stoeckl *et al.* 2007). As Ross Garnaut has said in regard to bio-sequestration, these markets offer options that could 'favourably transform the economic prospects of large parts of remote rural Australia' (Garnaut 2008).

The WALFA project injects \$1.0 million per annum in private funds into a region that mostly lacks private investment. And that investment is attached to contractual obligations to deliver well-specified and measurable services. It provides an opportunity for local people to design and operate businesses responsive to demands, standards and price signals originating outside the region, and so deliver on the repeated demands that Indigenous enterprise connect to the mainstream. The significance of the project runs much deeper than a single success story because it illustrates alternatives to failed approaches in emerging options for Indigenous people to realise financial benefits from land ownership.

Features of the project that have been recognised as contributing to success in Indigenous economic development and employment programs in other contexts (see Northern Territory Government (2005) and associated publications) include:

- role models

  The project is planned, authorised and supervised by senior Indigenous people whose talents have been recognised in both Indigenous and non-Indigenous society.

- Indigenous 'ownership', including Indigenous managed and controlled workplaces, or workplaces which respect Indigenous people

  The WALFA workplace is on Indigenous land and the culture of the workplace built on Indigenous norms.

- focus on alignment of opportunity, interest, aspiration, capacity and commitment

  It is difficult to imagine a project that better aligns opportunity (new markets based on ownership of land and control over resources), interest (fire management as a cornerstone of Indigenous life), aspiration (to achieve recognition of Indigenous knowledge and skills and to apply them to generate economic benefits), capacity (drawing on Indigenous knowledge and skills that are unrecognised or discounted in other situations) and commitment (demonstrated by the creation of the Ranger groups and long-term investment of effort and resources in creation of the infrastructure and institutions for regional fire management).

- partnership with other levels of government, Land Councils, private sector and community members

  WALFA has involved formal (contractual) and informal partnerships among the Northern Territory and Australian Governments, the Northern Land Council, Indigenous regional resource management bodies, Indigenous CDEP organisations, a major private sector investor and a diverse cross section of the region's people.

- realistic timeframes and long-term view

  The project was initiated in 1997 with modest goals, but because the vision was clear, was able to connect to new opportunities as they emerged. The contractual commitment with the private interest – DLNG – is long term, yet provides opportunity for review and renegotiation.

- adequate resources, realistic timeframes and effective communications and promotions of relevant programs

  Resources have come from a range of sources offering funding over various and often short time spans, but success has depended on two long-term commitments: namely that from the Tropical Savannas Management CRC underwriting the research and development work over a decade, and the private industry commitment used for operational funding, which will run for 17 years.

- training linked to a job outcome

  All WALFA training is linked directly to tasks that will be required long term and where work is available immediately, but also promote skills transfer to other forms of paid work.

- mentoring to support ongoing employment and business outcomes

  To date, mentoring has been confined to developing planning and operational skills, which are important for employment, but there has been no opportunity for mentoring in skills more directly connected to business development.

- innovation to take advantage of comparative advantages in:
  - Indigenous land ownership
  - Indigenous cultural knowledge
  - youthful demographic of the Northern Territory Indigenous population.

  WALFA is built fundamentally on opportunities deriving from land ownership and Indigenous knowledge in land management. It requires the participation of young people in the sometimes physically arduous work, but also provides opportunities for youth to develop new skills. The project actively seeks to engage youth through participation in on-country walks and other activities attractive to this age group.

- the need for quality data and research

  The project began with a strong applied research focus, which has been maintained throughout and continues with work to refine methods and measure outcomes.

The project is particularly strong in Indigenous 'ownership', consequent alignment with Indigenous interest, aspiration and commitment, participation of role models and mentors, emphasis on partnerships and orderly development over long timeframes. One of the reasons for strong Indigenous ownership is the role that is played by Indigenous knowledge in design and management of the project.

But this complex, potentially restrictive view of what works in community development in Indigenous settings in northern Australia is based on present circumstances. With experience in ventures such as WALFA – backed by improved educational and vocational programs – the range of development options available will be increased and the array of special conditions that need to be satisfied to enable success will be reduced in number and complexity. It will not be necessary for all businesses to reinvent approaches to all of these issues, once paths have been cut by ventures with a strong cultural fit and coherent support to overcome obstacles.

We do not suggest that interaction with carbon markets through fire management or otherwise should be seen as a goal for all remote communities. But the extraordinary alignment of opportunity, knowledge and commitment for Indigenous people to make – and be seen to make – a core contribution to economically and environmentally critical national goals, by facilitating transition to a carbon-constrained economy surely warrants serious attention and coherent effort.

NEW INSTITUTIONS

Perhaps the most challenging issues for Indigenous engagement with markets will be to build institutions that support reliable collective delivery and promote equitable distribution of benefits to participants and related interests (Whitehead *et al.* 2008). We have argued that these issues are resolvable, but they will require special effort from Indigenous organisations, groups and individuals.

To date, constructing institutions for regional scale delivery of land and resource management outcomes have been somewhat piecemeal and incremental. WALFA's success was not built on one or a few clever decisions. It was built on a long series of foundational actions by many individuals (e.g. Chapter 11; Box 11.1). Arguably, Indigenous institutions such as the Caring for Country Ranger groups, which were built over more than a decade, have been more successful than many other community-based arrangements in non-Indigenous society in driving positive change in management of land and natural resources and in empowering local groups to take leading roles.

But, if advancements in institutional design are to keep pace with national and international climate change agendas, it will be necessary to go well beyond an incremental pace. And as we have shown in previous chapters, real progress will require complementary action on many fronts by a number of different sectors. Among the most important are Indigenous organisations, governments, the philanthropic sector and private industry, as discussed below.

Indigenous organisations

Indigenous organisations should:

- develop collaborations capable of working over large areas
- address overtly the issues of 'property rights', through decisions about eligibility to participate and rules for sharing benefits
- add comprehensive dispute resolution mechanisms
- engage with governments, industry purchasers of credits, and philanthropic groups engaged in community development and/or environmental issues about their expectations for products and arrangements to validate performance and quality
- consider relationships between and within Indigenous organisations to avoid conflicts of interest and associated risk.

Robust commercial performance will require replacement of the role of committed and interested individuals, who have driven the incremental advances, with new institutions in the Indigenous domain establishing rights to participate in delivery and to share in benefits. Roles of traditional owners will often include planning and related high-level (strategic) decision making, but may or may not include active participation in on-ground work. Other Indigenous people may adopt roles equivalent to *djungai* (see Chapter 5), and so play important roles in planning. Many more will be involved in on-ground delivery of plans. Clarifying the array of roles, eligibility to take them on and the share of benefits they attract are important issues for Indigenous groups to resolve.

Above the operational level, the design of cooperatives or looser collaborations needed to realise economies of scale and manage risk is an important issue. Optimising the trade-off between the benefits of large spatial scale and the number and diversity of participants will be an important consideration. It may be challenging in some contexts to optimise the spatial configuration of project areas while also aligning compatible objectives in land management.

Processes for building 'beyond-WALFA' governance systems have begun (late 2008) through NAILSMA on behalf of mainland north Australian Indigenous Land Councils. With support from the Australian Government, consultations involving Indigenous natural resource management organisations, governments, philanthropic organisations, business and researchers in governance issues are planned. Those discussions will explore a broad range of options for engaging with the carbon economy, but will draw particularly on collective experience of the fire work. The insights of those with a long association with the project will be exploited to test ideas. It will be critically important that the Australian Government is kept well informed or, preferably, participates at a senior level to ensure that options are not lost in the detail of the CPRS design and the positions taken to international climate change negotiations.

Governments

Governments should:

- re-imagine the role of remote Australia in management of the national greenhouse gas emissions profile
- invest in science for understanding and exploiting the role of natural systems in carbon dynamics, especially the interaction between bio-sequestration and land-management practice
- commit to policy development in climate change that seeks multiple benefits including closing the gap of Indigenous disadvantage, revitalisation of regions, biodiversity conservation and improved water management
- clarify the roles of the Australian and other levels of government as purchasers of environmental services
- cooperate with other levels of government to define and develop systems to validate products.

In policy development, which is centred on the CPRS, the Australian Government is yet to seriously pursue the ideas put forward by Garnaut (2008) regarding options for use of the managed systems – that include much of the Australian land area – as carbon sinks, and hence to develop the skills and practices (such as fire management) that are necessary to sustain such a vision. Nor has there been much evidence – aside from the commitment to support Indigenous involvement in two segments of the large suite of opportunities – of a capacity to connect climate change responses to wider agendas. We consider that the WALFA project demonstrates that in north Australia, there is scope to 'transform regional economies' (Garnaut 2008) without compromising the integrity of the CPRS, and achieving national mitigation targets at lower cost.

To date, the policy focus on the land use, land-use change and forestry sectors and associated modelling has been principally negative: depending on continuing and perhaps deeper constraints on land clearing and hence reducing opportunities for some forms of economic development. For example, modelling by the Australian Treasury on the costs of the CPRS assumed linear declines in levels of land clearing (Commonwealth of Australia 2008). Although well-considered controls over land clearing are welcome, it is inconceivable that the positive should not also be embraced, and capacity developed to turn huge land areas to effective carbon sinks while producing biodiversity benefits and social benefits for the severely disadvantaged in the regions. There are technical issues to be resolved, but these appear considerably less challenging than, for example, relatively immature technologies such as carbon capture and storage linked to coal use that have already been endorsed by governments and industry (Baldwin 2008).

Philanthropic sector

The philanthropic sector should:

- clarify goals in community development and/or environmental protection and of willingness to accept 'mixed' models that include commercial sale of some products, especially carbon credits
- recognise the importance of social benefits
- clarify the standards in corporate governance that are expected

- specify the standards in measurement and validation of environmental services or social benefits.

Private philanthropic, donor-funded organisations are increasingly involved in support for community development and environmental management in northern Australia. It will be important that the expectations of such groups are well understood if they are to be involved in partnerships with Indigenous people. We have argued that the responses needed to reduce emissions, and to do so on a commercial footing, are closely congruent with those needed to deliver improved protection of biodiversity. But we also recognise that as techniques improve and ambitions increase, situations will emerge in which some trade-offs among commercial performance, social benefits and environmental outcomes will need to be considered. Many non-government organisations are also committed to equitable access to benefits of development activity and have experience in models for governance to manage competing interests in resource use. Engaging such groups in development of governance systems will be important both to ensure that their expectations can be met and to take advantage of their experience.

Private industry

Private industry should:

- support the development of robust business models
- clarify of the nature and scale of products sought
- design the standards in measurement and validation required.

Carbon markets are reasonably well-established internationally and in parts of southern Australia, but understanding is relatively weak in the north. Markets in other environmental services, including biodiversity credits, are even less understood. It will be particularly important that Indigenous groups be exposed to dialogue between government and potential purchasers of credits so that they clearly understand the standards that will be mandated and/or demanded by purchasers, and can design their businesses accordingly.

CONCLUSIONS

Independent of GHG emissions, fire management is one of the most pressing land-management issues for northern Australia. Important ecosystems such as the sandstone heaths and *Allosyncarpia* forests (Russell-Smith *et al.* 1998; Russell-Smith and Bowman 1992) and individual plant species, such as the native pine *Callitris intratropica* (Bowman and Panton 1991; Price and Bowman 1994; Bowman *et al.* 2001), are under threat from adverse fire regimes. A wide array of fauna (Franklin *et al.* 2005; Chapter 8) have suffered declines in distribution and abundance even in the most widespread forest and woodland types that dominate huge areas of the north. Although less well-documented, other attributes of land condition are also damaged by adverse fire regimes. Erosion may be increased (Russell-Smith *et al.* 2006) or production compromised (Dyer *et al.* 2001; Dyer and Stafford Smith 2003). Human health is also affected (Johnson *et al.* 2002).

Managing pervasive processes such as fire in vast, sparsely populated landscapes that are mostly funded in proportion to population – rather than land area or the demands of acknowledged problems – requires a contribution from much of the resident population (Whitehead 1999). But when that population is mostly socioeconomically disadvantaged, and has access to services and infrastructure severely limited in range and quality, achieving mobilisation can be difficult. Challenges are compounded by frequent shifts in Indigenous affairs, regional development, welfare and environmental management policies that sometimes conflict with each other (Whitehead *et al.* 2002).

The importance of the funding base potentially available from emissions markets with consistent rules and standards is much greater than dollar value alone might suggest. Value is greatly enhanced by the opportunities for regional communities to respond to properly regulated long-term markets. This stable base can underpin interactions with shorter-term government programs or emerging, and correspondingly less predictable, markets for other products, including other ecosystem services.

It is also important to emphasise that it is not only Indigenous groups who need to build capability in delivering environmental services at large spatial scales. Government agencies, including managers of the public conservation estate, have much to learn from the WALFA example (Chapters 4 and 10). In Kakadu National Park, accommodation of Indigenous burning with formal management practice has taken decades and remains incomplete, despite occurring within a formal joint management arrangement. WALFA, and its match of formal requirements of the Northern Territory's Bushfires Council to Indigenous objectives in fire use and control of delivery, has evolved much more quickly: arguably because the targets and accountabilities are clear and readily measured.

WALFA and the Garnaut review (Garnaut 2008) point to options to address in tandem an array of environmental challenges while promoting social and economic development in severely disadvantaged regions of northern Australia.

Will we be agile enough to grasp the opportunity?

REFERENCES

ABS (Australian Bureau of Statistics) (2008a) 2006 *Census Community Profiles by Location.* <www.censusdata.abs.gov.au>.

ABS (2008b) *Population Characteristics, Aboriginal and Torres Strait Islander Australians, 2006.* Catalogue Number 4713.0. <http://www.abs.gov.au>.

Access Economics (2008) *Access Economics Business Outlook, September Q 2008.* Access Economics, Canberra.

Altman JC (2004) Economic development and Indigenous Australia: Contestations over property, institutions and ideology. *Australian Journal of Agricultural and Resource Economics* **18(3)**, 513–34.

Altman JC (2006) 'In search of an outstations policy for Indigenous Australians'. CAEPR Working Paper 34/2006. Centre for Aboriginal Economic Policy Research, Australian National University, Canberra.

Baldwin S (2008) 'Carbon capture and storage'. NSW Parliamentary Library Briefing Paper No. 2/08. NSW Parliamentary Library Research Service, Sydney.

Bowman DJMS, Price O, Whitehead PJ and Walsh A (2001) The 'wilderness effect' and the decline of *Callitris intratropica* on the Arnhem Land Plateau, northern Australia. *Australian Journal of Botany* **49**, 665–672.

Bowman DMJS and Panton WJ (1993) Decline of *Callitris intratropica* R.T. Baker and H.G. Smith in the Northern Territory: implications for pre- and post-European colonization fire regimes. *Journal of Biogeography* **20**, 373–381.

Burgess CP, Johnston FJ, Bowman DMJS, Whitehead PJ and O'Dea K (2005) Healthy land: healthy people? Exploring the health benefits of Indigenous land management. *Australian New Zealand Journal of Public Health* **29(2)**, 117–122.

Commonwealth of Australia (2008) *Australia's Low Pollution Future: The Economics of Climate Change Mitigation.* Department of the Treasury, Canberra.

Dyer R, Jacklyn P, Partridge I, Russell-Smith J and Williams RJ (Eds) (2001) *Savanna burning: understanding and using fire in northern Australia*. Tropical Savannas Cooperative Research Centre, Darwin.

Dyer R and Stafford Smith M (2003) Ecological and economic assessment of prescribed burning impacts in semi-arid pastoral lands of northern Australia. *International Journal of Wildland Fire* **12(4)**, 403–413.

Franklin DC, Whitehead PJ, Pardon G, Matthews J, McMahon P and McIntyre D (2005) Geographic patterns and correlates of the decline of granivorous birds in northern Australia. *Wildlife Research* **32(5)**, 399–408.

Garnaut R (2008) 'Garnaut climate change review: final report'. Cambridge University Press, Melbourne.

Gerritsen R (2008) Constraining indigenous livelihoods and adaptation to climate change in SE Arnhem Land, Australia. *International expert group meeting on Indigenous peoples and climate change, Darwin, Australia, April 2–4, 2008*. <http://www.un.org/esa/socdev/unpfii/documents/EGM_cs08_Gerritsen.doc>.

Johnston FH, Kavanagh AM, Bowman DMJS and Scott RK (2002) Exposure to bushfire smoke and asthma: an ecological study. *Medical Journal of Australia* **176(11)**, 535–538.

Kakadu Region Social Impact Study (1997) 'Community action plan. Report of the Study Advisory Group'. Supervising Scientist, Canberra.

MAC (2007). 'Annual report 2006/7'. Maningrida Arts and Culture, Maningrida, NT.

Mansfield B (2008) *Privatization: Property and the Remaking of Nature-society Relations*. Wiley-Blackwell, London.

Northern Land Council (2006) *Celebrating Ten Years of Caring for Country*. Northern Land Council, Darwin.

Northern Territory Government (2005) 'Northern Territory Indigenous economic development strategy'. Northern Territory Government: Darwin. <http://www.ied.nt.gov.au/docs/ied_strategy.pdf>.

Northern Territory Government (2008) 'Northern Territory Government outstations policy discussion paper'. Department of the Chief Minister, Darwin. <http://www.action.nt.gov.au/outstations/docs/Outstations_Discussion_Paper.pdf>.

Price O and Bowman DMJS (1994) Fire-stick forestry: a matrix model in support of skilful fire management of *Callitris intratropica* R.T. Baker by north Australian Aborigines. *Journal of Biogeography* **21**, 573–580.

Russell-Smith J and Bowman DMJS (1992) Conservation of monsoon rainforest isolates in the Northern Territory, Australia. *Biological Conservation* **59**, 51-63.

Russell-Smith J, Ryan PG, Klessa D, Waight G and Harwood R (1998) Fire regimes, fire-sensitive vegetation and fire management of the sandstone Arnhem plateau, monsoonal northern Australia. *Journal of Applied Ecology* **35**, 829-846.

Russell-Smith J, Yates C and Lynch B (2006) Fire regimes and soil erosion in north Australian hilly savannas. *International Journal of Wildland Fire* **15(4)**, 551–556.

Russell-Smith J, Yates CP, Whitehead P, Smith R, Craig R, Allan G, Thackway R, Frakes I, Cridland, Meyer CP and Gill AM (2007) Bushfires 'down under': patterns and implications of Australian landscape burning. *International Journal of Wildland Fire* **16**, 361–377.

Russell-Smith J, Murphy BP, Meyer CP, Cook GD, Maier S, Edwards AC, Schatz J and Brocklehurst P (2009) Improving estimates of savanna burning emissions for greenhouse accounting in northern Australia: limitations, challenges, applications. *International Journal of Wildland Fire* **18**, 1–18.

Stoeckl N, Stanley O, Brown V and Stoekl W (2007) 'Regional economic multipliers in Australia's tropical savanna'. Tropical Savannas Cooperative Research Centre, Darwin.

Taylor J (2006) 'Population and diversity: policy implications of emerging Indigenous demographic trends'. CAEPR Discussion Paper No. 283/2006. Centre for Aboriginal Economic Policy Research, Australian National University, Canberra.

Taylor J, Brown D and Bell M (2006) 'Scoping population dynamics and demographic accounting in arid and savanna Australia: methods, issues and outcomes'. Report of a scoping study for the Desert Knowledge and Tropical Savannas Cooperative Research Centres. Desert Knowledge Cooperative Research Centre, Alice Springs.

Vanstone A (2005) 'Beyond conspicuous compassion – Indigenous Australians deserve more than good intentions'. Address to Australia and New Zealand School of Government, 7 December, ANU, Canberra.

Whitehead PJ (1999) Is it time to fill the north's empty landscapes? *Savanna Links* **11**, 6–7.

Whitehead PJ and McGuffog T (1997) Fire and vegetation pattern in a tropical floodplain grassland: a description from the Mary River, monsoonal northern Australia and its implications for wetland management. In *Bushfires 97. Proceedings of the Australian Bushfires Conference, 8–10 July 1997.* (Eds BJ McKaige, RJ Williams and WM Waggitt) pp. 115–120. CSIRO, Darwin.

Whitehead PJ, Woinarski JCZ, Franklin D and Price O (2002) Landscape ecology, wildlife management and conservation in northern Australia: linking policy, practice and capability in regional planning. In *Landscape Ecology and Resource Management: Linking Theory with Practice.* (Eds J Bissonette and I Storch) pp. 227–259. Island Press, New York.

Whitehead PJ, Purdon P, Russell-Smith J, Cooke PM and Sutton S (2008) The management of climate change through prescribed savanna burning: emerging contributions from Indigenous people in northern Australia. *Public Administration and Development* **28**, 374–385.

Yibarbuk DM, Whitehead PJ, Russell-Smith J, Jackson D, Godjuwa C, Fisher A, Cooke P, Choquenot D and Bowman DJMS (2001) Fire ecology and Aboriginal land management in central Arnhem Land, northern Australia: a tradition of ecosystem management. *Journal of Biogeography* **28**, 325–344.

Yu P, Duncan ME and Gray W (2008) 'Northern Territory emergency response: report of the NTER Review Board'. October 2008. Australian Government: Canberra. <http://www.nterreview.gov.au/docs/report_nter_review/docs/Report_NTER_Review_October08.pdf>.

INDEX